AF846124

Technology, Style and Society

Technology, Style and Society

Contributions to the Innovations between the Alps and the Black Sea in Prehistory

Compiled and Edited by

Lolita Nikolova

BAR International Series 854
2000

British Archaeological Reports are published by John and Erica Hedges and by Archaeopress.

This volume has been published by:
John and Erica Hedges
British Archaeological Reports
7 Longworth Road
Oxford OX2 6RA
England
Tel/Fax +44 (0)1865 511560
E-mail: bar.hedges@lineone.net

Enquiries regarding the submission of manuscripts for future publication may be sent to the above address

BAR S854

Technology, Style and Society: Contributions to the Innovations between the Alps and the Black Sea in Prehistory

Volume Editor: John Hedges BSc(Hons), MA, MPhil, FSA, FSAScot, FRAI, MIFA

Printed in England by Biddles Ltd

ISBN 1 84171 136 5

All BAR titles available from:

Hadrian Books
122 Banbury Road
Oxford OX2 7BP
England

The current BAR catalogue with details of all titles in print, prices and means of payment, is available free from Hadrian Books

All volumes are distributed by Hadrian Books Ltd

TABLE OF CONTENTS

Section 2

Cultures, Ceramic Styles and Contacts in the Bronze and Iron Ages

Section 3

Archaeology of Ritual

Section 4

Case Studies

Preface

The idea of the subject promoted on the technology, style and society in Prehistory, was to examine a possibility of collective approach to the problems of the innovations, from different points of view. Fortunately, in the period between 1997 and 2000 I have the opportunity to receive important contributions, concerning the region under discussion.

In this approach to the innovations between the Northern Alps and the Black Sea the first common focus is on new aspects of interpretation of prehistoric archaeological record. The second primary purpose is publishing of new important data from some key microregions, clarifying the topic and opening new questions and problems. A special participation of the young generation should be mentioned. They are authors of some of the best pieces of work in the book.

The publishing of the *Technology, Style and Society* was possible because of the kind respond of Mr. John Hedges whom I due my special debt of gratitude. I would also like kindly to thank all the contributors who accepted my invitation to participate in that integrative project and have represented such fascinating studies on various, mostly innovative topics.

Dr. Lolita Nikolova
International University –
Karlovo
Bulgaria

Abbreviations

AOR	*Archeologicheski otkritiya i razkopki*
BCH	*Bulletin de Correspondance Hellénique*
GDA NBU	*Godishnik na Departament Arkheologiya, Nov Bulgarski Universitet*
GMP	*Godishnik na muzeite v Plovdiv*
GMPO	*Godishnik na muzeite v Plovdivski okrug*
GMSB	*Godishnik na muzeite ot Severna Bulgariya*
GNAMPl	*Godishnik na Narodniya Arkheologicheski Muzej Plovdiv*
GNM	*Godishnik na Narodniya Muzej*
GPlNBM	*Godishnik na Plovdivskata Narodna Biblioteka i Muzej*
GSU GGF	*Godishnik na Sofijski Universitet, Geologo-Geografski fakultet*
IBAD	*Izvestiya na Bulgarskoto Arkheologichesko druzhestvo*
IBAI	*Izvestiya na Bulgarskiya Arkheologicheski Institut*
IIMKn	*Izvestiya na Istoricheskiya Muzej Kyustendil*
IIMVT	*Izvestiya na Istoricheski muzej Veliko Turnovo*
INMV	*Izvestiya na Narodniya Muzej Varna*
IVAD	*Izvestiya na Varnenskoto Arkheologichesko Druzhestvo*
KSIA	*Kratkie soobshteniya Instituta Arkheologii*
MAA	*Macedoniae Acta Archaeologica*
NACSN	*North American Commission on Stratigraphic Nomenclature*
PBF	*Prähistorische Bronzefunden*
PJZ	*Praistorija Jugoslovenskih zemalia*
Poročilo	*Poročilo a raziskovanju paleolitika, neolitika in eneolitika v Sloveniji*
SCIVA	*Studii şi Cercetări de Istorie Veche şi Arheologie*
SZMMÉ	*Szolnok Megyei Múzeumi Évkőnyv*
ZNM	*Zbornik Narodnog Muzeja, Beograd*

Section 1

Techno-Economy, Ceramic Styles and Society in Neolithic and Copper Age

Chapter 1

Chipped Stone Assemblages from South Bulgaria and North-West Turkey

(Epipalaeolithic / Mesolithic and Neolithic)

Ivan Gatsov

Introduction

There are three basic factors that substantiate the topics of the present work. The first one concerns the insufficient investigations of prehistoric stone technology and typology, especially of the Bulgarian Holocene chipped stone assemblages. Although recently the research on this subject has been intensified, the area remains still unsatisfactorily studied from this point of view. The second one can be linked to the accessibility of the collected chipped stone materials. In order to bridge this gap the present article tries to clarify as much as possible the rather chaotic image of that side of the archaeological investigation. The third one explores the strategic geographical position of North-West Turkey (Özdoğan 1995: 53-55; 1998: 63-92) and South-West Bulgaria in the process of Neolithization of South-East and Central Europe. This territory is considered as a natural bridge linking prehistoric cultures. South Bulgaria and especially the area of Upper Thrace and the valleys of Struma and Iskar in the western part of the country took up the role of a nexus between Anatolia, the Balkans and Central Europe.

I would like to notice that in the present analysis I include only personally studied chipped stone assemblages from South-West Bulgaria and North-West Turkey. I did not have the occasion to study materials from Aegean or Greek Thrace.

The main purpose of this paper is to reveal the chronological range of the Neolithic chipped stone assemblages and to present their most characteristic techno-typological features. The results of the geo-chemical and use wear trace analyses will be published elsewhere.

Bulgaria

Territorial range

The chipped stone assemblages which constitute the subject matter of the present study, are registered in some regions in Bulgaria; especially they come from North -East and South-West. From northern Bulgaria this is the Dikilitash site - an area close to the town of Varna, in the North of the Bulgarian Black Sea coast. The other regions studied include Upper Thrace (or Bulgarian Thrace) in South-East Bulgaria and West Bulgaria. Geographically, the last one covers the area among the Danube in the North, the Iskar River and the Rila and Pirin Mountains in the East, the Belasitsa Mountain in the South, the Osogovska, Vlakhina, Malashevska, Ograzhden Mountains and the Timok River in the West. Western Bulgaria is divided into two parts: North-West Bulgaria, between the Timok, Iskar and Danube Rivers and the Stara Planina Mountains (the Balkans). South-West Bulgaria lies between the Stara Planina, Rila, Pirin, Belasitsa, Ograzhden, Malashevska and Osogovska Mountains.

History of research

During the last years, different aspects of chipped stone assemblages, which belong to the Pleistocene/ Holocene transition, Neolithic, Chalkolithic (Copper Age) and Bronze Age have grown as an important subject matter in Bulgarian prehistory. Unfortunately, the information about the stone technology and typology is still insufficient. Especially, this state of investigations is relevant to the assemblages from to the very end of the Pleistocene and the beginning of the Holocene. From archaeological point of view, the period of Pleistocene/ Holocene transition is less known in comparison to the Neolithic period in Bulgaria.

For the time being, the chipped stone material from the surface of the 'Dikilitash' site, near Bulgarian Black Sea coast, can be linked to the beginning of the Holocene.

Up to now the youngest absolute date for the Late Palaeolithic in Bulgaria comes from the Temnata cave, North Central Bulgaria. It refers to the second half of the fourteenth millennium BP (Bluszcz et al. 1992). In the last years one unpublished absolute date has been obtained from another site - the Palaeolithic site in the Kozarnika cave, North-West Bulgaria. This date can be related to the eleventh millennium BP (personal communication of N. Sirakov).

The old excavations of the Bulgarian Holocene prehistory did not give equal weight to flint assemblages in comparison to the rest of the archaeological material. Usually, there is a lack of reliable information about the classification and origin of flint raw material presented on various prehistoric sites. In the old research tradition the flint artefacts were hardly ever mentioned. There was a lack of specialized terminology and illustrations were sparse and insufficient. Because the method of sieving was not very often applied, the full reconstruction of some stages of the production sequence was not possible. Another problem is that sometimes there is no exact information concerning the spatial distribution of the assemblages. Recently a complete stratigraphic information can be obtained from the newly investigated prehistoric sites in South Bulgaria.

Pleistocene/Holocene transition

Chipped stone material from "Dikilitash" (North-East Bulgaria)

The Dikilitash Reserve. The area covers about 50 sq. km situated to the west of the town of Varna - in the northern part of the Bulgarian Black Sea coast. This collection, gathered by archaeologist A. Margos, contains more than 11,000 artefacts and was dated - on the base of their techno-typological characteristics - from the very beginning of the Holocene.

Neolithic chipped stone assemblages from South Bulgaria

Below the main prehistoric sites in the area under question with chipped stone assemblages investigated by the author is presented (***Map 1.1***).

Karanovo. A complex information of this site can be found in the volume edited by V. Nikolov and St. Hiller (1997). In this work I am using the material from their excavations.

Stara Zagora-Azmak. The site was totally excavated in the 1960s by G.Il. Georgiev. A detail information about chipped stone assemblages from this site was published in 1990s (Gatsov and Gurova 1998).

Rakitovo. Investigations of A. Raduncheva and V. Matsanova in 1974-1975 (Matsanova and Raduncheva 1995: 114-125).

Eleshnitsa. Excavations directed by A. Raduncheva and V. Nikolov, which were conducted in 1983-1985.

Kapitan Dimitrievo. The most recently excavated site by V. Nikolov and V. Matsanova. The investigations took place in 1998-1999 (Nikolov et al. 1999).

Chavdar. The investigations carried out by G.Il. Georgiev and K. Kunchev in the period between 1964 and 1980. Around 3,000 sq. m were excavated (Georgiev 1981: 61-109).

Galabnik. In 1977, a Neolithic settlement was discovered within the limits of the village of Galabnik, near Radomir town, 2 km to the north-west of the village. It covered an area of around 30,000 sq. m. In 1979 rescue excavations started, conducted by M. Chokhadzhiev. A year later, J. Pavuk joined the excavations. In 1979 two trenches were made, the first measuring 10 x 2.5 m, and the second 10 x 7.5 m. Additionally, a total area of 1,500 sq. m was surveyed by block excavations. In the center of the settlement a trench was excavated, measuring 20 x 20 m. The thickness of cultural layers down to the ground water, which appeared reached 4 m depth from the surface. On an area of 50 sq. m a sterile layer was documented. So far, four 'cultural-historical horizons' have been distinguished.

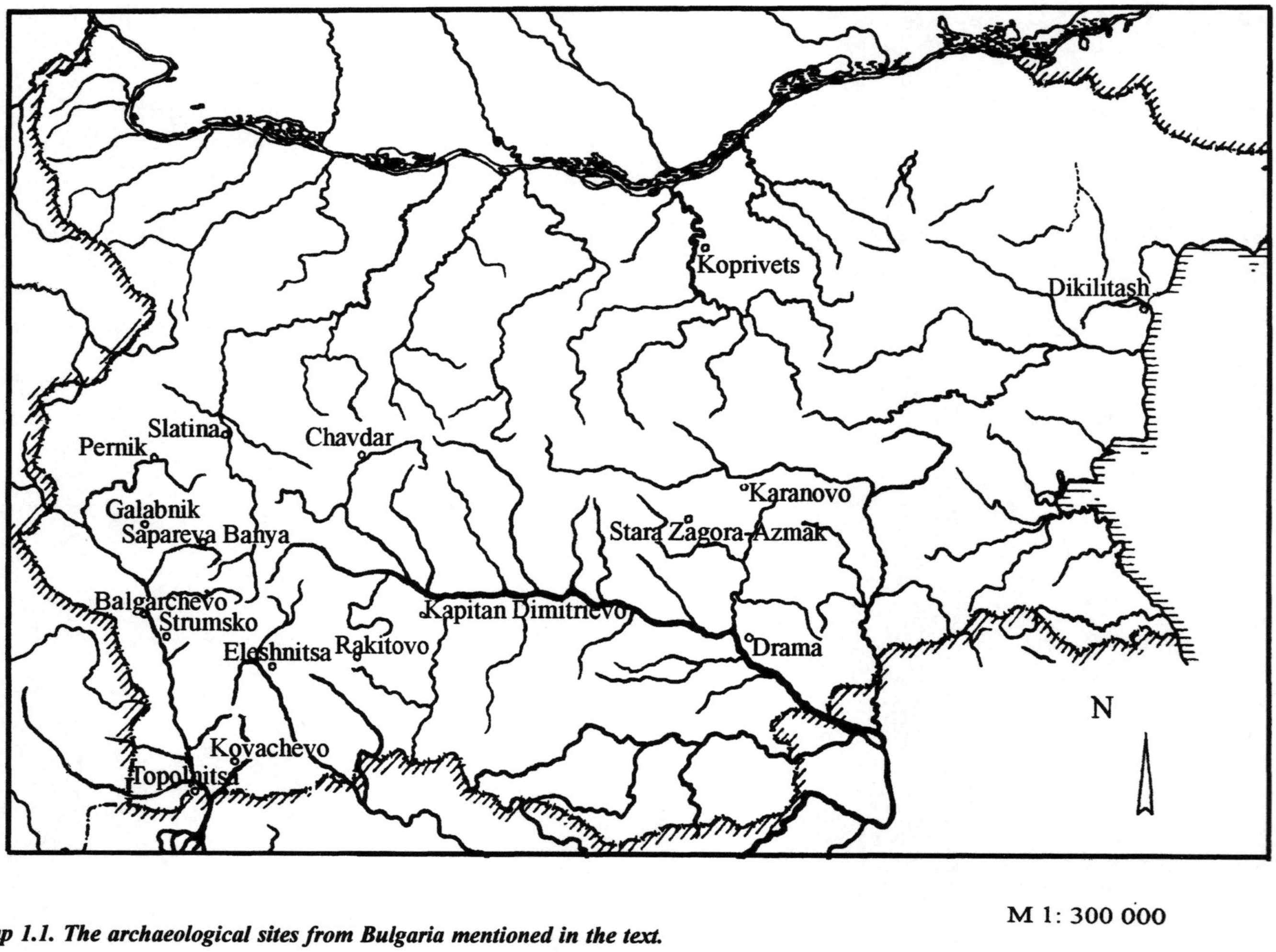

Map 1.1. The archaeological sites from Bulgaria mentioned in the text.

Pernik. In 1975 an Early Neolithic settlement was discovered in Pernik. Rescue excavations started in the same year and continued in 1976. In 1976 two trenches were made with dimensions: 4 x 4 m and 10 x 5 m respectively. The excavations revealed the existence of four building horizons in the first trench and two in the second. In both trenches there is no hiatus between the levels.

Sofia-Slatina. This site was discovered at the beginning of 1960s. The settlement is situated on the left bank of the river of Slatina, on the grounds of the modern housing area of Slatina in Sofia. In 1985 a part of the settlement was affected, and rescue excavations started. The last were directed by V. Nikolov (1992). The cultural level is up to a depth of c. 2.6 m and contains four building horizons from Early Neolithic. In 1985 the excavations focused on the earliest building horizon. From 1985 to 1989 in an area of 1,500 sq. m, the first horizon and part of the second one were explored. The excavations continued until 1998.

Topolnitsa. The settlement is located on the right bank of the river of Struma, 3 km South - East to the village of Topolnitsa and 11 km South to the town of Petrich. The excavations took place among 1980 – 1989. They were carried out by Y. Boyadzhiev, H. Todorova and I. Vajsov. The total settlement area is more than 15 000-sq. m, and the area of excavations conducted so far is around 800-sq. m. The excavations yielded two building horizons without any hiatus between them. In my work I use the excavation results until 1990.

Chronological range

From chronologically point of view, this work includes the assemblages from the Epipalaeolithic/Mesolithic periods and Neolithic sites. The chronological frames reliable are:

- The stage of Monochrome Neolithic (or Early Neolithic IA) – 6300 BC/6200 BC – 6000 BC/5900 BC
-The stage of Early Neolithic (or Early Neolithic IB-III) – 6000 BC/5900 BC – 5500 BC/5450 BC
-The stage of Middle Neolithic (or Late Neolithic I) – 5500 BC/5450 BC – 5250 BC/5200 BC
-The stage of Late Neolithic (or Late Neolithic II) – 5250 BC/5200 BC – 4900 BC/4850 BC (see Boyadziev 1995, 149-91; Todorova 1998: 28, 30; Nikolova 1998a).

Techno-typological characteristic of the chipped stone assemblages

Pleistocene/Holocene transition

Chipped stone collection from Dikilitash, the Bulgarian Black Sea shore

The collection from Dikilitash contains more than 350 cores, around 430 microliths and more than 10,000 flakes, blades and fragments (Gatsov 1984: 135-50; 1985: 471-74). The cores were classified on the base of number of platforms, number, shape and location of flaking surfaces. Different types of cores was divided in this way. Items with changing orientation have been prevailed among them. The 'plate' concretions and discoid cores were also detached. By the last, the circumference was served as a platform. The blank was detached from one or both surfaces. In most of the cases all surfaces were processed (Gatsov 1993). The final results are usually cores in shape, as polyhedrons. The presence of splintered technique was registered. It can be also noticed a completely lack of conical cores. The group of mircoliths tools can be characterized by backed bladelets, segments, triangles, trapezes, truncated bladelets. It can be also noticed the presence of splintered technique and completely lack of conical cores (***Figures 1.1: 1- 40; 1.2: 1- 45***).

On the base of main techno-typological features, the chipped stone collection from Dikilitash could be related to the Epi-Tardi-Gravetian tradition in the Iron Gate region of the Danube. It can be additionally maintained that these characteristic features differ from the chipped stone assemblages in the area of the North-West Pontic. In the last almost 20 years it has been proved that in the above mentioned region the Mesolithic assemblages are characterized by conical core processing techniques and the appearance of a blade technology has been established (Koen 1992).

It can be concluded that totally different core processing and typology of retouched tools and character of blank are characteristic of the Dikilitash collection. The main techno-typological features of this material in all respects are completely different from those from the Early Neolithic flint assemblages in South Bulgaria presented below. Up to now there is some information concerning the earliest phases (before white-painted horizon) of the Bulgarian Neolithic. This is the period of monochrome one. In North-East Bulgaria (Koprivets site) and South-West Bulgaria (Krajnitsi) chipped stone material from monochrome (before white-painted pottery) stage was studied. At Krajnici few pieces of flakes were found only. The collection from Koprivets (Popov 1996) can be characterized by more or less big flakes and flakes tools. The last are mainly notch tools. The raw material is of local origin, very low quality, extremely hard for knapping. The blades are practically missing. The material from Koprivets is completely different from those from Dikilitash (Zlateva 1997), as well as from the South Bulgarian macroblade industry. The last belongs to the white-painted Neolithic horizon in South Bulgaria. On the whole, at Koprivets mostly flakes and some kinds of flake tools present the flint artefacts from the monochrome period represented there. As a rule, these artefacts are made of a raw material of very low quality. But it can be stressed again that the information concerning the stone technology from the Monochrome Neolithic period in present Bulgarian lands is still initial.

Chipped stone assemblages from the Early Neolithic in South Bulgaria

Up to now, in South-East Bulgaria (Upper Thrace) and South-West Bulgaria the appearance of white-painted ceramics has been registered in the earlier Neolithic settlements. In the above-mentioned area the first Neolithic chipped stone assemblages belong to the phase or horizon of white-painted ceramic (Nikolov 1998a). These assemblages can be related to the sites of Karanovo I and Karanovo II, Stara Zagora-Azmak (Early Neolithic layer), Rakitovo, Eleshnitsa, Galabnik I, Pernik, earlier Slatina, Kovachevo, Kapitan Dimitrievo

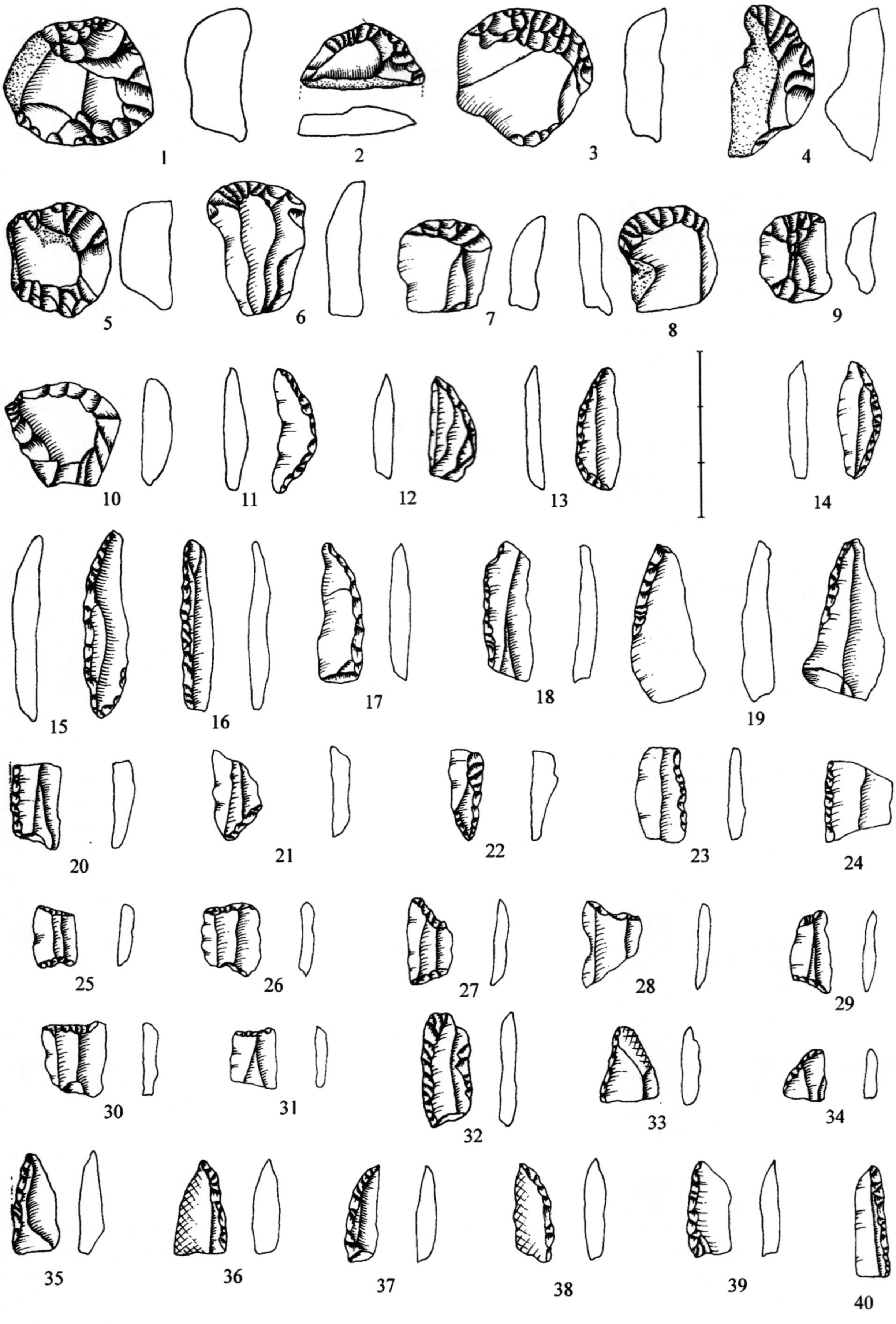

Figure 1.1. Dikilitas (North-East Bulgaria). Microtools.

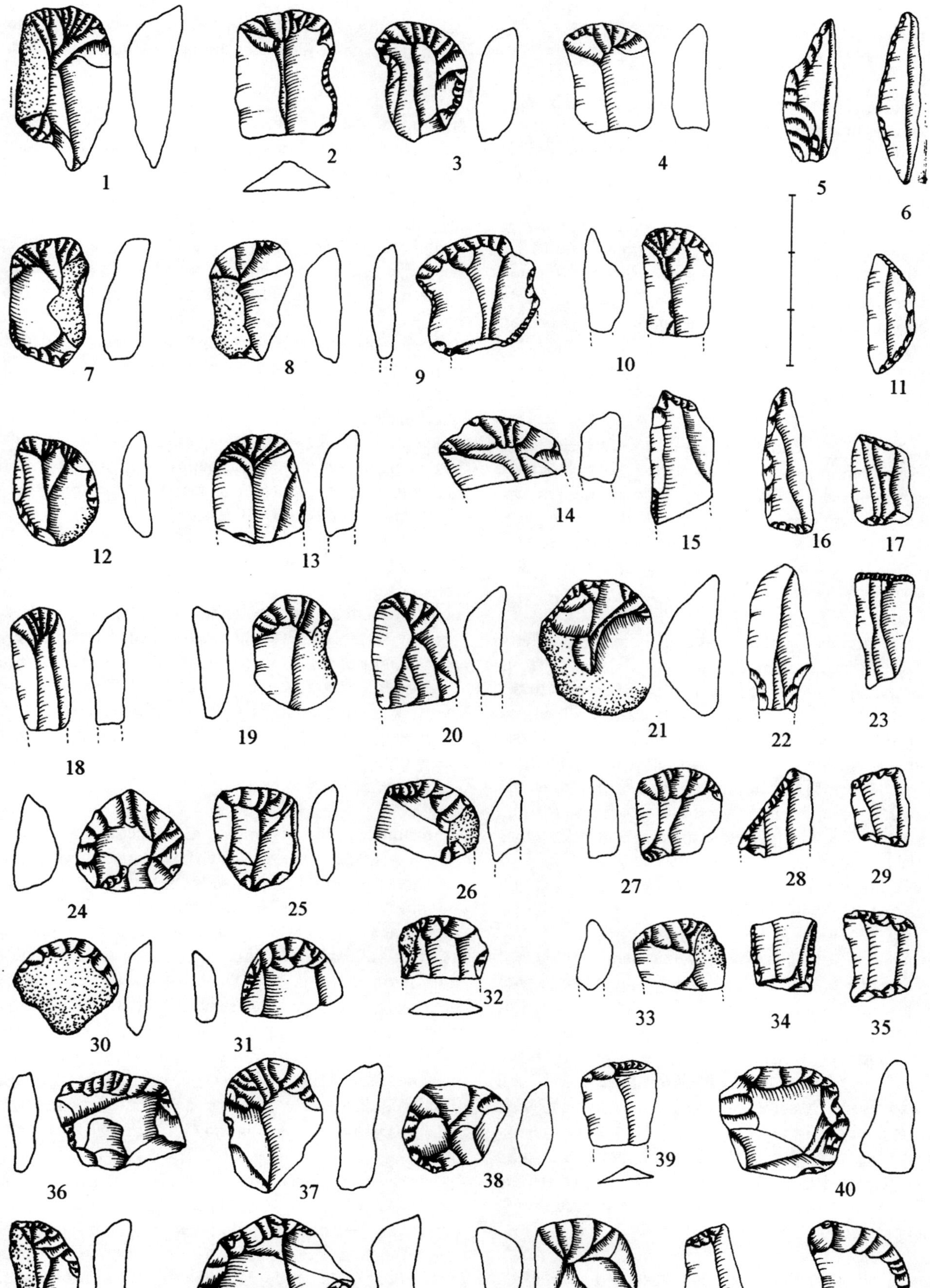

Figure 1.2. Dikilitash (North-East Bulgaria). Microtools.

(Early Neolithic layer) – ***Map 1.1;*** for the problematic of Neolithic period in South Bulgaria see Nikolov 1992: 5-163; 1996: 1-8; 1998a; Nikolova 1998a.

The main characteristic feature of the chipped stone assemblages belonging to this period is a technology for producing macroblades with or without high semi-steep retouch. The dimensions of blank and retouched blades suggest that for their producing, macroblade cores with a length no less than 12-14 cm were needed. This suggests, as it was noticed above, the dimensions of the macroblade cores should be min. 12-14 cm. As a rule, this macrocores were shaped from raw material of very high quality with yellow, yellow-reddish and yellow and brawn color with or without spots, very good for knapping. More detail characteristics of that type of raw material were elaborated by V. Kurchatov and published recently (Gatsov and Kurčatov 1997: 213-27).

It is also interesting to underline that at all early Neolithic sites in above mentioned region (including Turkish Trace) cores for macroblades from this type of raw material were not found. They were not found at all excavated areas of the Early Neolithic sites presented above, and also in all investigated Early Neolithic layer at Stara Yagora-Azmak tell (Gatsov and Gurova 1998: 7-21).

The lack of corresponding cores and concretions in the Early Neolithic settlement, trimming flakes, cortical flakes suggests that the macroblades were made elsewhere and brought to the areas under the study. For the time being, it can be considered as working hypothesisthat an initial preparation of those macrocores and receiving a blank had taken place off these sites (Voytek 1993: 55). An additional reason of such an assertion is the sizes and weight of the macrocores. All this suggests existing of workshops close to the source or outcrops of this type of raw material. Namely the totally lack of macrocores cortical flakes, blades and trimming flakes should be taken as a proof for this opinion.

But the problem with the place of macrocores and unretouched blades (***Figure 1.10: 9***) in the core processing is linked with the appearance of retouched tools at the excavated areas of Early Neolithic sites. The question is whether the blanks or ready-made tools were transported to the sites or were made in the workshops (***Figure 1.6: 1***). Unfortunately, it is still impossible to resolve this problem because the sieving method has not been applied. To answer this question it must be expected the results from the investigation of Drama in Upper Thrace and Kovachevo in South-West Bulgaria, where the results of sieving of culture layers excavated hopefully will clarify this problem.

From typological point if view, the main feature of all chipped stone assemblages is a typological monotony. Tool flakes are single examples. In all the collections the major typological groups consist of retouched blades. The others are presented by blades with marginal retouch, perforators and alternated perforators on blades, retouched truncations, end-scrapers on blades, usually with retouches on the edges. The splintered technique was also applied (***Figures 1.3: 1-8; 1.4: 1-12; 1.5: 3, 6, 8, 13-15, 17; 1.7: 1-17; 1.8: 1-3, 9; 1.9: 1-14; 1.10: 1-8, 11, 12; 1.11: 1-19; 1.13: 12-14***). About this high semi-steep or steep, undulated blade retouch it can be observed a tendency widths of the blank to have been reduced, namely by using this type of retouches (Gatsov, Kurchatov and Gurova 1999).

In most of the cases the blades were knapping from single platform cores, only exceptionally from double platforms ones. The blades and blade fragments are with a non-cortical dorsal pattern, parallel side edges and a trapezoidal section. The specimens were detached from cores in advanced phase of exploitation. Most of them are proximal fragments with a regular shape and straight or very slightly convex profile. The blade breaking takes place after retouching. Main part of the blades and blade fragments was connected with the same stage of core processing. This proposes an existing of specializing groups whose aim was receiving and selecting standardized macroblades only. This fact reveals a standardized blade production characteristic of the Early Neolithic society in the area under study, on the one hand. On the other hand, all this suggests existing of groups with certain degree of specialization and perhaps of initial degree of social stratification in the Early Neolithic period (Gatsov 1999b).

The blades with high, semi steep or steep slightly undulated retouch present the most characteristic and diagnostic type contrasting to abrupt, denticulate or notched retouches. This typological monotony could be explained by the relatively easy morphological transformations of the retouched blades. The type of raw material, the type of blank, functional requirements is main reason for these transformations. The subsistence activities of Early Neolithic society as a rule restricted the typological diversities of the chipped stone industries in the above mentioned region.

Unfortunately, the investigations concerning the sources of above described type raw material have not been undertaken in Thrace yet, on the one hand. On the other hand, this kind of flint is registered in shape of big blades. The last were occurred at all the sites chronologically relating to the first half of sixth millennium BC in South Bulgaria. According to the working hypothesis here, the raw material derived from the sources in the region of Saint Ilia hills in Thrace. This is situated around 30 km east of the Neolithic settlement of Karanovo. More information concerning geological and geochemical research in the area concerning Neolithic and Chalkolithic assemblages up to now are to be published elsewhere (Gatsov, Kurchatov and Gurova 1999).

As it was mentioned above, up to now the earliest appearance of macroblade technology on the Bulgarian territory can be related to the white painted horizon in South-East and South-West Bulgaria (***Figure 1.10: 9***). This macroblade technology was occurred in shape of macro blade, with or without retouch. From chronologically point of view, this is a period between 6000 BC and 5500 BC/5450 BC (***Figure 1.8: 10-18***).

Chipped stone assemblages from the Late Neolithic in South Bulgaria

The most important characteristic about Late Neolithic chipped stone assemblages was a very important changes in the basic

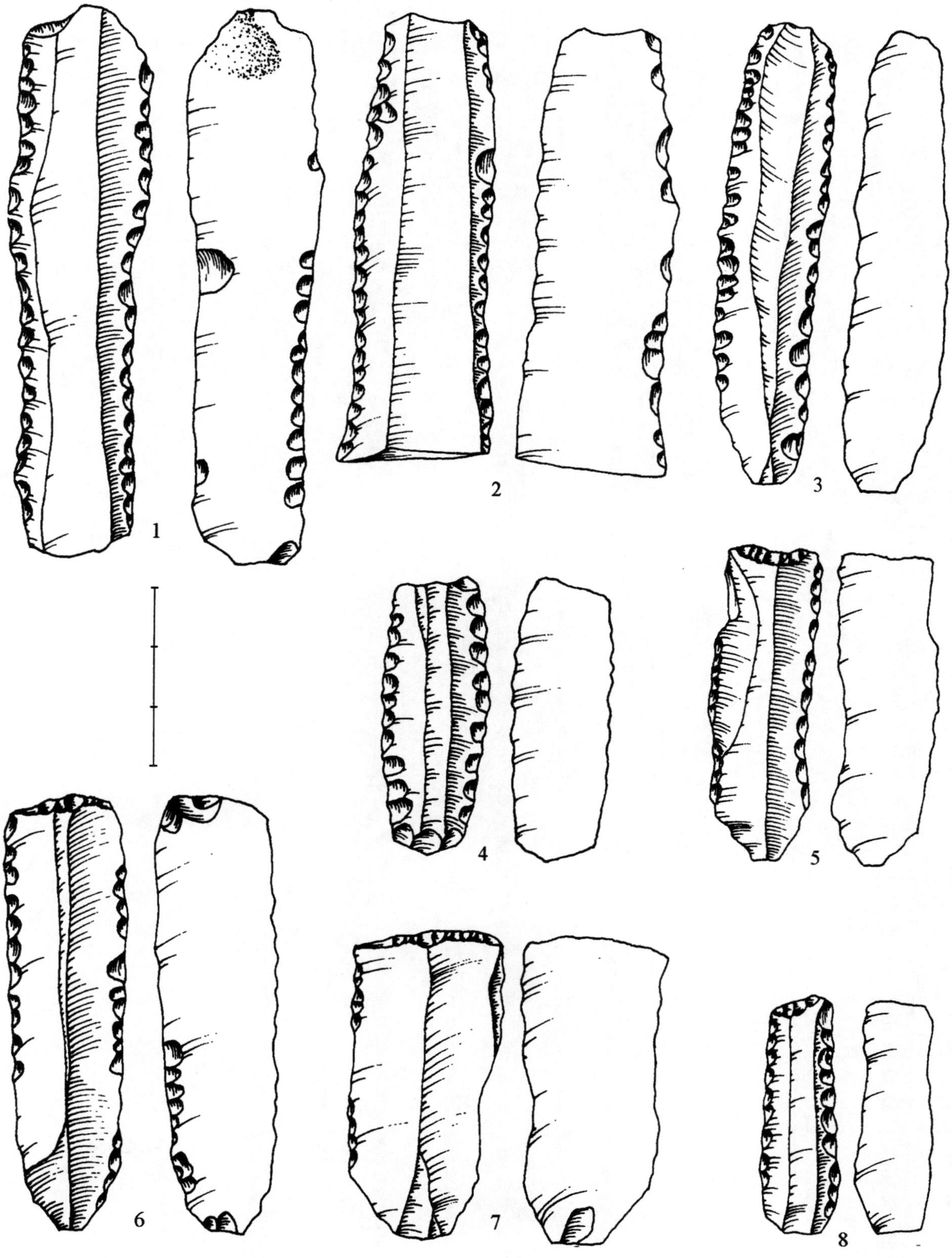

Figure 1.3. Chavdar. Retouched tools.

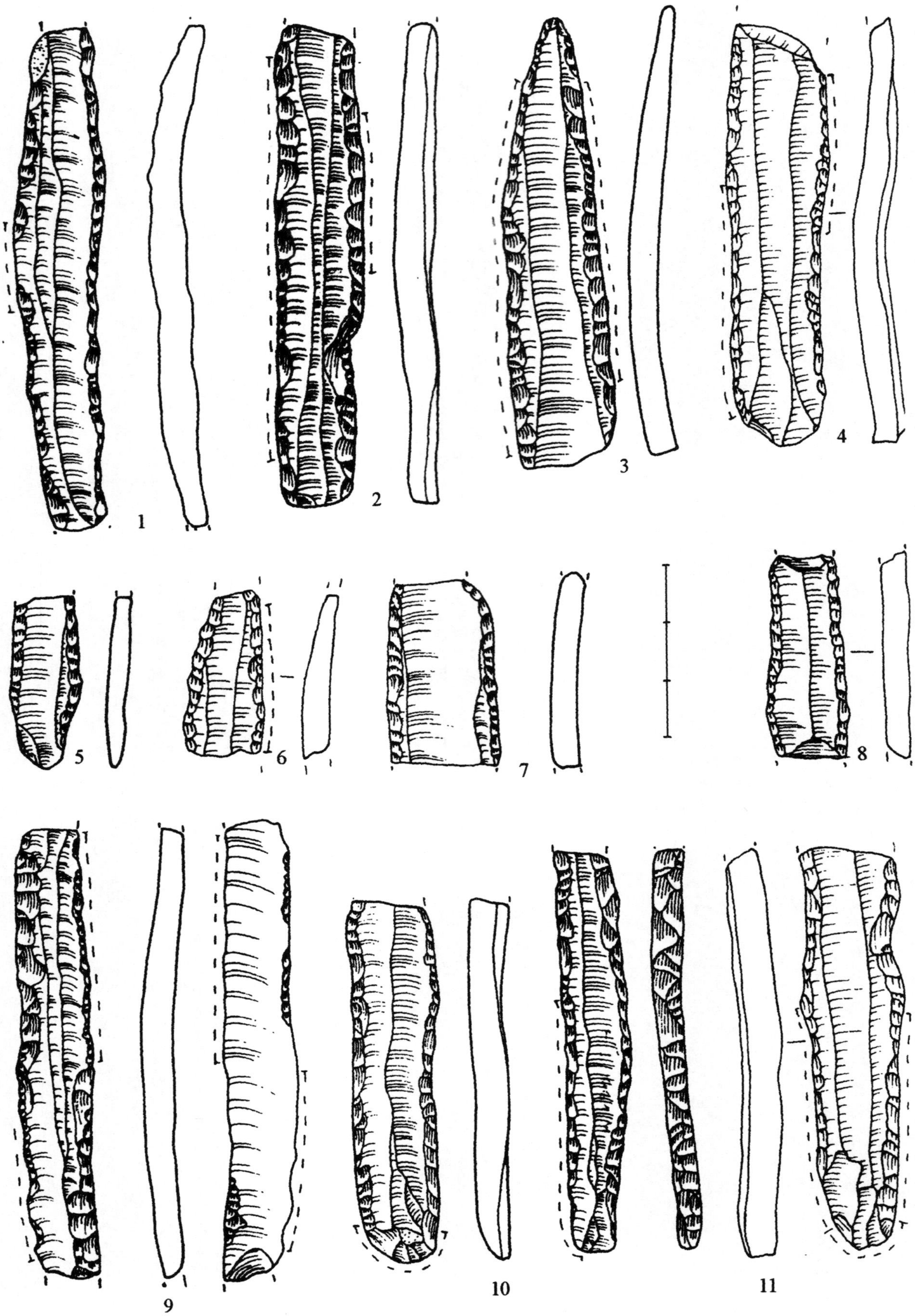

Figure 1.4. Karanovo I. Retouched tools.

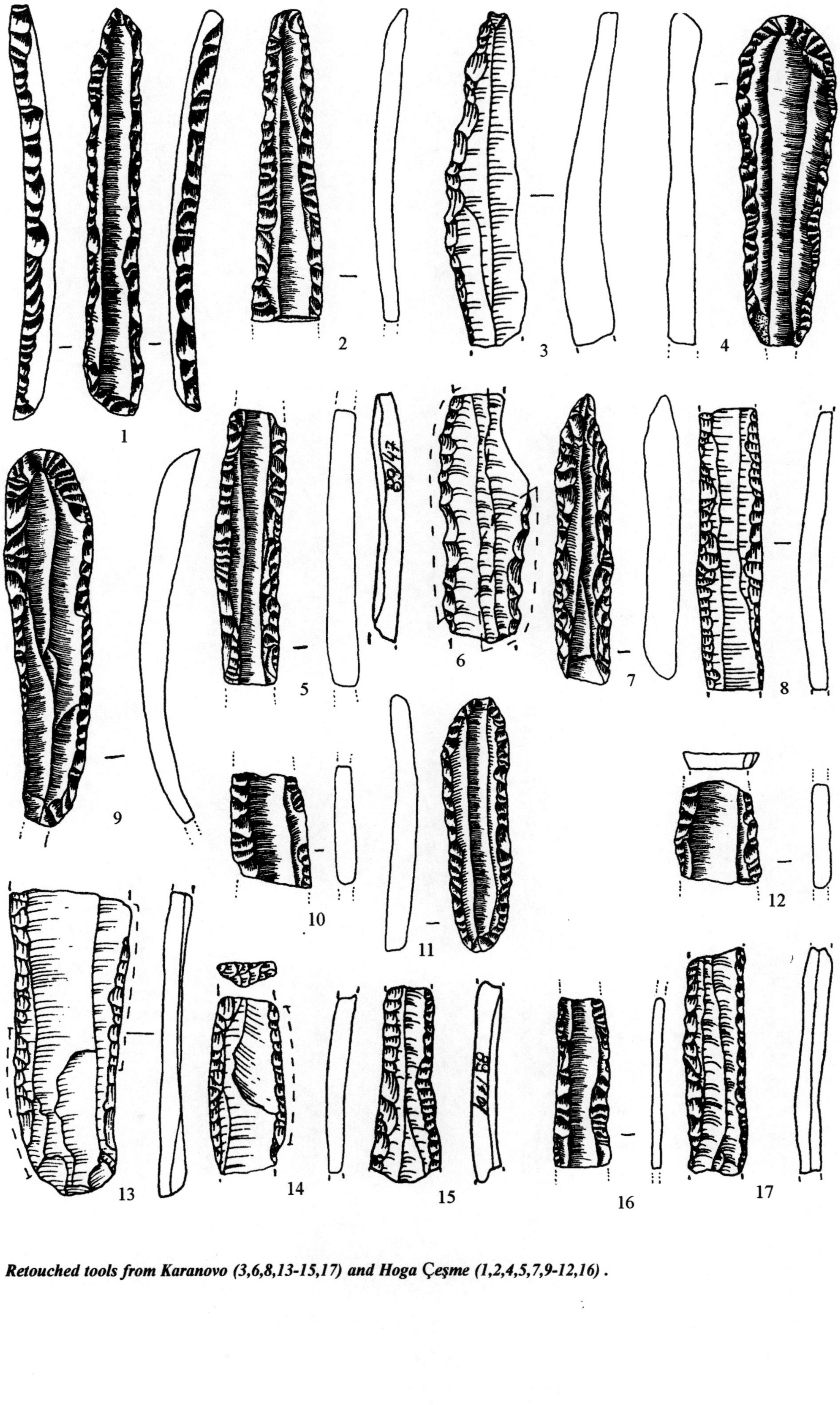

Figure 1.5. Retouched tools from Karanovo (3,6,8,13-15,17) and Hoga Çeşme (1,2,4,5,7,9-12,16).

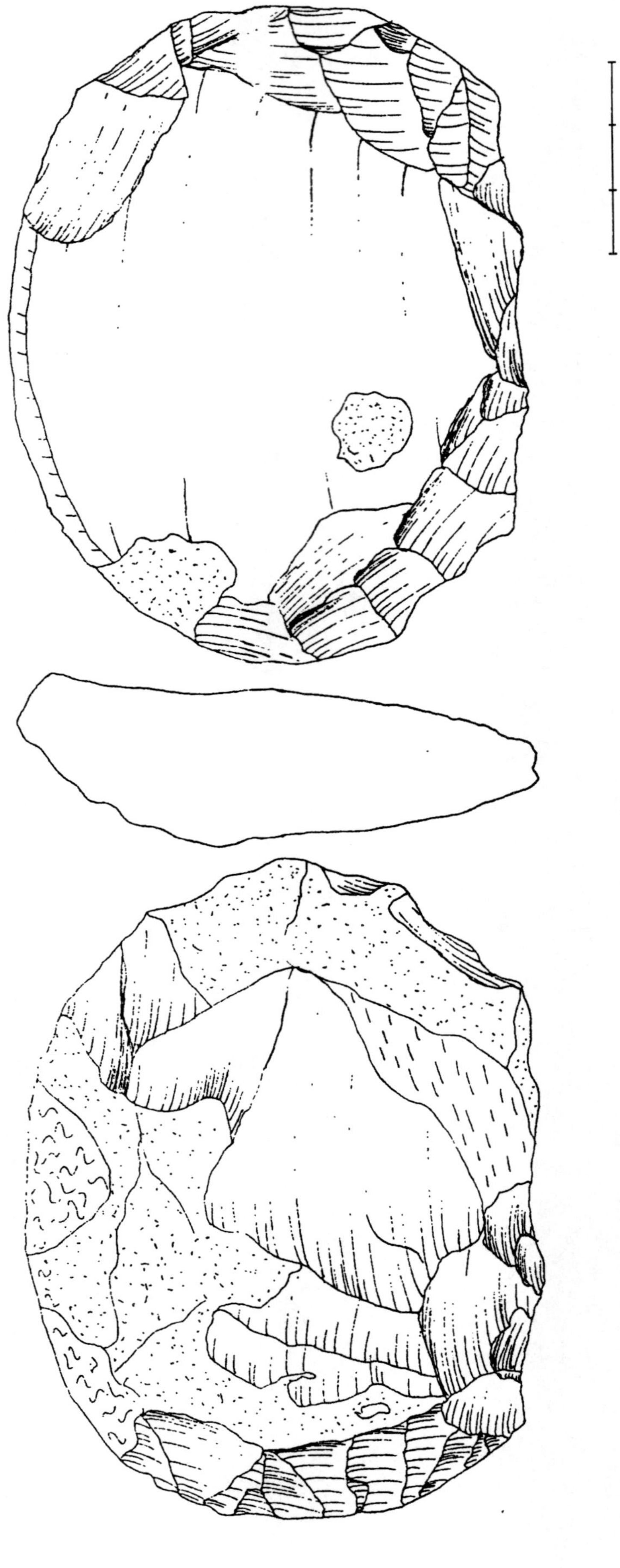

Figure 1.6. Stara Zagora-Azmak. A macro end-scraper.

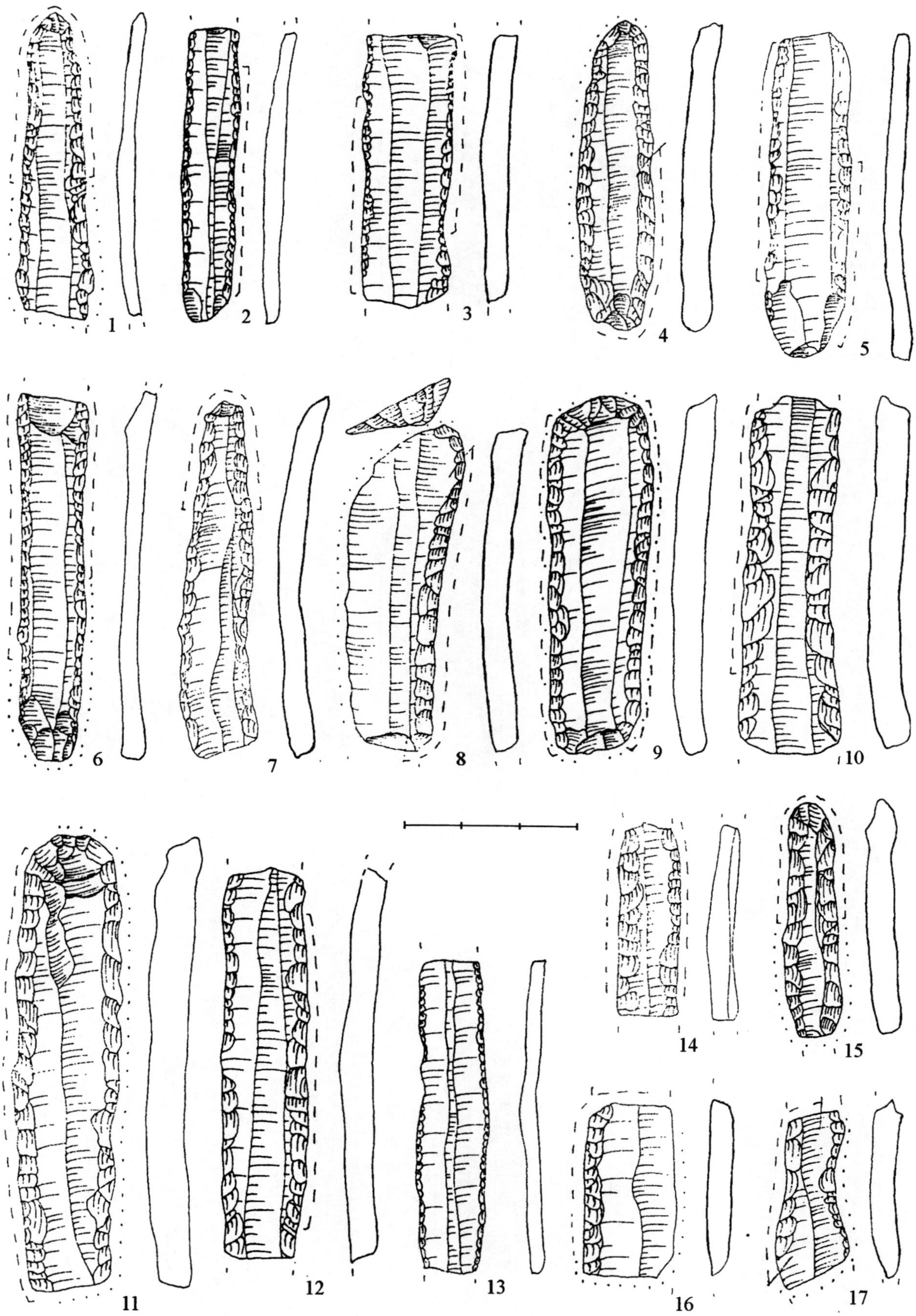

Figure 1.7. Stara Zagora-Azmak. Retouched tools.

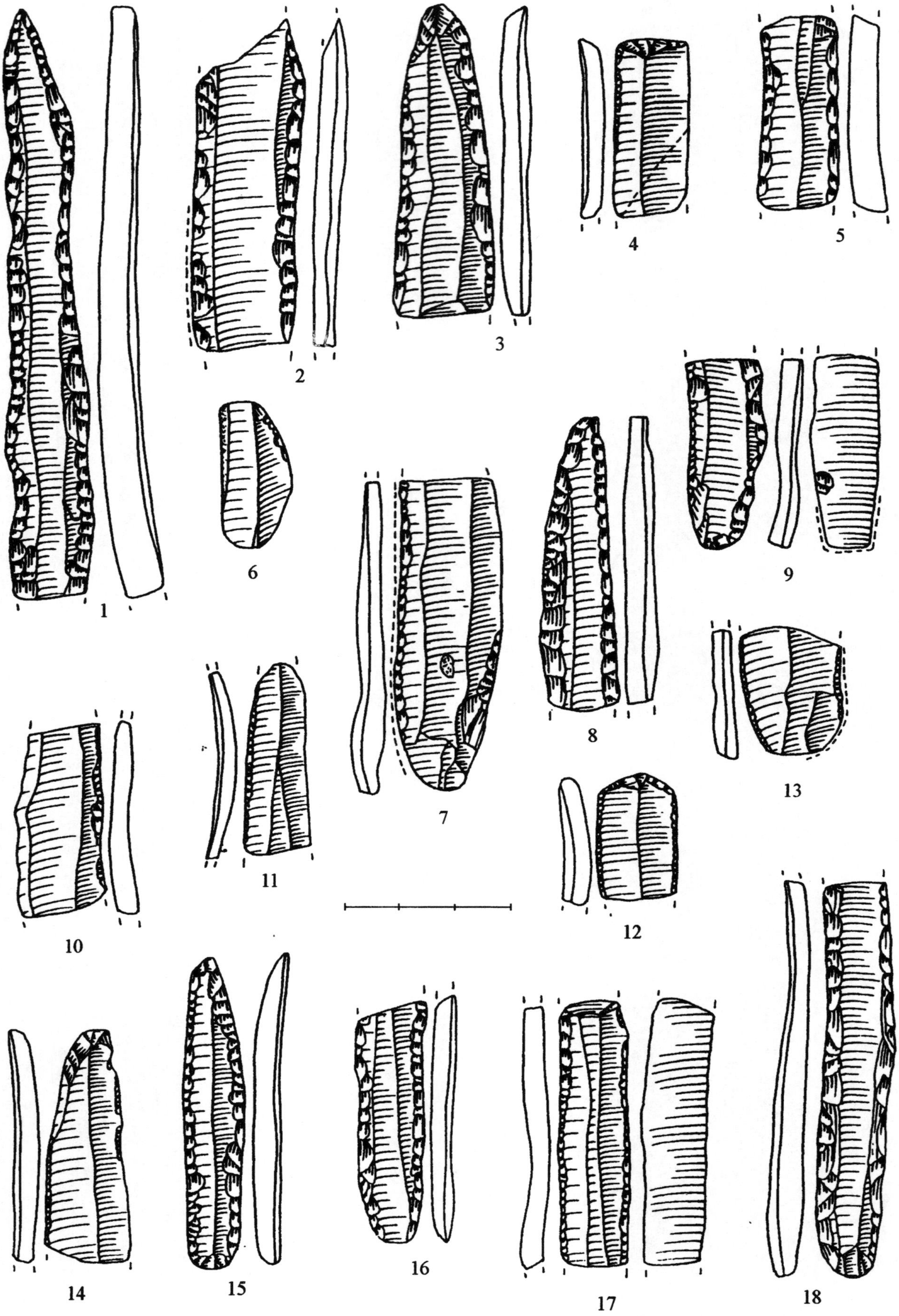

Figure 1.8. Retouched tools from Rakitovo (1-3, 9), Balgarchevo (4-8), Sapareva Banya I (10-12, 15), Sapareva Banya II (13), Sapareva Banya III (14,17) and Sapareva Banya IV-VII (16,18).

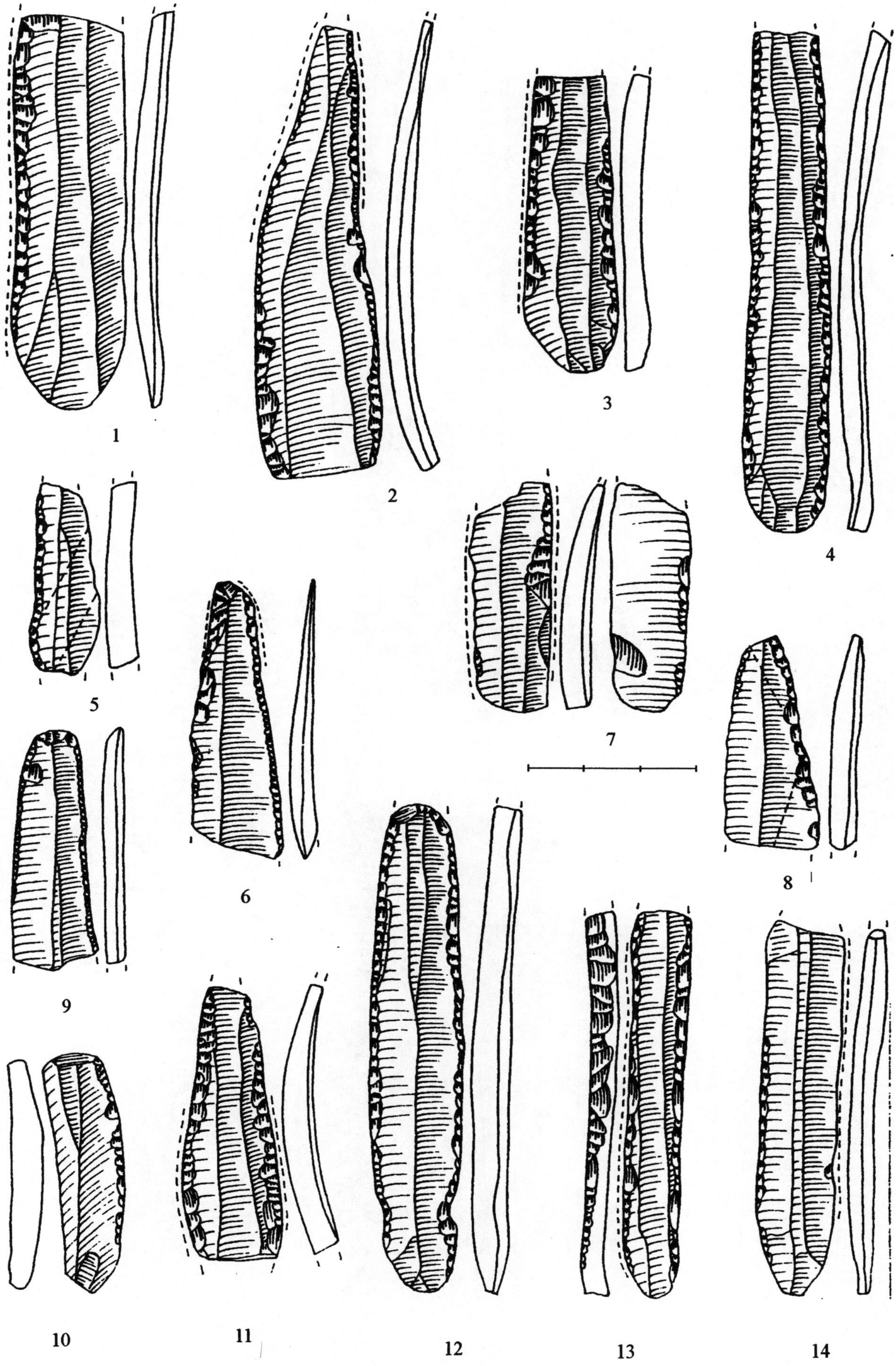

Figure 1.9. Retouched tools from Galabnik I (1-8) and Galabnik III (9-14).

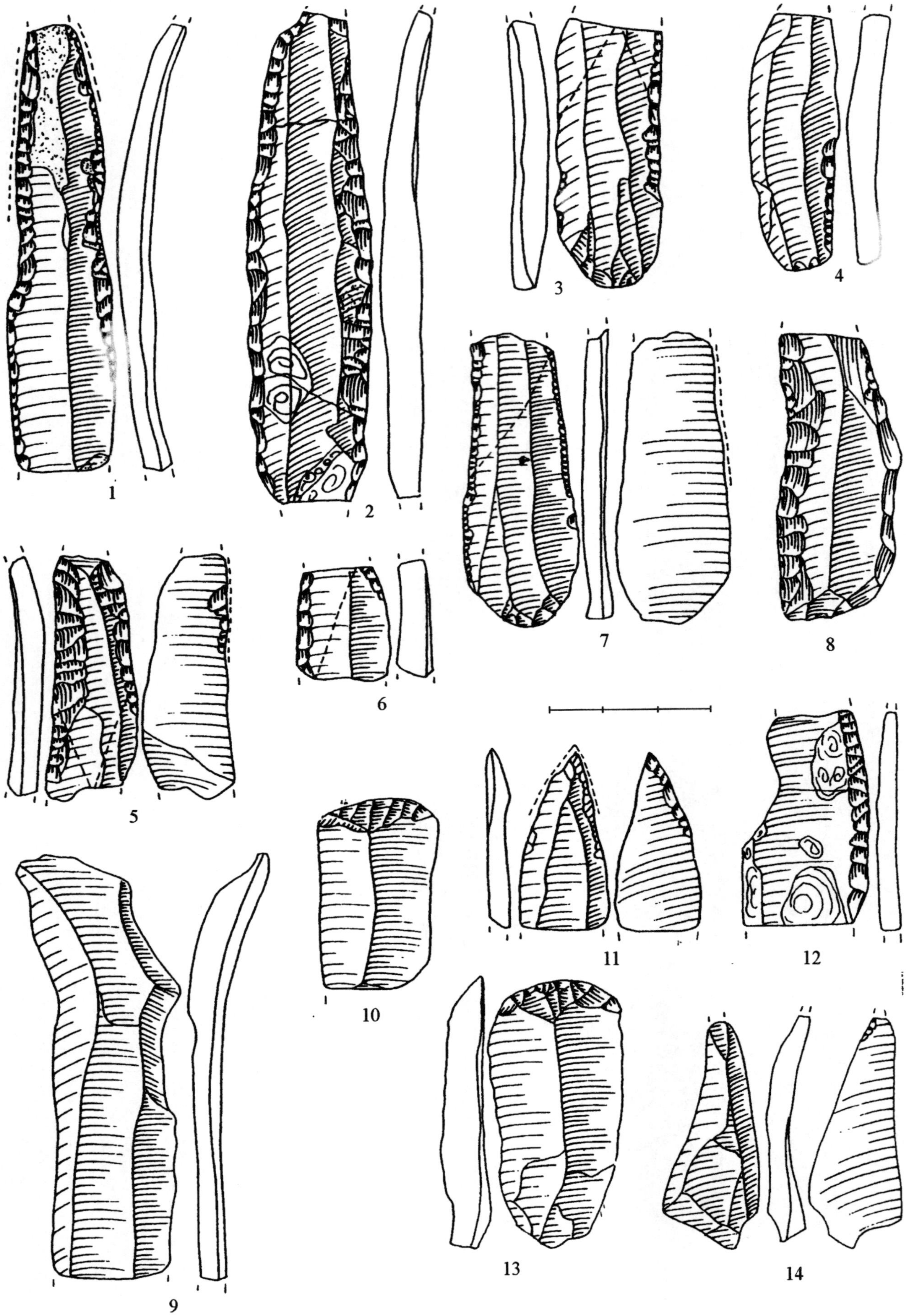

Figure 1.10. Retouched tools from Pernik (1-3,5,7-9,11,12), Galabnik II (4,6) and Strumsko (10, 13-14).

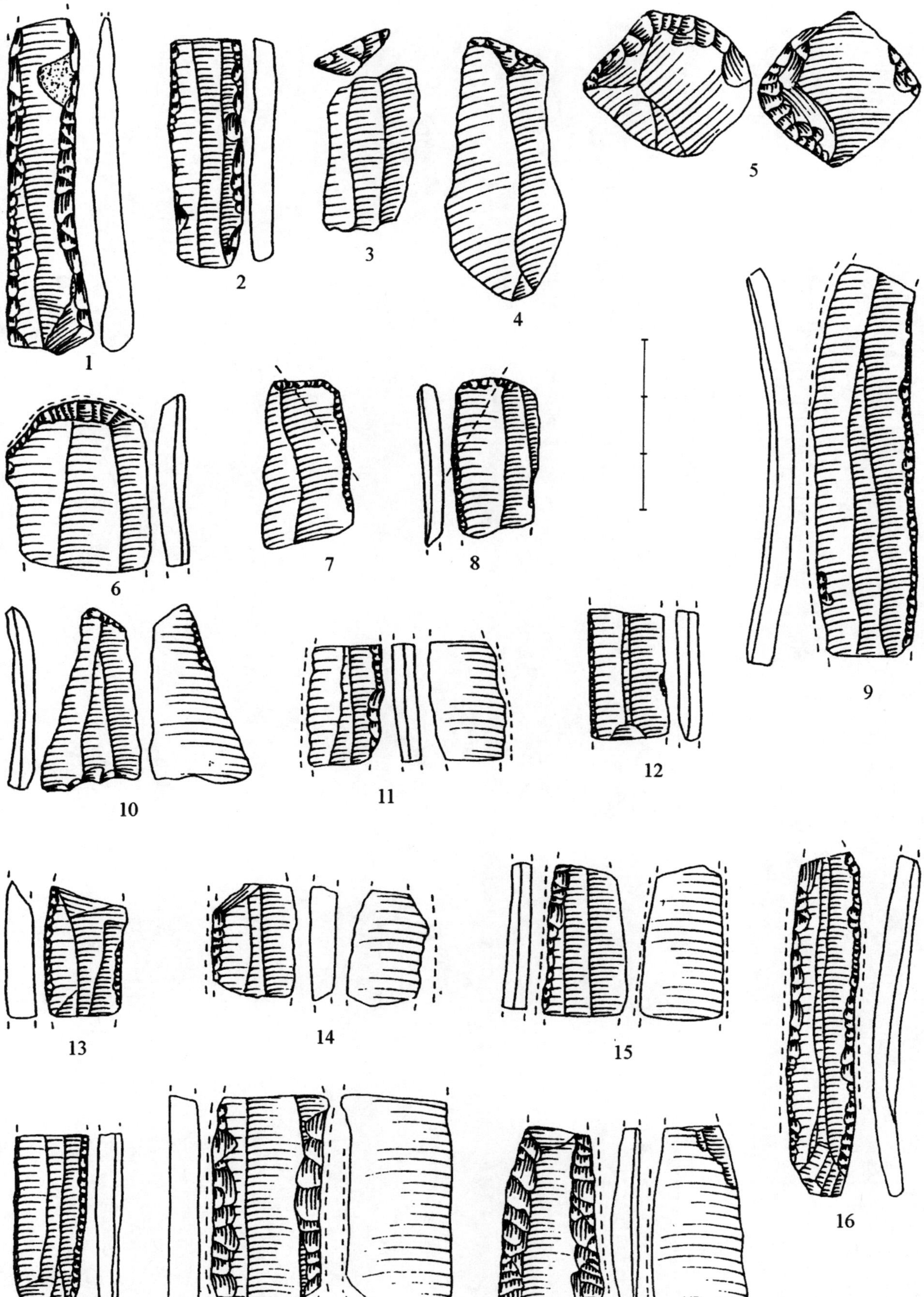

Figure 1.11. Retouched tools from Slatina IV (1-5) and Slatina I (6-19).

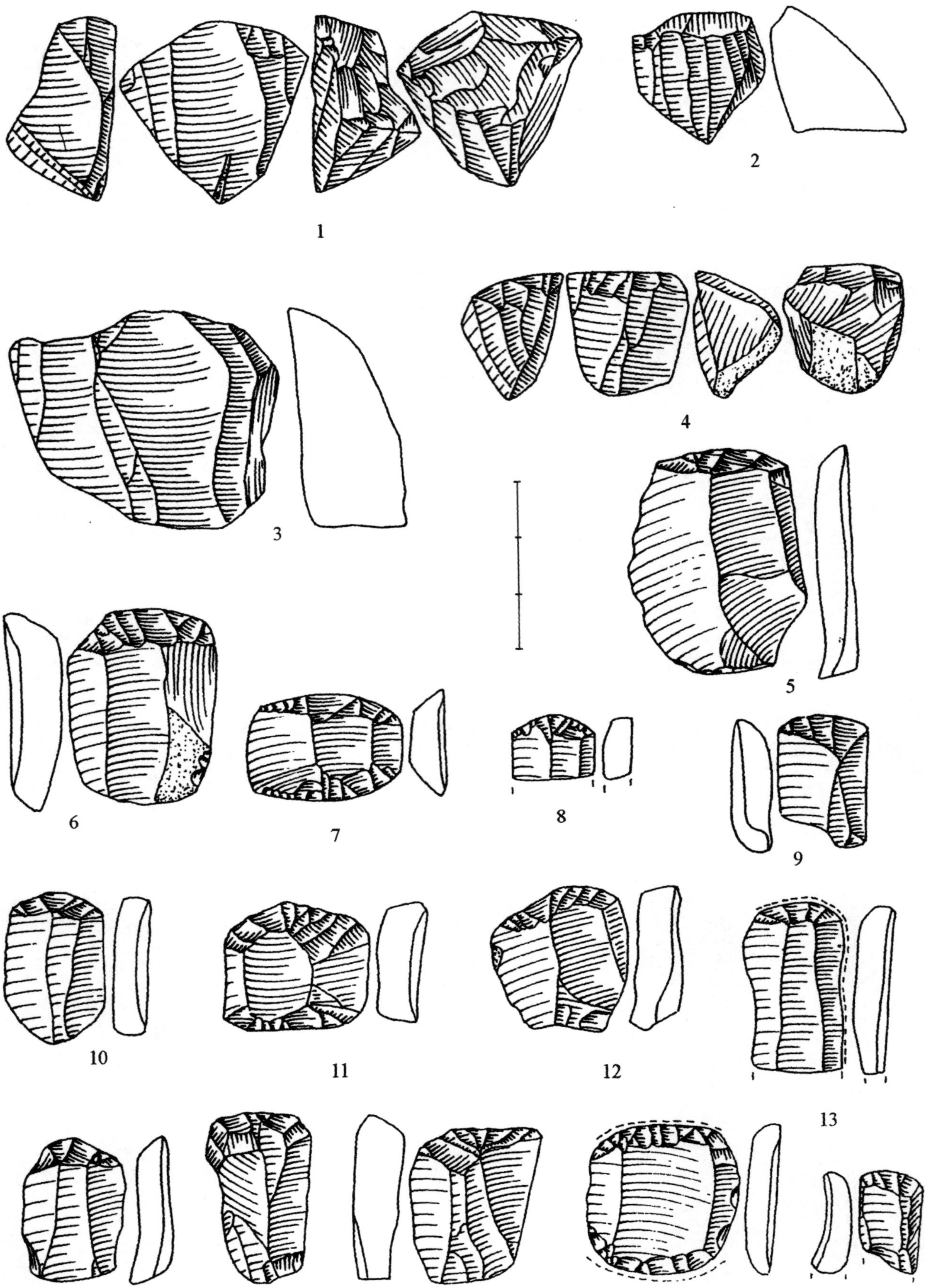

Figure 1.12. Topolnitsa: Cores (1-4) and retouched tools (5-18).

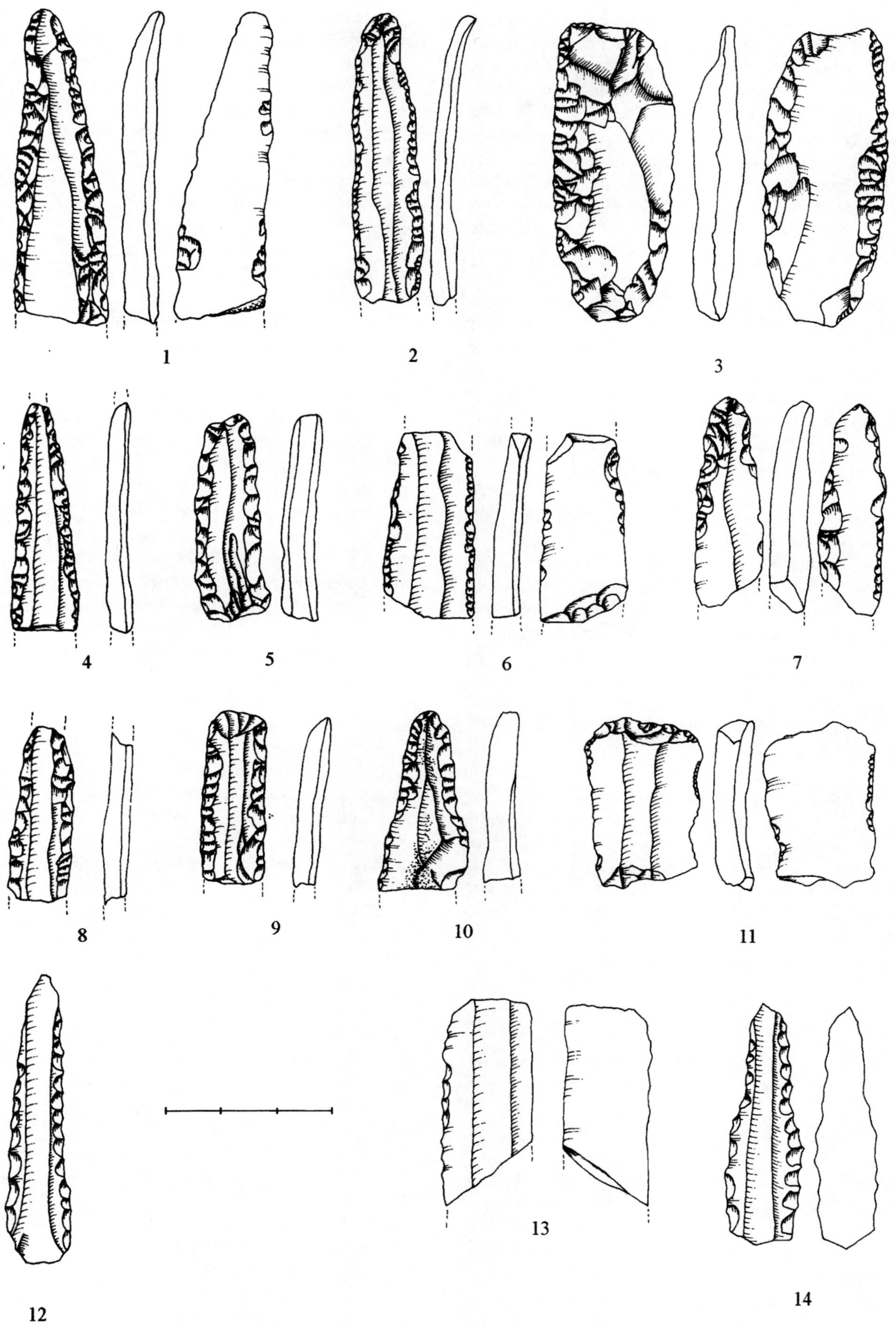

Figure 1.13. Kapitan Dimitrievo: Neolithic period (1,2), Early Neolithic (4,5,8), Late Neolithic (6,7,9-11) and Late Chalkolithic (3); Chavdar (12-14).

raw material and blank structure. Artefacts with smaller dimentions from low quality flint have beeen registrated. In Late Neolithic chipped stone assemblage the frequency of flakes also increased. In South Bulgaria, for the time being, these changes can be seen at Kapitan Dimitrievo site (Gatsov 1999a: 115-23) – ***Figure 1.13: 1-11***. The excavations there were recently renewed and published. The recent sequence documented contains materials from the Neolithic to the Chalkolithic periods: Early Neolithic and Late Neolithic layers, and a Late Chalkolithic layer. In view of that fact perhaps there were significant changes related especially to the Early and Late Neolithic.

At Kapitan Dimitrievo, in the Early Neolithic layer the tools from macroblades with high semi-steep retouch was the most typical feature concerning chipped stone material. They were made of blade knapped from macrocores from the same high quality yellow to yelow brownish raw material. The artefacts from this type and from this layer can be compared with those from South-East Bulgaria, such as Karanovo I and II, Stara Zagora-Azmak (Early Neolithic layer), etc.

During the Late Neolithic, on the territory of South-West Bulgaria changes in the stone technology and typology can be detected in the collections from Bulgarchevo (***Figure 1.8: 4-8***), Strumsko (***Figure 1.10: 10, 13, 14***), Damyanitsa and Topolnitsa (***Figure 1.12: 1-18***). These significant changes began in the flint technology and in the system of supplying with raw material. The South-West Bulgarian assemblages are characterized by presence of microlitic cores, and flake-blade cores with changed orientation in the final phase of exploitation. The macroblade technology typical of the Early Neolithic is absent.

Up to now, it can be concluded that in the Late Neolithic period on the territory of South-East and South-West Bulgaria some important changes in the technology and typology of lithic artefacts have been registered. These typological and raw material declinations are accompanied by replacement of the Early Neolithic by the Late Neolithic cultural units. They also contribute to better understanding and building explanatory models of cultural development and changes.

North-West Turkey

The stage of researches of chipped stone assemblages in North-West Turkey

Territorial range

In this report the chipped stone assemblages from Turkish Thrace and North-West Anatolia have been analyzed (Gatsov 1999b).

This territory includes a region around the Marmara Sea, Eastern Thrace and parts of Anatolia from the Aegean to the Black Sea. Eastern or Turkish Thrace lies to the east of the Meric/Evros River. This region belongs geographically to the Pontic and the Aegean zone.

This work aims to present the first results of the analyses of chipped stone artifacts from Nord-West Turkey. The assemblages are collected or investigated from two regions – North-West Anatolia and the European part of Turkey (***Map 1.2***).

Until now the techno-typological studies have been focused on the Epipaleolithic/Mesolithic collections from the Turkish Black Sea dune sites – Ağaçli, Gümüşdere, Domali. At the present stage of researches it is important to stress, that these assemblages are of particular interest for the investigations of the local Epipalaeolithic/Mesolithic cultures in the above mentioned area.

Epipalaeolithic sites in the Black Sea Region

For the time being, Epipalaeolithic/Mesolithic collections were collected mainly by M. Özdoğan from the surfaces fossillized sand dunes along the low terraces of the Black Sea. The research leaded by him started in the middle of 1970s and continued until 1980. Further bellow it will be showed the main techno-typological features concerning flint stone assemblages from Ağaçli, Gümüşdere, Domali. These assemblages are very important for the investigations of the local Epipalaeolithic / Mesolithic cultures in the above mentioned area.

It can be noticed that part of the stone material is mixed with Middle and Upper Palaeolithic artefacts. On the base of their typology and distinct patination, these artifacts were excluded from the collections. Up to now the Epipalaeolithic chipped stone assemblages have been investigated from Ağaçli and Gümüşdere on the European side of the Black Sea and Gümüşdere on the Asian side (Gatsov and Özdoğan 1994: 97-120).

Ağaçli. The most prolific site is located 35 km north – north-west of Istanbul. The extension of the dune area is about 1.5 km along the coast and 1 km inland.

Gümüşdere. The site is 27 km north of Istanbul.

Domali. It is situated on the Asian side of the region, c. 45 km north-east of Istanbul.

All sites are located close to the Black Sea shore.

Eastern Thrace and the Southern Marmara region researches concerning the Neolithic sites in Eastern Thrace and the South Marmara Region

In the area of Eastern or Turkish Thrace there have been processed chipped stone assemblages from Hoca Çeşme and Aşaği Pinar. I would like to notice that the work with the material from Pendik and Fikir Tepe and some others collections have not been finished yet (Özdoğan 1983: 401-411.; 1989: 201-215). Because of this reason I will present the first results from Hoca Çeşme and Aşaği Pinar only.

Hoca Çeşme. It is a small mound covering an area of 80 x 70 m, and situated in the area of Meric/Maritsa/ Delta, around 5 km inland from the Agean coast and 5 km east of the Enez District center. The excavation were carried out by M. Özdoğan from 1990 to 1992.

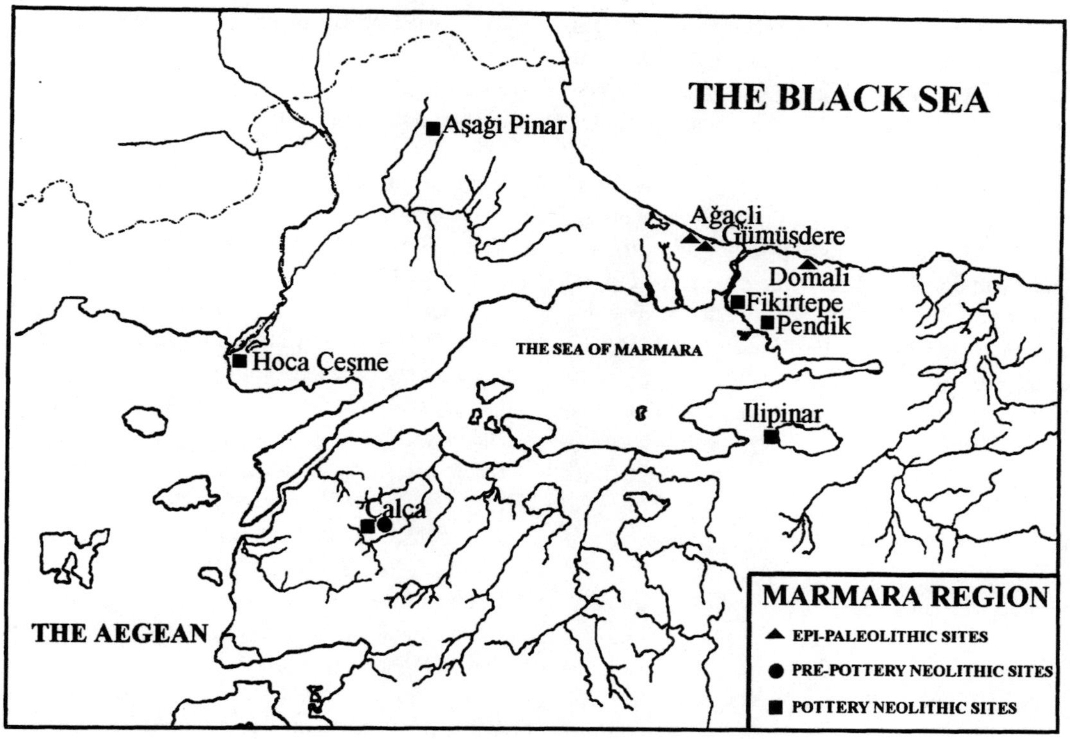

Map 1.2. Epipalaeolithic, Pre-Pottery and Pottery Neolithic sites in North-West Turkey (after Özdoğan and Gatsov 1998).

Aşaği Pinar. The Neolithic site is located in Turkish Thrace at the southern part of the town of Kirklareli and around 50 km directly from Turkish-Bulgarian border. The excavations started in 1993, directed by M. Özdoğan and H. Parzinger.

Up to now the approximately synchronizations with Karanovo phases are as follows:

Aşaği Pinar /AP/ 1 with Karanovo V, AP 2 to Middle and Late Karanovo IV, AP 3 with Early Karanovo IV, AP 4 with Karanovo III, AP 5 with Karanovo II/III transition and AP 6 with Karanovo II. The average calibrated dates for Aşaği Pinar are: Aşaği Pinar 3 – 5080 BC – 4900 BC, Aşaği Pinar 4 – 5250 BC – 5080BC, Aşaği Pinar 5 – 5350 BC – 5250 BC.

The Aceramic Neolithic period in Western Anatolia

From this area an analysis of the flint chipped stone assemblages from Çalca and Muslu have been processed. The sites were recovered by M. Özdoğan in 1988 and 1989. Both the sites yielded exceptional lithic artefacts and very interesting samples of raw material like flint and obsidian.

Çalca. It is located in the Can district. The chipped stone artifacts were collected by M. Özdoğan from one surface – 250 x 150m.

Muslu. It is situated in Manyas district. The material was gathered by M. Özdoğan in an area 300 x 300 m.

The Neolithic period in Western Anatolia

Illipinar. The site is located in the plain of Iznik Lake, around 2 km west of the lake and above 1,500 m south of the town of Orhangazi (Roodenberg et al. 1989-1990). The archaeological investigations had started in 1987, carried out by J. Roodenberg.

The following phases have been distinguished with their chronological equivalent: VB – 5450 BC – 5500 BC; VA – 5600 BC – 5500 BC, VI – 5675 BC – 5600 BC; VII – 5725 BC – 5675 BC; VIII – 5800 BC – 5725 BC; IX – 5875 BC – 5800 BC; X – 6000 BC – 5875 BC.

Techno-typological features of the chipped stone assemblages

Epipalaeolithic/Mesolithic transition. Ağaçli, Gümüşdere, Domali. Techno-typological features of the chipped stone assemblages

As main features of the technology and the most common core type was a single platform with semi-rounded flaking surfaces for blades and bladelets. End-scrapers on flakes belong to the group of retouched tools including specimens with microlith dimensions, backed pieces, perforators, retouched blades. Among the retouched tools end-scrapers prevail, usually with a length up to 2cm. Backed pieces, blades and flakes, are presented significantly of the above mentioned chipped stone assemblages. Most typically, the core processing techniques were linked with detaching of single platform conical cores. This type of exploitation was very common for chipped stone collections from the very end the Pleistocene and the beginning of Holocene. Up to now, the materials from Ağaçli, Gümüşdere and Domali has been considered as an evidence of the existing of local Epipalaeolithic cultures in that part of the Black Sea shore at the end of the Pleistocene and at the beginning of Holocene (Koen 1992) – ***Figure 1.14: 1-21***.

The other interesting techno-typological features of these collections shown are very significant technological differences as regards to the Epipalaeolithic assemblages from the Bulgarian Black Sea shore and the Early Neolithic ones in Thrace. At the present stage of investigations, the materials described above from Dikilitash could be linked to the Epi-Tardi-Gravetian tradition in the Iron Gates region of the Danube River (Gatsov 1993). The base of this hypothesis is on the one hand, 'plate' cores and cores with multiple changed of orientations have been registered. On the other hand, a very significant technological gap between Bulgarian and Turkish Black Sea Epipalaeolithic collections and Thracian Early Neolithic ones can be detected in relation to the Thracian Early Neolithic sites. As far as the Thracian Early Neolithic assemblages and especially from Upper Thrace concerns, the most characteristic distinguishing feature is the presence of macroblades from a raw material of a very high quality.

Techno-typological features of the chipped stone assemblages

Neolithic chipped stone industries from Hoca Çeşme

The assemblages which have been processed, are characterized of a relatively low level of technology, and typological monotony. The main reason of that is perhaps a homogenous raw material structure and almost the single type of blank as flakes. Of more than 3,000 pieces only c. 250 are blade fragments. The quantity of retouched tools is less than 130. Among them tool fragments are more than 15%. The group of retouched tools includes the retouched flakes, some types of end-scrapers, blade fragment with a partial marginal retouch. The changed in the raw material, blank and typological structures have not been registered in the particular units. The raw material is presented by microcrystaline quartz, chert, few quartz with local origin. The described types of raw material are not good for knapping. A few pieces of obsidian in shape of bladelet fragment have been found.

It is worth underlining that in phase II of Hoca Çeşme few macro blades with high semi steep and steep retouch appear (***Figure 1.5: 1, 2, 4, 5, 7, 9-12, 16***). The blank was obtained from same type of raw material characteristic of the Early Neolithic macro blade assemblages from Upper Thrace. Liked to South-East and South-West Bulgarian Early Neolithic sites, the lack of cores, trimming and cortical flakes and blades from the same raw material suggest the blades were brought off the excavated area. That kind of raw material, the type of retouch and the size of the blades are similar to those from the Early Neolithic horizon in Bulgarian Thrace. It can be stress that the second phase from Hoca Çeşme corresponds chronologically to the Early Neolithic period in Upper (Bulgarian) Thrace. Namely in this area the described technology can be registered from the Early to the Middle Neolithic in the region – between 6000 BC and 5500 BC/5450 BC.

Among the stone materials at Hoca Çeşme the cores and core fragments, trimming flakes, cortical flakes, and chips from the quarz and chert were distinguished. Perhaps, the knapping of local kind of raw material took place in the excavated area during the all four phases of this site.

Neolithic chipped stone assemblages from Aşaği Pinar

The site is still in process of research (Benecke 1998: 172-179; Gatsov 1998: 1-6; Parzinger and Özdoğan 1995: 5-29; Özdoğan et al. 1997: 2-11), so below it is to be presented some preliminary conclusions only.

The receiving of blank was based on the exploitation of small flint nodules and plate concretions, whose length is usually between 5-7cm.Very typical of this category are plate concretions for blades and bladelets. As a rule, the striking surfaces are situated on the narrowest side. During the excavations the artifacts with cortex like cortical flakes and blades, trimming flakes and blades, and waste have been found. It suggests that the exploitation of cores were on the spot, in the excavated area. The other interesting feature concerning the Aşaği Pinar assemblages is that no considerable changes in the artefact dimensions have been observed. Most of the blades and bladelets were obtained using soft hammer stone and punch (Binder and Perles 1990: 257-283). The following types like end-scrapers, perforators and alternated perforators, retouched blades, retouched flakes, splintered pieces and truncations also correspond more or less to the classical definition for these tools.

As a very characteristic concerning typological stricture it can be shown the occurrence of microlithes. To these group is related consistence of end-scrapers, perforators and alternated perforators, segments and retouched bladelets (***Figures 1.15: 1-23***).

Microlithic tools. This group includes micro end-scrapers, micro-perforators and segments. It can be stress that there are no geometrical microlithes such as trapezes and triangular.

Microend-scrapers. To this group belong tools with a diameter up to 25 mm. The tool fronts were done usually by semi-steep or steep retouch. Very often the specimens are on cortical flakes or with traces of cortex on the dorsal surfaces.

Micro perforators and micro alternated perforators. Most of the tools have a length from 12 mm to 22 mm, rarely, between 23 mm and 25 mm. None are smaller than 11 mm. Segments and truncation were made on blades or bladelets. They were registered in single examples in the flint assemblages.

About the link among the raw material and the type of retouched tool it is worth noticing another interesting feature. The last concerns namely the relation of microtools and raw material. Mostly the micro end-scrapers was made from yellow flint of high quality, with or without spots. At the same time, most of the micro perforators and alternated perforators were shaped from chert-striped brown and black or striped brown and gray raw material. This kind of raw material is with homogenous structures, with less or more inclusions, with shell break surface and is not very good for knapping. These tools were used for fabrication of malachite beads (Özdoğan and Parzinger 1998, personal information). The last are with holes responding to the diameter of the microperforator points and also beads with semi finished holes. This fact proves that the beadwork shops were operating at least in phase III and IV (Calley and Grace 1988: 69-89).

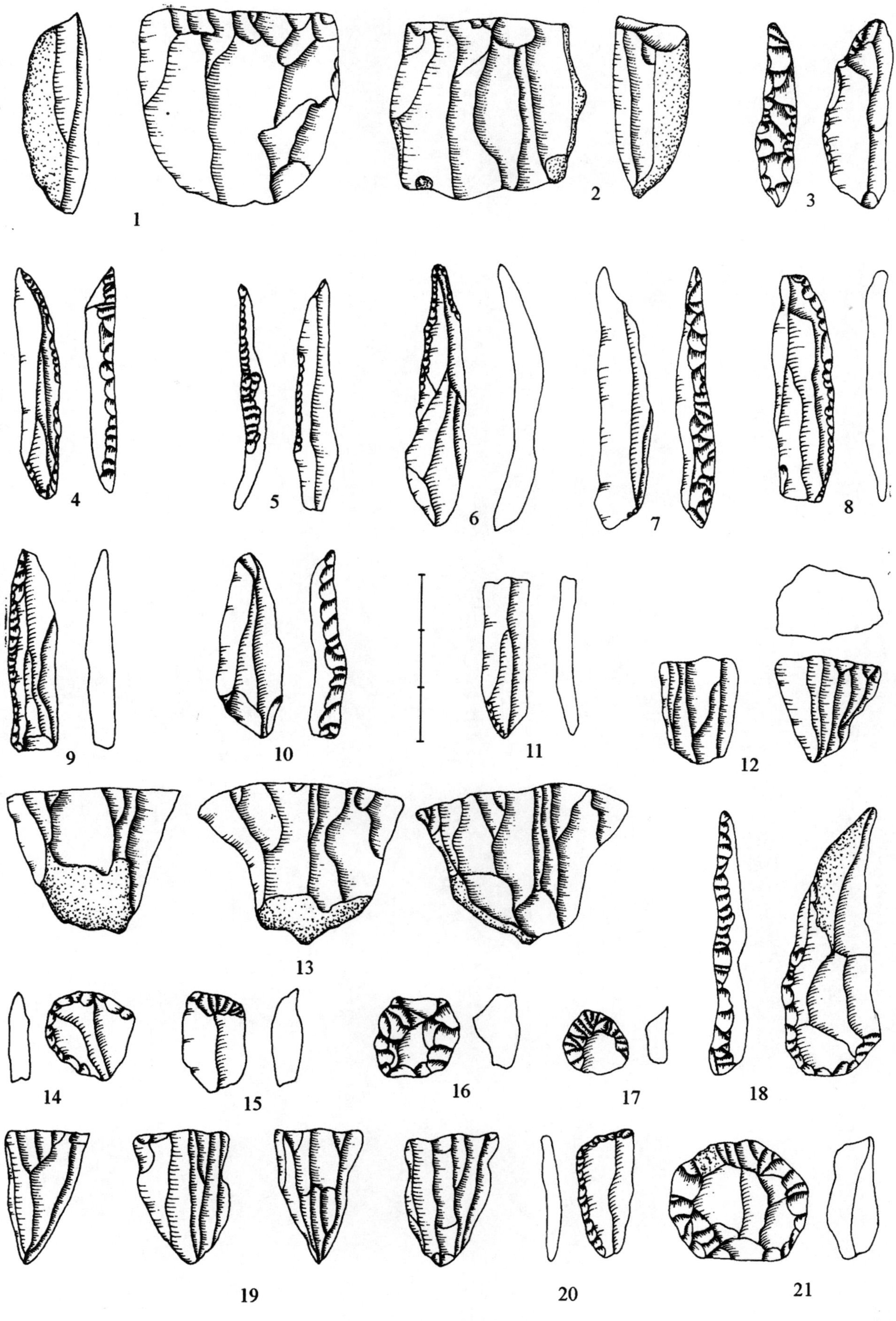

Figure 1.14. Epipalaeolithic sites from Turkey. Ağaçli: Cores (1,2,13,19) and retouched tools (4-11,14-18,20,21); Gümüşdere (3).

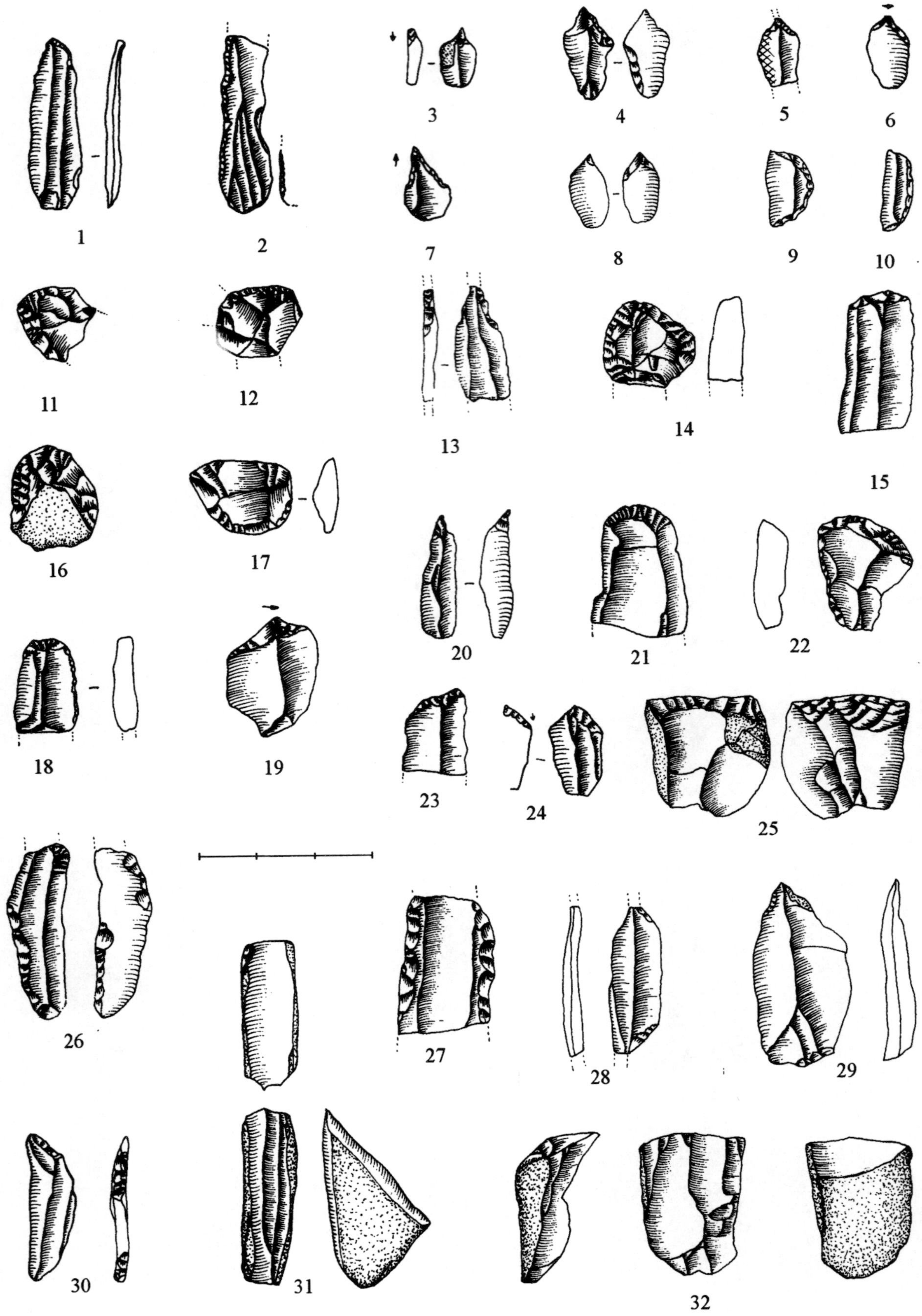

Figure 1.15. Aşaği Pinar: Retouched tools (1-30) and cores (31-32).

Very interesting is also the occurrence of single blades with high semi-steep retouches in assemblages in the assemblages from periods IV and V at the site. As it was mentioned above, this type of retouched blades is very diagnostic of the Early Neolithic period in Thrace. Accordingly, at Aşaği Pinar it might be expected appearance of earlier Neolithic stone materials.

The Aceramic Neolithic period in Western Anatolia

Chipped stone assemblages from Çalca and Muslu (Özdoğan and Gatsov 1998: 209-232.). The material from this site was published recently. That why further below it will present only some main techno-typological features concerning these collections.

The chipped stone assemblages from Çalca and Muslu concerned around 3,000 artefacts unevenly distributed in two assemblages – Çalca and Muslu (***Figure 1.16: 1-7***). The presence of cores and trimming flakes, cortical flakes and chips suggests that the flint raw materials were used on the spot. The core processing included adapting of platforms, striking surfaces and sides. On the base of the sizes of trimming flakes retouched tools, flakes and blades it suggests that these cores had been at least 12-15 cm long. It should be stressed that macro tools as end-scrapers have been registered. From functional point of view, the last could be referred to the category of the mining tools.

To the group of retouched tools the end-scrapers, scrapers and notched tools were related mainly. The obsidian artefacts are presented by single examples in shape of microcore for blades and bladelets (Cauvin 1994: 15-22; Özdoğan 1994: 423-31).

These sites yielded very interesting material in a very characteristic context. Then, they where related to pre-pottery ones. Of course the evidence from both the sites is still very premature. The suggestions need further confirmation from the excavations, especially from sites Kececayir and Kabakli, province of Eskisehir, which were recovered recently by T. Efe.

Neolithic sites from West Anatolia

Chipped stone industries from Illipinar

As the others chipped stone assemblages in North-West Turkey – those from Illipinar are still in process (Roodenberg, et al. 1989-1990, 61-144; Roodenberg 1995). Further bellow it will be presented only some preliminary information about the earliest phases, X and IX. The main features concerning these collections can be stressed as follows. By the raw material the very big quantity of samples have been registered. The last are more than 50.The samples were determined among more than 3,600 artefacts from all the phases. The samples are different kinds of flint (microcrystaline quartz) with different color. As a rule, they are with a good break surface, semi transparent or not transparent, more or less homogenous structures. The raw material distinguished at this site is good for knapping. During the surveys in the surrounding areas a similar kind of raw material (except obsidian) was found. The last is around 10% in phases X and IX and getting less in the next phases.

Among the debitage in X and IX phases the flakes prevail. Main part of the tools was also done on massive flakes (***Figures 1.17: 1-16; 1.18: 1-6; 1.19: 1***). These are big high end-scrapers or semicircular or circular more or less flat end-scrapers, massive retouched flakes, perforators The sizes of the blank and the retouched tools and suggest the existence of cores, which should be at least around 15 cm long. These massive cores as well as pre-cores were not registered in the area under the study.

Simultaneously the presence of relatively small single platform cores was established. These are examples with flat or semi rounded striking surface for blades and bladelets. It can be underline that the width of this type of blank and the scars of the width on the flaking surface is the same.

In the case of Illipinar, it is possible to suppose two types of core processing. The first one is links with exploration of massive cores somewhere outside of the excavated area. The second can be related with detaching on spot of single platform semi conical cores for narrow blades. In this case punch or haft hammer stones were used. According to wear traces analysis, in phases X and IX mostly blades without retouches were used (personal information of M. Gurova).

Conclusions

Summarizing the basic technical and typological feature of the above mentioned assemblages it the area under the study, some general points can be drawn.

The first is the problem with Pleistocene / Holocene transition. Up to now, a main difficulty in this field is a lack of stratified sites in the area under investigation. To resolve the problem at some extend, intensive paleoecological research also requires.

In Bulgarian lands, from technological and typological point of view, the material from Dikilitash has been related to the Epi-Tardi-Gravettien tradition of the Iron Gates assemblages. In North-West Turkey up to now the Epipalaeolithic / Mesolithic collections from the Black Sea shore like Ağaçli, Gümüşdere and Domali have been considered as a prove of existing of local Epipalaeolithic / Mesolithic culture in this area.

Another problem faces the techno-typological gap between the Dikilitash material and monochrome stone collection from Koprivets.

The technological gap also exists between the materials from Dikilitash, Ağaçli, Gümüsdere and Domali and the assemblages belonging to the white-painted horizon in Thrace (Bulgarian and Turkish parts).

The similar differences in the technological and typological characteristics exist among Koprivets and the stone collections from white-painted Neolithic in South Bulgaria.

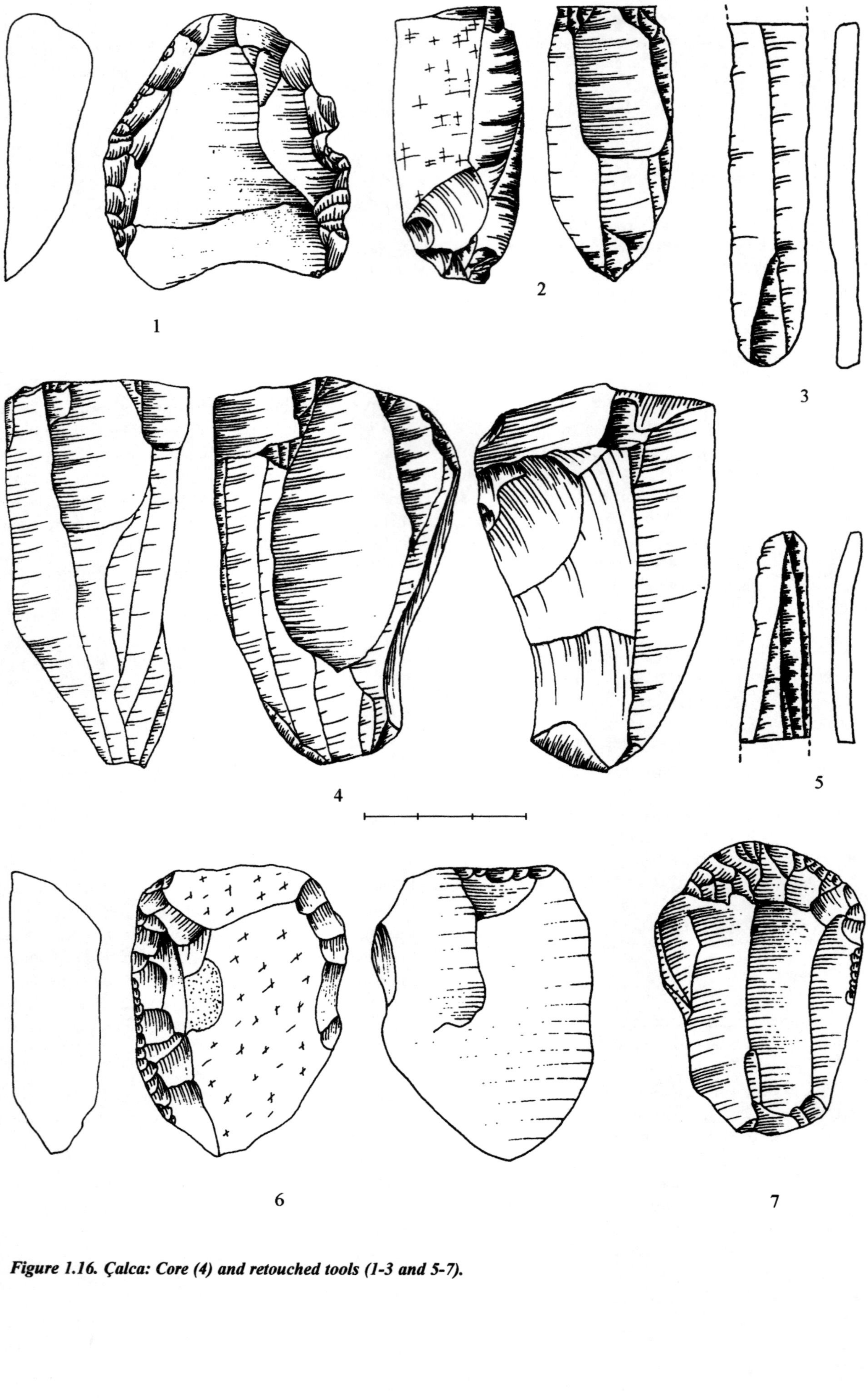

Figure 1.16. Çalca: Core (4) and retouched tools (1-3 and 5-7).

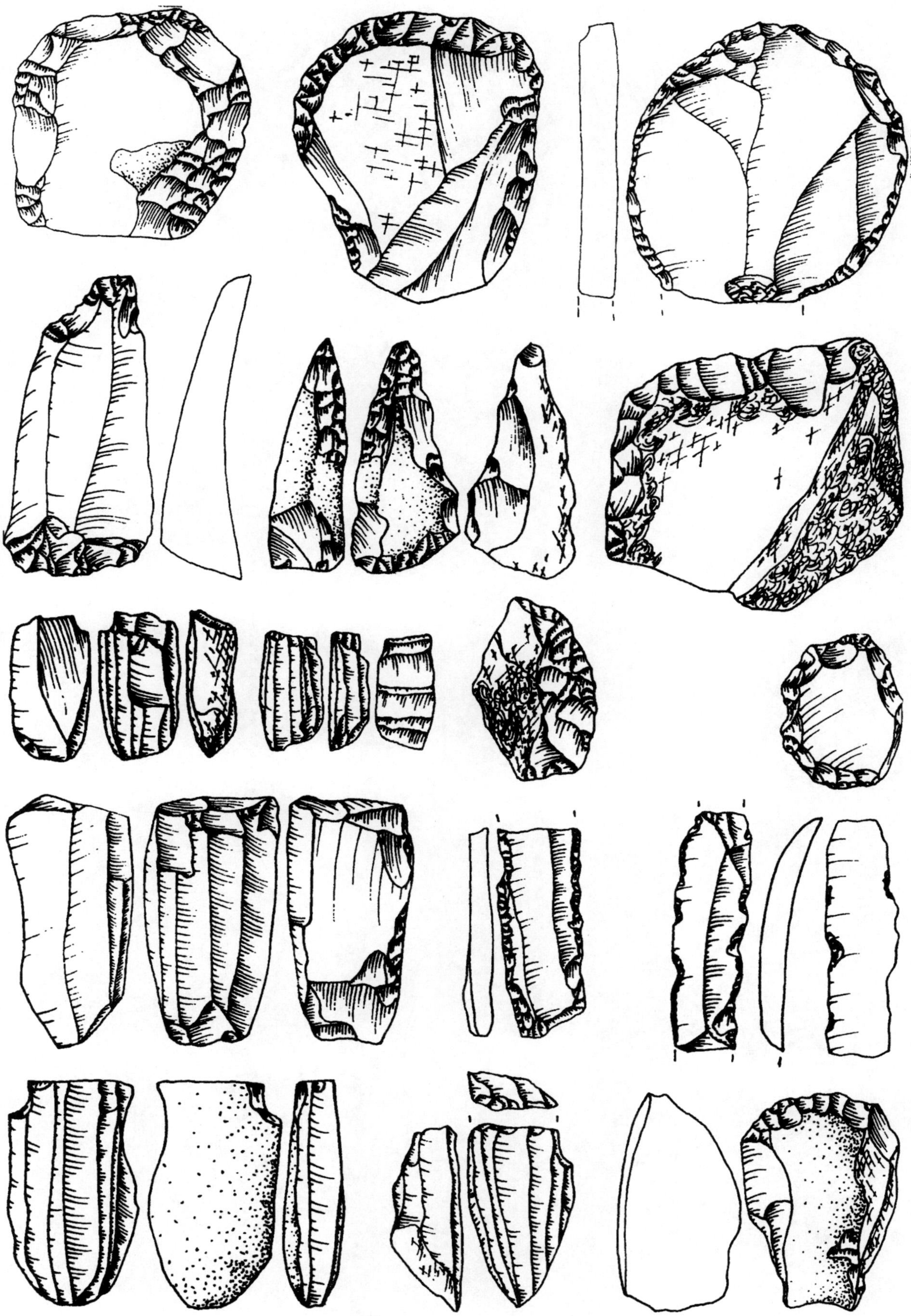

Figure 1.17. Illipinar: Cores (7,8,11,14,15) and retouched tools (1-6,9,10,12,13,16).

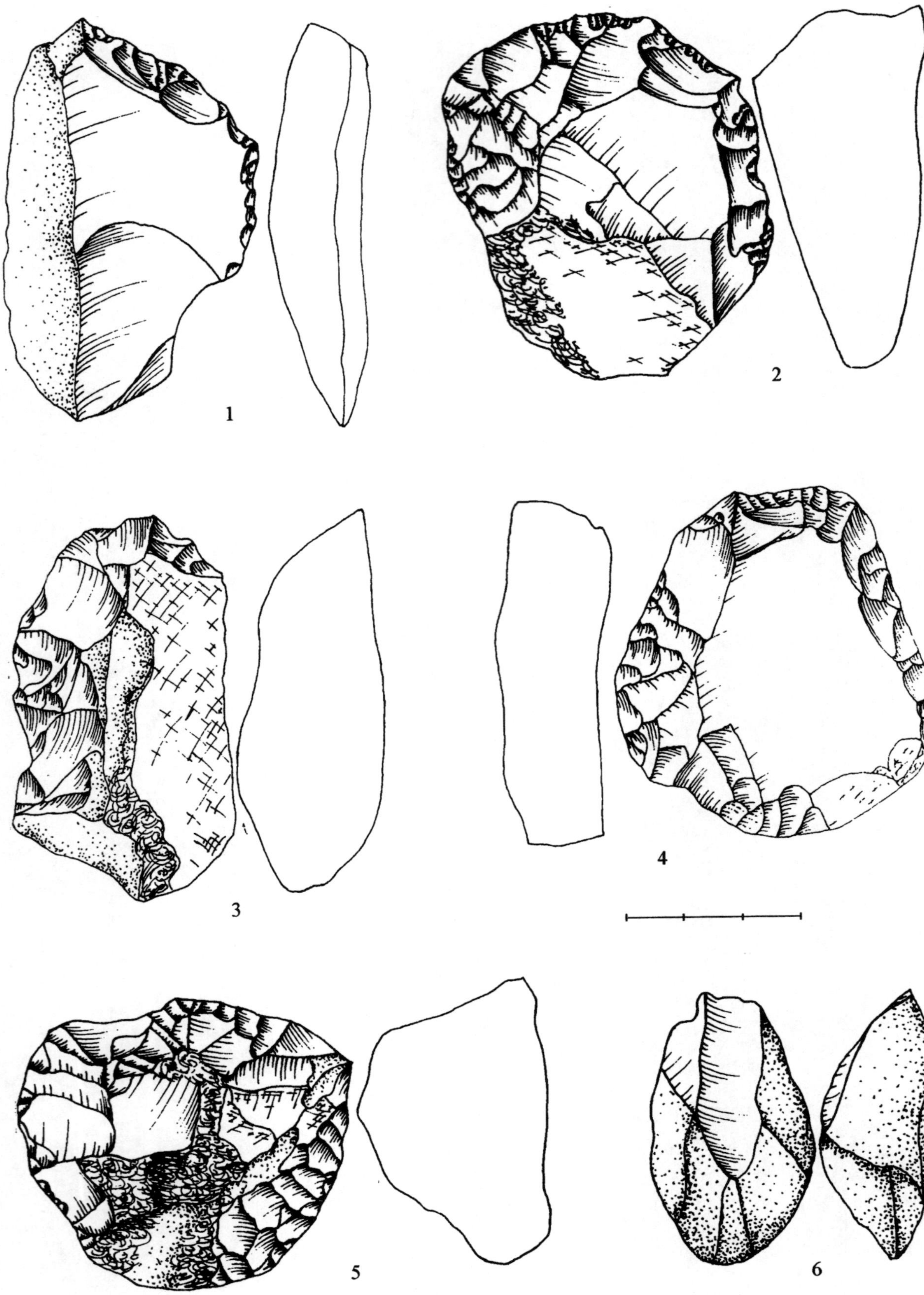

Figure 1.18. Illipinar: Retouched tools.

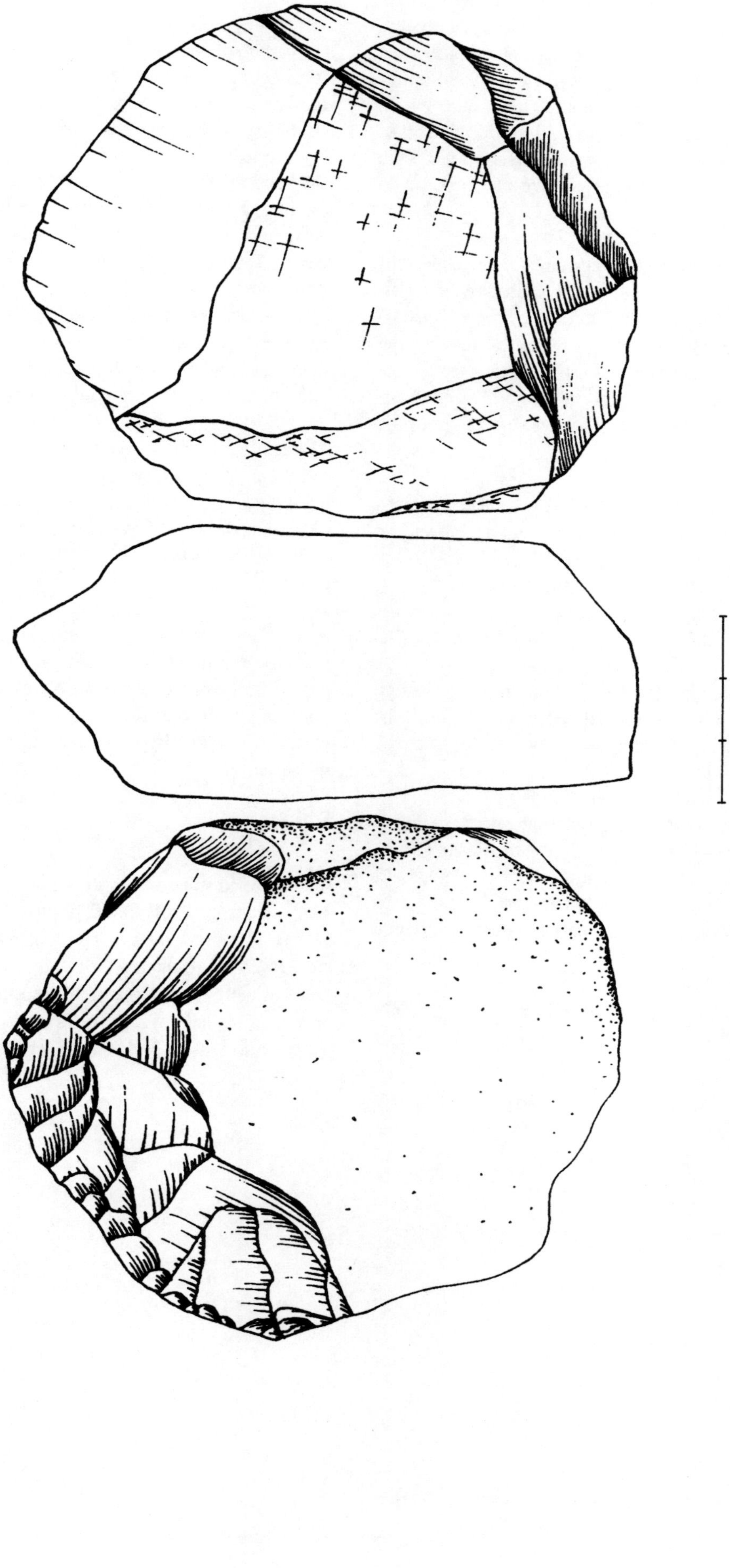

Figure 1.19. Illipinar: Macro end-scraper.

For the time being, in Upper Thrace the first Neolithic chipped stone assemblages belong to the phase or horizon of white-painted ceramics. The main distinguished feature of these assemblages is the macroblade technology and the type of a raw material. At all excavated areas in South Bulgaria blade cores, trimming flakes and cortical flakes have not been documented. This fact suggests an initial preparation of those macrocores and knapping the blades had taken place off the sites. Similar situation can be seen at Hoca Çeşme. In phase two, which corresponds chronologically to the Early Neolithic period in Upper (Bulgarian) Thrace macroblades have been found. Like in Upper Thrace, they were made elsewhere and brought to the site.

Perhaps during the Early Neolithic period in Thrace the operation chain in the area under the study includes the existing of the workshops close to the outcrops of the above-descried type of raw material. Because of lack of seaving the other ring of this chain – the place of tools-making is still not clear. But having in mind my work in Drama, South Bulgaria, I suggest that the ready-made tools were brought to the sites.

About the typological structure of the all assemblages from the white painted Early Neolithic horizon it can be showed its typological monotony (Kaiser and Voytek, 1983: 323-53). Most of the tools are retouched blades with high, semi steep or steep retouch slightly undulated. They can be considered as a most characteristic and diagnostic type in all collections.

As a rule the blades are linked with the same advanced stage of core exploitation. These standardized blade suggests of existing of groups with certain degree of specialization in the Early Neolithic period in Thrace.[1]

Another problem is the root of this macrotechnology, which took place in the first half of sixth millennium BC in Thrace and in my opinion also in Greek Thrace (Gatsov 1993). At this stage of investigation it can be considered Anatolia as a fatherland of this technology. But the archaeological evidence is still not enough.

Last but not least is the problem of disappearance of the technology under discussion. As it was mentioned, a typological declination and changes in raw material supplying have been registered in Late Neolithic period in the region. Simultaneously, these changes are registered with the replacement of the Early Neolithic by the Late Neolithic changes in cultural units.

The other phenomenon is the very high degree of microlitization in all periods of Aşaği Pinar, Turkish Thrace. Up to now the microlithes were considered as a main characteristic technological feature of the Mesolithic. At the Neolithic Aşaği Pinar around one thousand micro tools – mainly micro perforator and micro end-scrapers were found. The micro perforators have been registered with malachite bead, including semi finished ones. After the recovering of M.Özdoğan and H. Parzinger it is sure that workshops functioned at this site. Similar micro perforators are found in Anatolia (Calley and Grace 1988: 69-89). Having in mind the distance in the space and in the time, it is more probably at this stage of research to consider this type of tool first of all to have been a result of some functional acquirements (Calley and Grace 1988: 69-89).

About the pre-pottery Neolithic period up to now, the material from Çalca and Muslu from West Anatolia has been related only. But the question remains open. It is absolutely necessary to confirm the hypothesis argued by excavations, especially at the archaeological sites of Kececayir and Kabakli, province of Eskisehir.

As my last conclusion, I can say that the aim of this report has been to present the first results of stone technological investigation in South Bulgaria and North-West Anatolia. But these conclusions need to be completed not only by further similar research but also by use-wear trace analysis, geological and other research.

Acknowledgments

The chipped stone assemblages were kindly offered to me, for analysis, from Prof. Dr.V.Nikolov, Prof. Dr.Ozdogan, Prof. Dr. H. Parzinger, Dr. J. Roodenberg. I would like to thank also to Dr. N. Balkan - Atli and Dr. D. Binder for consultations as well. My work and this report also have been possible owing to the help of my Bulgarian students Peppa and Hristo Kantardjiev., New Bulgarian University, Sofia.

[1] On the subject of specialization in Chalkolithic period in Bulgaria see Manolakakis 1996: 119 –23.

Chapter 2

Knapping Technology vs. Pattern of Raw Material Supply

(On the Example of Early Neolithic Flint Assemblages from the Orlovets and Koprivets sites, Northern Bulgaria)

Tsoni Tsonev

In autumn 1997 a limited fieldwalking survey was completed together with sampling for archaebotanical analysis and ^{14}C dating of the Early Neolithic site of Orlovets. It combined the efforts of P. Stanev (excavator of the Orlovets site), Dr. J. Chapman, Dr. Ts. Popova (palaeobotanist) and Dr. Ts. Tsonev. In the present study I am using data from my own survey of the area and analysis of the lithic assemblagespart, as well as of the field survey of R. Zlateva (1997), some results of her analysis on the flint assemblage from the Koprivets site, which are available in her MSc Dissertation (unpublished).

Topography of the sites

The Early Neolithic site Orlovets is situated in the Ada Kuzu Locality, about 2.5 km northeast of Orlovets Village. It was found on the second, unflooded terrace of the left tributary to the Baninski Lom River. The Neolithic site spreads onto the terraces to the south and northeast. The attitude is about 300 m. The area surrounding the village belongs to the main water shed that divides the basins of the rivers of Jantra and Russenski Lom. The terraces are formed by Apt marls covered by thick layer of Quaternary loess. Gray, forest soil with mean thickness about 50-60 cm. developed above them (Fotakieva et al. 1976; Stanev et al. 1994; Stanev 1995). The area of the village Orlovets lies on the border between the forest-steppe zone of the Danube lowland and the lower mountain deciduous area of the central Balkan range. The landscape surrounding the site is characterized by slightly elevated heights and shallow valleys above which the locality of Orlovets dominates.

The Koprivets site is located on a higher terrace covered by thick loess layer which is slightly inclined towards the catchment area of three tributaries of the Russenski Lom River. On a straight line the Koprivets site lies 10 km northeast of Orlovets site. Since 1989 it was excavated by V. Popov and I. Vajsov (Popov and Vajsov 1992; 1996).

The present study is confined only to the earliest assemblages of both sites, which belong to the so-called Monochrome Neolithic period. A technical-typological study of the assemblage No 1, trench B from Koprivets will be presented with a stress on some particular features. The earliest assemblage from Orlovets, trench B_{22}, which consists only of few artefacts is also presented. Despite its number it shows particularities that are in contrast to that from neighboring Koprivets site.

Technical and typological description of the flint assemblage No1, trench B from Koprivets

General structure of the assemblage is as follows (in the brackets the number of the examples is given):

1. Cores (36).
2. Fragments of cores and flaked nodules (8).
3. End-scrapers (5).
4. Burins (1).
5. Retouched blades (2).
6. Blades with retouched notches (1).
7. Blades with unretouched notches (3).
8. Retouched flakes (12).
9. Flakes with retouched notches (5).
10. Flakes with unretouched notches (5).
11. Borers (2).
12. Splintered pieces (2).
13. Side-scraper (1).

Cores

There are 16 single platform cores, 8 double platform cores, 5 microcores and 6 cores with a changed orientation of flaking. All cores are of small dimensions- maximum 4 - 5 cm. in diameter. The flint concretions were collected from the alluvial deposits and gravel of the nearby river valley. The dimensions of these concretions are approximately the same size as those of the cores.

Five out of sixteen *single platform cores* are for flakes while the rest are cores for bladelets or bladelets and flakes (***Figure 2.1:** 3*). Only one of them is in an initial state of exploitation. All the other cores are in advanced and final stage of exploitation. The form of the cores varies between rectangular, trapezoidal, cylindrical, triangular. Four cores have an irregular form. Most often the flaking surface is flat and wide. Only in two cases it is convex. In two cases the striking platform is cortical; on one core it is a natural surface. In four cases the platform is formed by a single blow; one core - with prepared platform.

Traces of reutilization are observed on four cores.

Double platform cores number 9 artefacts (***Figure 2.1:** 1*). All of them are in final or advanced stage of exploitation. The form of the cores is distributed in the following way: two cylindrical cores, one trapezoidal, one oval, one irregular and the rest are pyramidal or semi-pyramidal. The form of the flaking surface is predominantly flat. On one cylindrical core the flaking surfaces are convex and go round the lateral sides. The striking surfaces are formed mainly by a single blow or are prepared. Traces of reutilization are visible on one core.

Cores with changed orientation. They number 6 artefacts. Five of them are made for producing flakes and the other one - for flakes and blades. Only one core is in initial stage of exploitation. The rest are in final or advanced stage of flaking. The general form of the cores is irregular. Only one of them has a spheroid form.

Microcores. Their number is 5. One of them is a core made on a thick flake. With one exception, all the cores were designed for obtaining bladelets. All of them are in a final state of exploitation. Two cores have irregular form, 2 - cylindrical, 1- pyramidal.

Fragments of cores and flaked nodules

There are 8 examples. One concretion and one fragment of core bear traces of flaking. The rest were made on thick flakes and after the initial flaking discarded.

End-scrapers

There are 5 examples. Two of them are made on blades (***Figure 2.1: 5, 9***) detached from single platform cores. One of them bears traces of cortex on its distal part. The retouches at the fronts are simple, uniserial. The three other end-scrapers are made on flakes (***Figure 2.1: 2, 4, 7***). They have semi-oval form with fronts made on ventral sides of the flakes by inverse, simple retouch. The third end-scraper has a circular form. This form, however, comes from the rejuvenation flake itself detached from a core platform rather than is intentionally sought shape of the tool.

Burins

There is one example. It is a dihedral, one side burin made on a proximal fragment of a blade (***Figure 2.1: 6***).

Retouched blades

There are 2 examples. The blades are formed by small, marginal retouch.

Blades with retouched notches

There is one example. It reminds to some extent an 'Étranglée' form with asymmetrically placed notches retouched by tiny, marginal retouch (***Figure 2.1: 10***).

Blades with unretouched notches

There are 3 examples. This series of artefacts consists of blades with single notches and notches grouped into series. They can be situated either on one of the lateral edges or on both (***Figure 2.2: 2***).

Retouched flakes

There are 12 example. These are flakes with marginal, simple retouch (***Figure 2.2: 3***). One of them has alternate retouch (***Figure 2.2: 1***).

Flakes with retouched notches

There are 5 examples. Just as in the case of the blades the notches appear as single ones or grouped in series. They are covered with a tiny, marginal retouch.

Flakes with unretouched notches

There are 2 examples. They show features similar to the artefacts made on blades.

Borers

There are 2 examples. The first one is formed by steep retouches converging at the tip of a flake. The tip of the second one is formed unilaterally by a steep retouch (***Figure 2.1: 8***).

Splintered pieces

There are 2 examples. There is a flake with traces of splinter retouch on its dorsal side. The other one is a flake with thinned lateral side by a bifacial splinter retouch (***Figure 2.1: 11***).

The earliest assemblage from trench B_{22} at Orlovets

The earliest flint assemblage from the Orlovets site comes from the third cultural horizon of the trench B_{22}. The series of artefacts is small. There are two flakes of flint variety known from the close vicinity of the Koprivets site. The rest of the artefacts are made of flint variety that is characteristic of the outcrops lined along the Turnovo-Lovec anticline. The local pebble flints were gathered from the alluvial deposits along Jantra, Rahovets, Bohot, Negovanka and Russenski Lom Rivers. The last deposit constitute the main source of raw materials for the assemblages known from the Koprivets site. Contrary to this, the raw materials that dominate in the assemblages from Orlovets come from the primary sources of the Cretaceous limestone strata of the Lovec-Turnovo anticline (Цанков 1946, Sirakov and Tsonev 1995). Generally, the imported flints come from larger nodules suited for production of big blades unknown so far from the Koprivets Early Neolithic assemblage.

The series of blades from the Orlovets site consists of 10 blades and fragments of blades. Three fragments of blades are burnt. Because they are almost destroyed their features are not considered here. The rest of the series consists of two intact blades, three proximal (***Figure 2.2: 8***), one mesial (***Figure 2.2: 7***) and one distal fragments. They all are without cortex on their dorsal sides. The butts are formed by a single blow. Blade profiles are convex in four cases and straight in three ones. They are made of imported flints.

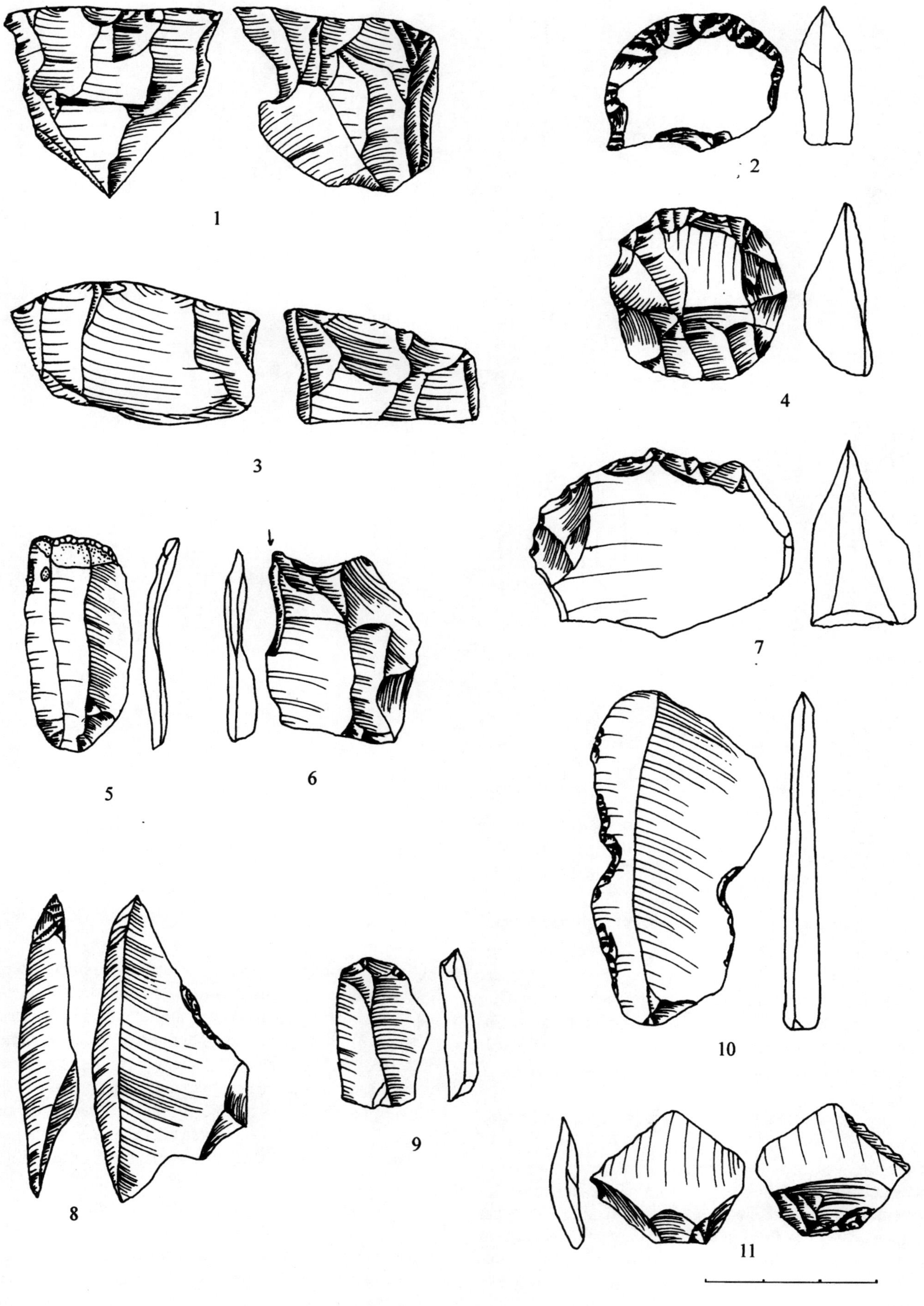

Figure 2.1. Flint artefacts from the Koprivets site (after R. Zlateva).

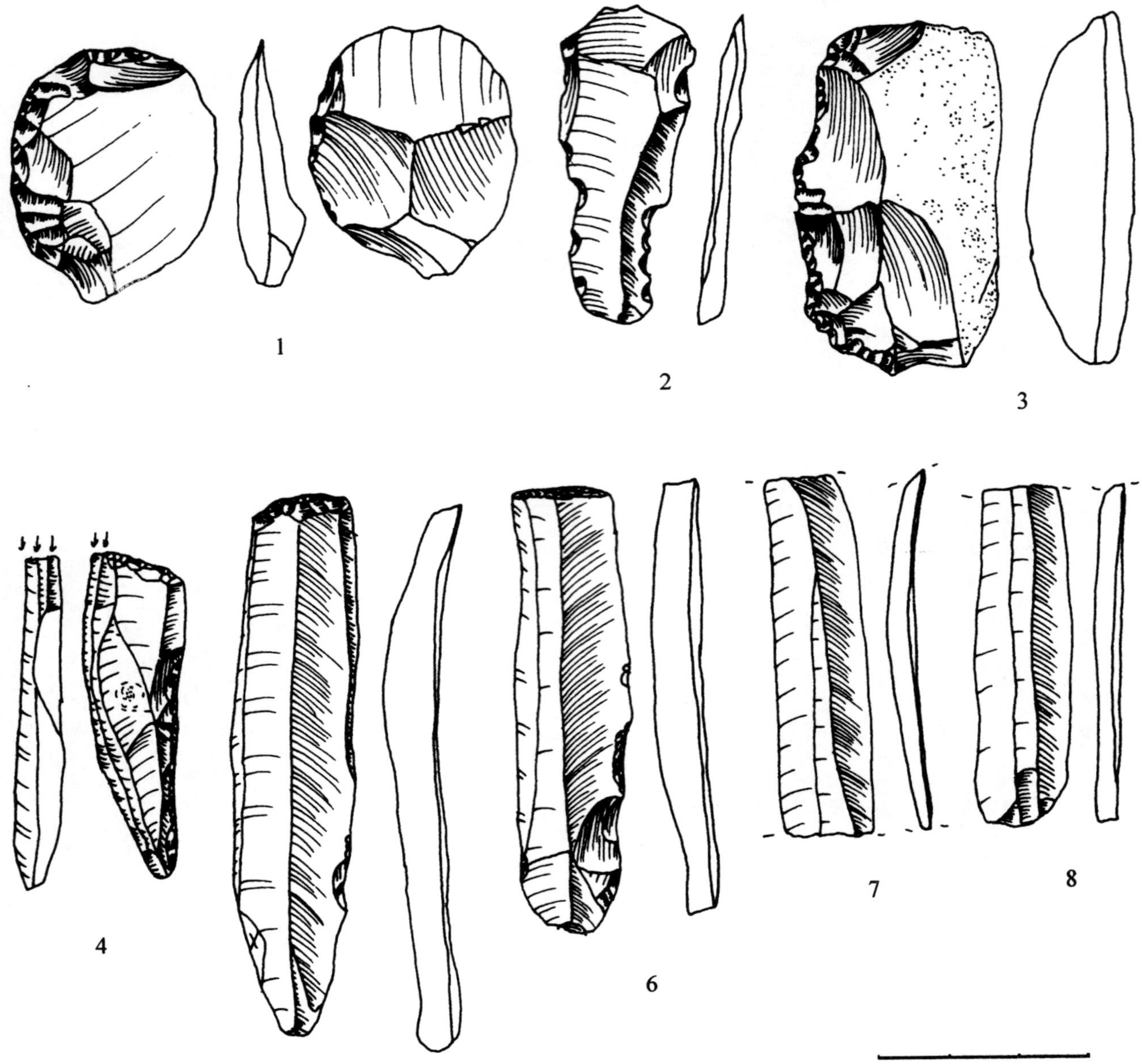

Figure 2.2. Flint artefacts from the Koprivets (1-3) and Orlovets (4-8). 1-3 after R. Zlateva.

The group of the retouched tools consists of two end-scrapers on blades and one truncated burin on a retouched blade (flint variety known from Koprivets). The fronts of the end-scrapers are formed by lamellar-convergent retouch (***Figure 2.2: 5, 6***). The truncated burin is presented in (***Figure 2.2: 4***).

Acquisition Patterns at Koprivets and Orlovets

The procurement of raw materials during the earliest occupations at both sites is in sharp contrast to one another. As it was mentioned above the region is characterized by two types of flint outcrops. The first one are secondary, alluvial deposits along the rivers mentioned. There were derived small, pebble cores with traces of rolled cortex that constitute the majority of the flint supply for the inhabitants of Koprivets. The second type of outcrops lie further to the South in the Cretaceous limestone strata of Lovec-Turnovo anticline. These are larger nodules of better quality suitable for production of big blades. And it was that these blades were exchanged through a distant trade within the region. Once established both systems of local and regional exchange patterns remained stable throughout the whole Neolithic. For example, the late Neolithic assemblage from Koprivets has more varied structure including flints from these distant sources. But the changes are not significant and show a conservative, closed system of raw material supply. The late Neolithic structure of raw materials in Orlovets also shows some increase of the local flints but the general pattern remains the same as in the earlier assemblages.

Comparison of the assemblages

Because large scale excavations were not possible at both sites, spatial patterning may contribute to some extent to the assemblage differences. But their similar topography, the excavation methods applied and the structure of raw materials that does not change throughout time suggest that sampling bias has little assistance to the contrast observed between the two assemblages.

In general, chipped stone production at Koprivets can be characterized as close to expedient in nature. On the average few flakes were removed from a pebble core and they would be used or discarded. Little or no attention was given to core preparation, recycling or maintaining cores. Manufacture techniques included both soft and hard hammers.

The various core types described above confirm this hypothesis. For example, among all cores there are few pyramidal ones with wedge-shaped bases. These cores are generally associated with maximum yield of blanks. Instead in the Koprivets assemblage cores with irregular, cylindrical and rectangular forms are almost equally presented.

The majority of cores are in advanced or final stage of exploitation. This explains to some extent their amorphous form. On the other hand the relatively small size of the nodules also contributes to this stage of discard. Pebble cores cannot be maintained long enough in their most efficient form and they were discarded after a second platform or changed orientation of flaking had been made to extend their use.

The microcores and one third of the single platform cores were designed for obtaining bladelets and small blades. Because sieving of the sediments removed was not applied during the excavations at Koprivets and Orlovets the microlithic aspect of the Early Neolithic technology is not possible to be examined. But it should be stressed that a considerable part of the cores were designed for obtaining small blades and bladelets.

There are significant typological differences between the two assemblages. In the earliest Orlovets assemblage no cores were found. In the middle and late Neolithic assemblages only a small number of cores (local flint known from Koprivets) were uncovered. Contrary to this the typological structure of the Koprivets assemblage is dominated by various types of retouched flakes and small and medium size blades while the Orlovets assemblage is a blade oriented industry. The size of blades from Orlovets vary between medium and big. They served as blanks for tool manufacture. From technological point of view these blanks come from large cores with elaborate initial preparation.

The blades bear no traces of cortex. Their butts are carefully prepared and the blades were removed by using soft hammer from slightly convex flaking platforms. Only two tool forms were found in the Orlovets assemblage but their general view and manufacture sets them in contrast to those from Koprivets. The truncated burin is similar to those known from the Epipalaeolithic tradition in the Eastern Balkans (Sirakov et al. 1994). Burins (most in an atypical form) rarely occur in the Early Neolithic context and almost disappear during the Neolithic. Generally, they miss in the lithic inventories till the Middle Eneolithic.

The dihedral burin from the Koprivets assemblage is close to an atypical form. This only shows that, in general, burins are not common artefacts during the Neolithic. This fact of rare occurrence of burins mostly in an atypical form is confirmed by the typological list of Neolithic artefacts created by I. Gatsov (1992) on the materials from Western Bulgarian Early Neolithic assemblages. There he mentioned only a single subtype - flat burin.

Although end-scrapers are common in the Early Neolithic they also show some controversy. According to I. Gatsov (1992), in the earliest phases of the Western Bulgarian Early Neolithic assemblages end-scrapers do not occur. This fact, however, cannot be observed in the case of the Early Neolithic assemblages in the central Northern Bulgaria. As it can be seen this type is well presented in both inventories studied. The end-scrapers from Orlovets are made on big elongated blades from imported flint. Their fronts are slightly convex, almost straight and covered by semi-convergent, partly lamellar retouch. The technology of manufacture of the blanks shows careful preparation of cores that leaves no traces of cortex. The section of the blades is trapezoidal and the profile is slightly concave, almost straight in one case. These tools also have small notches covered with tiny retouch. The Koprivets assemblage has two end-scrapers on blades. These tools, however, are made on shorter blades which come from small cores without careful preparation. There are traces of cortex on the dorsal side of one of the artefacts and both tools show some irregularity of the form of the blank which sets them in sharp contrast to the blades from Orlovets. Their fronts are covered only by simple retouch which is opposite to the well-prepared fronts of the end-scrapers from Orlovets. The other three end-scrapers are made on small flakes with irregular form. Two of them are atypical end-scrapers made on the ventral sides of flakes. The third one is round in form. In both assemblages notched tools made on flakes and blades are well represented.

From typological point of view the Koprivets and Orlovets assemblages find some parallels with Early Neolithic sites in Romania and Serbia. The former featured by the irregularity of the blanks and somewhat occasional retouch has some common traits with the lithic inventory from the Lepenski Vir I (Kozlowski and Kozlowski 1984). 'Étranglée' forms of blades and notched tools show similarities with Romanian assemblages of the Starcevo-Cris complex (Paunescu 1970). The latter characterized by long, regular blanks oftenly with notches also points to the Starcevo-Cris complex, north of the Danube. On the other hand, both Koprivets and Orlovets differ from Early Neolithic in Thrace - Karanovo I, II (Gatsov and Kurčatov 1997) and in Western Bulgaria (Gatsov1992). Koprivets differs because of the lack of big, regular blades and Orlovets - lack of high, steep retouch that covers the big blades from the Karanovo I and II assemblages and those from the sites in Western Bulgaria.

Conclusions

The question is whether the formal variation of the Early Neolithic assemblages has cultural meaning or other factors underlie it. If we take into account the fact that the nature of stone artefacts is much more conservative (do not change quickly in time) than the fast altering styles of pottery the first explanation should be excluded. The possible explanation of the sharp contrast of knapping techniques of two closely situated in space Early Neolithic assemblages may be linked to two factors: patterns of raw material supply and topography of the settlement. The topography of Koprivets suggests some isolation because it is situated in the small, upper valley of the Russenski Lom River. The valley is closed to the South and has no direct access to the large Jantra valley. The near vicinity of the Koprivets site is rich of secondary deposits of pebble flints.

Contrary to this no flint outcrops have been found in the immediate vicinity of the Orlovets site. But it lies close to the large Jantra valley that is widely opened to the north and south. It seems that the inhabitants of Orlovets profited much better from the exchange of high quality flints than their neighbours from Koprivets. This explains why in Orlovets in the middle and upper Neolithic assemblages there were found only few small cores (mainly flints known from Koprivets) while in Koprivets the number of cores of local flints is considerably greater. The small dimensions of the flint nodules in Koprivets impose some restrictions on knapping technology. The initial preparation cannot be done properly because it reduces significantly the dimensions of the cores. These cores cannot be maintained long enough in the most efficient state of exploitation. After initial detachment of a series of blades these cores needed rejuvenation or changing orientation. This was followed by short exploitation after which they were discarded.

Some of the discarded cores were occasionally picked up again and exploited additionally. But the re-utilization could only be short and after detachment of few flakes or blades the cores were discarded again. This is in sharp contrast to the long blades from Orlovets. They come from large, carefully prepared cores. Such cores are not found in Orlovets and Koprivets. Most probably the blades were made in workshops close to the outcrops of high quality flints (Cretaceous limestone along the Lovec-Turnovo anticline) and along the Jantra valley brought into the site.

This shows that as early as Early Neolithic there existed two patterns of raw material supply. Long blades of high quality flints were valued considerably. They were produced and distributed on long distances (mainly following the larger valleys of the rivers). Not all Neolithic communities profited from this exchange. Secondary deposits of raw materials with less quality flints were preferred by some communities to the imported flints. The high quality flints and valued blades were not attractive enough for them. Perhaps, application of fewer efforts and lower cost played an important role in their decision.

These mechanisms of raw material supply are stationery (does not change with time). The middle and late Neolithic assemblages from both sites show little deviation from these patterns.

CHAPTER 3

SALT USE IN THE EARLY AND MIDDLE NEOLITHIC OF THE BALKAN PENINSULA

Nenad Tasić

Introduction

The Neolithic period was regarded as a calm and peaceful period of human adaptation to the new conditions, sometimes as a dismal struggle for survival for an individual and a community as well. However the general impression on the Neolithic period was lacking a distinct character, and the studies were somehow limited on small groups and small territories, which miraculously, especially in the region of the 'Fertile Crescent', produced huge migrations which inhabited large territories. The discovery of Neolithic and even pre pottery Neolithic cities such as Çatal Hüyük, Jericho or Çayönü demonstrated the ability and readiness of Neolithic man to live in densly packed towns, with excellent urbanism. And of course, the study of everyday life, subsistence patterns, exchange or barter of exceptionally praised commodities such as obsidian, Spondylus, flint across almost the entire Neolithic world, has lead us to start perceiving the complex organization of this early societies. I hope that the following hypothesis can add a few new elements to the study of the dynamic nature of the Neolithic world.

At the end of the Ice Age, the Balkan Peninsula was inhabited by bands of hunters and gatherers who left rather few traces. The exception is of course the Lepenski Vir culture with sedentary way of life and it's exquisite material culture. During the period of the climatic optimum, some time around 6200 BC, the first groups of stockbreeders and agriculturists started inhabiting this region. Their settlements have been ascertained in Southeast Europe over a vast area: from Greece in the south to Hungary in the north. Further development of Neolithic communities gave rise to hundreds of settlements and a rather dense population pattern. This "cultural boom" and expansion of agriculture and stockbreeding occurred during the first centuries of 6th millennia BC. The cause of the change in subsistence pattern and the process of this epochal shift from hunting and gathering to agriculture and food production are still a controversial issue among archaeologists. Two main hypotheses exist which tend to explain and find origins of this change. The one traces the source of this development to the regions of Anatolia, Levant and Mesopotamia, where the climate and resources offered the most favorable conditions, and the other one searches for local origins of the Neolithic revolution. It is not my intention here to argue for either one of the hypotheses, but to establish a link, useful for the further investigation of the Early Neolithic in the Balkans and the process of Neolithization. This possibly useful link could be salt.

During the history of mankind numerous relationships, alliances and animosities have depended on salt. Salt was the cause for waging many a war. One example, among many, could be the animosity between Autariates and Ardiei. One seasonal salt spring existed in the bordering region between these two Ilyrian tribes, and as Strabo informs us - it was the constant cause for war (Papazoglu 1969: 71-72). Salt was the source of wealth, and a cause of the rise or destruction of many a state. Although it is still a basic strategic commodity, only recently we have come passing over it, at least in everyday life.

The salt value

In spite of its character as the most necessity commodity (yet so perishable from archaeological records), salt has been taken for granted and overlooked in discussions about the problems of the genesis of the Neolithic. The importance of trade and exchange in the Neolithic period has been discussed amply in archaeological literature, often dealing with obsidian, flint, sulfur, bitumen, Spondylus, cinnabar etc. However, proper attention has been paid neither to the use of salt nor to the trade and exchange of this indispensable commodity. The most probable explanation for this is its above mentioned perishable nature and a relative abundance of salt in the regions of primary Neolithization. There is more than one use of salt that could be vital to the Neolithic economies. Salt is used in human and animal diet, food preservation, curing and preserving of hides and probably for ritual and medical purposes.

Saline and places of salt trade and exchange have usually left salt-related toponomy behind, such as Salzburg, Salona, Salisbury, Hallstatt, Hallein, Halle, Tuzla or Tuz Gölö. For example, there are more than twenty places throughout Serbia, Romania, Bulgaria and Macedonia named Slatina, and some of them are connected with Neolithic sites.

The importance of salt as a vital commodity has already been recognized by Gordon Childe, who tended to explain the golden finds from Merseburg by the exchange and "commercial importance" of salt (Childe 1929: 244). There are of course numerous papers on the salt-exploitation in Hallstatt, Hallein and Halle. However, these sites belong to a much younger epoch of prehistory. Paradoxically, the significance of salt has been neglected in studying the cultures of the Neolithic – the time of sedentarization of human populations, of domestication of animals and cultivation of plants. Except for brief notes concerning salt trade (e.g. Sherratt 1997; Anati 1962; Runnels and Van Andel 1988: 91; Gimbutas 1991: 72) there have been no attempts to explain the connection between salt use and the Early Neolithic. Most experts agree that the need for free salt, for human and animal diet, started at the beginning of the Neolithic period. (Nenquin 1961: 11, 126, 156; Adshead 1992: 26). However, that statement is usually the end of the discussion concerning salt in the Neolithic. The methodological obstacles seem too high for drawing any useful conclusions. The evidence of exploitation of salt in Neolithic is either totally lacking or circumstantial - such is the case of polished stone axes found in rock-salt mines and the occasional finds of Neolithic pottery close to the installations for artificial evaporation (Nenquin 1961: 127, 128.).

Studying the Early Neolithic of Southeast Europe and its connections with the Near-Eastern Neolithic, mapping the sites and superimposing them on the maps of salines (according to the presently known deposits, marshes or brines, and relying on the toponomy research), one comes to the conclusion there is a close interrelation between Early Neolithic settlements and salines. Therefore, I am arguing that the settlement pattern in the zone of primary Neolithization and its penetration/diffusion into Southeast Europe is to be seen as closely connected with salt-rich soils and salines.

Firstly, we must ascertain that salt was really in use during the Neolithic. It was often discussed whether the need for (or addiction to) salt is physiological or psychological or perhaps merely a cultural phenomenon. Both humans and animals need salt, which is essential to blood, nerve impulses and heart action (the sodium content governs the osmosis of intercellular fluid; with potassium, sodium plays a major role in depolarizing the nervous and muscular fiber; with magnesium, sodium regulates the activity of the acetylcholin and therefore influences intestinal peristalsis). Current medical estimations are that we need 3 g of salt a day or roughly a kilogram a year. There is a reasonable confusion in figures when it comes to the annual quantity of salt per capita. Different authors give different estimations - from 1 kg to 7.2 kg a year. It is also true that the salt intake in all modern and as well in most traditional societies exceeded these quantities. It is also a medical fact that if one's diet contains sufficient quantities of raw or roasted meat, fish (or even insects) there is no physiological necessity to ingest additional quantities of salt. This was probably the case with hunting-gathering societies. On the other hand when a population lives mostly on cereal, vegetable or boiled meat - diets require supplements of salt. And that is exactly the case in the transitional phase between hunting-gathering subsistence strategies and early agriculture (Adshead 1992: 7). Finally, there are many cultural implications of salt use. The one, which seems very important, is that the use of salt, as everyone would agree, makes important distinctions in gastronomy. Perhaps the key word is the differentiation between those who use salt and those who do not. In the region of Balkan Peninsula where groups of agriculturists, emerging from the south, encountered groups of hunters-gatherers that inhabited the area, and probably did not regularly added salt to their diet. As we are all aware, one of the strongest taboos which had survived until our days are associated with food. This differentiation, among others of course, must have been very important one in Neolithic as well. Them and us!

The use of salt in animal husbandry is even more important than its use in human nutrition, because wild species provide sufficient salt intake from large areas of pastures. Once they were taken captive and habituated to the sedentary way of life, which happened during the Early Neolithic, extra portions of salt must have been introduced into their diet. This could be achieved either by grazing them on salt-rich pastures, which was probably the case in the Near East, or in cases of salt-starved soils, such as occur in the central Balkans, by adding imported blocks of salt. Some confusion in figures also occurs when the salt intake in animal husbandry is discussed. There are estimates that cows, especially milking cows, require up to 36 kg salt per year, and there are other estimates that only 10.3 kg of salt is required, and more recent tests of the consumption by horses gave an average consumption of 2.7 kg of salt per year. These figures could help us estimate the actual amounts of salt needed for animal husbandry. The amounts used for these purposes during the Neolithic are probably much lower, but most certainly represent a significant expenditure in salt for any community.

J. Nenquin presented the history of salt production in Western Europe. His particular field of investigation was the territory of France, Germany, Austria and England where he found almost 30 sites, which could have been used for salt production during the Neolithic period. The firm evidence according to Nenquin is "briquetage" the term used for large deposits of ceramic fragments. These assemblages were sometimes regarded as the pottery-workshops, but it is now certain that they represent works for the boiling of brine (Nenquin 1961, passim). For the assumption that salt was in use in the Neolithic times in the region for Southeast Europe, we have few direct and abundant circumstantial evidence. The direct evidence is the site of Lunca – Poiana Slatini in the Siret valley, where Middle-Neolithic pottery was clearly in context with a saline that was in use until very recently (Dumitroaia 1987: 253-258). Another one is Gornja Tuzla with its cone-shaped ceramic ware that has been linked with salt production (Čović 1961, Chapman J. 1988: 12). Gornja Tuzla is located a few kilometers from Tuzla, an important rock-salt mine in Bosnia, and is also in an extremely salt-rich region.

Another set of evidence which could confirm that salt was in use during the Neolithic could be the settlement pattern in the valleys of the Salzach and the Saalach. According to J. Nenquin, it seems a good indication that the presence of brine springs was the feature important enough to attract the first farmers and stockbreeders to inhabit these inhospitable surroundings (Nenquin 1961: 52).

Furthermore, there are numerous descriptions in ethnographic research of "primitive" salt production requiring only a source of salt and fuel, or in some cases merely manual labor for collecting surface salt from naturally evaporated inland salt lakes and marshes. No sophisticated technology, which might have been an obstacle for early agriculturists and pastorals, was involved. There is one suitable example from the ethnological literature describing the salt-production installation from the coast of Guinea at the beginning of 18th century. W. Bosman had reported "earthen pots, which they set ten or twelve next another; thus making two rows, being all cemented together with clay, as if they had been done by a bricklayer, and under the pots is something like a furnace of fire" (after Multhauf 1978: 21). This seems to have been the main salt-producing technique, except for the surface collection and mine works, for a long time – from the Bronze Age until the advance of more refined techniques during the Roman period. In some parts of Europe this method, and even more "primitive" methods such as pouring brine directly over hot coals, as was done at the end of 19th century in Bukovina, survived for much longer (Multhauf 1978: 20). In the regions where surface salt was not available there are different techniques, which do not involve mining, but have been efficient for different populations around the world. As early as the 3rd millenium B.C. the Chinese have already discovered two main techniques: salt-pan evaporation technique and evaporation of sea-water mixed with algae. In maritime regions it is not unusual that salt is obtained by burning and boiling algae and other salt-plants. Burning of salt-plants could have been the source of salt in saline regions where salt domes do not occur, before the invention of evaporation technique. The plants are simply burned and the ashes are used as salt. "Another method is to put grate over glowing embers, and to pile salt-plants on this grate. The juice of the plants falls into the ashes, and assembles into lumps, which are used as salt" (Nenquin 1961: 116)

There are many techniques of salt production that may have been applied in the Early Neolithic period. But one that is particularly interesting for our research comes from East Central Africa, from the so-called Kingdom of Bunyoro dating from the beginning of the second millennium AD, because it could be easily applied to the conditions and technology of the Early Neolithic period. The salt production technique is rather original and involves spreading the earth, which is constantly being reused or recycled, over exposed surfaces of the salt-impregnated deposits. The surfaces from which the salt is produced are called ebibuga, meaning 'salt-gardens'. They spread the earth every morning and scrape up each evening, so that it absorbs salty moisture from the ground, which is in turn evaporated by the sun, thus over a number of days gradually increasing the salt-content of the earth. The salt is then removed by filtering, and the process starts over again. Another remarkable information from Bunyoro is that exclusively women produce the salt (Connah 1991: 479-94).

One more mining technique is worth mentioning, which could have been used in Neolithic. The tools for salt mining are typologically not very different from Neolithic polished stone axes. The implements of that shape could be used to loosen the salt-block and also to perform rudimentary mining-operations. Later in prehistory, namely during the Bronze and Iron Ages, the following working-sequence in salt-mining could be established: shafts were sunk into deposits of salt from the place where salt appeared on the surface; to loosen the blocks of rock-salt stone and wooden wedges were used. Ancient miners followed the salt-banks sometimes making larger galleries in the places where good quality salt appeared. Wooden beams were used as a support. Salt-blocks were loosened by making a vertical groove in the "salt-face" dimensions about 80 x 15 x 15 cm. Then semicircular grooves were cut and salt-blocks were broken along these grooves, mounted on wooden handles shoveled into leather rucksacks and taken out by the means of wooden ladders. The pinewood torches gave light to the miners. It is rather unclear how the ventilation was achieved. There are few assumptions involving fanning with a screen of branches; difference in temperature; existence of parallel shafts (Nenquin 1978: 118-19).

The other set of evidence speaking in favor of the use of salt is the location of the earliest Neolithic sites. They are concentrated in the Zagros Mountains, the Levant and Anatolia – regions extremely rich in easily available salt. The example of Çatal Hüyük, the first Neolithic urban center, is more than self-explanatory. It is situated in the Konya Plain, several tens of kilometres away from what has remained of Tuz Gölö, a salt lake that evaporates during summer draughts leaving crusts of salt on the ground. During the Neolithic, the lake was bigger and the salt marshes extended as far as the very site. The Konya Plain in central Anatolia seems to have been densely populated during the Early Neolithic and Chalcolithic periods. Aşikli Hüyük, and Koşh Hüyük are situated not far from Hadji Bektaş (near Kayseri), the third largest inland salt source in the Ottoman Empire (Adshead 1992). Yet one more explicit example is the famous site of Jericho, which lies in the vicinity of the Dead Sea. Numerous sites in salt-rich regions may support the hypothesis that salt was used in food preparation and conservation and also in animal husbandry.

The salt in the Early and Middle Neolithic Balkans

The main Early and Middle Neolithic culture, or rather a cultural phenomenon, in the inland Balkans was the so-called Starčevo-Körös-Criş complex, which covered almost 500,000 sq. km (***Map 3.1***). The most important feature of this cultural complex was coarse and painted pottery, according to which this period can be chronologically divided into a fine sequence of phases and further distinguished from the following Late Neolithic period. The heart of this culture was the central Balkan area. As the ***Map 3.2*** shows, this region is extremely salt-starved. However, according to the dozens of contemporaneous Neolithic sites ascertained so far, it seems to have been rather densely inhabited. But, if we look at the ***Map 3.1*** showing Starčevo-Körös-Criş territory, we will find out that the outskirts of the complex are marked by salines. In the east, there is the salt-rich area of the Siret and Prut valleys, and Transylvania with numerous Starčevo culture sites. An excellent example of the abundance of salt in Transylvania is the note from a visitor at the beginning of 19th century who reported that rock salt broke the surface in so many places "that 90 watchmen were required to prevent the theft of this royal property". It was also very frequent that houses in Transylvania had cellars dug into the salt. And amid of all that opulence lies Gura Baciului, representative of the earliest of the Starčevo culture phases, with its astonishing pottery and rich cultural deposit, with imported Spondylus, obsidian and opal finds (Lazarovici and Maxim 1995: 154; 156; 158), and very interesting and early dominance of domestic breeds of cattle

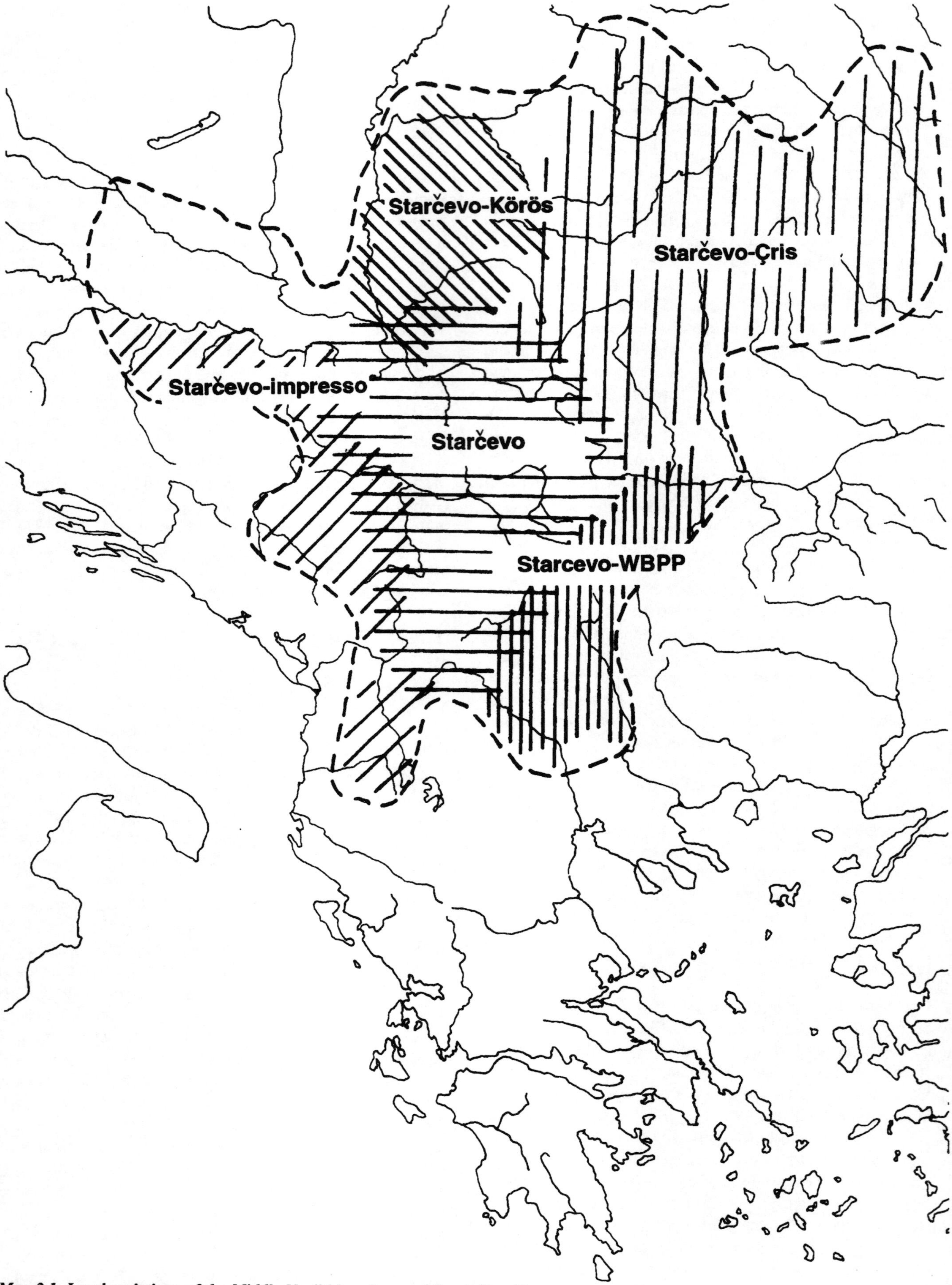

Map 3.1. Local variations of the Middle Neolithic cultures of South-East Europe.

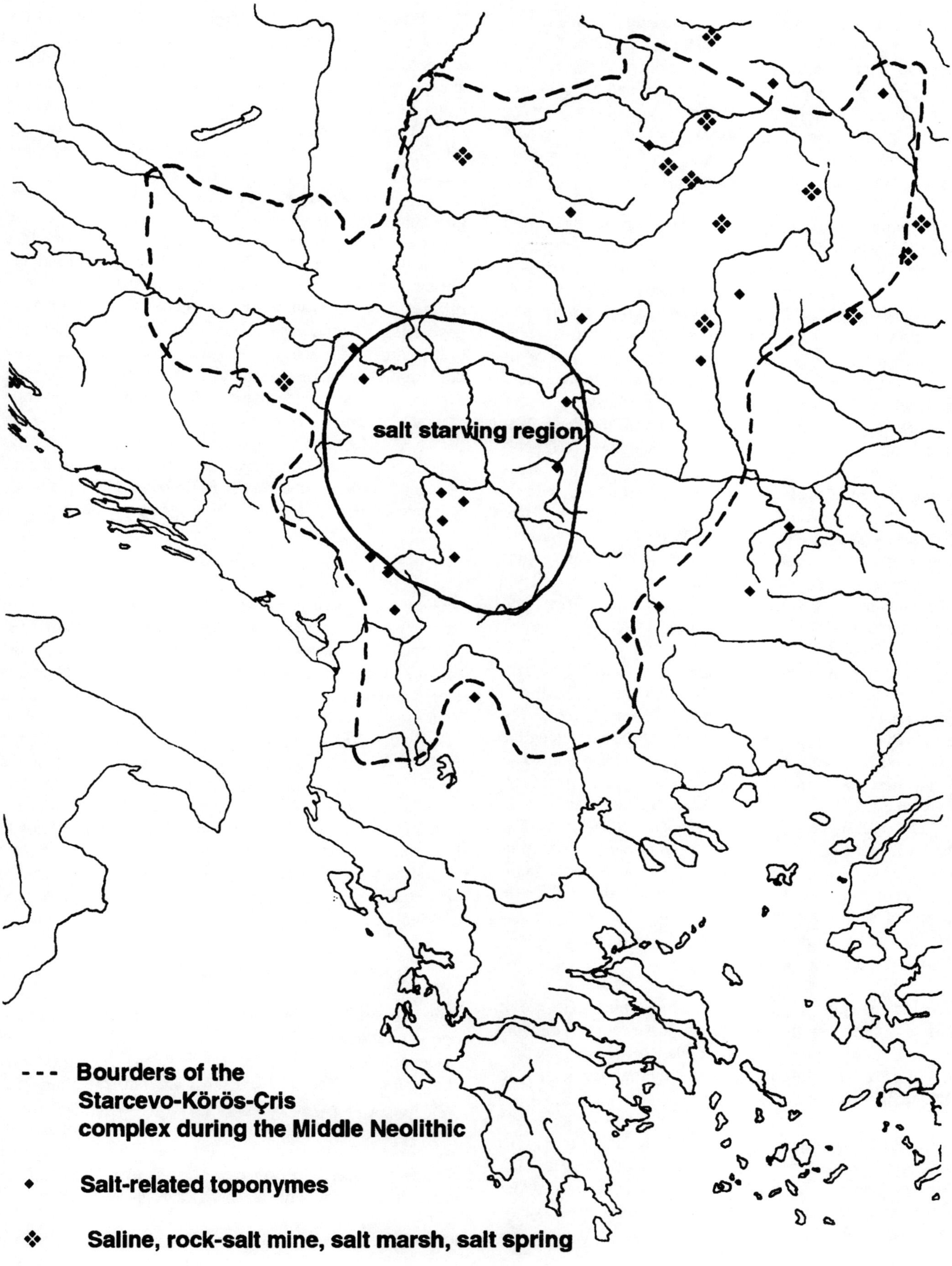

Map 3.2. Salt starving region.

over sheep and goat. Since salt was so easily acquirable, it is my opinion that it would be a mistake to ask for further evidence such as briquetage or evaporation pans before concluding that the wealth of Gura Baciului had to do something with salt. Even if there was a rock salt mine in vicinity, the Neolithic digs could hardly be confirmed because the salt has the quality of "regenerating" i.e. forming new crystals thus hiding the evidence of early mining.

In the north – not far from Oroshaza, between the rivers Tisza, Körös and Mures – there is a "depressed area under the water during the autumn, and in the summer covered with a 1-3-cm-thick layer of crystallized salt mixed with mud" (Chapman V. 1974: 31). This situation is ideal for salt extraction using one of the traditional evaporation techniques. Not far from this site the Early-Neolithic site of Szarvas is situated, as well as an agglomeration of more than forty Neolithic sites. This area was the northeastern border of the Starčevo culture, where it mixed with the local Neolithic Körös culture thus forming the Starčevo-Körös culture. In the west, the rock salt-mine near Gornja Tuzla marked the border of the Starčevo culture. The southern border is marked with several toponymic manes like Tuzi, Tuz (tuz = salt in Turkish) and Sllanice in southeastern Montenegro and in eastern Albania. Whether Starčevo sites in that area had any connection with inland or coastal salines, known to have existed near Valona and Dyrrhachium later in history, is still unclear (Adshead 1992).

The other important fact, according to archaeological records of household inventory, is that a striking mixture of cultures in salt-rich regions of the Balkan Peninsula can be observed. Such is the case with Starčevo-Körös sites along the Körös and Mures rivers, Starčevo-Criş in Transylvania and in the Siret and Prut valleys, Starčevo-impresso and Starčevo-Vinča in Gornja Tuzla, and also Starčevo-Impresso in Albania (Kolsh) and along the Yugoslav-Albanian border. Such a mixture of cultures could even suggest that no monopoly over salines existed in this period, and that the common interest in the important commodity brought different groups of people together. There is one interesting ethnological note by Le Page du Pratz from 1758, which speaks in favor of previous hypothesis:

'The Indians come a great way off to this place (Great Salt Spring) to hunt in the winter and make salt. Before the French trucked coppers with them, they made upon the spot pots of earth for this operation: and they returned home with salt and dry provisions'.

However, it is now quite clear that the Neolithic sites in salt-rich regions of the Balkan Peninsula, such as Gura Baciului, Gornja Tuzla and Szarvas appear to have been wealthier then other contemporaneous sites. The abundance of pottery, ornamental techniques and motifs is striking.

The dating of the archaeological material from Gornja Tuzla, which is the westernmost point of the Starčevo culture, could perhaps help us understand the pattern of inhabitation of the region. The earliest phase Gornja Tuzla VIb belongs to the Middle Neolithic of Central Balkan (phase MNCB II after N.N. Tasić or Starčevo IIb after D. Arandjelović-Garašanin). The painted decoration points to the close relationship with the south Central Balkan: the painted zigzag and rectilinear ornaments which sometimes form triangles, thin vertical lines with net painted beneath the rim and extremely high proportion of painted pottery almost 20%. For analogies I would note Rudnik in Metochia for painted pottery and also the elaborate technique and combining of different techniques in decorating coarse ware and Anzabegovo II for its arcade barbotine. This similarity could suggest that there was a single swift movement at one point and that Gornja Tuzla was established as a station from which the Starčevo culture later spread to the North. It was soon after the first phase of occupation of Gornja Tuzla, during the Middle Neolithic of the central Balkans (MNCB II after N.N.Tasić or Starčevo IIb after D. Arandjelović-Garašanin), that this region became very densely populated, which may be related to the discovery of a new salt-source there.

Conclusions

The introduction of salt in the archaeological interpretation, as an exchange item that has to be acquired on a regular basis, could help us explain the impressive accuracy in the rhythm of change in material culture at both ends of the Starčevo-Körös-Criş complex over a vast area of the Balkan Peninsula – from south-eastern Albania to northern Romania. We could also follow the trade and exchange routes and further investigate the control over the salines and perhaps the initial accumulation of wealth, which could lead to early social stratification. I do not imply here that salt was the only impetus to migrations or the only source of wealth in Neolithic communities, but I certainly do suggest that it should be taken very seriously into consideration when discussing the questions of human past dating back as early as the Neolithic.

Acknowledgments

This paper was written during the research stay in Heidelberg, at the Institute für Ur- und Frühgeschichte, which was made possible by generous grant by the Alexander von Humboldt Foundation. I am most grateful to Milutin Garašanin, Joseph Maran, Blagoje Govedarica and Stasa Babić for encouragement and comments on my thinking and writing. However, I remain solely responsible for the ideas and arguments presented in this paper. I wish to dedicate this paper to the memory of my teacher Dragoslav Srejovic.

CHAPTER 4

HOUSEHOLD VARIATION AND MEAT SHARING IN NEOLITHIC (SPATIAL DIMENSIONS OF THE FAUNAL REMAINS FROM OPOVO, YUGOSLAVIA)

Nerissa Russel

Opovo Ugar-Bajbuk is a Late Neolthic Vinča culture site in the Vojvodina region of Yugoslavia. From 1983-1989, an International team directed by Ruth Tringham and Bogdan Brukner conducted excavations at the site. A 16 x 20 meters block was excavated to sterile soil, revealing three building horizons containing houses, pits, ovens, and a general scatter of debris in between that I term 'midden' (Brukner 1990, Tringham et al. 1992, Tringham et al. 1985).

The faunal remains from Opovo were recorded and studied by the author (Russel 1993, in press). In brief, on the basis of the analysis of 46,594 fragments of animal bone (ca. two-thirds the total excavated) Opovo is unusual among Vinèa sites in that approximately 70% of the fauna is wild, mostly red deer and wild boar. Among the domestic animals, cattle predominate, with small amounts of pig, sheep, goat, and dog. There is little change in the fauna through time at the site, and the total occupation span appears to be only 100-200 years. I will therefore not divide my discussion by building horizon, but concentrate on the individual features.

In this paper, I explore the significance of the spatial distribution of the fauna at Opovo. Since the vast majority of the bones were not plotted, I cannot use such traditional means of spatial analysis as most forms of nearest neighbor analysis, etc. In any case, I feel that there is little point to such analyses, or to density mapping (which could be done at the level of the quadrate) at a site such as Opovo, where there is a clearly visible structure to the site (houses, pits) that is of obvious significance to the bone distributions. A preliminary density analysis based on several categories of material from the first three excavation seasons tends to confirm that this accomplishes little beyond confirming the location of the features (Miracle 1986). Rather, I examine the fauna in relation to these structural contexts. ***Figure 4.1*** provides a summary view of the features referred to here.

Some faunal analysts have noted variation in the composition of faunal assemblages from various contexts within a site, and have discussed the biasing effects that this can have in trying to reconstruct the overall economy of the group in question (Higham 1968; Meadow 1975, 1978, 1983). These differences between contexts are also potential sources of information about the social structure of the group studied, and it is to this that I address myself here. For the fauna, it is particularly the pits that have the potential to yield information about differences between social entities, since the houses contain very little bone. I will examine the differences between the various houses and pits with a view to distinguishing the activities of different households. In particular, I will attempt to determine whether meat was shared among households, and whether questions through several techniques to reduce problems related to limitations of particular techniques (Carr 1991, Rigaud and Simek 1991).

Houses

Five houses ot parts of houses were excavated at Opovo. These are numbered 2-6 because a concentration of house rubble once called House 1 was later determined to lack any actual structural remains. All these houses, as is usual in the later phases of the Vinča culture, were burnt down, probably deliberately, at the end of their life cycle. Only a small corner of House 4 lay within the excavation block, so I will no discuss it here. Most of the fairly limited bone found inside the houses was probably secondarily deposited in them subsequent to their destruction. Certainly most bone inside the houses when they burned would have been calcined from the high temperatures of the flames that fired and often vitrified the clay walls. Indeed a higher percentage of bone is calcined or partially calcined in the houses (5.8% as opposed to 2.8% outside houses), but the vast majority is unburned ot burned to a lesser degree likely to be the result of other processes such as cooking.

Therefore most of the bone must have been deposited in the houses subsequent to their burning and the end of their use lives as houses. It is thus not surprising that the species distributions of bone in the houses generally resemble those of the site as a whole; faunally, the houses are not really distinct from the middens. This means that the bone within them cannot be used in any straightforward way to address differences in household consumption.

House 5 stands out in terms of the amount of fauna in the various houses (*see **Table 4.1***). Given that the houses all seem to be approximately the same size and that only about half of House 6 falls within the excavation block (*see **Figure 4.1***), Houses 2, 3, and 6 seem to have about the same amount of bone in them, while House 5 has substantially less. This may

House	# of fragments
House 2	991
House 3	951
House 5	180
House 6	454

Table 4.1. Number of bone fragments in the houses

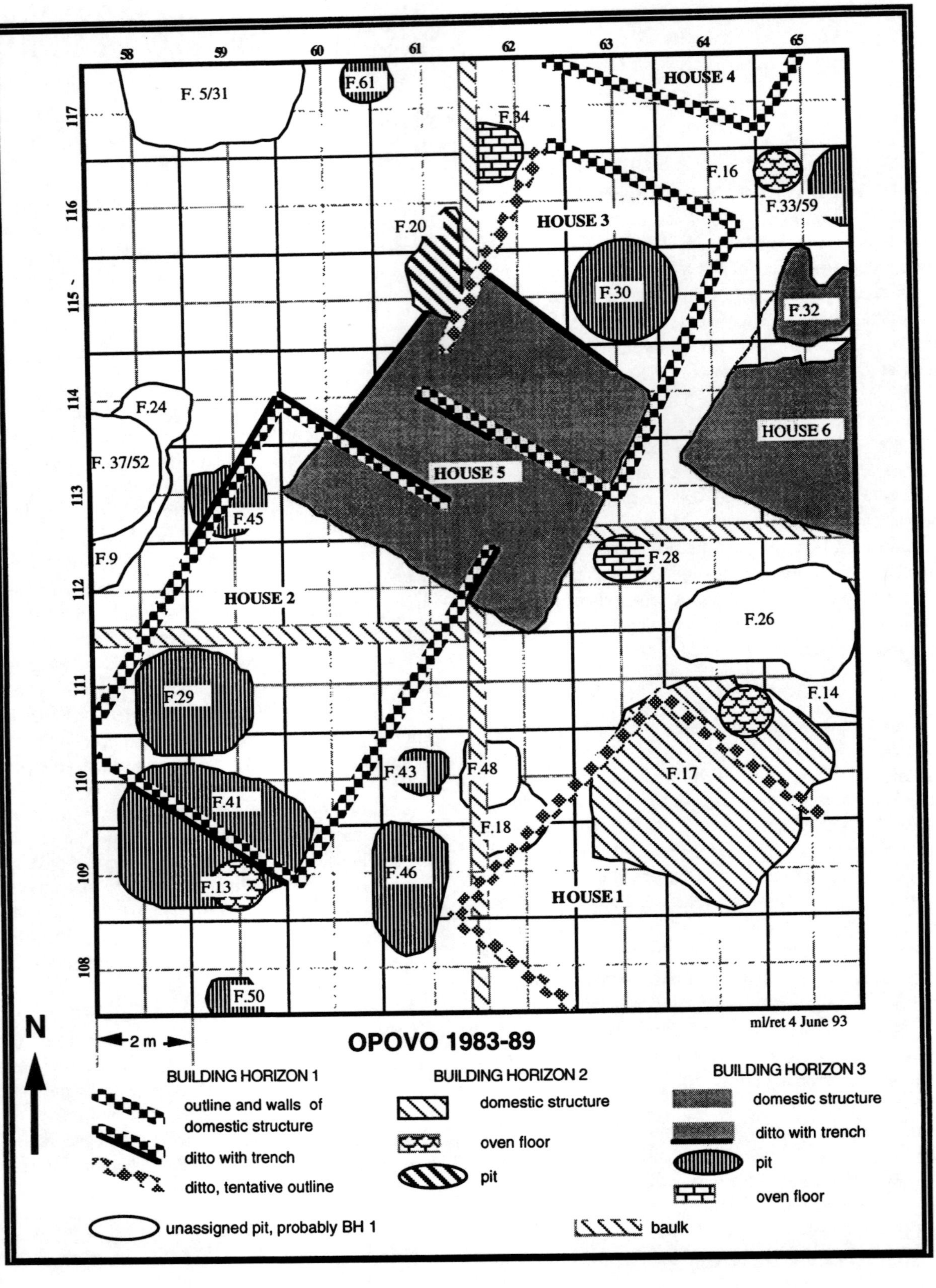

Figure 4.1. Palimpsest of Opovo features from the three building horizons.

indicate a more rapid sealing of this house after its destruction. In any case, it seems to have received much less use as a garbage dump than the other houses. This is particularly intriguing given that it is the earliest house excavated.

Pits

The Opovo pits show a distinctively patterned structure that begs for a consideration of their function beyond simple repositories of garbage. While we tend to think of discarded materials as 'just trash', ethnographic studies demonstrate that even garbage, or particular categories of it, can be laden with meaning and require special care in its disposal (Moore 1981, 1982, 1986). Legge (1989) has suggested that Neolithic pits may often have been filled in a ritual context. Refuse from feasts may be treated differently either because of the symbolic value that it carries (Davenport 1986) or because the large amount of garbage generated by feasts is more troublesome (Maltby 1985). Richards and Thomas (1984) have coined the term 'structured deposition' to refer to situations in which the remains of rituals are deposited in patterned ways that may themselves carry symbolic meaning. While the materials themselves may seem no different from everyday 'garbage', the spatial patterning and treatment of these remains is distinctive (*see* also Chapman 1996, Hill 1996).

Bones that appear to derive from the remains of feasting at Opovo are found differentially in pits as opposed to houses and midden (Russel 1993, in press). The lower levels of most of the larger Opovo pits (a meter and more) are filled with general 'garbage', including large amounts of bone along with charcoal, pot sherds and so on. These garbage levels, or the upper parts of them, generally also contain multiple layers of ash. Above this are often found one or more levels of *in situ* burning (burnt soil, ash), and these are in turn usually capped by a layer of burnt house rubble.

Bones that may result from feasting (roasting, carved, groups of similar pieces from different individuals) are found primarily in the ashy garbage levels and the *in situ* burning layers. Pits with evidence of feasting are: features 26, 30 (the bony deposits in the upper levels), 5/31, 41, 46, 50 and somewhat more tenuously features 18/48, 43, and 45. Not all the bone in the pits was deposited directly after feasting, however. Gnawed and digested bones (i.e., fragments that would have been voided in dog feces) are actually more common in the pits than elsewhere (Russel 1993), so some, perhaps most of the bone in pits likely results from mundane cleaning. Such grain as there is at Opovo is found almost exclusively in the pits (Tringham et al. 1992). If this is not simply a matter of differential preservation, it may be that the limited amount of grain was saved largely for feasting. Perhaps the only reason for growing it at all in the heavy soils and marshy environs of the site was because some grain product (bread, gruel, beer) was ritually necessary for feasts (Katz and Voigt 1986). The levels if *in situ* burning can be seen as intended to reduce the volume ot smell of garbage as the pits get full, or as a ritual of purification marking the end of their use, just as the houses were also burned at the end of their use lives. One can at least imagine a scenario where feasting remains are placed in garbage pits, which are then finished off with ritual burning, perhaps even including the burning and dismantlement of a house. Close study of the architectural remains at Opovo indicates that the houses were burned deliberately, and probably individually (Stevanović 1996, Tringham et al. 1992). A favored explanation is that houses were destroyed when the household head died, or when the house was otherwise perceived as being at the end of its life span. It may be that at least some of the feasting at Opovo occurred in the context of a ritual of house destruction.

Species distribution by pit

Figures 4.2-4.14 plot the distribution of the taxa of dietary significance in all the Opovo pits having appreciable amounts of faunal material, quantified with diagnostic zones (Warson 1979). The taxa are arranged with domestic species on top and wild on the bottom, bridged by the pigs, which are chiefly wild but include some domestic. It can be seen that there is substantial variation among the pits. A rather naïve way to interpret this would be to assume that each pit accurately reflects the animal production (hunted and herded) of a given household over a certain period of time. This assumes that each household deposits its refuse in one pit at a time, that the pit is not used by other households, and that there is no sharing of meat (at least not while still on the bone) between households. In this case, there are some households with very little domestic stock subsisting largely on wild animals for the meat portion of their diet, while others (features 29, 9/37/52) have substantial amounts of cattle. Features 20 (with small sample size) has sheep/goat but no other domestic species (the pigs are all wild), possibly representing a young household or one recovering from a disastrous herd reduction, concentrating on small stock until they can trade up to cattle.

Some of this may in fact represent actual household circumstances, but it is likely that the situation is more complicated. Given the ratio of pits to houses, more than one must have been associated with each house, although it is hard to know if more than one was open simultaneously. The larger pits probably contain at least a few years' worth of garbage, while some of the smaller ones may have filled in only a few months, based on seasonality estimates of the bone from the various levels of the pits. Features 5/31, 30 and 41, in the year-round category, all have essentially 'typical' species distributions, i.e., resembling that for the site as a whole. Features 9/37/52 (building horizon 2 or 3) on the other hand, has a fairly long period of use and respectable sample general, however, it would seem that in the long term there is little difference in the meat available to different households, either because all have similar herds (or there is a communal village herd) and hunt about the same amount, or because meat is shared so as to even out production differences between households. On the other hand, these large pits may be receiving the refuse from more than one household. Four pits, all from building horizon 3 (the earliest), have almost entirely wild fauna with notably higher than usual representation or roe deer (45, 46, 50, 59). The two that have seasonality data (45 and 46) appear to have been used only in the fall (or more generously, late summer – early winter). Feature 18/48, probably from the succeeding building horizon 2, is somewhat similar, although with a bit more domestic fauna. This could reflect the hunting of crop

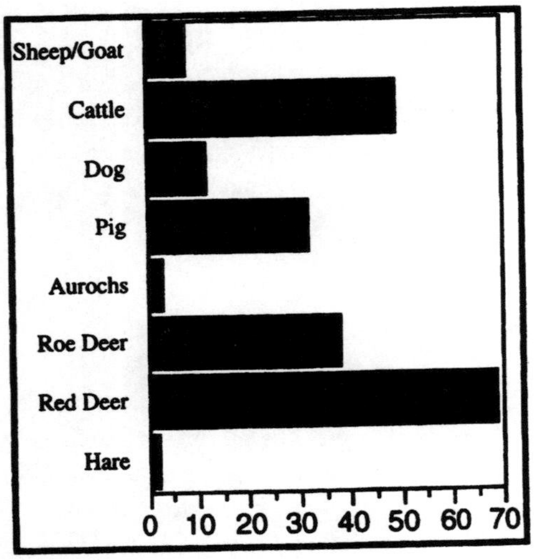

Figure 4.2. Distribution of taxa in feature 5/31 (DZ).

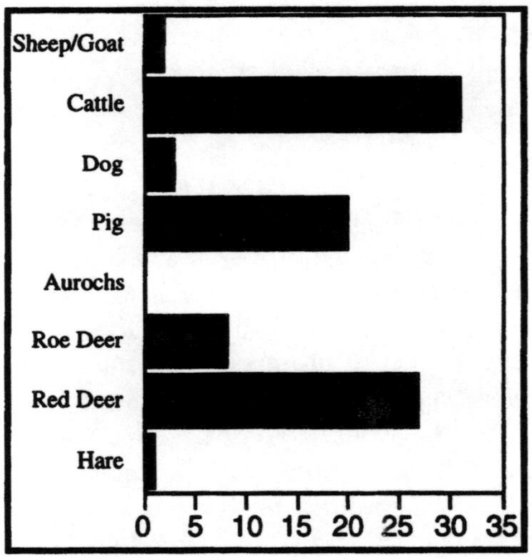

Figure 4.3. Distribution of taxa in feature 5/37/52 (DZ).

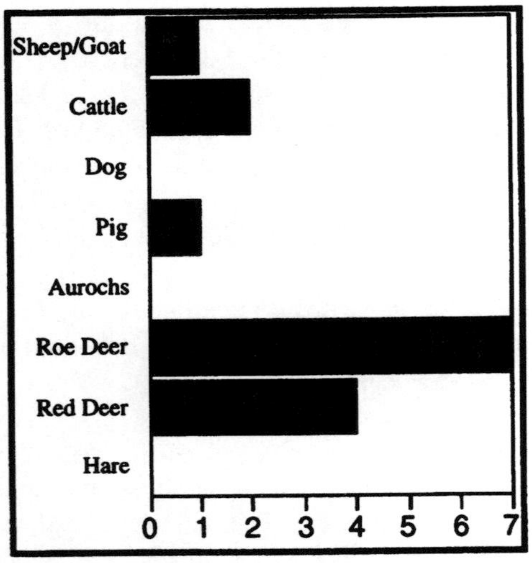

Figure 4.4. Distribution of taxa in feature 18/48 (DZ).

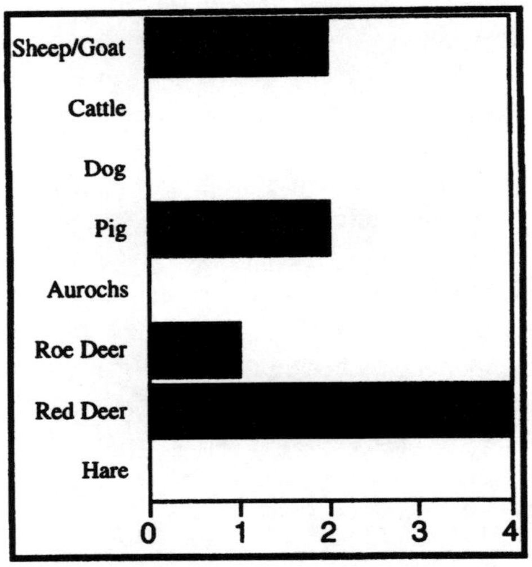

Figure 4.5. Distribution of taxa in feature 20 (DZ).

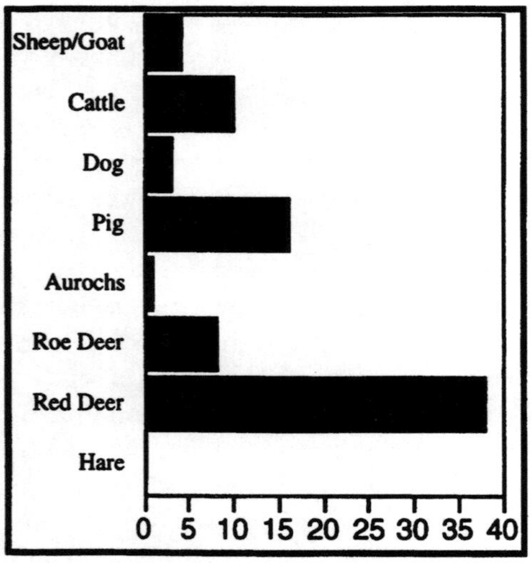

Figure 4.6. Distribution of taxa in feature 26 (DZ).

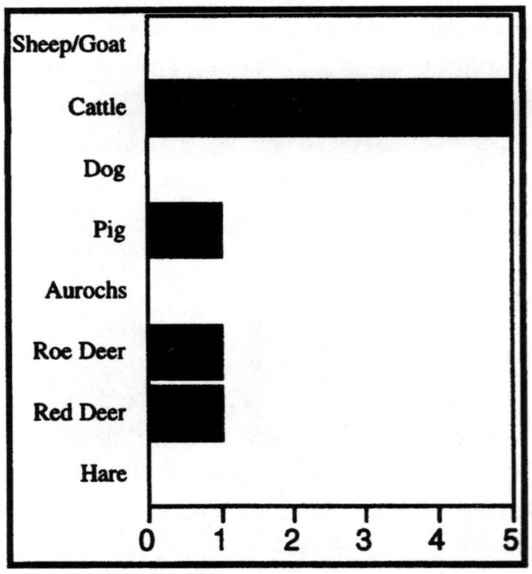

Figure 4.7. Distribution of taxa in feature 29 (DZ).

Table 4.2. Densities per cubic meter of major taxa in pits (DZ).

	F. 5/31	F. 9/37/52	F. 18/48	F. 20	F. 26	F. 29	F. 30	F. 41	F. 45	F. 46	F. 50	F. 59	F. 61
Sheep/Goat	.56	.27	.52	10.67	.75	0	.23	1.26	0	0	0	0	1.14
Cattle	3.45	4.11	1.04	0	1.88	5.0	1.28	2.78	.37	.34	0	.58	1.14
Pig	2.25	4.24	.52	10.67	3.01	1.0	1.05	3.37	1.46	2.38	5.71	3.19	9.14
Roe Deer	2.68	1.06	3.63	5.33	1.51	1.0	.82	3.31	2.93	4.08	6.86	4.06	6.86
Red Deer	4.86	3.58	2.08	21.33	7.15	1.0	2.57	4.96	3.66	3.74	5.71	4.64	28.57

raiders at harvest time as a major activity in these households. Features 61 (building horizon 3) is also heavily populated by wild fauna, but mostly red deer.

The density of diagnostic zones per cubic meter (***Table 4.2***) suggests that the pit relatively high in roe deer are actually low in other fauna.[1] This tends to confirm the crop raiding hypothesis: fewer animals were killed at these times, while those that were are crop raiders, indicating that most labor was devoted to agriculture and other tasks at this time of year. On the other hand, the high values of red deer in feature 61 appear to reflect genuinely high rates of hunting. Features 9/37/52 and 29 do have higher cattle densities than elsewhere, but not dramatically so. As these two pits do not contain identifiable feasting remains (*see* above), they may more closely reflect an everyday, household mode of consumption.

To the extent that materials in the pits represent the remains of feasting, they do not reflect the day-to-day productive output of single households, but rather the amount of meat (and other food/goods) that could be amassed through some combination of household production, kin ties, and exchange relationships by the host(s). It is also possible that some feasts may have been hosted by the whole village for visiting neighbors. It is likely that the remains of feasts are more apt to be deposited in pits than everyday garbage, of only because the larger amounts of refuse generated would be more troublesome: attracting flies, generating unpleasant odors, and getting in the way (Kramer 1982, Maltby 1985). Of we assume that most feasts are hosted by individual households, then their remains still be an index of that household's socioeconomic power (ability to marshal resources), but not if its direct production.

Meat Sharing

There is very little variation in the distribution of elements among the pits (Russell 1993), with the major exception of the red deer assembling from features 41, which is strikingly dominated by mandibles in contrast to the other pits. Since this is the largest sample, there is a good chance that this is more than random variation. A predominance of mandibles, which are quite dense (Binford and Bertram 1977, Lyman 1984), may be indicative of heavy attribution. However, the postcranial elements in feature 41 do not support this: the dense distal humerus is not dominant and softer parts such as the pelvis and distal radius, and to a lesser extent the distal femur and proximal tibia, are rather well represented. Thus there may have been a genuine concentration of mandibles, probably with tongues, in the household associated with this pit. This could indicate that a choice part was given to a high status household, possibly for use in feasting. Perhaps more likely, it could indicate that an elderly person with bad teeth resided in this household, and was given a particularly soft piece of meat by friends and relatives. If this is indeed the case, it suggests that the taphonomic history of the pits is such that differential meat distribution (flow of good cuts to particular households) could indeed be detected, at least in the larger pits, and thus that with this one exception such differential distribution did not occur.

One way to seek evidence of meat sharing is to examine the co-occurrences of species and elements in the pits through correlation matrices. This is based on the assumptions that each pit primarily contains the refuse of one household (although there may have been more than one pit per household, either simultaneously or in sequence), that extensive symmetrical sharing will tend to produce an even distribution of species across the pits, and that sharing is likely to result in clustering of adjacent elements as joints of meat are circulated, whereas if an animal is consumed entirely within one household there will be no such association. Thus clustering of certain species would indicate either a lack of sharing (with associations produced through variations in household herd size and in hunting skill ot frequency) or asymmetrical sharing with meat flowing only to some households.

Table 4.3 is a correlation matrix of the major taxa among the 13 Opovo garbage pits (features 5/31, 9/37/52, 18/48, 20, 26, 29, 30, 41, 45, 46, 50, 59, 61). All counts are based on diagnostic zones. Such correlation matrices are by nature sym-

[1]The samples sizes are so small in feature 20 that one should probably not give much credence to the density values, which are highly divergent from the other pits.

Taxon (Count)	Sheep/ Goat	Cattle	Dog	Pig	Aurochs	Roe Deer	Red Deer	Hare
Sheep/Goat (39)	1.00	.779	.842	.935	.719	.901	.876	.623
Cattle (155)	.779	1.00	.941	.892	.791	.823	.892	.950
Dog (32)	.842	.941	1.00	.903	.898	.894	.964	.899
Pig (166)	.935	.892	.903	1.00	.775	.933	.949	.767
Aurochs (8)	.719	.791	.898	.755	1.00	.794	.840	.762
Roe Deer (166)	.901	.823	.894	.933	.794	1.00	.900	.755
Red Deer (307)	.876	.892	.964	.949	.840	.900	1.00	.799
Hare (4)	.623	.950	.899	.767	.762	.755	.799	1.00

Table 4.3. Correlation matrix for taxa in pits.

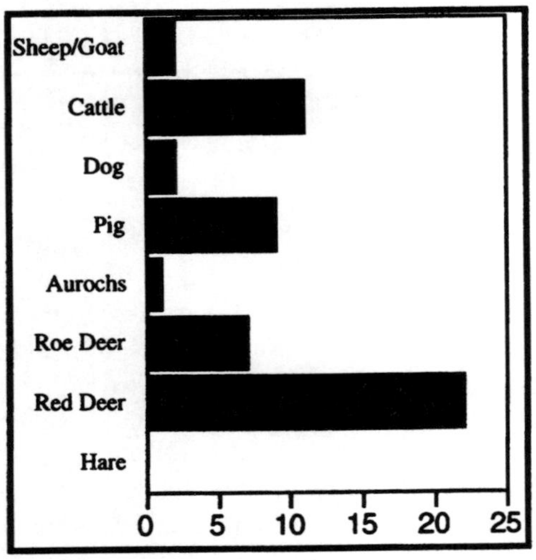

Figure 4.8. Distribution of taxa in feature 30 (DZ).

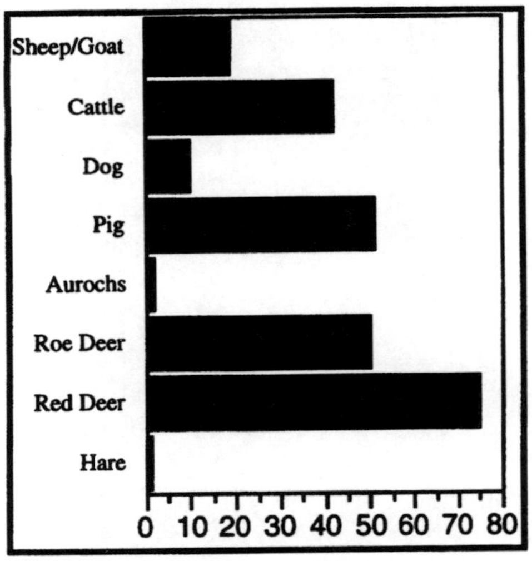

Figure 4.9. Distribution of taxa in feature 41 (DZ).

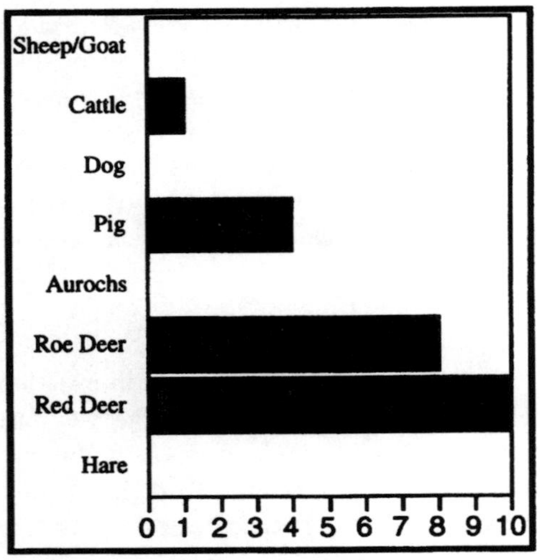

Figure 4.10. Distribution of taxa in feature 45 (DZ).

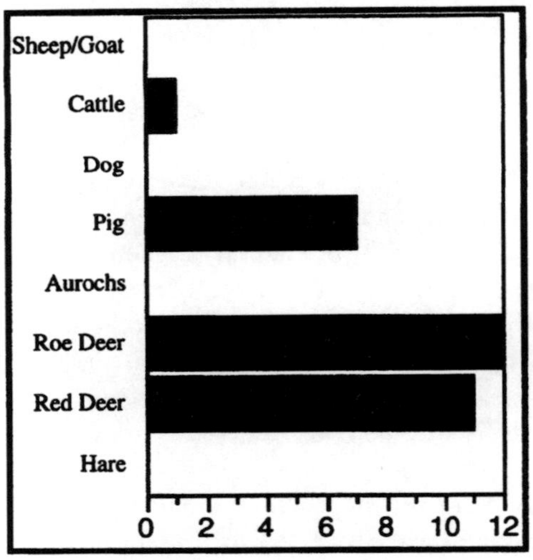

Figure 4.11. Distribution of taxa in feature 46 (DZ).

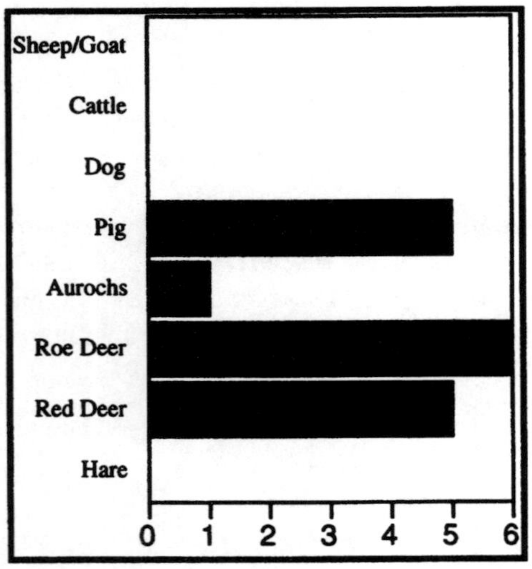

Figure 4.12. Distribution of taxa in feature 50 (DZ).

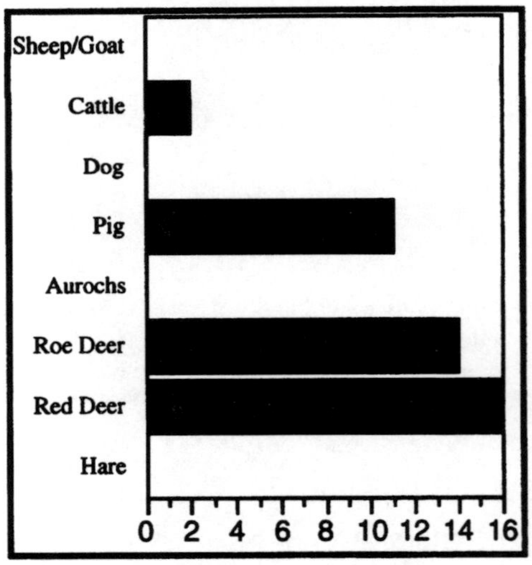

Figure 4.13. Distribution of taxa in feature 59 (DZ).

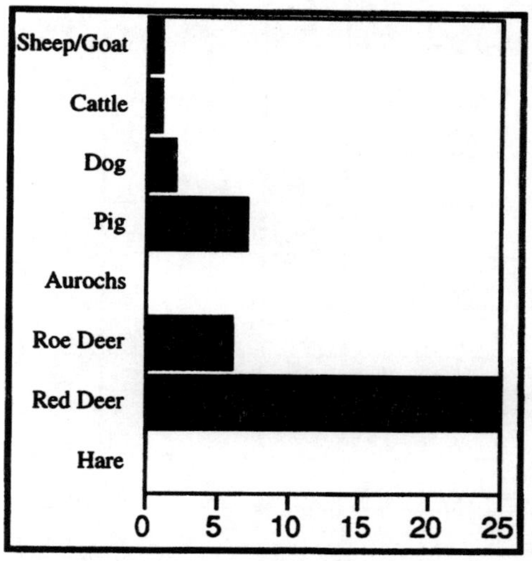

Figure 4.14. Distribution of taxa in feature 61 (DZ).

metrical around the diagonal axis of autocorrelations. I have nevertheless included both sets of values to make it easy to read the values associated with a category by following a single row or column.

It is striking that all the correlations in ***Table 4.3*** are quite high. This indicates a rather even distribution, with taxa occurring in similar proportions in most pits. This is a somewhat different picture than emerged from the visual comparison of species profiles by pit in ***Figures 4.2-4.14*** above (although less divergent from the density figures), and tends to suggest that symmetrical sharing was taking place. There does seem to be some clustering of wild species, which might indicate that these are not shared. Sharing the meat of domestic but not wild animals is contrary to what one would expect based on ethnographic analogy and our understanding of property relations, which would predict if anything the opposite. However, it is likely that these values reflect the smaller, apparently shorter term pits with high proportions of wild species noted above, and thus may be detecting a seasonal phenomenon rather than lack of sharing. A general problem with correlation matrices is that the more abundant groups will tend to correlate simply because they are more numerous. This is very likely a factor here, although it does not appear to be operating to any substantial degree in the elements matrices below, where the differences are smaller.

Tables 4.4-4.7 show correlation matrices for the four major species (cattle, roe deer, red deer, pig) for those pits in which they are represented by more than 2-3 specimens. These tables are restricted to the major meat-bearing bones. The whole bones is treated as the likely unit of circulation given stone tool technology, so for the long bones, the greater of the proximal or distal counts has been used. Likewise, for the radius and ulna, the greater count is used to represent this area, assuming that they would have moved as a unit. We can examine two sets of correlations to seek evidence of sharing: those between rights and lefts of the same elements, and those of adjacent elements. On the whole, if animals are consumed within a single household, the right and left specimens of particular elements should be equally abundant in each pit and thus highly correlated (it is hard to imagine why one side would suffer differential attrition). On the other hand, separating an animal into right and left sides is one obvious way to divide it (Tilley 1981), so if meat is being shared it is likely that the two hind legs, for instance, will end up in different pits. Given that we are not dealing with huge numbers of each taxon per pit (such that sharing might even this out over time), this would be reflected by a lower correlation between right and left of the same element. Thus we see that on the whole, right/left correlations are fairly low for cattle, red deer and pigs, but higher for roe deer, suggesting that the larger animals are being shared to a greater extent. For red deer, the correlations are lower for the rear than the fore limb, indicating that the rear legs are more likely to be distributed. Pigs show high (non-sharing) correlations for the pelvis and tibia, lower elsewhere, suggesting that the ham region and fore limbs are more often shared.

Second, if meat is shared, we would expect the animal to be divided into joints, such that adjacent bones that travel together in a unit of shared meat will be more highly correlated than those that are separated into different pieces. Thus we see that for cattle, right and left scapulae show low correlations with right and left human, respectively, and similarly for the pelvis and femora. The humeri and radii/ulnae are more highly correlated, while right femora and tibiae are correlated relatively highly, but left are not. This seems to indicate that meat was distributed, and that joints for sharing were separated at the shoulder and hip, but not normally at the elbow, and perhaps more variable at the knee. Roe deer, on the other hand, show fairly high correlations at most joints, and many quite high correlations among non-adjacent body parts, again suggesting a low degree of sharing. Red deer show rather ambiguous levels of correlation among adjacent body parts in the forelimb, and low values between pelves and femora as well as femora and tibiae. Again, this way indicate that rear legs circulated more, and apparently were divided at the knee more consistently than with cattle. Pigs show rather ambiguous levels of correlation at all joints, with the scapulae and humeri and the femora and tibiae highly correlated than the humeri and radii/ulnae and the pelves and femora. This suggests that some sharing occurred, but less frequently than cattle and red deer. This could reflect differences in size or perhaps different treatment of wild and domestic pigs, which are here combined, although most are wild. There is also a weak indication that the shoulder was included with the humerus, while the fore limb was separated at the elbow, while the hind limb was separated at the hip but not (or not always) at the knee, somewhat in contraction to the right/left correlation.

Kathy Prizer's (1994) conjoining study of some of the red deer bones from Opovo tends to support the inference offered above. Prizer was able to find only seven conjoining pairs across pits, suggesting that meat sharing occurred but was not extensive. There is no pattern of meat in general or particularly choice pieces flowing to particular pits (and hence perhaps households). Although the sample of conjoining pairs in not large, there are more cross-pit hind limb conjoins, deer hind limbs circulated more than fore limbs (Prizer 1994: 45).

Element (Count)	R. Mand.	L. Mand.	R. Scap.	L. Scap.	R. Hum.	L. Hum.	R. R/U	L. R/U	R Pelvis	L. Pelvis	R. Femur	L. Femur	R. Tibia	L. Tibia
Right Mandible (8)	1.00	.227	-.583	.579	.375	-.272	.444	.514	-.025	.408	.134	.748	.639	.172
Left Mandible (12)	.227	1.00	.583	-.208	.631	.357	.874	.450	.945	-.436	.052	-.238	.773	.345
Right Scapula (11)	-.583	.583	1.00	-.468	-.063	.612	.250	-.086	.700	-.408	.200	-.919	.221	.258
Left Scapula (4)	.579	-.208	-.468	1.00	-.200	.327	.134	.642	-.512	.873	-.071	.327	.118	.674
Right Humerus (7)	.375	.631	-.063	-.200	1.00	-.102	.792	.514	.590	-.612	-.535	.408	.332	.029
Left Humerus (3)	-.272	.357	.612	.327	-.102	1.00	.408	.560	.241	.167	-.218	-.667	.060	.890
Right Radius/ Ulna (8)	.444	.874	.250	.134	.792	.408	1.00	.800	.713	-.272	-.312	.068	.639	.554
Left Radius/ Ulna (9)	.514	.450	-.086	.642	.514	.560	.800	1.00	.177	.210	-.504	.210	.329	.826
Right Pelvis (7)	-.025	.945	.700	-.512	.590	.241	.713	.177	1.00	-.662	.079	-.361	.630	.101
Left Pelvis (3)	.408	-.436	-.408	.873	-.612	.167	-.272	.210	-.662	1.00	.327	.167	.060	.421
Right Femur (4)	.134	.052	.200	-.071	-.535	-.218	-.312	-.504	.079	.327	1.00	-.218	.512	-.245
Left Femur (3)	.748	-.238	-.919	.327	.408	-.667	.068	.210	-.361	.167	-.218	1.00	.060	-.281
Right Tibia (8)	.639	.773	.221	.118	.332	.060	.639	.329	.630	.060	.512	.060	1.00	.237
Left Tibia (12)	.172	.345	.258	.674	.029	.890	.554	.826	.101	.421	-.245	-.281	.237	1.00

Table 4.4. Correlation matrix for cattle elements in pits.

Element (Count)	R. Mand.	L. Mand.	R. Scap.	L. Scap.	R. Hum.	L. Hum.	R. R/U	L. R/U	R Pelvis	L. Pelvis	R. Femur	L. Femur	R. Tibia	L. Tibia
Right Mandible (18)	1.00	.111	.375	.494	.627	.706	.575	.185	.116	.297	.770	.892	.257	.582
Left Mandible (127	.111	1.00	.559	.733	.670	.279	.816	.789	.857	.732	.626	-.037	.777	.484
Right Scapula (11)	.375	.559	1.00	.753	.661	.527	.685	.683	.619	.651	.655	.238	.655	.333
Left Scapula (10)	.494	.733	.753	1.00	.942	.806	.915	.490	.827	.937	.926	.386	.926	.496
Right Humerus (9)	.627	.670	.661	.942	1.00	.754	.943	.422	.748	.813	.944	.488	.812	.553
Left Humerus (14)	.706	.279	.527	.806	.754	1.00	.643	.104	.459	.764	.856	.663	.701	.357
Right Radius/ Ulna (13)	.575	.816	.685	.915	.943	.643	1.00	.669	.853	.832	.917	.348	.853	.537
Left Radius/ Ulna (10)	.185	.789	.683	.490	.422	.104	.669	1.00	.683	.488	.447	-.098	.547	.293
Right Pelvis (2)	.116	.857	.619	.827	.748	.459	.853	.683	1.00	.905	.655	-.143	.946	.143
Left Pelvis (5)	.297	.732	.651	.937	.813	.764	.832	.488	.905	1.00	.800	.143	.994	.238
Right Femur (2)	.770	.626	.655	.926	.944	.856	.917	.447	.655	.800	1.00	.655	.778	.655
Left Femur (1)	.892	-.037	.238	.386	.488	.663	.348	-.098	-.143	.143	.655	1.00	.073	.714
Right Tibia (6)	.257	.777	.655	.926	.812	.701	.853	.547	.943	.994	.778	.073	1.00	.218
Left Tibia (6)	.582	.484	.333	.496	.553	.357	.537	.293	.143	.238	.655	.714	.218	1.00

Table 4.5. Correlation matrix for roe deer elements in pits.

Element (Count)	R. Mand.	L. Mand.	R. Scap.	L. Scap.	R. Hum.	L. Hum.	R. R/U	L. R/U	R Pelvis	L. Pelvis	R. Femur	L. Femur	R. Tibia	L. Tibia
Right Mandible (20)	1.00	.950	.668	.517	-.433	.210	.474	.651	.570	.687	.029	.773	.647	-.281
Left Mandible (27)	.950	1.00	.801	.575	-.252	.451	.656	.830	.494	.663	-.029	.883	.594	-.343
Right Scapula (15)	.668	.801	1.00	.835	.312	.765	.936	.933	.533	.792	.277	.813	.688	.089
Left Scapula (20)	.517	.575	.835	1.00	.263	.602	.705	.766	.684	.838	.696	.520	.622	.207
Right Humerus (17)	-.433	-.252	.312	.263	1.00	.554	.420	.190	-.012	.072	.279	-.081	-.076	.480
Left Humerus (12)	.210	.451	.765	.602	.554	1.00	.791	.791	.044	.272	.091	.612	.287	-.102
Right Radius/ Ulna (18)	.474	.656	.936	.705	.420	.791	1.00	.878	.413	.645	.137	.710	.595	.138
Left Radius/ Ulna (19)	.651	.830	.933	.766	.190	.791	.878	1.00	.373	.646	.210	.850	.502	-.145
Right Pelvis (16)	.570	.494	.533	.684	-.012	.044	.413	.373	1.00	.782	.464	.160	.464	.122
Left Pelvis (12)	.687	.663	.792	.838	.072	.272	.645	.646	.782	1.00	.562	.583	.811	.381
Right Femur (13)	.029	-.029	.277	.696	.279	.091	.137	.210	.464	.562	1.00	-.067	.250	.539
Left Femur (9)	.773	.883	.813	.520	-.081	.612	.710	.850	.160	.583	-.067	1.00	.672	-.161
Right Tibia (16)	.647	.594	.688	.622	-.076	.287	.595	.502	.464	.811	.250	.672	1.00	.393
Left Tibia (25)	-.281	-.343	.089	.207	.480	-.102	.138	-.145	.122	.381	.539	-.161	.393	1.00

Table 4.6. Correlation matrix for red deer elements in pits.

Element (Count)	R. Mand.	L. Mand.	R. Scap.	L. Scap.	R. Hum.	L. Hum.	R. R/U	L. R/U	R Pelvis	L. Pelvis	R. Femur	L. Femur	R. Tibia	L. Tibia
Right Mandible (18)	1.00	.268	.914	.010	.743	-.028	.284	.469	.558	.351	.932	.692	.692	.914
Left Mandible (15)	.268	1.00	.371	.308	.493	.422	.466	.628	.788	.778	.033	.778	.289	.308
Right Scapula (7)	.914	.371	1.00	.395	.525	.166	.404	.669	.445	.194	.866	.828	.686	.798
Left Scapula (7)	.010	.308	.395	1.00	-.359	.544	.219	.516	-.214	-.300	.035	.450	.024	-.143
Right Humerus (6)	.743	.493	.525	-.359	1.00	-.078	.114	.189	.812	.754	.570	.545	.389	.746
Left Humerus (6)	-.028	.422	.166	.544	-.078	1.00	.098	.485	.139	.050	.098	.200	.200	.071
Right Radius/ Ulna (11)	.284	.466	.404	.219	.114	.098	1.00	.394	.339	.170	.238	.488	.293	.219
Left Radius/ Ulna (14)	.469	.628	.669	.516	.189	.485	.394	1.00	.562	.361	.394	.808	.808	.593
Right Pelvis (3)	.558	.788	.445	-.214	.812	.139	.339	.562	1.00	.933	.339	.696	.603	.708
Left Pelvis (6)	.351	.778	.194	-.300	.754	.050	.170	.361	.933	1.00	.073	.547	.348	.476
Right Femur (1)	.932	.033	.866	.035	.570	.098	.238	.394	.339	.073	1.00	.488	.683	.866
Left Femur (3)	.692	.778	.828	.450	.545	.200	.488	.808	.696	.547	.488	1.00	.600	.639
Right Tibia (10)	.692	.289	.686	.024	.389	.200	.293	.808	.608	.348	.683	.600	1.00	.876
Left Tibia (7)	.914	.308	.798	-.143	.746	.071	,219	.593	.708	.476	.866	.639	.876	1.00

Table 4.7. Correlation matrix for pit elements in pits.

Discussion

It is difficult to separate the effects of differential access to both wild and domestic species and of sharing, particularly since both have been somewhat obscured by site formation processes (not all bones were deposited directly in pits). The small number of households represented in the excavated area also poses problems. There is a certain amount of evidence that meat sharing occurred to a fairly substantial extent at Opovo, chiefly involving in larger species. Both wild and domestic species were shared, although they were divided differently. This may in part reflect differences in body size and meat distribution over the body among the species, but may also result from symbolic association of certain species and their body parts or differing approaches to the sharing of the meat of wild and domestic animals or different species. For instance wild animals may have been shared routinely while domestic animals were shared only in the context of feasts, when different rules of distribution applied, or symbolic associations of particular species may have determined whether they were shared before or after cooking. On the whole, it appears that sharing was symmetrical. There is little sign of good cuts flowing to one household, suggesting that prestige was gained through distribution rather than possession.

Since there appears to be a size effect in sharing, with meat sharing much less common among sheep-size animals (judging from roe deer element distributions), we may best able to examine variability in herd size and hunting skill/frequency among the sheep/goat and roe deer. Apart from what may be seasonal variation seen in some of the pits used for shorter lengths of their nature there are probably the taxa that would be likely to vary least among households. Roe deer may have been hunted chiefly as crop raiders rather than actively sought out. Their relatively even distribution may at least indicate that all households represented in the excavation area did cultivate their own fields. Sheep and goats are relatively small, cheep livestock that would be relatively easily accessible even to poorer households. One would expect more variation in the number of cattle per households, but any such variability has been obscured by the apparently extensive sharing of beef.

For the red deer, since there is a suggestion that the rear limbs may have been shared more than the from part of the body (or else the front half was treated as a unit), we may perhaps be able to locate the better or more frequent hunters by identifying the pits where the fore limbs predominate: 5/31 and to some extent 41, from building horizons 2 or 3. Since these are both large and elaborate pits with evidence of feasting, it is quite possible that the pattern here does not reflect everyday subsistence differences, but the ability of these households to collect game for feasts, which probably involved manipulating kinship and exchange partnership networks as well as hunting to amass the necessary food.

The possibility that some pits, which appear to be seasonally restricted in their use, indicate different patterns of faunal exploitation during the late summer and autumn from the rest of the year provides clues to problems of labor scheduling faced by the Opovo dwellers. It would appear that agricultural work, including protecting the harvest from marauding roe deer in particular, was paramount at this time. The two such pits with seasonally data covered a minimum span of September-December (feature 45) and August-September (feature 46). The autumn killing of both wild and some domestic species seen in the seasonality profile for the site as a whole (Russel 1993) must either have occurred after the harvest was in or have been carried out by only some households.

Acknowledgments

I would like to thank Ruth Tringham and Bogdan Brukner for enabling my participation in the Opovo Archaeological project. Svetlana Blazicc provided me with space and support during one phase of the analysis at the Institute for the Protection of Nature in Novi Sad. I received help in the analysis from several able students: Heide Arteaga, Kathleen Bolen, Sheelagh Frame, Kirstin Karipides, Lisa Lindsey, Preston Miracle, Andrew Singer, Martin Vidal, and Oren Zaidel.

The Opovo Archaeological project as a whole was supported by grants from the Joint U.S. – Yugoslavian Institution; and the University Research Expeditions Program of the University of California. In addition, I received support for this project from the Wenner-Gren Foundation, the International Research and Exchange Board (IREX), the Department of Education Fulbright-Hays Doctoral Dissertation Research Abroad Program, the Mellon Foundation (through the Center for Slavic and East European Studies at the University of California at Berkeley), the Lower Fund of the Department of Anthropology at U.C. Berkeley, the Humanities Graduate Research Grants and the University Predoctoral Humanities Fellowship of U.C. Berkeley.

Chapter 5

Kapitan Dimitrievo Tell Sequence and Cultural Characteristics

(Brief Report in the Light of the Sounding Excavations in 1998-1999)

Vassil Nikolov

Introduction

The tell of Kapitan Dimitrievo, known in the literature as Banyata tell, is located 1.5 km west of the village of Kapitan Dimitrievo, Peshtera District - the westernmost part of Thrace. It is situated on a terrace sloped to the east, probably over very low natural hill, and has height of 13 m. The tell is oval shaped and has diameter 140 x 140 m at the bases.

The earliest archaeological investigations at the Kapitan Dimitrievo tell were directed by Petur Detev in 1947-1948 who excavated two trenches. The first one was at the eastern half of the tell, 240 sq. m, reaching sterile sediments at a depth of 12 m. Thick cultural layers from Neolithic and Chalkolithic were documented. The second one was on the northwestern periphery of the tell, 108 sq. m, reaching the depth of 7 m. The cultural layer was mainly from the Bronze Age.

P. Detev published relatively short report of these excavations (1950a). The analysis of the published by P. Detev material and his observations has provided sufficient evidence of the character and the periodization of the prehistoric material culture of the Kapitan Dimitrievo tell and highlighted its importance for the investigation of Neolithic, Chalkolithic and Bronze Ages in western parts of Upper Thrace. However, the information was limited and required new investigations of the tell.

The archeological investigations on the Kapitan Dimitrievo tell in 1998-1999 consisted of four small trenches in different parts of the cultural layers. They encircled the main trench of P. Detev. The aim was to uncover the Neolithic layer. It was a reason three trenches (soundings 1, 2 and 3) to be dug over the lower, peripheral areas of the tell. The last trench (No. 4) was situated at the highest part of the tell, where the Chalkolithic layers were investigated. The information collected relates to all Neolithic periods (the sixth millennium BC) and the second half of the Chalkolithic (the second half of the fifth millennium BC).

Stratigraphy

Trench 1

It was made on the eastern periphery of the tell in 1998. The total area was 15 sq. m. The sterile sediments were reached at a depth of 2.25 m. The lowest layer in the trench was homogenous almost without building remains, 1.00-1.20 m thick. It was characterized by white painted pottery over red slip attributed to Karanovo I culture. The overlaying cultural layer was re-deposited in Antiquity; without building remains. The pottery belongs to Late Neolithic and Late Chalkolithic. The thickness of the layer varied between 0.30 m and 1 m. The third cultural layer was represented by thin clay floor of a house and three deep pits. It dates from the Late Iron Age (6th-5th BC). The uppermost layer includes an oven for building pottery from Classical Antiquity.

Trench II

It was situated over the steep northwestern slope of the tell (10 sq. m) and was investigated in 1998-1999. It has been not finished. The main layer in the trench belongs to Karanovo I culture, c. 3.80 m thick. Several building phases have been documented. The earliest two remain unexplored. They consist of building horizons with burnt plasters, documented by trial trenches reaching natural sediments. The building phases of two superseding houses followed, earlier of which was burnt (marked as No. 2 and No. 6 respectively). Along with the characteristic white painted pottery, in the lower house a polychrome (brown and white painted) sherd was found. Uppermost levels consisted of white ash spots of open hearths. Along with white painted pottery a polychrome painted with light brown and white sherd occurred there.

[1]The recent excavations have been organized by a team from the Archaeological Institute with Museum at Bulgarian Academy of Sciences, in cooperation with the Historical Museum in Peshtera Town. They were carried out during two archaeological seasons (from 13 July to 7 August 1998 and from 29 August to 17 September 1999) and directed by V. Nikolov. V. Matsanova (Pazardzik), T. Stefanova, B. Bozhilov, and K. Bachvarov (Sofia) participated in the team. Several scholars joined the excavations for a short time – M. Gurova (1998), V. Petrova (1999) and V. Vandova (Kjustendil) – as well as A. Chokhadzhiev (V. Turnovo) and V. Skerletova (1999, Sofia, Vienna), students in Archaeology. Dipl. Eng. P. Valev set the plan-square working net. E. Marinova (Sofia, Bonn) collected samples for palaeobotanical analyses. D. Pavlov, Director of the Historical Museum in Peshtera Town, rendered considerable support to the organization of the excavation. Of special importance was the assistance of M. Tao from Tokai University, Haratsuka, Japan, who provided a part of the excavation expenses. The team of the investigations thanks the villagers of Kapitan Dimitrievo who supported the fieldwork.

A month after finishing the campaign in 1999 at Kapitan Dimitrievo, the results of the excavations in 1998-1999 were published in a small monograph in Bulgarian (Nikolov et al. 1999). In this chapter I am taking the advantage of the kind invitation of the editor of the book L. Nikolova briefly to present the main results of the excavations to a vaster circle of scholars abroad.

House 1

It was situated c. 1.20 m above the sterile base of the tell. The house was oriented approximately in the directions of four cardinal points. In the trench a part of the western wall and parts of the basic inner equipment have been documented. The oven was situated in front of the western wall, perhaps in front of its central part. Its base was approximately horse-shoe-shaped with podium and opening to the east. Next to oven's northern side there was a grain grinding stone attached to its wall, with a place for flour. Five round or oval grain storage-places were documented built in situ over the clay floor. In front of the oven and the grain grinding place there was the front part of another base of oven - podium and dome belonging to it. The massive clay body was almost in customary position. Broken stone from the base filling were documented behind the equipment. Unlike the former oven, the podium of the latter one was burnt to red and cracked. The equipment was recovered over well preserved part of clay ceiling. The construction of the oven and the ceiling were documented over red sherds of the walls of the grain storages, as well as over well burnt sherds of a vessel; another part of the vessel remained just near the remains described. The stratigraphic data suggested that the whole construction of oven parts and of ceiling fell from some height and covered the grain storage and the vessel over clay floor. The construction was a base of podium of the oven of the upper house (No. 1) and didn't not lie in situ in front of the oven of the lower house. Accordingly, the only explanation of the situation was that it fell from the second floor of the same house. The floor was made of rammed yellow clay (that layer was not investigated). Among the building's ruins relatively numerous sherds were found. In one of the grain storages there was a stone adz, and in another one a roe horn, diaphasa of cattle, stone adz and small anthropomorphic figurine. The figurine is made of yellow clay, unburnt, but hard because of the high temperature of the house burning. It is female with massive lower part, steatopygia stressed and schematically represented upper part with two small bulges-arms.

House 2

It was recovered c. 1.80 m above the sterile sediments. The house was oriented approximately in the directions of four cardinal points. Only part of its western wall and a small part of the main inner equipment, an oven and grain storage, occurred in the trench. The oven was found in front of the western wall, probably in front of its central part. Its base had approximately horse-shoe shape with a podium and related opening to southeast. In the area of the oven 15 cm thick clay floor of the house was documented. The second cultural layer in the trench belongs to the Karanovo III-IV period, but it was documented over small area only, c. 60 cm thick. There were no architectural remains.

Trench III

It was situated at a higher level on the south slope and, 15 sq. m, dug in 1998-1999. The excavations revealed the Late Neolithic layer. They stopped reaching down the level of the Early Neolithic.

Two cultural stages have been documented in the Later Neolithic layer. The lower layer was c. 80 cm thick. There are no architectural remains documented, but numerous white and whitish levels have been uncovered. The considerable amount of sherds and finds registered there infers that the layer was part of a vast area of dumping ground. The ceramic complex has the most important features of Karanovo III-IV period in Thrace. In this layer a big refuse pit was cut in. After being filled up, the disposition of garbage and ash continued on the same place, covering the pit. The garbage heap was 2.5 m thick. Whole filling of the pit and the heap over consisted of a big amount of sherds and finds from Late Neolithic. The ceramic complex has peculiar features named Kapitan Dimitrievo.

Trench IV

It was made over the highest part of the tell in 1999, 7 sq. m, and reaches a depth of c. 3.5 m. All cultural debris under the destroyed layer related to the Late Chalkolithic. The remains of that period were not completely investigated. At the lowest level (at a depth of 3.5 m) there were traces of house floor and oven. At a depth of 2.10 m, there was also badly damaged remains of oven. At a depth of c. 0.80 m a seriously damaged hearth made of well burnt clay was documented. A big refuse pit began at a depth of 1.80 m in the southwest part of the trench and reached its bottom. The pottery shows mainly the characteristics of Karanovo VI culture, but at least in the uppermost stratum there were some elements characteristic of the Central Balkans (Krivodol).

Therefore, in 1998 and 1999 the succeeding Neolithic periods Karanovo I, Karanovo III-IV and Kapitan Dimitrievo, as well as Late Chalkolithic were investigated at the tell of Kapitan Dimitrievo. There is no new information concerning Early Chalkolithic layer excavated by P. Detev. The data concerning the stratigraphy and periodization of the cultural layers at Kapitan Dimitrievo tell confirm the actualized picture of the Neolithic cultural appearance in Thrace (Nikolov 1998a).

Neolithic

Ceramic sequence

Ceramic complex of Karanovo I period

The following characteristic shapes are documented (***Figure 5.1***).

Tulip vessels. They are with prolonged body over high pedestal cylindrical or slightly conical base. Most of outer walls of the vessels are decorated with white.

Plates. They represent a big part of the ceramic complex. They have a shape of a truncated cone turned downwards, flat base or ring-like low base. Some of the exterior and interior walls of the vessels are ornamented with white paint.

Bowls. The bowls are frequently encountered with shapes that demonstrate a big variety: with rounded semispherical body, S-profiled, with semispherical body and profiled rim of the orifice shaping short neck, biconical with rounded biconical body and short conical neck, or with point-biconical body and

Figure 5.1. Kapitan Dimitrievo tell. Earlz Neolithic, Karanovo I period. Ceramic vessels.

short neck. A big part of the exterior walls of the vessels are ornamented with white paint.

Pot-like vessels. They have semispherical body and short cylindrical neck. The bases are usually low ring-like. The outer walls of the vessels of that sort are commonly decorated with white paint.

Closed pot-like vessels. According to their neck and body, those vessels can be attributed to four series: with high neck and flattened to spherical body, with short and higher cylindrical neck and spherical body, with short neck and prolonged body, with spherical body, high neck and "Fussboden". The base in most cases is flat, but sometimes low ring-like base characterize the vessels. Some vessels are ornamented with white paint, with channels or shallow incisions. As a whole, the pottery from the layer investigated represents the main characteristics of the ceramic complex of Karanovo I period in Thrace, but the characteristic peculiarities of the western parts are documented clearly as well.

Ceramic complex of Karanovo III-IV period

The following shapes are among the distinctive features of that period (***Figure 5.2***).

The bowls with inverted orifice. The vessels have open two-parted body. The lower part has a shape of a truncated cone turned downwards. The upper one has slightly conical shape. The short under rim part is shaped by thickening of well represented carination and gradually becomes thinner in direction to the orifice. Hence, in the upper part of the vessels there is biconical outer and rounded inner side of the profile.

Bowls. The vessels are with comparatively deep, wide and slightly closed body: rounded and without flattening, whose widest part is approximately the middle of the height, with cylindrical and conical body, with biconical body; the medium edge is high, under the orifice.

Plates. The vessels with open conical body turned down, the walls are always not flatted, the rim is straight.

Flaring plates with thickened rim. The vessels are with segmentation profile. They lie over flat base or over four cylindrical 'feet'. There is incised ornamentation of parallels oblique lines over the inner wall of the thickened orifice part.

Vessels (plates) for food preparing over embers. They have truncated turned down body with wide flat base. The walls are thin. The outer surface is very rough, with non-smoothed folds of modeling over mould. The orifice is thickened underneath. The inner surface is very smooth. The vessels sometimes have two opposite triangle handles, which raise from the rim.

Tub-like containers. They have truncated cone turned down with wide flat bottom. The orifice is wide profiles outside and as exception it is ornamented with oblique channels.

Jugs. The vessels with pear-like body and high neck, with one vertical stick-like handle with button.

Cylindrical cups with a high handle. The bottom is flat. The handle is vertical stick-like with cylindrical button, which has slightly mushroom-like widening above.

Closed pot-like vessels. They have spherical body with lower or higher cylindrical neck.

Lids. They are with cylindrical and conical body with equal cut and slightly unflatted upper part.

The ceramic complex from the first half of the Late Neolithic in Thrace is the only one with stratigraphic data from its western part. It provides the main evidence for distribution of the Karanovo III-IV culture over the whole territory in Thrace (Nikolov 1998) including the upper catchment area of the Maritsa River.

Ceramic complex of Kapitan Dimitrievo period

The following ceramic shapes are significant of that period (***Figures 5.3-5.5***).

Bowls with inclined orifice. They have two-part body with stressed turning point. The lower part has a shape of truncated cone turned down and the upper one is conical. The short orifice part is shaped by thickening of well-stressed turning point and gradually thinning in directing of the orifice. The thinned rim is often stressed outside by shallow horizontal channeling. In most cases the filed between the rim stressed and the turning point is covered with parallel vertical and oblique channels.

Bowls. There are several series distinguished: with almost cylindrical upper part, with unflatted body and slightly profiled rim, as well as biconical with short upper part.

Cylindrical cups. They have a vertical stick-like handle with button.

Tub-like containers. Vessels with truncated wide body, flat bottom, straight walls and turned outside rim.

Jugs. Vessels with pear-like body and high neck, with one vertical handle with button.

The closed pot-like vessels. The body is spherical and slightly biconical, with shorter or higher slightly conical neck. The distinct vessels have two arch-shaped handle attached at the high neck oppositely placed.

Open pot-like vessels. They have prolong rounded and little biconical body with straight orifice. In some cases those vessels are ornamented with bitumen.

Plates. They have a shape of truncated cone, with slightly unflatted walls. Significant of the period are those vessels, which walls are covered from both sides with decoration of incised lines, sometimes encrusted with white matter. Some of the bowls consist of high pedestal base, as parts of high conical pedestal bases have been found ornamented outside.

There are bottoms of two plates found, with densely incised and encrusted with white decoration of Karanovo IV type.

Plates with thicken orifice. They have a body-shape of wide truncated cone with flat base and as an exception with four 'feet'. Parts of the vessels have thicken from inside orifice stressed, sometimes ornamented with parallels incised lines.

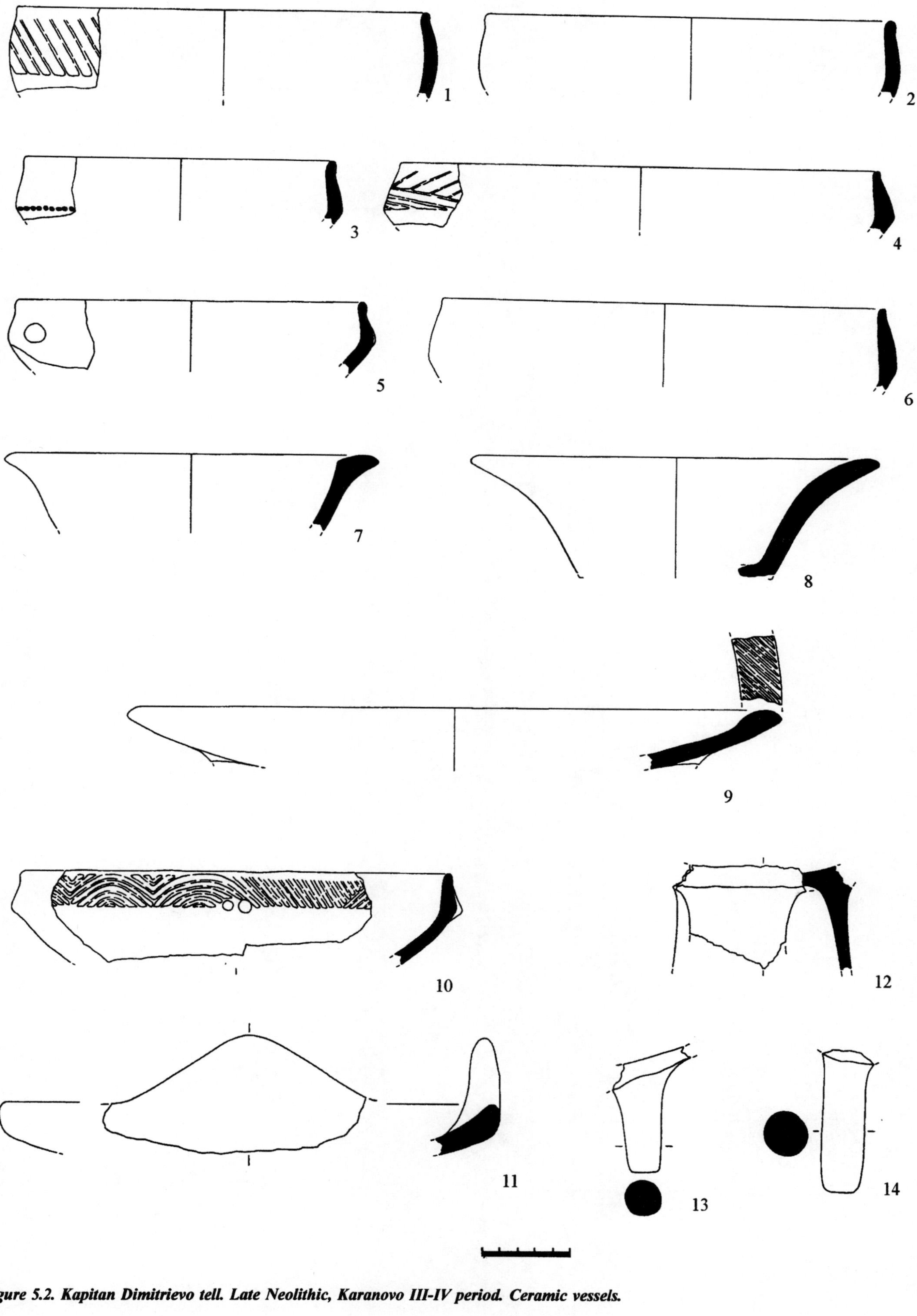

Figure 5.2. Kapitan Dimitrievo tell. Late Neolithic, Karanovo III-IV period. Ceramic vessels.

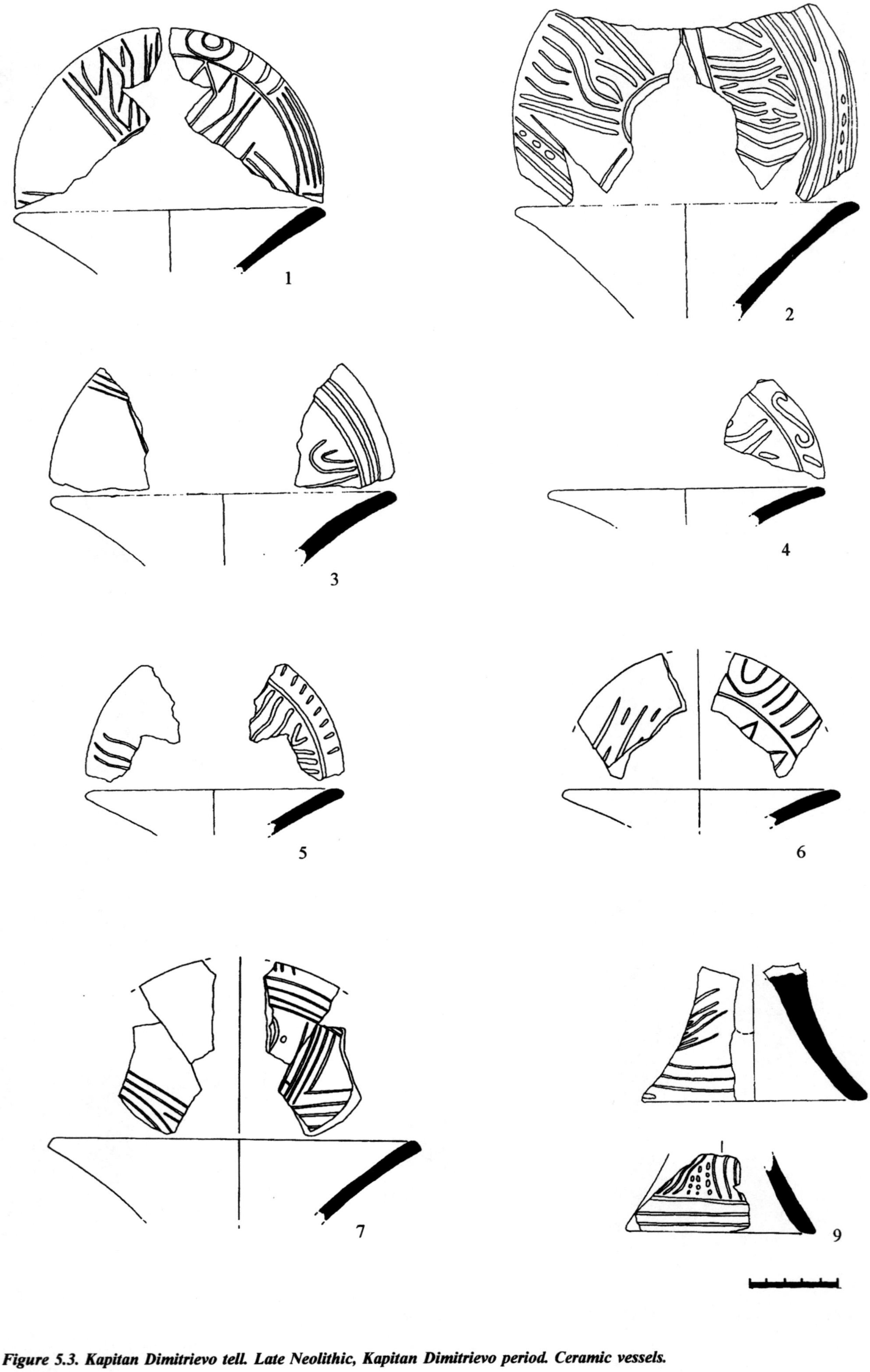

Figure 5.3. Kapitan Dimitrievo tell. Late Neolithic, Kapitan Dimitrievo period. Ceramic vessels.

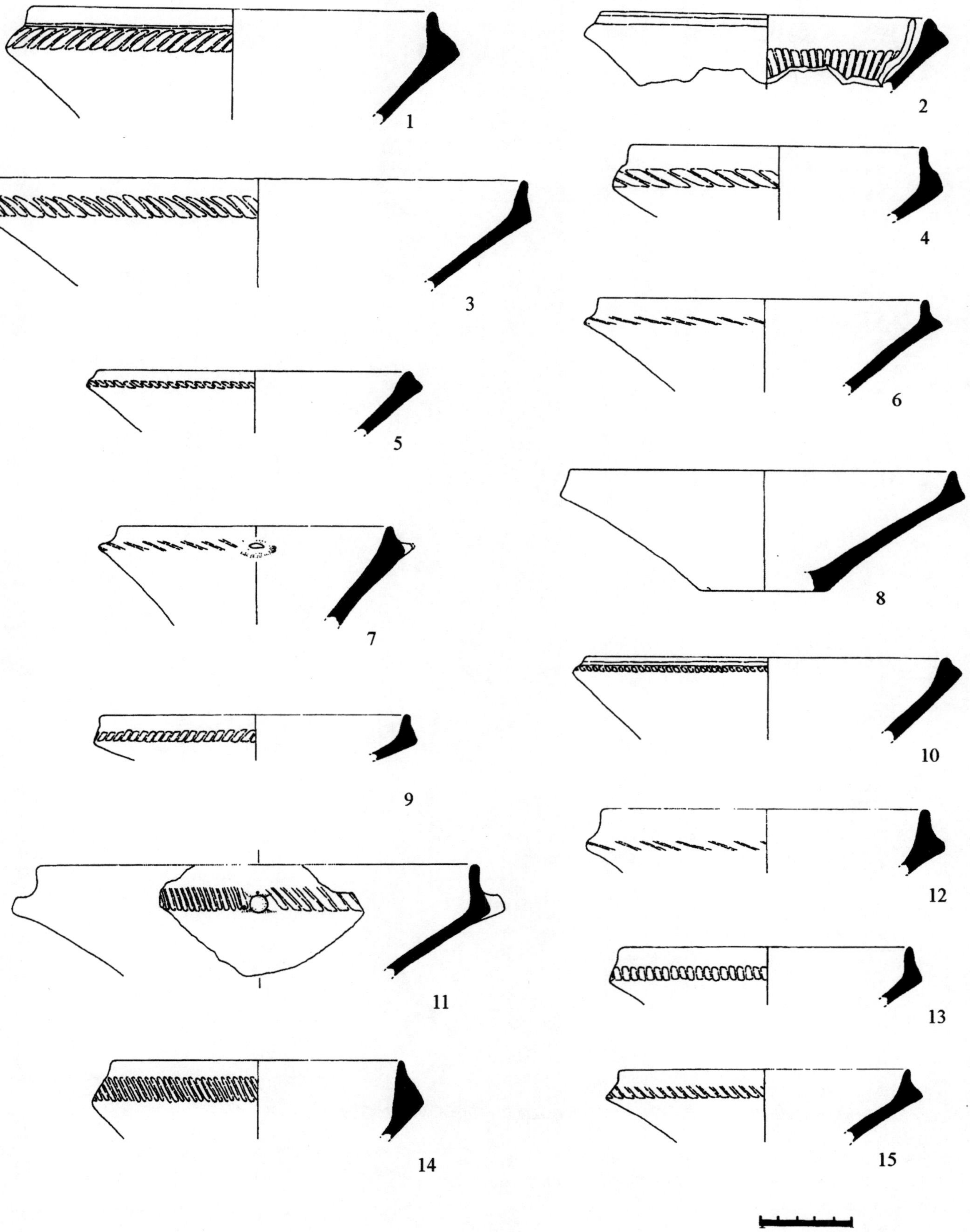

Figure 5.4. Kapitan Dimitrievo tell. Late Neolithic, Kapitan Dimitrievo period. Ceramic vessels.

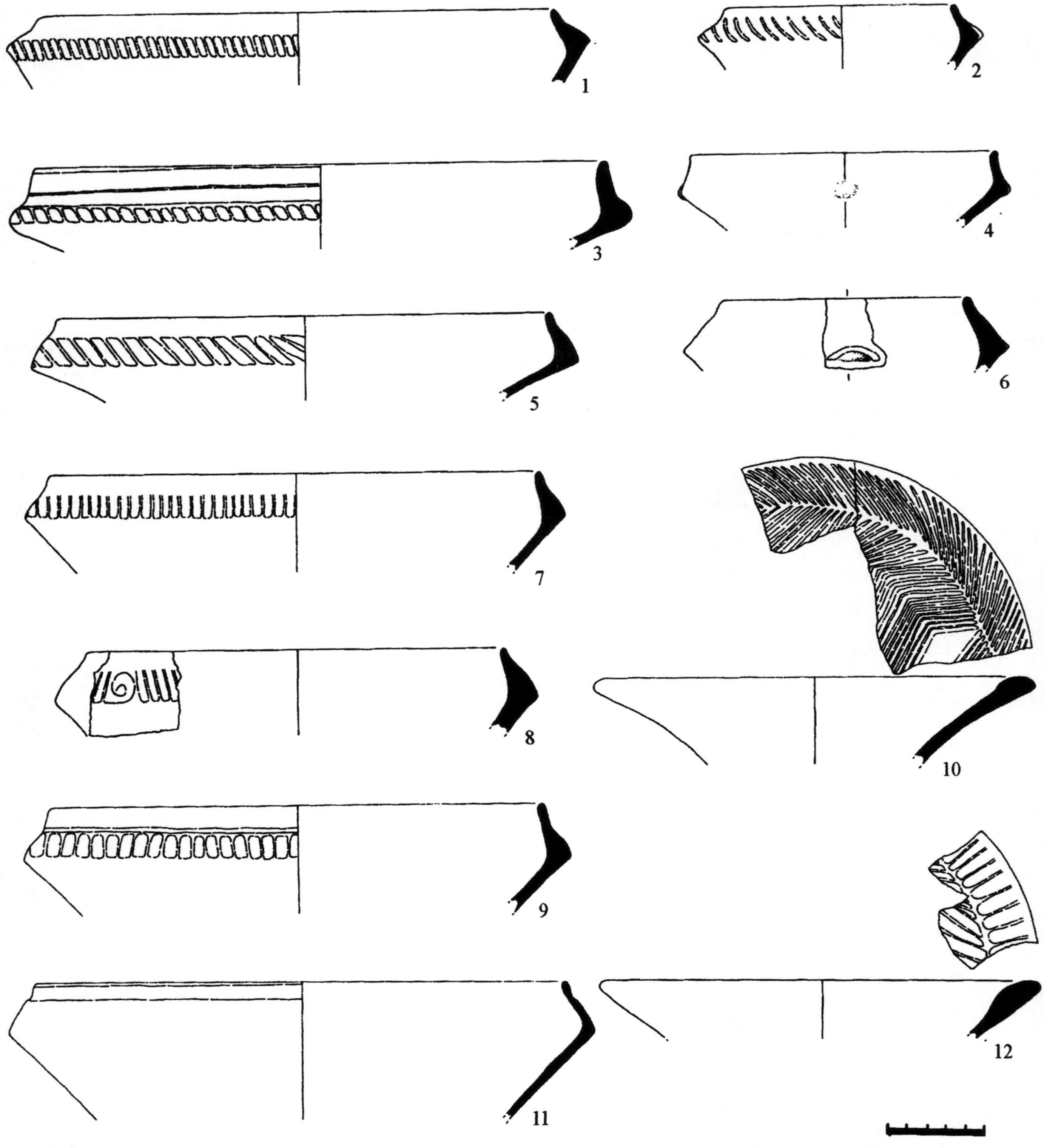

Figure 5.5. Kapitan Dimitrievo tell. Late Neolithic, Kapitan Dimitrievo period. Ceramic vessels.

Vessels with completely incised outer and inner walls with thin lines and sometimes white encrusted ornamentation belong to that group.

The rest of the vessels have gradually thickened orifice part, which sometimes is stressed outside with one or two horizontal channels and covered with parallels oblique cut channels.

Vessels (plates) for food preparing over embers. They have a shape of truncated cone, probably with rounded wide bottom. The walls are thin. The outer surface is rough, with non-smooth folds with traces of mould modeling. The walls are thin. The orifice is little thickened underneath. The inner surface is very well smoothed. In some cases the vessels have two opposite triangle handles with a hole starting from the rim.

Lids. They have conical body with little convex upper surface.

The ceramic complex of Kapitan Dimitrievo is a newly defined Neolithic appearance, which perhaps take place in the western parts of Thrace only. It represents a clear tradition of the previous complex Karanovo III-IV and consists of elements bearing in mind the pottery of the synchronous culture of the Karanovo IV in the eastern parts of Thrace, as well as of elements characteristic of Eastern Macedonia and the Cen-

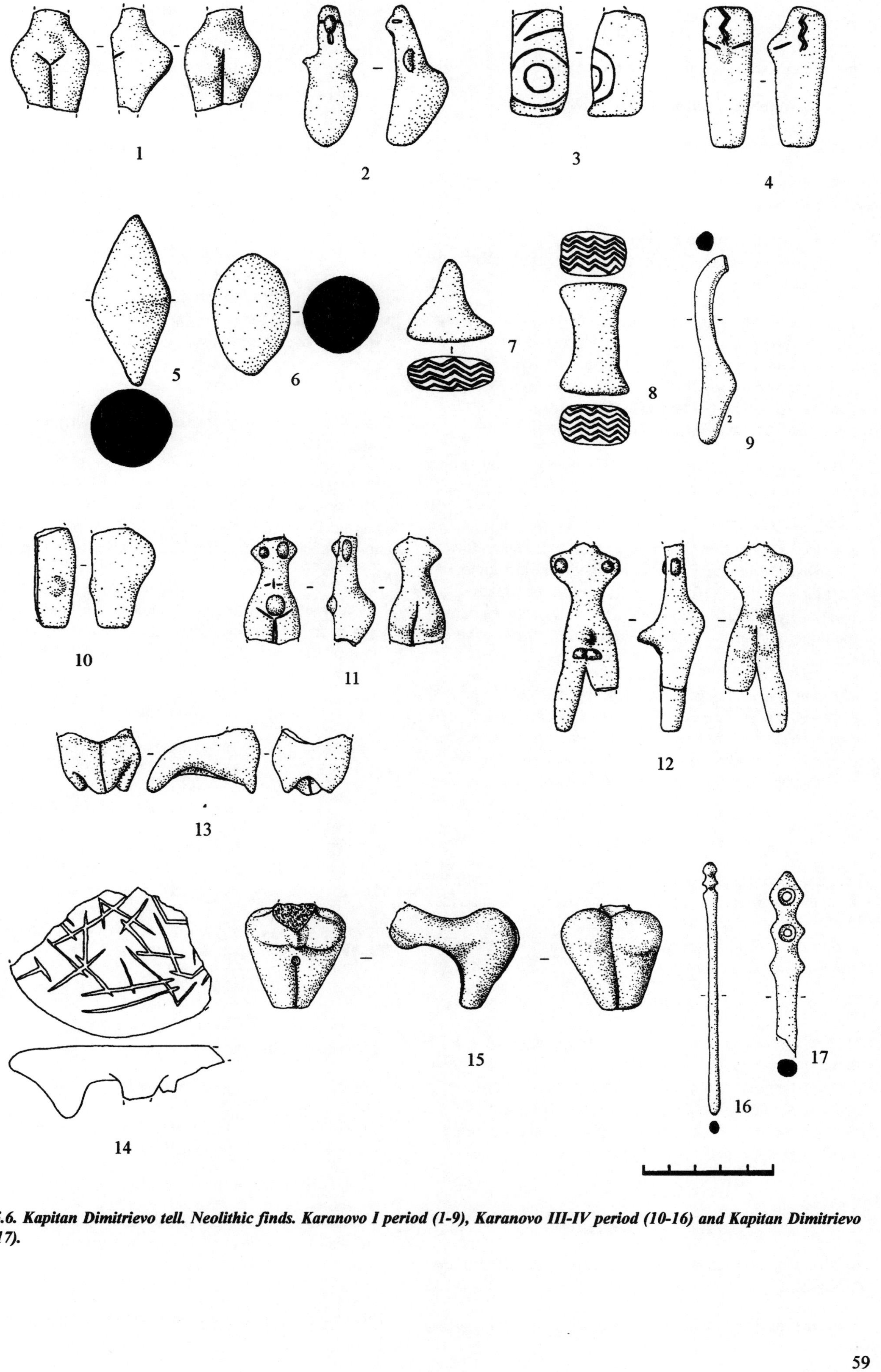

Figure 5.6. Kapitan Dimitrievo tell. Neolithic finds. Karanovo I period (1-9), Karanovo III-IV period (10-16) and Kapitan Dimitrievo period (17).

tral Balkan zone.

Other finds

Of the numerous finds only few are present here (***Figure 5.6***).

Ceramic anthropomorphic figurines

Standing figurines. They include:

A. Female figurines with massive, separately modeled and close-fitting legs.

B. Female figurines made of one piece of clay. Vertical incised lines distinguish the legs, or the last are compactly modeled.

C. Hermaphrodite (a figure with female and male gender elements represented). The legs are separated and conical modeled. The waist is narrowed, the arms are represented by short horizontal bulges. Additionally attached clay bulges mark the breast. A conical bulge represents phallus. Underneath, two flattened clay balls attached designate testicles.

The sitting figurines. They are with separately modeled legs bent in the knees. The fertility triangle is sometimes marked.

Inclined figurines. There is one inclined forward figurine. The legs are conical, modeled separately and fitted. The buttock is very stressed. The upper part is inclined forward almost at right angle with regard to the legs and is represented schematically. A round hole depicts the vulva starting behind at the upper part of the buttock where traces of red paint are occur.

Anthropomorphic heads. They have cylindrical shape and triangle relief nose. The eyes are marked by oblique cuts. The crown is equally cut and represented as conical bulge pointed backwards. From both sides of the head and above the nose of one of the figurines there are by two vertical zigzag incised lines.

Ceramic finds of pintadera sort

There are three finds discovered, one of which is compact. Two of them are characterized by prolonged oval-face surface and narrowed approximately conical handle. The third one is double-sided - Slatina type. Over the surface of all three finds there are parallel zigzag lines.

Cult tables

The cult tables are triangle and rarer quadrangular, with entered or raising above the table basin. Among the rest of the finds there are fragments of bracelets of Spondylus, fragments of marble and ceramic bracelets, fragment of stone ring, a part of bell hook anthropomorphically shaped, beads - ceramic and stone, decorative bone pins etc.

Late Chalkolithic

Ceramic characteristics

The following ceramic shapes are among the significant ones (***Figure 5.7***).

Cups. Biconical cups with two symmetrical vertical handles. The ornamentation is mainly of graphite, positive painted over the upper part of the body walls and the handles.

Wide-open plates with inverted thickened rim. The ornamentation is mainly of graphite, positive, on the inner wall. In some cases the graphite decoration is in combination with red distempered paint.

Wide-open to medium open high plates with everted orifice. When there is ornamentation, it is graphite, positive, mostly over the inner wall. The outer orifice part is sometimes ornamented as well.

Bowls. Part of those vessels have deep rounded body with straight rim. The body is often covered with incised or relief decoration, rarely with graphite. The rest of the vessels are biconical, with stressed carination. The decorations is positioned over the turning point (channels or shell-incised impressions) or over the upper part of the body (graphite or mixed graphite and distemper painted).

Pot-like vessels. They are deep vessels with little unflatted walls and compact body.

Closed pot-like vessels. They have biconical body and short cylindrical neck, with one handle. One other group has spheroid body with two vertical handles ornamented with a graphite positive decoration painted over the wall of the upper part of the vessel.

The comparative-typological analysis allows the ceramic complex from sounding IV to have been attributed to the third phase of the Late Chalkolithic (end of the fifth millennium BC).

Other Finds

Of numerous finds only some will be represented below (***Figure 5.8***).

Ceramic anthropomorphic figurines

Standing figurines. A. They are made of a piece of clay with compact modeled legs, distinguished with narrow hollow. B. They are with separately modeled legs fitted.

Sitting figurines. The figurines are separately modeled with legs fitted. The buttock is very stressed.

Anthropomorphic handle of a vessel

The indications of the anthropomorphic head are stressed with white and red distempered paint.

‚Cross-idols'

The ceramic finds with conical and flattened body with four opposite bulges (Matsanova 1994).

Anthropomorphic and zoomorphic vessel's handle

The handle represents an animal head and straight aside arms. It is decorated with graphite and distempered paint.

Cult tables

There are three kinds. A. Four-sided, with equal upper surface

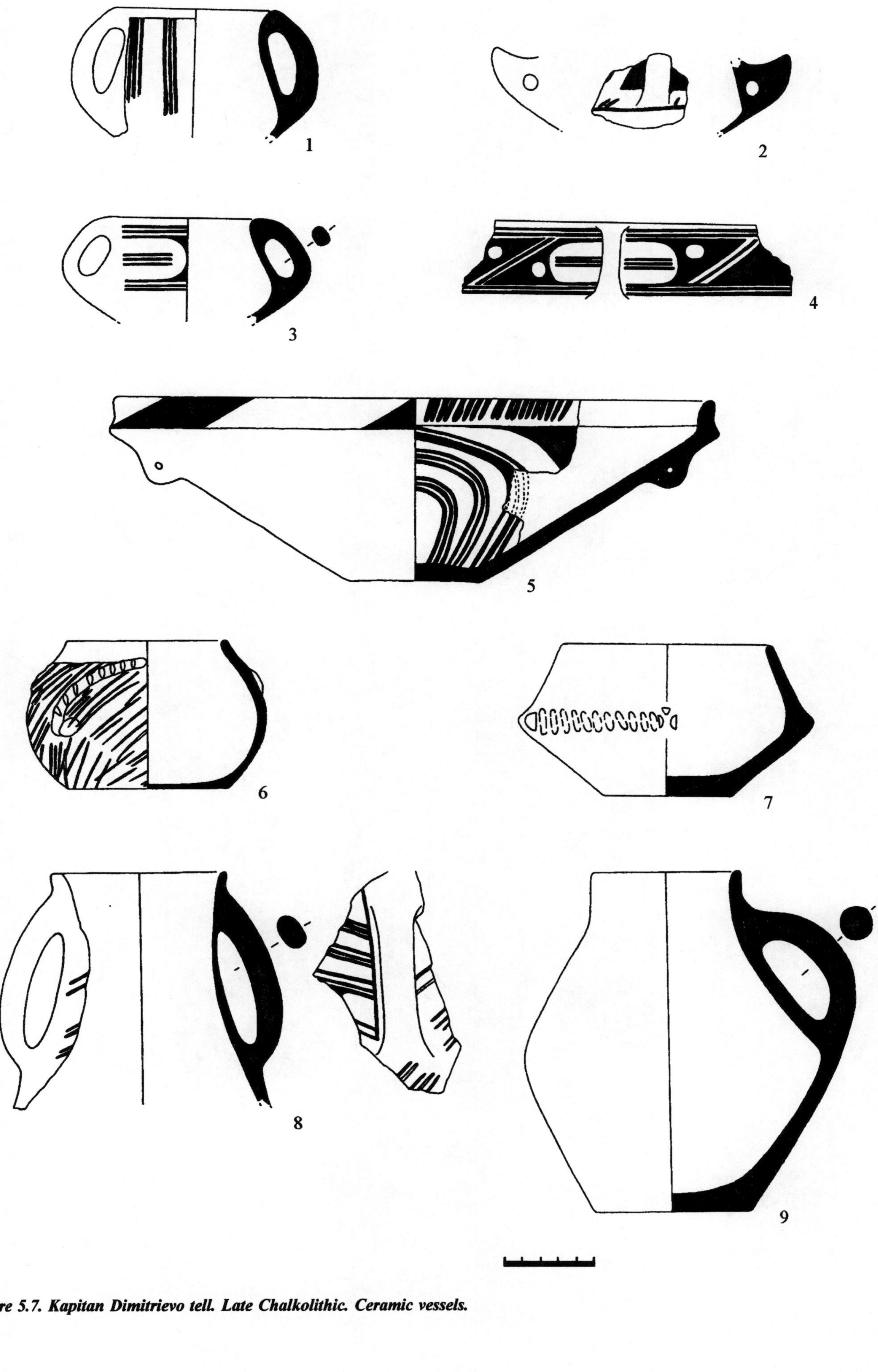

Figure 5.7. Kapitan Dimitrievo tell. Late Chalkolithic. Ceramic vessels.

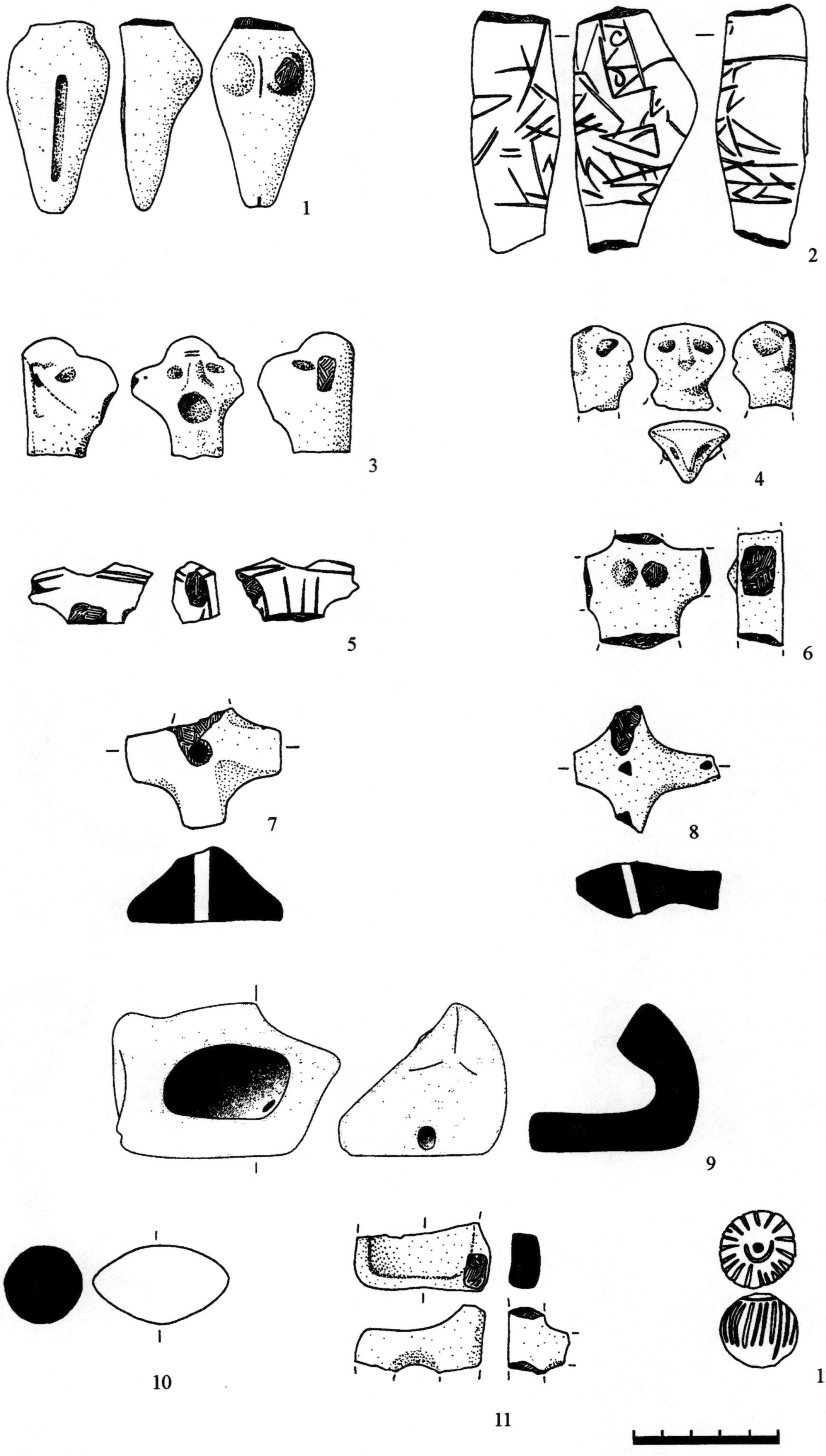

Figure 5.8. Kapitan Dimitrievo tell. Finds from the Late Chalkolithic layer.

placed over four feet. Over the front side there are two almond-like eyes. There is an instance when the object is completely covered with red paint, and the upper part of the ‚face‘ is additionally decorated with white painted lines. B. Four-sided, perhaps with quadrangular entered basin placed

over four ‚feet‘. They are decorated with incised lines or with chess-like ornament. C. Round in plan, with entered round basin, placed over four ‚feet‘.

Ceramic model of oven

Conical bulges modeled are modeled aside of the front side and at the point of the cupola.

A Ceramic chair (throne)

It has rectangular shape, four short legs, little concave upper surface and perhaps arch-shaped back. Among the rest of the finds there are copper ring, ceramic beads, fragments of Spondylus bracelets, knuckle-bones with smoothed walls.

Discussion and conclusions

The trial investigations at Kapitan Dimitrievo tell in 1998-1999 show that the cultural layers studied provide a solid base for constructing the stratigraphic sequence, periodization scheme and chronological system of a later prehistoric period in the western parts of Upper Thrace. In other words, Kapitan Dimitrievo tell may be considered as the second important after Karanovo tell focal point for the Neolithic and Chalkolithic in Thrace (the sixth and the fifth millennia BC).

According to the character of the material culture, in the Neolithic layer of the tell, there are three succeeding each other cultural and chronological periods documented.

The Early Neolithic period Karanovo I represents a long development during the first and the beginning of the second half of the sixth millennium BC. The documented for the time span studied maximum thickness of the cultural layer (3.80 m) is respectful. A peculiarity of that period is the production and using of ceramic vessels with white painted ornamentation. In the later stages some vessels with dark and polychrome decoration occur. Despite some local particularities, the white painted ornamentation is included in the common characteristics of the Thracian ceramic complex Karanovo I. The dark and polychrome decoration is a sign of cultural contacts with Starèevo culture complex in the Central Balkans. P. Detev left us a documented in the earlier layer big building, which combined with the information from trench II, constitutes sound evidence for the high level building skills of the early farmers. This concerns mainly the remains of the two-floor house - the earliest for the time being building of that sort in Southeast Europe.

The Late Neolithic Karanovo III-IV culture indicates longer development - probably it takes place during the third and beginning of the last quarter of the sixth millennium BC (these data correspond to those already known from the rest of Thrace). The maximum documented thickness of that layer of the tell is c. 60-70 cm. A special distinctive feature of the material culture of that time is the production and using of ceramic vessels with dark (mostly gray and black) surface. Similarly to the previous period, the vessels are often of biconical shapes and the painted decorations disappear. Instead the channel pottery occurs.

The ceramic complex obtained all the characteristics of the Karanovo III-IV period, which is registered over the whole Thrace. For the time being, there are no architectural remains, but it should be stressed the iconographically rich types of anthropomorphic plastics including a figure of hermaphrodite.

Perhaps the Late Neolithic period of Kapitan Dimitrievo has longer development during most of the last quarter of the sixth and the very beginning of the fifth millennium BC. For the time being, the material remains from the only one refuse pit are registered and the garbage heaped at the same place. The period of Kapitan Dimitrievo represents a cultural phenomenon, which differs from the culture of the other parts of Thrace covering the same time-span. As indicator of the cultural processes, the ceramic complex consists of elements reminding the period of Karanovo IV in the eastern parts of Thrace – incised ornamentation and its structure, number of basic ceramic shapes. Also there are features whose origin points out to the Central Balkan area (high holed bases, some ceramic shapes). Of special interest are cultural contacts with the Aegean documented, through the Mesta valley (ornamentation with bitumen). The pottery data are additionally confirmed by the Truncülariopsis trunculus shell whose provenance is undoubtedly Mediterranean.

The Early Chalkolithic layer exists according to P. Detev's documentation, though the new excavations do not prove it. Its ceramic complex consists of Thracian, as well as of Central Balkan elements.

Obviously the Late Chalkolithic period of the settlement has long development during the second half of the fifth millennium BC. The ceramic complex is still not well studied but, it is clear, it consists of the main characteristics of the Karanovo VI period in Thrace.

Permanently or at least in some chronological phases a considerable part in it have the Central Balkan elements (pot-like vessels with two opposite handles, red distempered paint in the ornamentation). The data on the architectural remains, for the time being, include hearths and ovens only.

The tell of Kapitan Dimitrievo reveals traces of Bronze Age settlement but this period is still not investigated in the course of the renewed excavations.

At the end I would like to stress once again that the scientific interest in Kapitan Dimitrievo tell is defined by the importance of that tell to provide a stratigraphic and chronological source of meaningful information not only about the western parts of Thrace, but about a larger region. The last is likely to include Thrace and the Central Balkan region, as well as the Aegean area. The tell bridges the long existing gap of our knowledge that is provided by the investigation of such a prehistoric monument, which can give a possibility to establish the cultural and chronological relation between Thrace, the Central Balkans and the Aegean in the sixth and fifth millennium BC. The new trial excavations in 1998 and 1999 documented this site, while considerable effort is required to conduct large-scale excavations in future.

CHAPTER 6

TOWARDS THE NEOLITHIC AND COPPER AGE DEVELOPMENT IN THE BALKANS

(DATA FROM THE STRYAMA VALLEY)

Lolita Nikolova

In this study the Neolithic and Copper Age development in the Upper Stryama valley will be analyzed to approach the nature of the innovations in earlier Prehistory. It has been initiated by a publication of basic Neolithic and Late Copper Age record (Nikolova 1998a; Nikolova and Zakharieva 1994).

The diachronic view on the Neolithic and Copper Age cultures

The upper Stryama valley is one of the newly archaeologically investigated microregions in the Balkans. Located in the heart of the Balkans, for the long time that region was a white spot in the prehistoric investigations. Just at the beginning of 1990s, the discovering and initial investigation of several prehistoric sites on both sides of the Upper Stryama River considerably increased the fragmentary information left by P. Detev. On the one hand, the new archaeological record has documented intensive contacts of the population from the region under discussion with the western Central Balkans. On the other hand, it has posed important new research problems. For the topic of the innovations, that region can be interpreted as an instance of interaction model represented, where the geographic border is the western Sredna Gora Mountains dividing both parts of South Bulgaria, but being the active communication line interconnected both regions as well. Accordingly, the Upper Stryama in Prehistory became not a periphery of the richly developed Thrace, but an integrated and important part of the southern Balkan prehistoric cultural system.

The Upper Stryama valley includes the Karlovo hollow located at the foot of southern slopes of the central Stara Planina Mountains and bordered from south, east and west from the Sredna Gora Mountains. It is more open towards the Upper Maritsa valley through the Stryama drainage. The second micro-region is the Hissar hollow situated to the south of the Sushtinska Sredna Gora Mountains.

The position of the Upper Stryama valley in the Balkan culture system is geographically determined – the Stara Planina Mountains divide the region from the Danube basin, as well as the Sredna Gora Mountains from the Iskur and the Struma basins and partially from Eastern Upper Thrace. The northern and western geographical borders were also cultural ones, but as the archaeological material shows that statement is only partially true. First of all to the west, through the uppermost Stryama drainage, the population could deeply enter the western Sredna Gora Mountains where by contemporary Koprivshtitsa it was possible the population to reach the upper Topolnitsa basin and from there to go into the Iskur and Struma valleys. The tells from Chelopech, as well as from Chavdar in Zlatitsa-Pirdop hollow indicate strong contacts with Thrace, so the upper Stryama valley was in a position to connect both the micro regions in the Central Balkans, despite the geographic mountain barrier.

We have a system of cultural elements that confirms the active interaction function of the Upper Stryama valley from early Neolithic onwards.

There is a multiple system of geographical preconditions that made it possible the Upper Stryama to become one of the earliest occupied by agriculturist-stockbreeders areas in Upper Thrace. The sort of soil, the presence of arable areas near the main river of Stryama, as well as numerous streams and smaller rivers coming from Stara Planina Mountains, the hunting resources in the Stara Planina and Sredna Gora Mountains and only part of the favorable geographical background for the prehistoric population occupation there. In addition, the moderate climate, with winter rains and not very hot summers should be stressed.

Of special interest are the data on Early Neolithic from the Upper Stryama valley, as they show the region was actively consisted of dense and dynamic settlement structure and of population with active cultural contacts with the neighbor regions. At the same time, the population followed cultural standards, which had selectively relation to the innovations in the Balkans.

Neolithic

Early Neolithic

The contemporary investigations in the Balkan Neolithic outline several problem directions:

1. Grounding of chronological horizons including numerous interrelated sites from close and distant regions, through comparative stratigraphy, typological analyses and context interpretation of the radiocarbon dates – for instance common radiocarbon chronology of the Neolithic (Nikolova 1998a). However, there are missing detail comparative analyses even of the basic Early Neolithic sites in the Balkans.

2. Analysis of the mechanism of the Neolithization as a graduate process of development of agricultural and stockbreeding

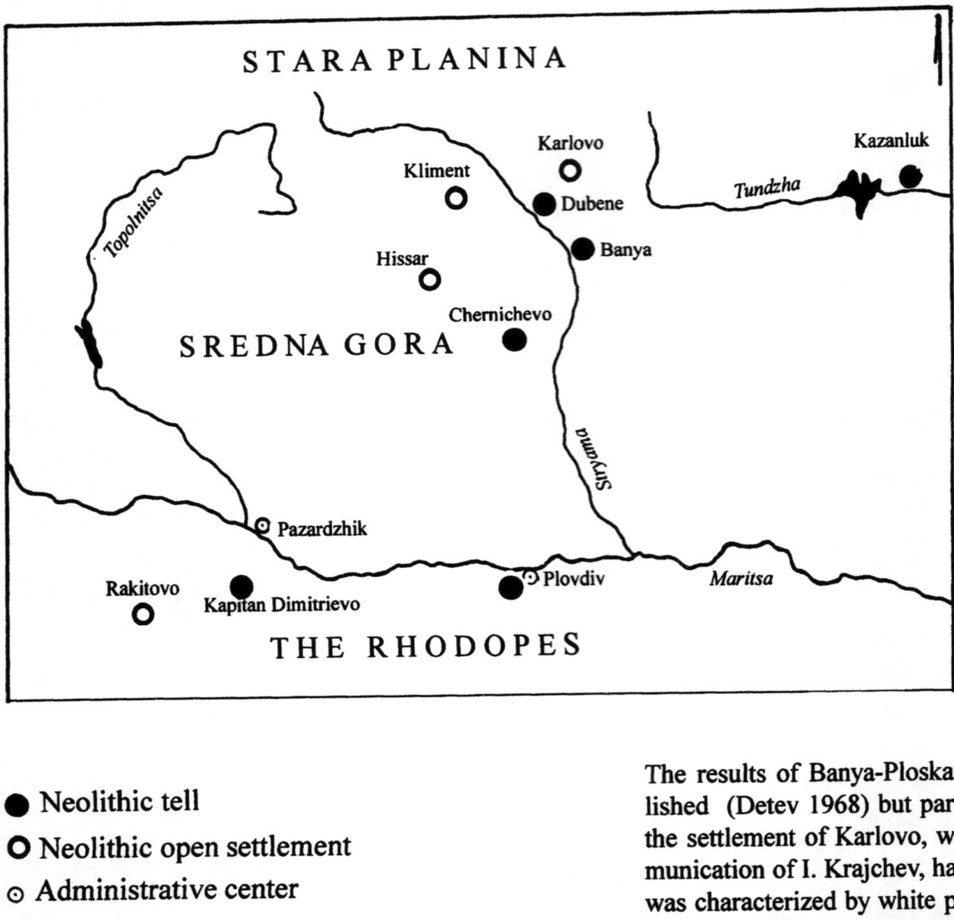

Figure 6.1. The basic Neolithic sites in Western Upper Thrace.

structures, including an adaptation of the Mesolithic population, especially in the Lower Danube, a colonization of fertile localities of the new population from Anatolia and a re-colonization of other regions from segmented social structures of the initial Early Neolithic substrate, and / or following migration groups (Nikolova 1998a with ref.; Nikolov 1986; see Willis and Bennett 1994 for the renewed discussion of the interrelated problems if the distribution of the agriculture on the one hand, and the distribution of the population and languages, on the other hand).

3.Analysis of the micro-social structures and a development of the concept of I. Hodder of 'domus' from theoretical point of view, for instance, through micro structural archaeological analyses such as the house of Sofia-Slatina (Nikolov 1992), modeling of the Balkan Neolithic village (Chapman 1989), analysis of the mechanism of cultural contacts between close and distant social groups (see for instance Nikolov 1986), etc.

4. Intensification of the micro-regional analyses, which would give a possibility of extraction of new common tendencies and argumentation of the peculiarities of the Neolithic development in the Balkans.

5. Macro-regional comparative investigations for an argumentation of different models of Neolithization, including parallels with neighbor regions of Eurasia (Nikolova 1998a).

Those problems in peculiar way reflect on the investigation of the Early Neolithic in the Upper Stryama valley. Its initial stage is connected with the drillings of P. Detev on the tell of Banya (Ploskata Mogila) and Chernichevo in 1960s and 1970s. The results of Banya-Ploskata Mogila excavations are published (Detev 1968) but partially. It should be also stressed the settlement of Karlovo, which according to popular communication of I. Krajchev, had cultural layer of 1-1.50 m and was characterized by white painted pottery (Krajchev 1970). In 1980s N. Madzhev excavated on the periphery of the Banya-Ploskata Mogila documenting two horizons from Late Neolithic (Karanovo III culture). At the beginning of 1990s two settlements have been documented and partially excavated – Dubene-Pishtikova Mogila and Kliment-Banyata (Nikolova 1994; 1998a; Nikolova and Madzhev 1992; Nikolova and Madjev1993).

Therefore, for the time being, 5 settlements are known in the Upper Stryama valley from Early Neolithic – three tells and two open thin-leveled settlements (***Figure 6.1***). The only stratigraphic data originate from Dubene-Pishtikova Mogila and Kliment-Banyata. Perhaps the tell of Chernichevo was investigated over the biggest area, but there are no available data to approve that statement. Unfortunately, the diary of the excavator consists of incomplete excavation information.

Approach to the investigation of the Early Neolithic ceramic style

Karanovo I culture

For this study, a special attention will be paid to the ceramic material of the Chernichevo tell preserved at the Archaeological Museum in Hissar Town. It originates from excavations of P. Detev but remained without stratigraphic data. According to the ceramic data, the tell belongs to Early Neolithic (Karanovo I culture), Late Neolithic (Karanovo III culture), Copper Age (Maritsa and Karanovo VI cultures) and to Early Bronze III (Yunatsite III culture). The predominated part of the ceramics belongs to Early Neolithic. Only few are Late Neolithic and Copper Age sherds. Several whole vessels from Early Bronze III indicate that probably in the time of P.

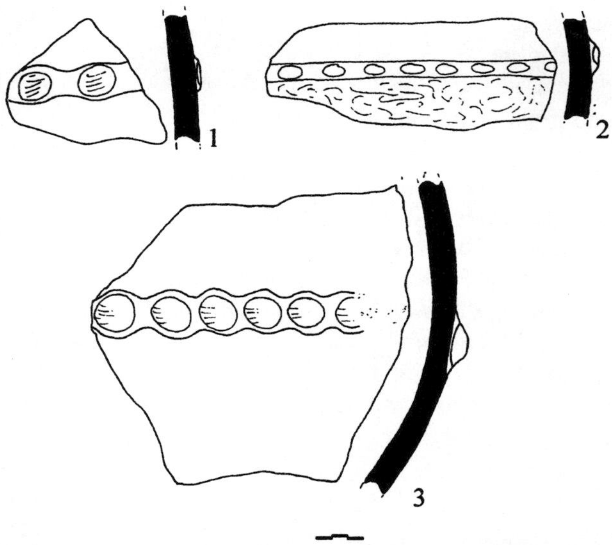

Figure 6.2. Chernichevo tell. Coarse pottery. Early Neolithic. Karanovo I culture.

Detev's excavations there was a layer preserved from that period there.

The Early Neolithic ceramic style of the Karanovo I culture at Chernichevo defines typical characteristics of that culture. Characteristic are the brown smoothed and slip pottery, with variants: brown, brown and red, dark brown and black or gray; brown burnished pottery; red burnished pottery, plain or white painted, as well as black burnished pottery.

As the ceramics are mostly very fragmented, some morphological characteristics can be defined only (***Figures 6.2-6.9***). The big storage vessels are represented by spherical pithoi and short-necked vessels with rounded body, in some cases with a relief ornamentation. Impresso-pottery has been documented in several variants. Vertical corded handles characterize the pots. They are with spherical body and flat bottom and belong to the brown smoothed and brown burnished group of ceramics. Another typical shape are the S-profiled or with inclined profile red burnished bowls with a white painted ornament. Other type widely distributed is the deep conical bowls, which are with different bases – pedestal, thicken or flat ones. The black burnished pottery is more rarely distributed characterizing spherical bowls with corded vertical handled. The pedestal bases, which belong to conical bowls, tulip-vessels and other shapes are massive or divided into two or three parts.

Some attributes of that pottery can be statistically elaborated. However, it should be kept in mind, the statistics are based not on all the pottery excavated but only on one preserved at the Archaeological Museum in Hissar Town. But it should be expected P. Detev rescued the best samples. However, there is almost completely missing dark painted pottery preserved.

Of 33 fragments of vessels with corded handles 19 belong to pots of brown smoothed pottery or it is 58%. 5 samples belong to brown burnished pots, or it is 15%. Another 6 vessels are pots or big bowls of black burnished pottery, or 18%. There are also 3 sherds of red white painted bowls with corded handled (9%). It seems that ratio may represent real relation between the different types of vessels with corded handled.

The conical deep bowls are monochromic with red slip or painted with white, as well as with light brown color. In ***Figure 6.6: 1*** there is a sherd of plate with light brown surface, biscuit with coarse sand which walls are covered with slip of fine clay. Typical of the Early Neolithic layer are the spherical closed bowls with equal cut or S-shaped orifice and thickness of the walls c. 0.5 cm. Wide distributed net ornament occurs in combination with curvilinear and dot ornament. The painted fragments in ***Figure 6.7*** are with small net and are comparable with the pottery from the second (later) layer at Dubene-Pishtikova Mogila tell.

For instance, the small net is documented at a depths of 1.30-1.45 m there. In this layer the ornament of big net dominates, which is emblematic of the earliest documented at Dubene-Pishtikova Mogila phase (at a depth of 1.90-2.10 m from the surface, in sounding No. 2) – ***Figure 6.11*** (see also Nikolova 1998a). At Pishtikova Mogila, the net ornament has typical position under the rim, but it is distributed as main body ornament with thickened bordering lines as well (***Figure 6: 8***). At the second layer of that tell a conical bowl with a linear ornament over the outer and net-like over the inner wall is documented (***Figure 6.11: 6***).

The preserved cultural layer at Dubene-Pishtikova Mogila is c. 2.20-2.50 m (see in details in Nikolova and Madzhev 1992). In the light of the recent evidence, three levels have been documented from the Early Neolithic, synchronous with the early and late stage of the Karanovo I culture in Upper Thrace. The lower two levels are characterized by white painted pottery, while in the upper level dark red painted ornament occurs but the white painted remains typical. However, the excavated area is very limited.

Kliment-Banyata is one-level settlement with thickness of the cultural layer up to 1.00-1.50 m (with the pits). The last was completely destroyed before the excavations. Because of the acidity soil, the very fragmented ceramics were discovered without slip over the walls almost in all cases. Because of that fact it was not documented the characteristic painted pottery of the late stage of the Karanovo I culture in Western Upper Thrace.

The pots with vertical corded handles characterize the Kliment-Banyata ceramic complex, along with conical deep plates, spherical bowls and barbotine-ceramics. Typical of the Early Neolithic cult table with stamped ornament was discovered fragmented in the destroyed cultural layer (Nikolova 1998a; ***Figure 6.12***). In the same layer a fragmented knife of obsidian was discovered (Nikolova and Madjev 1993; Tsonev 1995), which indicate possible organized system of trade, perhaps with the Aegean cultural circle. Kliment-Banyata possibly marks point in this system, as the obsidian was a result of indirect contacts along the Maritsa River. Chronologically, the site marks the end of the Early Neolithic in the Upper Stryama valley and maybe it is synchronous with Dobrinishte from the Struma basin (Nikolova 1998a; Nikolov and Radeva 1992) dated from the end of the Early Neolithic III. The radiocarbon dates from the latter site are valid for the end of the

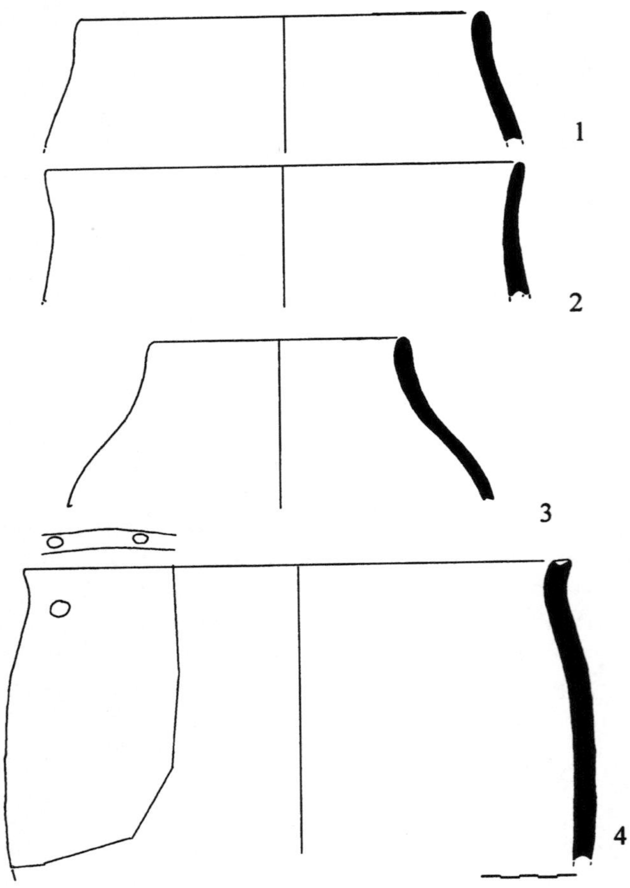

Figure 6.3. Chernichevo tell. Coarse and technomic pottery. Early Neolithic. Karanovo I culture.

Early Neolithic in the Balkans, dated from 5500-5450 BC (Nikolova 1998a).

Interrelations

The analysis of the interrelation of the Karanovo I ceramic style from the upper Stryama valley concerns several sites of primary importance for the Early Neolithic in the southern Balkans – Kapitan Dimitrievo in western Upper Thrace; Chavdar in Zlatitsa-Pirdop lowland, Slatina in Sofia Field, Gulubnik in the Upper Struma valley, as well as Kovachevo I1-4 from the middle Struma valley. The pottery from other important sites like Karanovo, Kazanluk, Eleshnitsa, etc. is unpublished or preliminary published.

In the first place the problem of the lower chronological borderline of the Early Neolithic settlements in the Upper Stryama valley occurs. But that problem faces the general question of the sequence of the settlements of the Karanovo I culture. According to V. Nikolov, Eleshnitsa and Slatina 4 represent the earliest Karanovo I culture and they precede Chavdar. On its hand, the ^{14}C dates from Chavdar 6 can very satisfying argue a belonging of the site to the very early stage of the Karanovo I culture. Some parallels in the pottery between Chavdar 6 (Georgiev 1981: figs. 47-48) and Slatina 4 (Nikolov 1992: figs. 61-66), on the one hand, and between Slatina 4 and Gulubnik 1, on the other hand (Nikolov 1992: fig. 65: 2; cp. Pavuk and Čochadžiev 1984: fig. 7[1]) indicate Chavdar 7-6 – Slatina 5-4 – Gulubnik 7-6 horizon (cp. ***Charts 6.1-6.4***).

To the south, to that horizon can be also attributed Eleshnitsa 1-2 and probably Kovachevo I2, as Kovachevo I1 is connected with Vaksevo and Nevestino traditions (Čohadžiev and Genadieva 1998: figs. 1-5). The peculiar dot ornamentation indicates the last sites preceded chronologically Karanovo I culture and perhaps dated from Early Neolithic IB or very early IIA (after Nikolova 1998a). But especially at Vaksevo there are some good parallels with Kovachevo 1-2 (Čohadžiev and Genadieva 1996: figs. 3; 4: 1; cp. Pernitcheva 1990: plate 11; 12: 12), so its chronological upper borderline may correspond to EN IIA. Morphological and ornament elements like conical deep plates and holed bases, net ornament, motif of oblique parallel lines, linear ornament over the holed base etc. are in fact Proto-Karanovo elements, according to the data from possibly earliest Karanovo at Kovachevo 1 (Pernitcheva 1990: figs. 1-12).

As far as Eleshnitsa concerns, according to V. Nikolov, it is phase Eleshnitsa that characterized the earliest Karanovo I

[1]Pavuk J. and Čochadžiev M. 1984 Neolitische Tellsiedlung bei Gălăbnik in Westbulgarien. *Slovenska Arheologia* 32, 195-228.

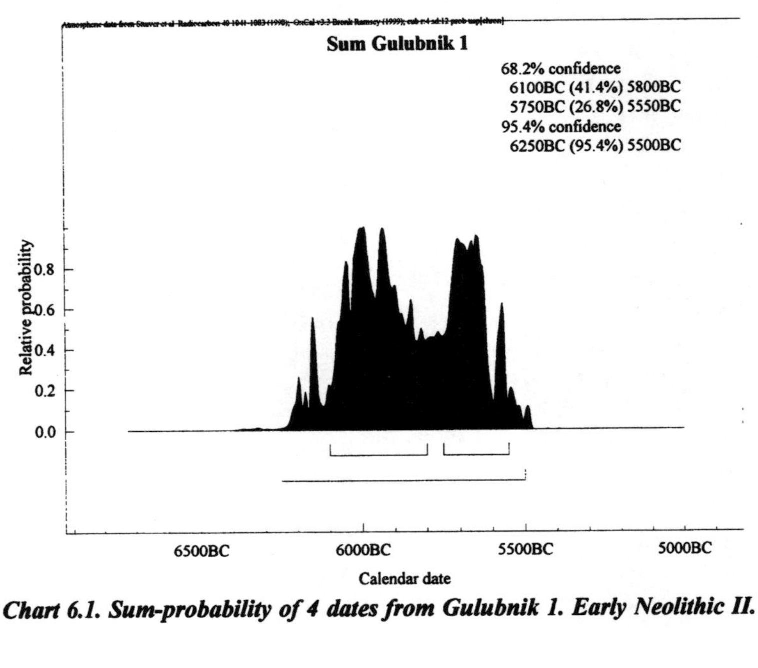

Chart 6.1. Sum-probability of 4 dates from Gulubnik 1. Early Neolithic II.

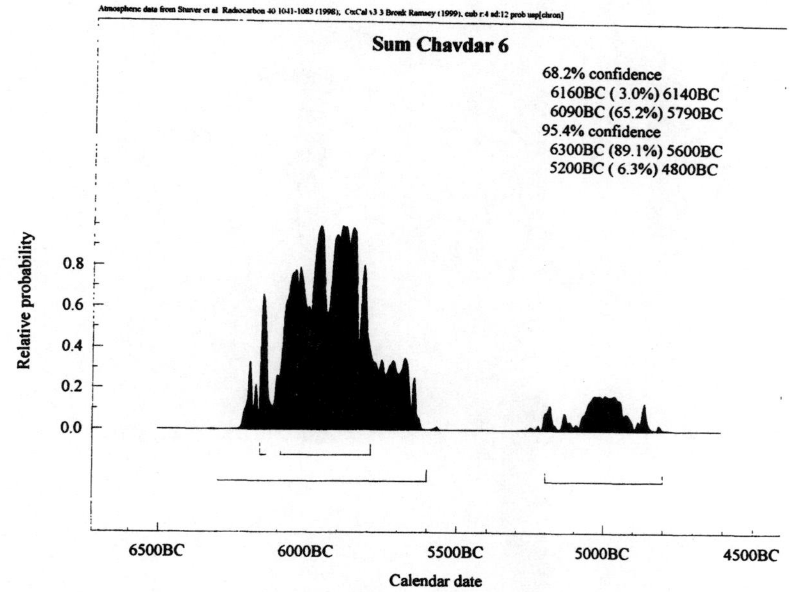

Chart 6.2. Sum-probability of 9 dates from Chavdar 6. Early Neolithic II.

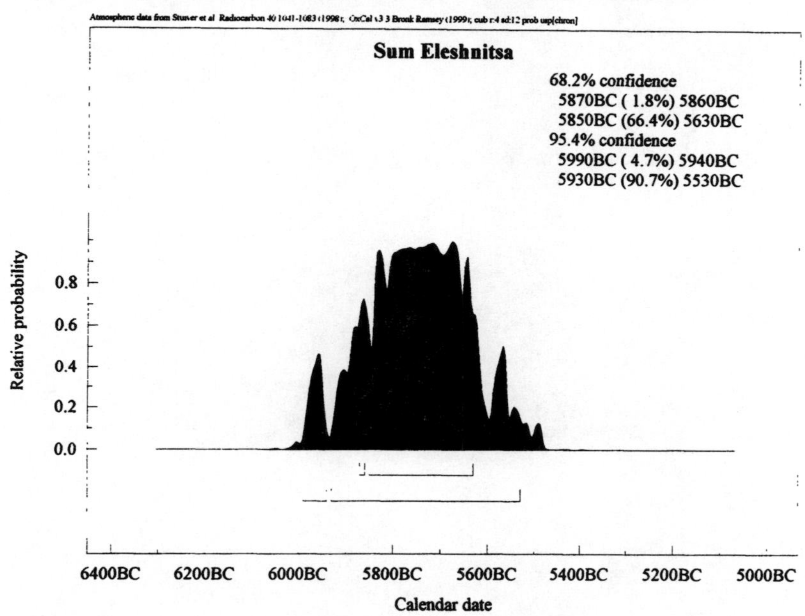

Chart 6.3. Sum-probability of 8 dates from Eleshnitsa 2. Early Neolithic II.

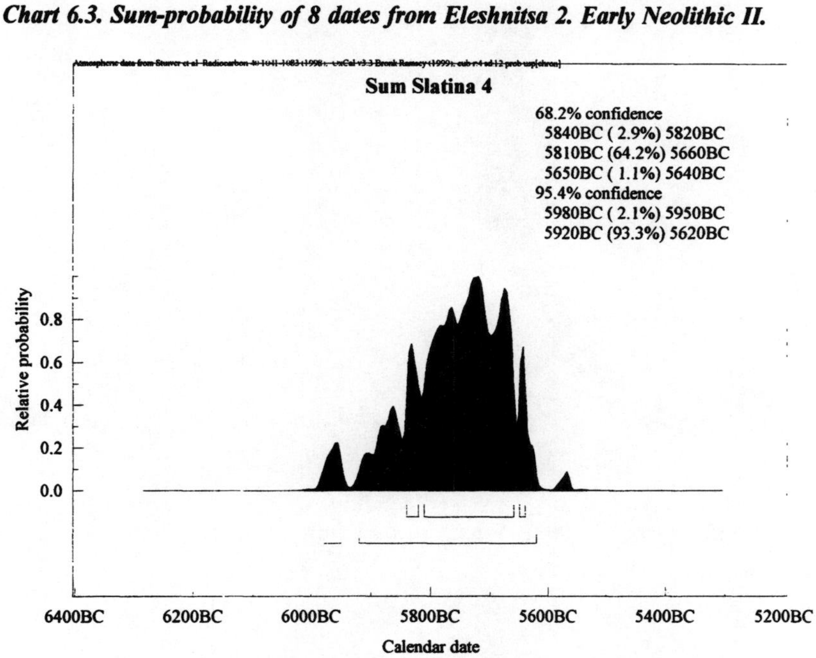

Chart 6.4. Sum-probability of 12 dates from Slatina 4. Early Neolithic II.

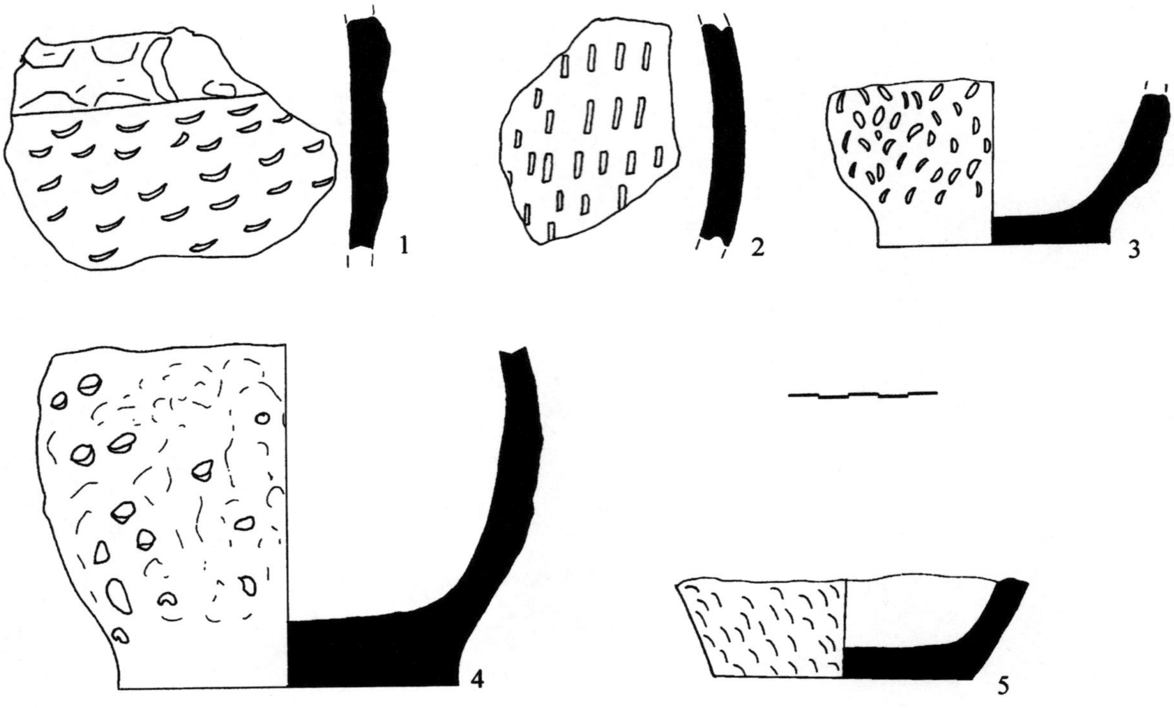

Figure 6.4. Chernichevo tell. Impresso-pottery. Early Neolithic. Karanovo I culture.

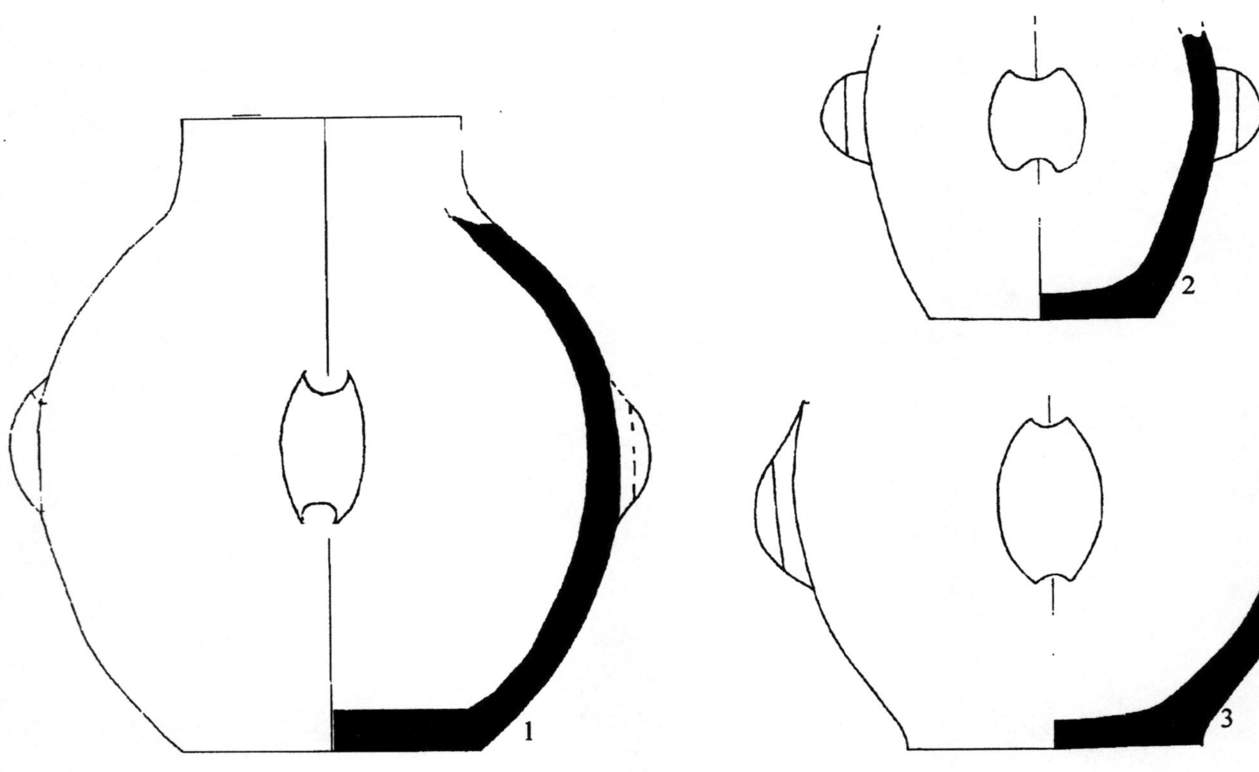

Figure 6.5 Chernichevo tell. Technomic pottery.Early Neolithic. Karanovo I culture.

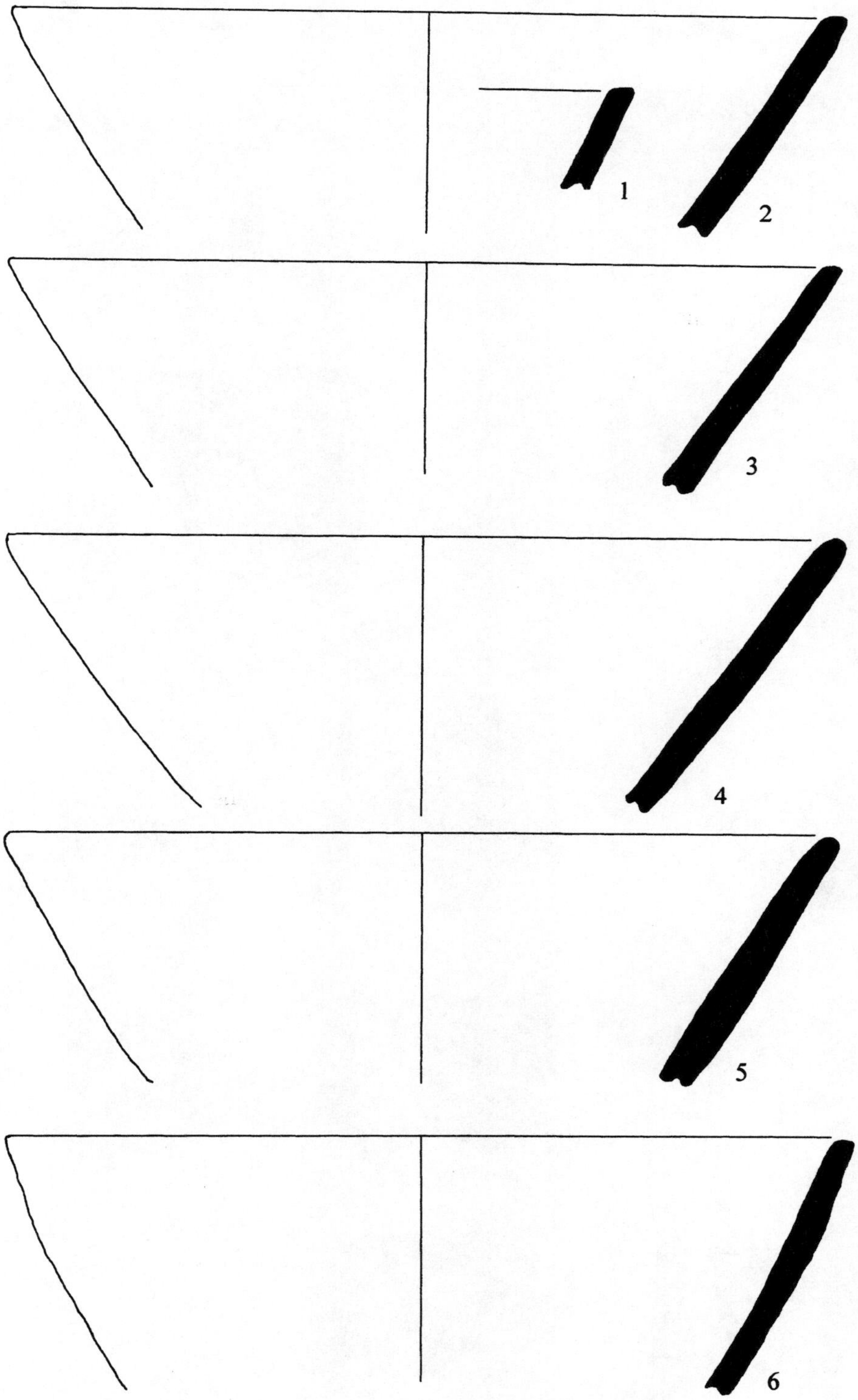

Figure 6.6 Chernichevo tell. Technomic and fine pottery.Early Neolithic. Karanovo I culture.

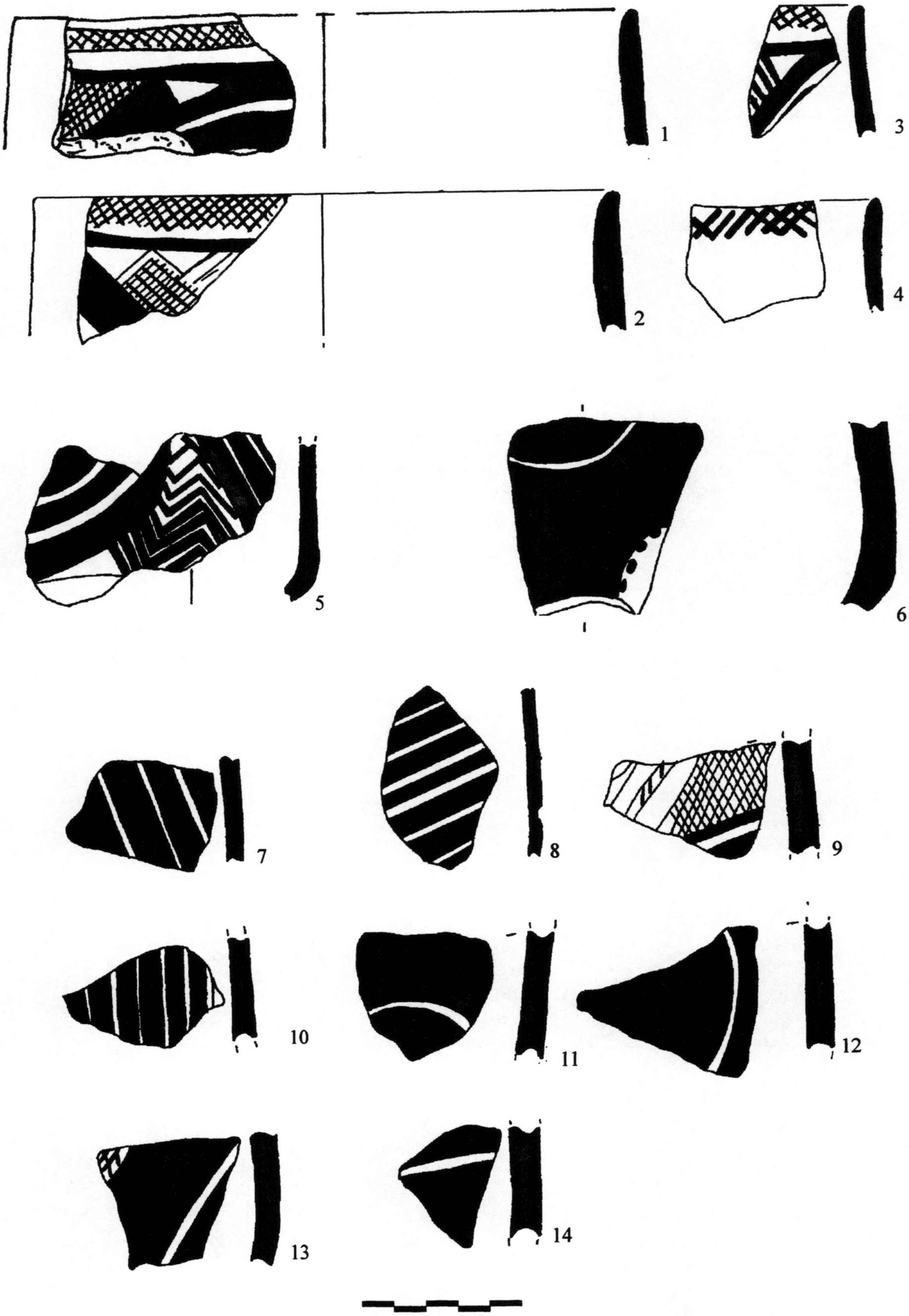

Figure 6.7. Chernichevo tell. White-painted pottery. Early Neolithic. Karanovo I culture.

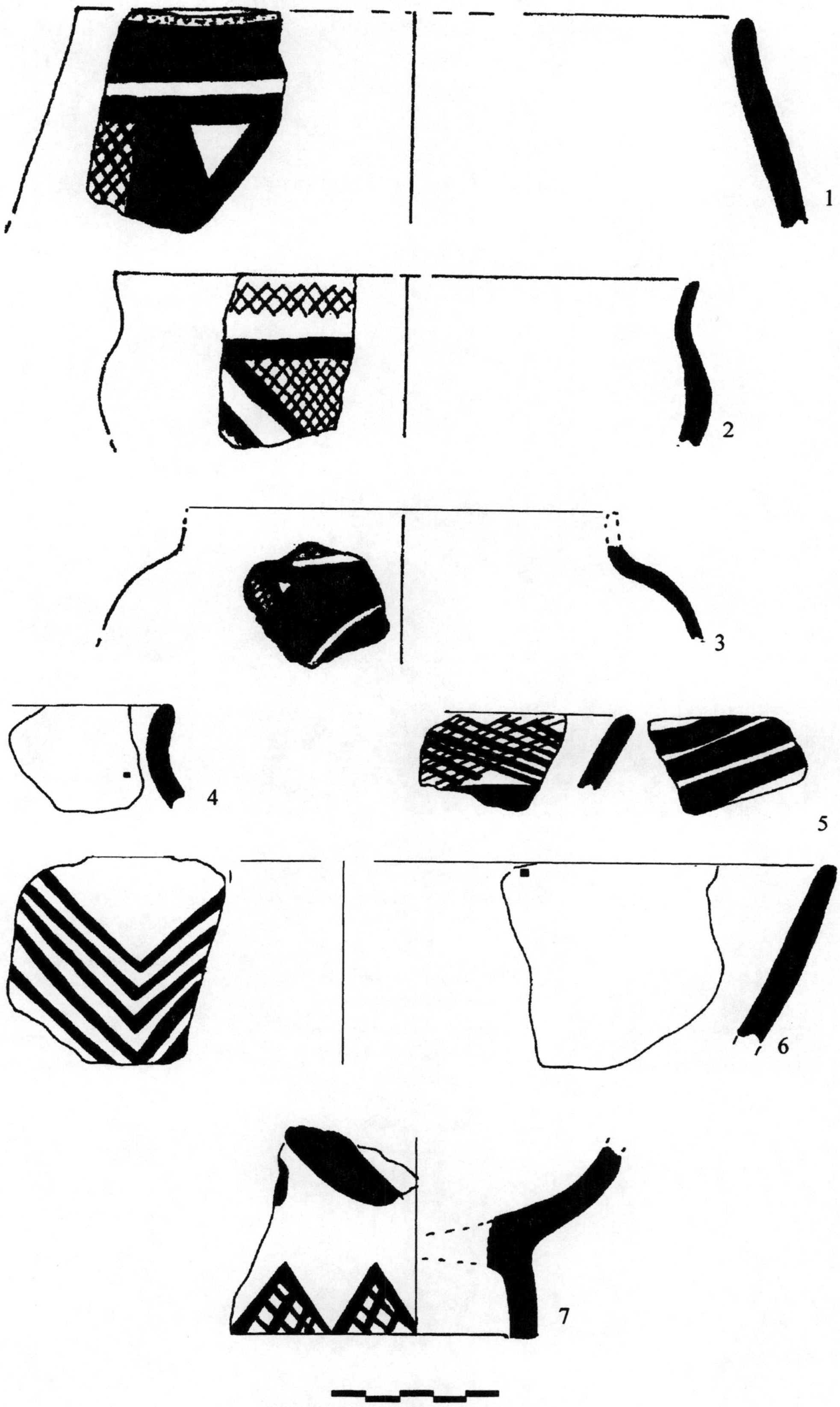

Figure 6.8. Chernichevo tell. White-painted pottery. Early Neolithic. Karanovo I culture.

Figure 6.9. Chernichevo tell. White-painted pottery. Early Neolithic. Karanovo I culture.

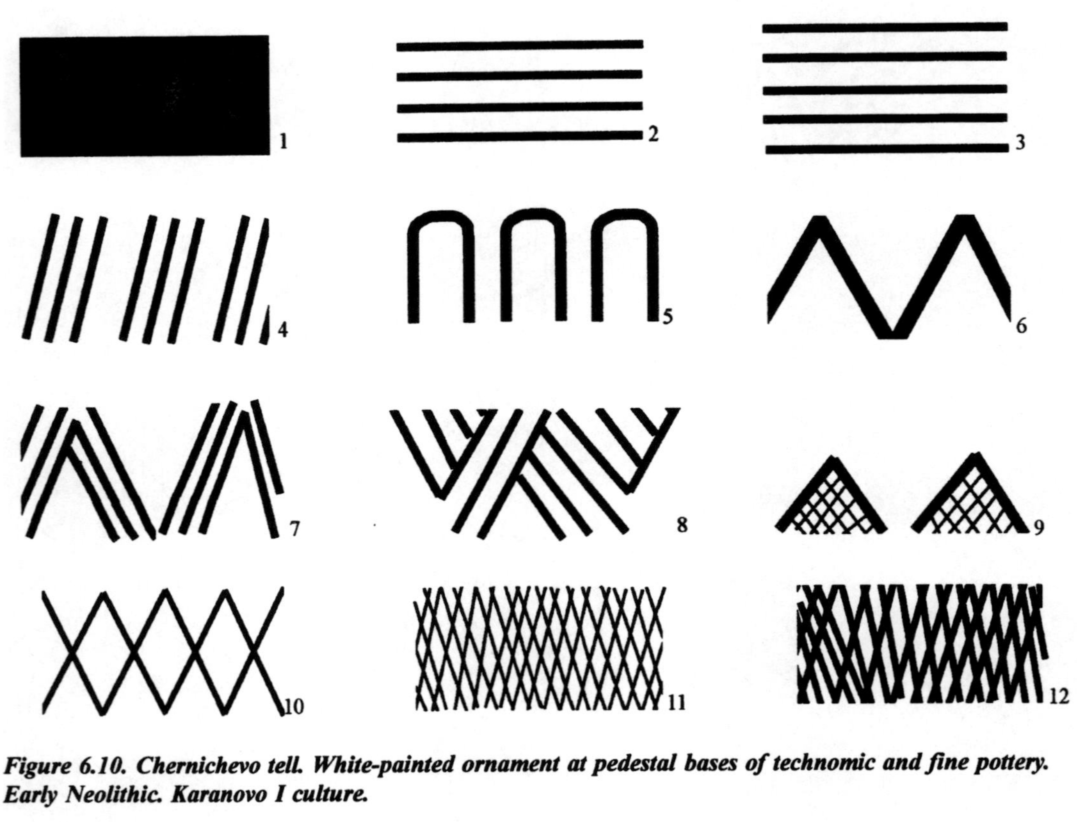

Figure 6.10. Chernichevo tell. White-painted ornament at pedestal bases of technomic and fine pottery. Early Neolithic. Karanovo I culture.

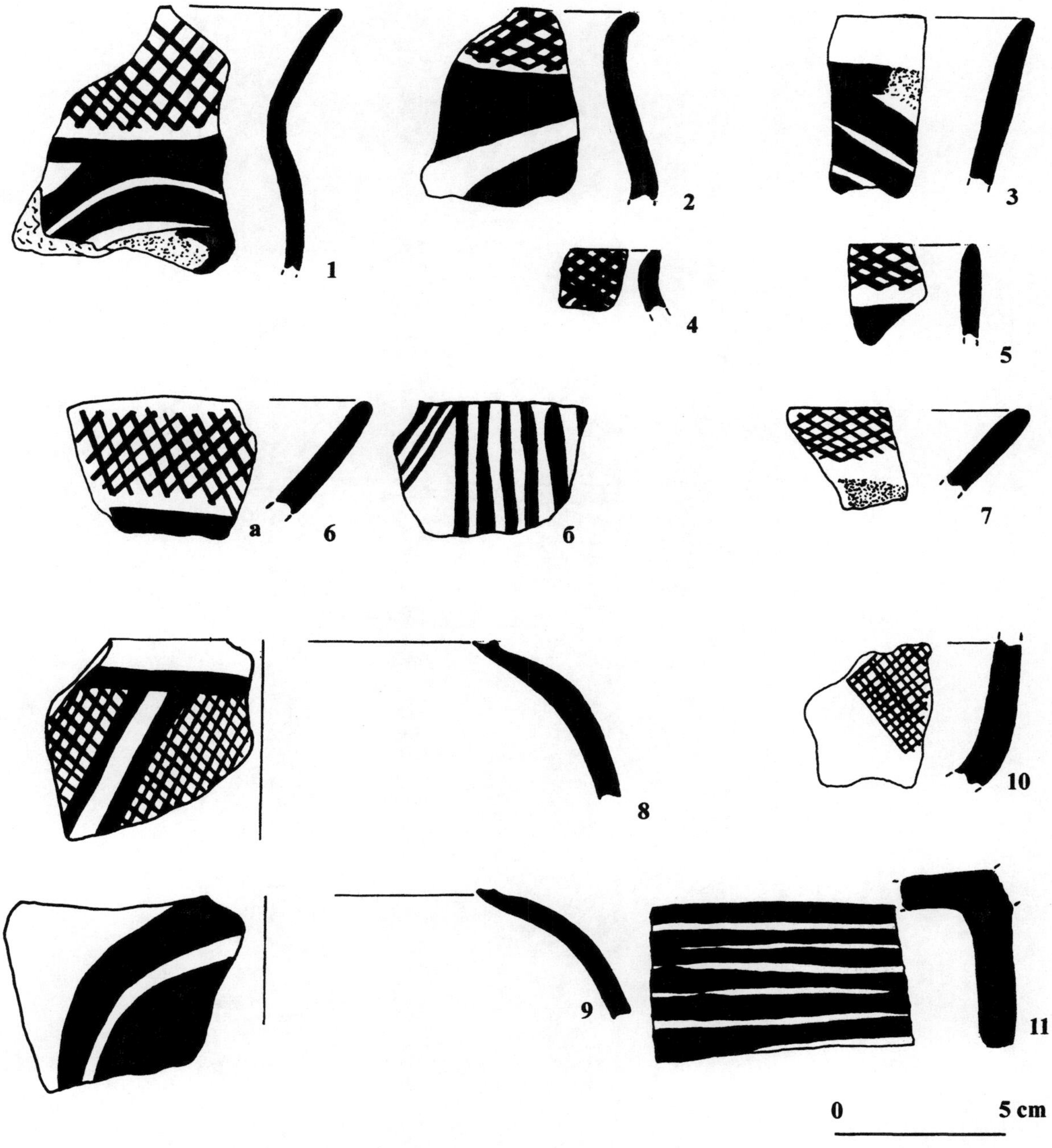

Figure 6.11. Dubene-Pishtikova Mogila tell. White-painted pottery. Early Neolithic. Karanovo I culture.

culture in its southwest variant (1996). But the pottery of that site is preliminary published. V. Nikolov suggests as idiosyncratic element of the Eleshnitsa phase the biconical bowls, which existence at Kovachevo 2 may indicate synchronism between Eleshnitsa 2 and Kovachevo 2 (Nikolov 1996: fig. 2d; cp. Pernitcheva 1990: plate 8: 9). However, the problem is complicated as there is no dark painted pottery documented at Eleshnitsa and Slatina 4 on the one hand. On the other hand, dark painted pottery characterizes already Chavdar 7. This is the reason V. Nikolov to believe Eleshnitsa and Slatina 4 preceded Chavdar (Nikolov 1992; 1996). But in the light of the ^{14}C dates Chavdar 6 seems to be much earlier if we compare only 68.2% probability values. There are two hypotheses more: 1. Only the earliest values of the ^{14}C dates from Eleshnitsa 2 and Slatina 4 with 95.4% probability, correspond to the possible historical chronology of the beginning of the site occupation and they are comparable with 68.2% probability values of Chavdar 6. In this case the contemporaneously between earliest Eleshnitsa, Slatina and Chavdar can be assumes, dated from c. 6000/5900 BC. 2. Chavdar 7-6 preceded Eleshnitsa and Slatina 4. We need more dates to decide that problem but the dates from classical Starčevo support early chronology of the Chavdar, respectively of emergence of dark painted pottery is very early appearance in the Central Balkans. All three hypotheses argued excluded any possibility Slatina 4 to be synchronized with Chavdar 4 as it is suggested by Ya. Boyadzhiev (in Bailey and Panayotov 1995: 163). According to the last author, the wood sample should be correlated with grain samples as given the date of the building of the house and the latter – the end of house's occupation. If fact sampled posts may belong to reconstruction of the house and to be near the age of the grains or the dates may not rep-

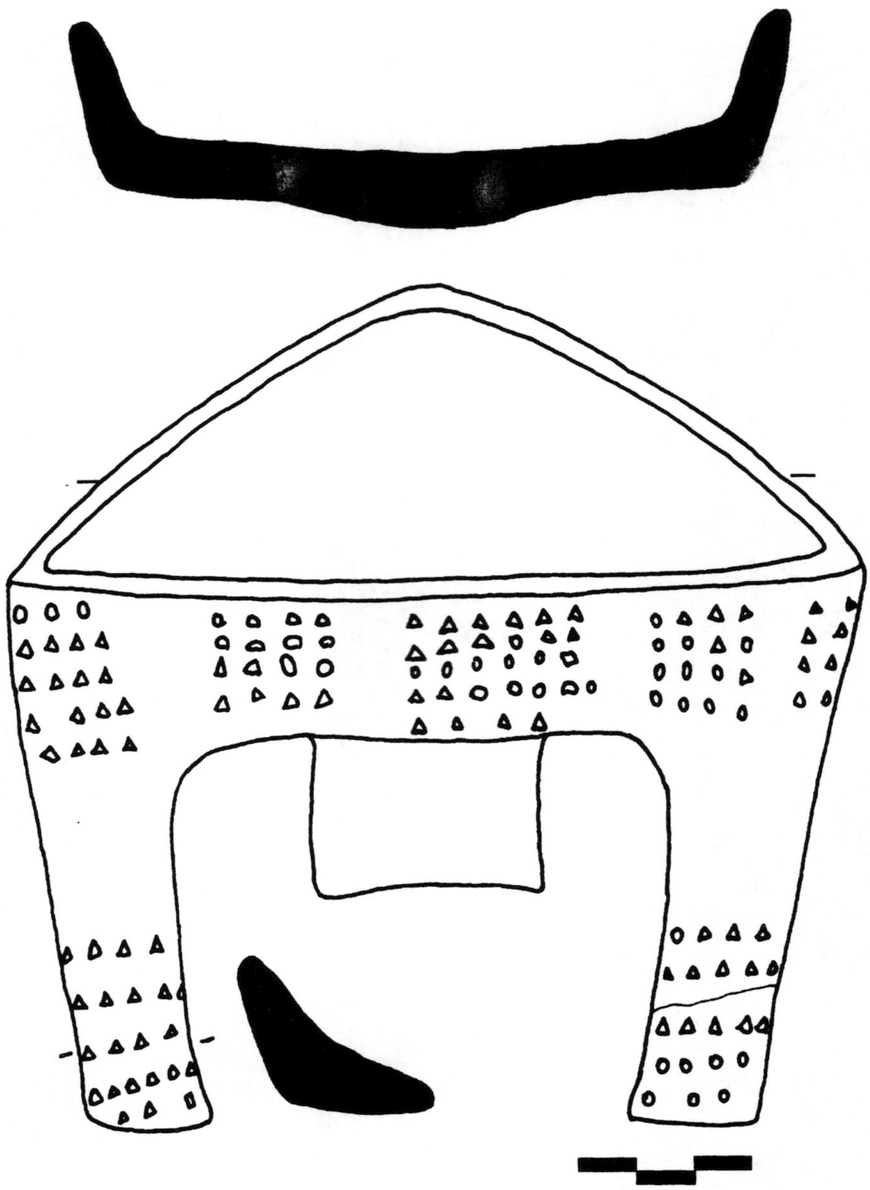

Figure 6.12. Kliment-Banyata. Settlement find. Early Neolithic. Karanovo I culture.

resent the real historical chronology, which is also very possible. The c. 5750 BC is accepted as terminus post quem of the end of Slatina 4 in my in Nikolova 1998a (p. 118), assuming a long occupation of the house.

The periodization of the Karanovo I culture is one of the complicated problems as the biggest excavated sites have remained without published data on the one hand. One the other hand, the ceramic style of the white-painted pottery is rather conservative that requires the using of statistic methods (Nikolov 1996: 1 with references cited there). That is the reason V. Nikolov to suggest two stages of the Karanovo I culture development - first stage (initial) that developed only in the Struma valley and second stage divided on the base of the existence of dark-painted sherds in Upper Thrace as imports from the "Central Balkans" (western parts of that region, according to my cultural and geographical terminology). But later the author suggested that there are polychrome pottery from Slatina 4 (Nikolov 1998: 205). Early dating of the horizon including dark painted pottery is not contradicted by the dates from early Starčevo (Nikolova 1998a). However, the polychrome and dark painted pottery to the west of Chavdar area – in Upper Thrace – is a rarely distributed in compare to the total predominance of the white painted pottery. But it is clear, it cannot be a criterion for the inner periodization of the Karanovo I culture in Upper Thrace.

The earlier levels from Dubene-Pishtikova Mogila can be attributed to the first cultural horizon mentioned above. The similarity between the Chernichevo white painted ceramics and Chavdar 6 sherds also may indicate the Upper Stryama valley belong to the area of the earlier Karanovo culture.

In the periodization of V. Nikolov (1996) the second stage of the Karanovo I culture is divided into two phases with regard to its southwestern variant (Rakitovo and Dobrinishte phases). Unfortunately, the pottery from Rakitovo is still not published. By the excavations only two levels have been documented (Mazanova 1966: 105) but it cannot be excluded completely destroyed levels as well. Some of the cult object from Rakitovo well parallel with the Chavdar 5 (a figurine, cp. Georgiev 1981: fig. 56; Mazanova 1996: table 4: 1), Chavdar 5-2 (figurines, cp. Georgiev 1981: figs. 55, 57; Mazanova 1996: tables 2: 2-4; 3), Chavdar 4-3 (tripods, cp. Georgiev 1981: fig. 53; Mazanova 1996: table 10). Those parallels may indicate the synchronism between Rakitovo 1-2 and Chavdar 5-4/3. We need a richer archaeological record from all microregions of the Karanovo I culture that may contribute to more detail regional chronology.

According to the comparative analysis, the last stage of Karanovo I in Western Upper Thrace is contemporary with Karanovo II and Karanovo II-III (after V. Nikolov). At two tells in the Upper Stryama valley there is a stratigraphic sequence Karanovo I – Karanovo III, according to the ceramic material (Banya-Ploskata Mogila and Chernichevo).

In the light of the recent evidence, the latest Karanovo I culture includes later Early Neolithic levels at Kapitan Dimitrievo, Kurdzhali and Krumovgrad, later EN Chernichevo and Dubene-Pishtikova Mogila, Kliment-Banyata, the site of Dobrinishte, etc. At Dubene-Pishtikova Mogila the layer up to 1.20 (?) m depth can be attributed to the latest stage including probably the uppermost levels of the tell preserved. The closest comparative data to the west originate from Chavdar 3-2. Kliment-Banyata belongs to the last stage being perhaps synchronous or postdating partially (?) Dubene-Pishtikova Mogila. The stamped motif over the triangle table has the earliest parallels at Krajnitsi (Tchohadjiev and Bakamska 1990: fig. 25: 1) and it is typical of the Early Neolithic including the western Rhodopes, where at Rakitovo is documented similar table (Mazanova 1996: table 10). According to the orifice, both are very characteristic of the later Karanovo I culture, with numerous parallels – Eleshnitsa (Nikolov and Maslarov 1987: fig. 6), Gradeshnitsa (Nikolov B. 1974: 6, 8). To the relative chronology of Kliment-Banyata correspond the prolong pots with vertical corded handles with parallels at Karanovo II. Of the scanty published pottery from Chavdar 3-2 the pot with channeling zigzag over the body should be stressed (Georgiev 1981: fig. 33b), as well as the triangle cult table (Georgiev 1981: fig. 53c) from the third horizon. G. Georgiev relates Chavdar 1 to Karanovo II-III (1981). K. Kunchev (1995) assumes that the first level is related to Karanovo III. On his hand, V. Nikolov (1998a: 144) indirectly assumes, that the horizon is related to Karanovo III-IV. Additionally, the latest EN Chernichevo occupation may also correspond to the end of the Karanovo I culture because of some Karanovo II parallels in the pottery.

For the time being, one of the most expressive for Karanovo I chronology dates are those from Karanovo (Karanovo I, 6000-5530 BC [Nikolova 1998a: 127]), which generally correspond to the common chronology of the Early Neolithic in the Balkans (Nikolova 1998a: chart 4; 1999: chart 3.1). For the chronology of the end of the Karanovo I culture in Western Upper Thrace there are indirect data (^{14}C dates and archaeomagnetic samples [Nikolova 1998a; Kovacheva and Toshkov 1991]). On the one hand, the radiocarbon chronology of the Karanovo II is 5750-5520 BC (Nikolova 1998a). On the other hand, the dates for the end of Dobrinishte are very precise and by 33,9% confidence that date the end of the site from between 5530 and 5480 BC. Then, the existence of despite minimal diachronism confirms the thesis of the archaeomagnetic investigation that the samples from Karanovo II originate from earlier than Dobrinishte level, as well as the possibility the last site to be contemporary with Karanovo II-III. Unfortunately, Kremenik is known by a scanty published material. V. Nikolov assumes very late dating of Kremenik 4, which is confirmed by two of the ^{14}C dates (Nikolova 1998a: 127). But at the same time, Bln-2550 has calibrated values between 5570 and 5380 BC, that includes the end of the Early Neolithic. That is a reason the Kliment-Banyata settlement to have been synchronized with a late phase if the Early Neolithic development at Kremenik (Nikolova 1998a: 108), Early Neolithic III. To that synchronization line can be added Bulgarchevo 1 (Pernicheva 1995).

In the light of the recent data, it can be concluded that the end of Karanovo I in Western Upper Thrace corresponds to Karanovo II-III culture in Eastern Upper Thrace.

Discussion

Returning back to the problem directions defined above, now I will summarize how the new investigations of the Early Neolithic in the Upper Stryama basin reflect on them.

The stratigraphic data from Dubene-Pishtikova Mogila and typological data from Chernichevo give a possibility to reconstruct a chronological horizon that connects Western Upper Thrace with the Sofia field and the Upper Struma valley, on the one hand, and with the Upper Maritsa valley, on the other hand (Kapitan Dimitrievo). Unfortunately, there are missing radiocarbon data from the Upper Stryama valley and the Early Neolithic in that region is included in the common chronological span – mainly the first half of the 6th millennium BC confirmed by the chronology of Chavdar (Nikolova 1998a). In the later stage of the Early Neolithic the Upper Stryama valley followed the cultural development in the Upper Maritsa and Upper Tundzha valleys, while the Zlatitsa-Pirdop area was the westernmost frontier of the strongly influenced by Kremikovtsi group.

The absence of evidence for the Mesolithic in the Upper Stryama valley gives an advantage to the migration theory of genesis of the Karanovo I culture in Western Upper Thrace. It should be stressed that there is missing evidence from Early Neolithic IA characterized by monochromic pottery. But period is still very poor investigated in Southwestern Bulgaria documented by one site only. So, the statement in the literature that it has been very well investigated is incorrect.

In the Upper Stryama, numerous parallels between the pottery and early Chavdar may indicate migration in the Upper Stryama valley from Zlatitsa-Pirdop area. At the same time, it should be stressed the identical ceramic style in the Upper Stryama valley and in the Upper Maritsa valley (Kapitan Dimitrievo). So, for Western Upper Thrace a segmentation of already settled down in the Balkans population is thinkable. It should be stressed there was no accidental settling down, but a well-planned colonization and contemporary grounding of several settlements. The territory preferred along the

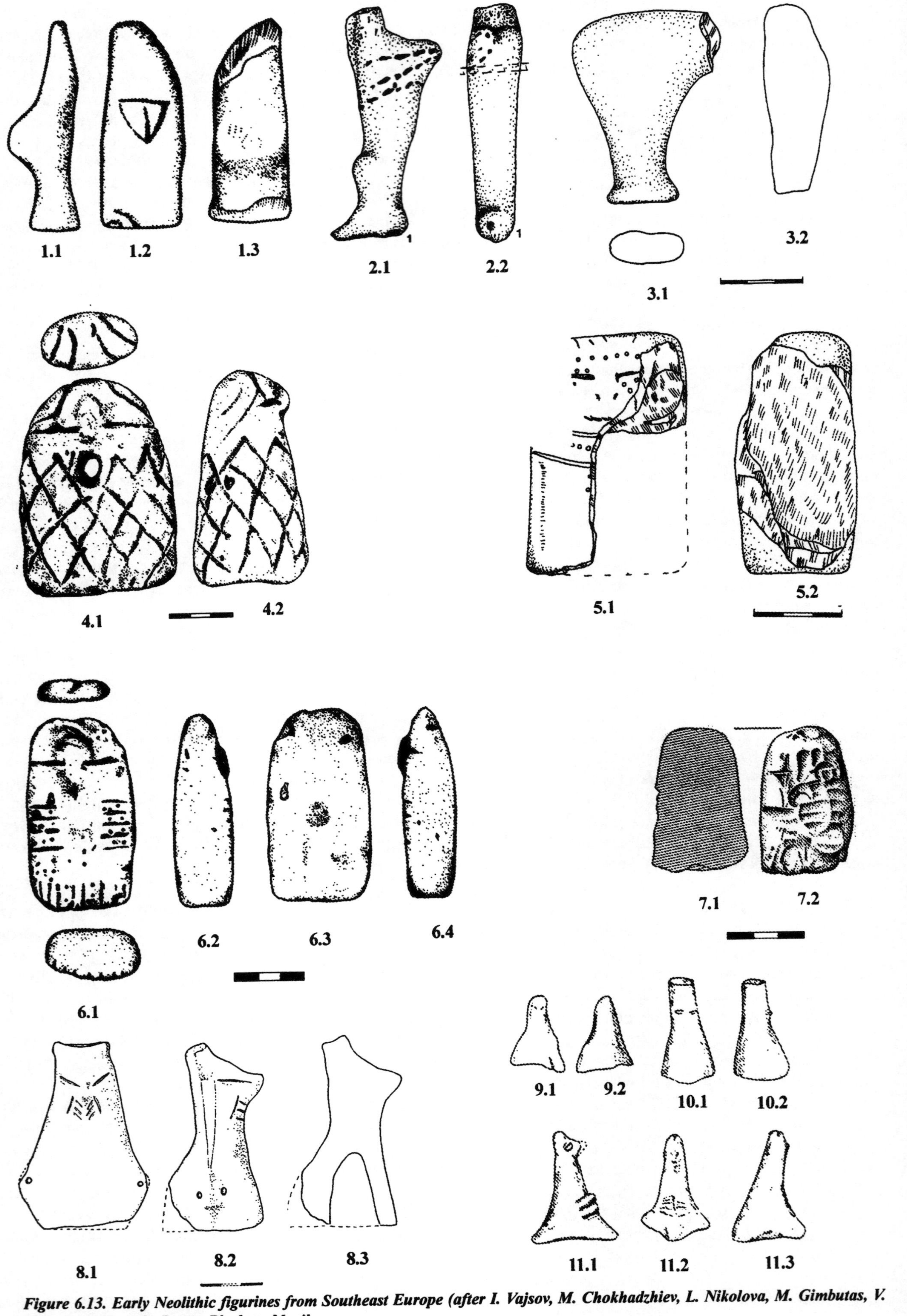

Figure 6.13. Early Neolithic figurines from Southeast Europe (after I. Vajsov, M. Chokhadzhiev, L. Nikolova, M. Gimbutas, V. Mazanova, etc.). 8: Banya-Ploskata Mogila.

Stryama River, confirms the thesis of dominating of the agricultural and stockbreeding structure of the Early Neolithic social and economic structures in the region.

The figurine from Banya-Ploskata Mogila

At least in the History Museum in Karlovo there are missing materials from the excavations of P. Detev on Banya tell. From that tell in the museum a fragment of a bowl with ochre surface and linear ornament is preserved, as well as a female figurine (***Figure 6.13:*** *8*; Nikolova 1998a: fig. 9). The comparative chronology allows the figurine to be attributed to the lower layer of Banya tell, which was one of the biggest settlements in the region in the Early Neolithic.

The figurine from Banya-Ploskata Mogila (No. 1) has no exact published analogies, but it consists of common elements for Early Neolithic figurines in the Balkans, especially characteristic of Western Bulgaria. More west, similar is the figurine from Gorubintsi-Rug Bair in the upper basin of the Vardar River in Macedonia (Sanev et al. 1975: No. 144), which is unfortunately, with broken lower part. Perhaps both represent one and the same type of standing figurines with trapezoidal vertical section and prolong upper part represented the head, as well as with corded handles symbolically representing the arms. Similar syncretism is known from the vessels, for instance from Rakitovo (Mazanova 1996: table 6). At the same Early Neolithic site a figurine with trapezoidal verticals section is discovered, a variant of the Banya-Ploskata Mogila type, without arms represented, but with a breast of conical applications (Mazanova 1996: table 4:1). Common element between the figurines from Rakitovo and Banya are the slightly flatten and prolong backwards end of the head, with horizontal rectangular section. That is a typical element of the figurine art in the Early Neolithic in the Central Balkans, with parallels again in the Upper Struma (Chokhadzhiev M. 1981: 76:3). To the north, in the area of Starčevo-Gradeshnitsa-Criş complex, the column-like anthropomorphic figurines with widen lower part is documented at settlement of Beneşova, along with other figurines of common of the Balkans types represented in ***Figure 6.13: 1-2, 5-6*** (Lazarovici 1979: table 10: 1-3, 9-11, 13-16, 25).

For the functional interpretation of the Early Neolithic figurine from Banya, of importance are two elements – representation of synthetic image, combining human and bird elements, as well as a transformation of the arms in cord handles. On the one hand, the figurine represents standing posture, with a possibility it to be placed standing, but at the same time it can be hanged. A female gender is represented as the figurine in the lower part is widened and in the upper part the head is shaped as a bun of hair (?). In contrast to Starčevo tradition, by which with incised lines is represented long hair with bang (Karmanski 1991: No. 4), the hair is not a object of naturalistic representation in the Struma-Stryama-Upper Maritsa basins, according to the recent data.

In the Indo-European religion, the roots of which are in the Balkan-Anatolian Early Prehistoric cultures, the cult of the hearth is usually connected with a goddess (for instance Hestia). But the house and especially the hearth are possible to have been more complicated mythological, as in the Vedian religion the personification of the fire was Agni (Nikolova 1985). In this context the mixed anthropomorphic and zoomorphic elements in the Early prehistoric plastic in the Balkans have mythological roots. Those elements can be connected with their function as elements of the house cult, having in mind the protecting rule as protectors of the house. At the figurine from Banya the sign of M is incised under the nose.

Together with the schematic representation of the mouse, it increases the function mentioned and has parallels in numerous prehistoric cultures. The upper shaping of the head is possible to be connected with representing of ritual hairstyle or hat. At the same type some figurines from Early Neolithic in the Central Balkans, are without clear gender indications (***Figure 6.13: 4-6***). That fact confirms the hypothesis, that through the figurines mythological ideas were presented, in which the reality and the fantastic were mixed (for instance, the combination of anthropomorphic and zoomorphic elements) or only the anthropomorphic idea was stressed without clear definition of the gender. The transmission of the anthropomorphic idea over the ceramic vessels also indicates that it was connected with the protecting function in that early period of the development of the society, which in peculiar way was opposed to the nature because of its purpose to preserve itself, as well as to develop the social environment. It is possible the opening of figurine from Banya to have been functioned (to make it lighter), but it cannot be excluded that it has symbolic meaning, representing the mother womb (as the anthropomorphic vessels). In this context, it is not possible functions as protecting and giving fertility to be opposed, because they complement one another. Common element of both cults is using of hanging idols, as it is not only the figurine from Banya, but from Pernik 2 where a standing column-like figurine with opening is discovered (***Figure 6.13: 4***). M. Chokhadzhiev interpreted it as a loom weight (1981: 73-74). Despite the possibility the incised ornament to represent hair or long dress, it should be stressed the rhomb ornament, which is a basic mythological symbol.

It seems the figurine was multi functional. It was used as standing and as hanging. In the first case it is possible it to represent singer participant in a ritual, but in the second one to have protecting function connected with the cult of local household. The dominating protected function is confirmed by the presence of similar Early Neolithic miniature figurines-amulets, which are only 2-3 cm high. Examples are known from Thessaly (Achilleion) – ***Figure 6.13: 9-11***. Two of the figurines from Achilleion IV (Gimbutas et al. 1989: fig. 7.23; ***Figure 6.13: 10-11***) are discovered in the context of house and altar and have elements close to Balkan figurine art, on the one hand. On the other hand, those amulets are possible to document manners of distribution of the common elements of the prehistoric figurines between close and distant cultures – through contacts among the people and through an exchange.

The standing figurines with unarticulated lower part are very popular in the Early Neolithic. M. Chokhadzhiev (1981) and I. Vajsov (1984) published number of finds from the upper Struma and Sofia region (***Figure 6.13: 1, 4, 6***). Later, V.

Matsanova represented a new archaeological regional model of ritual in the western Rhodopes (Rakitovo) in the Early Neolithic with several types of anthropomorphic figurines (Mazanova 1996: table 4: 1; ***Figure 6.13: 5***). Which have close parallels in Thrace and especially in western Central Balkans. In some cases the legs are articulated (***Figure 6.13: 2***).

Common types of figurines probably document the existence of similar rituals in different regions in the Central Balkans. At the same time they have parallels in Starčevo-Gradeshnitsa-Criş complex, and some of them in Körös culture in central Europe, as well as to the south in the Aegean. Accordingly, along with the ceramic contact-data, the Early Neolithic figurines are main record of the cultural contacts between the close and distant cultures in the Carpathians-Balkan-Aegean region. It should be stressed that during the Neolithic, as well as in the other prehistoric epochs, the Balkans were not a unique appearance and cannot be artificially cut from the other regions.

The question appears to which extent the figurine from Banya-Ploskata Mogila represents a household cult in the egalitarian society, or it represents a tendency towards an increased social stratification in the Neolithic society in the period of the so-called Big Man. Different elements of the Early Neolithic culture system can be interpreted as a proof of existence of horizontal, as well as elements of vertical stratification in the Balkans. From hierarchy in the micro and macro settlement structure up to difference in the burial ritual, the Early Neolithic Balkan structures represent a new level of stratification, which initial elements can be traced to Lepenski Vir culture from Mesolithic.

In the context of that treatment of the problem, the clay figurines probably also represent multilevel social strategies. In prospective plan, on the cemeteries from Hamangia culture rich burials with figurines occurred (Vajsov 1992). Similar hierarchy without doubt existed on the settlements. The Balkan Neolithic figurines are characterized by exactly defined cultural peculiarities showing that they were an important element in the cultural identification of the different population, despite the presence of common stylistic elements. The fact the similar elements are discovered in different societies – in this case Karlovo hollow and Western Bulgaria – proofs their social determination connected not only with one and the same cults but also with one and the same social levels.

The assumption that the figurine from Banya belongs to élite family is confirmed by the fact that it is a record for contacts with more distant areas, in this case from Western Bulgaria. The last can be individual, but the representative for the local élite, which aim was resolving the inter-'tribe' relations with the neighbor 'tribes'.

According to that hypothesis, the cultural contacts in prehistory were socially determined, connected with certain social levels. By absence of metal trade, in the Early Neolithic of peculiar significance were the ideological contacts between the élite levels of different close and distant regions. That fact may explain the similarity in the figurine art and the other objects of the cult between the east and the west Central Balkans despite of the mountain geographic barriers. However, the common ceramic style in the Early Neolithic II between the Sofia field and Karlovo hollow being an integrative part of the Karanovo I culture indicate the problem of unification of the culture is very complex.

Considerations

The earliest material from Early Neolithic in the upper Stryama valley is characterized by the white-painted pottery. Those ceramics are very close to the pottery from the earlier Slatina horizons in Sofia field (Nikolov 1992: figs. 56-66). But the closest parallels in the west originate from Chavdar from Zlatitsa-Pirdop hollow (Georgiev 1981; Nikolov 1994; Kunchev 1995). The similarity in the pottery indicates that there was an active communicating line through the western Sredna Gora Mountains passes. Moreover, the question is more complicated because, for the time being, it is not clear the directions of diffusions of the white-painted pottery style into Upper Thrace.

In chronological plan, there is no evidence in Upper Thrace for a distribution of the monochromic ceramics there. In addition, in Upper Thrace, according to the recent data, there are also no data on the earliest horizon of the white-painted pottery in the Balkans known by the western Central Balkans and the western Lower Danube and the southern Middle Danube. At least two possible theoretical explanation models exist.

According to the first model, the stylistic difference in the earliest Balkan ceramic styles was not chronologically determined. So, what we think that it is later appearance of the white painted pottery in Upper Thrace is in fact contemporary with the monochromic and early white painted ceramic style in the rest of the Balkans.

The second model accepted in my study from 1998 (Nikolova 1998a) suggests that for the time being, in Thrace is not documented the monochromic and the earliest white painted pottery style known from neighbor Balkan regions. That model does not explain the cultural development as a diffusion of the innovations in distinct regions but it takes into considerations the possibility in Upper Thrace the earliest phases of the Early Neolithic not to have been investigated.

Palaeoecological evidence indicates that the first farmers impact upon the forested landscape, which predominated, was negligible (Willis and Bennet 1994: 327), but in the record base for that statement there is in fact a deficiency from Upper Thrace.

The migration model for Upper Thrace possibly applied is only an element of multi-aspect process of the Neolithization in the Balkans including a local development of the Mesolithic substrate known from the Danube riverside along Iron Gates (Lepenski Vir culture). For the investigation of the innovations in the Balkans of importance is the anthropological view on the glacialization of the Lepenski Vir population as a possible local evolution influenced by the diet and the local social environment (Mikić 1990).

Late Neolithic

The periods of the Late Neolithic I (Karanovo III culture), Late Neolithic II (Karanovo IV culture [Kaloyanovets]) and Early Copper Age (Maritsa culture) a still not well investigated in the Upper Stryama Valley.

In Late Neolithic I the tells continue to characterize the prehistoric culture there, but it is not clear the chronological interrelation between the Early and Late Neolithic levels at Banya and Chernichevo tells. Also, based to the terrain observations there is a small tell to the north of the Sarovka locality, Dubene-Leshtaka, where the buttoned handled indicate that it is possible the site to have been a segmentation of the Banya-Ploskata mogila population. However, for the time being, it is not clear the precise chronology of the foundation of that tell.

Along with the tell type, in the town of Hissar there is an open settlement documented from Late Neolithic I, but the material published is very poor.

Recently, V. Nikolov has discussed the later Neolithic culture material in Western Upper Thrace in some detail, in his doctor science dissertation (1998a). It is very important for the study of the innovations in the ceramics style as it shows the transition period documented between different ceramic styles, especially between Karanovo II and Karanovo III, as well as between Karanovo III and Karanovo IV. In my study from 1998 the so-called stage Karanovo II-III (after V. Nikolov) was excepted as a late stage of the Karanovo II culture and the so-called Karanovo III-IV culture as late stage of the Karanovo III culture (Nikolova 1998a).

The basic site of the Late Neolithic I in Western Upper Thrace is Plovdiv-Yasa Tepe tell. By the excavations of P. Detev in later 1940s – 1950s, it was 4 m high and 150 m in diameter. In the course of the initial investigation in its northeastern part two cultural layers were documented – lower, 2.5 m thick, and upper – 1 m thick but the later layer in fact consisted of secondary posited Late Neolithic vessels while the earlier later belongs to Maritsa culture (Early Copper Age). By the later excavations three levels from Late Neolithic I were excavated and P. Detev (1968a) assumed it was possible a hiatus to have exited between the levels as a result of flooding. In my opinion, Late Neolithic is represented on the tell by different stages of the Karanovo III culture.

Some of the published pottery completely parallels with Karanovo III pottery from G. Georgiev excavations. According to V. Nikolov, only the latest Late Neolithic level belongs unconditionally to the so-called Karanovo III-IV (Nikolov 1998a: 139), but on the other hand, among all the Neolithic material there is nothing that can be attributed only to Karanovo III. Then a conclusion follows that Plovdiv-Yasa Tepe belongs to the ceramic complex Karanovo III-IV in Thrace and to its classical stage then (Nikolov 1998a: 141).

Two of the sites discussed by V. Nikolov occupy the Upper Stryama valley – Hissar, the open settlement, and Banya-Ploskata Mogila.

The site of Hissar is well dated by P. Detev to Late Neolithic and paralleled with Plovdiv-Yasa Tepe. According to V. Nikolov, the flat lid with arched shaped handle is characteristic of Karanovo IV, so the settlement indicate a development of older traditions [Karanovo III-IV] in the period of the Karanovo IV in the astern areas of Thrace (1998a: 142). But we have no complete published information for any site from Karanovo III or Karanovo IV culture in whole Upper Thrace. It is even better to say those cultures are known very fragmentary. The material from Hissar parallels with Plovdiv-Yasa Tepe I and therefore, Hissar belongs to the Karanovo III culture.

The second site is that of Banya-Ploskata Mogila. P. Detev published limited ceramics from Late Neolithic. According to V. Nikolov, the jug certainly belongs to the so-called Karanovo III-IV culture (***Figure 6.14: 1***). But that jug, together with a fragmented idol (***Figure 6.14: 2***) originates from the excavations of N. Madzhev who documented two settlement levels. The pottery from those excavations is not published. So, in the light of those data the jug can be dated only generally to the Karanovo III culture.

The analysis of the other sites to the west of Upper Stryama by V. Nikolov is based on the theoretical precondition such as: 1. The material parallels with Plovdiv-Yasa Tepe I, then, it is the so-called Karanovo III-IV. 2. There are some elements of the Karanovo IV cultures but they are not many, so the traditions of the Karanovo III-IV continued in Western Upper Thrace and westward in the Zlatitsa-Pirdop lowland in the period of the Karanovo IV in eastern Upper Thrace.

However, the data about Karanovo IV culture in Western Upper Thrace have been increased. After Kalugerovo, known before, recently Dubene-Popovka II (***Figure 16.14: 3***) and Kapitan Dimitrievo (Nikolov, chapter 5) clearly document Karanovo IV appearance in Western Upper Thrace. Even more. Kapitan Dimitrievo documented the Late Neolithic II in Western Upper Thrace in a new light. A peculiar ceramic style occurs there close but not identical to Karanovo IV, known for instance from Nova Zagora-Hlebozavoda. There is only one vessel close to that style – from Popovka II, near the village of Dubene.

Until 1998 it was the only vessel from Western Upper Thrace together with Kalugerovo documenting Karanovo IV period in that region. Now, along with Dubene-Popovka II and Kapitan Dimitrievo, all these finds can be defined as Kapitan Dimitrievo occurrence (culture?), which is close but not identical with Karanovo IV in eastern Upper Thrace. For the time being, it is not clear if Karanovo IV and Kapitan Dimitrievo are two variants of one and the same complex (culture?) or there are two cultures with similar style but different structure.

That suggestion is very important for the relative chronology, as well as for the cultural processes with regard to the problems of the innovations and their distribution in the Central Balkans. Because what looks archaic and retardation in the

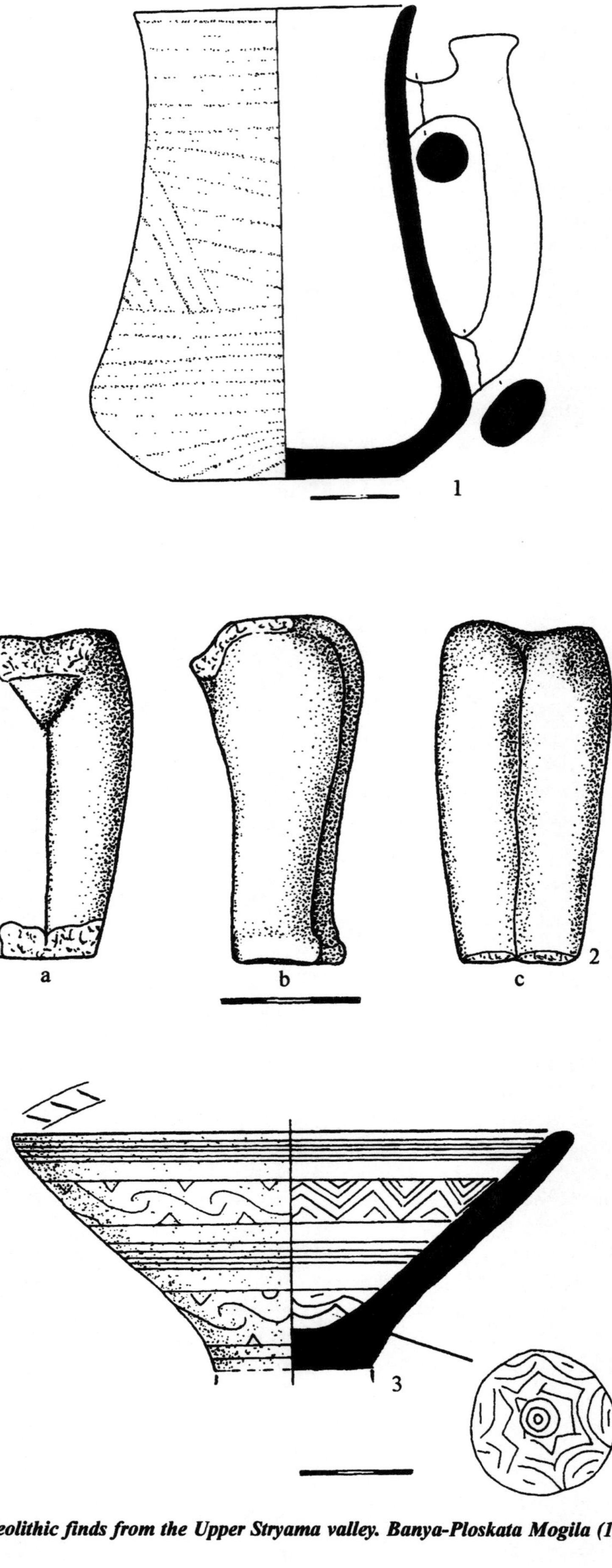

Figure 6.14. Late Neolithic finds from the Upper Stryama valley. Banya-Ploskata Mogila (1-2), Dubene-Popovka II (3).

development of Western Upper Thrace, is in fact contemporary development characterized by cultural integration including a formation of peculiarities but in the context of the innovation of white-encrusted ceramic style, widely distributed.

Copper Age

Early Copper Age

Sherds with deeply incised encrusted geometric ornament from Chernichevo tell indicate there was a settlement of the Maritsa culture on the tell. But they are only few sherds and only generally can be dated to the Early Copper Age. To the last period I can also attribute two accidental finds from Karlovo lowland –one-handled jug and an almost completely preserved anthropomorphic figurine (***Figures 6.15-6.17***).

The anthropomorphic figurine is an accidental find and has close parallels in Northwestern Bulgaria (Belotintsi). The pottery discovered with the last find is characteristic of the Gradeshnitsa culture (Alexandrov 1981: figs. 5 d-f; 6; pp. 36, 37). Indirectly, the relative chronology – possible later Early Copper Age is confirmed by the fact that in Late Copper Age, in the area of the Karanovo VI culture the female figurines with holes at the base have been distributed.

The jug is with high neck, profiled turning point and a handled attached to the rim and the upper part of the body. The closest parallels of the jug published can be find at Sofia-Poduene house pottery (Todorova 1990: plate 3: 3-4).

B. Brukner (1991: 205) already revised the given by H. Todorova relative chronology of the house. His opinion is it corresponds to Vinča D1 but not to Vinča D2 phase. That revision is very important as the Vinča D2 already may correspond to the Krivodol- Sălcuţa-Bubanj Ia complex represented by Dyakovo. From this point of view, Sofia-Poduene house can be attributed to the latest stage of the Gradeshnitsa-Slatino-Dikili Tash complex. The poor investigation of that region and the limited excavations at Poduene cannot answer important questions. For instance, if the house pottery discussed is an example of possible population moving from the Velika Morava valley into the Sofia field in the period of the great transformations of the late Vinča culture or it is only a strong influence.

The chronology of the Early Copper Age in western Bulgaria is divided into three phases by H. Todorova (1986: 12-27) and S. Chokhadzhiev (1997; see the synchronization table in chapter 8), but the sequence argued by both authors are rather different. According to the former author, to the first phase belong the upper horizons of Brenitsa tell, as well as it represented in Devetaki cave, Ikhtiman, Bukhovo and Vodna near Lovech Town. The last three sites are not published. As far as the Brenitsa concerns, it is the excavator who dated Brenitsa 3-6 from the beginning of the Early Eneolithic (Nikolov B. 1986) but the preliminary pottery (Todorova 1986: Drawing 45) published indicates later Gradeshnitsa shapes as well.

Based on well investigated micro region of Slatino area in the middle Struma valley, S. Chokhadzhiev (1997: 61) well argues the diachronous sequence of Slatino multi-level settlement. According to it, Slatino 2 belongs to the first phase of the Early Eneolithic with an assumption, that the very beginning of that period, respectively of the Gradeshnitsa-Slatino-Dikili Tash complex is represented at Kuro I and II in the Slatino micro region.

Stratigraphically and typologically, the second phase of the Early Copper Age is represented at Slatino 3-4, while the horizons 5-6 document the last phase of the Early Copper Age, respectively of the Gradeshnitsa-Slatino-Dikili Tash complex according to S. Chokhadzhiev. That author is correct dating the whole three horizons of Gradeshnitsa to the third phase of the Gradeshnitsa-Slatino-Dikili Tash complex but not to the second phase, postulated without any arguments elsewhere (Todorova 1986: 125-126). Gradeshnitsa C precedes Dyakovo in chronological table of M. Garašanin (1998: table 1). That correctly given chronology by S. Chokhadzhiev is confirmed by Sofia-Poduene evidence. Among the pottery there are parallels with Early Copper Gradeshnitsa C. So, the question that occurs is if the last site follows chronologically Gradeshnitsa, respectively Pernik-Krepostta (one of the nearest sites investigated from Early Copper Age), or it is parallel with Slatino 6 and Gradeshnitsa C. The positive answer assumed indicates that the end of the Gradeshnitsa-Slatino-Dikili Tash is synchronous with Vinča D1. In the light of the above suggestions, the jug from Karlovo lowland can be dated to the end of the Early Copper Age in the Balkans and synchronized to the west with the latest phase of the Gradeshnitsa-Slatino-Dikili Tash complex and Vinča D1.

Despite the absence of direct parallels, the cultural belonging of the Karlovo jug to the Maritsa culture area seems clear. But that suggestion is based mainly on the territorial position of the Karlovo valley – on the periphery of the Upper Thrace, while the jug as well as the encrusted sherds completely correspond to the Gradeshnitsa-Slatino-Dikili Tash complex as well. There are also few sherds with deeply incised encrusted pottery from Chernichevo tell that can be attributed to the Maritsa culture, but that style characterize also the complex mentioned.

With regard to more precise chronological relation to the other sites from Maritsa culture, it should be stressed the published record base is not very rich. It is worth mentioning that there is no ceramic definition of the defined up to now different phases (see Leshtakov 1997: 126 sq. with the references cited there), with the exception of the so-called fourth phase, which, as the excavations at Drama showed, is probably very early Karanovo VI culture. With regard to Western Upper Thrace, the statement that at the Yunatsite tell there are two horizons of Maritsa culture belonging to the so-called second and third phase at a depth of 8.30-9 m have no single argument (Todorova 1986: 66). So, it is not a surprise that the last author has mixed the evidence from two tells near Mudrets, publishing it as belonging to one and the same site (Leshtakov 1997: 127 and personal information). As far as Yunatsite concerns, V. Mikov published the Copper Age material very commonly and the sherds with incised ornament under discussion (Mikov 1995 [1939]: fig. 15) are attributed by the excavator even to layer B. Later he compared them with sherds

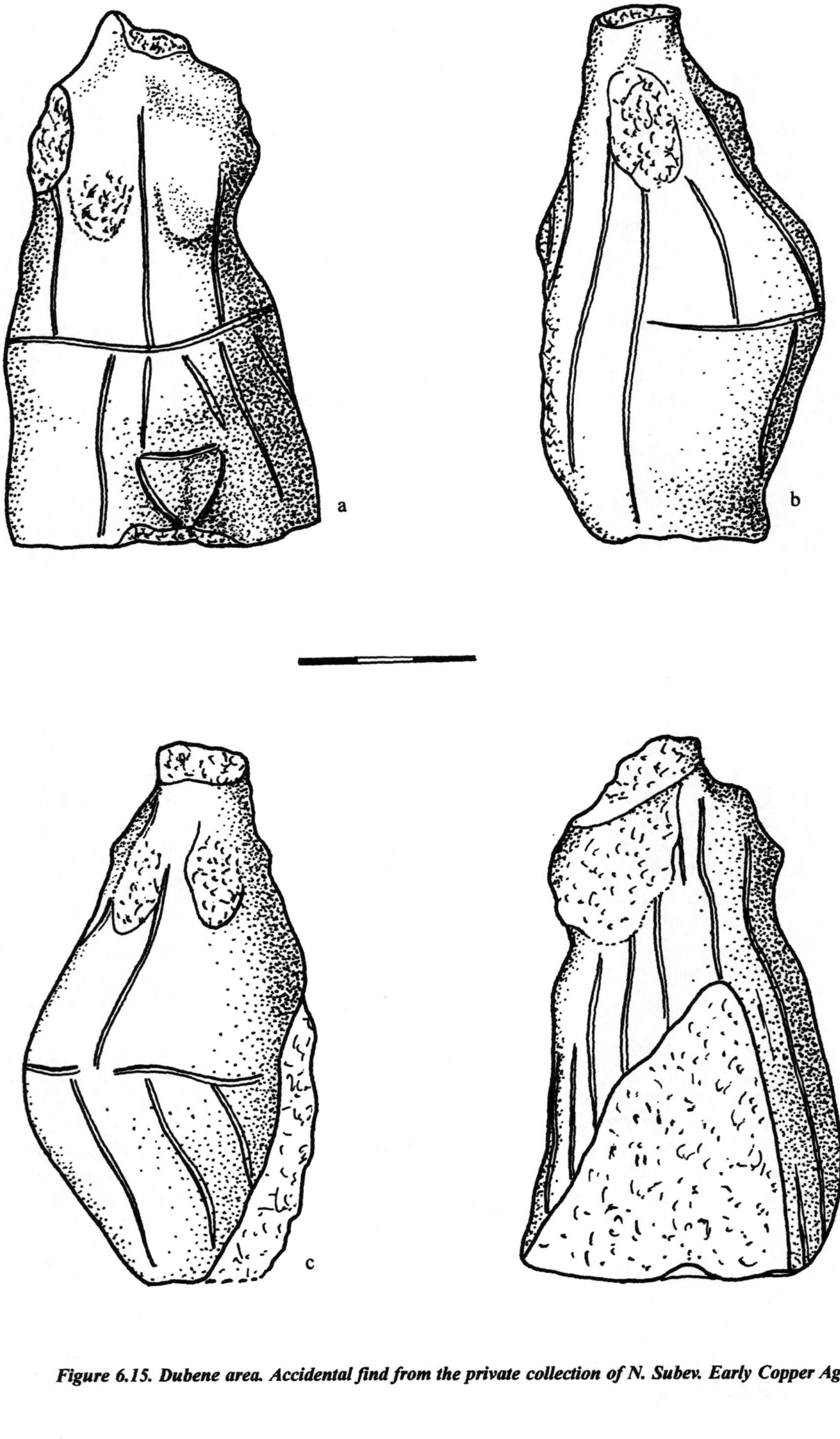

Figure 6.15. Dubene area. Accidental find from the private collection of N. Subev. Early Copper Age.

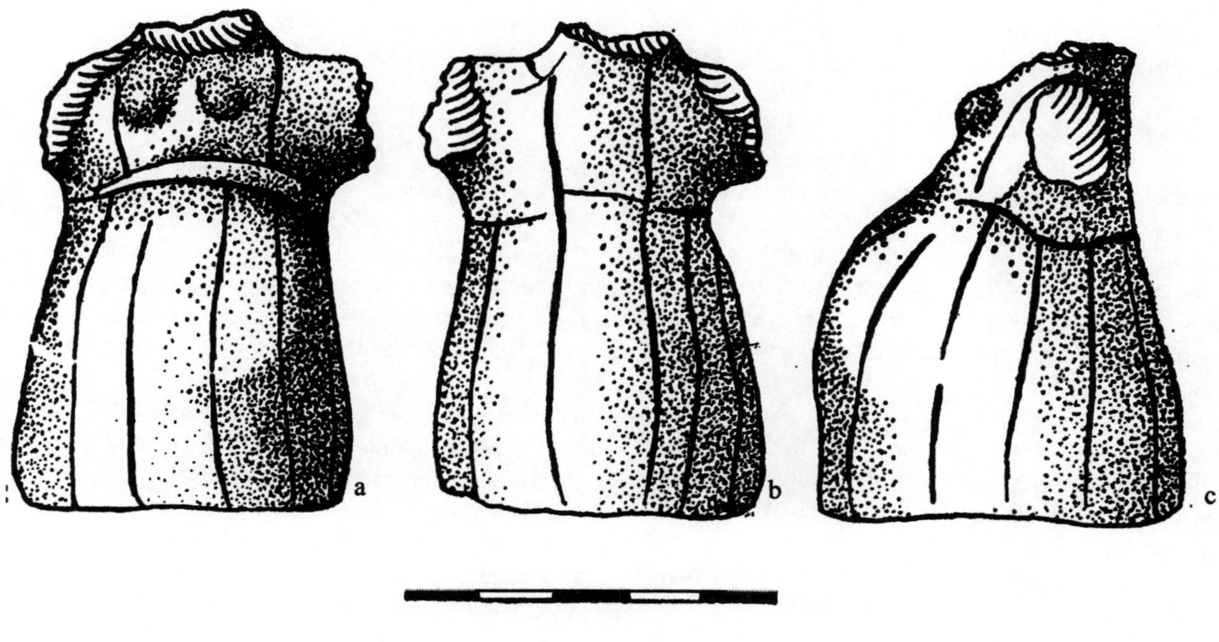

Figure 6.16. Settlement find from Northwestern Bulgaria (after G. Alexandrov).

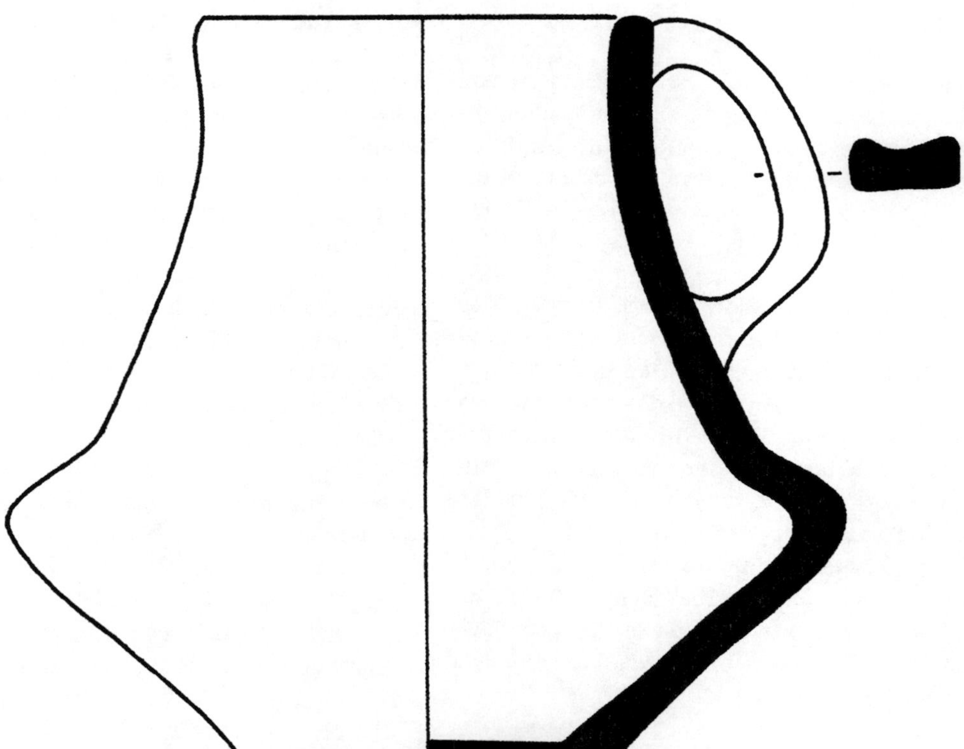

Figure 6.17. Accidental find from Karlovo region. Historical Museum in Karlovo. Early Copper Age.

from Plovdiv-Yasa Tepe (Detev 1948: 2). But we have no certain stratigraphic data for any concluding suggestions. Also, as the recent Kapitan Dimitrievo excavations showed, there was a smooth transition between Karanovo V and Karanovo VI in Western Upper Thrace. So, it is possible some Maritsa culture elements to have occurred in the earliest Karanovo VI or it may at the base of the tell there was some Maritsa layer but without clear, for the time being, precise chronology.

The most important site of the Maritsa culture in Western Upper Thrace is Plovdiv-Yasa Tepe excavated by P. Detev. There are two layers documented, from later Neolithic and from Copper Age, as well as some finds from destroyed Early Bronze levels.

In the course of the excavations in the 1950s, the stratigraphic observation of the Copper Age levels at Plovdiv-Yasa Tepe II suggests that in the earlier levels (at a height of 1.20 – 2 m above the natural terrain) the white encrusted pottery is much bigger. Of middle range is the white encrusted in combination with red ochre painted and graphite, white the graphite painted pottery characterizes the upper levels from Copper Age, at a 'height of 2-3.5 m' above the natural level (Detev 1959: 47). This is the only stratigraphic observation mentioned by the author with regard to the peculiarities of the Copper Age pottery. The absence of the stratigraphic context of the most of the vessels published gives a possibility the site to be dated generally to the Maritsa culture, probably to its early and late stage.

In this context, the very important excavations from Harmanli micro region in the Maritsa drainage should be mentioned - the excavations of K. Leshtakov at Luda Reka and Dervishov Odzhak sites (1997). There are thin-level settlements of the early Maritsa culture and the first precisely published material of that culture in Upper Thrace.

It is possible the site of Dervishov Odzhak to have been grounded before Luda Reka but both sites represent the earliest stage of the Maritsa culture in Upper Thrace. However, in the light of the comparative analysis – the encrusted pottery predominating in combination with red painted / encrusted and rare graphite painted ceramics – the characteristics of the Dervishov Odzhak correspond to the Plovdiv-Yasa Tepe IIA as the same is the characteristics of the ceramic style, after P. Detev. But it is not clear whether at Plovdiv-Yasa Tepe is represented the earliest or (more probably) there is the second phase of the Maritsa culture. In the light of the recent evidence, stylistically, three phases of the Maritsa culture in Thrace can be assumed as follows:

1. First stage. Deep white encrusted pottery with rare ochre painted / graphite and graphite ornamented and Karanovo IV elements preserved.
2. Second stage. Flourishing of the deep white encrusted pottery in combination with ochre painted / graphite and graphite ornamented.
3. Third stage. Increased graphite ornamented in combination with deep white encrusted pottery with rare ochre painted / graphite.

All of them belong to the Early Copper Age and the correlation with the different sites will be discussed elsewhere. The figurine and the jug from Dubene village area and from Karlovo Museum respectively, perhaps belong to the third stage of the Maritsa culture.

Late Copper Age

A Late Copper Age settlement was localized and sounding excavated on the Sarovka locality near the village of Dubene. The culture layer is c. 1 m thick, but no equipment was documented an the excavated area is very limited. The pottery dates the settlement from the later phase of the Karanovo VI culture (Nikolova and Zakharieva 1994; ***Figures 6.18-6.20***).

There is no periodization of the Karanovo VI culture argued that is based on long sequence from Upper Thrace, published precisely. In the stratigraphic-typological periodization of the Karanovo VI-Kodzhadermen-Gumelniţa complex it is assumed three phases of the development of that complex, or two-stage division (see Erdoğu, chapter 11). Yunatsite tell represents a long sequence of the Karanovo VI culture in Upper Thrace, but that layer is still not completely investigated by the renewed excavations. According to V. Mikov (1995 [1939]: 135-36), the Copper Age cultural layer there is 2.80 m thick (at a depth of 6.20 - 9 m).

In Eastern Upper Thrace, Bikovo tell (cp. Detev 1954; Georgiev 1961) represents the longest stratigraphic sequence of the Karanovo VI culture in this region up to 7 m thick, according to G. Georgiev. The stratigraphic data are mainly about later levels, in which late Karanovo VI culture was documented. In the literature is written about "Sălcuţa III" amphora-like vessel with shell incised ornament documented. This one or similar vessel originates from a depth of 2 m, according to P. Detev (1954: 174, fig. 48). From the "Eneolithic rubbish" originates a necked spherical with two horizontal disk-like handles and possible false-corded ornamentation discovered at a depth of 1.5 m with parallel at Zaminets C (Nikolov 1975: 88). On its side, the two-high handle cups is discovered at a depth of 4 m (Detev 1954: fig. 47:1, table 3). However, a typical Karanovo VI biconical bowl was found at a depth of 3 m.

These data may be indicate that at Bikovo there is at least 2.5 m thick layer of later Karanovo VI culture, that can be synchronized with later Sălcuţa II but not with Sălcuţa III (for Sălcuţa III see Nikolova 1999 and the references therein). The shell-incised ornament is popular in later Karanovo VI culture known for instance from Dolnoslav (Fol and Lichardus 1988: Fig. 44), which tell is also dated from later Karanovo VI culture. In combination, shell-ornament, false-corded and two-handled cups characterize the later Karanovo VI culture at Stara Zagora – Miniralni Bani. The handled amphora vessels are not unknown in the area of the Gumelniţa culture. They occur in latest Karanovo VI culture context at Yunatsite as well (Mazanova 1992: plate 5: 1), so it can be expected influence from North into Sliven micro region, as well as from the Central Balkans.

According to the newly excavated Kapitan Dimitrievo tell (trench 4), there is at least c. 3 m thick layer from Karanovo VI culture and at least 4 levels documented at a depth of 0.80

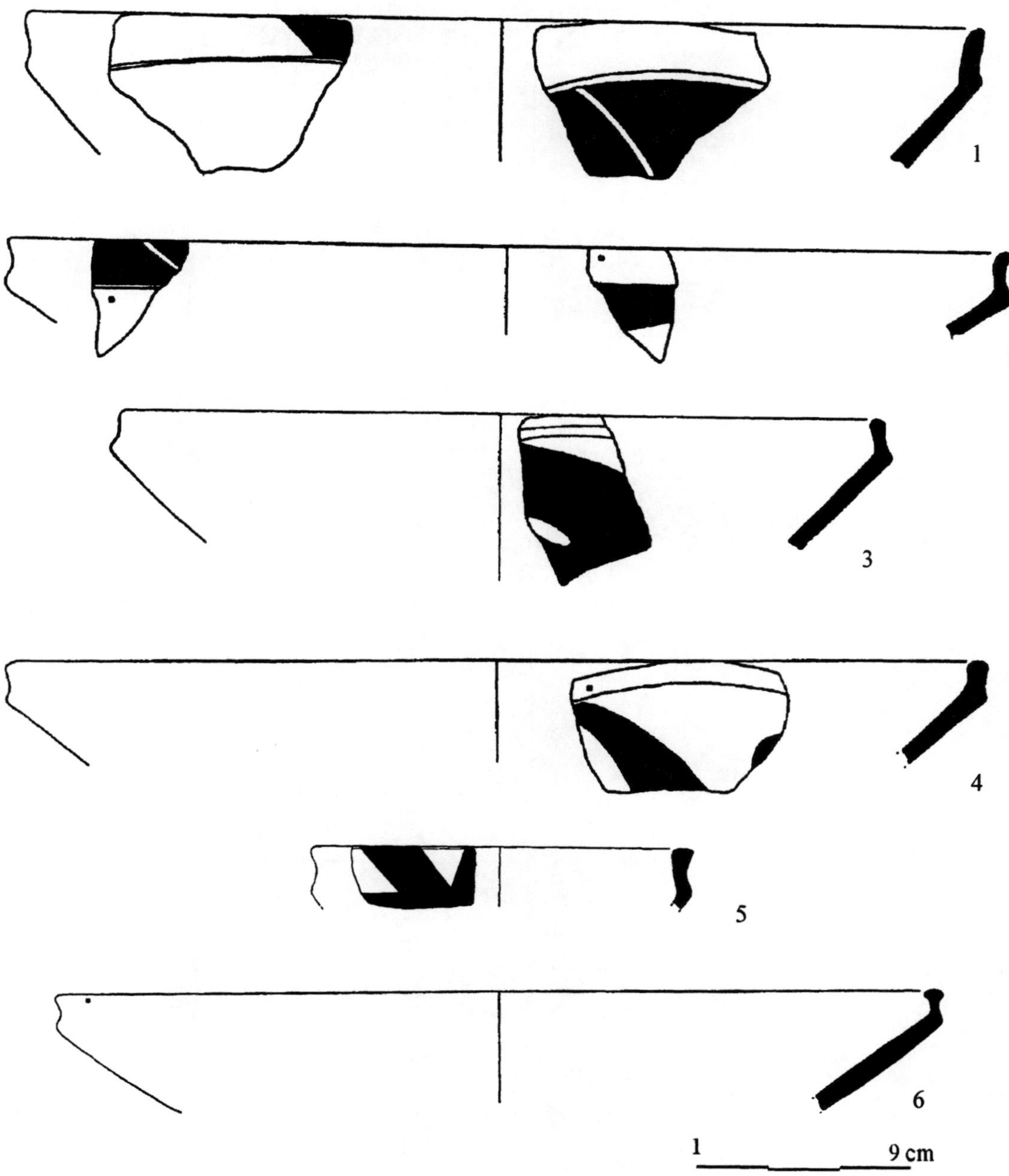

Figure 6.18. Dubene-Sarovka I. Late Copper pottery. Karanovo VI culture.

m, 1.80 m, 2.10 m and 3.5 m (Nikolov in Nikolov et al. 1999: 22). The trench was dug from the level of 99 m above sea level or at a depth of 1 m with regard to the highest point of the tell. P. Detev publishing his excavations mentions the upper layer from Copper Age was excavated at height of 5 m to 12 m from the ground level (Detev 1950a: 7). As the tell was 12 m high, the Copper Age layer was excavated immediately under the surface. But in the light of the renewed excavations on the tell at a depth of c. 5. 50 m there is already Late Neolithic layer. Also, according to V. Nikolov, at the base of the trench 4 there was already trace of Maritsa culture (1999: 20-22), which is confirmed by the pottery, published by V. Matsanova (in Nikolov et al. 1999: tables 1-9). But P. Detev wrote at a height of 6 m (i.e. at a depth of 6 m) there was an askoi discovered near a hearth (Detev 1950a: fig. 15), which is typical of the later Karanovo VI – Gumelniţa – Varna complex. It looks the askoi was discovered somewhere on the slope or on the periphery or in mixed context and its stratigraphic context cannot be accepted as reliable.

Comparing Bikovo with Yunatsite and Kapitan Dimitrievo, the latter two sites only partially covered stratigraphically the layering of the Karanovo VI features at the tells of Thrace, roughly its half duration. For the time being, the settlement of Dubene-Sarovka (***Figure 6.18-6.20***) corresponds to the classic (second stage) of the Karanovo VI culture, while the latest horizons at Yunatsite – to the third (last stage) of the Karanovo VI culture. It seems the middle levels from Karanovo VI culture at Kapitan Dimitrievo (see Matsanova in Nikolov et al. 1999; Nikolov, chapter 5) were contemporary to Dubene-Sarovka I.

In fact, Kapitan Dimitrievo, Yunatsite and Dubene-Sarovka I are the only well documented settlements of the Karanovo VI culture in Western Upper Thrace. On those sites, along with the standard ceramic style of the Karanovo VI-Gumelniţa-Varna complex, some elements occur, which are typical of Krivodol-Sălcuţa-Bubanj complex. They were a result of the cultural contacts or very strong influence in the period of de-

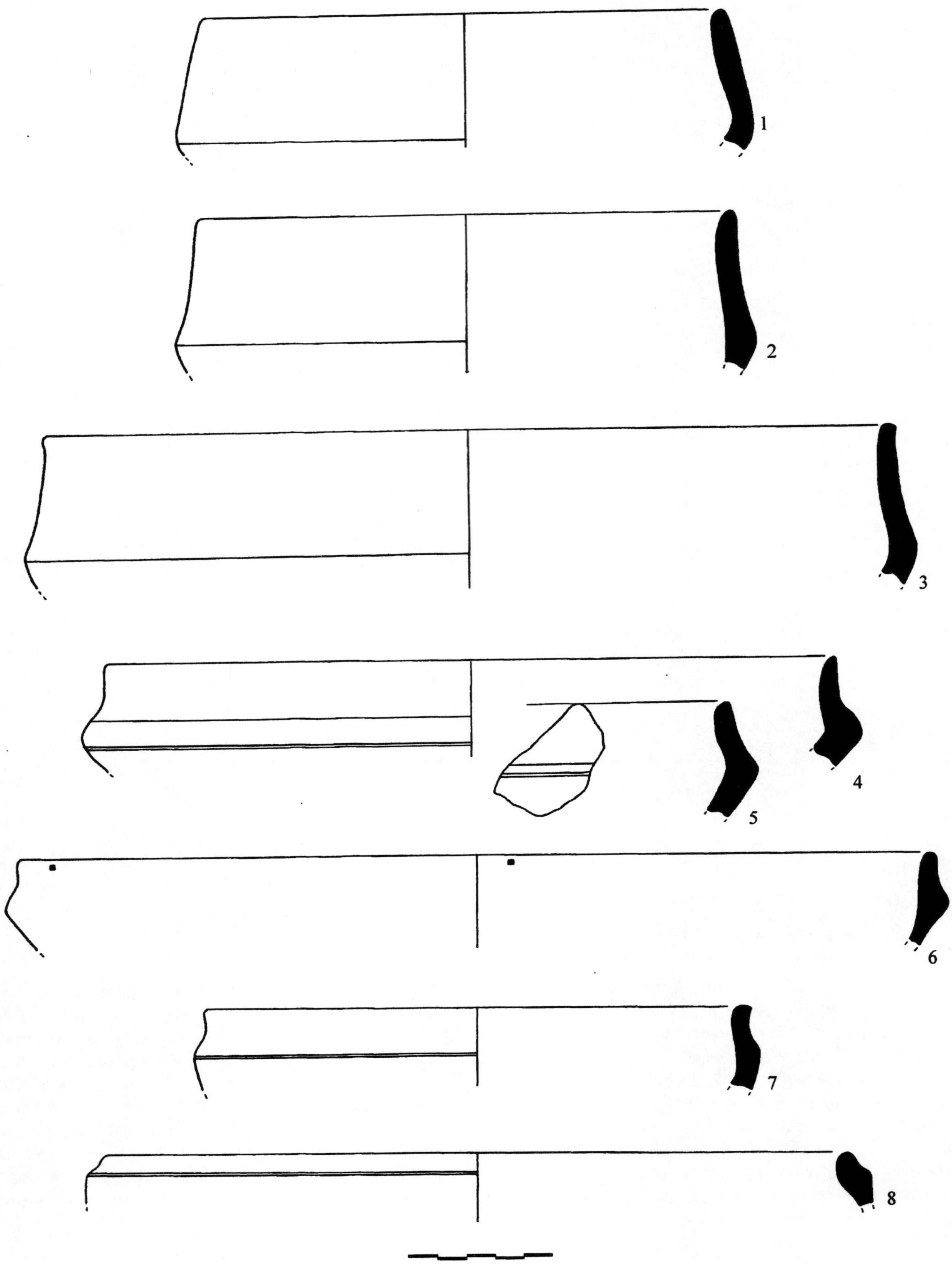

Figure 6.19. Dubene-Sarovka I. Late Copper pottery. Karanovo VI culture.

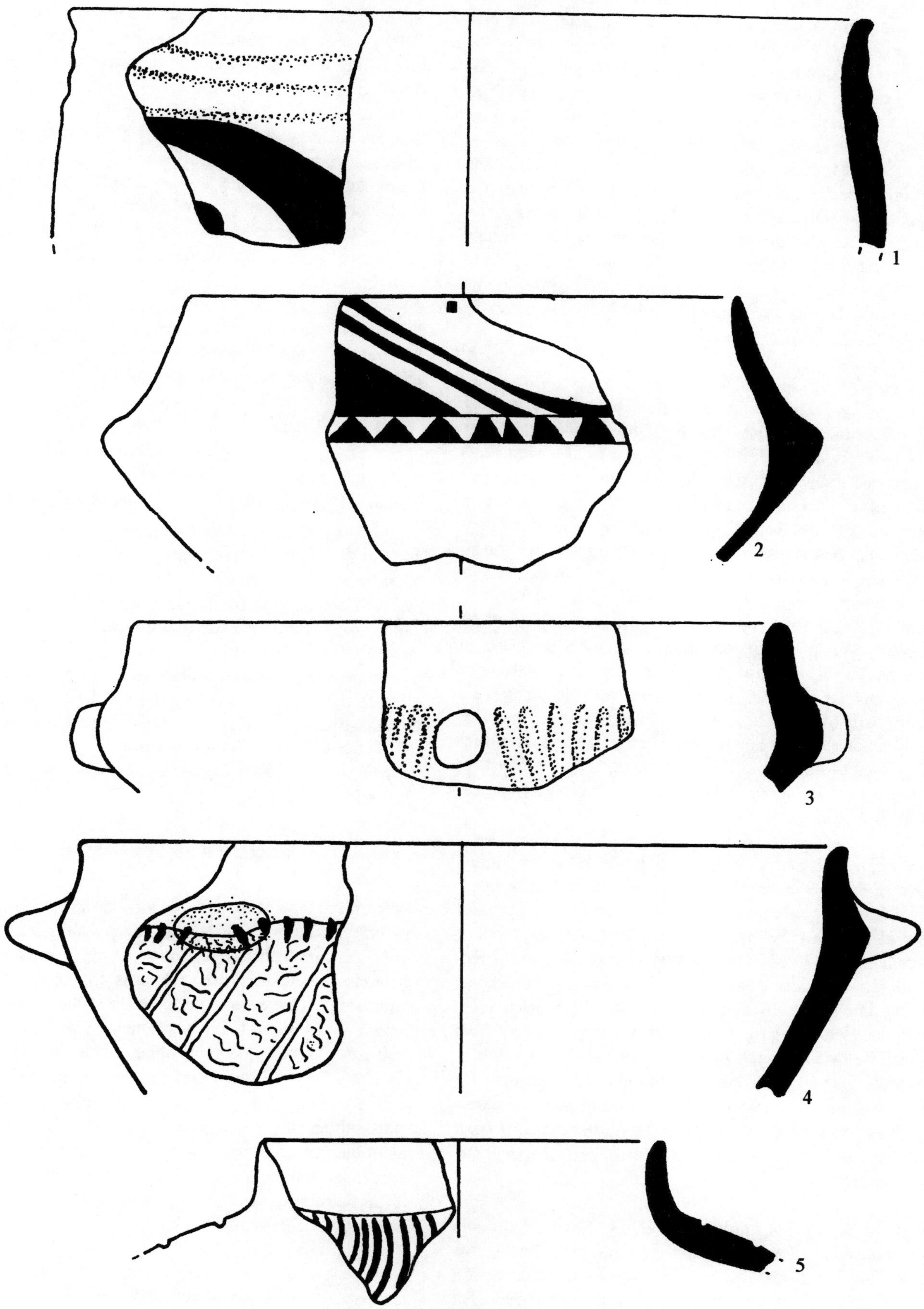

Figure 6.20. Dubene-Sarovka I. Late Copper pottery. Karanovo VI culture.

creasing of the ceramic style of the Karanovo VI-Gumelniţa-Varna complex (Yunatsite). Recently, south interrelations have been argued, based on the parallels of the pottery between the Karanovo VI culture and later Sitagroi III (Leshtakov, chapter 6). The existence of very late phase at Sitagroi was argued based on parallels with later Krivodol-Sălcuţa-Bubanj complex and ^{14}C dates in Nikolova 1999. As far as Yunatsite concerns, the excavations of the earlier levels will answer the question of the precise chronological comparison between the settlements of both the sites – in which cases the contacts are synchronous or diachronous. But the communication bridge between the upper Maritsa valley and the Northern Aegean in the Late Copper Age is a real model, which is confirmed by the Kapitan Dimitrievo evidence as well (Matsanova in Nikolov et al. 1999).

In the context of the local social environment of the upper Maritsa valley, the population of Dubene-Sarovka had another lines of contacts – those with the Upper Tundzha valley (Gabarevo), as well as through the Stryama River with Eastern Upper Thrace. Unfortunately, there are no Karanovo VI settlements discovered in Plovdiv area and one of the closest known up to now is Rupkite-Kaleto. Among the pottery published there are some sherds dated from the so-called Transitional (post-Karanovo VI) period in Upper Thrace. In fact there is typical Karanovo VI sherd with parallels at Dubene-Sarovka (Georgieva in Doumas and La Rossa 1997: fig. 9: 3). The red 'painting' is not something strange for the classical Karanovo VI culture. Also, as N. Todorova and V. Matsanova show in chapter 26, limestone pieces and organic material are characteristic of the latest Karanovo VI. As far as the so-called snail-shell concerns, it is not clear whether it was such a mixture or kind of limestone. But, morphologically the sherds in fig. 9:5-6 may belong to Karanovo VI, as the parallel of the handle in ***Figure 6.20: 4*** shows.

However, there is a bowl with red and white distemper painted shown in fig. 9: 9 in the discussed publication. Similar linear composition of concentric red-and-white bands characterizes Vaksevo-Skaleto (Chokhadzhiev, chapter 8, ***Figure 8.8: 11***), where it is discovered in the context of graphite painted pottery, along with false-corded of type Šuplevac ceramics, for the first time in Southwestern Bulgaria have been discovered. Those parallels indicate, on the first place, the pottery from Rupkite-Kalevo was incorrectly typologically distinguished. Secondly, if the bowls indicate contemporary phase with Vaksevo-Skaleto, it may post-date Yunatsite being synchronous with Sălcuţa III, but not with the *Scheibenhenkel horizon* named as 'Transition period' (in fact Final Copper II) by the excavator.

Moreover, we do not know the later Karanovo VI ceramic style at all in details – with exception of the published below pottery from Yunatsite (Todorova and Matsanova, chapter 26), so very important is the whole ceramic complex discovered at Rupkite-Kaleto, which is missing in the cited work.

As far as the western contacts of the Dubene-Sarovka community concerns the bowl in ***Figure 6.20: 3*** may indicate western influence, known for instance from Bagachina (Bonev and Alexandrov 1996: fig. 33: 1). The graphite-ornamented bowls in ***Figure 6.18*** are also common types of the Karanovo VI-Gumelniţa-Varna and Krivodol-Sălcuţa-Bubanj complexes. They indicate the common tendencies in the ceramic production, along with the bowls with the inverted and twisted rim (Nikolova and Zakharieva 1994) with later parallels at Vaksevo-Skaleto (Chokhadzhiev, chapter 8, ***Figure 8.5: 1-4***). The biconical bowl with thickened carination (***Figure 6.20: 2***) is also shared shape as it is documented at Vaksevo-Skaleto (Chokhadzhiev, chapter 8, ***Figure 8.6: 11***). However, the parallels between both sites are diachronous as Dubene-Sarovka belongs to the classic but not to the late stage of the Karanovo VI culture, in the light of the recent evidence, contemporary with Telish II.

For the time being, there are no certain arguments that the community of the Dubene-Sarovka was in direct contact with the Late Copper population in the Osum basin in Northern Bulgaria.

Considerations

Despite the Copper Age is still not well investigated in the Upper Stryama valley, there are important new evidence, which for the first time document the both cultures – Maritsa and Karanovo VI. The data of the Maritsa culture indicate active western contacts, through the western Sredna Gora Mountains, which continued the tradition of the cultural development from Late Neolithic. The settlement at Dubene-Sarovka I is with typical of later Karanovo VI culture ceramic style despite there is close typological similarity with Krivodol-Sălcuţa-Bubanj complex. The axe of Jászladány type (Mikov 1972) from Karlovo Region documents Final Copper I in the Upper Stryama valley, again in interrelations with Western Bulgaria.

Theoretical Applications

The leading type of lower wide tells in the Karlovo valley on the one hand, reflect the geographic preconditions for a long term occupation, but on the other hand, the existence of the dynamic settlement system which is comparable with highlands areas. That type of tells was introduced by the initial population from the Early Neolithic and was followed in the prehistoric development. In other words, in contrast to the Maritsa valley, the multilevel settlements did not develop as high tells. One of the explanation of the occurrence of large "flat" settlements is suggested for North Greece, where they might be characteristic of more wooden environment, "where rejuvenation of the forest and weeds precluded long-term use of the fields and, thus, long-term occupation of the villages" (Demoule and Perlčs 1993: 364).

In the case of Karlovo valley, the low tells indicate long-term occupation, but it cannot be also excluded that the environment was well timbered and the surrounded smaller arable area aroused periodical exhausting of the soil.

The other type of settlements – thin level occurred at the foothills of the Sredna Gora and Stara Planina Mountains in the Early Neolithic and along the Stryama River side in the Early

Bronze Age.

Another basic element of settlement pattern is the house. The Early Neolithic house in the Balkans occurs as wattle-and-daub, in contrast to Near East mud-brick, as a local invention (Treuil 1983). We have no discovered houses from Neolithic and Copper Age in Karlovo lowland, but the fragments of plasters indicate wattle-and-daub house. According to the other settlements excavated, the Neolithic house in the Balkans was mainly rectangular.

Technology-economy and the diffusion of the innovations

The imminent interrelation between the innovations and development of the technology in the prehistoric societies is indisputable. But there are difference between the technologies – they can be monopolized or can quickly diffuse from one area into another one (Kohl 1987: 22). The technology depends on the natural recourses and can accelerate the development of the frontier areas but also can lead to the overexploitation of the environment.

In the interaction process of the diffusion of the innovation in some cases it is difficult the donor and the recipient culture to be defined, because not all cores of the innovations are easily estimable. In this chapter the complicated problem with the distribution of the white-painted pottery exemplifies the statement. As far as the graphite ornamentation concerns, its wide distribution in Late Copper is connected with Early Copper initial development (see Chokhadzhiev, chapter 8).

Economy, exchange of goods and gifts

The distribution of the innovations can be analyzed as a process of distribution of technological, economic or informative (cultural) matters, in some cases interrelated with the gift exchange. Even in Aegean Bronze Age (Ugarit) the usual trade "still had the form of gift exchange, without fixed prices to be agreed upon, but with the exhortation: "send me much of it, and not little" (Burkett 1987: 49). From the first millennium BC another model is known. According to Herodotus the Phoenician merchants traded with Africa by a minimal form of contact, in fact avoiding direct touch (Burkett 1987: 48). That 'silent trade' on the marginal regions can be difficulty applied to the Stryama interaction model because the archaeological record indicates active communication between the communities and the individuals.

Generally, two main types of trade can be distinguished: close and distant. The trade or exchange can be direct or indirect, as well as the distant contacts can be subdivided into different types according to its character and the distant. Insofar as trade is a peaceful movement of goods among two and more societies (Schortman and Urban 1987: 49, and references therein), those characteristics can be also applied to the diffusion of the innovations through trade.

The distributed innovation through trade may result in increasing of the wealth of the distinct social segment(s) or of the social power. As possible account evidence (the so-called tokens [Budja 1998]) can be interpreted some small ceramic disks shaped from sherds which were distributed not the Neolithic Balkans, but also in continental Greece (Demoule and Perlčs 1993: 375).

Obsidian

In Early Neolithic the Upper Stryama valley was involved in the distribution of the obsidian artefacts. A small knife like obsidian was excavated from later Early Neolithic Kliment-Banyata. Unfortunately, the absence of chemical analysis gives a possibility that artefact to be included in the Mediterranean system of obsidian exchange only hypothetically. There are no data on such artefacts from the other prehistoric cultures documented there. Nevertheless, in the Balkans exception obsidian artefacts occur among different prehistoric cultures – for instance in Vinča culture (Liubcova, Parţa) in Romanian Banat (Comşa 1969: 21) that can be connected with the Carpathians resources. As rare artefacts the obsidian objects can be defined as a precious exotic item for the Balkan population obtained through long-distant contacts and possibly connected with the local elite of the Neolithic and Copper Age societies there.

River drainage as communication routes

In the prehistory of the Balkans the main communication lines for exchange of goods and other cultural contacts were the rivers. The Danube, Maritsa, Tundzha, Stryama, Iskur, Struma, Vit, Osum, Olt, Vardar navigable rivers can be at least stressed. Along the Black Sea coast and the Aegean, the coastal navigation was significant means of transmissions.

The contents of the exchange goods can be only partially reconstructed. First of all, an object of exchange were rarely distributed resource and precious metals of their products – like gold, silver, minerals, dyes, etc. The monopoly of salt production and distribution can be also expected, which N.N. Tasić has discussed in detail in chapter 3.

It is under question if there was intra or inter exchange of cereals and pulses between neighbor and distant population. J.-P. Demoule and C. Perlès (1993: 362) suggest that the existence of 'normal surplus' is not only a risk-minimizing device in Neolithic Greece in areas where annual yields fluctuate widely but it also "provides a means of exchange and may promote part-time craft specialization".

Ceramic style, society and innovations

In the context of the problem of the innovations the dichotomy tradition and innovation can be defined and that problem can be approached by a possibility an explanation model of a society versus tradition relations to be argued.

The main aspect of tradition in the prehistoric society is the ceramic style. I will live aside the technological aspects as the stone and flint implements, which have very slow development (in the context of the ceramic style). The innovation in the ceramic style means a general social opposition against the tradition. I can hypothesize it was the young generation which for some reasons in a rhythm of c. 200 years generally or partially change the ceramic style in the villages. Different simplified models can be suggested, but the distribution of the new ceramic styles usually in the Balkans includes vast territories.

Those innovation distributed styles can be summarized as follows:

1st sort of style. The first monochrome ceramic style. It characterizes the earliest Neolithic cultures.

2nd sort of style. The style of white, dark and polychrome painted pottery. The former characterizes Early Neolithic while the latter begins in the Early Neolithic and continues in the Late Neolithic. Accidentally, it has occurred in the Early Neolithic.

3rd sort of style. The first encrusted pottery style. It characterizes Late Neolithic and Early Copper Balkans cultures.

4th sort of style. The graphite painted style. It begins parallels with the 3rd style and continues into Later and earlier Final Copper Age in some regions in the Balkans.

5th sort of style. The second encrusted pottery style. It initiated in the Final Copper Age and continued in Early Bronze as general exception and successfully and plentifully developed in the Early Bronze II, decreasing in Early Bronze III.

6th sort of style. The second monochromic pottery style. It characterizes as exception Early Bronze II and plentifully develops in Early Bronze III.

7th sort of style. The third encrusted pottery style. It characterizes some cultures from Early Bronze III (Maros culture), Middle Bronze Age and from Late Bronze Age.

That classification is based only on the emblematic kind of the ornamentation (or its absence) and is defined to recognize a periodical change on the ceramic style as an opposition of the previously abundantly developed styles. We can add to specify the problem such other components as the morphology and the technology, but in my opinion the general ceramic innovation can be searched in the distinguished marker as the kind of the ornamentation or its absence and in the social relation to its temporal distribution.

To apply any social model to the cultural development in the Upper Stryama valley it is difficult because of the absence of the cemeteries investigated and the limited settlement data on the earlier Prehistory. Nevertheless, that micro region was an integration part of the Western Upper Thrace cultures, which were developed up to variant of bigger cultures or complexes or up to independent cultures argued basing on the peculiar ceramic style. In terms of the South Balkan prehistoric development the social evolution model can be argued according to data from such representative sites as Yunatsite, Karanovo, Ezero, Sitagroi, Dikili Tash, Slatina, Slatino, Kovachevo, Hoca Çe°me, Kirklareli etc., which represent data for different prehistoric periods.

The social evolution model may concern not only the evolution of the cultures but their transformations and rather visible changes up to formation of other cultures. That model can be applied to bigger prehistoric phases and stages of development, which can be traced over vast areas of the Balkans. For instance, they can be defines social horizons as:

1. The horizon of the Mesolithic (Epi-Paleolithic) societies of hunter-gatherers.
2. The horizon of the earliest agriculturists-stockbreeders from Early Neolithic.
3. The horizon of the developed (later) Neolithic and Early Copper Coppers as a stage of the development of the initial complex societies.
4. The horizon of the Late Copper cultures as a period of the flourishing of the complex societies.
5. The horizon of Final Copper cultures as a period of the transformative complex societies.
6. The horizon of Early Bronze I as a period of the evolution of the complex societies.
7. The horizon of the Early Bronze II and III cultures as period of emergence of the initial chiefdoms in the Balkans.

The horizons defined partially correspond to the sorts of the ceramic styles (see above). In the process of the inner developments of the different social horizon, as well as from one into another horizon, the innovations are of primary importance in the culture changes.

Regionally, there can be trace similar processes in the different parts of the Balkans without being very clear the interrelations between those processes. Good examples are the Karanovo I-II-III transition in eastern Upper Thrace and Starčevo-Vinča transition in the northwest of the Balkans. The former process seems to be smooth characterized by the change in the ceramic style from painted toward monochrome, channel and encrusted ceramics.

Unfortunately, there is no cemetery evidence from Late Neolithic from Upper Thrace and the limited settlement data published infer mainly a social homogeneity in the Neolithic in Upper Thrace. However, such a suggestion contrasts to the data from the other regions – especially from the area of the Vinča culture. Social-economic reasons aroused the evolution from Starčevo to Vinča culture according to some authors (Leković 1990).

The diachronic view on the prehistoric development in the Upper Stryama valley document a region where the innovations, exemplified by the ceramic style, were quickly and vastly spread. The ceramic changes and the settlement changes, however, can be explained by interaction and social model of local economic crises, but not with 'ethnical changes'. Of special interest above were the contacts with Western Bulgaria, documented by expressive new record.

CHAPTER 7

ANIMAL BONE FRAGMENTATION AND THE ORIGIN OF METALLURGY IN THE CENTRAL BALKANS

Haskel Greenfield

Introduction

The Balkans of southeast Europe (Bulgaria, Romania, and Serbia) is one of the acknowledged Old World centers for the autonomous development of metallurgical technology (e. g. Renfrew 1969; Glumac and Tringham 1990). From southeast Europe, metallurgy spreads to the rest of Europe setting the stage for the advent of the more complex societies of the Post-Neolithic (Eneolithic/Chalcolithic and Bronze Age).

European prehistorians have traditionally defined the end of the Neolithic and advent of subsequent prehistoric periods in terms of the dominant archaeologically-visible technology. The Neolithic is dominated by lithic technologies, while subsequent periods are characterized by the widespread use of metals. The subsequent prehistoric periods are named and defined in terms of the dominant metal found in contemporary archaeological deposits - Copper Age (also known as the Eneolithic or Chalcolithic – ca. 3500-2500 BC), Bronze Age (ca. 2500-1000 BC), and Iron Age (ca. 1000-0 BC).

It has traditionally been assumed that at the close of the Neolithic, Balkan economies rapidly move from a stone- to a metal-oriented tool economy. This assumption has been largely based upon the dramatic drop in frequencies of stone tools in post-Neolithic deposits. In Neolithic deposits, stone implements and waste are very common, while metal objects are extremely rare and limited to ritual or social functions. Beginning in the Late Neolithic (Vinča-Pločnik culture of Serbia, Boian of Romania, and Karanovo IV-V of Bulgaria), there is indisputable initial evidence for the mining, smelting, and distribution of metal objects (e.g. Jovanović 1980; Tringham and Glumac 1990). During the subsequent Eneolithic and Bronze Age, metal objects become more increasingly more common and functional, while stone implements quickly become relatively scarce. By the Late Bronze Age and Early Iron Age, significant quantities of metal tools appear in the archaeological record.

While most research concerned with the origins of metallurgy has relied upon the metal artifacts, this approach is fraught with a host of problems. The number of early metal tools from the earliest prehistoric periods (Neolithic, Copper, and Bronze Ages) is quite small. Almost certainly, the range of recovered metal tools does not reflect the full range then available (cf. Olsen 1988). Metal was such a precious commodity in antiquity that it was used and reused (Tylecote 1987, 1992). Metal was typically discarded only when there was too little to salvage, a condition that was relatively infrequent. Most metal was probably recycled - as attested by the paucity of simply discarded broken or worn tools. Most metal finds are worn, broken, or finished tools and weapons that were hidden and forgotten or lost, ritually deposited, or simply lost. Early metals are also chemically unstable so that they will decompose relatively rapidly under most conditions. "Their original rarity, recycling of worn tools, and decomposition prevent us from discovering all of the types of tools then in existence" (Olsen 1988).

In the absence of large quantities and varieties of metal objects in any site or region, it is difficult to investigate the origins of metallurgy and its effect upon early Balkan society. As a result, little evidence exists to determine exactly when the change-over from a stone- to a metal-oriented economy took place. Did it take place slowly or rapidly and was the spread of metallurgy a relatively uniform process? These are vital questions that must be answered before the question of causal priority can be established.

Previous research on the origins of metallurgy through zooarchaeology

In this paper, a non-traditional approach for investigating the origins of metallurgy in a region is proposed. Zooarchaeology, the study of animal bones from archaeological contexts, has proven to be a means of investigating this issue. Recently, Greenfield (1999a, 2000, n.d. a, n.d. b) attempted to identify the transition from a stone to a metallurgical technology in the subsistence economy through the analysis of butchering cut marks on animal bones. A series of tests were conducted to systematically be able to distinguish between stone and metal cut marks. The results were then applied to two prehistoric sites from the central Balkans (Petnica and Ljuljaci) in order to monitor the origins of metallurgy in a region. The results of the above research can be summarised as follows: The percentage of stone tool cut marks declines over time. The incidence of metal cutting implements is minimal prior to the Bronze Age. They are present in the later Neolithic, but in such small quantities that they can probably be attributed to the occasional drift of material into lower levels or misidentification. Metal cut marks begin to appear in some quantity at the end of the Neolithic (Vinča D culture). This is the period of earliest metallurgy in the Balkans. Large copper veins were mined in nearby eastern Serbia and metal axes and other implements appeared in sites throughout the region (Jovanović 1980). During the Eneolithic (Baden-Kostolac culture, 3300-2500 BC, calibrated), the frequencies of metal tools remain low (13%). The numbers of metal cut marks dramatically increases during the Bronze Age (2500-1000 BC, calibrated). This substantial increase coincides with the appearance of high tin-bronze tools (Branigan 1974; Coles and Harding 1979: Tylecote 1992). This would indicate that bronze tools are effective for butchering from early on in the Bronze Age, contrary to the belief that metal tools would only become effective butchering implements when high tin-bronzes are developed (e.g. Champion et al. 1984: 163). During the Bronze Age, there is a difference in proportions between high and low status sites. In high status sites, the incidence of bronze cutting tools is dramatically higher (84%) than in lower status sites (41%). Stone tools remain important at lower status sites.

The origins of metallurgy– a bone fragmentation hypothesis

In this paper, a different approach is taken to the identification of the moment when local economies shift from a stone to metallurgical technology. Another means of establishing whether metal or stone cutting tools are evident is the fragment size of the bones. About two decades ago, Louis Binford (1981: 104-5) proposed that introduction of metal tools and the abandonment of a stone tool technology into a region would result in changes in the nature of butchering. This concept can be rephrased into a series of hypotheses that can be tested by comparing bone assemblages from either side of the Neolithic-Bronze Age divide.

1. It is expected that animal bones from the post-Neolithic would be more fragmented than those from the Neolithic. There would be an increase in the frequencies of small bone fragments in assemblages. The logic behind such a hypothesis is that the nature of bone fragmentation as a byproduct of animal butchering would shift. Metal tools, particularly axes and heavy swords, could be used to chop bones through joints or bones, and break them up into smaller sizes in a more systematic series of size classes than with stone tools. In contrast, butchering with stone tool blades would require more careful disarticulation of limbs. Otherwise, the stone tool blade would quickly dull and need to be replaced or sharpened (a time consuming task). Sharpening of metal tools is a rapid task since it simply requires rubbing the edge with a whet stone. Bone fragmentation in the Neolithic would take place largely when the bone is smashed against an object or heavy hammer stones were used to break the bone.

2. The opposite hypothesis is that bone fragmentation would be less in the post-Neolithic than in the preceding Neolithic animal bone assemblages. Since metal knives have a greater tensile strength than stone blades (cf. Binford 1981:104-5), they can be inserted into areas and withstand stresses that stone knives cannot. This option would result in the need for fewer bones to be fragmented during butchering, since one of the goals of butchers is to minimize the damage to their tools.

3. An alternative to either of the previous extreme options is that there is no significant change or shift in bone fragmentation frequencies between the Neolithic and post-Neolithic assemblages. This option recognizes that bones, in particular long bones, would be fragmented in bone Neolithic and post-Neolithic assemblages since the cooking technology in both periods is similar. Roasting of meat was probably of a minor significance based on the low frequencies of burnt bones in sites from both time periods (Greenfield 1986b, 1991). Most meat was probably cooked (boiled) in pots. In order to fit the meat (particularly of large mammal such as cattle, deer, etc.) into ceramic pots, most of which were not very large, the meat either had to be stripped off the bone or the bone broken to a size smaller than the pot. As a result, the major shift in fragmentation would probably take place between pre-ceramic and ceramic using cultures.

In the rest of this paper, I will test these options using data collected from archaeological sites in the central Balkans.

Data

The faunal assemblages from 11 prehistoric archaeological sites were examined to test the above hypothesis - Blagotin (Greenfield 2000a, n.d. c), Stragari (Greenfield n.d. d), Opovo, Novačka Ćuprija, Crkvina, Ljuljaci, Sarina Medja, Vrbica, Vecina Mala, Livade (Greenfield 1986a,1986b, 1986c, 1996), and Kadica Brdo (Greenfield n.d. e).

The sites have occupations that range from the Early Neolithic to the Early Iron Age. Since the introduction of a functional metallurgical technology is hypothesized to have taken place sometime during either the Late Neolithic, Eneolithic and Early Bronze Age, the scope of the data base must extend it into the periods before and after the proposed point in time. Therefore, data from Early Neolithic (before), and Late Bronze Age and Early Iron Age (after) are included. As a result, assemblages from the Early Neolithic (when there is little doubt that only lithics were employed in butchering activities) through the Iron Age (when there is unambiguous evidence for the use of metal tools in butchering activities) will be examined.

The faunal samples were excavated with a variety of recovery strategies. Most of the sites (N=7) were excavated relatively carefully by hand-collection (Stragari, Opovo, Ljuljaci, Sarina Medja, Vrbica, Vecina Mala, Livade). Initial analysis of the distribution of elements indicated that it was mostly the smaller bone elements were lost. Relatively representative bone distributions for the larger bone elements of the middle and larger mammalian forms were found (Greenfield 1986a, 1986b, 1986c,1996). Two sites were excavated with a combination of careful hand-collection and water-sieving of remains in a floatation tank (Blagotin, Kadica Brdo). Two of the sites were excavated where all of the sediments were either dry or wet sieved (Novačka Ćuprija, Crkvina). Theoretically, the differences in collection strategy may affect the quality of the inter-assemblage comparison (cf. Greenfield 1991; Casteel 1973; Payne 1972). But there does not appear to be any pattern showing that sieved assemblages produce significantly larger numbers of the smallest fragments (***Table 7.1***). For example, Blagotin, which was a mixture of hand-collection and water-sieving, produced in all periods the highest percentages of fragments that were less than a quarter in size (including the only a splinter category). As Clason and Prummel (1977) noted, comparisons of sieved and careful hand-collected assemblages most dramatically affect the smallest taxa (or the smallest bones of larger taxa). This problem has been largely obviated by using only collection with evidence of careful hand-collection, or more systematic collection procedures, and by utilizing only the data from the medium and larger mammals from each site (see below).

Methodology

All faunal remains from each site was identified to finest possible taxonomic category over the course of nearly 20 years of field work in the area. In addition to taxon, all specimens were also analysed for a variety of categories (see Greenfield 1986b), including the size of the fragment relative to its original size. A series of categories were established prior to analysis that enabled the analyst to rapidly assign size categories (e.g. whole; equal or greater than three-quarters, etc.) to each specimen. Over the course of the analysis, additional categories, such as splinter, were added, and only one of the original categories were deleted (half). Hence, the information for some of the categories (e.g. less than one-quarter and splinter) is merged in some lines of ***Table 7.1***, where the cell is empty (e.g. Stragari).

One of the major problems that exists in Balkan and other zooarchaeological collections is that they were often analysed by different analysts who employed different analytical criteria. This problem does not exist with the assemblages employed in this study. One of the advantages of working in a region over a long period of time is that there is a great deal of comparability between assemblages. Such comparability is essential for comparing assemblages across time and space. The analytical categories and their definitions have been published for a long time (e.g.Greenfield 1986b), however it is disheartening to see that few other zooarchaeologists have employed them or published their own categories and criteria.

In order to avoid inflating the effect of differential fragmentation between cranial and post-cranial bones on the samples, only post-cranial remains were used in the analysis. Also, only the remains from identified large and medium mammal taxa were used (e.g. *Bos primigenius*, *Bos taurus*, *Canis familiaris*, *Canis lupus*, *Capra hircus*, *Capreolus capreolus*, *Cervus elaphus*, *Equus caballus*, *Ovis aries*, *Sus scrofa* dom., *Sus scrofa* fer., *Ursus arctus*). This avoids some of the differential effect of identifiability upon fragmentation.

Results

As I conducted the field analysis of the various bone assemblages, I always felt (on a subjective level) that the size of bones in Neolithic levels or sites was larger than those in the succeeding post-Neolithic levels or sites. It is interesting to observe from the data in ***Table 7.1*** that this is not the case. There does not appear to be any overall pattern from the Early Neolithic through the Early Iron Age in terms of the size of the bones. Bone fragmentation does not appear to covary with time, nor with the expected introduction of metallurgy. The degree of fragmentation, therefore, must be related to variables other than the introduction of a metallurgical subsistence technology.

The percentage of whole bones in an assemblage is commonly cited as a variable useful for monitoring the effectiveness of bone recovery in unsieved assemblages (e.g. Casteel 1973; Payne 1972; Clason and Prummel 1977; Greenfield 1986b). However, it does not appear to covary with collection methodology. It varies from a high of 31.04% at Novačka Ćuprija (an entirely sieved sample) to a low of 4.17% (an entirely unsieved sample). The same can be said about the percentage of fragments that are less than a quarter (including splinters). Small fragments range from a high of 51.45% at Blagotin (a mixed recovery sample) to a low of 19.51% at Crkvina (an entirely sieved collection).

If recovery and butchering technology are not the cause of the fragmentation pattern, then other explanations must be sought. One proposed above (hypothesis c) sought to explain the constancy in fragmentation size between the Neolithic and post-Neolithic as a function of the cooking of meat in ceramic vessels. Vessel size, rather than butchering technology, would determine the size of the fragments. Another explanation may lie in the various depositional conditions that attend to each assemblage. This variable can be investigated only if the differences in assemblage attrition (as reflected by fragmentation) could be measured against the conditions of deposition. This is an issue that is poorly reported in most excavations from this region (cf. Greenfield 1986b, 1991).

Conclusions

In conclusion, this essay attempted to monitor the origins of metallurgy though animal bone fragmentation. No temporal pattern in bone fragmentation could be established that corresponded with the advent of a metal butchering technology. Other variables, such as bone recovery methodology, were also investigated and shown not to be a factor in bone fragmentation. Whether samples were sieved (dry or wet) and/or hand-collected did not make a difference in the patterning. Overall, it would appear that bone fragmentation is probably governed by other variables, such as the local depositional conditions (cf. Gifford 1981; Lyman 1994; Meadow 1978) and/or fragmenting bones to fit them into ceramic vessels during the cooking process. In light of the conclusions made above, further research will be necessary before the temporal variation in bone fragmentation can be explained.

Site	Site code	Period	Period order	Collection methdology	Fragment Size Whole	%	= or >3/4	%	= or >1/2	%	1/2	%	= or < 1/4	%	>1/4	%	Splinter	%	Grand Total
Blagotin	BL	Early Neolithic	1	Hand-collected and floated	871	15,35%	304	5,36%	930	16,39%		0,00%	1775	31,29%	1732	30,53%	61	1,08%	5673
Stragari	ST	Middle Neolithic	2	Hand-collected	384	19,76%	159	8,18%	162	8,34%	2	0,10%	767	39,48%	469	24,14%			1943
Opovo	12	Late Neolithic	3	Hand-collected	101	22,39%	24	5,32%	40	8,87%	5	1,11%	120	26,61%	161	35,70%			451
Blagotin	BL	Eneolithic	4	Hand-collected and floated	17	12,32%	8	5,80%	16	11,59%		0,00%	26	18,84%	71	51,45%	0	0,00%	138
Novačka Ćuprija	5	Eneolithic-Late Bronze Age	4	Sieved and floated	365	31,04%	182	15,48%	112	9,52%	43	3,66%	219	18,62%	255	21,68%			1176
Crkvina	1	Early Bronze Age	5	Sieved and floated	22	26,83%	9	10,98%	17	20,73%	4	4,88%	14	17,07%	16	19,51%			82
Ljuljaci	20	Early-Middle Bronze Age	6	Hand-collected	220	17,92%	147	11,97%	144	11,73%	2	0,16%	375	30,54%	340	27,69%			1228
Sarina Medja	17	Late Bronze Age	7	Hand-collected	33	18,44%	18	10,06%	14	7,82%			61	34,08%	53	29,61%			179
Vecina Mala	18	Late Bronze Age	7	Hand-collected	3	13,04%	2	8,70%	5	21,74%			7	30,43%	6	26,09%			23
Vrbica	19	Late Bronze Age	7	Hand-collected	1	4,17%	3	12,50%	3	12,50%			11	45,83%	6	25,00%			24
Livade	10	Late Bronze Age	7	Hand-collected	68	9,94%	61	8,92%	86	12,57%	1	0,15%	237	34,65%	231	33,77%			684
Blagotin	BL	Early Iron Age	8	Hand-collected and floated	31	16,40%	10	5,29%	25	13,23%		0,00%	68	35,98%	53	28,04%	2	1,06%	189
Kadica Brdo	KB	Early Iron Age	8	Hand-collected, sieved and floated	427	18,91%	76	3,37%	322	14,26%	4	0,18%	636	28,17%	793	35,12%		0,00%	2258

Table 7.1. Fragment size of bones from major food taxa (large and medium-sized mammals) from archaeological sites in the central Balkans. Post-cranial remains only.

Chapter 8

Developement of the Graphite Ornamentation in the Copper Age

(In the Light of Evidence from the Struma Cultures)

Stefan Chokhadzhiev

Introduction

The pottery style (technology, morphology and ornamentation) is the most precise and chronologically variable indicator of the culture in Prehistory.

The graphite ornamentation characterizes the Copper Age in the Struma Valley. The use of graphite is already known from the Late Neolithic (Georgiev et al. 1986: 134), but it was used only as a slip component, resulting in appearance of a peculiar silver-like brilliance of the black burnish surface of the earthenware. The real use of the graphite for painting begins with emergence of the Copper Age cultures and continues up to the Final Copper Age (Kolarovo 1-2, chapter 12).

The graphite painted motifs from the Early, Middle and Late Copper Age is a subject of that study, especially the emergence, development and disappearance of the certain motifs and compositions, with regard to their interrelation with the different types and variants of pottery shapes.

The periodization of the Copper Age is elaborated elsewhere (Chokhadzhiev 1999; ***Table 8.1***). In this study the statistic definition of the Early Copper graphite motifs in Western Bulgaria and Northern Greece (Chokhadzhiev 1989: 93-95, tables 13-18 and 20; ***Figure 8.1***) is also used, including only graphite and graphite in combination with pricked, red painted and incised ornamentation. But it should be stressed that later Copper Age is still incompletely investigated there. The purpose of the investigation is the analysis of the emergence, development and the disappearance of the graphite motifs and compositions, as well as of characteristic types and variants ceramic shapes related to distinct phases and stages. The case study in that analysis is Vaksevo-Skaleto.

Characteristics of the graphite ornamentation

The basic characteristics of the graphite decoration in the Struma drainage are the positive ornamentation and its subdivision into one or two horizontal bands.

In the initial stages of the first phase of the Early Copper Age the graphite decoration is painted with wide and rough coats, 1-2 cm wide, which give no possibility of definition of certain motifs. But the technique of the painting with graphite was probably mastered and elaborated quickly, as at the end of the first and the beginning of the second phase of the Early Copper Age some masterpieces occur of painted graphite ornamentation. The common tendency in the positive graphite ornament is making them of thinner lines. In the Middle Copper Age and the beginning of the Late Copper Age the line width is 2-3 mm, and at the end of the Late Copper Age 1,5-2 mm. There is a process towards negative motif observed in the Struma drainage already in the second phase of the Early Copper Age. During the next, the third phase, there are also cases of negative graphite ornamentation, but that tendency did not develop and terminated its existence in the next stages of the Copper Age in the Struma drainage. The same is the picture in Northwestern Bulgaria (Mikov 1948: 26-62; Nikolov 1975; 1984). So, in all Western Bulgaria there is a common way of development of the graphite ornamentation, differing from the Northeastern Bulgaria and Thrace where in the Middle and Late Copper Age there is a tendency towards negative ornament, according to H. Todorova (1986: 102, 107, 111) – cp. chapters 5 and 6.

The second phase of the Early Copper is characterized by appearance of new shapes and decoration. Emergence of combined ornamentation characterizes that phase such as a graphite ornamentation in combination with pricks and incisions.

For the first time in this phase painted with graphite swastika is lied on the inner surface of open bowls' walls shaped as truncated cone. The swastika is of two main variants: with square in the center as a result of the elements' configuration of the four sectors. The opposite two sectors consist of similar motifs – the swastika becomes extremely popular in the third phase of the Early Copper Age and continues in the Middle and the bigger part of the Late Copper Age, which demonstrate numerous variants.

The curvilinear compositions, coming from the spiral, lied on the inner walls of the same open vessels, shaped as truncated cone, occur at the end of the Early Copper Age and continues until the end of the Late Copper Age (Mikov 1948: 38: fig. 23b; 42: fig. 28). The spiral, in its whole variety, is very popular during the third phase of the Early Copper Age, when it is lied on the walls of various ceramic shapes. In the Late Copper Age, as we will see later, its distribution is limited mainly over the type of open bowls above mentioned.

At the beginning of the third phase of the Early Copper Age

Figure 8.1. Copper Age graphite and combine ornaments from Western Bulgaria and Northern Greece.

PERIOD\ PHASE			STRUMA RIVER BASIN	WESTERN BULGARIA	EASTERN SERBIA AND MACEDONIA	OLTENIA
Late Chalkolithic	III a III b	(degradation)	**Vaksevo-St.Voda 6** **Kolarovo II**	Galatin II	Bubanj Hum Ib	Sălcuţa IV ?
			Kjustendil			
	II	(final)	**Vaksevo-"Skaleto"**	Krivodol IV-V Telish III Galatin I	Crnobuki I Šuplevac I	Sălcuţa III
	I	(flourishing)	**Negovanci** **Vaksevo 5** **Kolarovo I** **Sitagroi III B-C**	Krivodol I-III Zaminec Telish II		Sălcuţa II b-c
Middle Chalkolithic		(forming)	**Djakovo**		Bubanj Hum Ia ?	Sălcuţa II a
Early Chalkolithic	III		**Slatino 5-6** **Sitagroi III A** **Dikili Tash II C**	Gradeshnica Telish I	Bubanj Hum Ia-1 Vinča C	
	II		**Slatino 3-4** **Balgarchevo** **Strumsko** **Dikili Tash II B**	Borovan	Vinča B2/C	
	I		**Slatino 2** **Strumsko** **Dren** **Dikili Tash II A**	Brenica 6-3		

Table 8.1. Synchronization table of the Copper Age sequnce in Western Bulagria and neighbor regions.

the closed sharply biconical vessels of group 2 occur. Two variants were mostly distributed: 1. With almost equal upper and lower parts, and 2. With considerably shorter upper part. The motifs of group 24 (after the classification in Chokhadzhiev 1989) – opposite passing one another triangles – are more distributed over the outer surface of the rim of the bowls of the first variant. Almost identical composition occur by vessels from the Middle Copper Age, at the site of Dyakovo (***Figure 8.2: 1, 2***). We should stress on the development of that type of shapes – already at the end of the third phase there is a tendency towards a straight upper half, which becomes almost vertical. The same tendency is observed at Dyakovo (Čochadžiev 1984: 77, fig. 17: 1,2; 18: 4) where certainly standardization in the sizes and the shapes of that type of vessels has been observed. The whole composition consists of 12 triangles passing each another – by six in the space between two handles. The filling of each triangle comprises two or three bands perpendicular to one of its side. The bands consist of three or four thin lines. Almost identical configuration of the motifs decorates similar vessels in the next stage of the Copper Age – the first ("classical") phase of the Late Copper Age (***Figure 8.3: 7, 9***), but it is maybe their last using. The motifs of hanging solid triangles are rare distributed, lied on the rim of the same type of closed sharply biconical bowls. Obviously, they are more universal because they are distributed over generally different shape (group 5.118 in the common typology) known from Dyakovo. Similar to those described above compositions, consisted of motifs with a triangle shape, are documented by the bowls of the second variant (with shorter upper part), often distributed in the third phase of the Early Copper Age, and known up until the first phase of the Late Copper Age.

In the third phase of the Early Copper Age open profiled bowls of group 2.32-2.34 occur distributed in the Middle Copper and in the first phase of the Late Copper Age. The motifs of group 24 are again preferred for the ornamentation of those shapes – triangles opposite situated and passing each other.

The ornamentation of the outer walls of rims characterizes the bowls with motifs of group 24, while the Late Copper ones are with ornamented inner rim walls.

Along with the complex profiled bowls, chess motifs occur, which continue their distribution in the Middle Copper Age and probably at the beginning of the Late Copper Age (Raduncheva 1981: 20, fig. 9: 2).

Another chronological indicator for the beginning of the third phase of the Early Copper Age in the Struma drainage is the emergence of a specific shape of group 7 or the so-called 'fruit dish', which has comparably rich inner typology. The upper chronological borderline of their existence can be accepted the first phase of Late Copper Age represented at Pernik-Krepostta (Raduncheva 1981: 20: fig. 9: 5). The decoration with graphite ornament and especially over the whole visible surface characterizes those vessels only. The ornamentation is usually divided into two horizontal bands – under the rim and over the pedestal base respectively. For the decoration of the rims, motifs of group 24 are again used, but at the lower part – spiral motifs of group 28 are usually distributed.

One motif should be stressed especially, which demonstrate a preference for a definite place of painting over the vessel. That is 25.482 of the common typology appeared in the third phase of the Early Copper Age and distributed up to the Middle Copper Age. It is preferred for the decoration in the zones under or above the handles.

As it was mentioned above, the graphite decoration was lied on mostly in two horizontal bands. Such an arrangement of the ornaments is also distributed among the vessels with two symmetrical handles. They have been known in the Struma drainage since the time of Late Neolithic, but they become characteristic shape in the second half of the Early Copper Age and especially in the Middle and Late Copper Age. The upper ornamental band is lied on cylindrical and semi-cylindrical necks. It is documented in two variants: 1. Horizontal and vertical lines of group 23. 2. Motifs from group 27 of the common typology of the Early Copper Age shapes and decoration (Chokhadzhiev 1989: table 17; ***Figure 8.1***) – meanders, descendant of meanders, interconnected quadrangles etc. The motifs of the first variant appear to be very stable appearance during the time, because they are discovered over similar place in combination with incised and combined decoration from Middle (Čochadžiev 1984: figs. 15: 3; 17: 4; 18: 6) and Late Copper Age (***Figure 8.4: 2, 4, 6***). The motifs of the first variant are usually lied on the necks of vessels with specific proportions – the diameter of the rim is bigger or almost equal to the height of the vessel. In the second phase of the Early Copper Age it occurs by vessels with two symmetrical vertical handles.

The motifs of the second variant are more often distributed, characterized by rich inner typology. The lower band consists of three variants: 1. Motifs of spiral or concentric circles at the base of the vessels. 2. Motifs consisted of oblique cut lines, lied on, with channels, over the widest part of the vessels. 3. Motifs of hanging triangles. The longest distribution characterizes the second variant of motifs – from the second phase of the Early Copper Age, through the Middle Copper Age, until the first phase of the Late Copper Age. In that group, with the time, the spirals, spiral and meander and their motifs-deviations also gradually lost their significance.

In the Middle and the Late Copper Age the spiral and the spiral and meander continue to exist but mainly by the incised decoration, while by the graphite ornamentation the spiral breaks down and disappears (Chokhadzhiev 1991: 31, 32; ***Figure 8.7***). In the beginning of the Late Copper Age distinct elements of the spiral remain only – short curvilinear lines, arches etc. In the second phase of the Late Copper Age, when the ochre painted ornament dominates over the graphite paints, the spiral and spiral and meander graphite ornamentation disappears. The motifs are defined by bands of 3-6 thin lines. In the last phase the graphite painting consists of thin 5 to 12 lines, which describe straight, oblique, seesaw, curvilinear and etc. lines.

An interesting exception from Dyakovo should be mentioned – a lid of vessel with graphite ornamentation. It is interpreted in different way. It was not discovered in the course of the excavations in 1980, but by surface observation in 1998. The lid is ornamented with graphite spiral and is the only vessels with such an ornamentation.

On the one hand, the specific shape preconditions a decoration of a spiral, which fact contradicts the opinion represented above, but on the other hand, it is a proof for close connections of the ceramic complex from Dyakovo with the Early Copper Age from the other sites on the region.

Among the mass material of the Copper Age from the basin in the Struma drainage, there are fragments with foreign motifs. They are outlined with thin sheaves, 8-12 in number, with 1 mm wide and they can be connected with Maritsa culture in Thrace.

In the last phase of the Late Copper Age (the so-called Post-Eneolithic) the graphite ornamentation disappears. The tendency towards limited using of graphite paint in the second half of the Late Copper cannot be connected not only with the social-ethnical changes characteristic of the whole Balkan Peninsula. In the region of the Struma drainage are known freely accessible resources of graphite. It is possible after thousand of years using of graphite, the replacement of graphite with ochre naturally to be required.

A case study: Vaksevo-Skaleto

The prehistoric site is located 2.2 km west of the village of Vaksevo, Kyustendil District. It is situated over hill, 609.2 m above see level, located on the right bank of the Eleshnitsa River. The hill is accessible from its east side only. At its west part it finishes with vertical rocks, whence the name "Skaleto" occurs. The locality is known as 'Ilindeno' as well (***Figure 8.5***).

The settlement was discovered for the first time by road tracing, which destroyed Late Middle Age cemetery. It required archaeological investigations in the destroyed sector, which

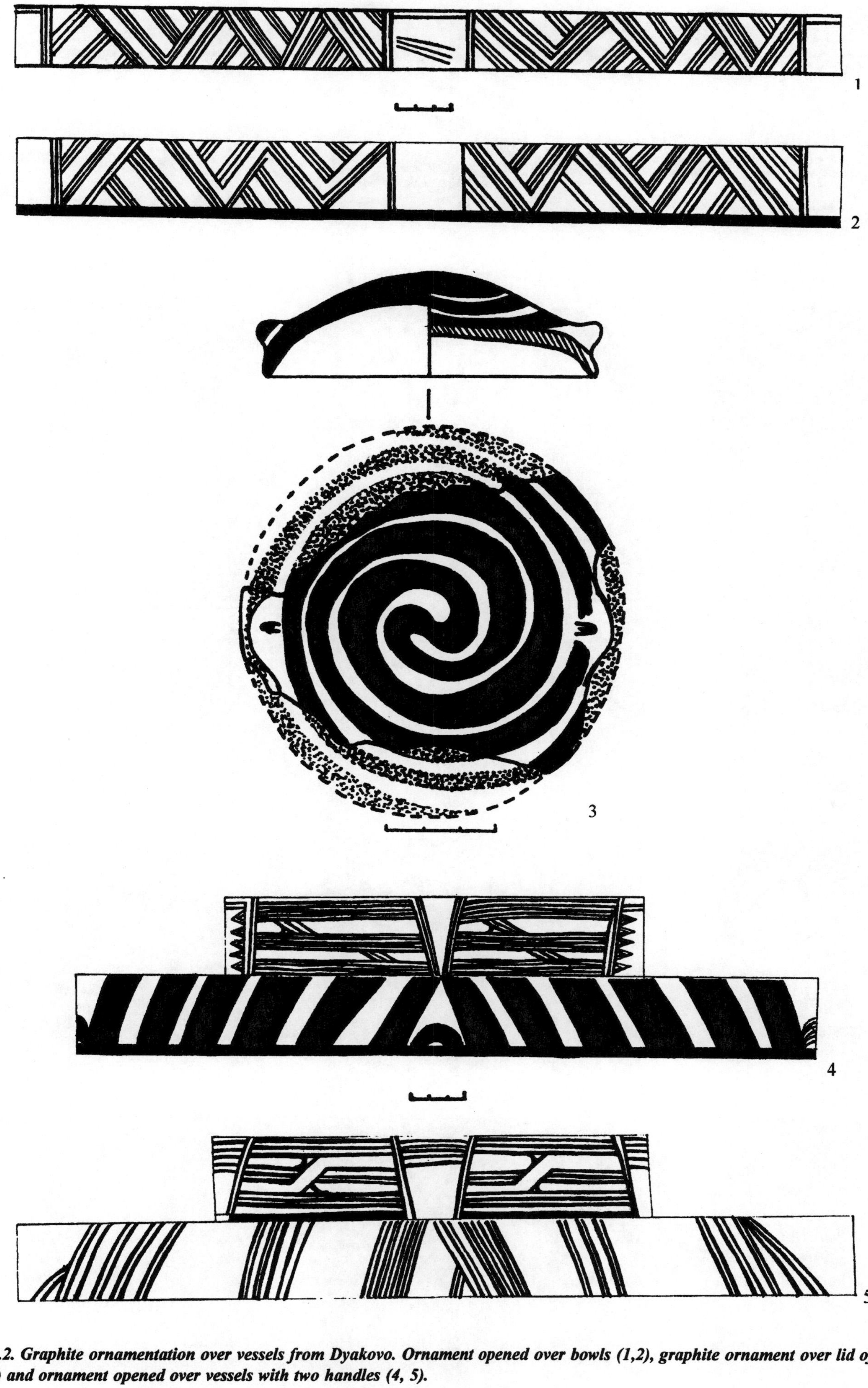

Figure 8.2. Graphite ornamentation over vessels from Dyakovo. Ornament opened over bowls (1,2), graphite ornament over lid of a vessel (3) and ornament opened over vessels with two handles (4, 5).

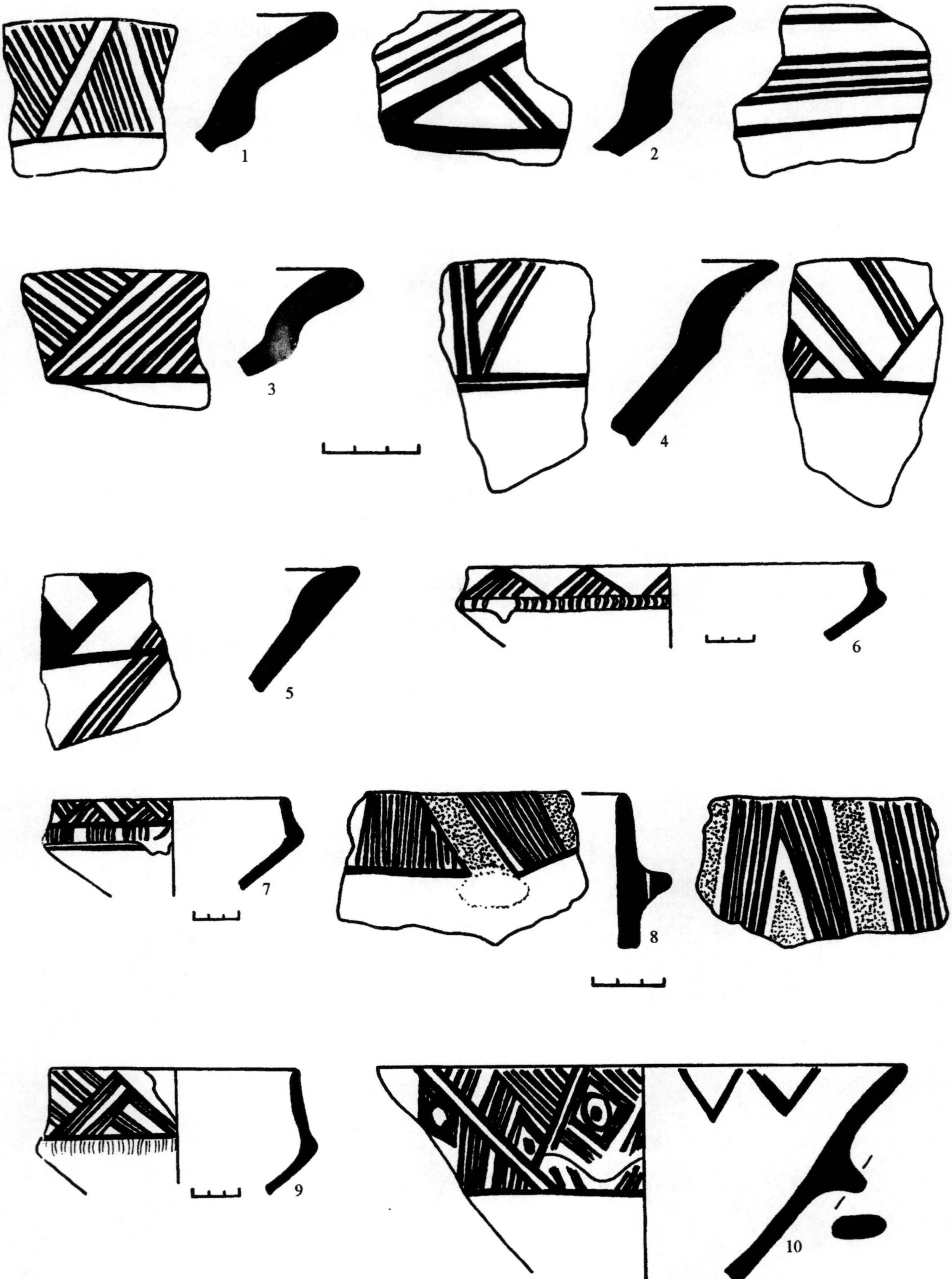

Figure 8.3. Graphite ornamentation from fifth horizon at Vaksevo-Studena Voda.

Figure 8.4. Graphite ornamentation from fifth horizon at Vaksevo-Studena Voda.

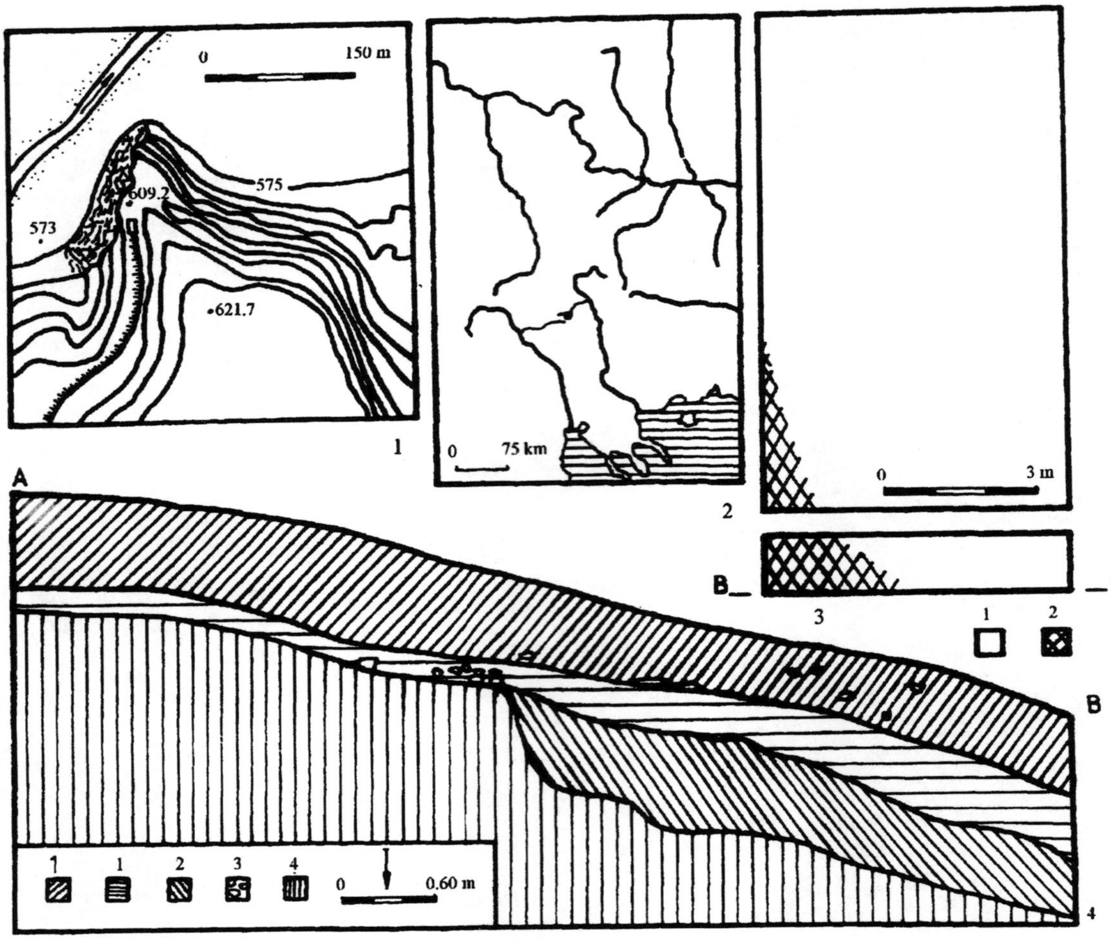

Figure 8.4. Vaksevo-Skaleto. Topographic plan of the locality (1), situation of Vaksevo village (2), excavation area (3 – 1, total area investigated; 2, traces of a prehistoric settlement), North profile (4 - 1, gray and brown soil; 2, gray and yellow soil; 3, brown soil; 4, stone; 5, continent).

lasted two seasons – in 1989 and 1990.

The common sloping of the locality is in east-west direction, weaker in north-south direction. Over the area investigated, 17 graves have been recovered. In whole drilling the natural sediments have been reached. The traces of prehistoric settlement have been documented in the west part of the excavated area. It is preserved as a thinner in north direction strip, but 0.60-1 m wide in north-south direction. The rest part in west direction was destroyed of the road. The settlement probably widens in south direction, but the terrain fells into private courtyards where the investigation was impossible.

The soil of the prehistoric settlement is brown. It has lower compactness and consists of a big quantity of very fragmented pottery. No architectural structures were preserved. The thickness of the culture layer is c. 0.40 m.

The fragmented pottery does not allow a precise typology of the pottery to be elaborated (***Figures 8.5-8.9***). It consists of 943 fragments which, generally speaking, define the characteristic elements of the Late Copper Age in the region.

The ceramic is made of paste with fine and coarse mineral mixture, 0.5 cm big. The color varies from gray and black to gray and brown and rusty brown.

According to wall thickness and the degree of smoothing, the pottery can be divided into two main categories – coarse and fine. In the first group, the thickness varies from 0.7 to 1.2 cm, and in the second it is c. 0.6 cm. The surface of the first group is well smoothed and often ornamented with distemper paint or graphite.

According to common shapes and the ratio between the height and the diameter of the rim, 8 groups of vessels can be distinguished.

The most distributed shapes are those of he first group – open vessels with a conical shape (***Figure 8.5: 1-5***). By that group the diameter of the rim is rather bigger than the height. They represent 20.96% of the total number ceramic sherds, as most numerous are the vessels with the thickened from the inner side rim, 10.3%. The variety of that group is due to the way of thickening of the orifice, the existence of spout or vertical handles, the existence of biconical edge, oval and concave from the inside profile.

The vessels of group 1 have the vastest distribution. They are known from numerous settlements of the Krivodol-Sălcuţa-Bubanj complex: Sadovets, Krivodol, Gnilyane, Galatin, Devetaki Cave, Kolarovo, Pernik, Šuplevac, Crnobuki, Dyakovo, Sălcuţa, Pločnik, Sitagroi III etc. (for these and par-

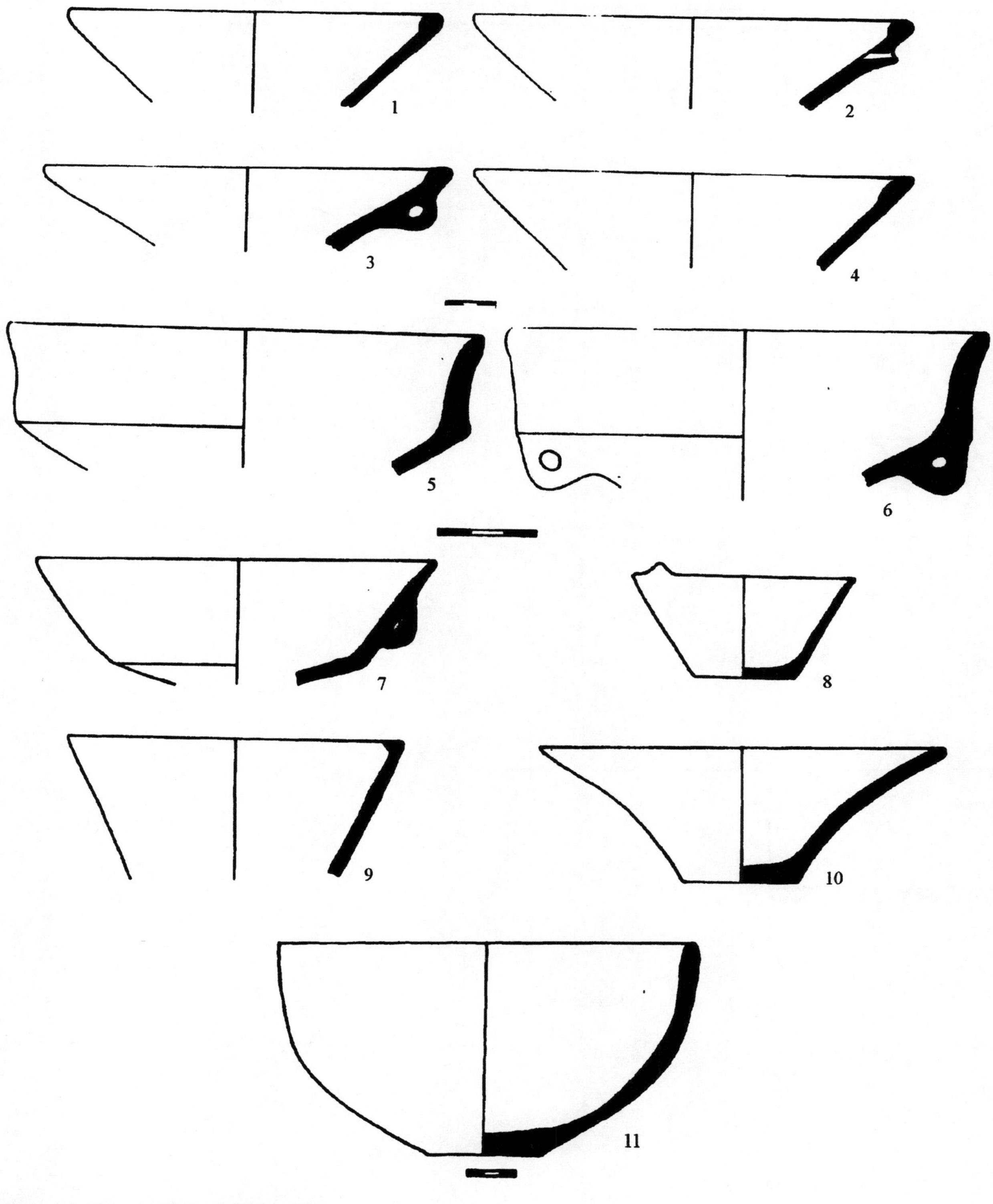

Figure 8.5. Vaksevo-Skaleto. Plain pottery.

allels given below see the references in Chokhadzhiev 1991). The pottery from the newly discovered site confims the assumed considerable chronological difference at Sitagroi III in previous my works, as Sitagroi IIIa is attributed to the beginning of Early Copper, and Sitagroi IIIb and IIIc to the Late Copper Age – see also chapters 6 and 26.

The vessels of the second group have the same ratio between the diameter of the orifice and the height, but the orifices recline inside. Comparatively often the shape shown in ***Figure 8.6: 1-2*** is distributed (4.6%). Similar vessels are known from Sălcuţa, Krivodol, Galatin, Kolarovo, Šuplevac, Crnobuki, Sitagroi III etc.

The third group includes vessels with biconical shape, by which the height is equal or bigger than the diameter of the orifice. The turning point of the vessels is rounded with different degree and is stressed with relief band with pits. The shapes are comparatively often distributed – 3.4% (***Figure 8.6: 4***) and 2% (***Figure 8.6: 5***). Similar vessels are known from Crnobuki, Pernik, Galatin and Dyakovo.

Typologically, the fourth group us most various. It includes vessels with S-profiles. The biggest is the variety of rim shap-

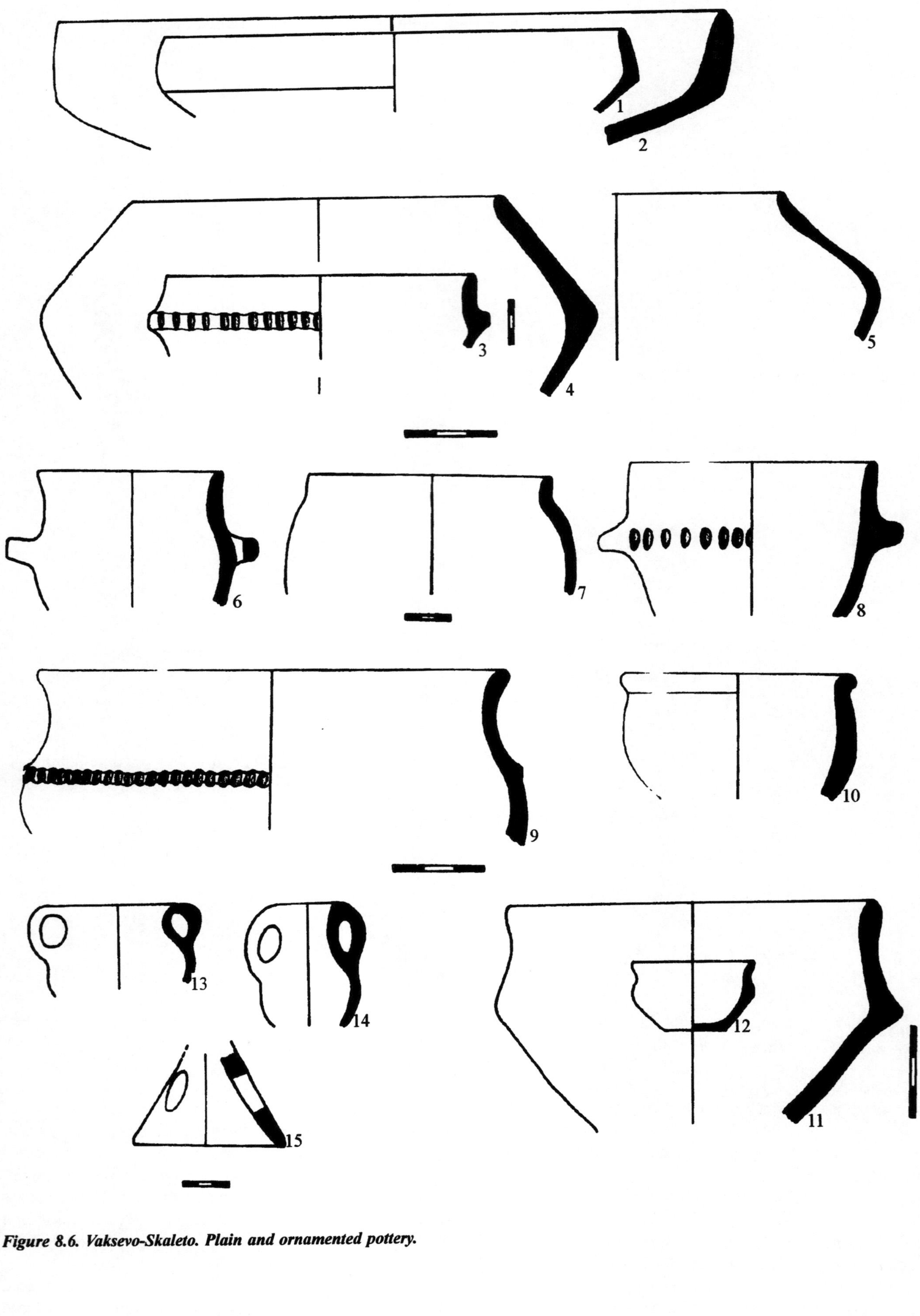

Figure 8.6. Vaksevo-Skaleto. Plain and ornamented pottery.

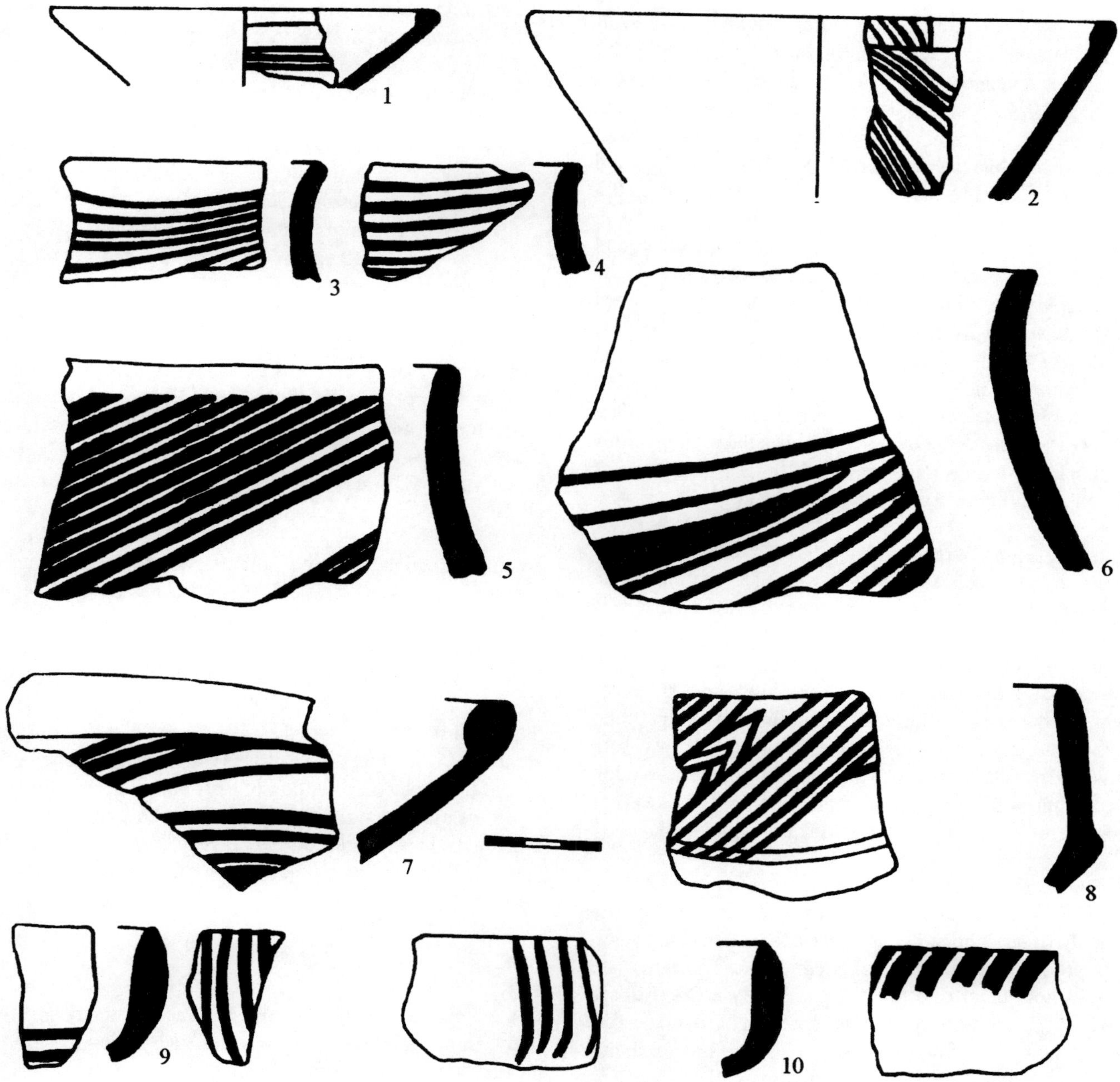

Figure 8.7. Vaksevo-Skaleto. Graphite ornamented pottery.

ing, which from thin and out-turned changes up to 'Lippenrand'. The upper part is usually thoroughly smoothed, but in all cases close to their round turning point there is a relief horizontal band or horizontal grooves.

Among the most distributed one is the vessel shape in ***Figure 8.6: 9***, 4.2% The vessels with 'Lippenrand' are considerably rarely distributed (***Figure 8.6: 10***). They are known from different settlements of the Krivodol-Sălcuța-Bubanj complex in Serbia. Most of the rest shapes have analogies mainly at Šuplevac and Crnobuki.

The fifth group consists of vessels with different sizes of the necks (***Figure 8.6: 6-8***). Those shapes are characteristic of whole Late Copper Age, despite at Vaksevo-Skaleto each of them is distributed comparably rarely – c. 0.5%. Analogies of those vessels are known from Sălcuța, Sadovets, Krivodol, Galatin, Devetaki Cave, Zaminets, Pernik, Dyakovo, Crnobuki, Šuplevac, Peklyuk, Gnilyane, numerous settlements in Serbia and Greece.

The sixth group includes characteristic of the Krivodol-Sălcuța-Bubanj vessels with two vertical handles started from and just below the rim (***Figure 8.6: 13-14***). The existence of that characteristic shape outlines the borders of the complex – Sălcuța, Krivodol, Sitagroi III, Šuplevac and Crnobuki.

It is not clear the function of two fragments of vessels, which have a flat edge from their inner side.

Fragments of possible vessels with high pedestal base (***Figure 8.6: 15***) were also discovered at Vaksevo-Skaleto, similar to those known from Slatino, Dyakovo, Sitagroi III, Galatin etc.

Bulges over the vessels are 4% of the total number sherds. Relief buds are comparatively rarely distributed – 1.6% and are located at the widest part of the vessels.

The ornamentation of the vessels varies (***Figures 8.8-8.9***). It includes ornamentation with graphite and distempered paint (red, white and yellow), as well as incised, dragging, pricked, and stamped decoration. The different techniques are very often in combination. The graphite ornamentation was mostly used (5.7%). The graphite lines are thin (0.15-0.2 cm), often painted in-groups by 4 to 7. In single cases the thickness is up to 0,6 cm. The sheaves outline straight lines, oblique and broken lines. The curvilinear motifs shaping spirals and spiral and meanders almost disappear, which are so characteristic of the Early Copper Age. Similar graphite ornamentation is known from the lower levels at Bubanj Hum in Serbia, which M. Garašanin (1958: 243) reasonably connects, along with the red and ochre painting, with Krivodol. The new investigation at Krivodol indicates that in the earlier horizons the graphite paints predominate while in the upper ones the white, yellow and red ochre paintings (Nikolov 1984). In fact, V. Mikov has stressed the affinity of paint to graphite yet (1948).

At Vaksevo, the use of red painting is comparably often, 1.7%. It is lied on the vessels after their burning. The motifs consist of wide bands, 1.3 cm, but painting over the whole surface is often distributed as well. It is considerable painting with red distemper in combination with contour of white paint, yellow, as well as with white, yellow and graphite, and with graphite and yellow encrustation. Such combinations of painting with graphite, red, white and yellow paints are known from the lower horizon at Galatin and the upper horizon at Krivodol. The painting with red paint is characteristic of the earlier phases at Crnobuki, Šuplevac and Bakarno Gumno II.

In incised ornamentation is comparably often distributed, 2.3%. The incised lines are 0.1-0.2 cm wide and outline motifs mainly of straight and broken lines. The drag ornamentation is frequently distributed as well, 1.9%. The drags are usually in horizontal direction. There is only a sherd documented with finger drawing. But the percent of barbotine earthenware is comparatively high, 1.6%. It is rare the combination of incised lines, pricked ornamentation and red paint or incised lines and horizontal pit rows. By this ornament the incised lines bordered rectangular field filled with another ornament.

The pricked ornamentation is not numerous but various. By this decoration there are horizontal rows of pricked pits and holes, rows of pricked triangles, rows of nails incisions etc. Such ornaments are known mainly from Crnobuki and Šuplevac.

Despite the least in number (0.4%), the cordware fragments again interconnects the site with Pelagonia – the settlement of Crnobuki and Šuplevac (***Figure 8.9: 10, 13-14***). It is possible the ornament known as 'Wickelschnur' to have been performed with the help of thin wire turned over round big ring stamping the motifs, but not with turned cord.

Female figurines represent the anthropomorphic plastic (***Figure 8.10: 1-5***). They are with arms straight aside and incised ornament. Similar figurines are known again from Crnobuki and Šuplevac. The zoomorphic plastic is represented by two fragments – one of a head of cattle and the another a part of body (***Figure 8.10: 6***).

Very numerous are the so-called handles with two or three cross-holes (***Figure 8.10: 7-8***). Such objects have been found at many sites of the Krivodol-Sălcuţa-Bubanj complex. And are characteristic of its inventory. In the light of recent evidence it can be assumed they were weights for vertical loom because they allow a maximum possibility for arrangement of the fibers.

To summarize, the location of the settlement near the watershed of the Eleshnitsa and Bregalnitsa Rivers explains the big number of analogies with sites from Pelagonia. At the same time, the settlements consist of all characteristic elements of the Late Copper Age settlement in Western Bulgaria. We think it is closest to Crnobuki I, Šuplevac I and Galatin 1. So, it can be attributed to the final Krivodol-Sălcuţa-Bubanj complex. At the same time, it is a certain point in the definition of the southeast direction of distribution of that complex. The ceramic material clearly indicates the evolution of the ornament characterized by decreasing of the graphite ornament tradition and its replacement by the distemper coloring ornament.

Conclusion

The analysis of the graphite ornament shows, that each phase of the Copper Age in the Struma drainage is characterized by peculiarities, which definitely must be taken into consideration by the periodization of the Copper Age cultures, on the one hand. On the other hand, at certain moment the definite motifs appear and continue to exist during almost all the epoch, while the others are very rarely used. Some motifs have universal function, while others have distinct place over the vessel. Without doubt, those differences are due to their specific semantics, which probably have been changing in the course of the time. During different stages of the Copper Age the position of certain motifs over the ceramic vessels often change. All these interconnections, tendencies and regularity in synchronous and in diachronous aspect require a complex approach by each interpretation of the graphite ornament.

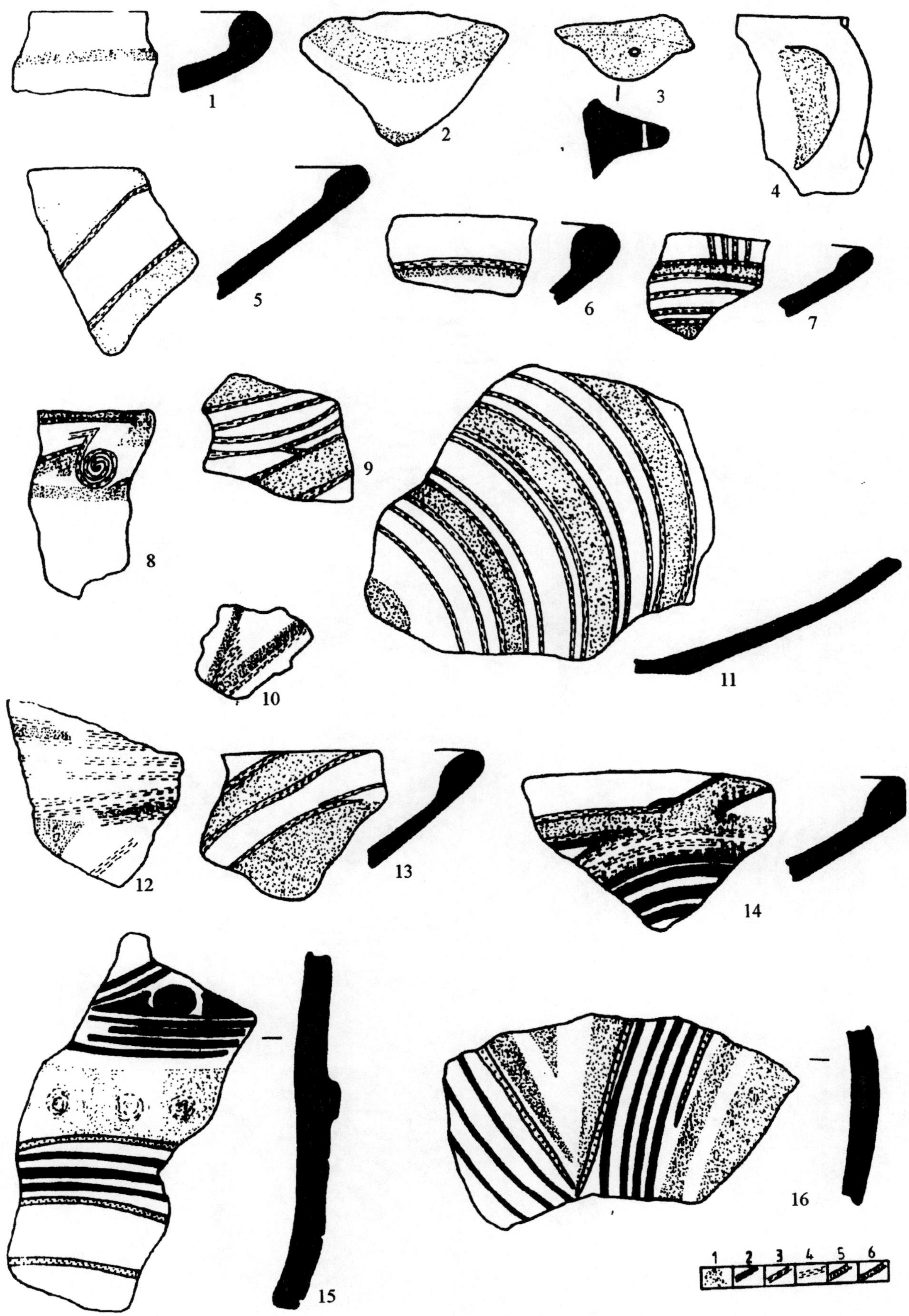

Figure 8.8. Vaksevo-Skaleto. Pottery ornamented with graphite and paints. Legend: 1. red paint; 2. graphite; 3. white paint; 4. yellow paint; 5. encrusted white matter; 6. encrusted yellow matter.

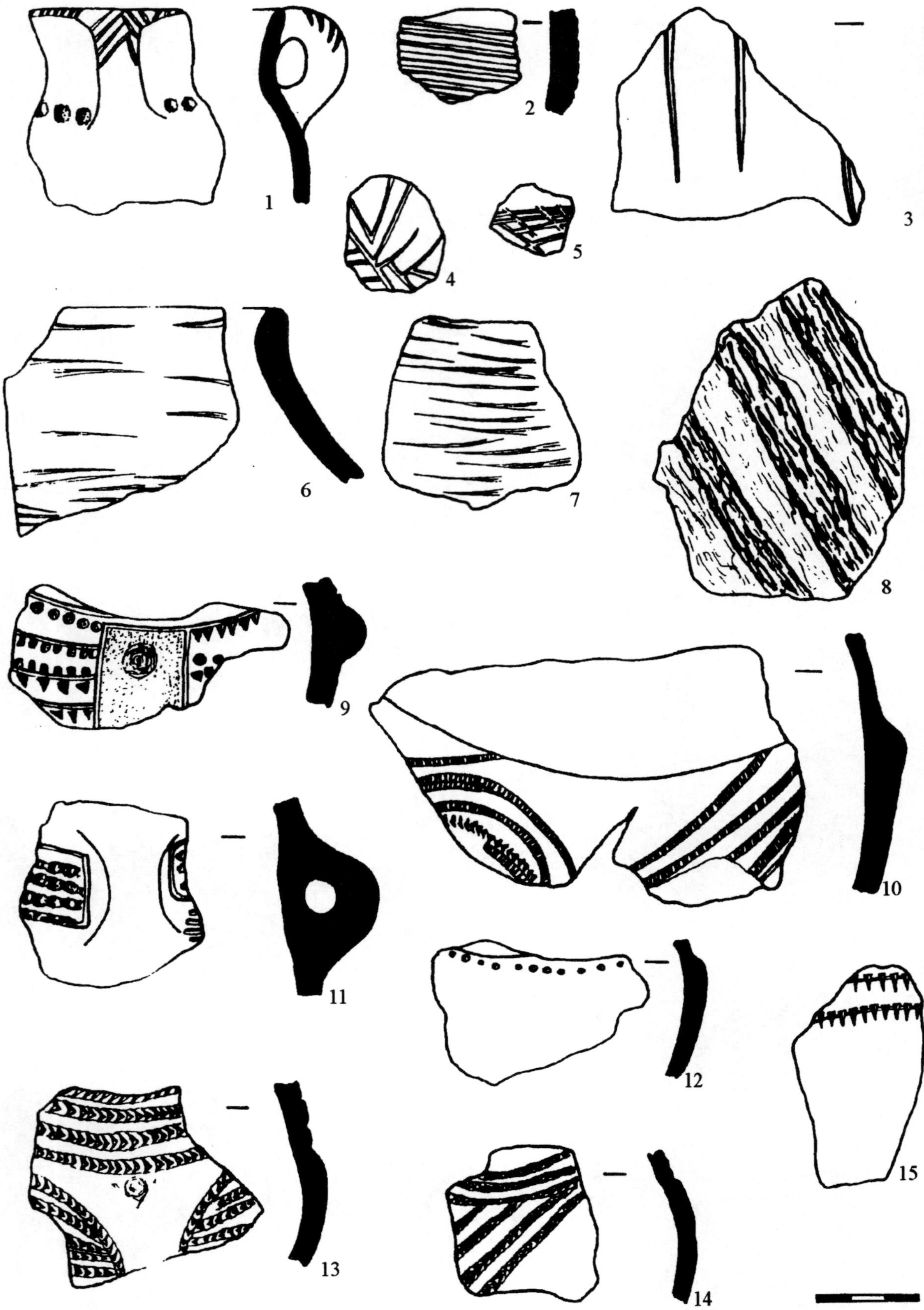

Figure 8.9. Pottery with incised (1-5), drag (6,7), barbotine (8) and pricked (9-15) ornament.

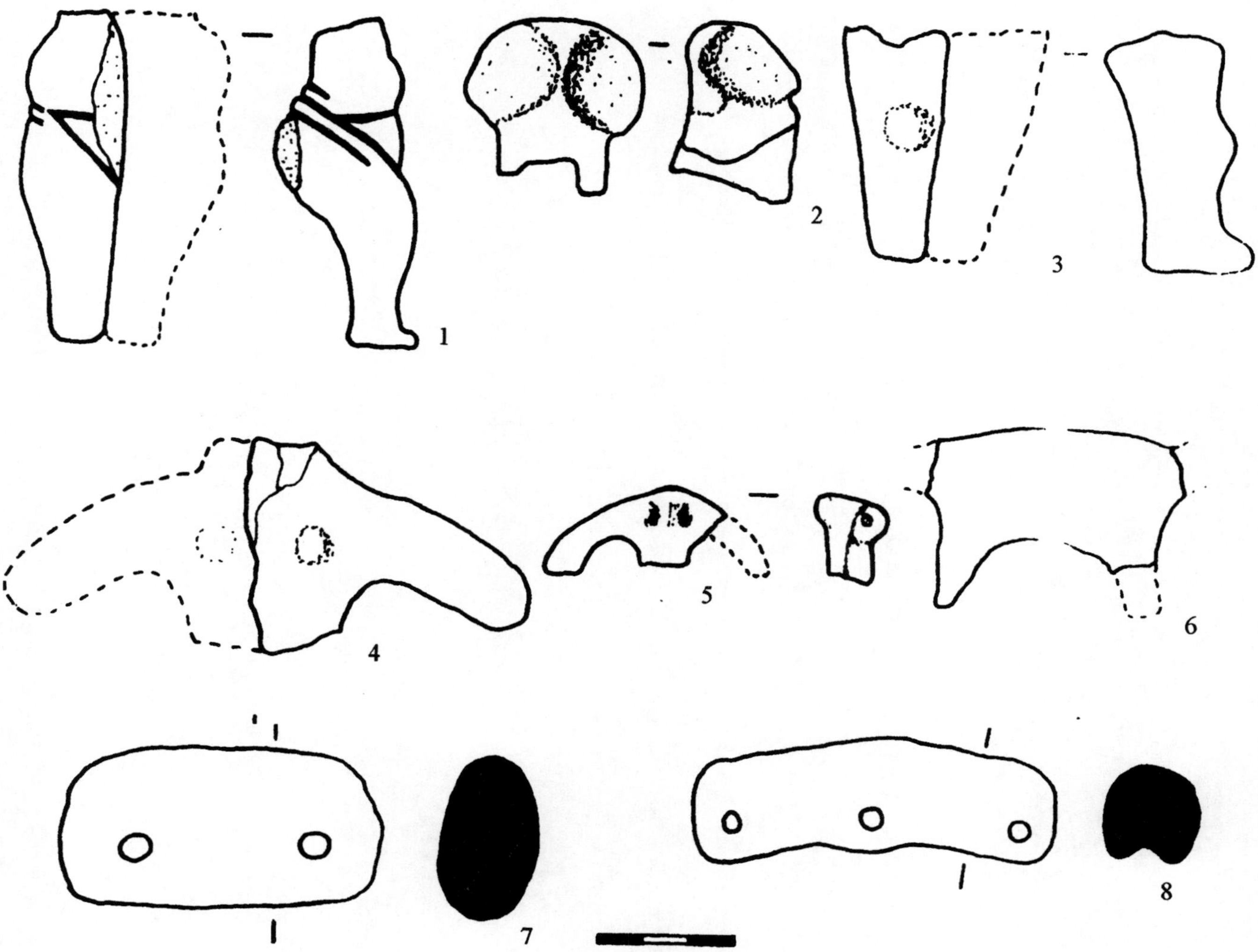

Figure 8.10. Anthropomorphic (1-5), zoomorphic (6) and other small finds (6-7).

Chapter 9

An Attempt for an Interpretation of One Peculiar Vessel from Dyakovo

(Contribution to the Spinning Process in Prehistory)

Alexander Chokhadzhiev

Introduction: The site of Dyakovo and the vessel

This study is an effort for giving an idea of intention of one vessel found at a prehistoric settlement near the village of Dyakovo, Kyustendil District (Southwestern Bulgaria) – ***Figure 9.1.***

The site is located in Gradishtito Locality, some 2.5 km northeast from the village of Dyakovo. Initially the settlement was dated to the Late Bronze Age (Velkov 1942: 209). In 1980s S. Chokhadzhiev provided rescue excavations. The results refer the site to an unknown up to then period of the Copper Age in the Struma Valley. Before these excavations the Copper Age in Western Bulgaria was divided into two periods – Early Chalkolithic (Gradeshnitsa culture) and Late Chalkolithic (Krivodol culture [Nikolov 1974: 21]). But the pottery from Dyakovo takes an intermediate position between the Early and the Late Chalkolithic pottery. The explorer refers the site a little bit earlier that the first known phase of Krivodol culture and according to his opinion it realizes the transition between the Early and Late Chalkolithic in Western Bulgaria (Čohadžiev 1984: 81; cp. Garašanin 1998). The vessel which is an object of this research (Čohadžiev 1984: Fig. 13: 2) is found in a house from the first building horizon in drilling 1. There are no absolute dates from this horizon, but there is one originated from a carbonized post of a dwelling from the second (later) horizon in sounding 3 – 4450 BC-4340 BC (Görsdorf and Bojadžiev 1996: 151). So, the discussed vessel could be only earlier that this dating if the date represents the real absolute chronological position of the level under discussion.

The vessel has an approximately cylindrical-conical shape, with strongly prolonged lower part. It is 21.4 cm high, the diameter at the rim is 23.2 cm, the maximal diameter is 26 cm and the diameter of the bottom 6 cm. The vessel has six horizontally drilled handles and one on the inner side of the bottom with a diameter of the hole – 0.9 cm. The shape of the vessel is unlike for the period. The six handles are rarity, and the interior handle is unique for our territory. These rare elements and especially their unusual combination ascent the question about the function of the vessel.

Interpretation

The explorer S. Chokhadzhiev mentions that the intention of the vessel under discussion remains unclear. Without having

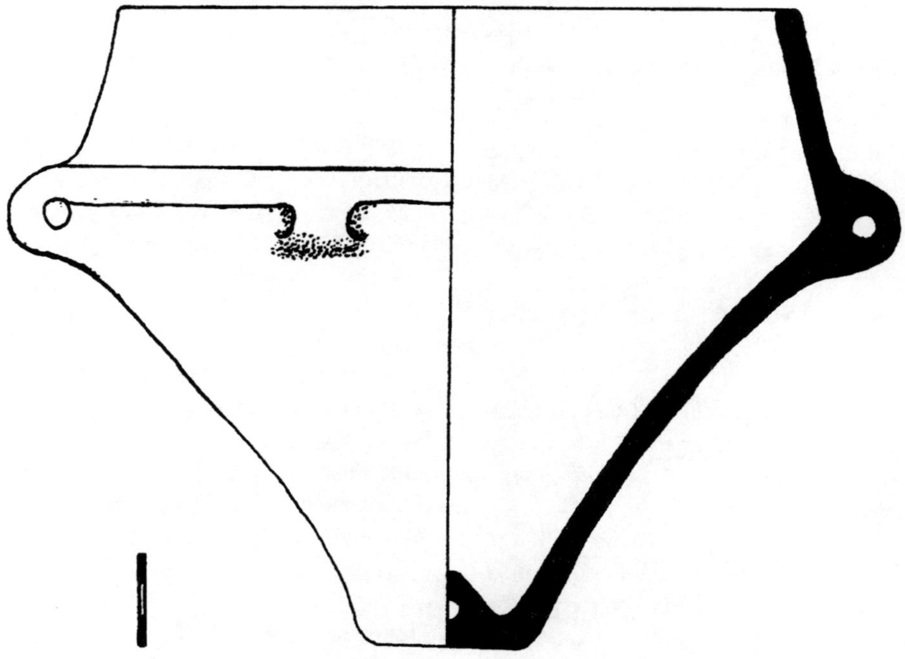

Figure 9.1. The vessel from Dyakovo. After Čohadžiev 1984: Fig. 13:2.

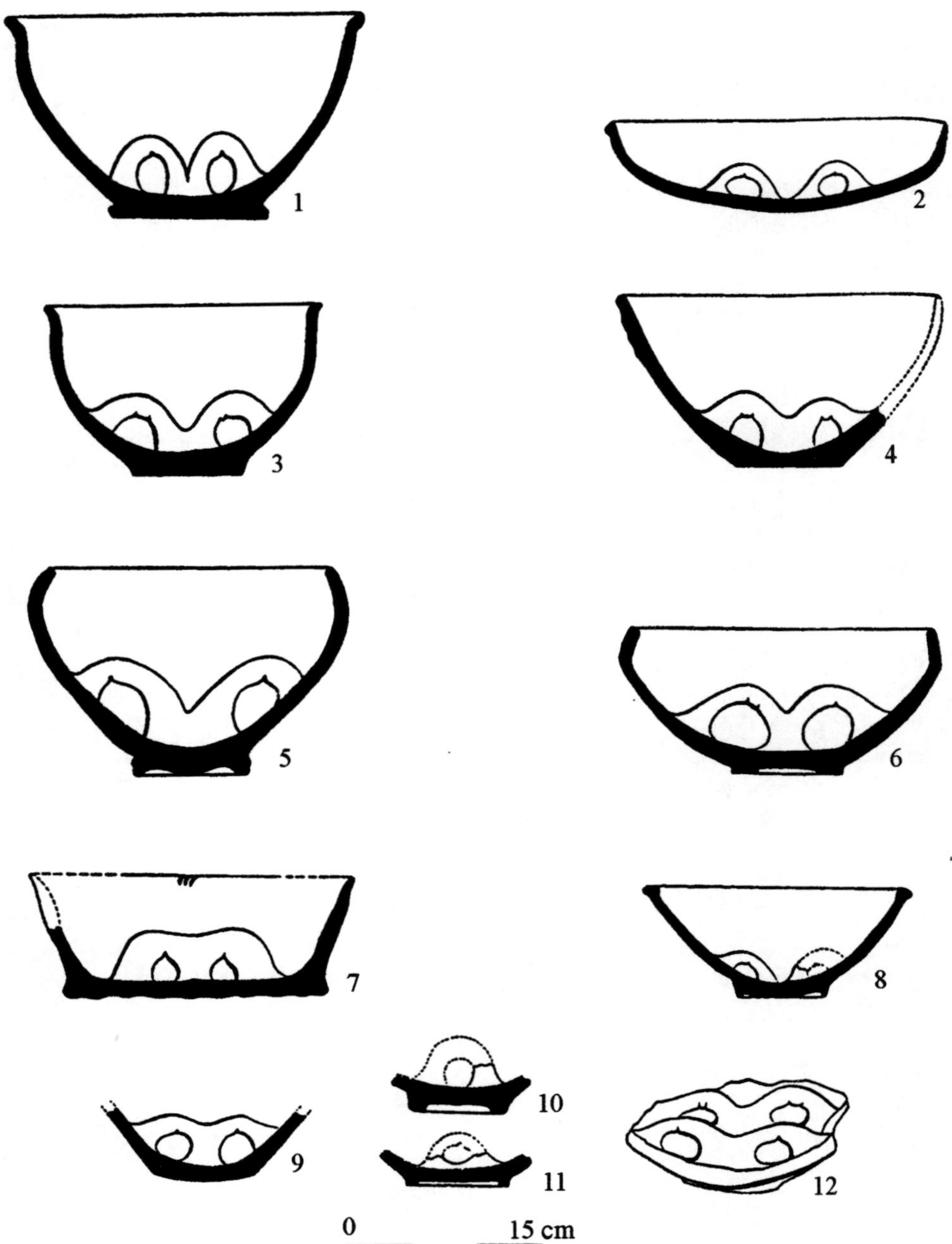

Figure 9.2. Spinning bowls from Palestine. After Dothan 1963: fig. 1

analogies in the in the Struma basin or surrounding areas, he cautiously expresses the hypothesis that the vessel was used for greasing wicks (1984: 74). On account of this lack of close comparative evidence, in the process of my study it becomes necessary to extend the territorial and chronological range of research, finding analogies in Egypt and Palestine from rather later period.

A group of pottery and stone vessels discovered there is characterized by interior handles (with one or more grooves on their undersides) and generally known as 'spinning-bowls' (Dothan 1963: 97; ***Figures 9.2-9.4***).

In Palestine, the oldest instance known with defined context comes from excavations at Tell-el-'Ajjul. Its layer could be dated to 14th-13rd century BC (***Figure 9.2: 1*** [Dothan 1963: 97: Fig. 1:1; Petrie 1932: Plate 27]). The oldest one belonging to level IX dates from the first half of the 14th century BC. At level VIII there were also found several spinning bowls – from the end of the 14th century BC (Fitzerald 1930: 20). One of them belongs to LBA II level (***Figure 9.2: 8***) and the others – to layer of Early Iron Age (***Figure 9.2: 5, 10-11*** [Dothan 1963: 99, Fig. 1: 5,8,10 and 11]). Two of them are unique for Palestine because of their single interior handle (***Figure 9.2: 10, 11***).

A group with numerous spinning-bowls from Early Iron Age originates from level IV at Beth Shean. All of them are well burnt, with flat or disk-like bases. All except one have two joined handles, grooved on their undersides (***Figure 9.2: 3,4*** [Dothan 1963: 99; Fig. 1: 3,4]). One pot base found there is unique with its four interior handles, each with several grooves (***Figure 9.2: 12*** [Dothan 1963: 99; Fig. 2: 12]).

Fragments of such bowls are also found at Tell Qasile (***Figure 9.3*** [Mazar 1951: 128, 139; Dothan 1963: Fig. 2]). A lower part of a spinning-bowl is discovered at level VIIA at Megido (1175/1163-1120 bc).

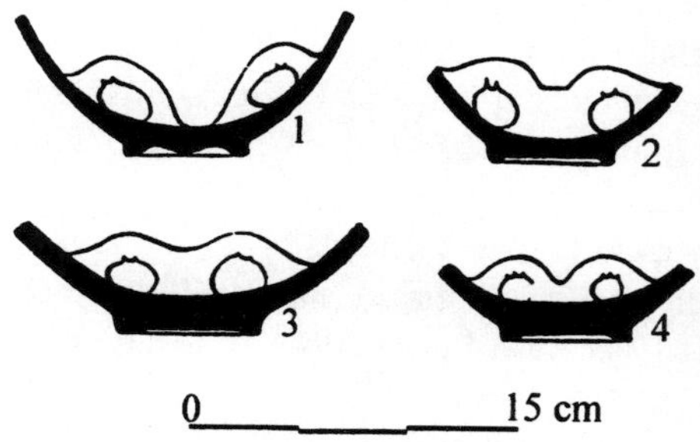

Figure 9.3. Spinning bowls from Qasile Tell.After Dothan 1963: Fig. 2.

A spinning bowl of a type differing from that ordinarily found in Palestine came to light during the excavations at Tell Jemeh, published initially by W. Petrie as a lid (1928: 22, Plate 61: 98M). The vessel is hand shaped and the base is flat and uneven at its underside (***Figure 9.2: 7***).

The oldest instances of spinning-bowls from Egypt were found during the excavations at Abu Ghalib (***Figure 9.4: 1***) and Kahum, both in a stratigraphic context of the 12th Dynasty (2000-1780 [Larsen 1941: 30-31]). But the most extensive group of spinning-bowls in Egypt are discovered at the city of el-Amarna (***Figure 9.4: 2-4, 6*** [Peef and Wooly 1923: 61, 137). The entire group dates to the time of Akhenaton (1370-1335 BC), and were found within the city itself and in the worked village to the east. Inside the city of el-Amarna, the spinning-bowls are both ceramic and stone (Dothan 1963: 101).

The workers' village at Deir el-Medineh also offers a large group of spinning-bowls (***Figure 9.4: 5,7-8*** [Nagel 1938: 183-88]). Two spinning-bowls made of gypsum were documented at Lisht (Dothan 1963: 103). They date to the 20th and 21st Dynasties (1200-945 BC).

There are several theories proposed for the use of this kind of vessels. It had been suggested they are lamps (Peef and Wooly 1923: 137), lids (Fitzgerald 1930: 26) and spinning-bowls (Dothan 1963: 105). The first two suggestions do not explain some peculiarities of these bowls. The presence of interior handle in combination with flat base makes the shape completely unsuitable for a lid. The handles never are at the outside surface of the bottom and this type of bowls is always found without any other pot, which lid they could be.

If we take the proposal for the lamps, the only way to explain the interior handle would be that the wick had been strapped in it; afterwards, the bowl had to have been filled with fat. After the solidification of the fat, the wick remains stiff, but it would be the same even without its lying in the handle. If it was used a liquid fat, the stiffing of the wick would be reason-

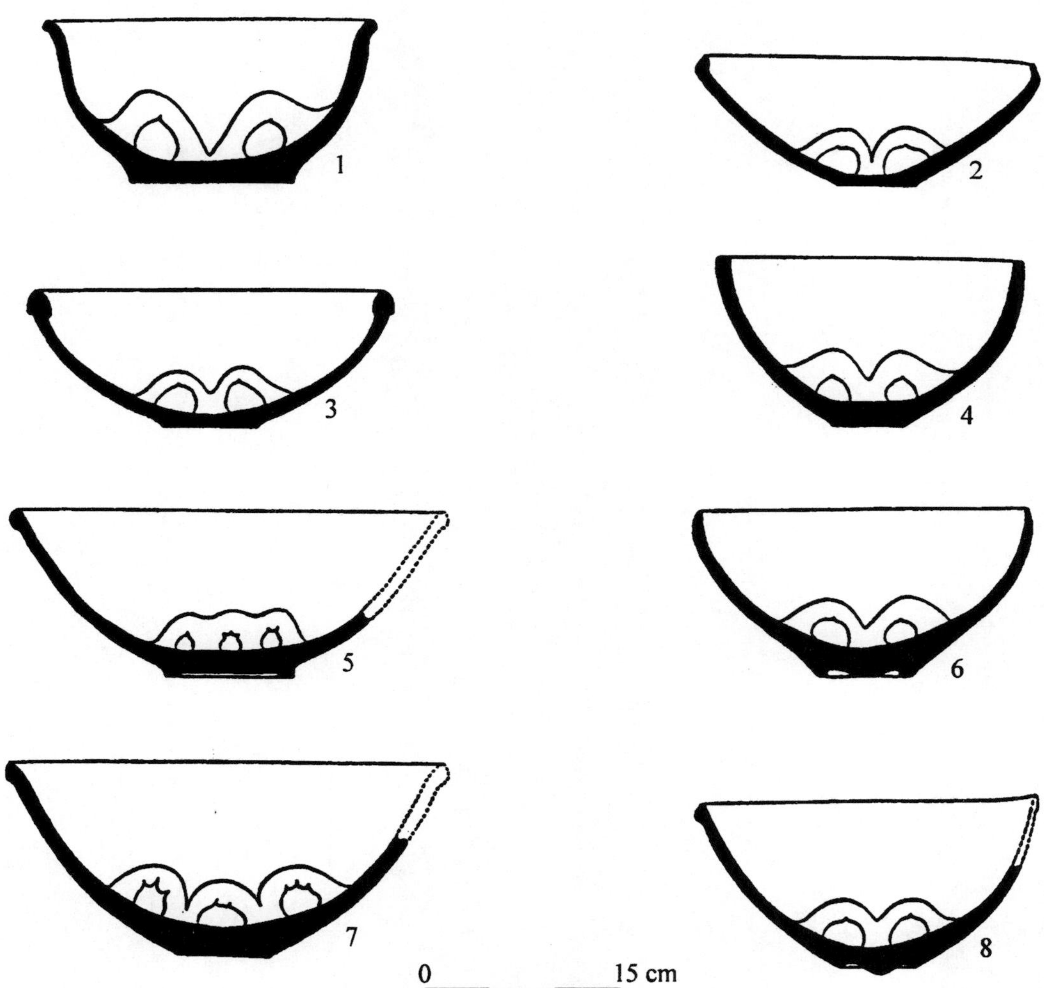

Figure 9.4. Spinning bowls from Egypt. After Dothan 1963: Fig. 3.

able but this in no way could explain the several interior handles in some cases, and the grooves on their undersides. These grooves are probably a result of continuos spinning motion, caused by the drawing of the thread through the 'ear' of the interior handle.

So, the most reliable theory is that this specific type of vessels was really used as spinning-bowls. To ascertain the hypothesis we have to turn to certain Egyptian wooden models of spinning and weaving workshops, and tomb wall paintings. These are the basic sources for revealing the ancient spinning and weaving, not only for Egypt, bur for the ancient world. If were have the necessary archaeological data, the conclusions we draw ca be transferred back for studying some textile techniques in prehistoric period.

The wall paintings of our interest are discovered in tombs in Beni Hasan, el-Bersheh and Thebes. Equally interesting are the models from Thebesh, Gereh and Beni Hasan.

The wall painting in the tomb of Kheti at Beni Hasan (Newberry 1893: Plate 13) dates from the 12th Dynasty (2000-1788 BC [***Figure 9.5: 1***]). There are shown the three main spinning techniques in ancient Egypt (Dothan 1963: 105):

1. The technique of the suspended spindle (***Figure 9.5: 1 left***) – the spinner draws the thread from the coil before him.
2. The technique of the supported spindle (***Figure 9.5: 1 center***) – the spinner draws thread from the bowl before him and rolls the spindle on his thigh.
3. The technique of the grasped spindle (***Figure 9.5: 1 right***) – the spinner draws thread which issues from a bowl before him, through a forked stick, down to a spindle which is revolved with both hands. This technique shows the doubling of several threads into one (Crowfood 1931: 19; Nagel 1938: 158; Larsen 1941: 31; Dothan 1963: 106).

On a wall painting of the tomb of Thuti-hetep at el-Bersheh (***Figure 9.5: 2***) are depicted two spinning women (Newberry 1895: 35, Plate 26). Behind each one, another woman is sitting winding a coil. Before the coil a shapeless mass is presented – probably an image of the textile raw material, which has to be prepared for spinning.

The scene from the tomb of Daga at Thebesh is also of our interest (***Figure 9.5: 3*** [De Garis Davies 1913: 34; Plate 37). The depicted spinner works with two spindles; the one is hanging from her left hand, while she holds the thread that she probably turns into motion by fingers. The other spindle she turns on her thigh. Both threads are drawn from a vessel behind her back. That points to the proposal of the two interior handles, which has to the inside the depicted bowl.

Similar scene we find in the tomb of Tuti-nefer - ***Figure 9.5: 4*** (18th dynasty – 1555-1335 BC [De Garis Davies 1929: 234]). An entire spinning and weaving establishment is depicted here located in the cellar of Thuti Nefer's house. The sitting spin-

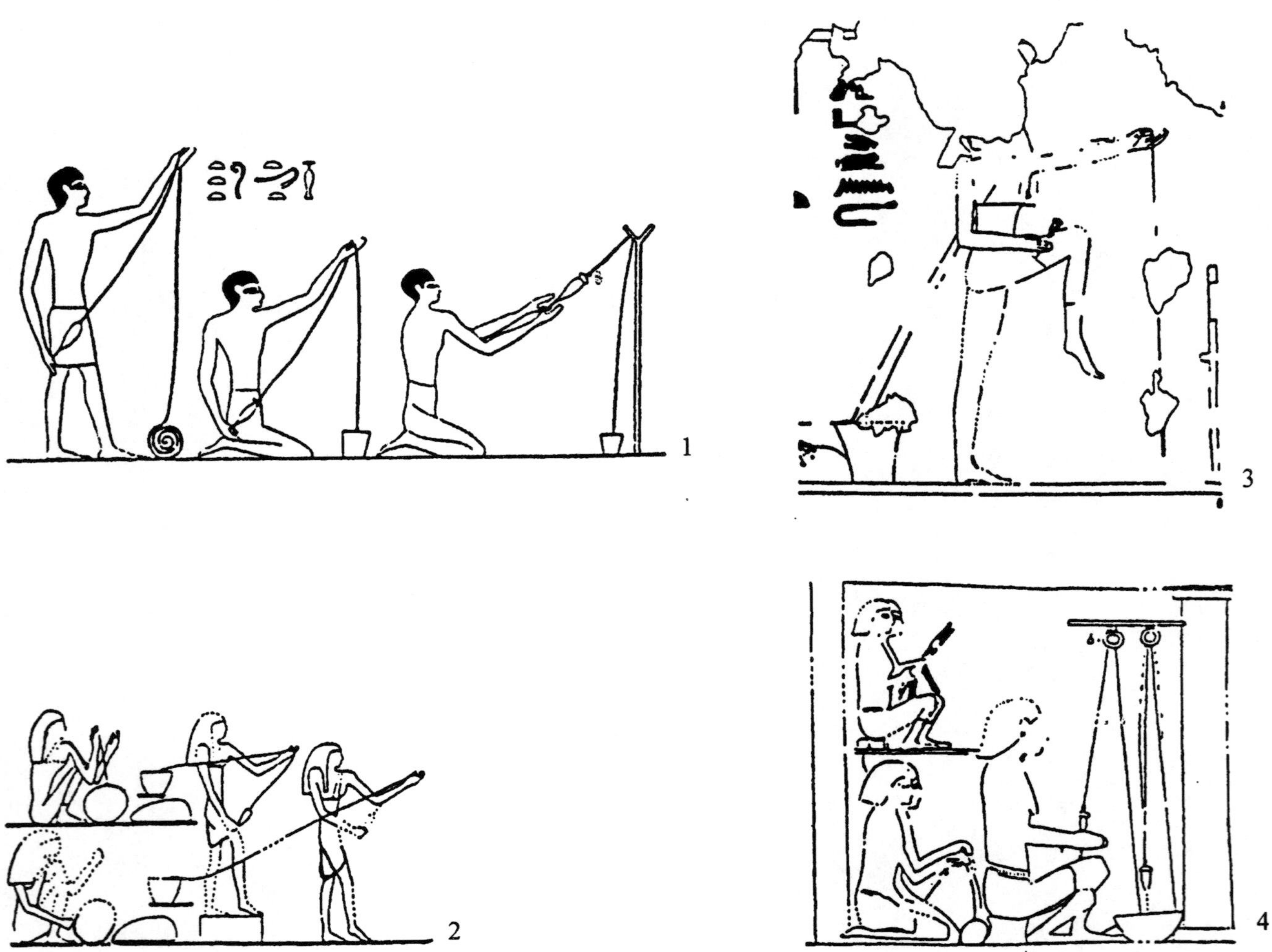

Figure 9.5. Spinning scenes from tomb wall painting from Egypt. After Dothan 1963: Figs. 4-7.

Figure 9.6.Drawing of a spinning bowl from the wooden model from the tomb of Meket Re' at Thebesh. After Dothan 1963: Fig. 8

Figure 9.7. Neo-Hittite relief from Maraş. After Bossert 1942: Plate 814.

ner holds the spindle with two hands. She draws the threads upwards, from a shallow bowl through two rings above. These rings make a longer spin possible (Dothan 1963: 109).

The most concrete information for the spinning process we find in wooden models of textile workshops.

The wooden model from the tomb of Meket-Rĕ′ at Thebesh was found quite intact, despite the delicacy of the threads (Winlock 1955: 301). The model shows in amazing details each stage of the process – preparing the threads, spinning and finally – weaving. It is evident that the spinning is done by means of a bowl and a spindle (***Figure 9.6***).

The principle of spinning by means of a vessel is known also from elsewhere, but without having enough data for the presence of spinning-bowls. For instance, on a Hittite grave stele from the end of the 8th century BC (Bossert 1942: 71, fig. 817) a woman is depicted seating on high-backed chair with a young man facing her. In her left hand she is holding a spindle with the thread coming out of rectangular box at her feet (***Figure 9.7***). Naturally from this relief we cannot draw conclusion that inside the depicted box there is an interior handle, but the principle the raw material to be in some kind of vessel is observed.

The Mayas know similar kinds of tools. For spinning cotton they used a special kind of vessel (***Figure 9.8***) with short pedestals and a spout, which plays the role of the interior handle of the spinning-bowls. The spinning vertebras that often are found around them confirm the conclusion that these vessels were used for spinning.

In a traditional Bulgarian culture the method of spinning wool from a basket is known, where one of the brakes of the bottom is used in the same way as the interior handle of a spinning bowls.

The sense of the spinning with the help of a bowl, basket or a box is in the wish of every spinner to prevent the threads of getting dirty, and rolling around. But in the spinning-bowls there is something more special – the interior handle.

How was spun with a spinning-bowl? The small size of the hole of the interior handle does not allow the spinning to be realized directly from distaff-full – the thread will be easy to

Figure 9.8. A Maya vessel for spinning cotton.

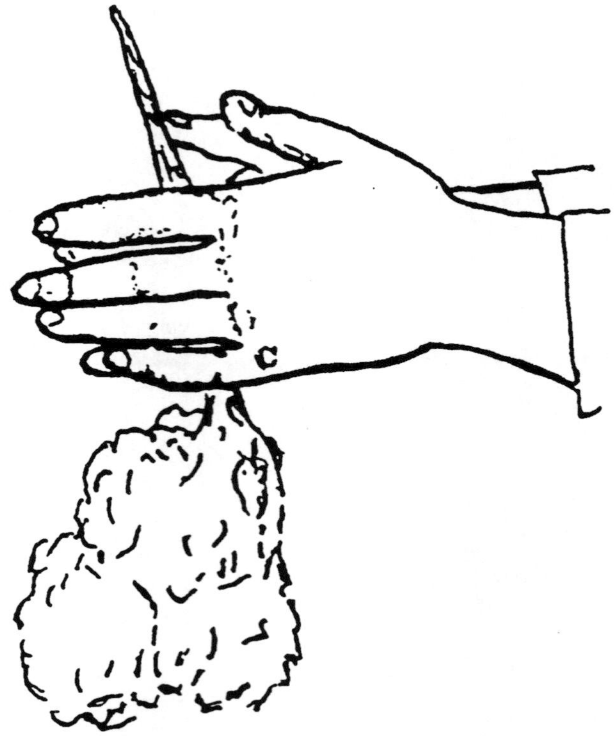

Figure 9.9. The method of rolling hoses.

tear. So, some kind of preliminary preparation was necessary. This preparation could be shaping the material (mainly wool) into 'hoses' by rolling it between the hands (***Figure 9.9***). Afterwards, these 'hoses' already winded in coils and moisturized for better result, were put in the spinning-bowl(s) and one of the ends was threaded into interior handle. Then, the thread was roved to yarn with the help of the spindle. There is another possible method of working with the spinning-bowl. After the traditional spinning from distaff-full, the received threads are winded each on a coil; two or more fibres are threaded in the hole of the interior handle, twisted and then doubled by spinning to a multi-layer yarn.

After all it was said above, let's turn back to the vessel of Dyakovo, that in fact is the subject of this research. Having in mind its interior handle, the only provable suggestion could be its use as a spinning-bowl.

Under these circumstances it occurs the question about what was the function of the six horizontally drilled handles on its surface, because there is no analogy among the spinning bowls. The following hypothesis is based only on logical arguments, but there is no evident data for any other interpretation.

In comparison between the diameter of the rim and the diameter of the bottom their correlation is approximately 4:1, and to the total height of the vessel, the diameter is related as 1:3,5. These proportions make the vessel too unstable of placed on the bottom. But the relations seem to be intentional.

There are known some vessels from the same period with even smaller bottom (in comparison with the rim), but some of them were put inside a pit on the floor. They do not have many handles. Opposing them, the vessel from Dyakovo has handles – even six in count, which makes its putting in a pit insensible. In this case it seems very possible the vessel to have been hung from the house's ceiling, tied through its handles to two parallel beams (***Figure 9.10***). This construction allows the vessel to sag on the force of the drawing of the thread. The weight of the vessel itself causes a force opposite in direction. In this way the two forces facilitate the process of drawing the thread.

One more characteristic of the vessel supports the given proposal for its use. If the coil has approximately the same diameter of the rim (a little bit smaller) there always will be some empty space over the interior handle that is of use for the good spinning and preserves the thread from the tearing (***Figure 9.11***).

Generalizing, we can say that spinning with a vessel is of use for the process because the vessel itself holds the coil still and do not allow it to roll around; the interior handles leads the thread and the small size of the hole presses it and makes it compact.

Finally, it can be said that the presented research approaches one but not the only possible interpretation of the function of the vessel from Dyakovo. Attaching it to the group of the spinning-bowls sets one very interesting question – about the advent of such 'tools' in the Balkans some 1500 years before their earliest appearance in Egypt and Palestine. Are they related or they do appear and develop independently from each other? But to resolve this problem it is possible only after discovering more material of that kind. For the time being, the Dyakovo's vessel remains unique for its time and territory.

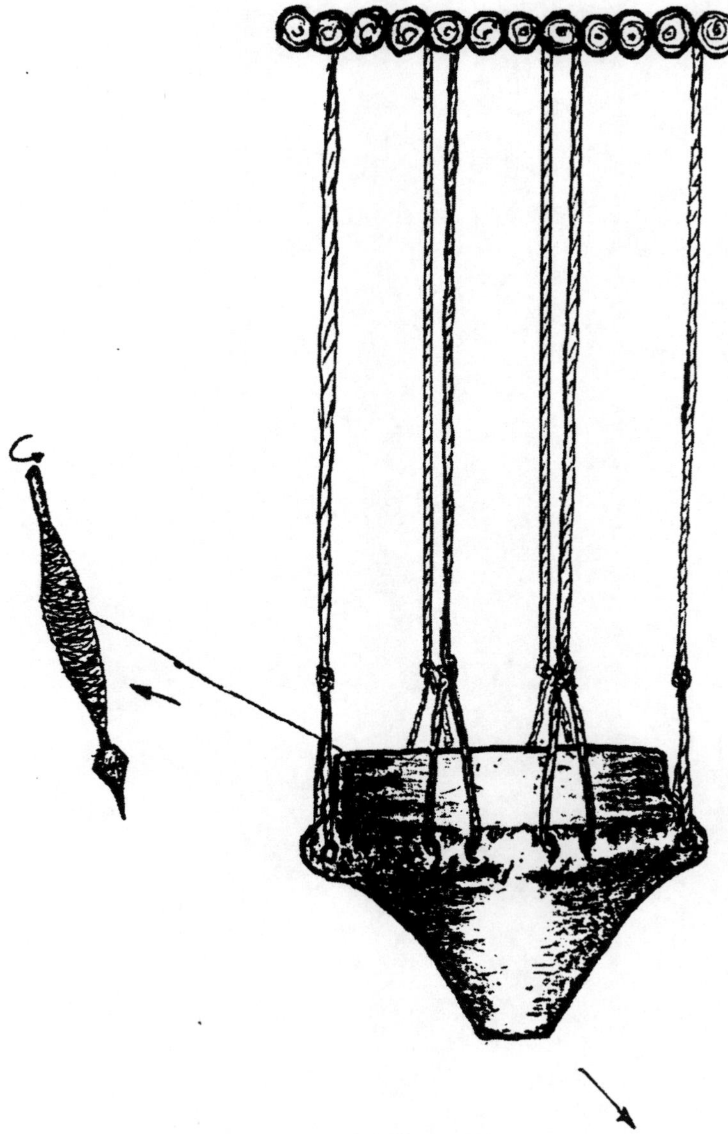

Figure 9.10. Reconstruction of hanging the vessel to the ceiling.

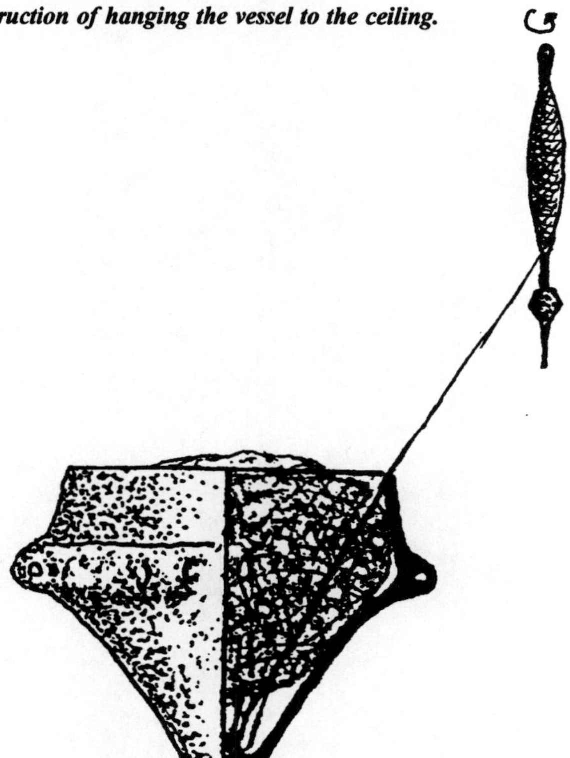

Figure 9.11. Spinning with the vessel from Dyakovo.

Chapter 10

Metal Finds from the Copper Age in the Middle Drainage of the Osum River

Martin Hristov

The excavations in the Middle Osum drainage in 1990s have revealed some new copper finds dated from the Copper Age. They originate from Barata Locality near Umarevtsi Village and from Devetaki Cave.

In this chapter the finds are systematized and interpreted. Some of them are published for the first time.

Copper finds from Umarevtsi-Barata

The archaeological excavations in 1990 included two drillings: sounding 1 (northern) and 2 (southern). Both are 5 x 2.5 m each, A and B plan-quadrate respectively. According to the excavator, the settlement is one-leveled, attributed to the Early Copper Gradeshnitsa culture (Stoyanov 1992a: 13).

In the course of the investigation in sounding 2, plan-quadrate A, a copper awl was discovered next to the so-called stone 'path', turned with its thicker part up. The awl is with a square cross-section, passing on round towards the point. The backside is flattened, wearing thin (***Figure 10.1:*** *1*).

A copper pin was discovered in the gallery there, 0.15 m north of the awl (Stoyanov 1992a: 9). It is with a sharp point, round cross-section and twisted at the backend (***Figure 10.1:*** *2*).

Copper finds from Devetaki Cave

The earliest excavations in Devetaki Cave were carried out in 1920s and renewed in 1950s, no more than rescue investigations. In 1992 they started again, but over a terrain, which bigger part had been considerably destroyed.

They continued till 1999 with a break in 1998. According to the material published, diaries and reports of the excavations, the layers investigated in course of the recent explorations date from Late Copper and Final Copper (the so-called Transition period). Among the artefacts there are some copper finds, discovered in a different context.

In 1992 under the vault near the cave opening, three copper artefacts have been discovered in house remains. According to the preliminary publication of the vertical stratigraphy, the house belongs to the 7th building level dated from the second phase of the Late Copper Age in Bulgaria (Stoyanov 1993).

The first of the finds is a copper awl with a rectangular cross-section along its whole length, and pointed front part (***Figure 10.1:*** *3*). The second find is an awl again, pointed at its front part. It is with a rectangular cross-section along its whole length and bigger that the first one (***Figure 10.1:*** *4*).

The last implement is fragmented and discovered together with its bone handle. The artefact is an awl, but a considerable part is missing. It has a rectangular section. It is preliminary published as consisted of body of rectangular in section wire with a blunt back part and sharp cutting side (Stoyanov 1993: 16). But the documentation and the autopsy of the find without doubt infer the awl was fragmented, which is visible in its graphic reconstruction (***Figure 10.5***).

The next two finds were discovered in the course of the excavations in 1994.The former, an awl (***Figure 10.1:*** *6*), has relatively precise stratigraphic data and the latter (probably also an awl [***Figure 10.1:*** *7*]) was discovered without an exact stratigraphic context.

The awl in ***Figure 10.1:*** ***6*** was discovered by the finishing of the investigation connected with the so-called podium of the Late Copper house, recovered in 1992. It was found by the excavations of the 'rib' near the northeastern corner of the podium and the eastern wall of the cave from the same building level. It has rectangular section along its whole length, and sharp point is better represented from one of the sides only.

The other one find (***Figure 10.1:*** *7*) is more massive. Its back part has rectangular cross-section, while pointed front part has almost ellipsoidal cross-section.

In 1996 the excavations in Devetaki cave was conducted on other place located under the first opening of the cave ceiling. The sounding is marked as No. 1. Several (sixth, up to 1999) building levels have been recovered, of which the first three counted from the top downwards are attributed by V. Gergov to the so-called Transitional period from Chalkolithic to the Bronze Age. The next three are attributed to Late Chalkolithic. In my opinion, the material culture at least of the third and fourth horizons is characteristic of Late Copper Age.

In the course of investigation of a house from the third horizon in sounding No. 1, two copper finds were recovered.

The first is a copper awl of not very big dimensions. It has square cross-section in its back part and round one in its front (working) part (***Figure 10.1: 8***). The second artefact is a copper fishhook. It is made of wire with rectangular section (***Figure 10.1: 9***).

Discussion and conclusions

The settlement of Umarevtsi-Barata excavated has been dated from the second phase of the Early Chalkolithic culture of Gradeshnitsa (Stoyanov 1991: 19) despite there is a hesitation regarding the relative chronology in later preliminary communication (Stoyanov 1992a: 13). The awl of that site has several parallels in the Copper Age. The copper pin, however, has parallels only among the finds from the Late Copper Age. For the time being, such pins are certainly known from graves Nos. 5 and 15 at Varna Late Copper cemetery (Ivanov and Avramova 1997: 33, 35). Perhaps similar pin originates from Late Copper later in Emenska cave, Veliko Turnovo District (Nikolova and Angelov 1961: fig. 8), Kodzhadermen tell (Popov 1918: 99, fig. 91), etc.

Two fishhooks similar to those from Devetaki cave originate from Ruse tell, perhaps from Late Copper layer (Georgiev and Angelov 1957: 69-70, fig. 30: 2; Chernykh 1978: table 19: 14, 15). There are such finds in Gumelniţa culture, stage B – from Gumelniţa, Vidra, etc. (Comşa 1978: fig. 4: 1-3). They are rather well distributed in Tripolie A and B1, according to the periodization of T.S. Passek – at Bernevo Luka, Luka-Vrublevetskaya, Soloncheno I and Novo Ruseshti (Ryndina 1971: fig. 5: 1-5; fig. 25: 8).

A part of fishhook is discovered at the second building level on the prehistoric site of Telish-Reditite, Pleven District (Gergov 1987: fig. 8e). One identical fishhook of the same type as that from Devetaki cave is unearthed by the excavations at Fărcaşele and dated from Vădastra II according to M. Nica (1970: fig. 10: 4).

According to the directions of turning of the upper end, which was attached to the halter, the fishhooks are divided into two types:

Type 1. The turn is just behind the trunk, and is parallel to the axe of the whole hook, as it is by the finds of Tripolie and Gumelniţa.

Type 2. The turn is not parallel to the axe of the hook, and is aside the trunk. The fishhooks from Fărcaşele and from Ruse tell exemplify that type. The find from the third horizon at Devetaki cave belongs to it.

The section of the copper wire of both types is with round and ellipsoidal section.

The parallels of the finds from the middle Osum drainage infer they all they are included in the context of the copper metallurgy in the Copper Age. The opinions concerning its initial development can be divided into two groups:

1. According to some writers, who treat that problem the beginning of the metallurgy is connected with a production of small, unmassive objects, as awls, pins, or adornments, and is dated mostly from Early and Middle Copper Age (Cernykh 1978: 276; Todorova 1986: 145; Ryndina 1993: 139).

2. According to second opinion, based on relatively new finds, in the third phase of the Early Copper Age in the area of Gradeshtitsa-Slatino-Dikili Tash complex massive implements have been produced yet. The last process continued in the Middle Copper Age, according to the data from the Struma drainage (Chokhadzhiev 1997: 18, figs. 51: 23; Čochadžiev 1998: 10-14).

As far as the metal finds from the middle drainage in the Osum River concern, some of them are the first metal finds with exact stratigraphy in that region. On the whole, they contribute to the development of the typology of awl and fishhook types in the Copper Age. The Early Copper metal awls indicate the innovation of metallurgy was spread over the southern middle areas of the Lower Danube as well.

Catalogue

1. *Awl*. Copper. Umarevtsi Village. Barata Locality. Lovech District. Length – 10.1 cm. Maximum thickness – 0.4 cm. Historical Museum Lovech. Inv. No. 1393. ***Figure 10.1: 1.***

2. *Pin*. Copper. Umarevtsi Village. Barata Locality. Lovech District. Length – 6.3 cm. Maximum thickness – 0.2 cm. Historical Museum Lovech. Inv. No. 1394. ***Figure 10.1: 2.***

3. *Awl*. Copper. Devetaki Cave. Lovech District. Length – 7.4 cm. Thickness – 0.5 cm. Historical Museum Lovech. Inv. No. 2548. ***Figure 10.1: 3.***

4. *Awl*. Copper. Devetaki Cave. Lovech District. Length – 5 cm. Width – 0.3 cm. Thickness – 0.3 cm. Historical Museum Lovech. Inv. No. 2597. ***Figure 10.1: 4.***

5. *Awl*. Copper. Bone handle preserved. Devetaki Cave. Lovech District. Thickness of the wire – 0.4 cm. Historical Museum Lovech. Inv. No. 2632. ***Figure 10.1: 5.***

6. *Awl*. Copper. Devetaki Cave. Lovech District. Length – 7.1 cm. Width – 0.3 cm. Historical Museum Lovech. Inv. No. 2816. ***Figure 10.1: 6.***

7. *Awl*. Copper. Devetaki Cave. Lovech District. Length – 7.2 cm. Width – 0.5 cm. Historical Museum Lovech. Inv. No. 2821. ***Figure 10.1: 7.***

8. *Awl*. Copper. Devetaki Cave. Lovech District. Length – 4 cm. Width – 0.2 cm. Historical Museum Lovech. Inv. No. 2909. ***Figure 10.1: 8.***

9. *Fishhook*. Copper. Devetaki Cave. Lovech District. Length 2.9 cm. Width of the point – 1 cm. Width of the wire – 0.4 cm. Thickness of the wire – 0.2 cm. Historical Museum Lovech. Inv. No. 2910. ***Figure 10.1: 9.***

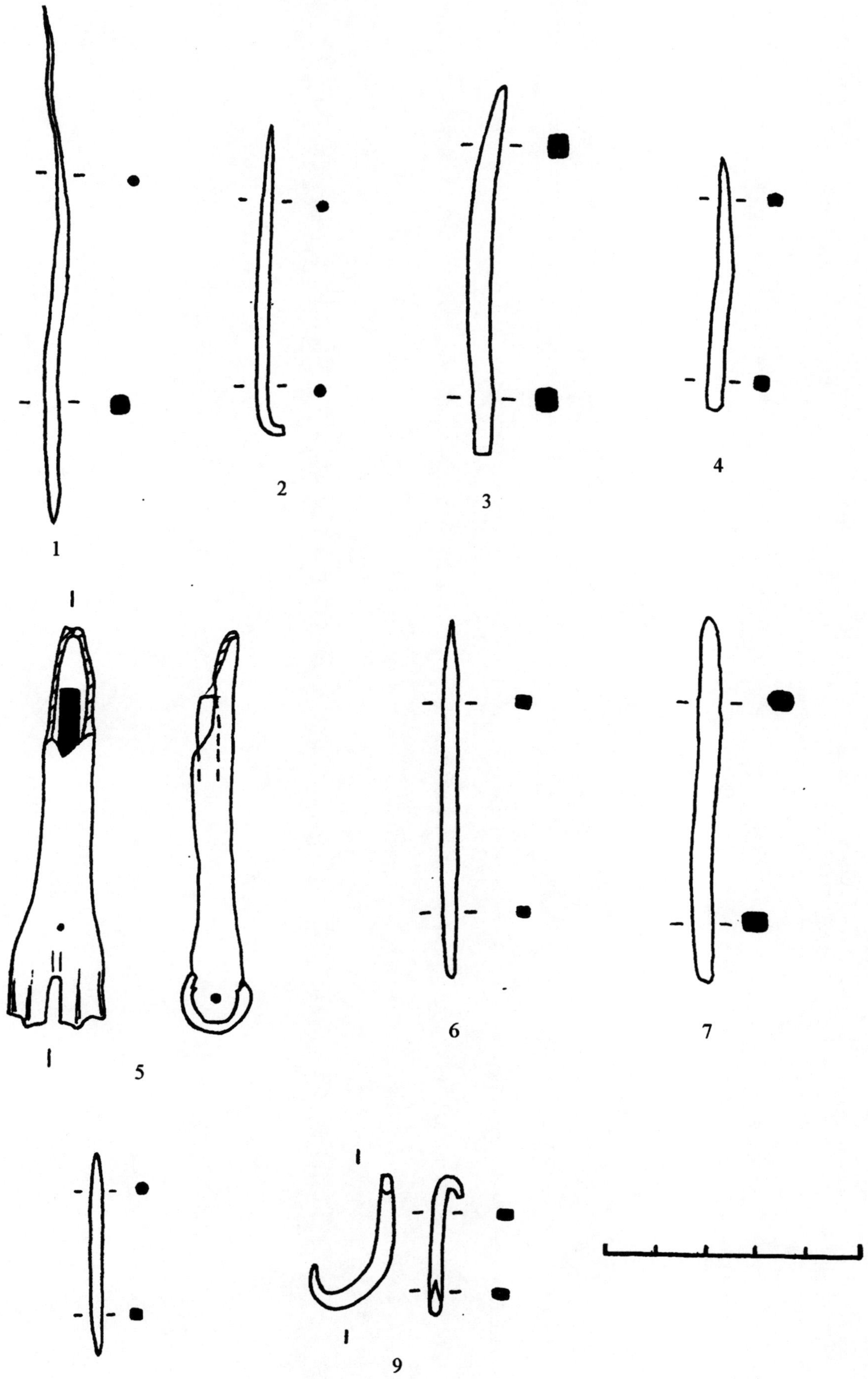

Figure 10.1. Copper Age finds from the Middle Osum valley.

CHAPTER 11

LATER COPPER AGE POTTERY FROM YUMURTA TEPE

Burçin Erdoğu

In 1995 an archaeological survey was initiated in the province of Edirne, Turkish Thrace, by the University of Thrace, Department of Archaeology. It has contributed extremely to our knowledge of the prehistory of this region and has thrown light on the relations of Turkish Thrace not only with the Balkans but also with Anatolia.

In the course of our survey, the Late Copper earthenware from Yumurta Tepe is particularly noteworthy. This group supplies a missing link in the pottery sequence of Eastern (Turkish) Thrace.

Location

Yumurta Tepe is located c. 12 km to the north of the district center of Havsa and c. 1 km east of the village of Haskoy in the province of Edirne. The mound is situated on the east bank of the Suloglu Stream, a tributary of the Ergene River. To the east of the stream is a high terrace on which the mound has formed. It is approximately 250 m in diameter and perhaps 1 m high.

The pottery

All the Yumurta Tepe pottery is handmade and unburnished or lightly burnished. There is a limited number of sherds with fine burnish. The firing temperature varied from low to fairly high. The Yumurta Tepe pottery divides into three main types. The common ware is the dark colored (50%). It is tempered with sand, grit and mica. Both interior and exterior surfaces are evenly smoothed. They are unburnished or they are less frequently lightly burnished. There are a few sherds coated with a dark colored slip on the exterior surface. Surface colors are in tones of pale black or grey. The reddish colored ware is rather common (45%). It has also tempered with sand, grit and mica. Both surfaces are smoothed, unburnished or lightly burnished. Surface colors are in variety of tones of reddish brown. Apart from these wares the course ware rarely occur (5%). The paste is rather course, including unbounded use of sand and grit temper. Mica and organic temper are also occur. Both surface are smoothed and sometimes both surfaces are left totally untreated. The surface colors are cream, light brown or red-

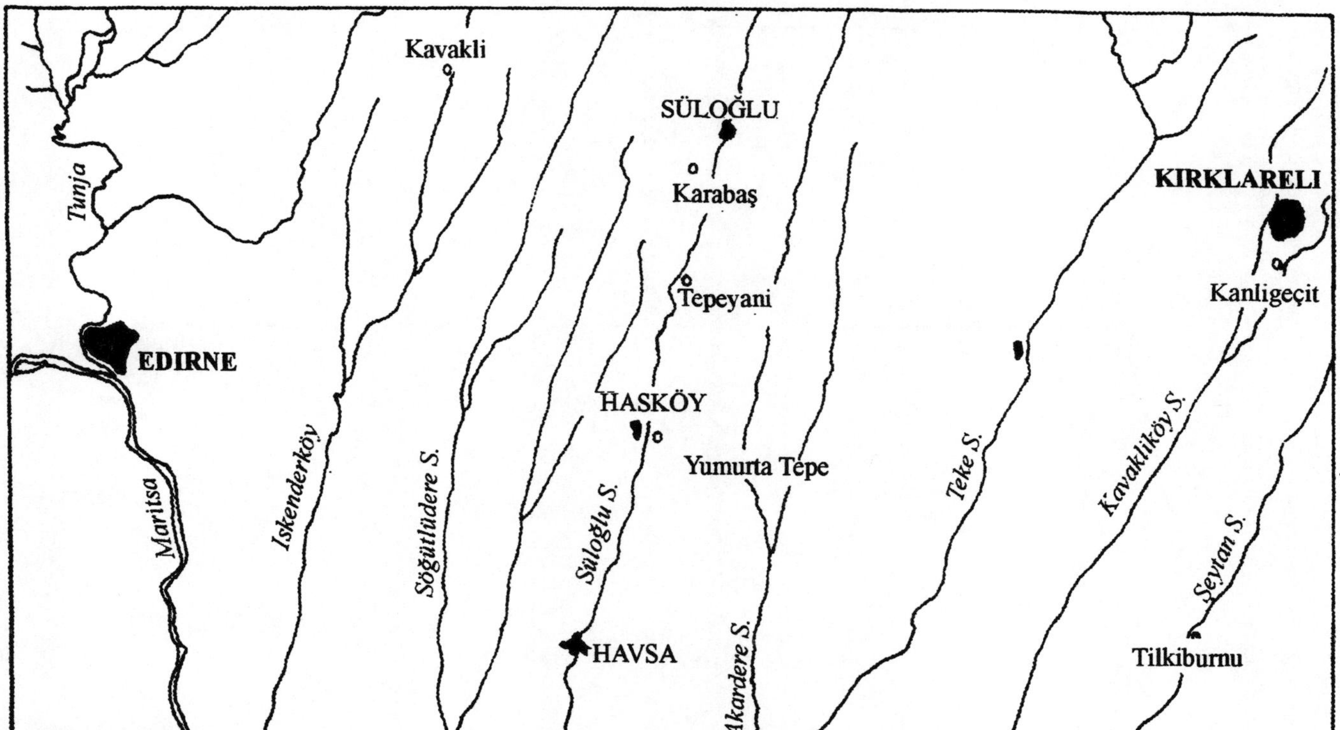

Figure 11.1. Map showing the Late Eneolithic sites in Turkish Thrace.

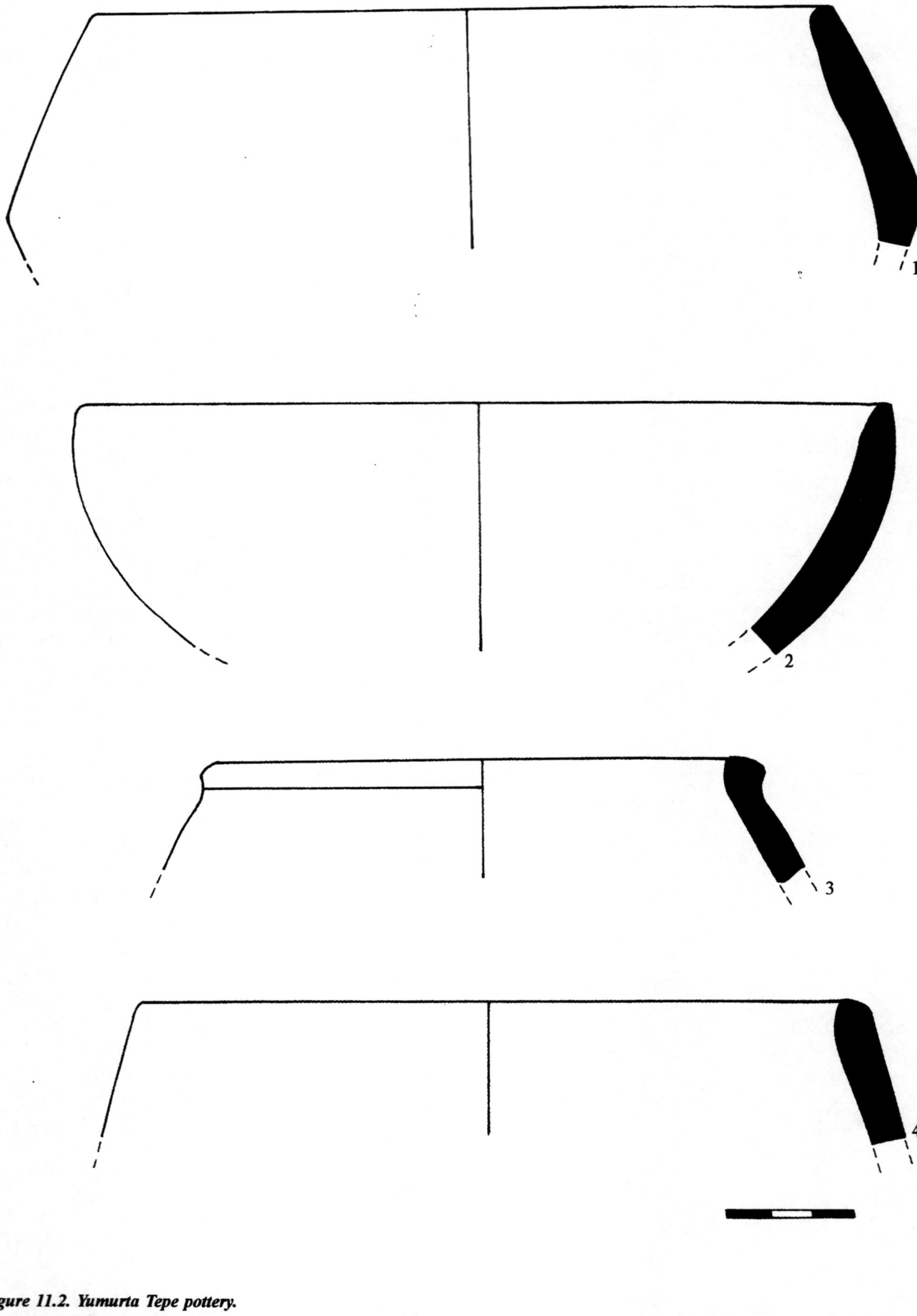

Figure 11.2. Yumurta Tepe pottery.

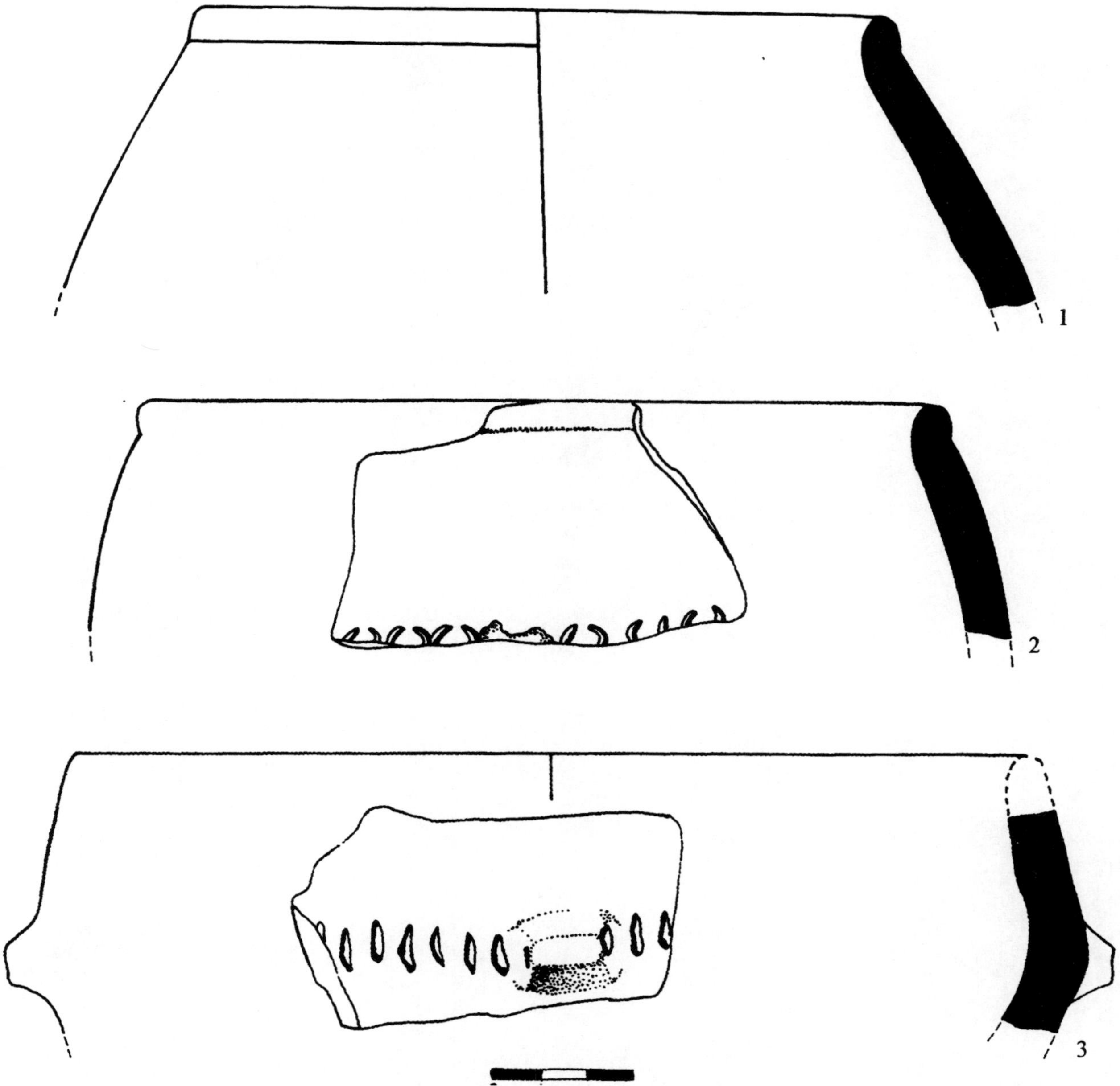

Figure 11.3. Yumurta Tepe pottery.

dish brown.

The repertoire of shapes is not rich. The most common shape is a bowl with incurved rim (***Figure 11.4: 1,3***). Curved sided bowls (***Figure 11.5: 1-4***) and straight sided bowls (***Figure 11.2: 2***) are also attested. Convex sided pots with bead rims are also common (***Figures 11.2: 3; 11.3: 1,2***). The other shapes are hole-mouth pots (***Figure 11.2: 4***) and closed pots with carination (***Figure 11.2: 1***). Sometimes knobs on the carination (***Figure 11.3: 3***). There is one example of S-shaped pot with large flat bottom and a vertical handle (***Figure 11.7: 7***). Clay ladles are also occur (***Figure 11.6: 1,2***). Bases are well developed being flat or concave (***Figure 11.4: 4,5***).

Various decorations techniques have been employed, such as channel, barbotine, incision and impressions (***Figures 11.6: 3,6; 11.7: 3-8***). The most common pattern for the channel and incision decorations, is the running vertical parallel lines (***Figures 11.6: 6; 11.7: 6***). Nail and stroke impression are much in use (***Figures 11.6: 3; 11.7: 3,7,8***). Shell-impression decoration also occurs (***Figure 11.7: 4,5***). Some of the sherds are decorated with concentric incised bands imitating the print of a cord (***Figure 11.7: 1,2***). There is only one sherd decorated with white paint (maybe burnt graphite decoration?), was found on the interior rim of a curved sided bowl (***Figure 11.5: 3***).

Discussion

The relation between Eastern Thrace and the Upper Thrace / the Northern Balkans, as well as Anatolia during the Later Copper Age period still pose a problem. Until now, only a few Late Copper material has been recovered in Eastern Thrace, and two sites, Tilkiburnu and Kanligeçit have been investigated.

Tilkiburnu is located c. 18 km to the south of the town of Kirklareli. A very restricted salvage operation took place by M. Özdoğan (1982a). 12 pots were found together in a pit and the excavator argues that all the pottery bears some re-

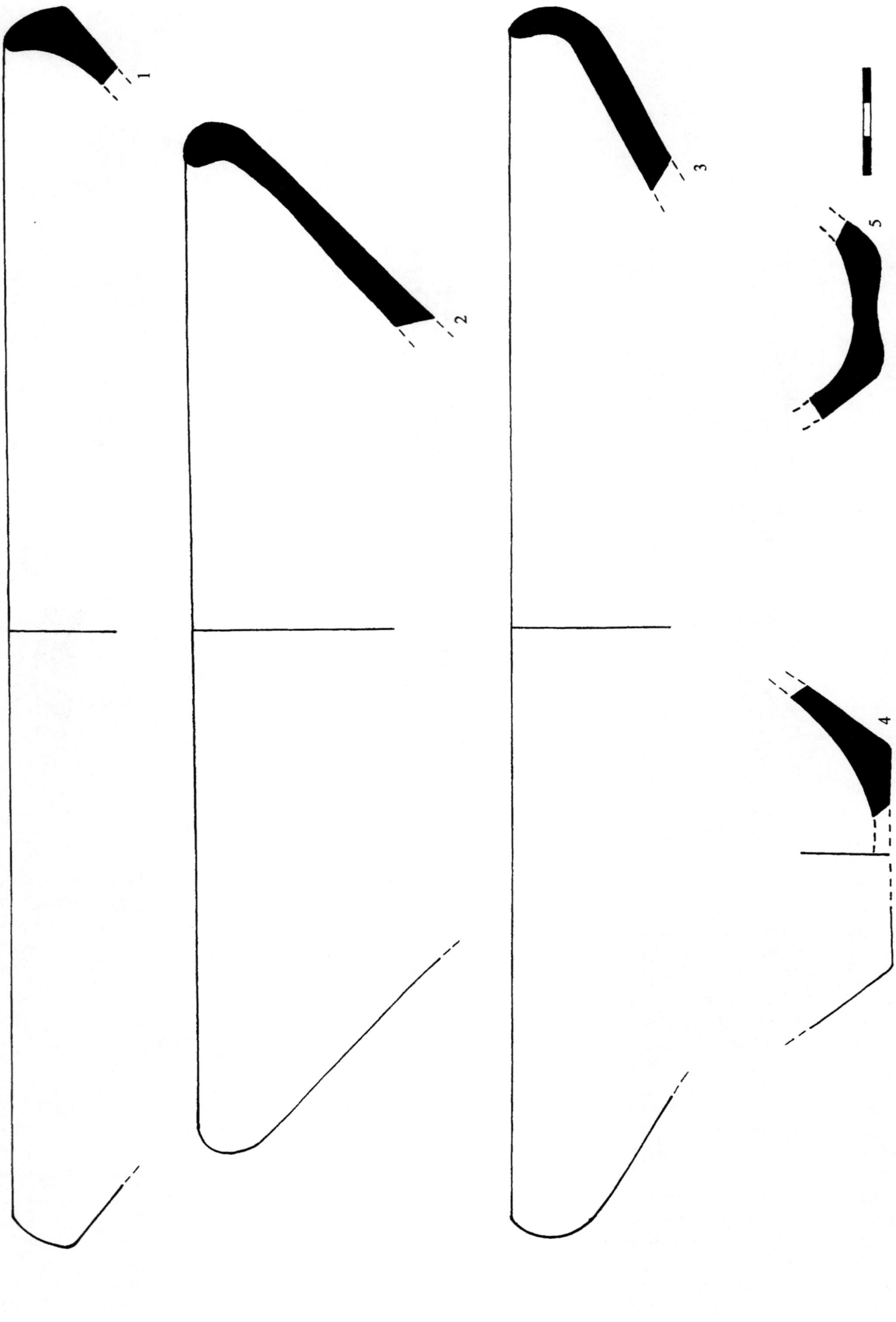

Figure 11.4. Yumurta Tepe pottery.

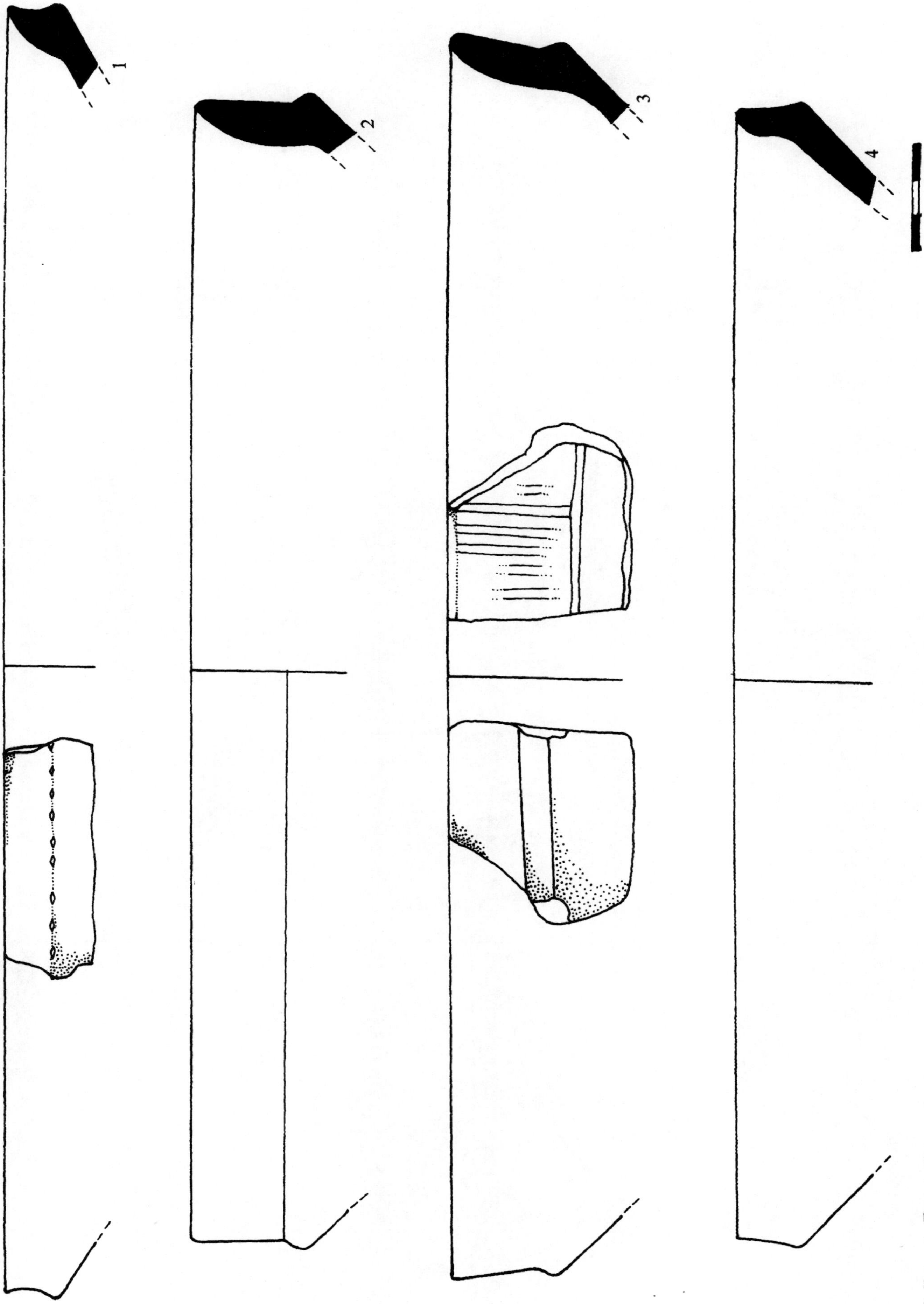

Figure 11.5. Yumurta Tepe pottery.

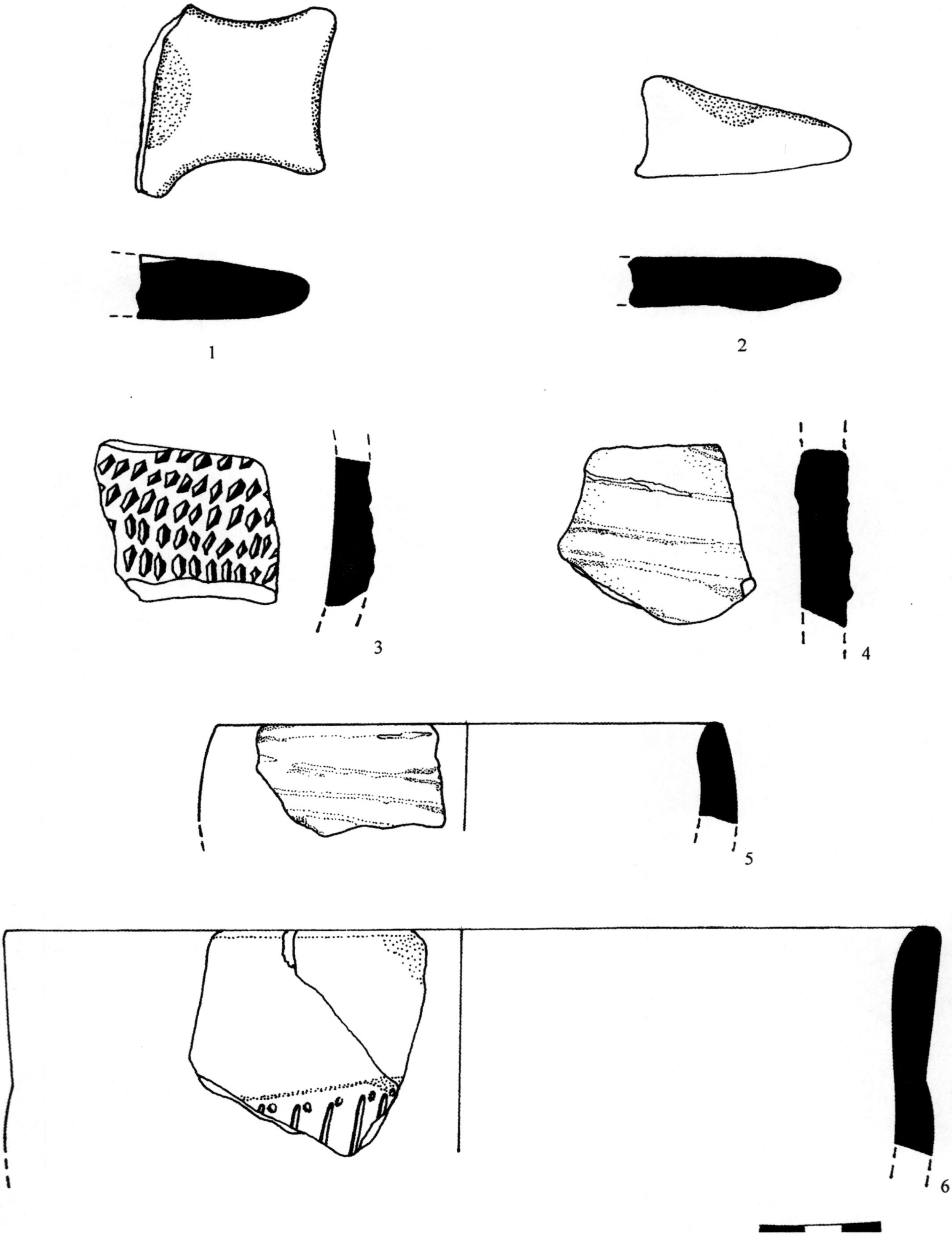

Figure 11.6. Yumurta Tepe pottery.

semblance to the Late Karanovo VI – Gumelniţa horizon (1982a: 11). However, some forms are similar to the Northwestern Anatolian Kumtepe Ib horizon.

Kanligeçit, the second investigated site in Turkish Thrace, is located c. 250 m to the south of the town of Kirklareli. The Later Copper Age material was recovered In two pits which were dug into virgin soil. Autopsy of the sherd material revealed some similarities with the Karanovo VI – Gumelniţa period of the Balkans. However, the most of the pottery is rather different that Tilkiburnu. Nevertheless, we have neither at Tilkiburnu nor at Kanligeçit any example of a classical Karanovo VI-Gumelniţa assemblage. All the characteristic pottery from Karanovo VI-Gumelniţa complex, like graphite painted decoration, extravagantly angled vessels, large lids, amphora-like vessels and all the other forms and decorative techniques are absent at both the sites. It is more reasonable to see Eastern Thrace sites as of basically local development (Erdoğu 1993: 28).

To turn now to the evidence from Yumurta Tepe. At Tilkiburnu (Özdoğan 1982a) there are excellent parallels for the dark colored and reddish colored unburnished of lightly burnished ware from Yumurta Tepe. The ware shows unmistakable parallels in the shapes as well. Our inverted rim bowls and curved sided bowls (***Figures 11.4: 1-3; 11.5: 1-4***) are again comparable with Tilkiburnu (Özdoğan 1982a: figs. 6: 2; 7: 2), as well as the convex sided pots with bead rims (***Figure 11.3: 1,2***; see Özdoğan 1982: fig. 6: 5,6). Our channel decorated sherd from Yumurta Tepe (***Figure 11.7: 6***) has strong affinities with Tilkiburnu channel decorated vessel (Özdoğan1982a: fig. 6: 7). At Tilkiburnu however, are absent nail, stroke and shell impressed decorations attested at Yumurta Tepe.

The pottery from Kanligeçit is hard to correlate exactly with Yumurta Tepe. The similar inverted rim bowls occur in Kanligeçit among the fine ware. However, there is no such fine ware in Yumurta Tepe. Only the clay ladles from Yumurta Tepe seem to have good parallels at Kanligeçit.

The Yumurta Tepe pottery shows rather good similarities with the pottery from the Balkans. At first sight, nearly all the parallels point to the last phase of the Karanovo VI-Gumelniţa complex and to the second and the third phase of the Krivodol-Sălcuţa-Bubanj complex. The similarity is especially apparent at the types of decoration, and to some extent also in the repertoire of the shapes. However, most of the shapes are different.

Nail and stroke impressions of Yumurta Tepe can be compared with sites such as Sălcuţa I-II (Berciu 1961: fig. 88: 8, 105: 4,5), Krivodol (Nikolov 1984: 53: 2), etc. For shell-impressed sherds from Yumurta Tepe similar examples can be found at sites such as Sălcuţa I (Berciu 1961: fig. 88: 5), Karanovo VI (Georgiev 1961: Plate 21: 7), Gumelniţa (Comş a 1976: fig. 18: 19), etc. The white painted decoration on the interior rim of the Yumurta Tepe curved sided bowl is reminiscence of a bowl with a graphite decoration from Sălcuţa II (Berciu 1961: fig. 120: 1).

Conclusions

To conclude I may sum up the main points of the Later Copper Age pottery material from Yumurta Tepe.

1. Decorative techniques and some forms of Yumurta Tepe are strongly reminiscence of the late phase of Karanovo VI-Gumelnişa and the second and the third phase of the Krivodol-Sălcuţa-Bubanj complex. However, many characteristic elements of the Later Copper Age Balkan cultures are absent at Yumurta Tepe, which fact can be explained by the suggestion that the pottery is a local variant.
2. Almost all of the ware groups noted at Yumurta Tepe seem to be presented at Tilkiburnu. But there are some forms and decorations that are absent at Tilkiburnu.
3. It is rather difficult parallels between Yumurta Tepe and Kanligeçit pottery to be found. Yumurta Tepe pottery was recovered at Karabas, Tepeyani and Kavakli during our survey in the province of Edirne (Erdoğu 1995: 97-98). At Tepeyani there was also found Kanligeçit pottery is earlier that the Yumurta Tepe pottery.
4. Without excavations, it is difficult to be argued, but it seems the two-phase system of the Karanovo VI-Gumelniţa horizon argued basing on the evidence from Drama (Fol et al. 1989: 92-94) should also be applied to Eastern Thrace.
5. The problem of the graphite pottery in Southeastern Thrace is too complicated to be dealt with here. We suggest that the graphite painted ceramics have much wider spreading to the north that in Turkish Thrace.

Catalogue of the pottery illustrated

Figure 11.2. 1. Medium fine fabric, reddish brown in color, interior lightly burnished, exterior smoothed only. 2. Course fabric; black in color; unburnished. 3. Medium fine fabric; reddish brown in color; lightly burnished. 4. Medium fine fabric; dark brown exterior, black interior; lightly burnished.

Figure 11.3. 1. Medium fine fabric; pale black in color; lightly burnished. 2. Medium fine fabric; reddish brown in color; lightly burnished on exterior, interior smoothed only; decorated with stroke impressions. Medium fine fabric; black in color, fine burnished on exterior, interior smoothed only; decorated with impression dots.

Figure 11.4. 1. Fine fabric; pale black in color; fine burnished on interior, exterior lightly burnished. 2. Medium fine fabric; pale black in color; lightly burnished. 3. Fine fabric; pale black in color; fine burnished on interior, exterior lightly burnished. 4. Medium fine fabric; pale black interior; lightly burnished. 5. Course fabric; cream in color; unburnished.

Figure 11.5. 1. Fine fabric; pale black in color; fine burnished interior; exterior lightly burnished. 2. Fine fabric; black in color; fine burnished on exterior; interior smoothed only. 3. Medium fine fabric; black in color; lightly burnished; decorated with mat white paint on interior. 4. Fine fabric; reddish brown in color; fine burnished on

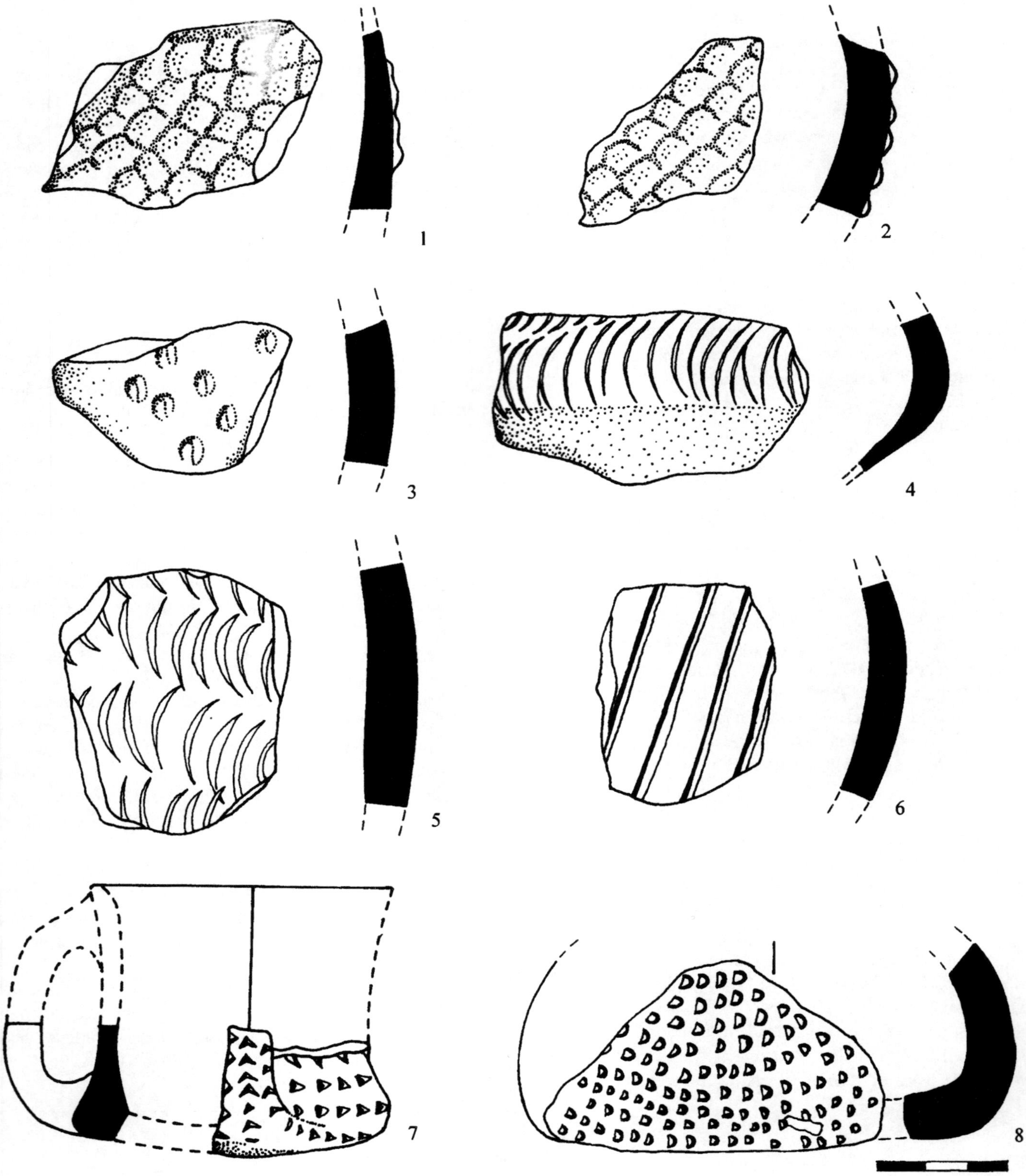

Figure 11.7. Yumurta Tepe pottery.

interior; exterior smoothed only.

Figure 11.6. 1. Medium fine fabric; grayish brown in color; unburnished. 2. Medium fine fabric; grayish brown in color; unburnished. 3. Medium fine fabric; reddish brown in color; unburnished; decorated with stroke impressions; decorated with stroke impressions. 4. Medium fine fabric; grayish brown in color; unburnished; decorated with barbotine. 5. Medium fine fabric; reddish cream in color; unburnished; decorated with barbotine. 6. Medium fine fabric; grayish brown exterior; grayish black interior; lightly burnished; decorated with shallow channels and impressed dots.

Figure 11.7. 1. Medium fine fabric; black in color; unburnished; decorated with concentrated incised bands imitating the print of cord. 2. Medium fine fabric; black in color; unburnished; decorated with concentrated incised bands imitating the print of cord. 3. Medium fine fabric; black in color; unburnished; decorated with nail impressions. 4. Medium fine fabric; grayish brown in color; unburnished; decorated with shell impressions. 6. Medium fine fabric; reddish brown in color; unburnished; decorated with shallow channels. 7. Medium fine fabric; grayish brown in color; unburnished; decorated with stroke impressions. 8. Medium fine fabric; reddish brown in color; unburnished; decorated with stroke impressions.

Chapter 12

The Final Copper Settlement of Kolarovo

Lilyana Pernicheva

The prehistoric settlement of Kolarovo is located in the Strumeshnitsa valley, the most southwestern part of Bulgaria The Strumeshnitsa River is a right tributary to the Struma River. The geographic region called Podgorie includes the edges of the sub-mountain areas of the Belasitsa Mountains whose slopes run down towards the Strumeshnitsa River bed. The site occupies that area of prolong flat watershed with a step-like slope in north-south direction towards the Strumeshnitsa River. Morphologically, the watershed belongs to the oldest geological background of Podgorie – the Young Pleistocene surfaces crossed in north-south direction by ravines of drained rivers and streams emptied into the Strumeshnitsa River. Deep ravines border the terrace where the site is located, but it evolves down to the north towards the flooded terrace of the river. The height of the terrace over the riverbed is c. 30 m. In south direction the terrace is bordered by the next higher step of the watershed with a ditch slightly visible over the surface. It is likely the ditch to have belonged to the prehistoric sites. The locality, called Musovitsa, is situated in c. 4 km north-northeast of the village of Kolarovo.

The settlement size was limited topographically, so it did not occupy a big area but c. 5-6 Decares. C. 300 m south of the ditch mentioned there were rare sherds of later Copper Age pottery, dispersed over the surface without traces of building remains. It is possible at some point the settlement to have occupied the higher part of the watershed in the south. It occupies almost flat terrain with slight slope to the north and does not belong to the typical highly located settlements. However, the site is naturally well defended from three sides and remains almost hidden for a view from the river valley. At the same time, there is a good view towards the surrounding environment in all directions. Its natural defense is not worse than the typical highly located settlements, characteristic of

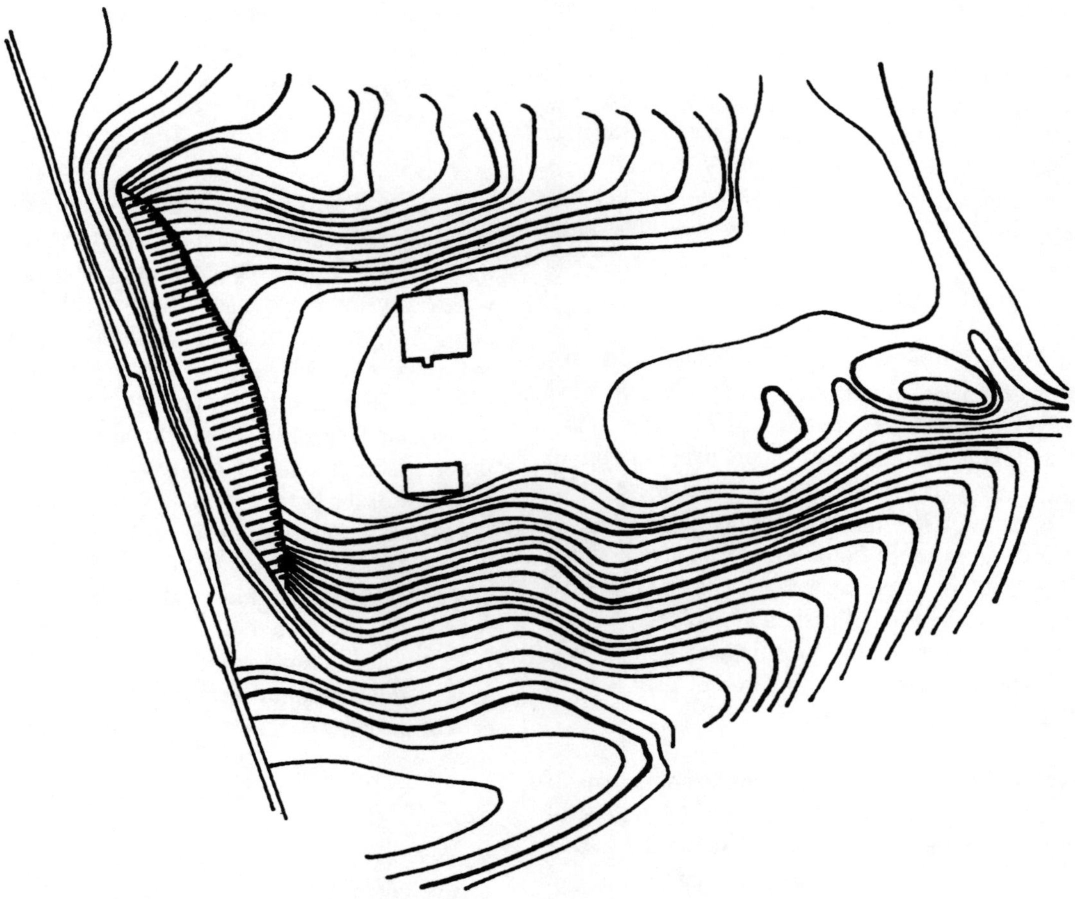

Plan 12.1. Kolarovo. Tachometric photo.

the Late and Final Copper Age in Western Bulgaria.

The area of the settlement was put under agricultural activity until 1960s, up to 0.15-0.20 m. The type of the soil is cinnamon, forest and clay, so that it was very difficult to be dug.

Stratigraphic situation

The archaeological investigation included two big trenches, oriented in the four cardinal points (***Plan 12.1***). Sounding A was situated on the northeast periphery of the terrace where the surface was concentrated with burnt plaster. Shaped as a square, it was 12 x 12 m, but later was widened c. 1 m to the east, up to the edge of the terrace. The total excavated area was 150 sq. m.

Sounding B was situated on the opposite west periphery of the terrace. A rough crumble stone was spread over the terrain of c. 70 sq. m. Initially, 4 x 10 m with long axis in direction north-south, that sounding was later widened 1-1.5 m to the west, following the slope terrain. It has total of 45 sq. m. Accordingly, the whole excavated areas reached 200 sq. m.

Sounding A

The culture layer in the east and the south part of the drilling is 0.50-0.65 m thick (without the pits). The later became thinner in the west and in the north of the excavated area, up to 0.30-0.40 m (***Plans 12.2-12.3***). Obviously, in this part the terrain was cultivated deeper and probably cleaned of building rests, as the bigger part of the culture layer was destroyed.

Up to 0.15-0.20 m depth from the surface a brown and gray layer was discovered, cultivated in the past. By the excavations it was very hard and did not contrast to the uncultivated layer. The cultivated layer was rich of burnt plaster and fragmented pottery, especially in the sectors of the burnt houses. In the northeast part of the drilling, partially under the north profile, the darker soil of pit has been already outlined in the upper layer (pit No. 2) – ***Plan 12.4***.

Extremely hard clayey layer of yellow and brown color followed in the biggest part of the drilling under the cultivated layer, 0.15 – 0.40 m thick. The soil had fine structure. The layer consisted of remains of one burnt settlement horizon. In the east and the south part the sector investigated there was a concentration of burnt plaster of houses (***Plans 12.2 and 12.6***). At the base of that layer house floors were discovered, which probably were slightly dug into the terrain. The floor level of the burnt horizon was very good marked by the bases of two pithoi in the southwest part of the sounding, as well as by a floor plaster in the south part, up to a depth of 0.40-0.50 m from the surface (***Plans 12.2 and 12.6***).

On the strength of the stratigraphic situation in some parts of the drilling it can be assumed the existence of later stage of occupation that chronologically followed the burnt horizon. To that stage perhaps can be attributed an equipment of two almost identical in size and shape, symmetrically situated plates, similar to steps situated at a distance of 0.60 m. They were located next to the pithoi in the southwest part of the drilling and had destroyed the layer of the burnt horizon (***Plan 12.6***). That structure borders to northeast a sector of gray and yellow clay, which differed generally from the layering of burnt houses, despite stratigraphically, it was located at the same depth. The areas with a concentration of plasters were also disturbed on other places by similar sectors of beaten lighter clay without traces of burning. Those sectors were unearthed just below the fallow land west of the stone structure mentioned (an area of 3 sq. m) and between two sectors of houses (an area of 13 sq. m [***Plan 12.6***]). It is possible they to have been a beaten clay areas for the later occupation of the settlement. Unfortunately, the layering of that stage mainly fell in the cultivated layer. If there was a hiatus between it and the burnt horizon, it could not be documented stratigraphically. A part of the archaeological material at the settlement showed a definite typological difference with that from the burnt horizon, and its quantity increased typologically in the cultivated layer, especially in the uppermost levels of some pits. That observation also indicated an existence of later (second) stage of occupation of the settlement, which was not well documented stratigraphically.

There are also data on earlier building stage of the burnt horizon, which was registered in the sectors with rests of houses, more precisely in their substructure. Under their floor levels, at a depth of 0.65 m there were two not very big dugouts, construction pits and ceramic sherds of small quantity have been documented. The dugouts in the substructures of the houses were filled with burnt plasters. Obviously, a later stage of their occupation has been recovered. Outside their sectors the culture layer finished at a depth of 0.30-0.40 m. The ceramics of the house substructures was rather fragmented and do not show considerable typological variety in compare to that from the burnt horizon.

Sounding B

At a depth of 0.20 m there was a stone structure documented in all the drilling, consisted of big crumble and plates (long up to 0.70 m) mainly of sandstone and gneiss. The stones were recovered without any order, mostly horizontally, rather compactly, welding in extremely hard clay mass, but without indication of masonry (***Plan 12.8***). Situated in south/southwest – north/northeast direction, the stone structure covered all the sounding and was unearthed up to 11 m. Its width in the sounding was c. 4.5 m. In the southwest part it reached the edge of the terrace, and part of the stones were dragged down over the south slope. In north direction the stone equipment did not reach the end of the terrace and it seemed to turn slightly to the east, following its outlines. The structures have been unearthed up to 0.60 m. In the lowest part of the equipment the stones shaped indivisible part with the clay and it looked at its base they were joined with the natural rocky terrain of the terrace. The last had the same geological structure of sandstone and gneiss.

The role and the chronology of that structure are not well known. Having in mind the numerous stone on the surface, it can be concluded that it continues on the north periphery of the settlement, as well as to the south. There was no archaeological material in the clay, which joined the stones. Small

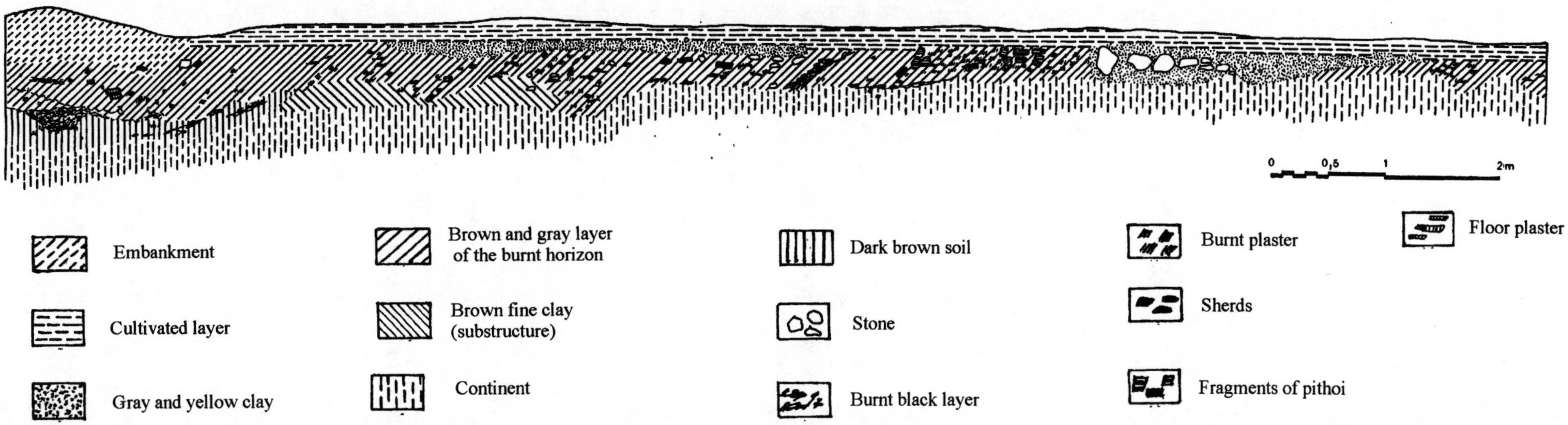

Plan 12.2. Sounding A. Southern profile.

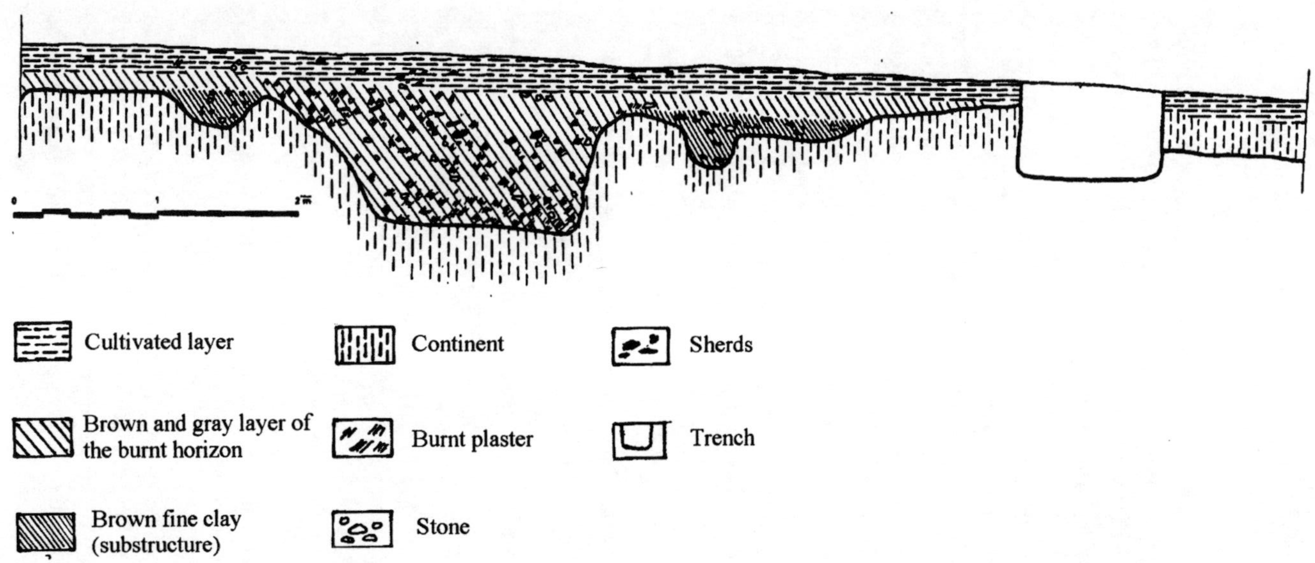

Plan 12.3. Sounding A. Western profile, sourthern part.

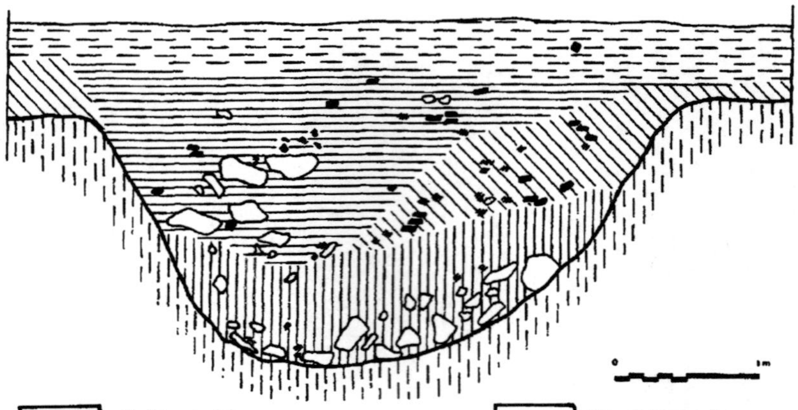

Plan 12.4. Kolarovo. Sounding A. Pit 2.

sectors with thin cultural layer were recovered in the neighborhood of the equipment in the southeast and the northwest sector of the sounding. The ceramic material of those sectors was scanty and corresponded to the material from sounding A. Perhaps a defense equipment was recovered but it remained unexplained the fact that at the opposite east edge of the terrace there were no traces of similar construction. Nevertheless, there were clear traces of earth embankment, which covered the east periphery of the burnt house in sounding A and have been well outlined in the south and the north periphery of the sounding (***Plans 12.2 and 12.5***). Despite there is no obvious argument to connect chronologically both structures, I think they are later that the burnt horizon and probably belong to the second stage of the settlement occupation. They probably defended and stabled the terrain from sliding down.

Houses' remains

The ruins of the houses were recovered as a kind of platforms, closely concentrated, both sectors, consisted of burnt plaster, with approximately rectangular shape in plan. The first house better outlined was situated on the very periphery of the terrace. Because of this fact the sounding was widened up to its edge (***Plan 12.6***). Probably a part of the houses was destroyed over the erosion slope. The dwelling's area was c. 8 m long in north-south direction. It had a maximum width of 6 m in its north part. The floor level of the house was recovered up to 0.45 m depth, i.e. the thickness of the burnt horizon together with the destruction in that sector was c. 0.25-0.30 m. To the northwest the habitation area was destroyed by the beaten clay layer mentioned above.

The floor of the house was hardly beaten and in some places it had a red color from the fire, with traces of black burnt layers. In the central part of the house a sector with plastered clay was documented, 2.20 x 2.60 m. The clay plaster, 3-5 cm thick, was placed directly over the terrain and was strongly cracked. Perhaps it covered a bigger area but there was no evidence of plastering of the whole house floor. The clay of the plaster was without any organic matters and consisted of a lot of sand. Fragments of similar, extremely hard clay floor plaster were documented among the destruction of the habitation and in its south part, as it is visible at the south profile of the drilling (***Plan 12.2***). It seems only the south part of the house floor was plastered and it obviously continued to the south. Under the well-documented floor level, in the substructure of the house the beaten floor was 15-30 cm thick with slightly concentrated plasters and ceramic fragments. It has rather homogenous red color from the burning, so it can be assumed it was preliminary burnt before erection of the house construction. No holes of posts have been excavated in that sector, which are traditional elements in the wattle-and-doubt constructions. Only on the west periphery of the house area, in the substructure of the houses three oval in plan pits were recovered, dug into the continent up to 0.18-0.30 m (***Plan 12.7***), but their depths from the house level were only 15-20 cm lower. The pits were situated in a row in north-south direction, in 0.30 and 1.10 m distance between (***Plan 12.7***, pits 11,12,13). The pits cover a total length of 4.5 m and maximum diameter of 1.10-1.20 m. Their floor were filled with lighter clay mixed with small fragments of plaster and in the higher level – with dark soil because of burning and small quantity of ceramic fragments. Pit 11was widened in its deepest part. Apparently, these were well-drained pits for placing of wooden posts in. Their belonging to the house described is not certain, because they were not visible at the level of the floor platforms. It is possible they to belong to the earliest building stage of the settlement.

A small part of beaten clay platform was recovered in the northernmost part of that building sector arising c. 10 cm above the floor. Around it there were several dislocated fragments of floor plaster of an oven. Perhaps there was also the place of the oven.

The displacement of the house level was from 15 cm north to

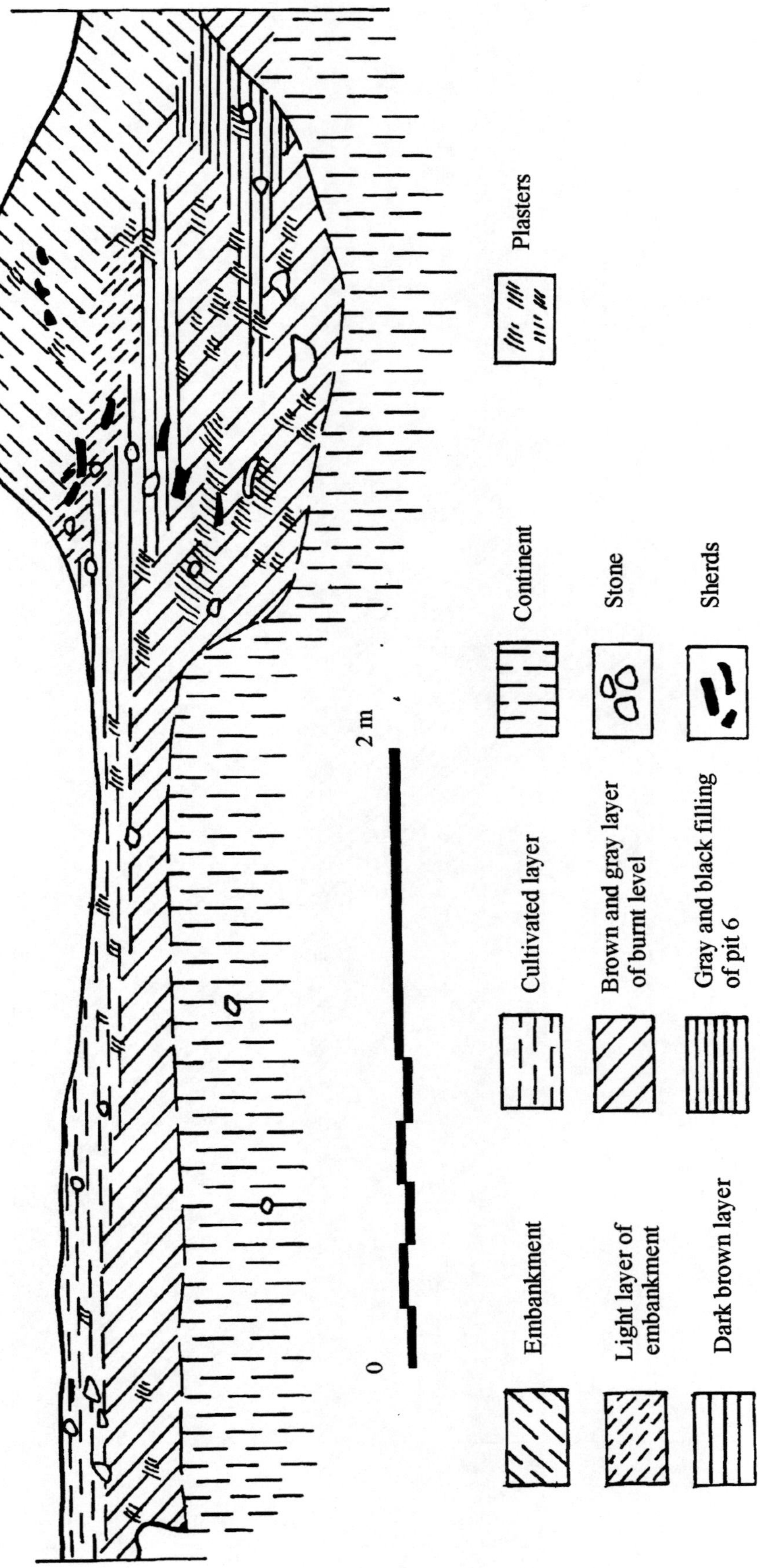

Plan 12.5.Sounding A. Northern profile, eastern part.

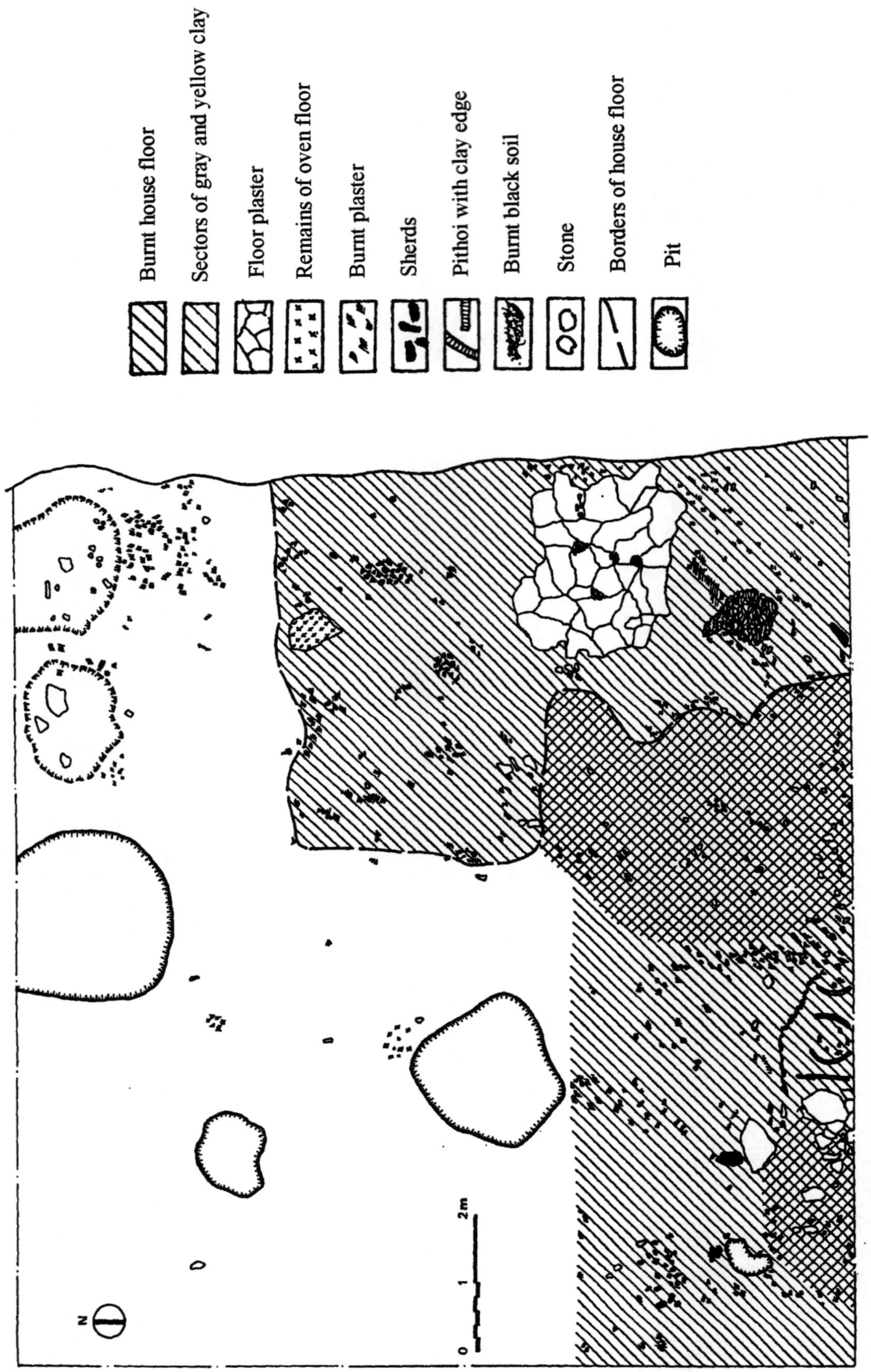

Plan 12.6.Sounding 6. Horizontal plan of the burnt horizon.

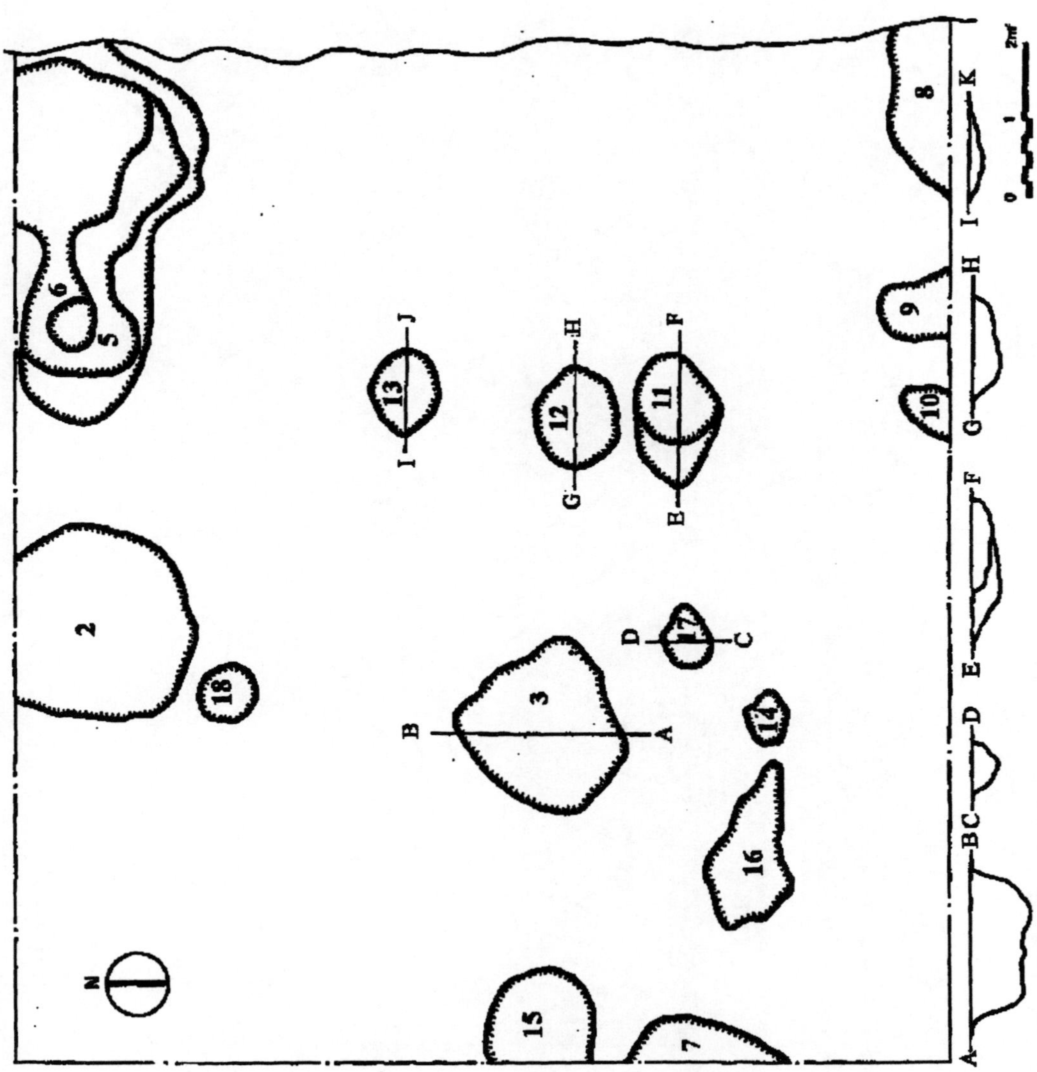

Plan 12.7. Sounding A. Pits' substructures.

Plan 12.8. Sounding B. Horizontal plan.

south. In its area there were found comparably well preserved ceramic material – three whole vessels, two cylindrical stands for vessels and other ceramic finds.

A part of a second building was recovered in the southwest part of the sounding. There was a concentration of burnt plasters covering an area 4 x 4 m. It also continued to south, outside the sounding (***Plans 12.2 and 12.6***). In north the habitation area broke down rather sharply there. That assumes a kind of artificial clearance of the terrain in the Antiquity or by the contemporary cultivation of the land there. It was also visible in the strongly thinned culture layer in the northwest part of the sounding (***Plan 12.3***). In that building sector, at the very end of the south profile there were documented the bases of two grain earthen pithoi situated 0.20 m under a thick culture layer with high concentration of plaster, and they themselves over big fragments of burnt destruction. The pithoi were located in a distance of 0.50 m. They have a diameter of the bottom 0.50 m and their walls were preserved up to height of 15-18 cm. The thickness of the walls was 1-2 cm. They were made of clay slightly mixed with organic matter and more sand. From the west that 'economic' sector was bordered by clay fence with a north-south direction preserved at the height of 5 cm, and 3 cm thick, without traces of wooden framework. On the bottom of the west pithos, a shallow bowl was broken in situ. It has a big diameter and was used obviously for covering of the pithos. Around the pithos there was visible a platform of plastered clay. The floor under that household equipment was beaten and burnt up to a depth of 15-20 m under the pithoi. In the area of the house there were again no postholes. In the northeast corner of the house platform, under the ruins of the house a pit was discovered with a diameter of 0.60 m, dug 30 cm into the continent filled with concentrated very burnt plaster and burnt wooden material. Identical but deeper dug pit was recovered c. 5 cm north out of the house platform. Probably it belonged to the same dwelling (***Plan 12.7, pits 17, 18***). Obviously, these pits were with construction role for some of the building stages of the house.

In the substructure of the houses, as well as in the continent outside the house platforms there were registered and investigated numerous pits (***Plan 12.7***). Their structure and way of filling constitutes their grouping as follows:

1. Construction pits mentioned, connected with the building of the houses (Nos. 11, 12, 13, 17, 18). The are comparatively shallow and filled with beaten clay, plaster and burnt soil.

2. Pits for clearance of the burnt ruins of the houses. Pits 7, 9, 10 and probably 15 belong to that type. They were situated next to the houses or under the house area. By color and structure of the soil they did not differ considerably from the layer of the burnt horizon. They were filled by the clearance of the terrain for new building activity, probably only once and mainly with burnt plaster, house inventory and a small quantity of ceramic fragments.

3. Puts with several different layers and sectors with a concentration of stone and gravel in. The layers of clay and of burnt soil or burnt organic material followed one after other. In some sectors there were a concentration of an archaeological material – fragmented pottery and cult objects. Most of them were found in those pits. It looks they were filled consequently, during different stages. It cannot be excluded they had a ritual function. For instance, the bottom of pit 2, next to the rock base of the terrace was covered mainly with broken stone plates. In the pit the layering of the burnt level differed very clearly from one much darker layer. In its upper part where was a concentration of pottery from the later stage, as well as two of the flat holed schematic figurines discussed below, were found there. Under that layer there was also flooring of broken stone, with clear aim demonstrated both layers to be distinguished. The upper part of the pit was sealed with a floor of big stone plates. In the pit there were well-preserved clay altars and other cult objects, all from the first stage of occupation of the settlement.

As a conclusion, it can be stressed the following points concerning the building technique, as the investigated area permitted. It seems the houses were slightly dug into the terrain. They had very solid clay beaten floors, partially or completely plastered with clay. The house had wattle-and-doubt structures but with limited using of the wooden material assumed. Perhaps because of the very hard terrain, only some basic posts were placed into in especially dug bog pits, additionally drained with fragments of plaster and clay. It is possible the wall to have been erected with the help of wooden tramps attached to the basic posts. It cannot be excluded a clay beaten technique in combination with wooden posts. That technique is already known from Krivodol-Sălcuţa-Bubanj complex's houses.

Ceramics

The ceramic material from the burnt building horizon, especially in the areas of the houses and under their destruction give a good base for dating of the first stage of the settlement. More complicate it is the selection and the typology of the vessels from the later occupation stage. Because of the thin cultural layer of the site and the obvious destruction of the upper stratum, in which were no traces of building remains, the characteristics of the later ceramic material could be elaborated based on some certain data only. In the cultivated layer of the settlement there was mixed pottery from both horizons, which made difficult the later one to be more precise characterized. Despite that fact the ceramic complexes from some pits document those changes, as well as the relatively big quantity of some new elements in the pottery registered almost in the cultivated layer and the mentioned pits only. At any rate, it looks that both chronological stages followed one after another and the analysis of the pottery to some extend gives a possibility of its distinguishing.

Structure of the material

In technological aspect the ceramics from the burnt horizon (the first stage) show no big variety. Characteristic is the presence of more or less mineral mixture in the clay – sand, rarely stone; organic mixture as exception. The fine paste without or with a lot of mixture and the vessels with rather rough structure are rare. There is a big contrast neither with regard to quantity of the clay paste and the kind of the mixture nor to the burning and the surface treatment. Accordingly, there is no difference between the distinct technological groups.

The predominated part of the vessels has a well-smoothed surface, rarely burnished. The surface covered with a slip is an exception. The brown color strongly predominates, with different variants – from red and brown to gray and brown, obviously depending on the intensity of the burning. The vessels have often brown and porous outer surface and black, better smoothed inner surface (for better pervious defending of the walls).

The clay paste of the thick-walled vessels (more than 1 cm) is mixed with rather big quantity of mineral inclusions (sand and small stone). The light brown outer surface is often in combination with black inner surface and inconstant burnt biscuit.

Mainly the vessels with a thickness of the walls 0.5-1 cm define the character of the ceramic complex (c. 85 %). That ceramic category includes almost all sorts of ceramic shapes. They are made of paste with rather small mineral mixture. The surface is well smoothed without burnishing. In some cases there is an additional slip of fine clay paste, mainly by the painted with graphite pottery and especially by the decorated with channels pottery. In this group the brown and the brown and black ceramics also predominates. The ratio is 4: 1 in compare to the pottery with gray and black surface. It is valid for the painted with graphite ceramics. The surface is more constantly burnt in compare to the thick-walled ceramics, but there are also vessels with variable biscuit.

The thin-walled ceramics (with up to 0.5 cm thick walls) in the first chronological stage of the occupation of the settlement is only c. 5 % of the whole complex. But only c. 2 % of the thin-walled vessels is made of fine paste. The rest includes sand mixture at different rate. In this category the black slip pottery is more often represented.

A specific technological group differs from the whole ceramic complex, which is well differentiated stratigraphically. This is the category, which is missing in the closed complexes of the two dwellings. It includes thin-walled vessels of clay paste mixed with quite a lot sand and small stone, with crumbly structure of worse quality. As a rule those vessels are with dark – black and gray – surface and dark slip, which easily comes off and often is fallen down. The inner structure and the biscuit are lighter – brown or gray and porous. That technological group is connected with specific shapes and ornaments, which will be discussed below.

To conclude, the traditional in the Neolithic and Copper Age in the area clay mixed with sand and rock, characterize the pottery of the ceramics from the Final Copper Age at Kolarovo. Vessels from clay mixed with 'shamot' (sandy crashed ceramics) are rare, and the clay paste with mollusk or snail shell crushed was not registered.

Morphology and typology

The ceramic shapes analyzed are divided into sorts according to their formal-functional characteristics – plates, bowls and pots, in correlation with the common criteria for their proportion as well. By some shapes are distinguished types and variants according to the shaping of the mouth and the body. In the cases when the shape of the fragments don't allow the proportion to have been defined, I use the term vessel, especially by some bowls and pots.

Plates (shallow bowls)

They are the leading shape in the ceramic complex of Kolarovo and have the typical character of the Late Copper Age. They are distributed among all technological categories, but predominate in the category of the technomic earthenware with medium thick walls and relatively good clay paste. The color is most often brown and red and brown, in rare cases they belong to the category of the dark friable pottery. The surface is well smoothed. In the substructure of the dwellings, which can be with some reserve attributed to their earlier occupation phase, there is earlier variant of gray and black plates with well smoothed surface.

Type 1. Conical plates with a ledge. The orifice has straight walls or is inverted. The height is 6-9 cm, and the diameter orifice - 19-32 cm. It is made of clay paste with a small quantity of sand, with comparatively good quality.

Variant A. Plates with very small cylindrical ledge, 2.5-3.5 cm high, with stressed turning point of the body (***Figure 12.1: 3, 4***). The rim is straight. It is rarely distributed variant, mainly in the earlier stage.

Variant B. Plates with rather low cylindrical ledge (1-2 cm) and rounded turning point (***Figure 12.1: 1, 2, 7, 8, 12, 16***). The cylindrical part is vertical or slightly inclined, thickened and rounded from the inner side. The rim is thinned and rounded.

Variant C. Plates with more or less inclined ledge (***Figure 12.1: 5, 6, 9, 10***). It differs very slightly from the previous variant (only by the inclination of the orifice). The turning point and the ledge are thickened in comparison to the lower part of the body. The biggest diameter is at the turning point. Some of them represent a remote variant, by which the orifice is only thickened and slightly inclined. The orifice rim is rounded and worn thin.

The plates of type 1 in some cases consist of attached to the lower body part conical stringholes lugs or knobs, as well as small horizontal handles (***Figure 12.1: 6, 16***). There are rare instances of vessels with graphite ornamentation.

Type 1 characterizes the first occupation stage of the settlement. A considerable part of them are found in the area of the house 1 and 2, in closed ceramic complexes (***Figure 12.1: 1, 3, 4, 6, 7, 8, 13, 14, 16***). There are probably later instances (***Figures 12.1: 9, 10***) discovered in pit 2.

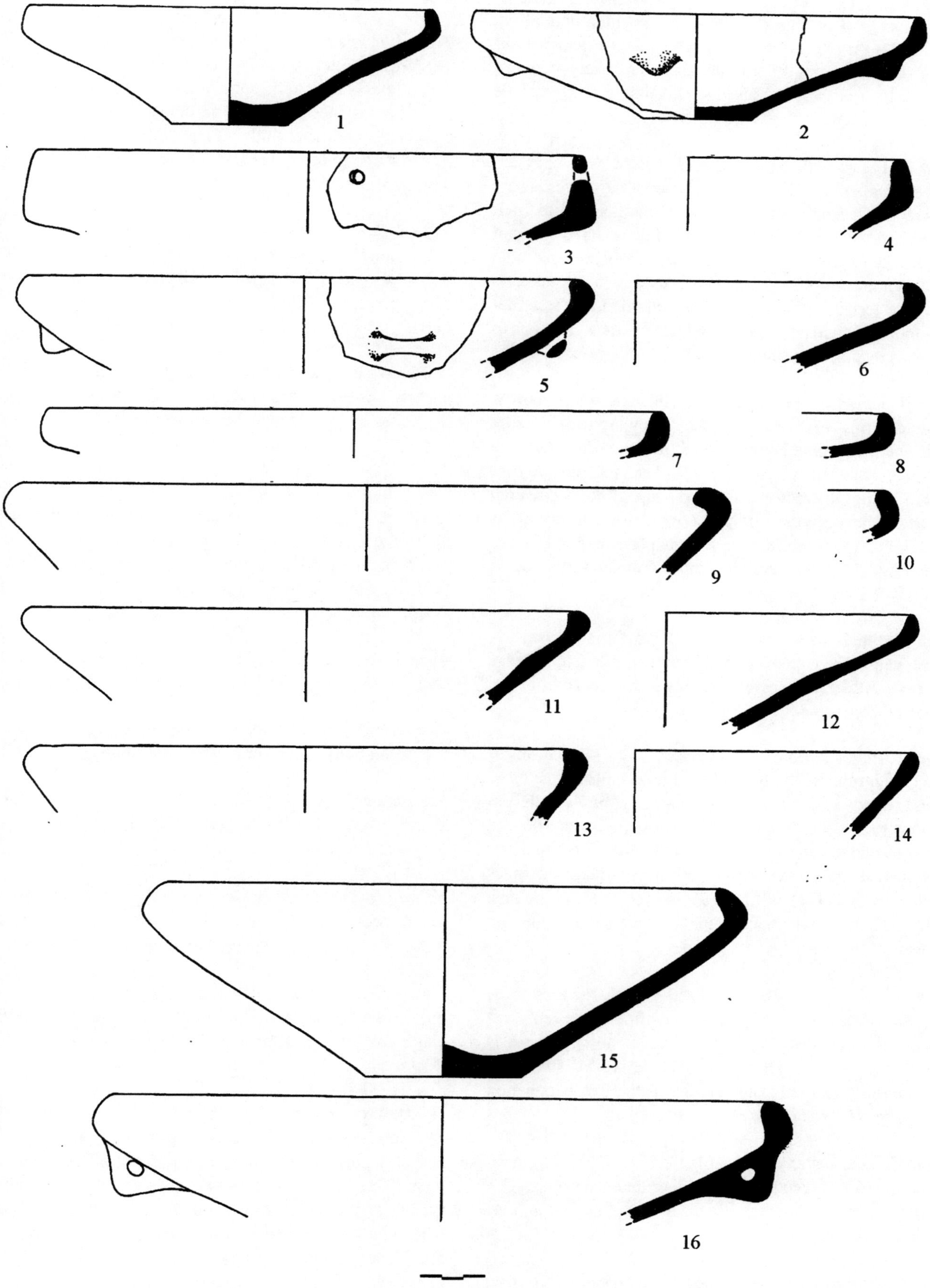

Figure 12.1. Conical plates. Type 1.

Type 2. Plates with twisted and thickened from the inner wall rim. They are usually wide open and shallow, but deeper shapes are also distributed. The body is conical, straight and rounded. That is most distributed shape and is documented in the first, as well as the second occupation stage of the settlement. According to the shaping of the rim, there are many variants, some of which are not sharply distinguishable. They can be clustered as follows:

Variant A. Plates with thickened and twisted rim. They are usually wide open, with 29-32 cm diameter of the orifice. There are horizontal stringholes lugs most often attached to their body (***Figure 12.2: 1-7, 14-16***). Some of plates have traces of graphite ornamentation (***Figure 12.2: 6***). They have medium grained structure of clay paste with brown or black surface. They characterize the first stage of occupation of the settlement – more of them originate from houses' areas. Perhaps to its second stage belong the instances in ***Figure 12.2: 14-16.***

Variant B. Similar to previous plates, but with more narrow and rounded rim (***Figure 12.2: 8-12***). They are usually more closed. There are sometimes traces of graphite ornamentation or they are ornamented with channels over the rim (***Figure 12.2: 8, 9***). They characterize the first stage of the occupation of the settlement, but they are probably distributed during its second stage as well. Exceptions are the plates exemplified in ***Figure 12.2: 13***, which are with more straight walls and in ***Figure 12.2: 17*** with horizontally cut rim.

Variant C. Plates with wide up to 3.5 cm flattened rim unstressed, which are with open in different way orifice (***Figure 12.3: 1-4***). All belong to the first stage for apart from the instance in ***Figure 12.3: 4***.

Variant D. Plates in comparably small size (13-25 cm in diameter), with rounded body, slightly thickened from the inner wall and rounded rim. They are comparably thin-walled, but with poor technological quality of producing. They belong definitely to later complexes at the settlement (***Figure 12.3: 5-9***). Some of them have tongue-like bulges from the outer side of the orifice (***Figure 12.3: 10***). Exception is the instance in ***Figure 12.3: 12***, which has conical stringhole lug from the inner wall of the orifice.

Variant E. Deep thick-walled plates with slightly rounded body, weakly profiled from the inner wall orifice and evenly cut rim. Their diameter is 22-26 cm. Despite the clay paste is with rather grained structure, the surface is well smoothed and the burning is good. They belong to the first occupation stage (***Figure 12.3: 13-15***).

Type 3. Conical plates with profiled orifice. Their stratigraphy is nor clear – it seems they were distributed in both stages. They have a relatively small diameter of the orifice, mostly 12-15 cm.

Variant A. The rim is with thickened outer wall, and worn thin inner wall (***Figure 12.4: 1***). That rare distributed variant was found only in the cultivated layer.

Variant B. Plates with profiled orifice and shoulders. Profiled upper part, with firmly open stressed from outside body with horizontal edge and slightly rounded inner wall (***Figure 12.4: 2-5***). The rim is rounded and worn thin.

Type 4. Conical straight walled bowls (***Figure 12.4: 7, 10-14***). The diameter of the orifice is 12-36 cm, the rim is rounded or straight. The body is sometimes slightly rounded. Almost all of the instances are found in the cultivated layer or in the complexes with later pottery. The plates in ***Figure 12.4: 11, 12*** are made in the technology of the later stage (friable clay with porous surface).

Type 5. Deep conical plates with proportions, which are close to those of bowls (***Figure 12.4: 6, 8, 9***). The orifice is straight or funnel-open (***Figure 12.4: 6***). The rim is rounded or straight. They are rare distributed, mainly in the cultivated layer, but outside the context of the later pottery.

If we have in mind the bases found, the plates have had flat or slightly concave bases (***Figure 12.11: 14-23***). It is possible the plates of type 5 to have been with pedestal base.

Bowls

Deep vessels represent that category, which height is less and approximately 2/3 of the maximum diameter. The last can be represented by different parts of the body. According their profiling the following 2 types can be distinguished, with different variants.

Type 1. Bowls with cylindrical upper part and conical lower part. They are very well distributed in the ceramic complex.

According to the proportion of the upper part and the profiling of the body, three variants are distributed.

Variant A. Bowls with high cylindrical part, straight and slightly widened orifice (***Figure 12.5: 1, 4, 7***). Almost all of them are found in the area of house 2.

Variant B. The upper shorter part is slightly inclined in (***Figure 12.5: 3, 6, 8, 11***). They are distributed in the first stage in the area of the houses, as well as in the cultivated layer.

Variant C. S-profiled bowls with short cylindrical neck, ornamented with channels. They have very fine structure of the clay paste and are distributed in the earlier occupation stage (***Figure 12.5: 2***).

The bowls of that type have as a rule thickened turning point and well profiled medium part (***Figure 12.5: 4, 7***). The diameter of the orifice of the smaller bowls is 13-18 cm, and of the bigger ones – 24-31 cm. Almost all of them have high technology of producing and in some cases well-burnished surface. The ornamentation of different channels is most often distributed, as well as pseudo-channels and incised ornament over the profiled part of the body (***Figure 12.12: 10, 12, 19***). In rare cases the channeling is in combination with graphite ornament. The bowls of variant A are not ornamented.

Type 2. Bowls with rounded body, non-profiled. The diam-

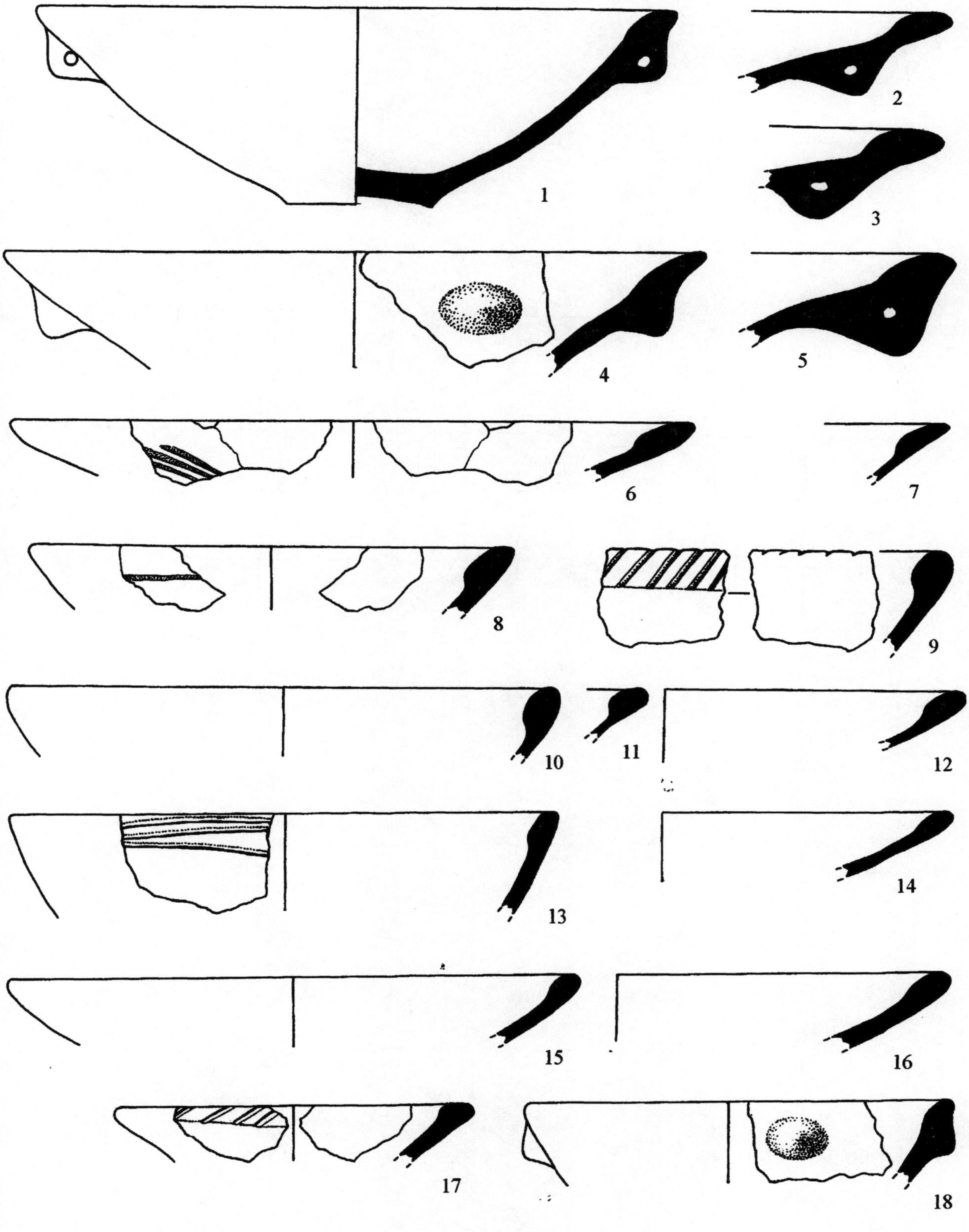

Figure 12.2. Conical plates. Type 2.

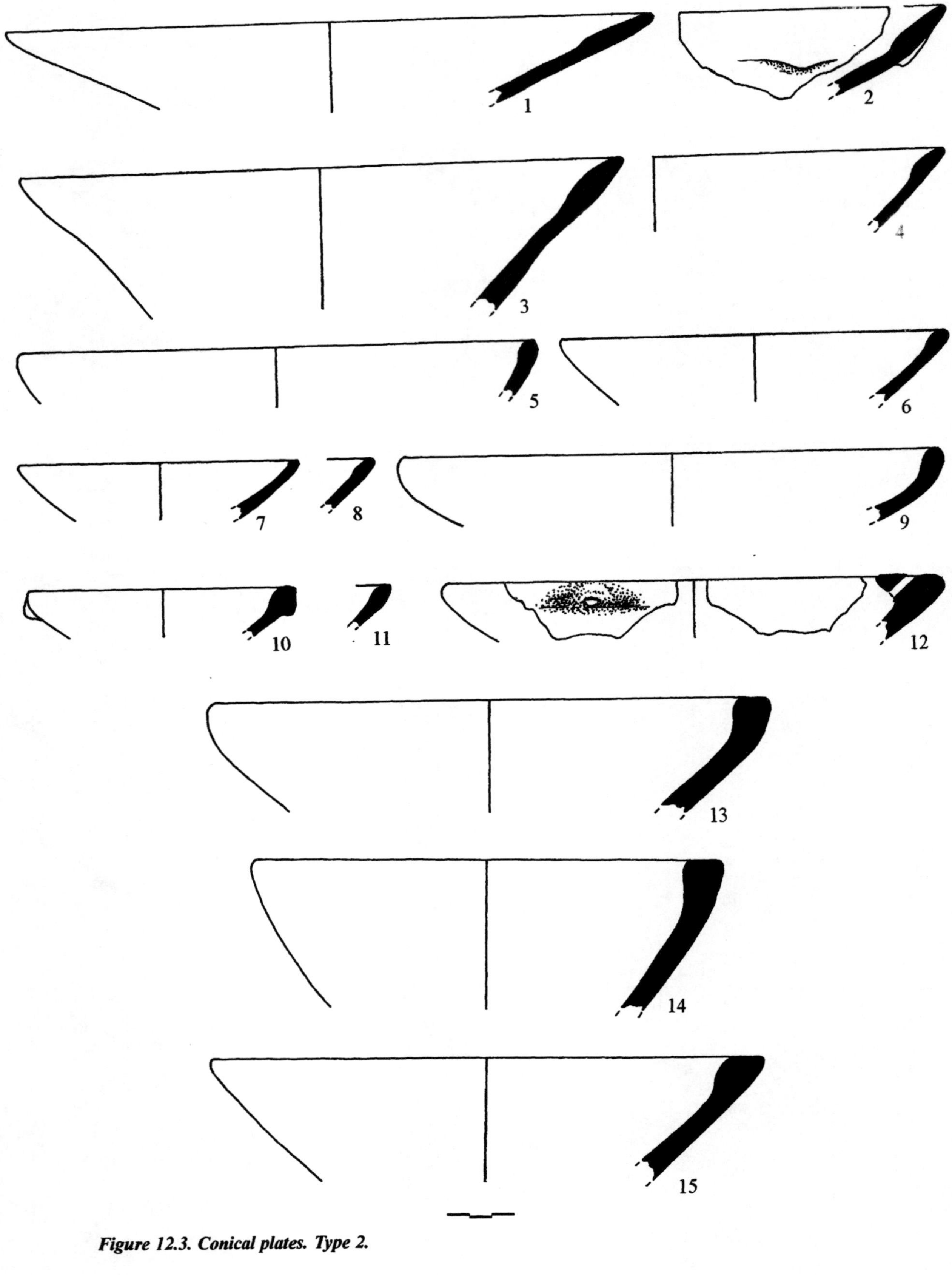

Figure 12.3. Conical plates. Type 2.

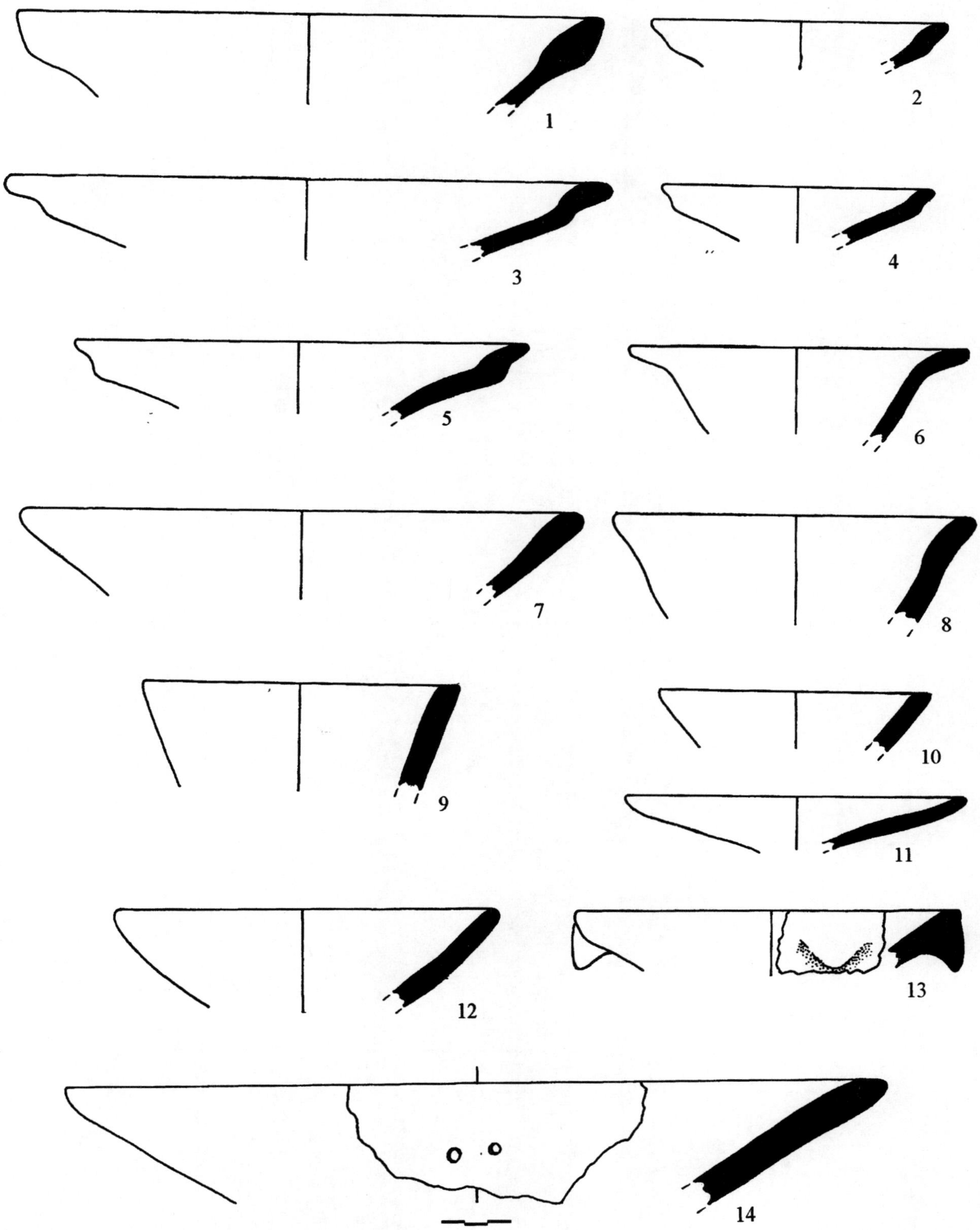

Figure 12.4. Conical plates. Types 3, 4 and 5.

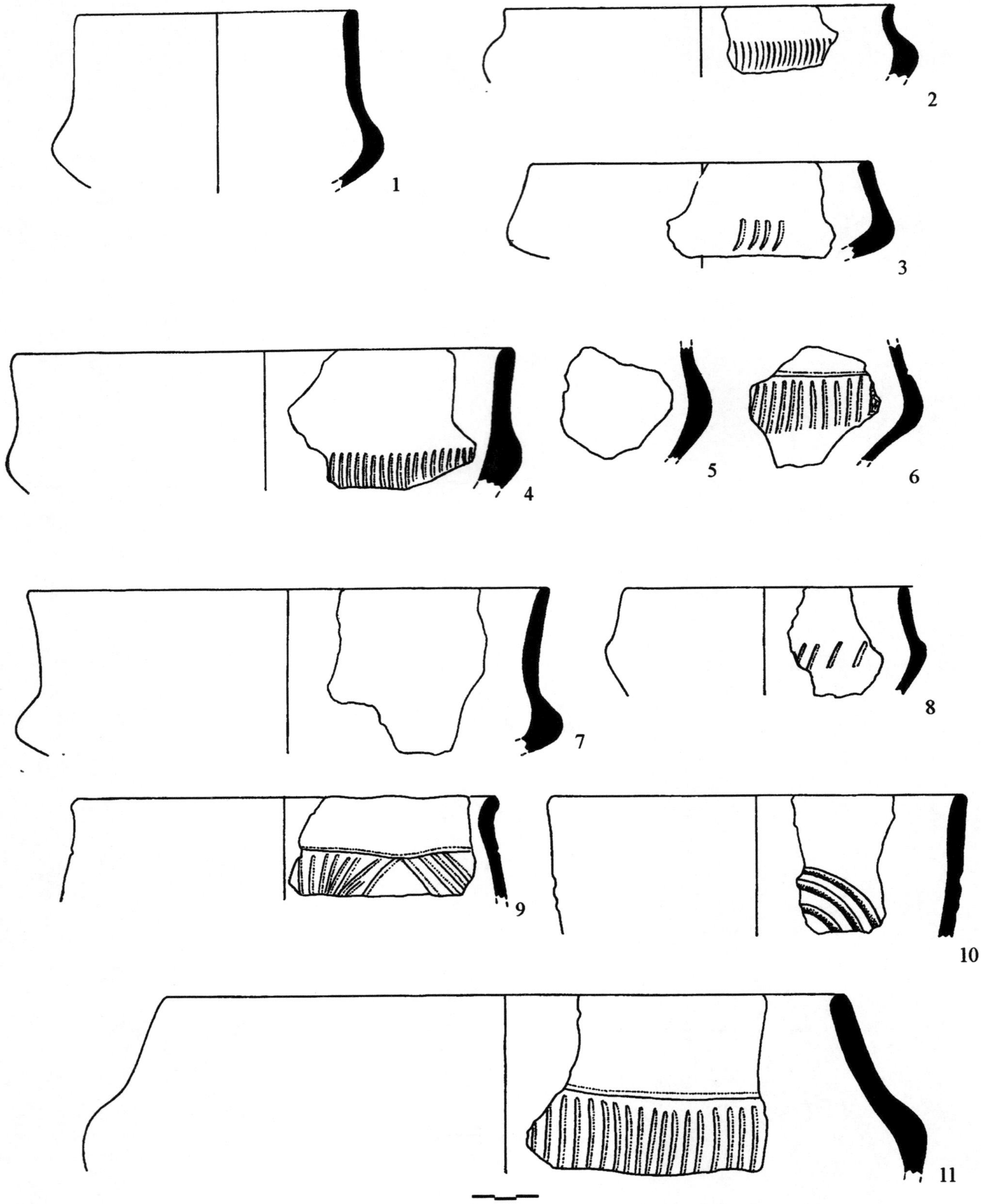

Figure 12.5. Cylindrical and conical bowls.

eter of the orifice is 20-30 cm. According to the degree of rounding of the body, there are 5 variants.

Variant A. Semispherical, with straight or slightly widened orifice (***Figure 12.6: 1-4, 8***). They are distributed in the category of thick-walled and technomic earthenware. The rim is rounded, slightly worn thin, and diameter of 27-30 cm. In some cases there are tongue-like knobs (***Figure 12.5: 1,8***).

Variant B. Semispherical bowls, with arched-profiled and slightly inclined in orifice. The last is sometimes thickened (***Figure 12.6: 6, 7, 9, 10***).

Variant C. Bowls with almost spherical body, closed orifice, with rounded or straight rim (***Figures 12.6: 11-14; 12.7: 1, 4, 7***). The orifice of some bowls is limited from outside with a horizontal groove. The orifice rim is rounded, in some cases with profiled outer and inner walls.

The semispherical and spherical bowls are not ornamented. In some cases they have an incised ornament and are with plastic bulge attached. Some of the bowls have small vertical handles (***Figures 12.6: 1, 8, 14, 15; 12.7: 1; 12.13: 1***).

Variant D. Bowls with more or less differentiated cylindrical neck (***Figures 12.7: 8, 10, 11, 14; 12.8: 1***). The neck is inclined in, straight or slightly out-turned. The orifice edge is rounded or slightly out-turned. Perhaps the body has had spherical or rounded biconical shape. The diameter of the orifice is 20-27 cm. They are distributed in the categories of the technomic and thick-walled pottery. They are rarely ornamented with pits-like elements under the rim or with pseudo-channels (***Figure 12.12: 16***). In some cases it is not possible by the fragments of the orifices preserved it to be distinguished if they belong to bowls or to deeper shapes of pot type (***Figure 12.7: 9, 12***).

Variant E. S-profiled bowls (***Figures 12.9: 1-18; 12.15: 1-4***). They have not only common shape but identical technology. Despite they are extremely thin-walled they are made of clay paste with rather big amount of sand, rarely mixed with 'shamot' (burnt clay fraction). Their surface is black or gray, rather well smoothed, burnished and covered with slip. In some cases the surface slip has fallen down because of the bad joining with the rough friable surface below. Those vessels are in comparatively small size (12-25 cm diameter of orifice). The profile of the body and the neck is rather various. They have slightly demonstrated or higher and stressed neck, in some cases funnel-open (***Figure 12.9: 10, 11***). The body is rounded or rounded to biconical. The orifice is shaped in various way – straight or slightly out-turned and inturned. The rim is rounded and worn thin, in some cases with profiled inner wall. They have no handles and most often they are not ornamented. They have rarely plastic ornament, pricked ornament or shallow incisions (***Figures 12.9: 2-4; 12.14: 4***). In one case there was documented a graphite under the rim band (***Figure 12.9: 1***).

A whole series of that type originates from Kolarovo. Their number, together with the ornamented, is more than 25. A predominated part was found in the upper part of pit 2, in some other pits, as well as in the cultivated layer. Their lacking in the closed complexes of the burnt houses indicate they are a certain shape of the second stage.

Pots

The data on that shape are not numerous, despite their number is big. Fragments of their necks are mainly documented and only one almost complete instance (***Figure 12.8: 16***). They have prolong-spherical or rounded biconical body, with a height rather bigger in compare to their maximum diameter and with small base (***Figures 12.8: 16, 17; 12.10: 13-17***). Perhaps they have been often provided with plastic handles-bulges or with small vertical handles. The fragments found of neck parts allow a selection according to their height and the orifice kind.

Type 1. Pots with high, almost cylindrical straight neck, which has a small diameter (8-13 cm). The body is rounded biconical, the rim is rounded (***Figure 12.8: 2, 3, 16, 17***).

Type 2. Pots with short cylindrical neck slightly stressed and little out-turned (***Figure 12.8: 4-7, 9, 10***). The rim is differently profiled – slightly thickened from the inside and worn thin; slightly out-turned, evenly cut; in both cases with a horizontal bulge, probably symmetrically placed from both sides (***Figure 12.8: 9, 10***). Probably the bodies of that type were prolong and symmetrical.

Type 3. Pots with strongly open funnel-shaped neck (***Figure 12.8: 8, 11-15***). It is possible a part of the fragments found to have been belonged to vessels with proportions of bowls. They probably have spherical body. The diameter of the orifice varies from 6.5 to 13 cm. They belong to the category of the technomic and thin-walled ceramics and have well smoothed to burnished gray or gray and black surface. One sherd has traces of white distempered paint over the black burnished surface (***Figure 12.8: 14***).

The pots from the first type are documented in the context of the first occupation stage of Kolarovo. The pots of the second type originate from the first, as well as from the second occupation stage. Of the type 3 pots with funnel-shaped necks there is one (***Figure 12.8: 8***) found in a house. The rest originate from pits and the cultivated layer.

Two-handled vessels

They are provided with vertical handles, starting from the orifice. According their proportion, they sherds may belong to pots or bowls. There is one two-handled small vessel originating from the area of house No. 1 (***Figure 12.10: 1***). Some sherds of orifices and handles (***Figure 12.10: 3-5, 7, 8***) are a proof that that shape was distributed in two variants – with arch-shaped handles and with oblique attached handles. The orifice parts are almost vertical, the rims are rounded, and the handles have circle or ellipsoidal section. In one case the vertical handle is ornamented with graphite motif (***Figure 12.12: 7***), which seems to characterize the first occupation stage of the settlement.

Some other sherds of handles should be mentioned despite

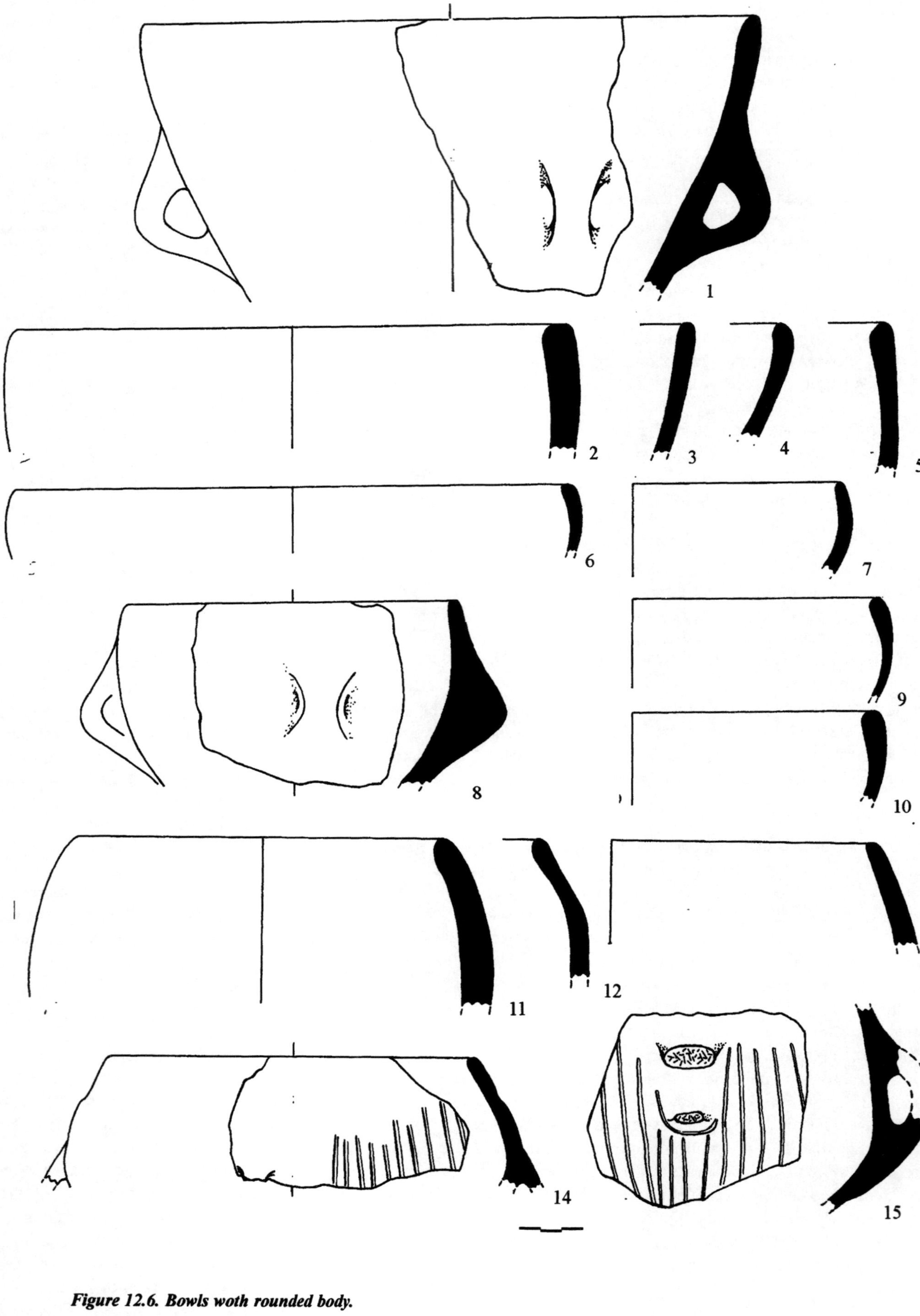

Figure 12.6. Bowls woth rounded body.

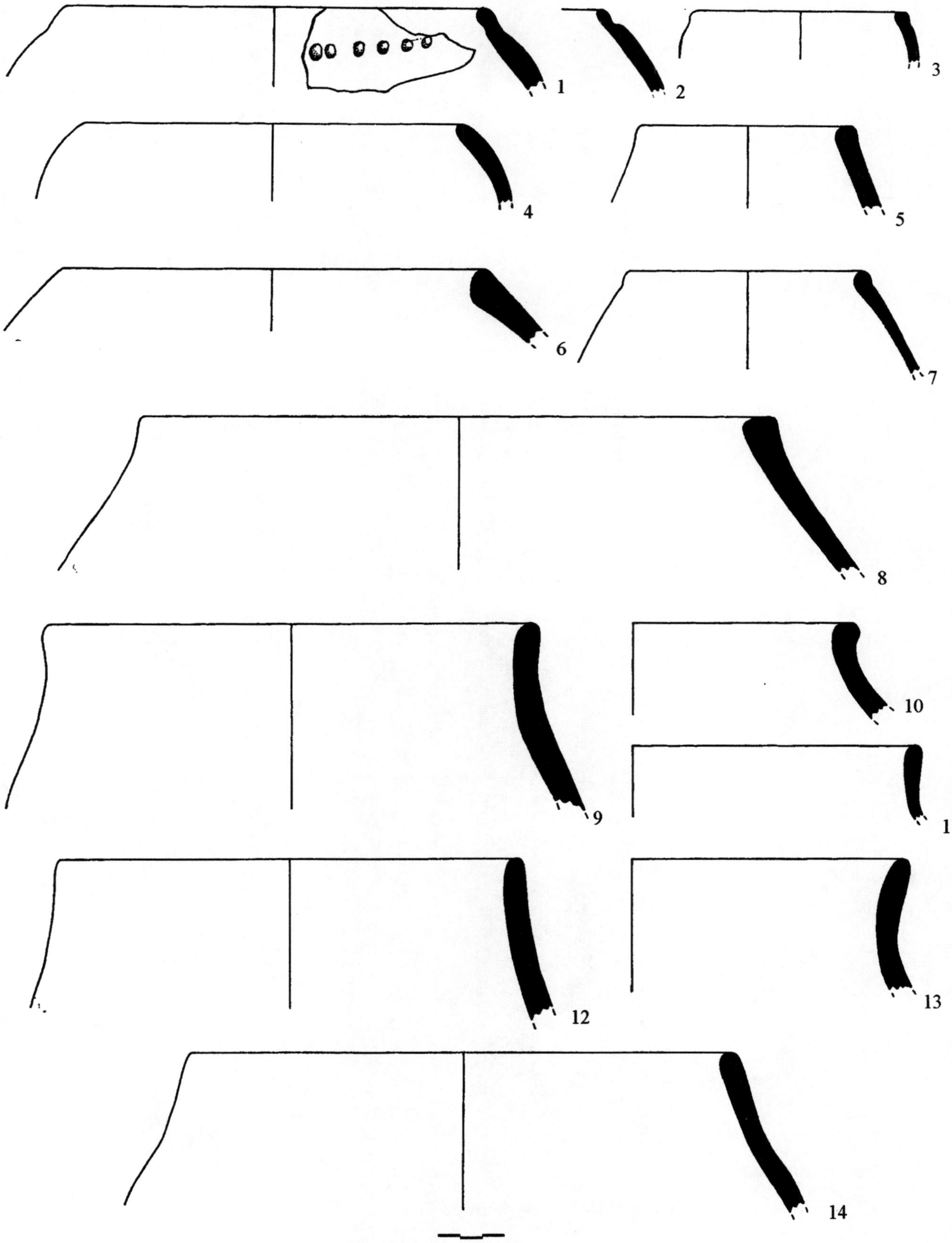

Figure 12.7. Bowls with spherical body and pots.

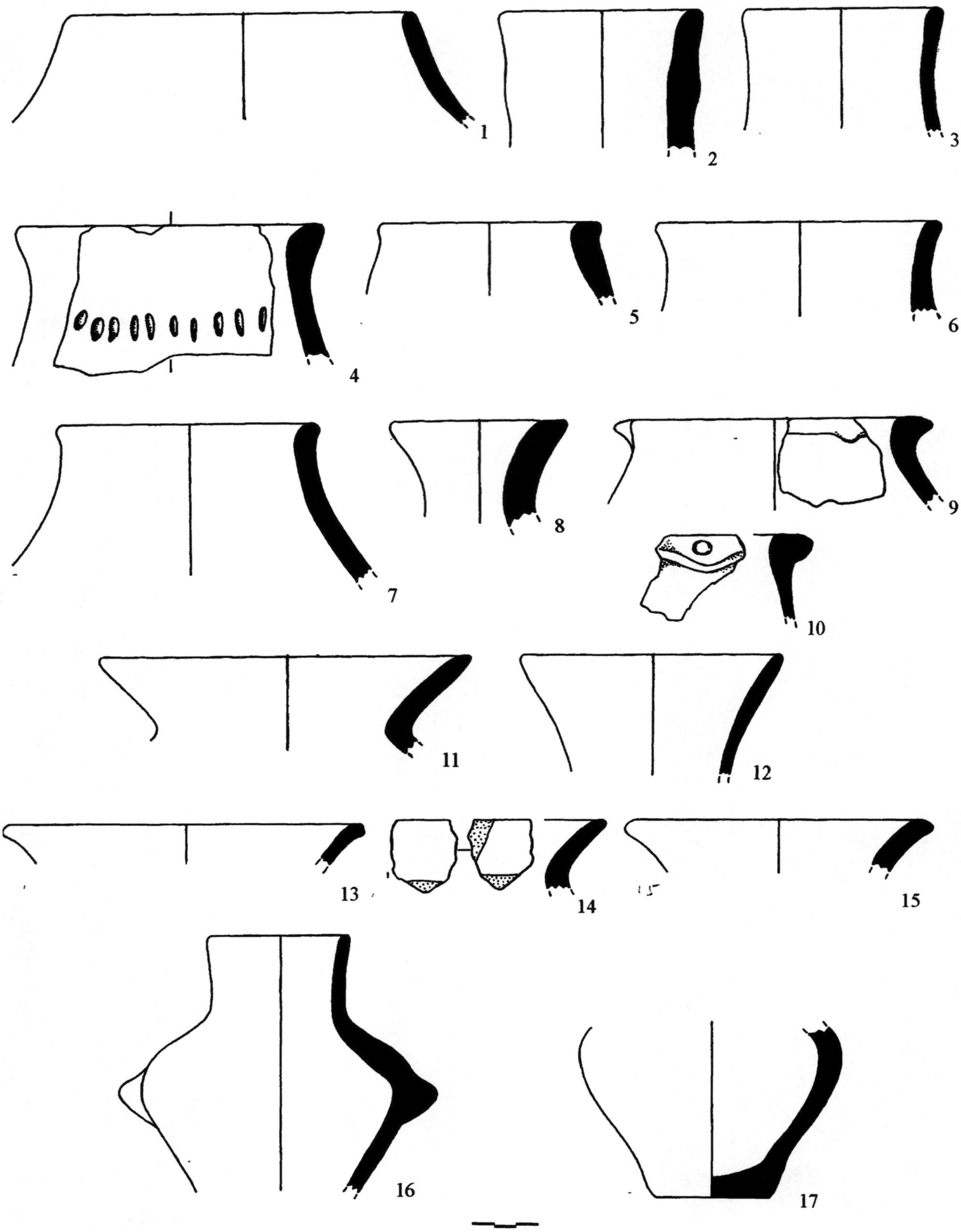

Figure 12.8. Pot. No. 14 painted with white distempered paint.

they do not belong to certain shape. They are single finds – from 3 to 6 sherds in whole ceramic complex:

- Fragments of wide vertical handles with rough structure of the clay paste (***Figure 12.11: 5, 6, 10***). They have flat or quadrangle section and originate from the cultivated layer;
- Small vertical handles with horn-like bulges (***Figure 12.10: 9-11***). They have no stratigraphic data;
- Vertical handles with inclined prolong profile (***Figure 12.11: 4***) found in the cultivated layer.
- Vertical handles with ellipsoidal section, with one or two prolong incised lines (***Figure 12.11: 1-3***). Two of them belong to the complex of the ceramics from stage 2 in pit 2.

Principally, the vertical handles attached to the body, as well as the tongue-like knobs, are slightly distributed in the ceramic complex (***Figures 12.10: 2; 12.11: 9, 11, 13***). Only in one case the tongue-like knob is decorated with pits (***Figure 12.15: 5***). The bottoms of the two-handled vessels are flat and unstressed, sometimes slightly concave, very rarely slightly profiled (***Figure 12.11: 15-23***). Single fragments of not very high pedestal bases have been found as well (***Figure 12.11: 14***).

Lids and stands and other ceramic finds

From the area of house 1 some whole or fragmented lids originate, one of which has traces of graphite ornament (***Figure 12.15: 16-18***). They are of two types.

Type 1. Semispherical or cylindrical and spherical, provided with vertical stringholes for hanging (***Figure 12.15: 16, 17***). ***Type 2***. Conical, with well-shaped handle in the upper part (***Figure 12.15: 18***).

To the ceramic complex of the burnt houses cylindrical stands for vessels should be added (two whole finds, unornamented) – ***Figure 12.19: 1, 2***. Fragments of ceramic strainers and miniature vessels were found in the area of the houses, as well as in some pits (***Figures 12.15: 19; 12.16: 1, 3***).

Ornament

The relative portion of the ornamented earthenware is not big, but with regards to the technique of the ornamentation used, there is a big variety.

Graphite ornament. Because of the wasted surface of the vessels there are often only traces of graphite, without a possibility the motifs to be reconstructed. The relative part towards the rest of the ornament is considerable. It was mostly painted over red and brown surface without slip or over specially prepared slip dark surface. The graphite ornamentation is always positive, as narrow strips. They shape simple motifs: vertical, parallel, filling wide fields, or all the surface; horizontal or oblique lines, as well as in combination with them, painted over the widest part of the vessel. The motifs painted with graphite are similar, even identical with the motifs of the ornaments with channels and pseudo-channels (***Figure 12.12: 1-11***). The plates of types 1 and 2 were mainly decorated with graphite ornament (sometimes over outside and inside walls), some lids, profiled bowls, thin-walled slip vessels and two-handled vessels (***Figures 12.2: 6, 8; 12.12: 1-12***). The graphite painting was rarely in combination with plastic ornament and channels (***Figure 12.12: 10, 12***). In the later occupation stage of the settlement there is a plastering with graphite of the surface over the walls of bowls with S-profile, or with one horizontal line at the neck base (***Figure 12.9: 1***).

White and red distempered painted ornament. It was laid on after vessel burning. Many sherds have on their inner or outer wall surface traces of white and tile-red paint, sometimes in combination of both. Because of its technology the paint is mostly fallen down and absolutely without possibilities to be reconstructed. It seems it was less used in compare to the graphite ornament. It was used in both stages at Kolarovo. According to the information of the sherds preserved, the motifs are lied on as horizontal strips or in painted geometrical fields (negatively ?) – ***Figures 12.8: 14; 12.13: 10***.

Channels. The number of the sherds in its classical shape of appearance is small. It is very often lied on in the technique of the pseudo-channeling sharing the place and the rhythm of the real channel. It mostly ornament the widest profiled part f the cylindrical and conical bowls as vertical pseudo-channeling – ***Figure 12.5: 2, 4, 6, 10, 11***. The typical channeling is distributed over burnished bowls with stressed turning point of the body (***Figure 12.12: 12. 15, 19, 22***) or over the profiled rim of bowls of type 2 (***Figure 12.2: 9, 13, 17***). As narrow and wide grooves (horizontal, vertical and in two cases concentric) lied on light brown vessels with burnished surface. The channeling is distributed in the complexes of the pottery from the second occupation stage of the settlement, or in the cultivated layer and obviously represent later variant of that ornament (***Figures 12.5: 5; 12.12: 13, 14, 17-18, 20, 21; 12.13: 1***).

Incised ornament. It is very distributed used with different techniques.

A. Incised in wide regular lines, which in many cases imitate exactly the motifs of the pseudo-channels over the profiling part of the cylindrical and conical bowls (***Figure 12.5: 8, 9***).

B. Incised vertical, horizontal and oblique lines, incised over wide fields of the vessels, mostly in combination with circle and drop-like pits (***Figure 12.13: 5-7, 19-22***), or with cylindrical bulges (***Figures 12.12: 14***).

C. Fine incised lines or shallow drags shaping sheaves, hatched triangles, single horizontal incisions or fields filled with oblique lines (***Figure 12.14: 4, 5, 8, 9, 12-15, 17-20***). That ornament is occasionally an imitation of the technique of dragging with hard brush and broom, but obviously a sharp instrument was used (***Figure 12.14: 7, 11, 16***). In rare cases the frames of the dragged fields are outlined with horizontal incised lines (***Figure 12.14: 14, 15, 17, 19***).

D. Short incised lines, which are always lied on regularly in

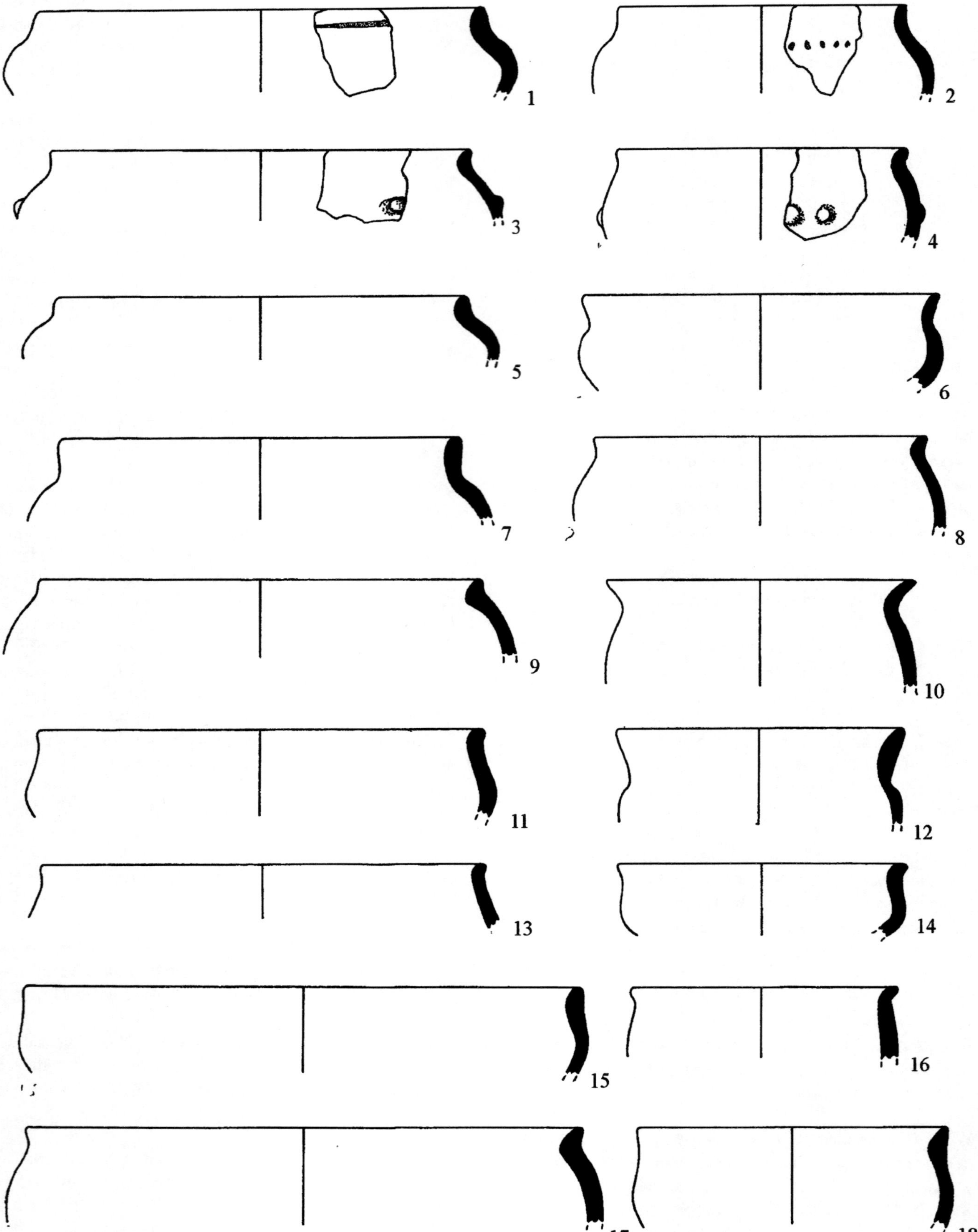

Figure 12.9. Bowls with S-profile.

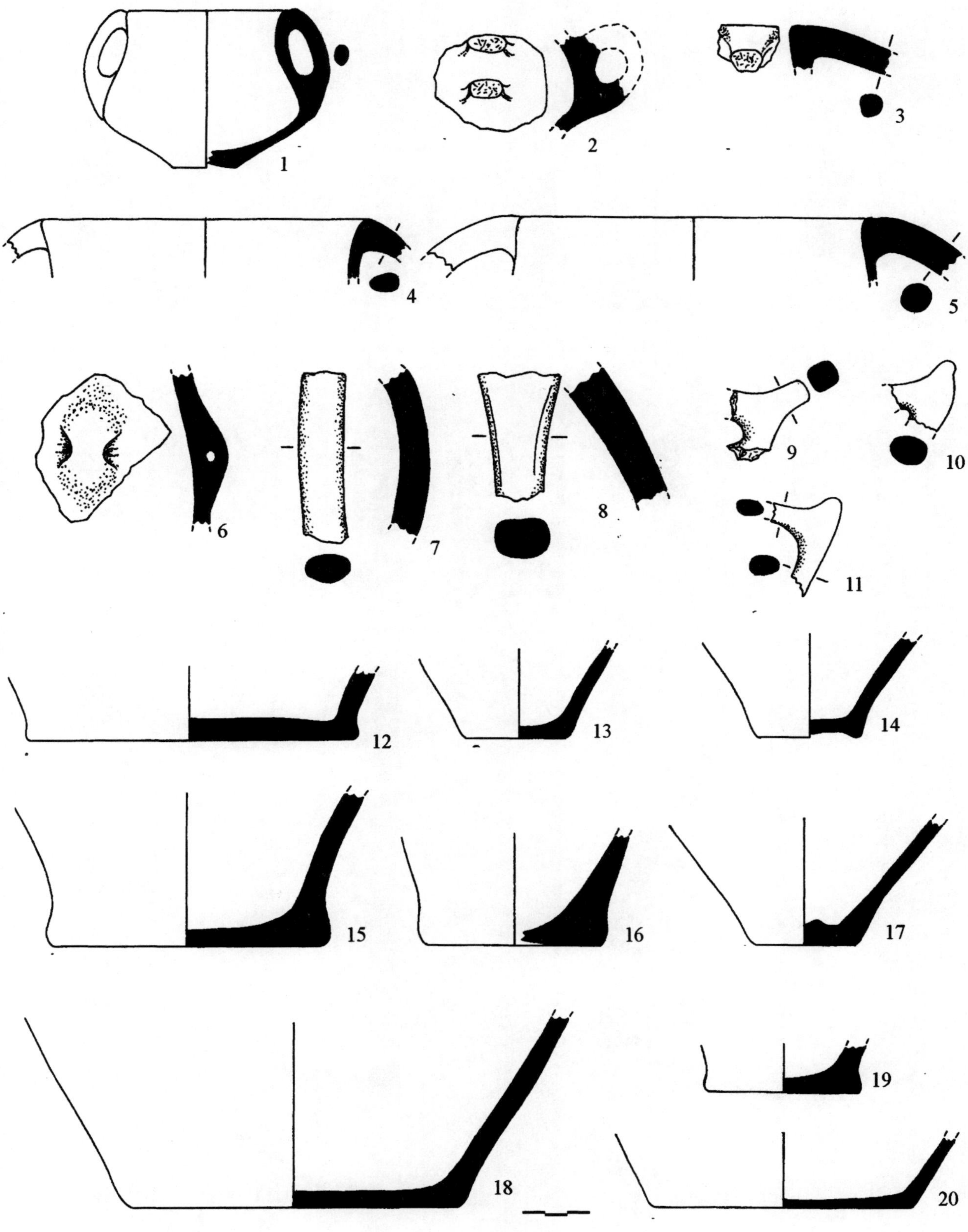

Figure 12.10. Amphorae-like vessels, fragments of vertical handles and bases of bowls and pots.

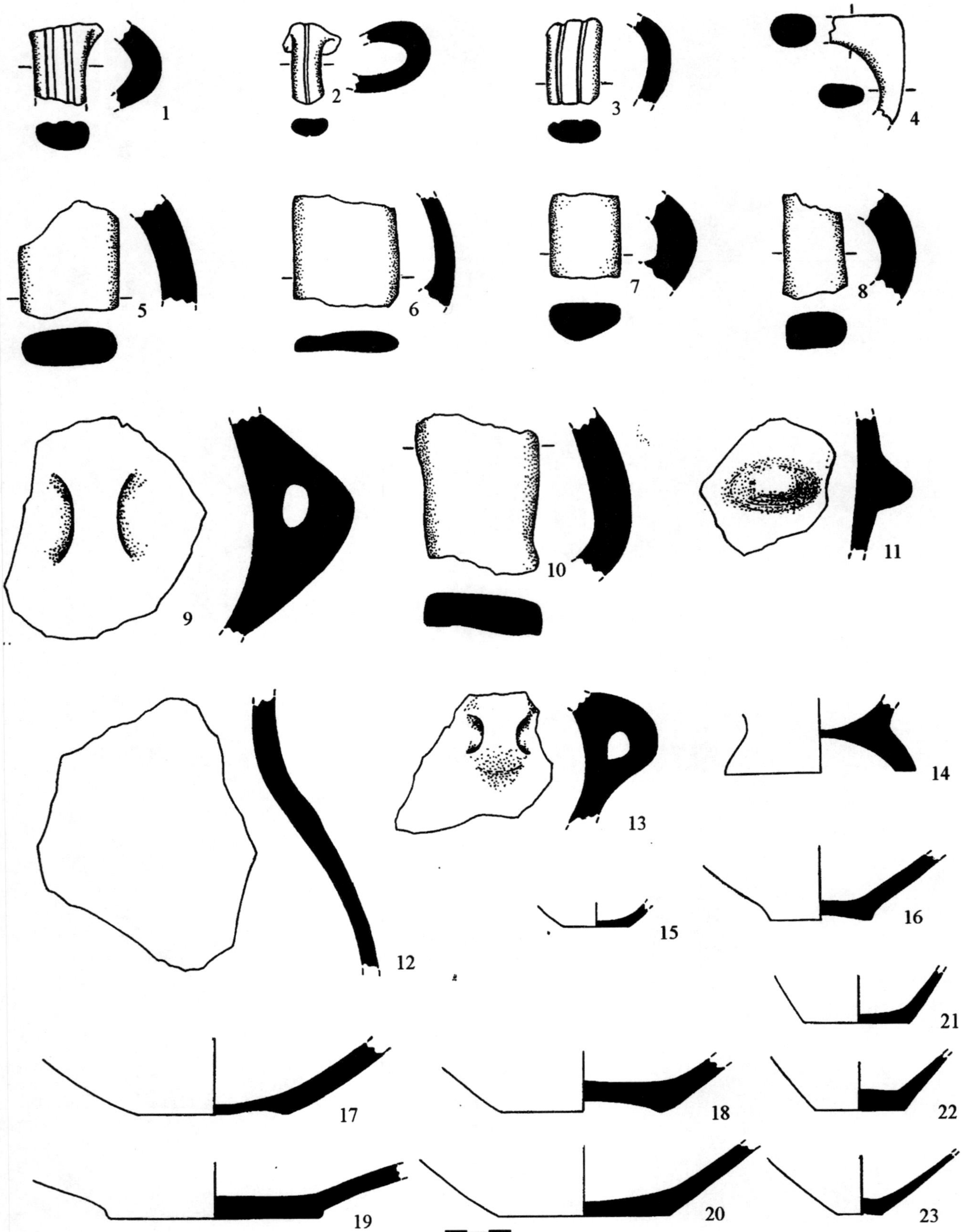

Figure 12.11. Fragments of vertical handles; bases of plates.

horizontal bands – over the body (***Figure 12.14: 10***), or at the base of the orifice part (***Figure 12.14: 4***).

The incised ornament of types a and b was distributed certainly at Kolarovo I, but it cannot be excluded its using in the later stage. The incised lines of type c and type d are typical of the late ceramic complexes and characterize the vessels with bad technology of the pottery.

Pricked ornament. It is distributed as circle and drop-like pits over the body or around handles of the vessels, often in combination with incised lines (***Figure 12.13: 3-9, 19***), or over the cylindrical relief bulges (***Figure 12.13: 14-16***). Perhaps it was pricked with holed stick that leaves circle or ellipsoidal stamp. Another variant is the ornament pricked with thin point at the neck base (***Figures 12.9: 2; 12.14: 6***). The last characterizes the later occupation stage.

Relief ornament. Short, cylindrical attached bulges, occasionally ornamented with circle pits, are often distributed in the ceramic complex, probably in the two occupation stages of Kolarovo (***Figures 12.9: 3, 4; 12.13: 11-16; 12.14: 9, 12***). It is interesting to note the note the relief band with pits attached horizontally at the body of the vessels is almost missing in the ceramic complex of both stages.

The instances of a combination of two or more ornamental techniques over one and the same vessel are not a lot. The incised ornament is sometimes combined with pricked one or with pits. Occasionally occurs a combination of graphite painting with channeling, as well as plastic bulges with pricks and pits (***Figures 12.12: 10-12; 12.13: 5-7,12-16, 19-22***).

There is missing a barbotine made with nail-like or other instrument, as well as the organized finger-like barbotine. The plastic pinched band is an exception (***Figure 12.15: 11***). In two cases pits are hollowed over tongue-like handles of vessels (***Figure 12.15: 5, 14***). Pits and incisions were documented over rims of a certain type of open deep bowls (***Figures 12.15: 5, 10; 12.14: 3***). Only one sherd of deep plate has a wave-like rim (***Figure 12.15: 6***).

Some fragments look foreign in the ceramic complex, according their technology and ornament. Two of them are of thin-walled vessels made of well purified clay paste with light gray surface and are with white encrusted ornament with net motifs in combination with cuts (***Figure 12.14: 1, 2***). The fragments of thick-walled vessels in ***Figure 12.15: 12, 13, 15*** differ with their high technology of burning up to cement hard biscuit and very smoothed surface. Their decoration with plastic bands with hollows us also not typical of the ceramic complex. Their presence may indicate Early Bronze appearance.

Implements

In the production inventory smoothed or polished stone implements are almost missing. There are only two fragments of adz-like polished implements with asymmetrically cut working end (***Figure 12.16: 7, 12***). Two more adz-like implements of *andécite* are coarsely made, with uneven, rough surface. By one of the finds almost is preserved the initial shape of splitting off and only the cutting end is additionally processed (***Figure 12.16: 10, 11***). There were found several such implements, used for powdering of mineral paints and only one stone smoother.

The limited quantity of smoothed stone implements might be due to their replacement from the copper implements. That fact is registered at other sites of the Final Copper Age. In the earlier layer of pit 2 in Kolarovo there were registered fragments of very roughly made thick-walled vessel with rounded base and clear traces of copper ore on the inner side of the body – a proof for use of that metal.

The bone material from Kolarovo is extremely limited and in bad condition. The reason of that is the acid character of the soil, in which it was almost completely destroyed. Over one holed bone traces of treatment only were documented.

Most numerous are flint and the quartz artefacts. The flint artefacts are 14. 13 represent prismatic retouched with one to three working ends. The use-wear trace analysis give the following data on their use – 7 of them were used as blades for sickle, two of them were implements for wood and a blade for sickle but later used for other operation as well. The comparatively rare distribution of the flint, which as a resource is not distributed in that region, is a reason flint and the obsidian implements to have a long-life of using, often with more than one function.

At Kolarovo for the first time the often use of quartz was documented, which replaced the flint for a production of the implements needed. About 60 quartz artefacts have traces of treatment and utilitarian use. As a resource the quartz splinters with rectangular, trapezoidal, triangle, semicircle and irregular shape were mainly used. Their length is between 3 and 7 cm. About half of the quartz splinters have traces of retouch, the rest were used without preliminary treatment. By many of them the working edge was shaped in process of breaking away of the stone from the nucleus. In the first view many of the quartz splinters looked as they were without traces of use, but the ware-use trace analysis gave a real data for their using. In the inventory the scrapers for leather treatment followed by the cutting implement edges (retouched and non-retouched splinters etc.).

The number of the bores and the implements for the treatment of the wood is small, which is due to the character of the material and for what the flint artefacts were used. A big number of the quartz splinters has traces of making nuclei. Some nuclei were also found (that resource material consists of 28 pieces). Typical of the quartz inventory are the pestle, the implements for retouching and the fragments for powdering of mineral paints (total of 11 pieces). A part of the quartz pestles have regular shape made after preliminary treatment – spherical and flatten spherical, with wide working platform over the whole periphery of the implement and with several working platforms. Some of them have traces of powdering of the grain, and others were used for powdering of mineral paints. There are also prolonged pestles with two working ends, as well as with irregular shape and with one working end only. It is interesting to note, after the breaking of those

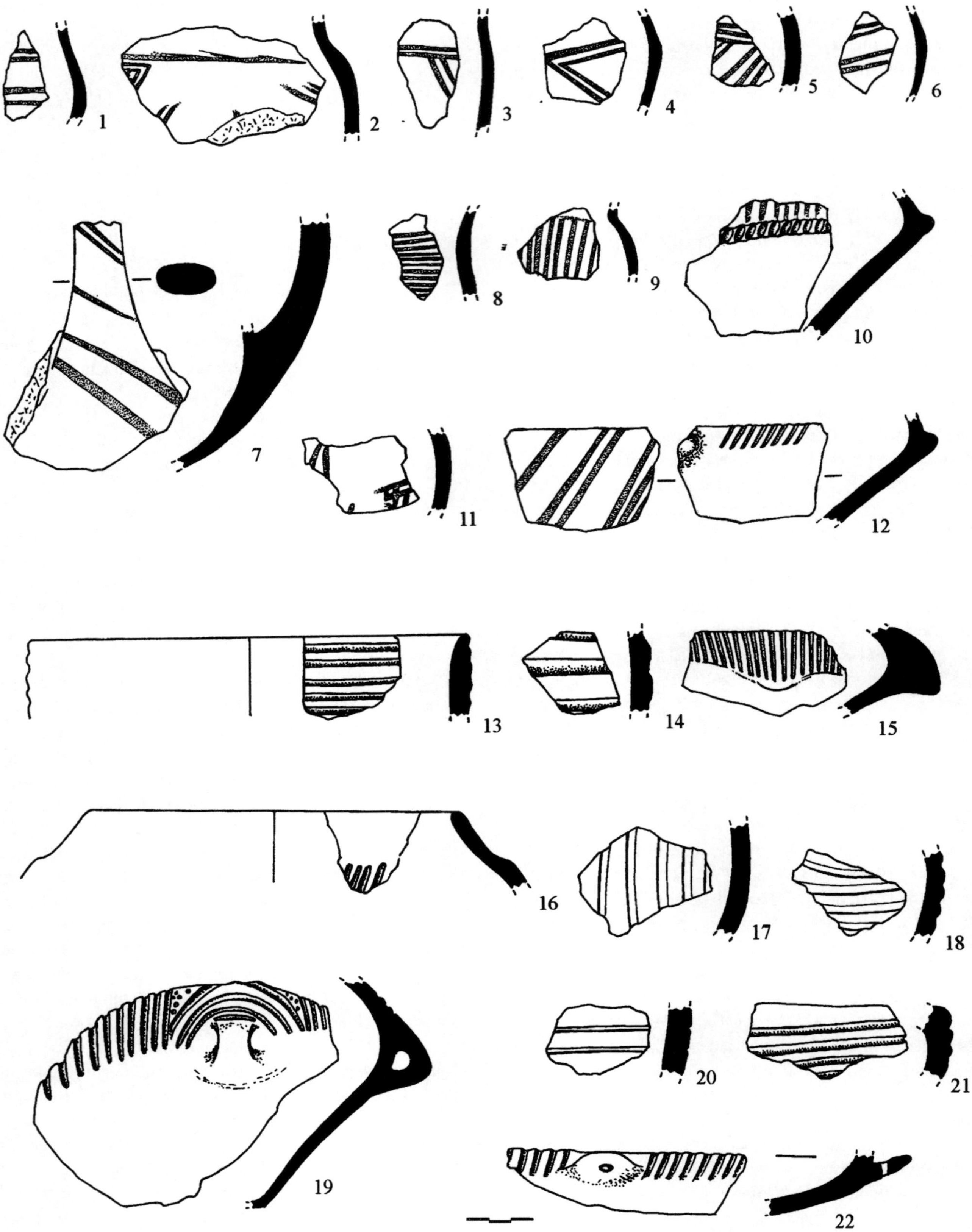

Figure 12.12. Graphite ornament (1-12) and channels and pseudo channels (13-22).

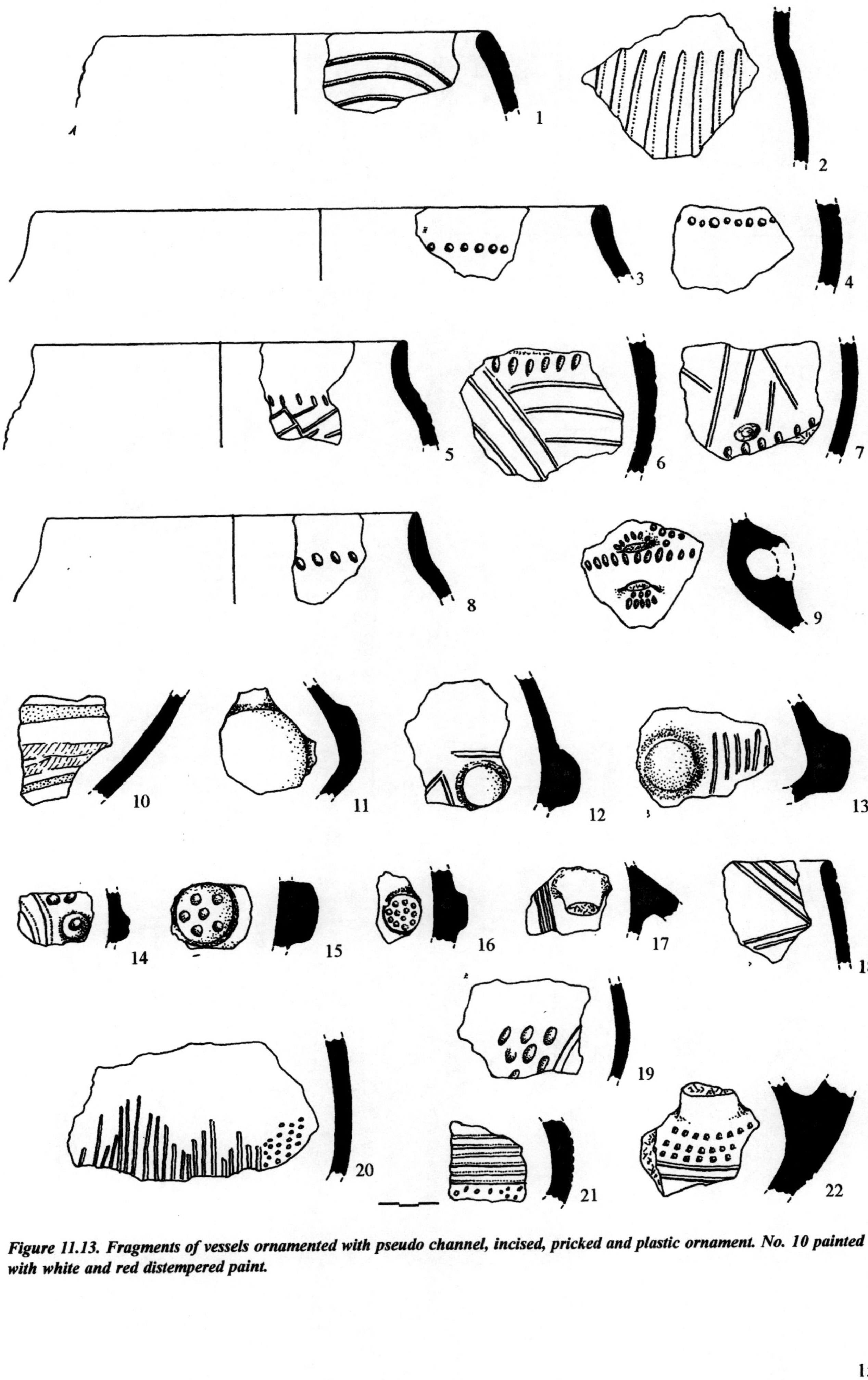

Figure 11.13. Fragments of vessels ornamented with pseudo channel, incised, pricked and plastic ornament. No. 10 painted with white and red distempered paint.

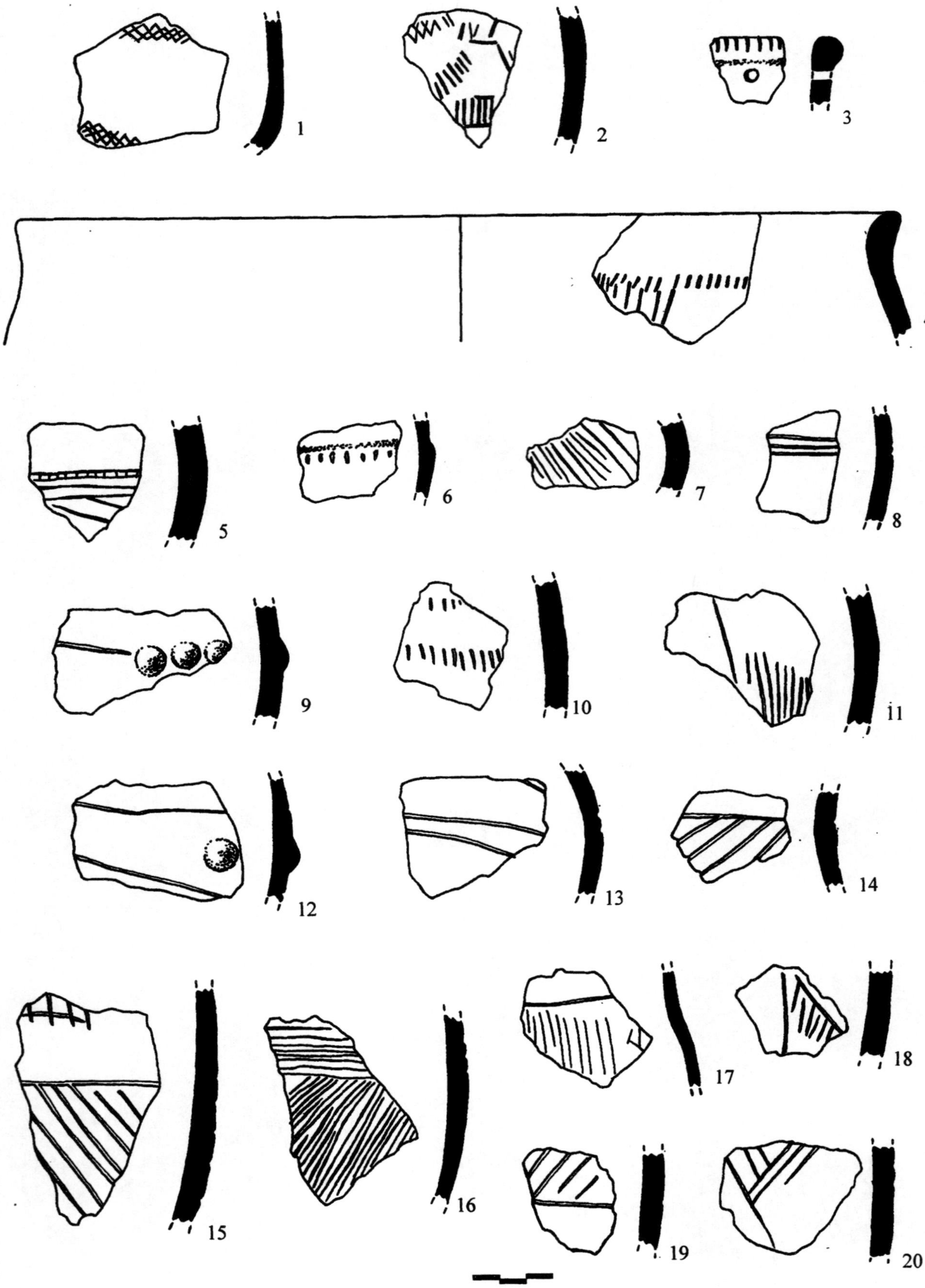

Figure 12.14. White encrusted sherds (1-2) and fragments with incised, drag and plastic ornament (3-20).

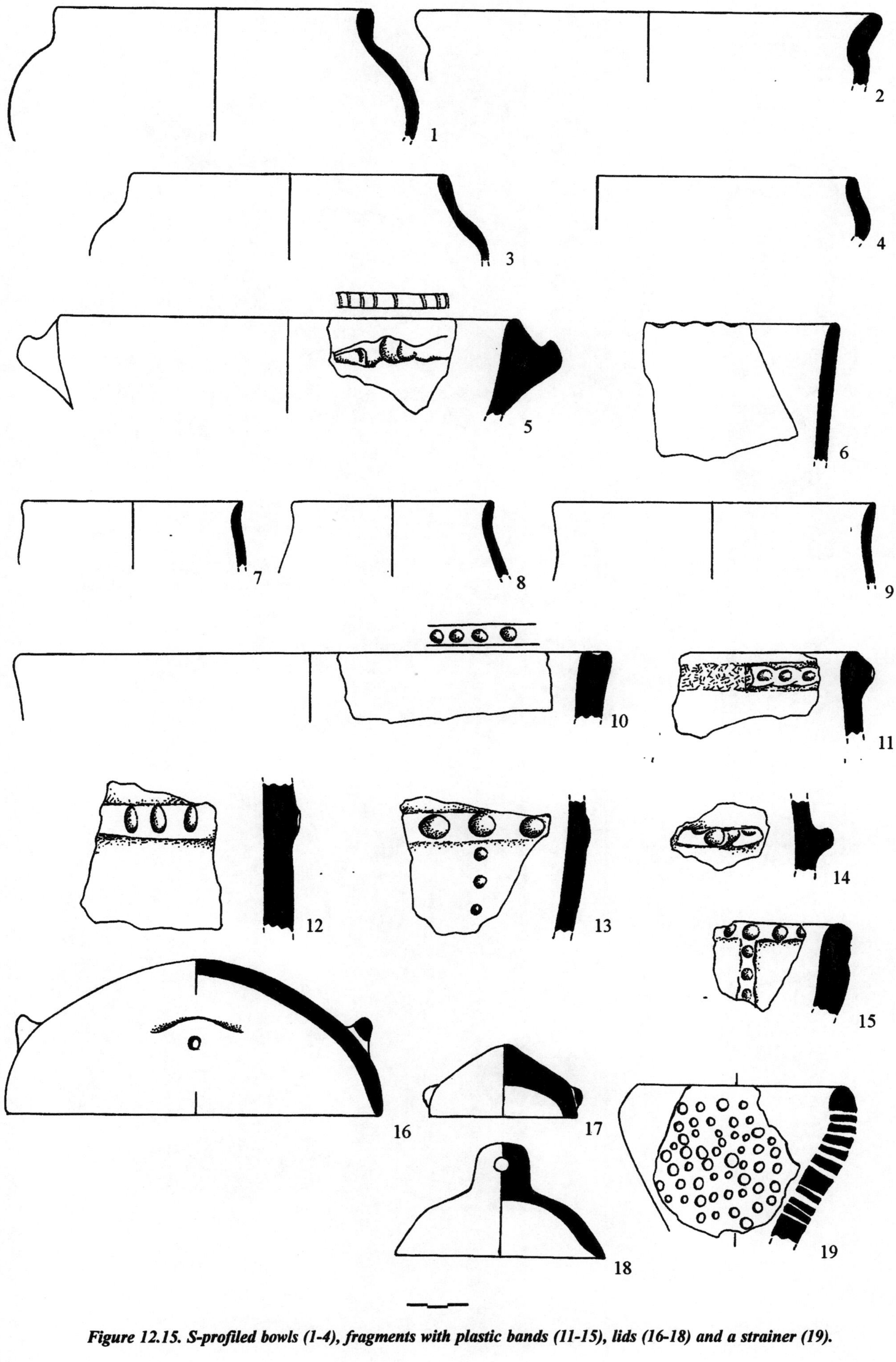

Figure 12.15. S-profiled bowls (1-4), fragments with plastic bands (11-15), lids (16-18) and a strainer (19).

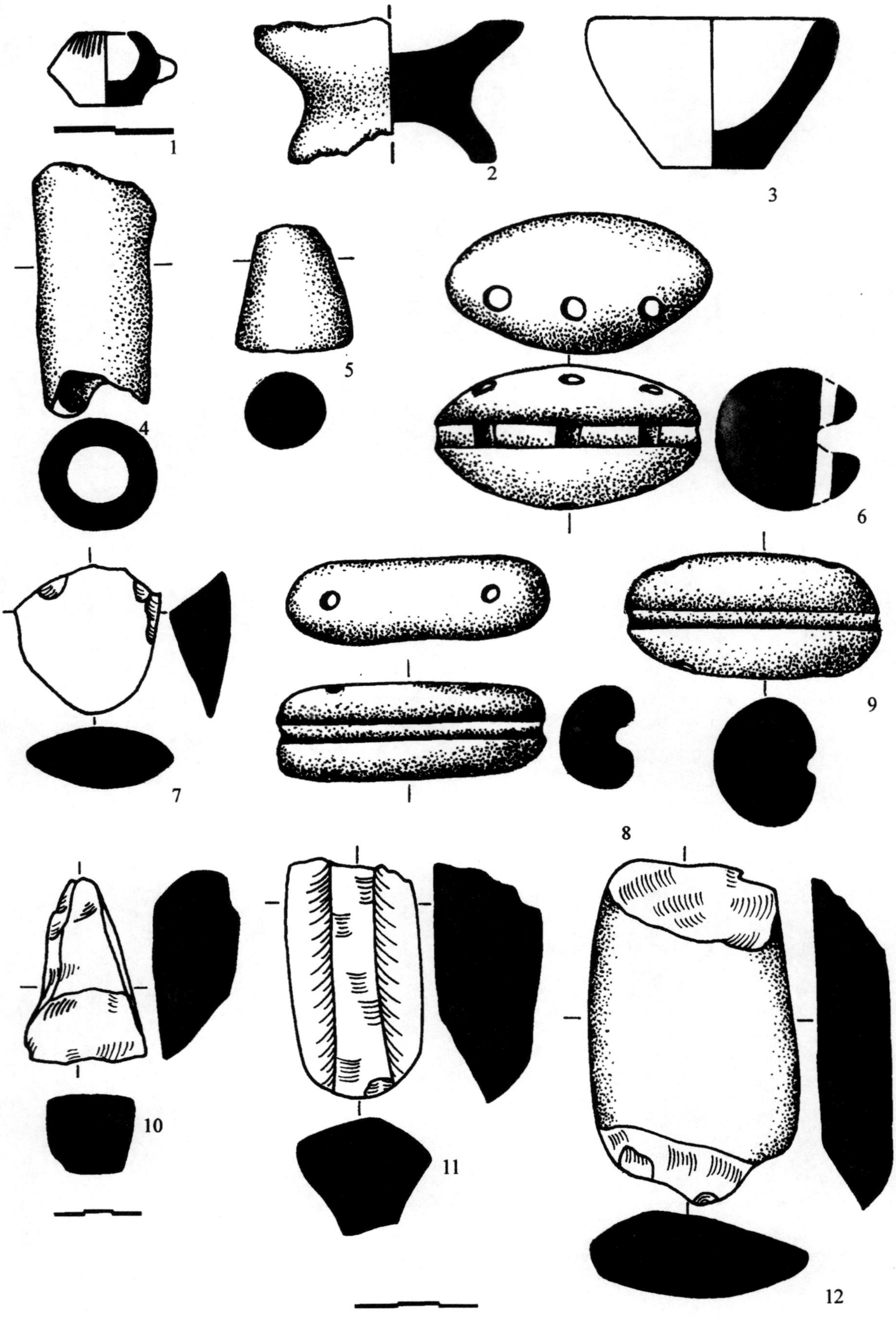

Figure 12.16. Small finds.

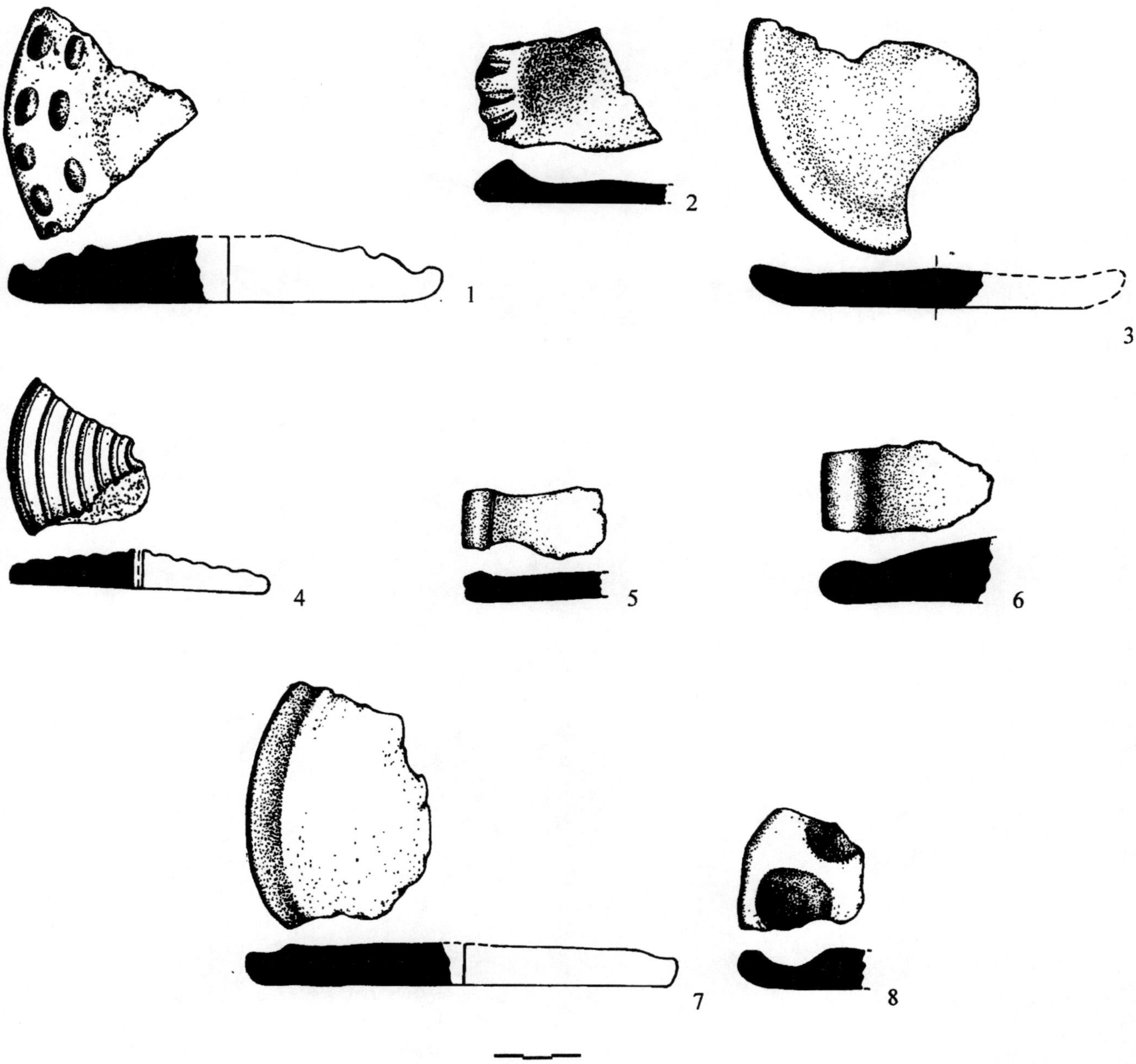

Figure 12.17. Ornamented clay plates and "podnitsi" (vessels for baking of bread).

implements the splinters were used as a resource for preparing of other implements or were directly used for other operations.

Therefore, because of the limited use of the flint resources, the whole complex of quartz and flint material was used very rationally. There is almost all kinds of implements of that sort – for wooden treatment, cutting, for treatment of leather, agricultural (sickle blades), pestle for grain. Agricultural implements of stone and bone were not found.

Clay household objects

Clay objects with spindle shape (handles for hanging of vessels?). Three whole and two fragmented finds represent that sort at Kolarovo. The length varies between 6.2 and 7 cm (***Figure 12.16: 6, 8, 9***). They have prolonged groove over the one side and two or three cross-holes. Characteristic are different variants for all stages of the Krivodol-Sălcuţa-Bubanj complex and are registered in the neighbor Šuplevac-Bakarno Gumno group in Pelagonia (see for instance Simoska and Sanev 1976: 152; Chokhadzhiev 1991: fiig. 11; Nikolov 1975: No. 106). At Kolarovo they were found in the area of house 2, but also in the fallow land and in pits 5 and 6.

Bread-baking bases ('podnitsi'). In this group are included conditionally all flat clay objects with circle shape in plan, without handles. They were very popular in the region in Neolithic and Copper Age. In the earlier periods they usually have bigger diameter, slightly profiled ledge and are unornamented (Pernicheva 1995: fig. 6: 198, 199; 1985: fig.

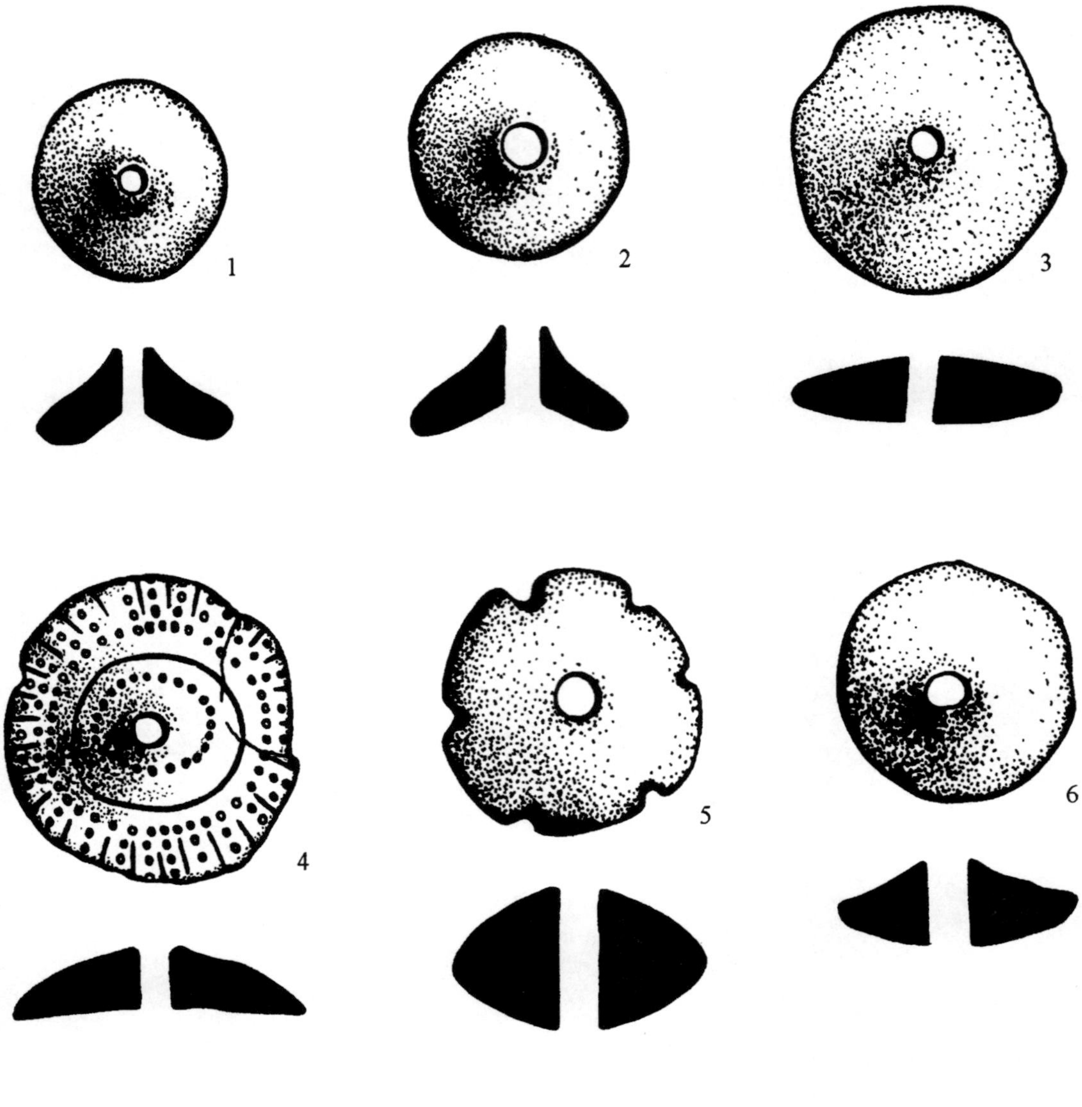

Figure 12.18. Clay whorls.

7). Perhaps not all similarly shaped finds can be treated as bread-baking bases. They have smaller size – diameter 8-17 cm and different profiling. A part of those finds are almost flat and with slightly profiled periphery, unornamented or ornamented with cuts over the rim (***Figure 12.17: 2, 3, 5, 7***). The last can be treated certainly as bread-baking bases. The rest four fragments are of clay plates which become considerably thinner towards the periphery and are decorated with pits with concentric incised circles (***Figure 12.17: 1, 4***) or are not ornamented (***Figure 12.17: 6***). It is possible they to have been holed as it shows the fragment in ***Figure 12.17: 4***. The function of those finds is unclear. More special is the fragment in ***Figure 12.17: 8***, which was probably covered with round pits and probably belongs to a kind of game. Similar objects with different modification characterize the Early Copper Age in that region (Chokhadzhiev 1997: fig. 196, 197).

Clay spindle whorls. They are the most distributed finds among the household inventory (11 intact and fragmented). Their diameter is between 3.3 cm and 5.5 cm. They are made carefully, conical, double conical and ellipsoidal, with symmetrical periphery and show a big variety in the cross sections (***Figure 12.18: 1-6***). The conical finds have slightly turned periphery. Only one find is decorated over whole surface with ornament of concentric strips in combination with oblique incisions on the periphery (***Figure 12.18: 6***). Another find has stressed thickened double conical shape, with holes on the periphery and can be interpreted as a weight of fishing net or loom weight (***Figure 12.18: 5***). The whorls were found mainly in the area of house 1, with exception of the finds in ***Figure 12.18: 5 and 6***, which were documented in the area of house 2. Ornamented and unornamented conical whorls are traditional finds at the settlements of the final Krivodol-Sălcuța-Bubanj complex (Bonev and Alexandrov 1996: fig. 42; Mikov 1948: fig. 37) and especially of the synchronous group in Pelagonia (Sanev et al. 1976: 424; Garašanin and Samoska 1976: plate 8). They are often distributed in the area of Rachmani culture as well (Weisshaar 1989: Table 63, 82, 91).

Clay lids. One of lids, a flat one, with small cylindrical handle (diameter 8.3 cm) was discovered on the floor in house 1 (***Figure 12.19: 3***). Mushroom-like find with a handle (height of 9 cm) was found in pit in a complex of early pottery. The function of those finds (Mikov 1948: fig. 36) is not clear, for the time being. Probably they were also lids.

The fragments of clay bases were found in different stratigraphical context – in pits, cultivated layer, in the area of houses. Their definition as typical of certain period of occupation of the settlement would be not correct.

To the single finds belongs a red ochre cone for ornamentation of vessels found in a pit from the early stage of Kolarovo.

Clay small altars. The altars discovered at Kolarovo are quadrangular with legs attached (***Figure 12.19: 5-7***). Two better preserved finds with documented in pit 3, in the complex of the pottery from the earlier stage (***Figure 12.19: 6, 7***). It should be stressed the bog size of one of them which preserved length is 15 cm and the height is 11,6 cm. It is decorated with incised ornament and pits under the rim and is provided with tongue bulges for hanging at the base (***Figure 12.19: 6***). The second altar of that type is with square base, 7 cm long, and 6 cm high. It is not ornamented. If we have in mind the proportion of the base it can be stressed that it also have had a square base. The relatively both depth of the basin along with the missing legs differ both from the other similar objects. The only similar altar but not identical is known from Zaminets (Nikolov 1975: fig. 40).

A relief representation of a hand over a fragment of a clay vessel (***Figure 12.20: 1***) *and a sign*. It is a rare find from that period. It is found in pit 4 and probably belonged to Kolarovo 1.

One unusual sign was documented over the clay vessel's wall. It is shallow incised isosceles triangle with small pits which mark the three angles from outside (***Figure 12.20: 2***).

Clay models of houses. Three sherds of house models give an information of two-sloping roof. They are identical by size and the degree of the slope and the edge of the fronton marked c. 5 cm long (***Figure 12.20: 3-5***). One of the fragments was found in the ruins of the oven in house 1 and probably had a role in the ritual connected with that house equipment. The other two have no clear stratigraphy.

Objects of cult

Clay zoomorphic figurines

Four fragments of bodies and heads of animal figurines were found (***Figure 12.21: 1-4***). Perhaps one of them represent deer (***Figure 12.21: 1***) and two other – ram and ox (***Figure 12.21: 2, 4***). The last two were found in the earlier occupation stage of the settlement (one of them in house 2). The other two were found near to the bottom of pit 2. There are not well elaborated, so they differ technologically of the characteristic of Southwest Bulgaria finds since the end of the Neolithic (Pernicheva 1993: figs. 1-3).

Anthropomorphic figurines

The fragments found of anthropomorphic figurines show variety in their typology documented at the site. Three types can be distinguished.

Type A. Sitting figurines. A sherd of an ischiatic part. It is well elaborated, found in the area of house 2. Height – 3,8 cm (***Figure 12.21: 6***). Obviously it is female figurine.

Type B. A leg of semi-sitting or sitting with straight legs in the lower part figurine. Height – 7 cm. The leg has elegant turning point, but without an artistic aspiration of representing anatomical peculiarities. It could belong to female of male figurine. The figurine originates from the area of house 1 (***Figure 12.21: 3***).

Type C. Standing figurines. A leg of such a figurine ornamented with oblique incisions (***Figure 12.21: 5***). It was probably holed indicated at the place of break. It was found also in the area of house 1.

The three fragments originate from figurines made of clay with different quantity of fine sand and well smoothed surface. They have light brown or brown and red color.

The anthropomorphic plastic of Krivodol-Sălcuţa-Bubanj complex includes standing and sitting anthropomorphic male and female figurines. With this respect, Kolarovo figurines are not exception and are characteristic of wide cultural area.

Type D. Flat clay figurines-busts. Four figurines of that type originate from Kolarovo. They are very schematic modeled and generally differ stylistically and to some extend technologically from the clay figurines found in the context of the material in the burnt houses (***Figure 12.22***).

The figurines are made of well-purified clay or of clay with fine sand regularly burned to brown or gray and brown. With exception of the ornamented figurine they all are with well-smoothed surface, almost burnished. They are equally flattened from both sides representing a body without head, with wide shoulders and slightly stressed widening in the lower part, under the waist. The torso finishes without the lower extremities. The arms are represented as short bulges outstraight from the shoulder, horizontally (***Figure 12.22: 2, 3***) or slightly turned down (***Figure 12.22: 1, 4***). They have no gender indication. The strongly widened shoulder part in the first view gives an idea of male presentations. The bodies are holed with narrow holes (2-4 mm) along whole height. In the openings there can be put a thin stick or cord for hanging, because they can be placed difficulty to stand. The heights of the whole figurines are: 3,8 cm, 5,2 cm, 5,8 cm. The widths of the shoulders are 4.3 cm, c. 4.3 cm, c. 6.5 cm, i.e. the maximum width is equal or bigger than the heights. The thickness of the figurines is 1.4-1.6 cm and it slightly varies in the different parts of the body and became thinner towards the periphery. Only one figurine has ornament. The last consists of incised lines over the shoulder and pelvis part and longer ones in the area of the breast and the stomach (***Figure 12.22: 2***). Two of the figurines are found in pit 2, in the complex of the pottery from the Kolarovo 2 (***Figure 12.22: 3, 4***). The figurine in table 22: 2 is found on the surface, and the fourth one is from the highest level of pit 6.

Figure 12.19. Ceramic stands (1-2), lids (3-4) and clay altars (5-7).

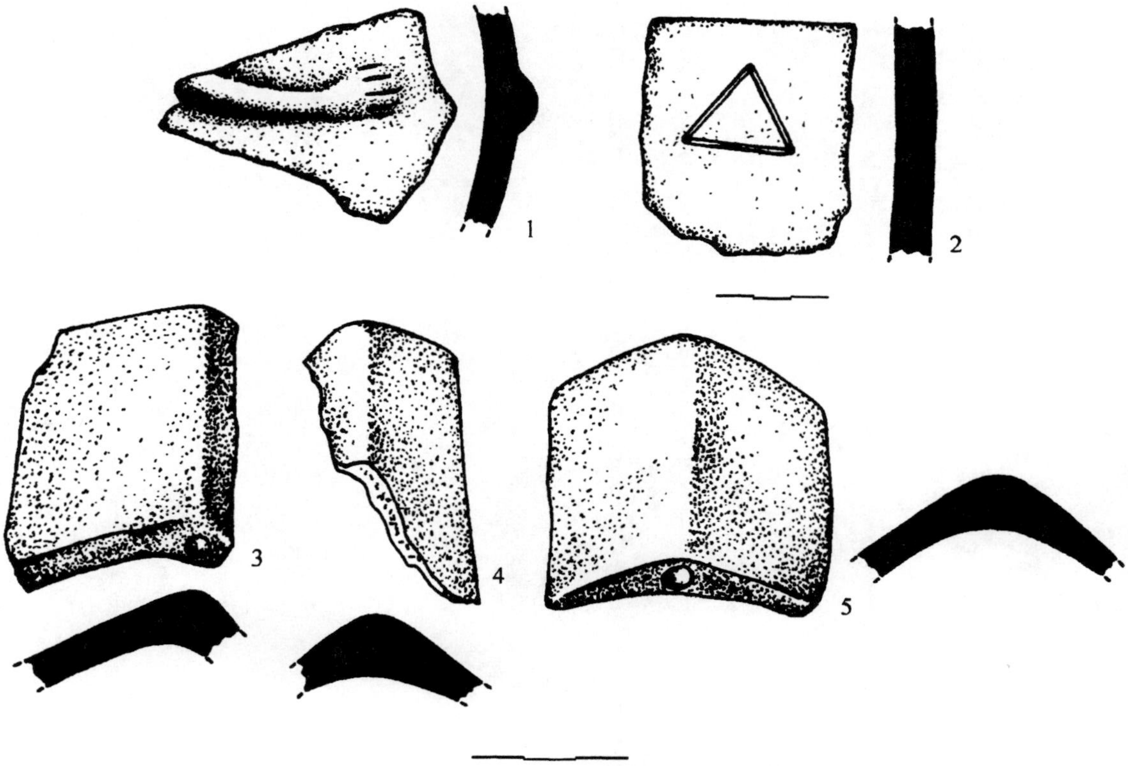

Figure 12.20. Fragments of cult vessels (1-2) and of models of houses (3-5)

That type of figurines has no parallels at the Late Copper settlements in Bulgaria including the area of the final Krivodol-Sălcuţa-Bubanj complex, with exception, probably of the figurine at Vaksevo (Chokhadzhiev 1991: fig. 11: 5). For the time being, such figurines have been not registered in the so-called Transition period in Bulgarian lands. Territorially, the closest parallels originate from Šuplevac in Pelagonia and one find from Bakarno Gumno in the same region (Kitanoski 1971: N. 7; Hauptmann 1967: fig. 6: 7-9). To the south, similar figurines but with less similar elements and mostly without prolong hole are known from Rachmani and Pevkakia (Weisshaar 1989: figs. 82, 83; Heurtley 1939: figs. K, L, M). It should be stressed, many figurines from Thessaly have identical prolong and cross-sections but it looks like that with exception of one figurine (from Zerelia) they were not holed (Weisshaar 1989: figs. 96: 20; 82; 83).

Territorially, very far are the parallels of those figurines from Cernavoda III culture (Morintz and Roman 1968: fig. 29: 13, 17, 18, 22, 23), which according to the periodization of L. Nikolova belong to Early Bronze I (1999: 317-323).

Most of the figurines mentioned are holed in the upper part only. A big part of them have marked despite schematically, breast, so the gender of Kolarovo finds cannot be certainly defined. Most probably, in this style were represented females, as well as males.

Typological and chronological definition and problems

The stage of Kolarovo I includes the basic part of the materials at the settlement – those of the burnt horizon, a lot of pits, but a part of unstratified materials at the settlement as well. Many parallels can be pointed out, indicating differently that stage of the settlements from the final Krivodol-Sălcuţa in the Central Balkans. Considerably limited similarities exist with the settlements from Bubanj-Hum culture I including comparatively closed region of the Morava basin (Stalio 1972), which shows chronological difference with that stage of Kolarovo (Garašanin 1998: 273).

A good number of sites from Krivodol culture in Northwestern Bulgaria show close or more distant typological connection with the material from Kolarovo I. Most important are Krivodol (Mikov 1948; Nikolov 1984); Zaminets (Nikolov 1975), Telish (Gergov 1992; 1994), Sadovets (Todorova 1968: fig. 3:5) and Bagachina (Bonev and Alexandrov 1996: fig. 36). Despite it is a question that has given rise to much controversy, the inner periodization of part of those sites is in general outline (Nikolova 1999: 17-21; Garašanin 1998: 268; Todorova 1986). It is clear that the ceramic material from Kolarovo, as well as a big part of the other household inventory point to a dating close to the final stages of the Copper Age of those settlements.

But it is under impression the whole complex of Kolarovo I shows some general differences in relation to the sites from late Krivodol culture. It is expressed best in the decreasing of the fine pottery style: the absence of some ceramic shapes; in the technology of surface treatment of the pottery; especially in the ornament technique and motifs. A big part of the motifs are ornamented with graphite and with different distempered paints characteristic of Krivodol culture are not repre-

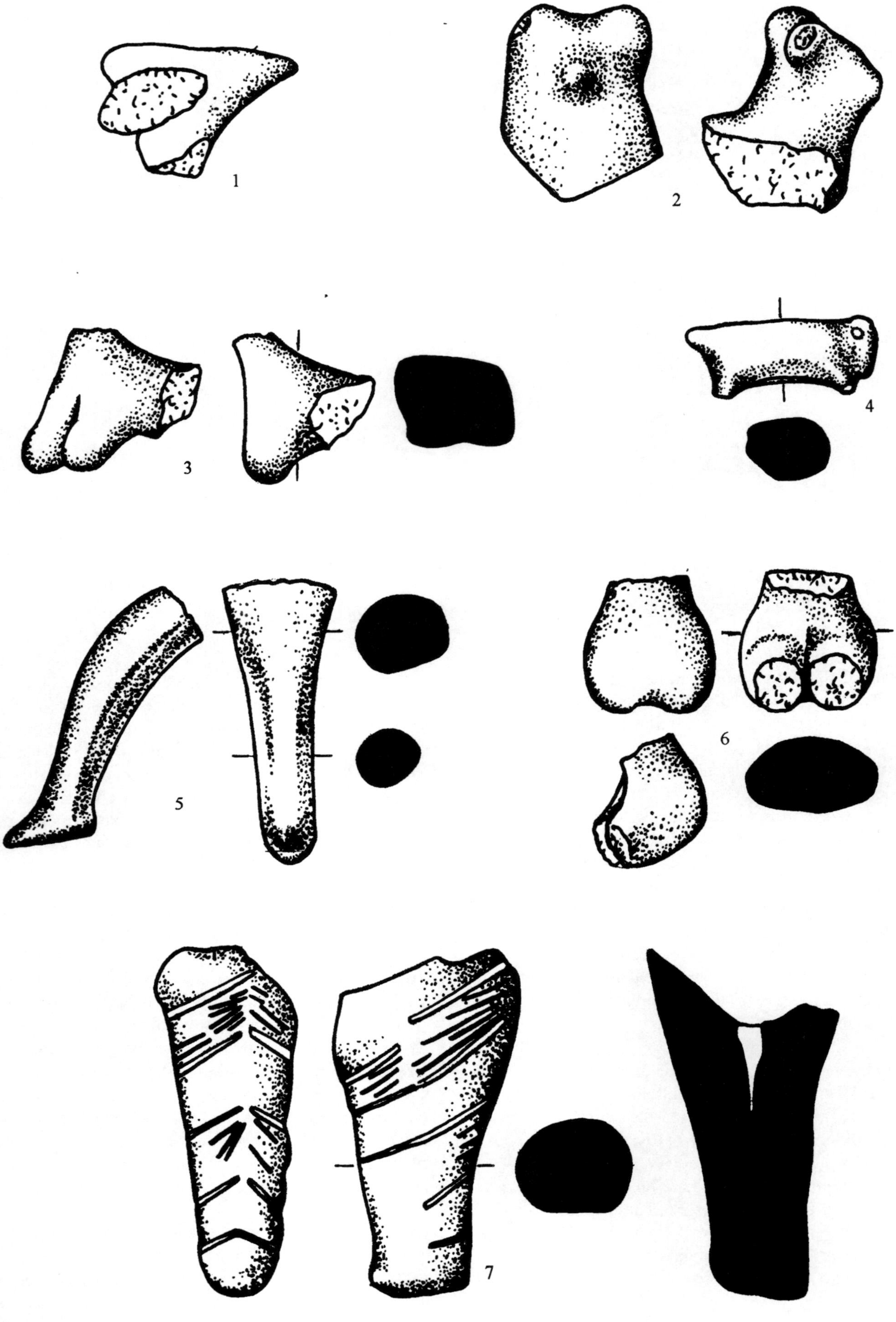

Figure 12.21. Zoomorphic plastic (1-4) and anthropomorphic plastic (5-7).

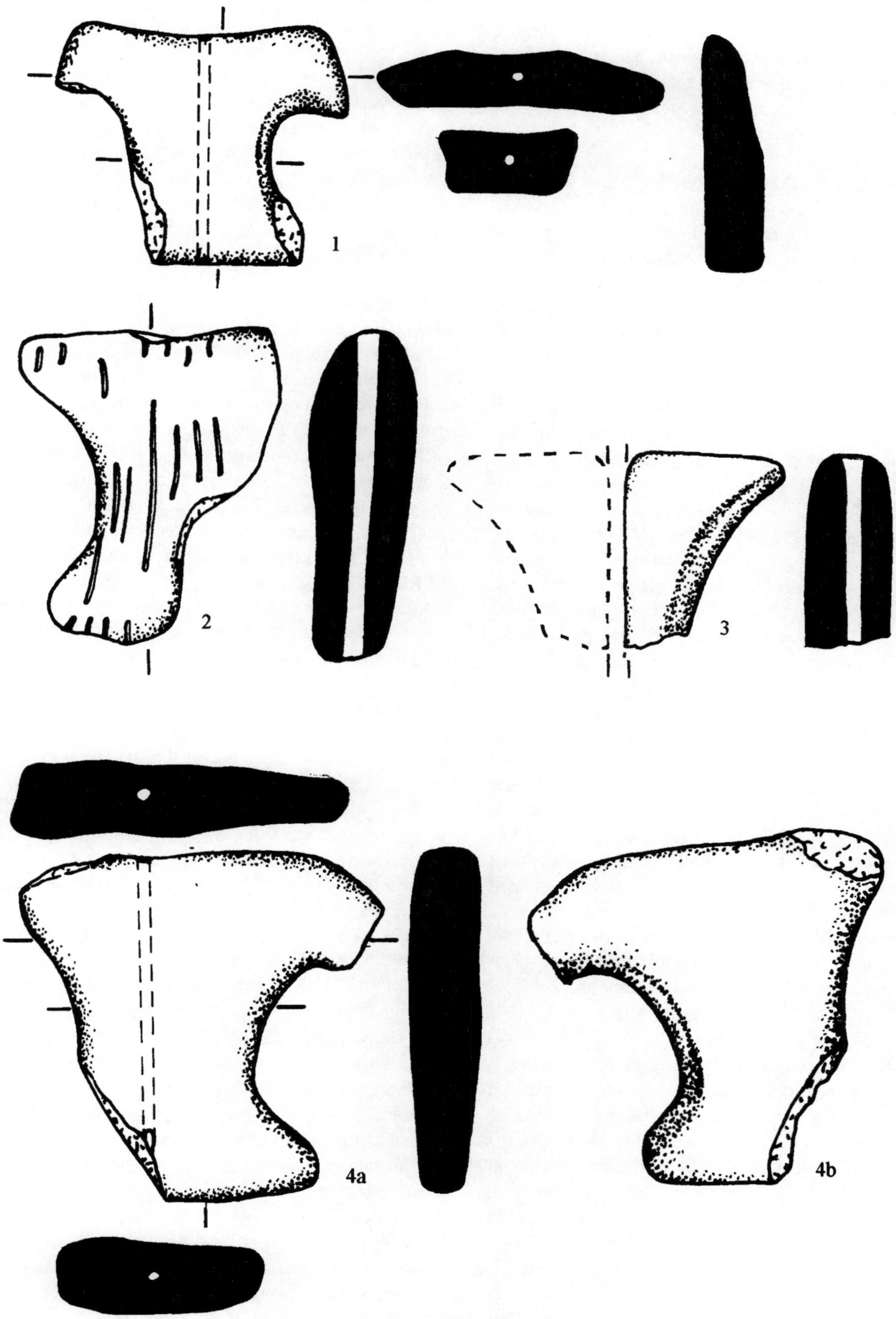

Figure 12.22. Kolarovo anthropomorphic figurines with holes.

sented at Kolarovo. There are missing there a specific pots and amphora-like vessels covered with different barbotine and incised ornament. The graphite ornament at Kolarovo is very simple, and the white and red painting after pottery burning is rare distributed, with very simple motifs. The painting with yellow color was not registered. But at Kolarovo the channeling dominates, pseudo-channeling, as well as different types of incised and pricked ornament.

The analysis of the Late Copper culture in Struma basin and especially in its south region is very complicated. The sites from that period are very rare. In the Late Copper complex of Pernik-Krepostta there are close analogies if some ceramic types from Kolarovo (Raduncheva 1981, figs. 14, 16: 2-8). Despite the materials there are unstratified. It can be assumed that there the late stage of Krivodol chronologically close to Kolarovo I.

Vaksevo from the Struma drainage is the closed territorially settlement dated by S. Chakhadzhiev from the final Krivodol-Sălcuța-Bubanj I (Chokhadzhiev 1991, chapter 8). At Vaksevo, in some cases there are identical with Kolarovo I shapes pottery and household inventory there - for instance, all the series of conical bowls, the profiles orifices of the pots, and probably some of the cult tables finds (Chokhadzhiev 1991: figs. 1, 2: 1-3, 3: 5-7). As far as other ceramic shapes concern – the profiled biconical bowls and vase-like vessels (Chokhadzhiev 1991: figs. 4: 5,6; 5: 13) and especially the ornamented techniques and motifs, there is a big difference between both settlements. The ceramics from Vaksevo is characterized with more various techniques of incision and especially pricked with different instruments decoration, with a presence of pseudo-corded and plastic pinched band. The graphite and painted with different distempered paints ornament is of much bigger quantity and considerably more varies in compare to Vaksevo I (Chokhadzhiev 1991: figs. 8, 9 and 10). The vessels with S-profiled at Vaksevo represent a different variant of the related to them shapes from Kolarovo I and II. The question arises if there is no local difference between both settlements, as Kolarovo is more isolated, aside from the main natural way along the Struma River. Bit more possible thee is a chronological difference, as at Vaksevo considerably more clearly contacts with Šuplevac-Bakarno Gumno in Pelagonia.

I think, for the time being the synchronization between Kolarovo and Šuplevac-Bakarno Gumno is under discussion (Garašanin 1998: 271-72). Despite the territorial closeness of Strumeshnitsa with Pelagonia to the west, the direct typological parallels of the settlements of that culture with Kolarovo material are very limited and concern mainly the anthropomorphic figurines with opening and secondary elements in the pottery (Garašanin and Simoska 1976: tables 5: 20; 6: 21; Sanev et al. 1976: No. 388; Simoska and Sanev 1976: 22; Kitanovski 1971: Nos. 1, 7, 9, 11; Hauptmann 1967: fig. 6: 7-9). According to the published material from Crnobuki there is a similar situation (Simoska et al. 1976: tables 1: 1; 3: 6, 8; 4: 6, 8, 9: 5: 1; 8), as well as with Karamani near Bitola (Simoska et al. 1977). Recently investigated Late Copper Age in Macedonia also give no significant comparative material with Kolarovo (Kolištrkoska-Nasteva 1993; Georgievski 1993: tables 8: 3; 11: 5; Kuzman 1996: Nos. 24, 25, 26-29). The comparison of the settlements of that cultural group with Kolarovo can be elaborated indirectly only and their earlier chronology (Garašanin and Simoska 1976: 20-21) looks more believable as synchronization with Kolarovo (Garašanin 1998: table 1).

Kolarovo I shows similar and even more indicated similarity with the classical phases of Sălcuța in Oltenia as with Late Copper Age settlements in Northwestern Bulgaria. Numerous parallels with the pottery from Sălcuța (phase II and III after D. Berciu) are obvious (Berciu 1961: figs. 89; 94; 96; 96; 101; 112 and especially fig. 93: 9 as the only parallel of the bowl from Kolarovo in ***Figure 12.3: 12***.

The Kolarovo II is characterized with the following leading new elements defined stratigraphically. In the technology of the pottery a group with poor structure occurred characterizing a whole series of differently profiled small S-vessels and slightly profiled conical bowls. With the same technological group the vessels with specific ornamental techniques are connected – pricked ornament, sloppy incised and drag motifs, short incisions, sheaves of lines and sloppy hatched triangles (***Figures 12.9; 2: 12, 14, 16; 12.14: 4-20***). Despite their unclear stratigraphic position, it seems with the same stage are connected the vessels with incisions over the rim (***Figures 12.15: 5; 12.14: 3***); fragments with band handles and arched-shaped vertical handles with prolong incised lines, but without information of the vessels to which they have belonged (***Figures 12.11: 1-3, 5, 6, 10***), fragments of plastic band (***Figure 12.15: 11, 14, 15***). It cannot be clearly documented to which extend the graphite painting was used, but it is certain its very simple applying over S-profiled vessels as simple band or graphite plastering. It seems the flat holed idols belong to Kolarovo II as well.

It is very significant that at Sălcuța IV (after D. Berciu) as new elements occur the same peculiarities as at Kolarovo II (Berciu 1961: 309-328). S-profiled vessels are identical (Berciu 1961: figs. 138: 8-11; 141), slightly profiled conical bowls (Berciu 1961: fig. 142), the technique and motifs of the late incised, drag and pricked ornament (Berciu 1961: figs. 136: 1, 3, 4; 137: 1, 2, 5); waved-shaped and rims with incisions (Berciu 1961: 139: 6), slightly stressed plastic band with hollows (Berciu 1961: fig. 140: 1, 7). With Sălcuța IV the changes in the clay paste are connected documented at Kolarovo II, but there is missing earthenware of shell-crashed compound. It should be also mentioned the absence of disk-like handles with application and of high handles (Berciu 1961: figs. 143-46). In the anthropomorphic plastic of Sălcuța there are almost no analogies with those from Kolarovo II (Berciu 1961: figs. 152, 153, 157). But the expressive parallels infer synchronization of Kolarovo and Sălcuța IV. By direct analogies (Roman 1971: figs. 18, 19) and indirectly it can be assumed the eventual synchronization with Băile Herculane I-II and with Ostrovul Corbului (Garašanin 1998; Nikolova 1999).

It is here necessary a comparison of Kolarovo II with Reburkovo and Galatin 2 from Northwestern Bulgaria (Vajsov 1992; Georgieva 1988; 1994). In relation of all investigated sites from the Final Copper Age in Northwestern Bulgaria the ceramic complex from Kolarovo has the closest analogies at Reburkovo. It is obvious the typological similarity of the S-profiled bowls, of multi variant conical bowls, as well as of a big part of the spherical bowls and pots with differently profiled necks (Georgieva 1994: figs. 2: 5, 6; 5; 6: 7-16; 7: 1-9; 8: 1-10). Despite the ornamentation generally differs there are some analogies in some distant ornamentation techniques of incised and pricked ornament, as well as of wide horizontal channels over some shapes of bowls (1994: figs. 5: 3; 11: 2-12). It is identical the technological group of slip pottery with crumbling structure of the biscuit. But on the whole a big part of the motifs of the pricked and incised ornament from Kolarovo II and Reburkovo differs. The minimal using of graphite ornament and the manner of painting and plastering of the surface with colored distempered paints on the latter side are related to a big extend to the same ornamentation manner at Kolarovo.

There are less typological parallels between Kolarovo II and Galatin 2. Parallels exist among the S-profiled bowls, the variants of the conical deep plates and the bowls with horizontal channels (Vajsov 1992; Georgieva 1988: figs. 17; 18: 5-8).

By the comparative analysis of Kolarovo with Sălcuţa and Galatin 2 most reliable is the synchronization Kolarovo II – Sălcuţa IV – Reburkovo – Băile Herculane II? (Garašanin1998). So, the Kolarovo I stage should belong to a little earlier stage, eventually end of Sălcuţa III or Sălcuţa III-IV.

There is a peculiarity in the development of the region of the Middle Struma south of Kresna, as well as in the valley of the Strumeshnitsa basin (Pernitcheva 1983: 25-29). In that region were registered no settlements from the developed phases of the Early Copper Age (Slatino-Dikili Tash culture), as well as from the earliest phases of the of variant Dyakovo of Krivodol culture. Obviously, in that period the south part of the middle Struma was a transition zone towards Northern Greece where at Dikili Tash and Sitagroi tells variants of those cultures developed in different phases (Dikili Tash IIA-C, Sitagroi IIIA-C). At both tell the final phase of the Copper Age is not documented. Then, the assumed synchronization of Sitagroi IIIC and with Sălcuţa II cannot be accepted (Demoule 1989: tables 2, 3).

White the settlements of Krivodol culture in Sofia field and in Northwestern Bulgaria are at different degree multi-level and usually situated over natural hills, to the south of Kyustendil field, along the Struma River and its tributaries, the settlements over terraces and are thin-leveled, as often their existence can be documented only by materials from destroyed surface layer of the terrain, without traces of house equipment. According to the data obtained, as Kolarovo, those settlements belong to the final stages of the Copper Age. The territorially closed settlements sometimes show typological differences between, which probably speak of different ethnical component. It is significant in this respect the second site from Vaksevo dated by S. Chokhadzhiev from the Transition period to the Bronze Age (Chokhadzhiev 1994). In the closed complex there are no even close parallels with Kolarovo I and II, as well as with the earlier Copper Age site of Vaksevo mentioned above.

It seems to me, similar situation exists on the settlements of Šuplevac-Bakarno Gumno in Pelagonia. Despite they are multilevel, it looks like the classical phases of Krivodol-Sălcuţa-Bubanj complex is not documented there (Garašanin1998: table 1). Despite their territorial closeness with the Strumeshnitsa valley, typologically they have less contact-data with Kolarovo as with Vaksevo. It is obvious, Kolarovo I and II has definite connection directly to the north and represents a late variant of Krivodol-Sălcuţa-Bubanj complex, being the southernmost sites of the final phases of that culture known up to now. In my opinion, it is still under discussion the synchronization of Šuplevac-Bakarno Gumno with Kolarovo I and II and some sites from Macedonia.

To conclude, based on the investigation of Kolarovo, the following conclusions and problems can be defined:

1. Kolarovo marks, for the time being, the southernmost borderline of the distribution f the Krivodol-Sălcuţa-Bubanj complex in its final phases. Of interest is the fact that while the pottery is closed typologically with north regions, the anthropomorphic plastic documented connects the site with Pelagonia and Thessaly.

2. The chronological stage documented at Kolarovo is under discussion with regard to its definition (see Problems 1992). In different way it is defined as Transition period (Post-Eneolithic and Proto-Bronze [Vajsov 1992; Todorova 1998]), and as Final Copper Age consisted of several sequent phases (Nikolova 1999: Table 3.2). As far as Kolarovo concerns, I believe it is a stage typological connected with final phases of the Copper Age in that region (Nikolova 1999).

3. It cannot be excluded in the Strumeshnitsa valley there to have been existed retardation of the Krivodol culture elements. It seems, the final of the Copper Age represented at Kolarovo I-II was not connected with total ethnical and cultural changes, a fact documented in the isolated region of the Central Rhodopes as well (Avramova 1981). There was possibly a process of graduated integration of already close ethnically population in the frames of the old ethnical and cultural population.

The location of the site on the borderline of two and even three ethnical and cultural regions explains the existence of elements of cultures located in west and south direction. It allows one to assume the existence of different interactions at one and the same time in this border geographical zone.

CHAPTER 13

LAW AND SOCIETY IN THE LATER COPPER AGE

Ana Raduncheva

Introduction

As all the ancient societies, the prehistoric population of the Central Balkans were ruled by and brought under the sway of the unwritten laws of the customary law. It defined the moral categories of good and evil, correct and wrong, moral and unmoral, permitted and prohibited. It also controlled the societal and personal behavior of everybody and provided for order and placidity. The customary law regulated the relations of the society to the property and power, as well as among the different institutions, 'civil' and religious, at all the hierarchical levels. It sanctioned and secured the community order and system, which on their side preconditioned its establishment.

It was also transmitted from generation to generation and probably was existed without considerable changes. The customary law determined relative closed status quo of the different community groups, as well as preserved its inner structure, identity and characteristic custom, psychological and moral peculiarities, which to some extend could differed from the same definitions of the surrounding communities. In any case one and the same traditional-law norms were under the obligation to all members belonging at least to communal formations closely blood-related.

In the Central Balkans the initial customary law was brought along with the first immigrant groups. The common homeland and the similar life preconditions where it was formed helped the elaboration of the comparatively close definitions of the norms of behavior. Without considerable changes it continued to exist up to the middle of the Copper Age when it began a process of rapid and serious community changes followed by changes in the behavior and establishment of new relations between the persons and the institutions. In this chapter the similarity and the difference in the law relations up to and after the middle of the Copper Age will be outlined.

The patriarchy

The background

Base on the results of the archaeological investigations of the prehistoric sites and the previous my theoretical research (Raduncheva 1986), we can certainly concluded that already in the Early Neolithic the immigrants brought in their new homeland and continued to live in the conditions of the patriarchal law (the patriarchy). The argument used in course of lot of decades as an evidence of the matriarchy structure of the prehistoric society was the presence of female presentations, which according to all the researches should represent the position of the mother as a personification of the really existed matriarchal society. According to our opinion, those presentations in fact argue the long lost female real power, so the canonical image was already filled with a semantics mostly of the carrying and delivering of new life. In this aspect it is very expressive a statuette from Çatal Hüyük in Anatolia which represents a woman delivering a male child (Mellaart 1967). The act of and the placing of the figurine in sanctuary show that the male power was long before sanctioned from the customary law and confirmed by the religion. There is also a presumption that it is difficult one to believe a female leading and control over mass migration assumed by the colonization of the Balkans. Therefore, it cannot be accepted the thesis about coincidentally existence of the patri and matrilocal prehistoric communities in the Bulgarian lands (H. Todorova).

So, we believe that in the Bulgarian lands the patriarchy is the only community formation with considerable development and changes after the middle of the Copper Age.

Arguments

Up to the middle of the Copper Age, the behavior slowly changed in the course of more than a Millennium. It can be assumed a relatively closed way of life based on the tribe organization that included blood-relative units. The tribe had no power to change and invade the neighbor or distant organizations. That 'organization-administrative' locking reflected clearly on the formation of what today is called local ethnocultural groups. The tribe organization and the followed norms of the customary law did not permit invading from outside on the local traditions of life – circumstance that secured the originality of the distant group. That statement can be confirmed by the fact the Early Neolithic pottery and other finds are peculiar in the different parts of Bulgaria – Upper Thrace, Northwest and Northeast Bulgaria. That statement can be extended for the other Balkan territories as well. The originality of the tribes is very expressive.

As we have stated elsewhere, in the second half of the Copper Age in Upper Thrace ran societal processes that provoked a strong expansion. That expansion finished with an establishment of the population traditions from the region under question over the neighbors, as well as with an organization of a strong tribe union. Its main characteristics were its organization on the territorial principal but not on the kinship.

It can be assumed at the beginning there was twofold system. On the one hand, the clan and tribe problems were judged in the blood-relative groups according to the old order. But when the problem concerned the new territory union and its functions, then the new laws were set. As we will see below that on the first view very elementary and easily performed change went to the deepest and serious societal changes. The first of them was the problem of the power.

According to the traditional-law forms of authority, the power in the prehistoric society was paired. It was divided between the chiefs and the priests but who were at once responsible for the law order. Having in mind that in the earliest periods of the human history the priest was a mediator between the man and the god, it looks naturally his power to have increased gradually. The time when the barriers of the Early Neolithic culture settled down in the Bulgarian lands is characterized by the higher, dominated power of the priests over the 'civil' institutions. Both the lines developed parallel and supplemented each other, but up to the middle of the Copper Age they were possible to be excised effectively only in the sphere of the local blood-relative unions.

In the time of the Late Copper Age, what can me defined as a time of great change, that structure of the power-ruling form preserved mostly by the governing and regulation of the life of the relative and tribal blood-relative units. It should not only regulate the following of the old laws of the customary law, but also of the new laws. In this period new centers of power were organized and consolidated – the organs of ruling of the tribe union. It was common and federative in character also including principal priests and chiefs. The materialization of that new structure can be seen in the emergence of Balkan cult centers and rock sanctuaries. However it can be assumed that highest power did not invade the life of the blood-relative groups. So, it continued the dualistic form of government.

The level reached of own law creation posed another important question – that of property. It can be argued the accumulation of the personal wealth and property in the households was much bigger than in the previous prehistoric periods. The inventory of the houses – despite of the limitation of their information – as well as of the cemeteries confirm that conclusion. There is diversity among the finds – from few unvalued finds to a bog gold adornment discovered in a building from the settlement near the village of Hotnitsa (Angelov 1959). The case with the burialgoods is even more eloquent. In the different graves of one and the same cemetery there are graves without inventory or only with one or two small and unvalued finds. But near them there are graves with such burialgoods, which include many and various vessels and other expensive finds – copper items, gold adornments, Spondylus adornments etc. That situation is indication, that at the end of the Copper Age it was a strong stratification in the population. If we add the results of the investigations of the Varna cemetery, the hierarchical and stratified structure of the society becomes clear. The difference in the personal and communal property should have been in direct relation with the hierarchical position of the individuals and their families.

The problems for re-decision in the Late Copper Age new social-political structure were those of the communal and tribe property. Especially, when it is posed the question of the possession of the important recourses of primary importance used to be the question of the property over the copper and gold mines, the graphite mines and of course of salt mines. According to the customary law, those whom the territory of the recourses belonged possessed them. By the agency of the centralization system, the property was taken in the hands of the rulers of the whole tribe union (for the copper ore mines see Raduncheva 1999). That influenced the change of the form of the realization and of the organization of the market. According to the tradition law each blood-kindred community choose the way, direction and the kind of trade. The quantity of the goods was also defined locally. In later Copper Age the craftsman's center produced certain quantity of goods for the market and defined their value according to the good equivalent established. When we add to that suggestion the fact that the new kind of recourse possession of the main productions, one goes to the conclusion, that the new society order allows the new leading institution of the tribe union to structure and control the market. The trade becomes considerably more centralized and regimented. At the same time the different kind of possession over the different goods goes to a possibility a difference in the good value to occur. We can certainly ascertain most expensive and farthest were distributed the goods produced by the supreme power – the big tribe union. For instance, the Karbuna treasure was made of copper, obtained from Ai Bunar mines near Stara Zagora Town (Chernykh 1978). It is known that the graphite obtained from Sveti Ilijski hills was distributed to the Mediterranean to the south and to the Carpathians to the north. As we stressed elsewhere (Raduncheva 1986) the salt, obtained from the Bulgarian Black Sea coast was the most important and standard good in the trade process. The new organization of property, trade and economic relation goes to a new degree of property and stratified status of the different levels of the society organization giving almost unlimited possibilities of centralization ruling. It can be assumed that in some cases – for instance in the mining process – the forced labor was used (Raduncheva 1999).

The emergence of new guard settlements confirms the mentioned above. The united territory of the newly established tribe union should be defended. It is very likely a guard defending function performed the tribes on the frontier territories. As most expressive instance are the Late Copper settlements along the Black Sea. Today they are under the water, but in the second half of the fifth millennium BC they were outposts of the tribe union, which met firstly the ones coming with peaceful or aggressive purpose. To that kind of sites it should be also added the frontier sites. So, the new societal formation had the function to defend the united territory.

In fact the question of inner and outer threat is not elaborated in the scholar literature. But having in mind the results of the archaeological research, we should stress there are no traces of destabilization and war destruction towards the end of the Copper Age. The rich cultural remains from that epoch allow one to assume the life was quite and such organized that it was possible a prospering society to have lived without troubles. Another argument confirming such statement is the fact that among the Late Copper finds there are no such classified as weapons for attack of defensive ones. The stone 'balls' and axe-hammers, as well as the flint and bone arrows are more hunting equipment that weapons. It is significant the fact that most often the last building levels of almost all the Late Copper settlements are poor or unfortified. It seems, towards the end of the epochs the population did not need them.

In the time of the second half of the Copper Age there was almost complete unification of the material and spirit culture in a vast geographical region of the Balkans. The innovation of the Late Copper ceramic style seems to have spread from the Thracian valley towards the other regions in the Balkans.

Along with the material culture, the religious belief was also unified, including distribution of similar three-sized figurines. At the same time sanctuaries' centers and high mountains rock secret places were ground and functioned, worshiped by population from vast areas in the Balkans. Following that kind of thinking, we could remind the organization of the king burial field near Varna Town clearly demonstrated the general change allowed the establishment of 'ruler' institution and distinguishing of a special place for burial ground, as well as of centralized system of worship. It can be assumed a process of increasing of the role and centralization of the spirit organization on the second half of the Copper Age and especially towards its end. The last dominated by taking decision important for the society despite as a structure it followed the level of structuring of the temporal power.

Conclusion

The changes in the social organization during the second half of the Copper Age correspond to drastic changes in the traditional-law norms. The meaning of the change is a transition from kinship decentralized society units (tribes) to centralized territorial unit (tribe union) and all corresponding secular, religious and controls institutions, which characterize the state as an institution. Of course, in later fifth millennium the steps are initial and timid. The territorial union in question more reminds the structure of the state of the Thracian known from the ancient record – federation of close territorially tribes. The long way to that change, in the course of thousands years of history, at the end of the fifth millenium BC was suddenly interrupted, leaving behind one shining way of development and the possibility of establishment of the earliest state in the world. Despite the disappearance of the highest developed in Europe at that time civilization there are some traces that occur considerably later in the social life of the settlers in Bulgarian lands.

CHAPTER 14

THE 4TH MILLENNIUM BC IN NORTHERN ITALY: WHEN AND WHERE WAS THE TRANSITION FROM NEOLITHIC TO ENEOLITHIC ?

NOTES FOR FURTHER DISCUSSION

Emanuela Montagnari Kokelj

Preface

In the last decade there have been many new discoveries and studies concerning the period defined as Eneolithic in northern Italy[1]. These developments have sensibly changed some aspects of the *scenario* reconstructed on the basis of previous data, to the point that a critical review would be necessary. It is obvious that such a review would require an open confrontation among scholars who usually deal with this period and the periods immediately preceding or following it. In this study I'll try to give a sketch of the main data now available and to indicate some issues that in my opinion are of particular interest for further discussion.

History of research

The history of research is a topic on its own and it is beyond my scope, but few basic data are necessary:

•Tthe term Eneolithic was introduced by G. Chierici in 1884, at the discovery of the Remedello cemetery, to indicate the period when metal objects were used besides lithic ones (Grifoni Cremonesi 1987: 16 sq.);

• This term is currently in use in Italian literature, together with Chalcolithic and Copper Age, with no semantic distinction, but from c. 1930s to 1960s Neo-Eneolithic was a common definition, clearly indicative of both theoretical and practical confusion;

• Remedello was first defined as a culture by O. Menghin (in 1931) and became The Eneolithic culture of northern Italy in the studies of P. Laviosa Zambotti (in 1939 and 1943) and M.O. Acanfora (in 1956) (Cocchi Genick and Grifoni Cremonesi 1989: 4-5), but the identification, in 1971, of collective burials in Lombardia, defined as the Civate culture by O. Cornaggia Castiglioni (1971a), and the subsequent discoveries of other funerary complexes, with different ritual aspects, again in Lombardia - Riparo Valtenesi /Manerba (Barfield 1983), in Valle d'Aosta - St. Martin-de-Corléans/ Aosta (Mezzena 1997) - and farther south, in Emilia - Spilamberto (Bagolini 1985) - they have clearly indicated that the Eneolithic of Northern Italy is much more articulated than previously thought (*see* for instance, Cornaggia Castiglioni 1971; Bagolini and Fasani 1982; Bagolini *et alii* 1987; Barfield 1988; Cazzella and Moscoloni 1992; Cazzella 1994);

• This almost exclusively funerary evidence contrasts with the evidence, only from settlements, relative to the cultures traditionally considered as representative of the final Neolithic (end of the 5th – beginning of the 4th millennium BC[2] – but see *infra* and *Appendix*) : the 3rd stage (Incised and Impressed style or Rivoli-Castelnuovo) of the Square Mouthed Pottery culture (Cultura dei Vasi a Bocca Quadrata = VBQ in the text), in eastern Lombardia, Trentino Alto Adige, Veneto and western Friuli; the Chassey-Lagozza culture, in western Lombardia, Piemonte, Liguria and farther south (Borrello 1984 – BUT the interaction of these cultures is actually very complicated, see *infra* - ***Figure 14.1***). This discrepancy in the nature of the evidence hampers the analysis of the transition, thus producing rather contrasting opinions as to how and when the passage should have occurred: in short, the majority of scholars place it around the middle of the 4th millennium BC (among the most recent literature *see* for instance, Barfield *et al.* 1995: 25; De Marinis and Pedrotti 1997: 289) or even later (Cremonesi and Vigliardi 1988: 308), while others anticipate it in the first centuries of the 4th millennium (Cazzella and Moscoloni 1992: 383).

I think that these divergences arise from the theoretical positions of the authors more than from the sets of data they use. A. Cazzella and M. Moscoloni, for instance, while stressing the apparent continuity at the turn of the 5th millennium, consider innovative, intertwined elements – such as the intensification of trans-alpine contacts, that might be related to an increase in stock-breeding and pastoralism as well as to the beginning of the exploitation of ore mines – as sufficiently indicative of the onset of the Eneolithic. They clearly accentuate the importance of processes of transformation versus their material results, in agreement with the opinion expressed elsewhere by A. Cazzella that 'Archaeologists like to find clear divisions between their phases, but this is not consistent with historical complexity. Even if social phenomena are more difficult to detect than the early forms of productive economy typical of the Neolithic, we may, one day, be able to speak of processes of 'chalcolithi-

[1] As "geographical" and "archaeological" boundaries of Northern Italy do not coincide, and prehistorians have interpreted them loosely, in this paper I have decided to include Liguria and to mention Toscana and Emilia-Romagna only in comparative terms.

[2] All dates are calibrated and usually presented with a 2σ level of confidence (see *Appendix*).

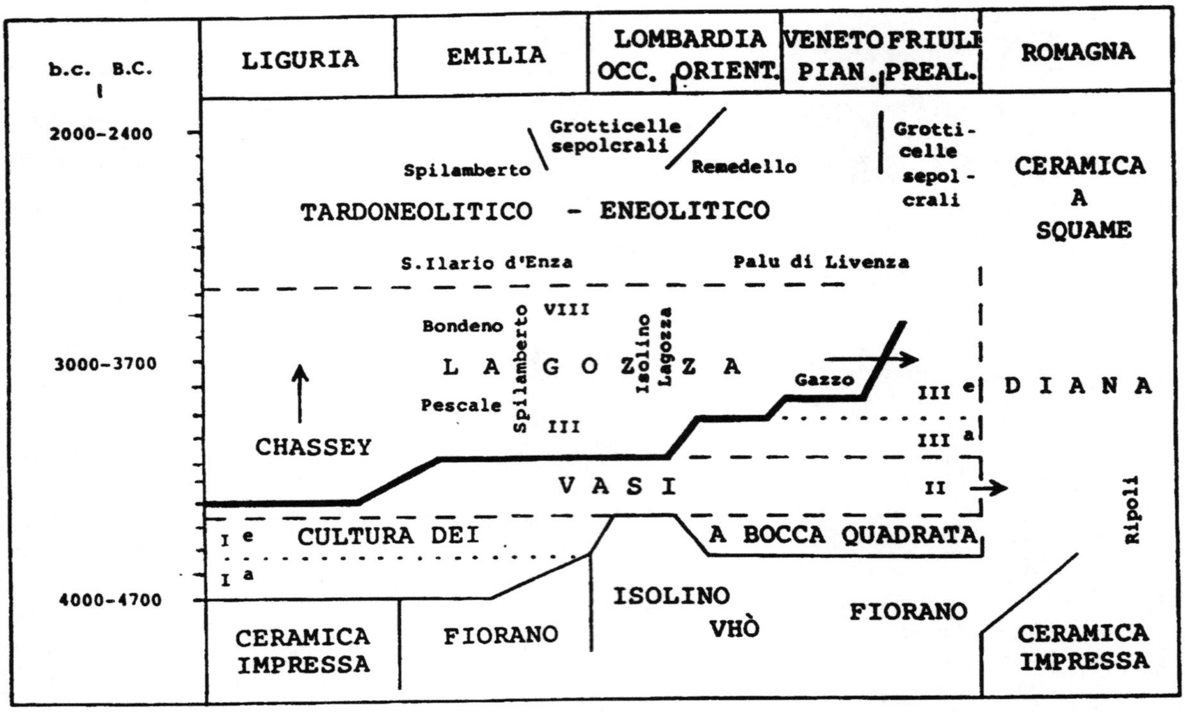

Figure 14.1. Chronological chart of northern Italian cultural developments from the 5th to the 3rd Millennium BC (from Bagolini, Barfield 1991).

sation' in Italy and adjacent areas' (Cazzella 1994: 13).

I basically agree with this position and think that an open confrontation on theoretical issues is particularly necessary now, in the light of the most recent acquisitions.

Recent data

In fact, in the last decade there have been many new discoveries and studies concerning the Eneolithic, often accompanied by new sets of ^{14}C dates, as well as re-examinations of complexes excavated in the past. These are not balanced equally in the different regions and this fact obviously makes a sensible difference.

Moving from east to west of Northern Italy the main new data are the following:

- Friuli-Venezia Giulia - Grotte Ciclami, Gallerie, Zingari, Mitreo, Ansa, Riparo Monrupino (Grotte Orso, Azzurra, Lonza and Gigante in preparation) (TS): critical re-analyses and exhaustive publications of materials from old excavations (*see* Montagnari Kokelj and Crismani 1997, also for previous literature);

- Veneto - Sovizzo (VI), 1994: discovery of a megalithic-funerary area with traces of Late Neolithic presence (Bianchin Citton 1995);

- Trentino Alto Adige - Senales/Giogo di Tisa (BZ), 1991: discovery of the "Similaun Man" - set of ^{14}C dates (Spindler *et al.*, 1994);

- Arco/Valle del Sarca (TN), 1989-1990: discovery of four decorated stelae (Bagolini, Lanzigher and Pedrotti 1992;

- Isera la Torretta (TN), 1990-1991: excavation of a stratigraphic open-air site attributed to the Late Neolithic-Eneolithic - set of ^{14}C dates (De Marinis and Pedrotti 1997);

- Lombardia - Remedello (BS): re-analysis of old documents and materials of the cemetery - set of ^{14}C dates (De Marinis and Pedrotti 1997);

- Riparo Cavallino/Monte Covolo (BS), 1990-1994: discovery and excavation of collective burials (Barfield *et al.* 1995);

- Piemonte-Valle d'Aosta - St. Martin-de-Corléans (AO): study of the megalithic funerary area excavated between 1969 and 1990 – set of ^{14}C dates (Mezzena 1997);

- Liguria - Arene Candide (SA): re-analysis of the deposit and materials excavated by L. Berbabò Brea and L. Cardini between 1940 and 1950 – set of ^{14}C dates (Maggi 1997);

- Research programme (in progress) on burial caves and rituals, land use in mountain areas and raw material procurement and working processes (Maggi and Del Lucchese 1988).

Analysis and results

It is obvious not only that new data resulting from recent, systematic field activities and new elaborations based on old evidence do not have the same weight, but also that these differences in quality have to be added to other differences, in quality and quantity, of available data in any analysis of the transition from the Late Neolithic to the Eneolithic.

A clear picture of the Italian situation up to the early 1990s is given, with rather different approaches, by A. Cazzella and M. Moscoloni in their volume of 1992 and by B. Bagolini and A.L. Pedrotti, L.H. Barfield and F. Nicolis in the chapter dedicated to norhern Italy of the *Atlas du Néolithique Européen* (Atlas [Bagolini and Pedrotti 1998]).

While I refer to this chapter for a complete review and for bibliography of sites not directly quoted in this text, I have extrapolated information relevant to the transitional phase (this means that I have taken into consideration sites with evidence of VBQ3, Chassey-Lagozza, Remedello and funerary traditions in caves and rock shelters). The most recent developments (see *supra*) are integrated to complete these data.

Moreover, as to radiocarbon dates, I have collated all dates listed by B. Bagolini and P. Biagi (1990), R. Skeates (1994) and again Bagolini and others (Atlas [Bagolini and Pedrotti 1998]) and updated them by adding those published after 1994 (*see Appendix*).

On this basis I have then tried to comply with the frequent claim to exclude from analysis all contexts that have raised, or may raise, controversy, by isolating the contexts which possess the main requisites of stratigraphical reliability, continuity of use and support of ^{14}C dates.

A comparative analysis of all these data (*see Appendix* for details) permit some basic observations:

• the number of sites that satisfy the requisites of reliable vertical stratigraphy and continuity from the Late Neolithic to the Eneolithic without, or almost without solution of continuity in estremely low: one site in Trentino, Isera - La Torretta, and one in Liguria , Arene Candide. Chronological data are available for these sites too, but only Arene Candide has a satisfactory set of dates for the whole sequence, while there are three dates for Isera that do not cross the 5th millennium BC.

In Trentino these dates of Isera 1 have been combined, on typological grounds, with other single dates from Fiavè I, Romagnano Loc, Acquaviva di Besenello and Riparo Gaban, so as to build up a chronological sequence that spans the whole 4th millennium (***Figure 14.2***). This sequence matches with the stratigraphical distinction of 5 phases proposed for Isera: Isera 1 = VBQ3; Isera 2 = VBQ3 with western influences (Chassey-Lagozza) and northern ones (Münchshöfen); Isera 3 = rather similar to Isera 2, with trans-alpine influences of the Altheim and Pfyn cultures; Isera 4 = characterized by the disappareance of western elements, while contacts with northern areas continue; Isera 5 = Eneolithic. Moreover, this sequence would support, and refine, the previous subdivision of the last VBQ stage into 2 parts, respectively without and with Chassey-Lagozza elements, and at the same time would cover the possible gap between the end of the VBQ culture and the beginning of the Eneolithic. But it is important to remind that the validity of such a subtle chrono-cultural articulation still has to be demonstred for sites other than Isera.

• Rivoli-Rocca, in Veneto, might also have been a site with requisites similar to those of the two previous sites, if the authors of the excavations themselves had not acknowledged that the presence of an Eneolithic level was quite dubious (Barfield, Bagolini 1976. 141).

As to the Late Neolithic, there are two dates, coming not from the stratified deposit (site D), but from an area of intersecting pits (site L, pits P-Q-R). One of the two dates of Rivoli falls only in the 5th millennium BC, the second goes well into the 4th; both samples come from pits without Chassey-Lagozza materials - but these are present in other, undated pits attributed to the same sub-phase Rivoli-Castelnuovo I, in 2 cases out of 4 even with elements of the older, Rivoli-Chiozza, phase (see *ibidem*, fig. 29 and 35 for provenance of elements and cultural attribution of the pits of provenance). This might imply the possibility of either anticipating the presence of Chassey-Lagozza elements or extending the duration of a 'pure' VBQ3 complex.

• The interaction of Chassey-Lagozza and VBQ is very complex (Bagolini, Barfield 1991; ***Figures 14.3-14.4***) and, besides Isera-La Torretta and Rivoli-Rocca, there are few other stratigraphically investigated sites where this interaction is documented: Castelnuovo di Teolo and Ronchetrin di Gazzo in Veneto; Breno, Isolino di Varese, Ostiglia and Rocca di Manerba in Lombardia. But none of them have radiocarbon dates: the only site that might have had all the requisites, Monte Covolo in Lombardia, has recently been critically reconsidered (*see infra*).

This means that the chronological position of mixed Chassey-Lagozza/VBQ contexts is still highly uncertain, and that they might be placed in the first half of the 4th millennium BC only if the dates of Isera 1 are accepted as the *terminus post quem* for the spread of Chassey-Lagozza influences from western regions eastwards.

Again, only the combination of dates from three, geographically distant sites - Isera-La Torretta in Trentino, Arene Candide in Liguria and Spilamberto/site VIII in Emilia – would support the parallelism between Chassey in Liguria and the early VBQ3, without influences from Chassey-Lagozza, at least in Trentino in the second half of the 5th millennium BC, on the one hand, and the later development of Lagozza in the first half of the 4th millennium, on the other.

• As to the chronological position of "pure" Lagozza contexts, the single ^{14}C date for the Lagozza occupation at Spilamberto/site VIII does not cross the middle of the 4th millennium BC, but the majority of the dates from the eponymus site, Lagozza di Besnate, do so. If the latter were reliable, a development of still Late Neolithic aspects contemporaneous with already Eneolithic ones - absolutely plausible in a period of cultural transition - might be expected. But such a coexistence might be artificial as well, since doubts can be raised about the reliability of the dates[3] and the use of different definitions such as 'Sub-Neolithic', 'Sub-Lagozza', 'First Eneolithic',

[3] Although these dates are regularly quoted in the literature, the fact that they were produced in the 1960ies by the laboratories of Rome and Pisa and that at least the latter is essentially unreliable suggest that the dates should be used very cautiously. These and other observations on chronology derive from constructive conversations with Claudio Arias, whom I wish to thank warmly.

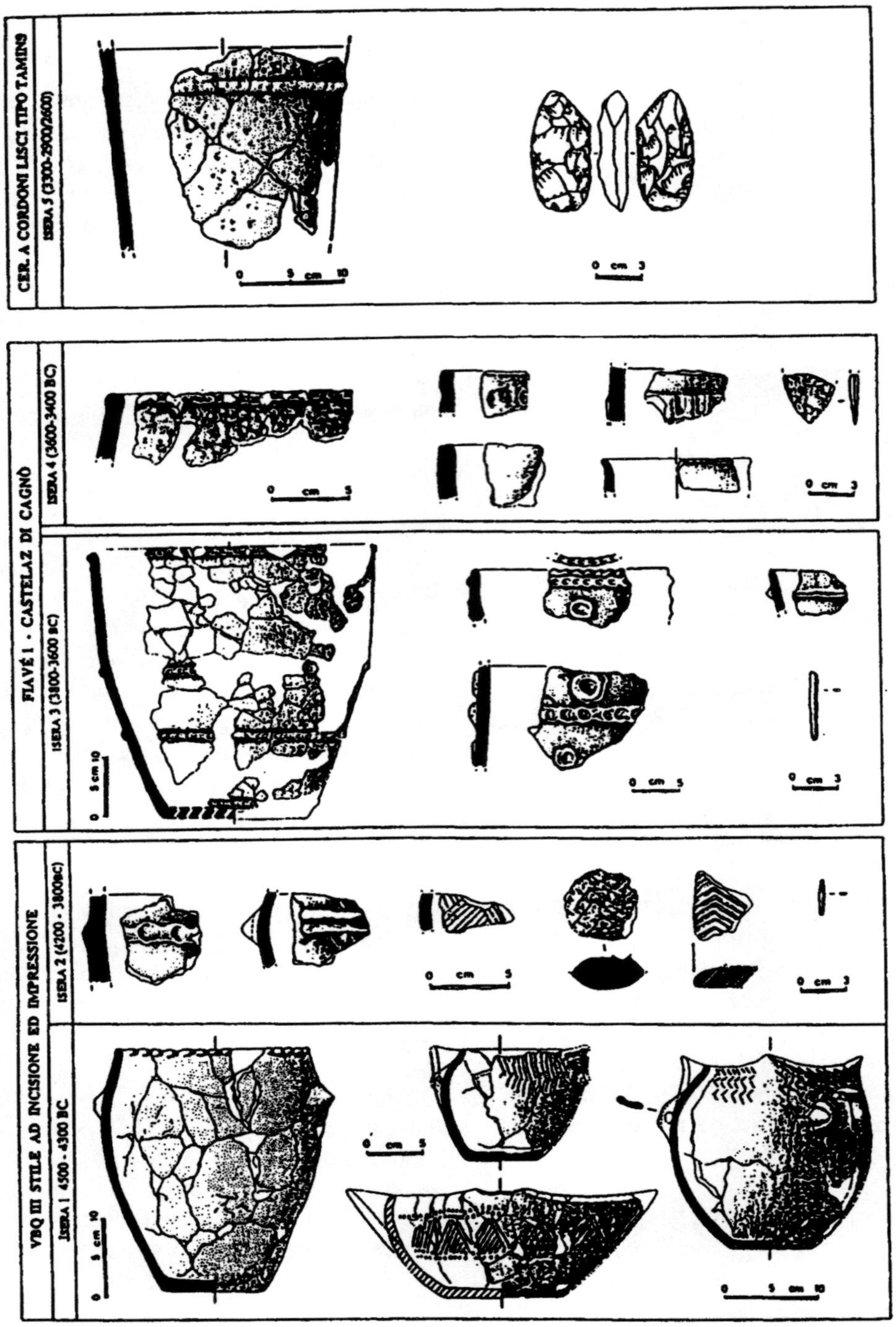

Figure 14.2. Chronological-typological sequence of Isera-La Torretta, Trentino (from De Marinis, Pedrotti 1997).

'Transitional Phase' (Cremonesi and Vigliardi 1988. 309-310) – for presumably similar contexts can also interfere[4].

• As to undoubtedly Eneolithic contexts - besides those of Arene Candide and Isera La Torretta (though the latter without ^{14}C dates), with Neolithic and post-Neolithic superimposed layers - horizontal stratigraphical data combined with a series of chronological ones are available only for the cemetery of Remedello (Lombardia, ***Figure 14.5***) and the megalithic-funerary area of St. Martin-de-Corléans (Valle d'Aosta).

[4] The article written by R. Peroni (1962-63) to present the results of the stratigraphical excavation of a famous rock shelter in Toscana, Romita di Asciano, is still illuminating to understand the complexity of the influences in the period of transition under study and the consequent inadequacy of a simple definition.

The recent ^{14}C dates of Remedello have contribuited to a new subdivision of the Eneolithic into 2 stages – 'Remedello 1' = 'Rame 1' (3400 - 2900/2800 BC) and 'Remedello 2' = 'Rame 2' (2900/2800 - 2400 BC), but the dates actually cover also the second half of the 3rd millennium and the whole 2nd millennium BC (*see Appendix*).

As to St. Martin-de-Corléans, there is a discrepancy between the sets of dates produced in the 1970ies and those produced in the 1990ies, with a shift from the 4th to the 3rd millennium which seems to be accepted as more coherent by the majority of scholars (see, for instance, De Marinis 1995: 220; Gallay 1995: 178-180; Pedrotti 1995: 265).

• Less marked discrepancies exist between the old and the

new dates obtained for Riparo Valtenesi/Manerba: two of them are slightly older (second half of the 4th millenium BC) than those of Riparo Cavallino, while the others, covering the first half of the 3rd millennium, are broadly coincident with those of the second rock shelter and of Remedello. The two sites are similar also in typology (cemetery), location (western Lombardia) and nature of investigation (stratigraphical excavations). These parameters should apply also to Buco della Sabbia/Civate (***Figure 14.6***) and, a part from location (Trentino), to Riparo Gaban, but their single dates - and the one from Ca' dei Grii, a cave in Lombardia with stratigraphical problems - would show a certain chronological *décalage* (for further discussion of this important point see *infra*).

Among all these contexts that exemplify the "funerary tradition", rather than the "culture", of collective burials in caves and rock shelters of the pre-alpine regions - as traditionally opposed to the individual burials of Remedello and Spilamberto in the Po Plain -, Riparo Cavallino occupies a particularly important position for two main reasons: the contribution that its recent excavations have given to understanding the funerary ritual as well as its connection, together with the close Riparo Persi, with the near site of Monte Covolo, the only connection between funerary area and settlement known so far in Northern Italy (Barfield *et alii* 1995. 26).

The settlement of Monte Covolo has long been considered fundamental for its sequence of "Recent" Neolithic - a mixture of Lagozza with VBQ3 and a decorative style called "Breno Nera", identified about 10 years ago at the eponymous site and also at Manerba, Nave and Rivoli-Rocca (Barfield *et al.* 1995: 24) - ; Early Eneolithic, characterized by the so called "White Ware"; Late Eneolithic, represented by Bell Beakers; Early Bronze Age of the Polada culture and later evidence. But problems of stratigraphy as well as of ^{14}C chronology - that imply further problems of relationship between settlement and funerary rock shelters - have been acknowledged by the author of the excavations himself (Barfield *et al.* 1995: 6-7, 46 and 25).

• To complete the *scenario* of funerary evidence, I have to quote also the collective burials in caves and rock shelters of Liguria[5] and the cemetery of individual graves of Spilamberto, to the south of the Po River in Emilia.

In Liguria a research programme on burial caves and rituals is in progress and new data are likely to be expected in the near future (*see supra*). As far as it is known at present (Maggi 1996), there are 28 caves/rock shelters used during the Eneolithic and/or Early Bronze Age, 16 of which with human remains. Besides Arene Candide (but without funerary evidence for this period), Grotta del Pertuso and Grotta di Prima Ciappa in val Frascarese would be the only two caves with ^{14}C dates: these dates cover together the second half of the 4th and a large part of the 3rd millennium BC.

As to Spilamberto (***Figure 14.7***), the cemetery – separated from the Late Neolithic settlement area (*see supra*) – has one single date that falls in the first half of the 3rd millennium, thus corresponding to part of the period of use of Remedello, the other necropolis of the Po Plain with individual graves. In contrast with Remedello, Spilamberto shows ritual aspects inflenced also by southern Italian cultures.

• To complete the whole *scenario* of reliable evidence attribuited to the Eneolithic, I still have to mention the very famous, casual discovery of the 'Ice Man' in the Similaun glacier (Alto Adige). This discovery has triggered off substantial re-analyses of specific categories of material culture, as well as a reconsideration of conceptual and chronological aspects, due to the good set of dates that place the death of this man in the second half of the 4th Millennium BC.

Some comments

Just the Similaun mummy is useful to underline one basic difference among the data mentioned so far. In this case ^{14}C dates refer to a single event, the death of a man, totally isolated from other events or sites. The dates of individual burials are again related to a single person, but usually within the context of a prolonged use of a cemetery. In the case of collective burials it is extremely difficult to ascertain the association between a date and a single person. In the case of a settlement, the connection is even looser, as a date might reflect one event, one action out of a continuous succession of events. A direct comparison among dates coming from so different contexts might consequently result in misleading conclusions.

Other differences - such as typology of sites (open-air sites, damp sites, caves etc.) and nature of samples (charcoal, wood, human or animal bones etc.) - have also to be considered as they imply peculiarities of cultural use and depositional and post-depositional conditions, on the one hand, differences in the parameters to be used for measurement and calibration, on the other.

Moreover, interpretative discrepancies may also interfere, in particular in real or apparent situations of chronological overlapping in moments of epochal changes.

On the grounds of what has been presented so far, as to the presence and distribution of sites that have all the parameters necessary to discriminate between reliable and unreliable contexts, Arene Candide in Liguria, on the one hand, Remedello in Lombardia and Isera-La Torretta in Trentino are important points of reference that MIGHT indicate a chronological *décalage* from west to east in the transition from late Neolithic to Eneolithic cultural aspects; BUT large areas of Northern Italy still do NOT have stratigraphical sites, either radiocarbon-dated or undated, to support a global reconstruction of the cultural evolution in the period under examination.

Nevertheless, it is clear that the association of stratigraphical reliability, continuity of use and ^{14}C dates support represents an ideal situation that is far from being frequent in Prehistory;

[5] The funerary caves/rock shelters of north-western Toscana should also be included in this review, but this has intentionally been avoided in order not to open a new area of discussion that would involve cultures of central and southern Italy as well.

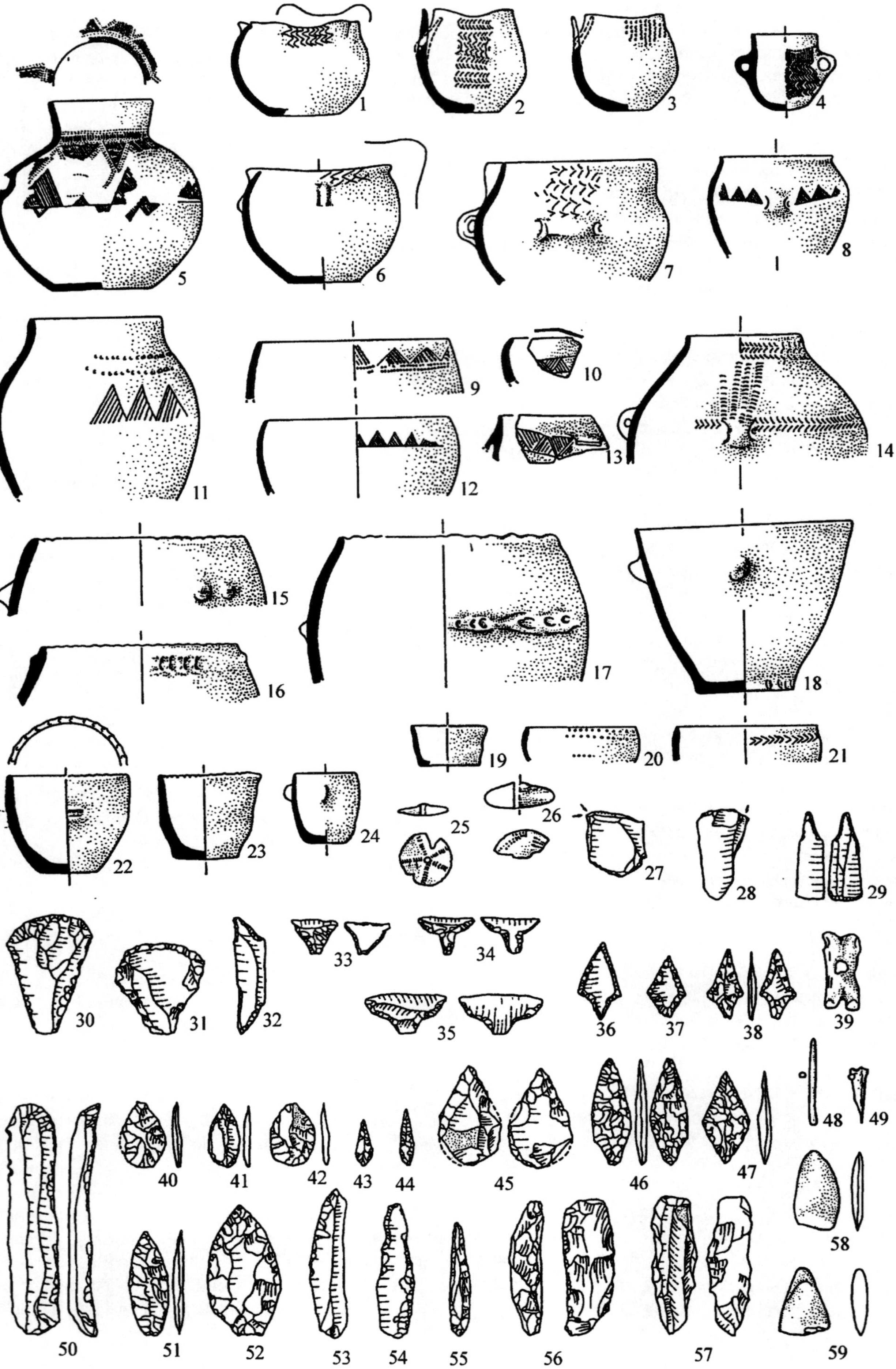

Figure 14.3. Main components of the the 3rd stage (Incised and Impressed style or Rivoli-Castelnuovo) of the Square Mouthed Pottery culture (from Atlas).

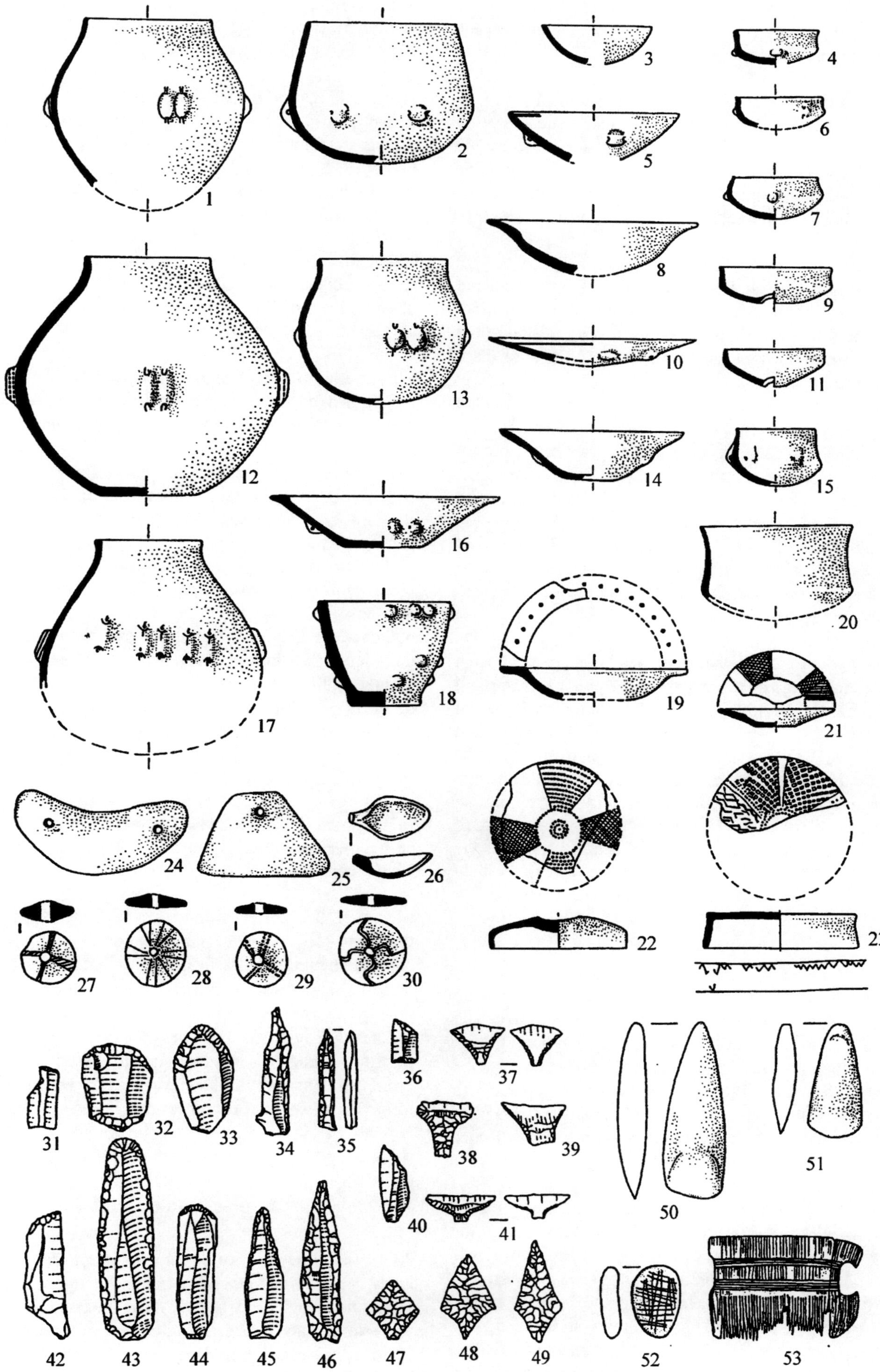

Figure 14.4. Main components of the Chassey-Lagozza culture (from Atlas).

very often it is only one of these parameters to be satisfied and even more often it is typological analysis the main instrument to trace cultural connections.

If we cannot but accept these objective limitations inherent in the archaeological evidence, at the same time we must be very careful not to stretch the results obtained at a specific site, or in a specific micro-region, to a macro-regional scale without critical re-evaluation: cultural sequences worked out in specific, limited territories should not be automatically generalized.

I have already stressed the incidence of interpretative discrepancies in particular in moments of cultural transition. Besides the differences in terminology directly related to the concept of Eneolithic (and consequently Late Neolithic), there are other differences concerning the basic concept of culture. The most patent example is given by Remedello itself, called a "culture" by some scholars, a "group" by others. But the same is true also for some regional groups of collective burials in caves and rock shelters: the one in Lombardia, for instance, was defined as Civate culture by O. Cornaggia Castiglioni at the discovery of the eponymous site, but this definition has later been dismissed by L. Barfield and others. The group of north-western Toscana is often, but not alway, referred to as Vecchiano *facies*, while no specific definition has been proposed for the group of Liguria.

In 1980s opportunities of confrontation were created by the 1985 national congress of Firenze (*see* Bagolini B., Barfield L.H., Cremonesi G. 1987) and by the international ones of Lazise-Verona, in 1980 (*see* Bagolini and Fasani 1982) and Viareggio, in 1987 (*see* Cremonesi and Vigliardi 1988). The problems of the transition from the Late Neolithic to the Eneolithic in Italy, in particular the scarcity af data and the contextual diversities, were focused by some scholars in their large-scale reviews, but the concept of culture as well as the parameters sufficient and necessary to choose between the label of Neolithic and that of Eneolithic were not deeply rediscussed.

Interpretative models

The relatively high increase in archaeological evidence in the 1990ies has still not produced any significant theoretical debate among Italian scholars, nor models of interpretation. I think of the methodological elaborations of foreign scholars, for instance C. Strahm (1988) and others, whose papers were collected and edited by J. Lichardus in Germany (*see* Bagolini and Barfield 1991) and by Voruz in Switzerland (*see* Roger 1995).

It might be interesting to notice that the "availability" model of M. Zvelebil and P. Rowley-Conwy – originally created for another fundamental period of transition in Prehistory, that from the Mesolithic to the Neolithic - was tentatively applied by J. Lewthwaite (1987) to Italy, by stretching the process over a very long period which actually includes the 4th millennium (Lewthwaite 1987: Fig. 2; ***Figure 14.8***). This proposal of application was made in very general terms and never put to the test: an attempt in this direction might be promising, but – if the time span is actually stretched so far – this model might clash with the "secondary products revolution" model specifically elaborated by A. Sherratt in 1981 (Sherratt 1997).

I have said that no models have yet been formulated for the Italian Eneolithic, but what has been produced is a research programme for the Late Neolithic and the Eneolithic, advanced first by B. Bagolini and L. Fasani (1982: 348 sq.), embraced later by others and presented by L. Barfield in this form: 'The material culture and burial traditions present conflicting and overlapping patterns of similarities which has led to the conclusion that, rather than seeking to define cultures or culture groups, a description of the patterns of the distribution of individual artifact types is all thet we can hope to achieve at present.' (Barfield 1988: 411).

This methodological approach is strictly related to the definition of culture, but in operative terms it actually proceeds in the "disarticulation" of a culture into its main components.

I see a link between this approach of culture disarticulation and the following statement made by B. Bagolini and L. Barfield about the Square Mouthed Pottery culture, the culture that unified almost the whole of Northern Italy in the first half of the 5th millennium and progressively reduced its presence to central-eastern regions due to the penetration of the Chassey-Lagozza culture (*see supra*): 'The square mouthed pot, once it had appeared, seems to have served as a powerful unifying symbol for the culture since we can see that it survived several centuries and through three radical changes in pottery design, which characterized the main sub-phases of the culture. Without the square mouthed form indeed, these stages might well have been identified as three totally different cultures' (Bagolini and Barfield 1991: 288).

In my opinion this passage underlines quite well the risks of an over-evaluation of typological comparisons independent from a contextual analysis.

Moreover, it is perfectly consistent with the idea underlying this study - but present also in other contributions of the same authors (in particular Barfield 1981) - that the search for, and trade in raw materials should have played a role of fundamental importance from the Neolithisation onwards. As to the spread of Chassey-Lagozza is concerned (***Figure 14.9***), they say: 'One possible scenario, or model, could be ethnic migration prompted perhaps by economic causes (or results) which are reflected in pattern of trade in raw materials. The western Alps are rich in raw materials for axe production and evidence for axe production is here abundant.' (*ibidem*, 292-293).

Application of models

As to the risks of typology independent from contextual analysis and to the possibility of using and testing the "trade in raw materials" model, I will make an example concerning the Trieste Karst (Venezia Giulia), an area that, for various reasons, has often been disregarded in general analyses of Northern Italy.

B. Bagolini and P. Biagi (1983: 201-202), M.A. Borrello

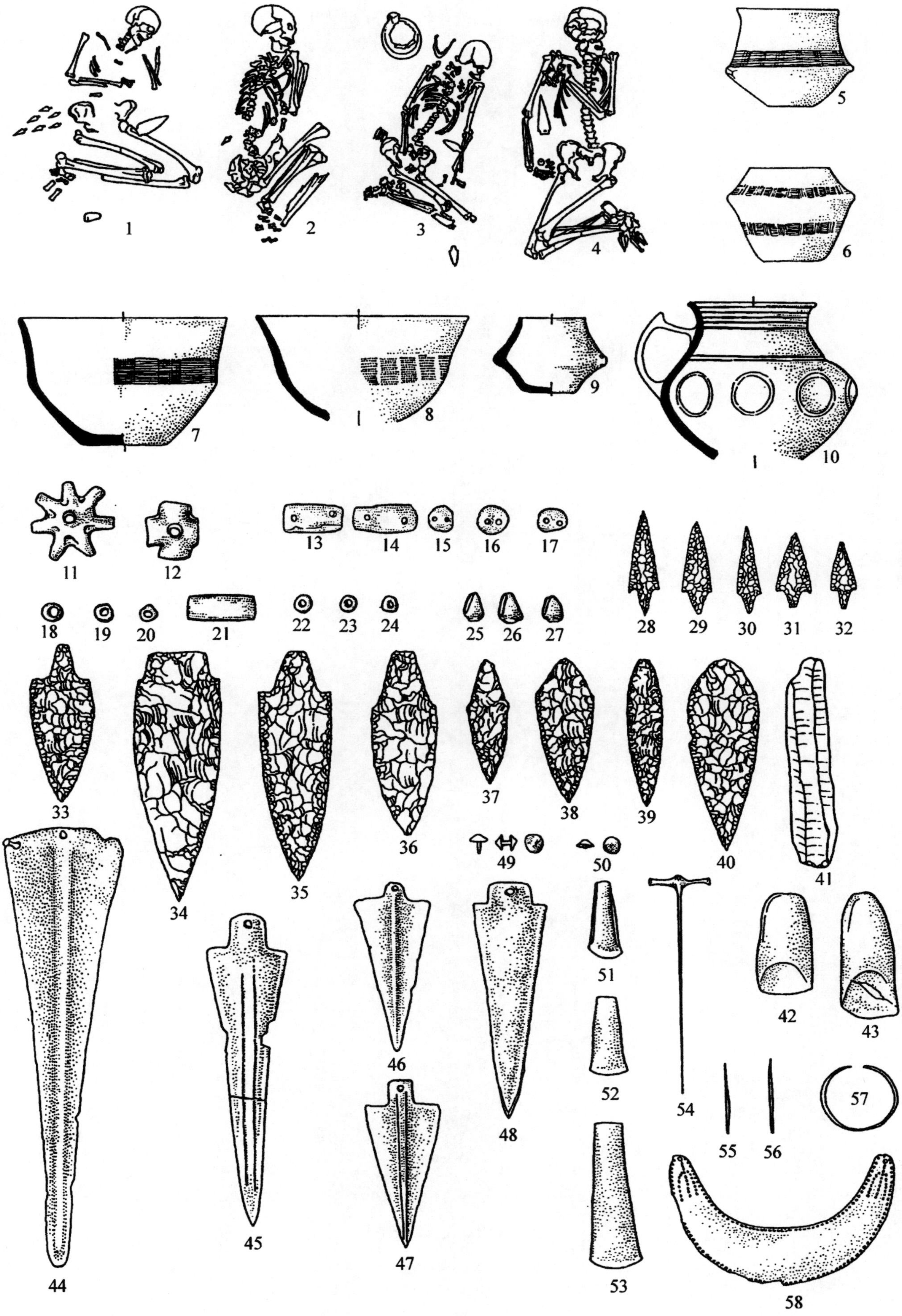

Figure 14.5. Main components of the Remedello and metope decorated fine ware group (from Atlas).

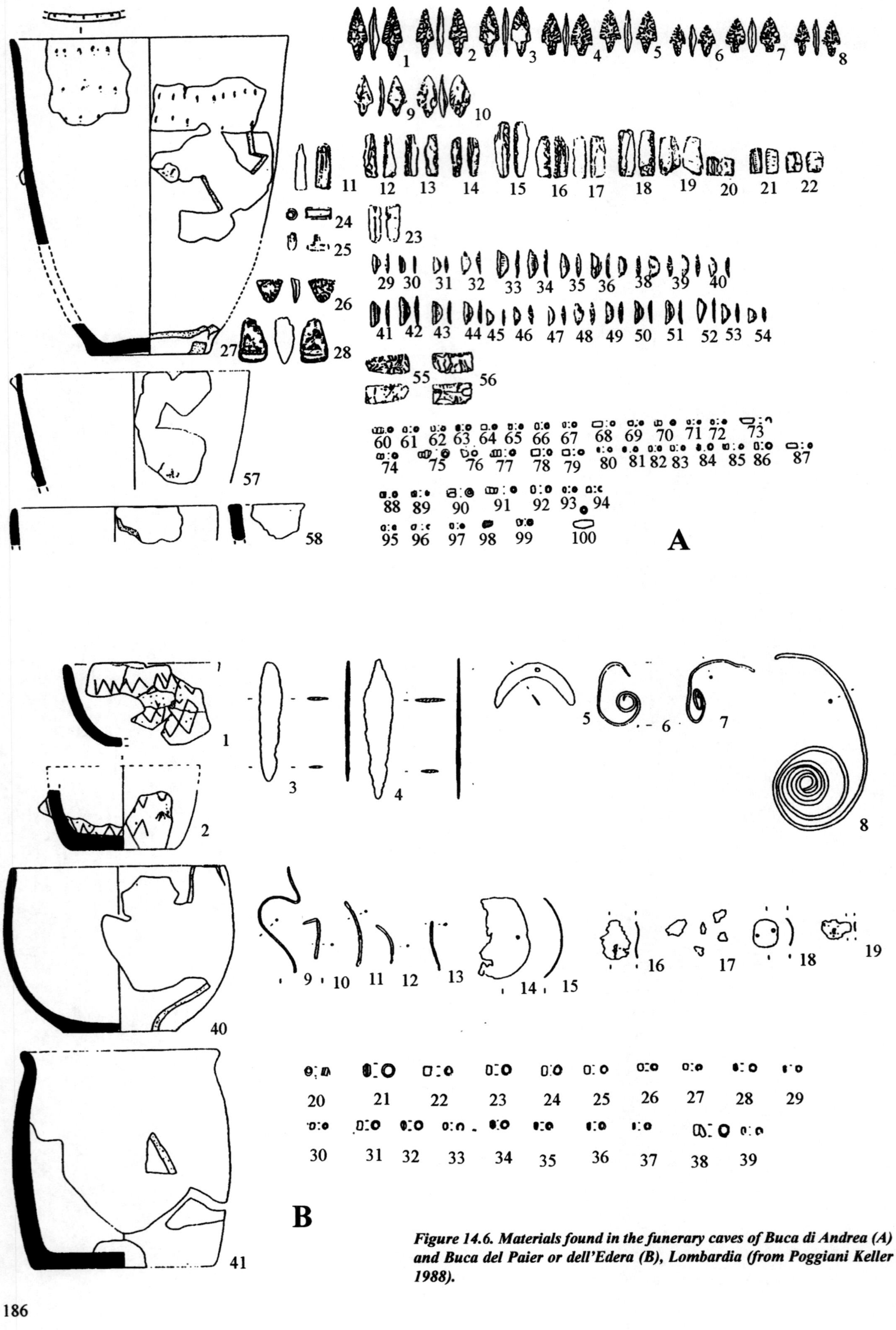

Figure 14.6. Materials found in the funerary caves of Buca di Andrea (A) and Buca del Paier or dell'Edera (B), Lombardia (from Poggiani Keller 1988).

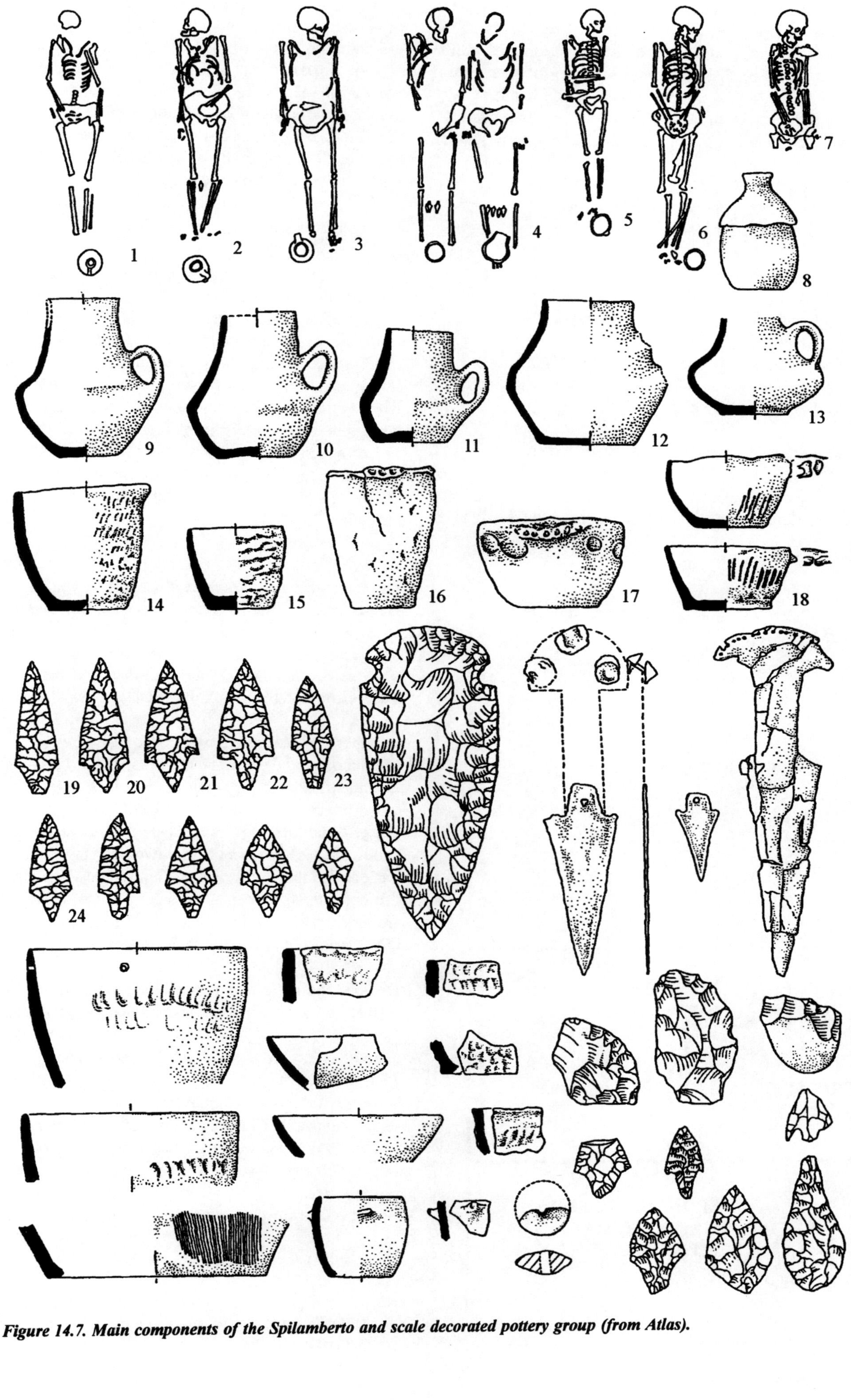

Figure 14.7. Main components of the Spilamberto and scale decorated pottery group (from Atlas).

(1984: 105-106) and also B. Bagolini and L. Barfield (1991: 292) mention this area, and Grotta delle Gallerie in particular, for the finds of square mouthed pots and Lagozza sherds. The materials of this cave have recently been re-examined (Gilli and Montagnari Kokelj 1994) and the number of sherds that might be referred to VBQ3 and Lagozza culture in this cavity, as well as in the others of the same territory, is extremely low. Moreover, focusing on the 'presumed' Lagozza pottery, it is interesting to notice that:

• These same types are well comparable also with non-decorated types of the Dalmatian Danilo culture (***Figure 14.10***);

• The final stages of this culture are still uncertain in terms of chronology and of relationship with the Hvar culture;

• The Karst maintained strong links, though with solutions of continuity, with the eastern Adriatic coast from the Neolithic till the Early Bronze Age and even later;

• The stratigraphy of the Grotta delle Gallerie is unreliable;

• No physico-chemical analysis of the pottery has been carried out so far, though the pottery seems to be different at a macroscopical analysis.

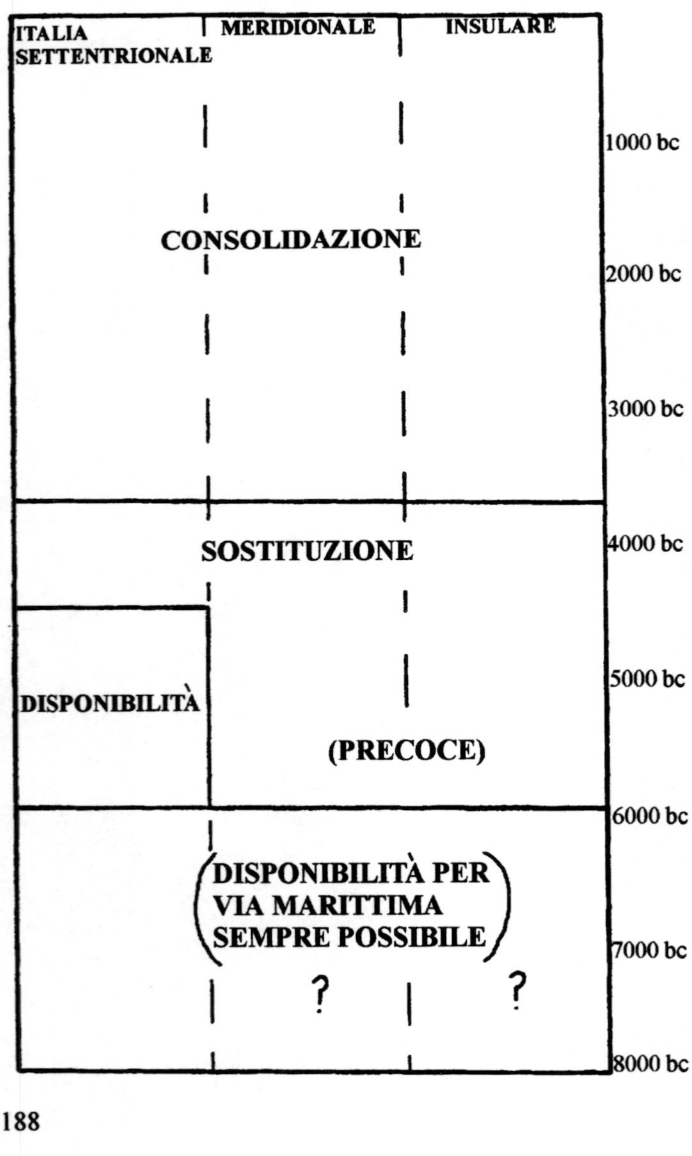

Figure 14.8. Attempt of application of the "availability" model of M. Zvelebil and P. Rowley-Conwy to the Neolithisation of Italy (from Lewthwaite 1987).

In my opinion, on these grounds any conclusion about a specific connotation of the Trieste Karst, marked by a stable penetration of western or eastern cultures, would be strongly biased. On the contrary, a role of mediation of this area between the eastern Adriatic coast and Northern Italy would be highly realistic if seen in the light of the 'trade in raw materials' model, in particular of an 'Adriatic trade axis' involving flint from the Lessini, Lipari obsidian, greenstones, *Spondylus* shell beads etc. (Thorpe *et al.* 1979; Barfield 1981; Bagolini and Barfield 1991; D'Amico *et al.* 1996).

Again on models and lines of research

To go back to general problems, the approach of isolating and studying individual artifact types, of disarticulating a culture into its main components, was applied in the 1980s to the Remedello cemetery itself, first with the enucleation of the metope decorated fine ware (Bagolini and Fasani 1982. 348). In the cemetery only 9 pots are representative of this style, while pottery is present on the whole in 15 out the 84 burials with grave goods (the cemetery is composed of 128 pit graves, 117 of which discovered between 1884 and 1886: all graves contain single inhumations, apart from 2 with two bodies (Cornaggia Castiglioni 1971; De Marinis, Pedrotti 1997. in particular 260-267).

All over Northern Italy 2 other pots decorated with a metope design come from Colombare di Negrar, in Veneto, and single sherds from 11 different sites (***Figure 14.11***). B. Bagolini and P. Biagi, who studied this pottery and its distribution, considered it as 'an intrusive element common to various facies of the same chronological horizon', that is, loosely, Eneolithic/ Early Bronze Age (Bagolini and Biagi 1988. 186).

A rather different position has been taken by R. De Marinis, the scholar who - on the grounds of the recovery of old documents and of his study of the two classes of materials which are considered those that identify the cemetery, i.e. lithic daggers and metal axes and daggers - has advanced the new subdivision of Remedello into two sub-stages. De Marinis has divided the 9 pots of Remedello into two groups and has attributed them, as well as the other typologically similar pots from different sites, to one or the other of the two sub-stages (De Marinis and Pedrotti 1997: 267 sq.).

Besides the metope decorated fine ware, other individual artifact types were isolated and studied in the 1980s.

As to pottery, four main categories of coarse ware were identified, as summarized by L. Barfield (1988: 413-415): pottery with rims with partial perforations or bosses (ceramica a fori non passanti – *see* also Gambari and Venturino Gambari 1985-1986) and pottery with finger pressed or plain cordons, apparently exclusive of Northern Italy; brushed decorated pottery (Besenstrich), present also in Toscana; scale decorated pottery (ceramica a squame), well represented in the

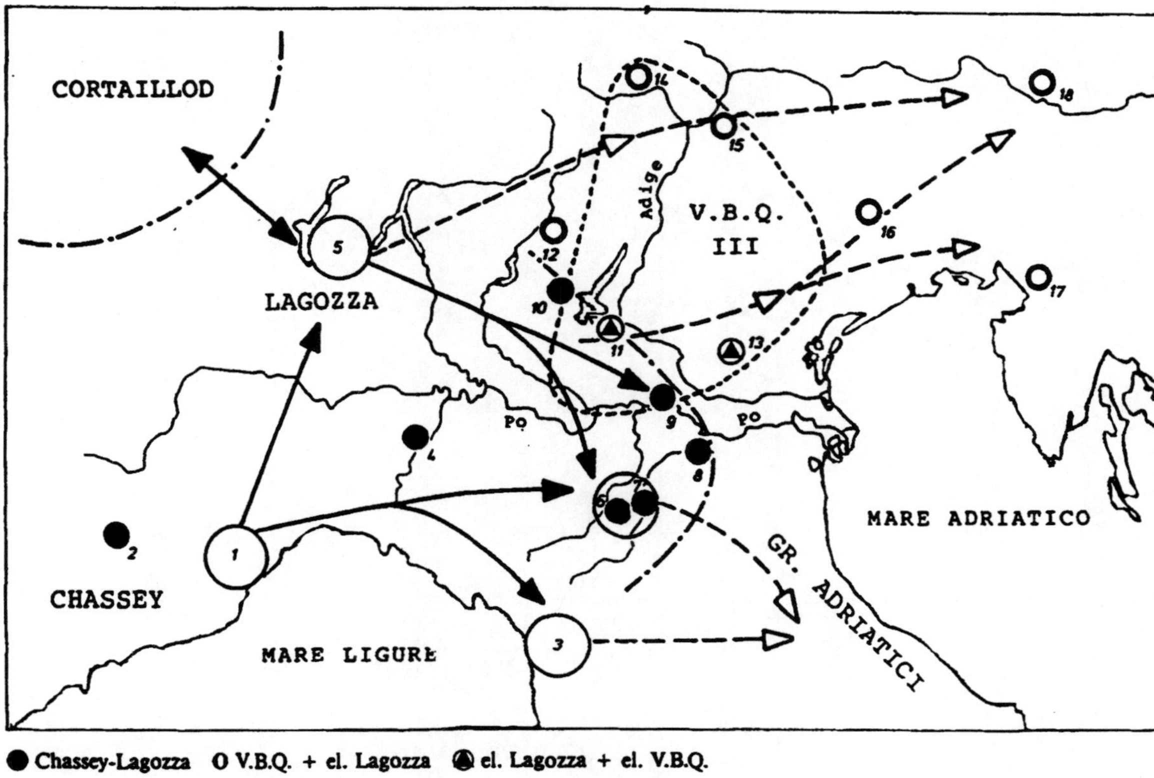

Figure 14. 9. Western penetration of northern Italy by the Chassey-Lagozza culture and its interaction with the Square Mouthed Pottery culture (from Bagolini, Barfield 1991).

Figure 14.10. Comparison among the "presumed" Lagozza pottery from Grotta delle Gallerie (1 – from Gilli, Montagnari Kokelj 1994) and Grotta dei Ciclami (2 - from Gilli, Montagnari Kokelj 1993) in the Trieste Karst, the Chassey-Lagozza pottery from La Romita di Asciano in Toscana (3 – from Peroni 1962-63) and pottery of the Dalmatian Danilo culture (4 – from Batovic 1979).

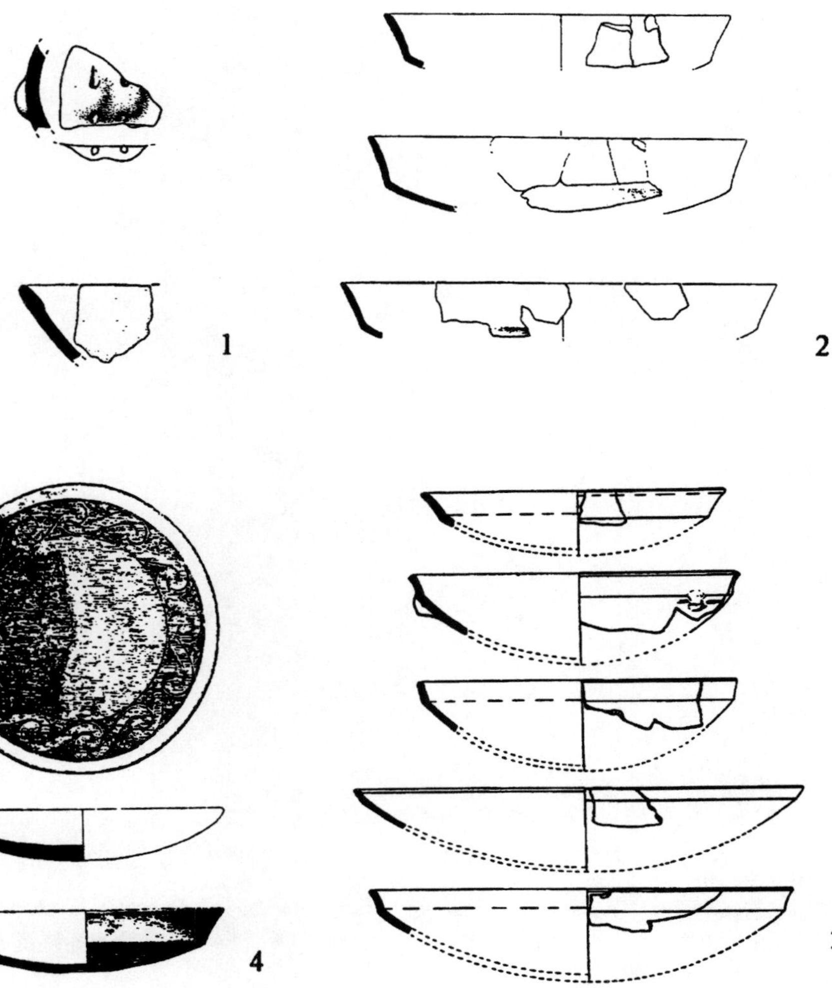

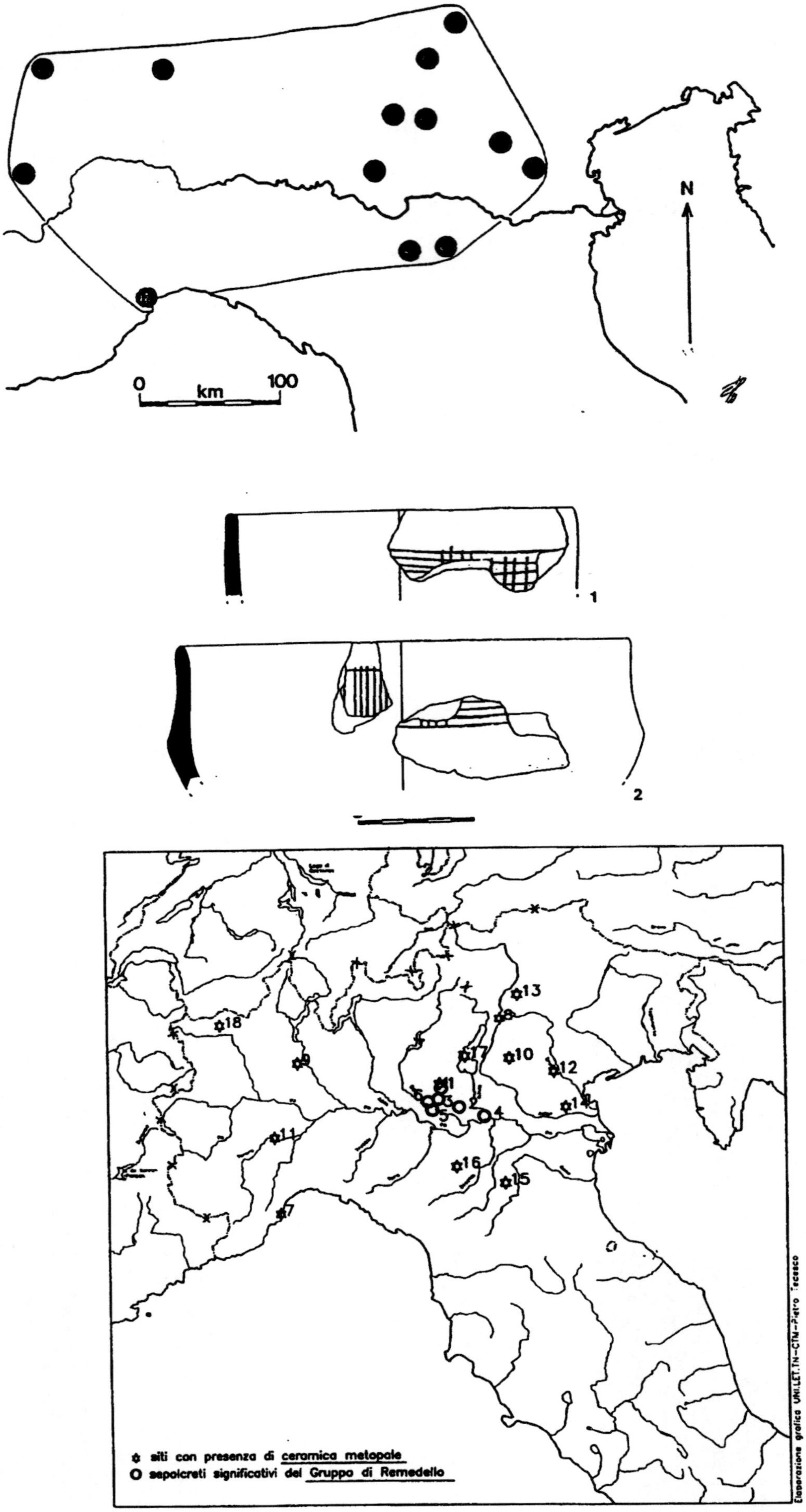

Figure 14.11. Distribution map of the sites with metope decorated potsherds (above – from Bagolini, Biagi 1988 – the examples are from the same source) integrated with the cemeteries of the Remedello group (below – from De Marinis, Pedrotti 1997).

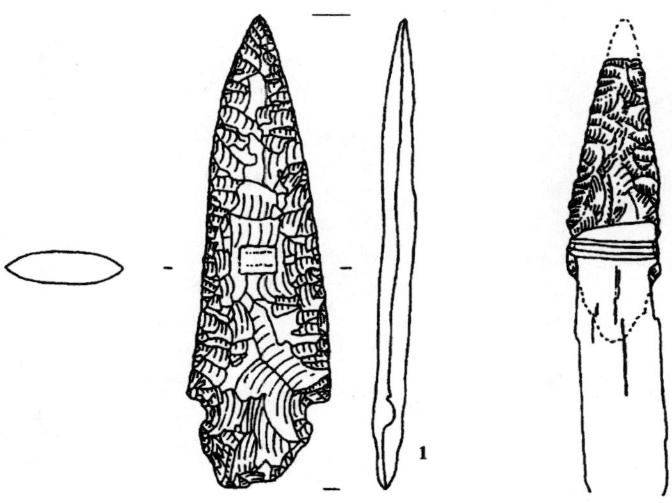
1 2

Figure 14.12. Comparison between the flint dagger from grave 97 of the Remedello cemetery (1) and the dagger of the Similaun Man (2 – above, from De Marinis, Pedrotti 1997).

Spilamberto cemetery (Emilia) (Bagolini 1985), apparently distributed south of the Po River in Northern Italy but scattered all over central and southern Italy (Bagolini and Cremonesi 1988). In his study L. Barfield could but admit that no clear relationship between artifact distribution and cultural tradition has yet emerged, in spite of the potentiality of this kind of analysis demonstrated by J. Hodder.

As to lithic industry, the dagger typical of the Remedello cemetery has been the object of two recent studies, still in progress. The preliminary results hint at two somehow different approaches. A basically morpho-typological approach has been adopted by R. De Marinis: he has analysed 50 instruments - 27 from graves and 9 out of context in the Remedello cemetery, plus 12 and 2 respectively from those of Fontanella Mantovana and Volongo – and on this basis has defined 18 different types of daggers. This typology has then been extended to the rest of Northern Italy (De Marinis and Pedrotti 1997: 267 sq.).

The reverse, that is a sample higher in number and larger in geographical terms, has been chosen as the basis for building a new typology by E. Mottes (1996), who has also strictly associated typological parameters with technological and functional ones. The last point seems particularly promising in the light of recent discoveries which confirm a basic difference between ritual, symbolic artifacts and functional ones: the dagger that the Similaun Man brought with him is a clear example of a tool destined to be used during his journey up into the mountains (***Figure 14.12***).

Other specific discoveries of Remedello daggers open up new perspectives also on chronology as well as on trade connections. Daggers of this typology found in Italy at Fiavè (Fiavè 1) and north of the Alps at Ergolding Fishergasse and Pestenacker (Altheim culture) would indicate that their production might have started well before their deposition as grave goods in the Remedello cemetery (De Marinis, Pedrotti 1997. 258, note 14). This evidence would strenghten the long time of use of these weapons/tools, not only after the span of time covered by the Eneolithic Remedello cemetery, but even before.

Moreover, these and other discoveries on both sides of the Alps represent a further proof of exchange/trade connections that would have started – or intensified to the level of archaeological visibility - in the second half of the 5th millennium and in particular in the 4th millennium (see also Pedrotti 1990).

Real objects and typological affinities document these connections in particular at Isera-La Torretta, the important site of Trentino where recent excavations have revealed a stratigraphical sequence covering the Late Neolithic and the Early Eneolithic (*see supra* and *Appendix*).

Among real objects there are a fragment of copper rod from Isera 2, interpreted as the likely result of exchanges with communities belonging to the Münchshöfen culture, and a fragment of copper awl from Isera 3, related to the Altheim and Pfyn cultures. What is rather intriguing is that the presence of these metal objects has virtually been ignored in the cultural attribution of these levels within a sequence where only the last horizon, Isera 5, is considered as Eneolithic. The reason, though not explicitly stated, might be that these presumably imported objects do not testify a local smelting activity[6]: on the contrary, a local production is documented by the presence of one crucible at Acquaviva di Besenello, a site paralleled with Isera 5 as well as with Remedello 1, and of a second one at Romagnano, layer Q[7].

Early metallurgy

This point reminds of the very interesting and controversial problem of the beginning of metallurgy, that has been the object of a recent, critical and acute review by L. Barfield (1996). He writes:

1. The definition of a Remedello culture is by no means easy and south of the Alps we do not have the clearer culture categories that exist at this time north of the Alps. Remedello is at best defined as an area of single burial which is to be contrasted with the area of collective burials along the Alpine fringe concentrated mainly in Lombardy known loosely as the Civate group.

2. The Remedello dagger type should not be considered as synonymous with the Remedello culture because of its much wider distribution, which is shown not only by the finds of metal blades themselves but also by the extensive representation of such daggers on statue stele in the southern Alps, between Sion and the Alto Adige, and in Liguria. Such representations comprise a greater number of daggers than the actual metal daggers themselves.

[6] My interpretation has been confirmed directly by the author, A.L. Pedrotti, whom I wish to thank for her always kind collaboration.

[7] A. Broglio and R. Perini (1995. 105) have recently attributed the ^{14}C date of 4810 ±50 to layer Q, in contrast with the majority of other scholars who attribute it to the deeeper layer R (see *Appendix*).

PHASE	DOCUMENTÉE PAR	STADE DE DÉVELOPPEMENT
Importation	Disque à bossette de Hornstaad. Poignard de Reute.	NÉOLITHIQUE
Début de la production autonome	Creusets et types de haches propres à la culture de Pfyn	CHALCOLITHIQUE
Extraction sur place et production métallurgique	Variétés de cuivre spécifiques et régionales ; creusets	CHALCOLITHIQUE
Production métallurgique étendue et intensive	Nombreux moules et riche inventaire d'objets en métal	MÉTALLIQUE

A

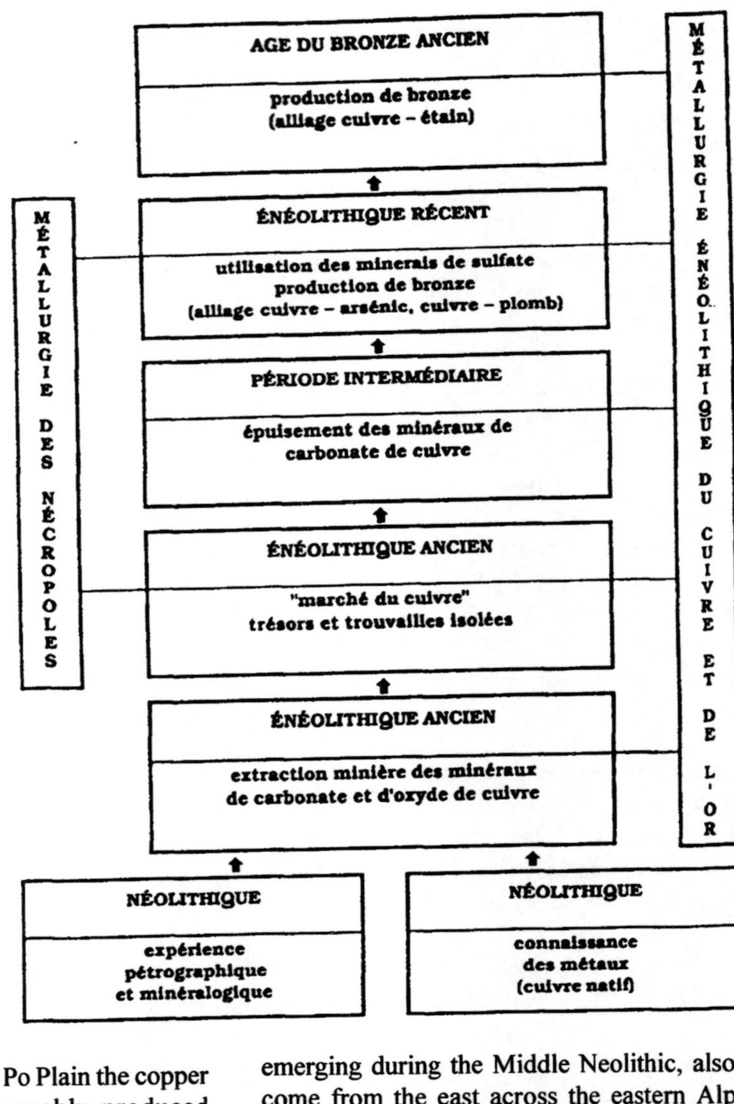

B

Figure 14.13: The models presented by B. Jovanović (A, Jovanović1991) and C. Strahm (B, Strahm 1991) to interpret the long process of acquisition of metallurgy.

3. Since there are no copper sources on the Po Plain the copper objects from Remedello were presumably produced somewhere along the Alpine fringe, where contemporary copper working has been demonstrated at sites in the Trentino. The daggers would therefore have been produced in the area of the Civate group and not Remedello. The same observation, about non local sources, can be made on the silver, stone, shell and flint artifacts from the site. (Barfield 1996: 69).

He adds: 'The first metallurgy in the north, if not already emerging during the Middle Neolithic, also appears to have come from the east across the eastern Alps in the form of Bocca Lorenza/Gurnitz axes, and other metal into the late VBQ culture with its markedly developed east alpine connections. Whether these first axes were just trade items, as the hoards might suggest, or whether at the same time they lead to the development of local metallurgy is not possible to tell at present. The fact that they are restricted to metal rich area of north west Italy may support the latter view.

A wider context for the spread of metal to northern Italy is the contemporary penetration of metal working into the Northern Alps (Pfyn, Mondsee) and its diffusion northwards into Silesia (Jordanova).' (Barfield 1996: 70).

I want to stress a few points of these passages:

• Metallurgy would have started in Northern Italy before the conventional limit of the Eneolithic;

• Metallurgy may mean imported objects and/or local production;

• There are sources of copper ores in the Alps, in Toscana and in Liguria, that are likely to have been exploited rather early: in Liguria, for instance, a wooden haft coming from old investigations in the mines of Libiola has been dated to 3500-2900 BC – and this single date does not necessarily mark the beginning of the mine culture, but a single event which might be a point within a presumably long process of use. In addition, preliminary results of new research at Monte Loreto/Sestri Levante indicate that these mines were in use around the same time if not earlier[8].

In my opinion, these important hypotheses and data are somehow in contrast with the following facts:

• There has been so far a prevalent interest in typological studies, mainly of heavy tools, i.e. axes and daggers (*see* for instance, Carancini 1993, 1996; De Marinis 1992; De Marinis and Pedrotti 1997[9]);

• Even when attention has been given to ore deposits and mining activities, a comparative, analytical study of their distribution *versus* the distribution of metal artifacts has not been accomplished;

• The analyses now available are almost exclusively spectrographic analyses of artifacts: analyses of copper ores or lead isotope analyses, trace element neutron activation analyses, x-rays diffraction ones etc. are still few (*see* for instance, Campana *et al.* 1996a).

I think that this situation is largely due not only to the lack of research oriented towards the understanding of the very beginning of metallurgy but also, again, to the lack of theoretical discussion on the parameters necessary and sufficient to define a context as Eneolithic.

There are many important studies of foreign scholars focusing on early metallurgy: I'll mention, for instance, those collected by Jean-Pierre Mohen in *Découverte du métal,* published in 1991 (*see* Mohen 1991), and those presented at the XIII International Congress of Prehistoric and Protohistoric Sciences, in 1996, at the Colloquium *Metallurgy: Origins and Technology*, edited by B. Bagolini and F. Lo Schiavo (*see* Barfield 1996).

Many scholars stress not only the long duration of the process of acquisition of metallury - B. Jovanovic (1991: 95), for instance, says that "*deux millénaires au minimum séparent la connaissance du cuivre des débuts de sa production métallurgique*" - but also the complex, intertwined technological and socio-economical steps of this process (*see* for instance, Ottaway 1996). B. Jovanović (1991: table 2) and C. Strahm (1991: 144) present two models to interpret this process (***Figure 14.13***): the models are not mutually exclusive and it would be very interesting to see whether one of them, or both in combination, may be applied to Northern Italy.

Funerary caves and rock shelters

Many authors also stress that the production of small objects - needles, awls, beads, ear-rings etc. - preceded that of heavy copper tools such as axes and daggers (see, for instance, Gale *et alii* 1991, in particular p. 52). I think that it would be important to reconsider from this point of view the presence of objects of this typology (see ***Figure 14.6***) often found – together with ornaments in stone, bone etc., usually largely prevailing in number - in collective graves in caves and rock shelters in the pre-alpine regions of Northern Italy (Barfield 1985; 1986; Poggiani Keller 1988; Maggi 1996; Nicolis 1996).

This presence has usually been considered as indicative of differences in ritual *versus* the cemeteries of the Po Plain, where metal heavy tools/weapons prevail, but it might also indicate differences in chronology[10].

These funerary caves and rock shelters have traditionally been attributed to the Eneolithic, though the early bronze age components have often been underlined (*see* in particular Maggi 1996 and Nicolis 1996) and occasionally also the Late Neolithic ones. These oldest elements have probably been underestimated: the meaning of their presence should be reconsidered in the light of the apparent absence of funerary evidence in the last phase of the VBQ culture and in the Lagozza culture – absence that might be due to an early adoption of some form of exposure or delayed burial (L. Barfield, 1986: 241; 1985: 24) – as well as of the long-lasting tradition of using caves and rock shelters for burying the dead (Bagolini and Grifoni Cremonesi 1994; Grifoni Cremonesi 1996). But also discontinuity and change in use – often in

[8] R. Maggi, whom I warmly thank, has communicated to me that a first radiocarbon date would be about 4700 BP.

[9] A direct comparison between the studies of G.L. Carancini and R.C. De Marinis reveal that there are important differences in the attribution of metal objects – and consequently of the graves where these are present – as to relative chronology. Also the metope decorated fine ware has a different chronological position: Carancini (1993. 136) considers this style older than his "first metallurgical phase of Remedello (= first horizon of diffused metallurgy = Final Eneolithic). The opinion of De Marinis has been presented in the text.

[10] In specific cases of human remains associated with traces of smelting activities – Acquaviva di Besenello (Angelini, Bagolini, Pasquali 1980) and La Vela/Valbusa (Fasani 1990) – the hypothesis of a metallurgist's grave might be a very suggestive one (see Mohen 1991).

connection with stabling activities (*see* for instance, Maggi 1996), that might reinforce the 'secondary product revolution' in progress – should be taken into consideration.

Radiocarbon dates are still too few - absolutely and in comparison with the number of funerary caves and rock shelters[11] - to support attributions based on typolgical analyses, but they cover a rather long span of time whose limits might reasonably be extended by new data (*see supra* and *Appendix*).

I'll make a last comment by referring again to Friuli-Venezia Giulia, the easternmost region of Northern Italy that has usually not been considered in general studies on funerary caves/rock shelters and early metallurgy. This is certainly largely due to the lack of stratigraphical deposits fully reliable, good publications of data, recent field research etc., but there are elements that might be re-evaluated and others emerging from recent re-analyses of old excavations (*see supra*):

- In the Trieste Karst there are at least one cave and two rock shelters where human bones were found in association with materials that can be attributed to the Eneolithic/Early Bronze Age (Montagnari Kokelj 1998); another similar site is located further north, in the Valli del Natisone (Bressan 1982);

- These territories, close to the present border with Slovenia, have many points of contact with presumably contemporaneous Slovene cultures, in particular the Ljubljana culture, where important metallurgical activities are documented (Korošec and Korošec 1969);

- The still rather uncertain inner chronological sequence of this culture does not allow to precise, unless with approximation, the beginning of this production, that nontheless should have been largely responsible for the dispersal of metal objects towards western regions. Mainly axes and daggers have in fact been found not only in Slovenia (Šinkovec 1995) but also in some Karst caves (Marchesetti 1889; Veluscek *in press*) and in a few open-air sites in the Friuli Plain (Moretti 1983);

- If we confront these finds with those of Remedello flint daggers - in some cases just in the same sites (Ljubljansko Barje/Slovenia, Grotta Tominz/Trieste Karst etc.), in others not in association with metal objects, but loosely in the same areas (Montagnari Kokelj 1990)[12] - the hypothesis that the Karst and the Friuli Plain might have continued to play, also after the Neolithic (*see supra*), a role of mediation in long-distance trade/exchange contacts seems strongly realistic.

Final comment

It is again very difficult to define the contexts involved in this mediation as Late Neolithic, Early or Late Eneolithic, Early Bronze Age, but in my opinion, besides new specific discoveries, also a general redefinition of the concept of Eneolithic in contemporaneous Italian literature would be of basic importance.

Acknowledgments

I am particularly grateful to my colleagues and friends who have read this paper and given me precious advice: Lawrence Barfield (University of Birmingham), Giorgio Stacul (University of Trieste), Elisabetta Mottes and Franco Nicolis (Ufficio Beni Archeologici - Provincia Autonoma di Trento) and Joseph Maran (University of Heidelberg - in his Institute most of my research was made in the best favourable conditions).

Due to problems of time not all their suggestions have been fully elaborated, but it is my intention to proceed further on this subject and take them into due consideration.

[11] A complete catalogue of the funerary caves and rock shelters of Northern Italy is still missing, but by combining the data collected by B. Bagolini (Bagolini *et alii* 1991), R. Maggi (1996) and me (research in progress – see Montagnari Kokelj 1998) one can estimate that their total number should be at least a hundred. It is consequently even more evident that the number of available radiocarbon dates is absolutely insufficient to build a general chronological framework.

[12] This possibly contemporaneous presence reminds me of the interesting idea of M. Séfériadès (1991) of a "competition" between flint and metal *prestige* items at the beginning of metallurgy.

APPENDIX

In recent years two important studies have been devoted to the chronology of the whole Italian peninsula, from the Paleolithic onwards (Skeates 1994), and of Northern Italy in the Neolithic and Eneolithic (Bagolini and Biagi 1990) respectively. A collection of ^{14}C dates has been included also in the *Atlas du Néolithique Européen* (Atlas [Bagolini and Pedrotti 1998]).

Although there is a broad coincidence among the lists of radiocarbon dates presented, in all of them there has been a selection of available dates due to different reasons, mainly of archaeological rather than archaeometrical nature. R. Skeates lists in detail his "junk dates", about 5% of the total number, and explains the reasons for rejection, quite often due to the overlooked disturbance of cave deposits and the presence of too large, and consequently useless, standard deviations (Skeates 1994: 271-272). B. Bagolini and P. Biagi are less explicit, but their choices are presumably related to their evaluation of the congruity of stratigraphy and/or cultural attribution of the contexts. Similar motivations should be at the basis of the choices made again by B. Bagolini together with A.L. Pedrotti, L. Barfield and F. Nicolis in their contributions to the chapter of the *Atlas du Néolithique Européen* concerning Northern Italy.

New measurements have become available in the last years, after these studies[13], and an update is consequently useful and particularly important for the period under examination. A critical review of all data would be appreciable, but it would require a combined archaeological and archaeometrical approach that goes beyond the limits of this paper. Therefore, I have decided to update the lists presented in the three studies just mentioned - taking the one by R. Skeates, the most recent and the most complete, as the basis – and to present the data as they are published.This obviously means that there may be discrepancies in the figures, depending on the different calibration methods used[14], but, again, a general recalibration would imply also a cross re-evaluation of all dated complexes. This operation might result in changes in the cultural attributions previously advanced, with consequent chain reactions on a wide scale.

I have decided to present a time span sensibly broader than the one under study, aiming at a clearer evaluation of continuity/discontinuity as well as geographical differences in the radiocarbon dates available to support cultural reconstructions. Then I have considered all cross "vetoes" and discarded unreliable dates: conclusions are discussed in the text[14].

Friuli-Venezia Giulia

SITE	LAYER	CULT. ATTRIB.	LAB. N.	UNCAL. BP	±	CAL. BC 1σ	CAL. BC 2σ	MATERIAL	REF.
Palù di Livenza	US 9			5395	90	4456-4001*			12
	ES 79			5305	90	4343-3984*		Wood	12
	US 6			3210	265	2198-830*		Charcoal	12
Grotta Mitreo	Base of layer 4	E/EBA	R-902	3820	50	2451 –2149	2460-2140	charcoal	1
	Layer 5	E/EBA	R-903α	3720	50	2269-2035	2308-1966	charcoal	1

Veneto

SITE	LAYER	CULT. ATTRIB.	LAB. N.	UNCAL. BP	±	CAL. BC 1σ	CAL. BC 2σ	MATERIAL	REF.
Rocca Rivoli	Site L,pits P-Q-R	VBQ3	Birm-617	5370	70	4339-4047	4360-4004	Animal bone	1, 2
	Site L,pits P-Q-R	VBQ3	Birm-616	5070	100	3990-3725	4214-3690	Animal bone	1, 2
	Pit L	VBQ3	Birm-103	5220	120	4406-3943			2
Mondeval De Sora			R-1938		40	3160 *			8
			R-1940		45	3010 *			8

[13] Some of the new ^{14}C dates have been included in the lists of the last draft of the Atlas – an update made necessary by the long time elapsed from its preparation to the publication – without remarkable changes in the text.

[14] I have indicated the references for each set of dates, in order to facilitate the comparison of accepted/rejected dates as well as to help finding the quotation of first-hand literature. The references are abbreviated as follows: 1 = Skeates 1994; 2 = Atlas (Bagolini and Pedrotti 1998); 3 = Bagolini, Biagi 1990; 4 = Barfield *et alii* 1995; 5 = Maggi 1997; 6 = Mezzena 1997; 7 = De Marinis, Pedrotti 1997; 8 = Guerreschi 1995; 9 = Campana *et alii* 1996; 10 = Maggi, Del Lucchese 1988; 11 = Campana *et alii* 1995; 12 = Vitri 1995. The symbol * indicates that information on the calibration methods used is incomplete.

Trentino-Alto Adige

SITE	LAYER	CULT. ATTRIB.	LAB. N.	UNCAL. BP	±	CAL. BC 1σ	CAL. BC 2σ	MATERIAL	REF.
Isera La Torretta	US 61	VBQ3-Isera 1	ETH-12494	5440	55		4445-4091	Animal bone	2, 7
	Us 75	VBQ3-Isera 1	ETH-12495	5580	65		4544-4264	Animal bone	2, 7
	Us 76	VBQ3-Isera 1	ETH - 12496	5570	55		4514-4332	Animal bone	2, 7
Fiavé I	Sett. IX str. F	Late N-Isera 3	Eth-12498	4950	55		3932-3640	Animal bone	7
Romagnano Loc III	R 4	Late N-Isera 4	R. 775	4810	50	3687-3524	3771-3384	Charcoal	1, 7
Völseraicha	Layer 3	VBQ3 + Chassey/Lagozza	Bln-2715	4780	100	3690-3380	3780-3350	Charcoal	1, 3
Similaun glacier	"Ice Man"	Rame I (ref. 7)	OxA-3371	4660	55	3510-3360	3622-3206	Human bone	1
			OxA-3372	4565	60	3371-3135	3501-3046	Human bone	1
			ETH-8342	4560	65	3371-3108	3503-3041		1
			ETH-8345	4555	35	3357-3141	3372-3104		1
			OxA-3373	4550	70	3369-3104	3502-3040	De-fatted human skin	1
			OxA-3419	4540	55	3359-3105	3491-3041	De-fatted human skin	1
			OxA-3420	4530	70	3361-3098	3497-2935	Human bone	1
			OxA-3375	4530	70	3361-3098	3497-2935	Human skin	1
			OxA-3421	4480	55	3337-3040	3360-2927	De-fatted human skin	1
			OxA-3374	4460	80	3338-2929	3370-2910	De-fatted human skin	1
			OxA-3376	4450	80	3335-2928	3370-2910	Human skin	1
Acquaviva di Besenello	Grave 1	Rame I-Isera 5	ETH-12497	4410	70		3332-2897	Human bones	7
Laas/Lasa	Colluvial humus	E	VRI-272	4350	100	3097-2900	3289-2699	Charcoal	1, 3
Tolerait		E	Birm-1005	4130	100	2888-2510	2920-2460	Charcoal	1, 3
Riparo Gaban	Layer C 5	E	Bln-1776	3985	50	2579-2464	2853-2399	Charcoal	1, 2, 3, 7

Lombardia

SITE	LAYER	CULT. ATTRIB.	LAB. N.	UNCAL. BP	±	CAL. BC 1σ	CAL. BC 2σ	MATERIAL	REF.
Mosio	Pit	VBQ3 + Chassey	Bln-2977	5110	50	3989-3818	4033-3790	Charcoal	1, 3
Lagozza di Besnate	Lower structure	(Chassey)Lagozza	R-338	4980	50	3906-3703	3950-3690	Wood	1, 2, 3
	Lower structure	(Chassey)Lagozza	Cross-check sample on R-78	4810	115	3773-3383	2931-3350	Wood	1
	Lake sediment	(Chassey)Lagozza	R-337	4805	50	3677-3522	3771-3383	Wood	1, 2, 3
	Structure	(Chassey)Lagozza	Pi-34	4794	90	3694-3384	3780-3360	Wood pile	1, 2, 3
	Lower structure	(Chassey)Lagozza	R-78	4735	50	3626-3380	3640-3370	Wood	1, 2, 3
	Lower structure	(Chassey)Lagozza	R-78 A	4580	50	3372-3146	3501-3100	Wood	1, 2, 3
Remedello Sotto	Grave 106	Rame I	ETH-12182	4485	60		3350-2927	Human bones	2, 4, 7
	Grave 34	Rame I	ETH – 12189	4400	70		3331-2888	Human bones	2, 4, 7
	Grave 75	Rame I	ETH-12183	4280	65		3078-2624	Human bones	2, 4, 7
	Grave 65	Rame I/II	ETH.121 2188	4185	70		2911-2578	Human bones	2, 4, 7
	Grave 40	Rame I/II	ETH-12185	4155	60		2885-2508	Human bones	2, 4, 7
	Grave BSII	Rame II	ETH-6196	4070	70	2711-2512	2884-2460	Human bones	1, 2, 7
	Grave 27	Rame II	ETH-12190	4025	65		2865-2341	Human bones	2, 4, 7
	Grave 73	Bell Beaker	ETH-12184	3800	60		2453-2037	Human bones	2, 4, 7
Riparo Valtenesi / Manerba	Grave 134	Funerary cave/rock shelter	Birm-1132	4420	90	3321-2920	3360-2890	Wood timbers	1, 2, 3
		Funerary cave/rock shelter	OxA-4550	4320	100	3350-2650*			4, 7
		Funerary cave/rock shelter	OxA-4595	4265	70	3050-2650*			4
		Funerary cave/rock shelter	OxA-4548	4105	65	2850-2450*			4
		Funerary cave/rock shelter	OxA-4547	4040	60	2850-2400*			4
Lagozzetta Besnate	Peatbog	EBA ?	R-336	4385	50	3093-2921	3302-2910	Wood from vessel	1
Riparo Cavallino / Monte Covolo		Funerary cave/rock shelter	OxA-5287	4245	65	3050-2650*			4
		Funerary cave/rock shelter	OxA-5288	4090	65	2850-2500*			4
		Funerary cave/rock shelter	OxA-5286	4060	60	2850-2450*			4
		Funerary cave/rock shelter	OxA-3746	3860	80	2500-2100*			4
Monte Covolo	Layer 20	Lagozza + VBQ	Birm-473	4790	210	3790-3350	4032-2927	Charcoal	1, 2
	Layer 13	Lagozza + VBQ	Birm-472	4240	190	3091-2580	3370-2340	Charcoal	1, 2, 3
	Layer 8	Bell Beaker	GrN-8013	4010	40	2583-2479	2854-2463	Charcoal	1, 2

Piemonte-Valle d'Aosta

SITE	LAYER	CULT. ATTRIB.	LAB. N.	UNCAL. BP	±	CAL. BC 1σ	CAL. BC 2σ	MATERIAL	REF.
Balm'Chanto	Layer 10	E	Bln – 3285	4090	70	2870-2506	2889-2470	Charcoal	1, 3
	Layer 6	E	Bln 2838	4010	60	2602-2468	2863-2402	Charcoal	1, 3
St. Martin Corléans	P12		F-78	5020	180	4000-3640	4240-3380	Wood charcoal	1
	P17		F-82	4910	200	3960-3386	4223-3137	Wood charcoal	1
	P16		F-81	4840	190	3904-3370	4034-3100	Wood charcoal	1
	P15		F-80	4800	170	3780-3370	3980-3100	Wood charcoal	1
	P7		F-77	4720	150	3690-3350	3897-3040	Wood charcoal	1
	P19		F-83	4450	180	3370-2910	3633-2615	Wood charcoal	1
	P13		F-79	4325	170	3305-2700	3494-2493	Wood charcoal	1
	Grave II – outside		UtC-1669	4220	110	2921-2637*		Wood timber	6
	Grave II – outside		UtC-1668	3960	60	2573-2457*		Charcoal ?	6
	Grave II – inside		UtC-1671	3870	60	2464-2283*			6
	Grave II – inside		UtC-1001	4000	100	2855-2460*			6
	Grave V - inside		UtC-1690	4000	100	2855-2460*		Cremated bones	6
	Grave IV – inside		UtC-1689	3540	60	1963-1777*		Human bones	6
	Grave IV – inside		UtC-1687	3510	60	1926-1749*			6
	Grave IV – inside		UtC-1688	3320	50	1682-1525*		Cremated bones	6
	Grave VII – inside		UtC-1693	3930	70	2564-2342*			6
	Grave VII – inside		UtC-1691	4170	80	2898-2625*			6
	Grave VII – inside		UtC-1692	3900	90	2559-2283*			6
	Grave III – inside		UtC-1686	3980	60	2581-2461*		Accumulated bones	6
	Grave IISE – inside		UtC-1670	3760	60	2289-2047*		Semi-burnt bones	6

Emilia-Romagna

SITE	LAYER	CULT. ATTRIB.	LAB. N.	UNCAL . BP	±	CAL. BC 1σ	CAL. BC 2σ	MATERIAL	REF.
Spilamberto	Site VIII	Chassey/ Lagozza	I-11817	4995	100	3960-3700	4000-3540	Charcoal	1
Spilamberto	Area X	E	I-11816	4195	95	2914-2619	3031-2500	Charcoal	1

Liguria

SITE	LAYER	CULT. ATTRIB.	LAB. N.	UNCAL. BP	±	CAL. BC 1σ	CAL. BC 2σ	MATERIAL	REF.
Arene Candide	Layers 9-10H	Chassey	Beta-60700	5320	70	4230-4060	4300.4000	Charcoal	5
	Layers 11-13H	Chassey	Beta-60692	5290	80	4210-4020	4300-3965	Charcoal	5
	Layer 17H	VBQ2	Beta-60694	5280	110	4230-3990	4340-3910	Charcoal	5
	Burial VI		GX-16962G	5260	135	4250-3960	4350-3780	Human bone	5
	Layer 13	Chassey	Beta-65306	5250	140	4260-3940	4350-3760	Charcoal	5
	Layers 12-13	Chassey	R-104	5075	45	3970-3789	3992-3781	Charcoal	1, 2, 3, 5
	Layer 8H		BetA 60691	5020	110	3940-3715	4040-3610	Charcoal	5
	Gandolfi Branch	E ?	Beta-56694	4720	120	3640-3370	3730-3110	Animal bone	5
	Layer 2	E	Beta-66550	4430	60	3130-2930	3310-2920	Charcoal	5
	Layer 5 H	E	Beta-48682	4320	80	3060-2880	3130-2650	Charcoal	5
Tana di Barletta	Lower hearth 3	E ?	HAR-8388	4980	100	3950-3690	3990-3533	Charcoal	1, 10
	Middle hearth 2	E	HAR-6435	3880	120	2564-2147	2860-2030	Charcoal	1, 3, 10
Bargone			S RR-3814			3700-3100*			11
Riparo Arma di Nasino	Layer VIII	Lagozza ?	R-332	4705	70	3622-3371	3650-3340	Charcoal	1
	Layer VII	E	R-261	4680	100	3623-3350	3690-3100	Charcoal	1
	Layer VIII		R-314α	4340	60	3035-2911	3257-2786	Charcoal	1
	Layer VIII		R-314	4275	65	2923-2784	3040-2668	Charcoal	1
	Layer VIII		R-262	4220	90	2917-2629	3040-2580	Charcoal	1
	Layer VI	Bell Beaker	R-309	4220	55	2911-2702	2922-2621	Charcoal	1
	Layer VI	Bell Beaker	R-309α	4110	55	2871-2585	2887-2497	Charcoal	1
	Layer VII		R-311	3765	70	2332-2040	2470-1970	Charcoal	1
	Layer V		R-308α	3690	60	2200-1974	2291-1892	Charcoal	1
	Layer VI	Bell Beaker	R-260	3680	95	2269-1924	2430-1776	Charcoal	1
	Layer VII	Bell Beaker ?	R-331α	3645	70	2139-1921	2273-1783	Charcoal	1
	Layer VI	Bell Beaker	R-329α	3600	50	2108-1886	2138-1777	Charcoal	1
	Layer V		R.308	3550	55	2013-1778	2128-1740	Charcoal	1
	Layer V		R-257	3525	65	2010-1747	2111-1665	Charcoal	1
	Layer VI	Bell Beaker	R-330	3490	50	1886-1741	2011-1664	Charcoal	1
	Layer VI	Bell Beaker	R-329	3430	50	1871-1662	1890-1606	Charcoal	1
	Layer V		R-258	3390	90	1865-1602	2007-1465	Charcoal	1
	Layer IV		R-307	3390	50	1744-1625	1879-1532	Charcoal	1
	Layer VI		R-259	3340	90	1740-1521	1883-1430	Charcoal	1
Grotta Pertusello	Layer III	Late N	R-122	4650	90	3606-3344	3640-3100	Charcoal	1
	Layer III	Late N / E ?	R-155	4390	70	3254-2918	3340-2900	Charcoal	1
Libiola	Copper Mine	E	Bln-3367	4610	50	3494-3343	3510-3135	Wooden tool	1, 3, 10, 11
	Copper	E	Gif-7213	4490	90	3352-2949	3497-2920	Wooden tool	1, 3, 10, 11

Section 2

Cultures, Ceramic Styles and Contacts in the Bronze and Iron Ages

CHAPTER 15

DUBENE-SAROVKA, YUNATSITE CULTURE AND THE INTEGRATIONS IN THE EARLY BRONZE BALKANS

Lolita Nikolova

Introduction

Dubene-Sarovka is a newly investigated Late Copper and Early Bronze tell in Western Upper Thrace (Nikolova 1999 with ref.; 2000a; ***Figure 15.1***). The Early Bronze Age levels belong to the Yunatsite culture in Western Upper Thrace (***Figure 15.2***). In this chapter the ceramic style of Dubene-Sarovka II will be a background to define the evolution the Yunatsite culture to which the site belongs, as well as its relation to the development of the Early Bronze I-III ceramic styles and different cultures in the Balkans.

Among the purposes of this analysis is a definition of some characteristics of the prehistoric innovations in the light of new evidence, well stratified. It will be shown that the innovations in the Early Bronze Balkans were spread as a result of multi-aspect integration of the neighbor and distant cultures. In Early Bronze I of special interest is the distribution of the channel and groove pottery, while in the Early Bronze II it is the encrusted pottery of different sorts that characterize many cultures in the Balkans. An innovative phenomenon in Early Bronze III is the distribution of the ceramic styles of plain pottery in several microregions of the Balkans.

The recent periodization of the Yunatsite culture is based on the Early Bronze stratigraphic sequence at the Yunatsite tell (layer B), Dubene-Sarovka II tell, Plovdiv-Nebet Tepe, Ognyanovo tell and Manole tell (Nikolova 2000a). The general background is the evolution of the ceramic style from an emblematic fluted and groove pottery (the first stage) with exception encrusted pottery, towards an emblematic encrusted pottery (the second stage) and a stage of pointed bottom cups

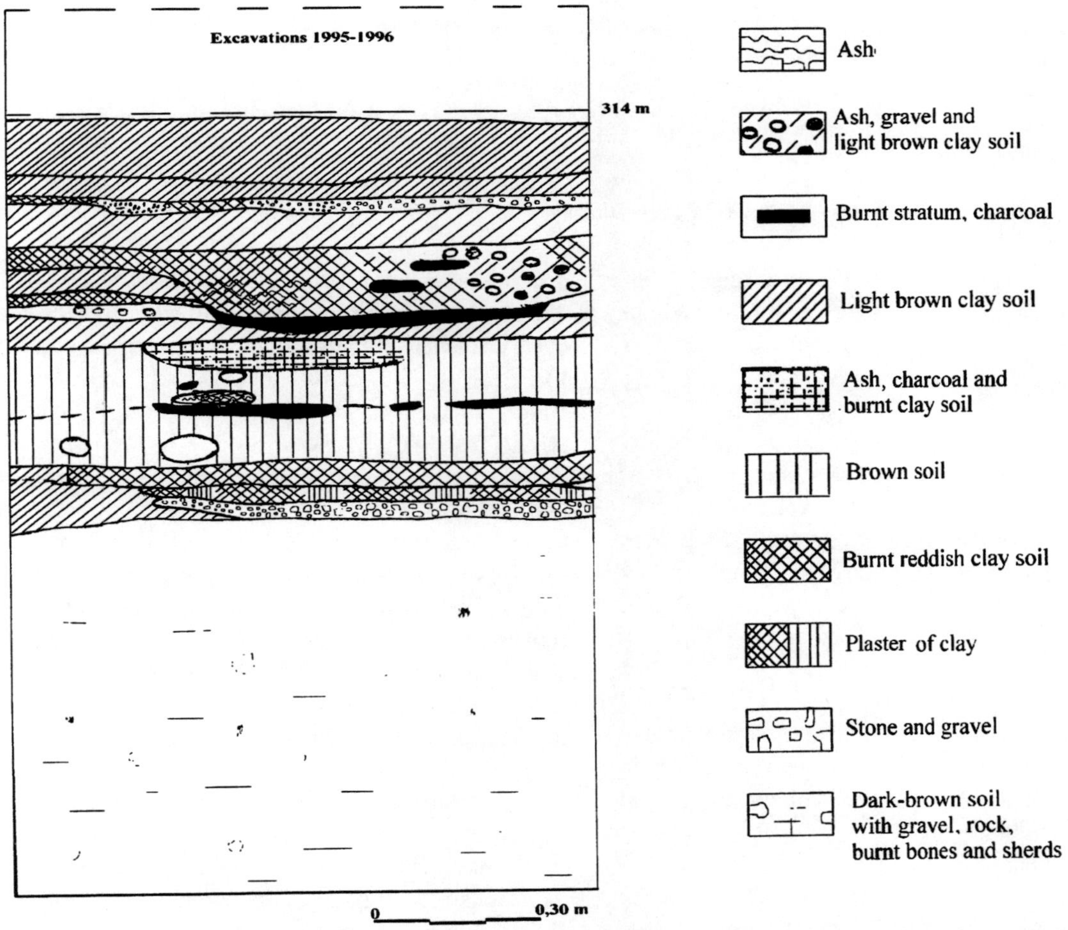

Figure 15.1. Western section of trench H18-4 at Dubene-Sarovka (1999).Geomorphologic description of the levels see in Kenderova, chapter 28, Figure 28.1.

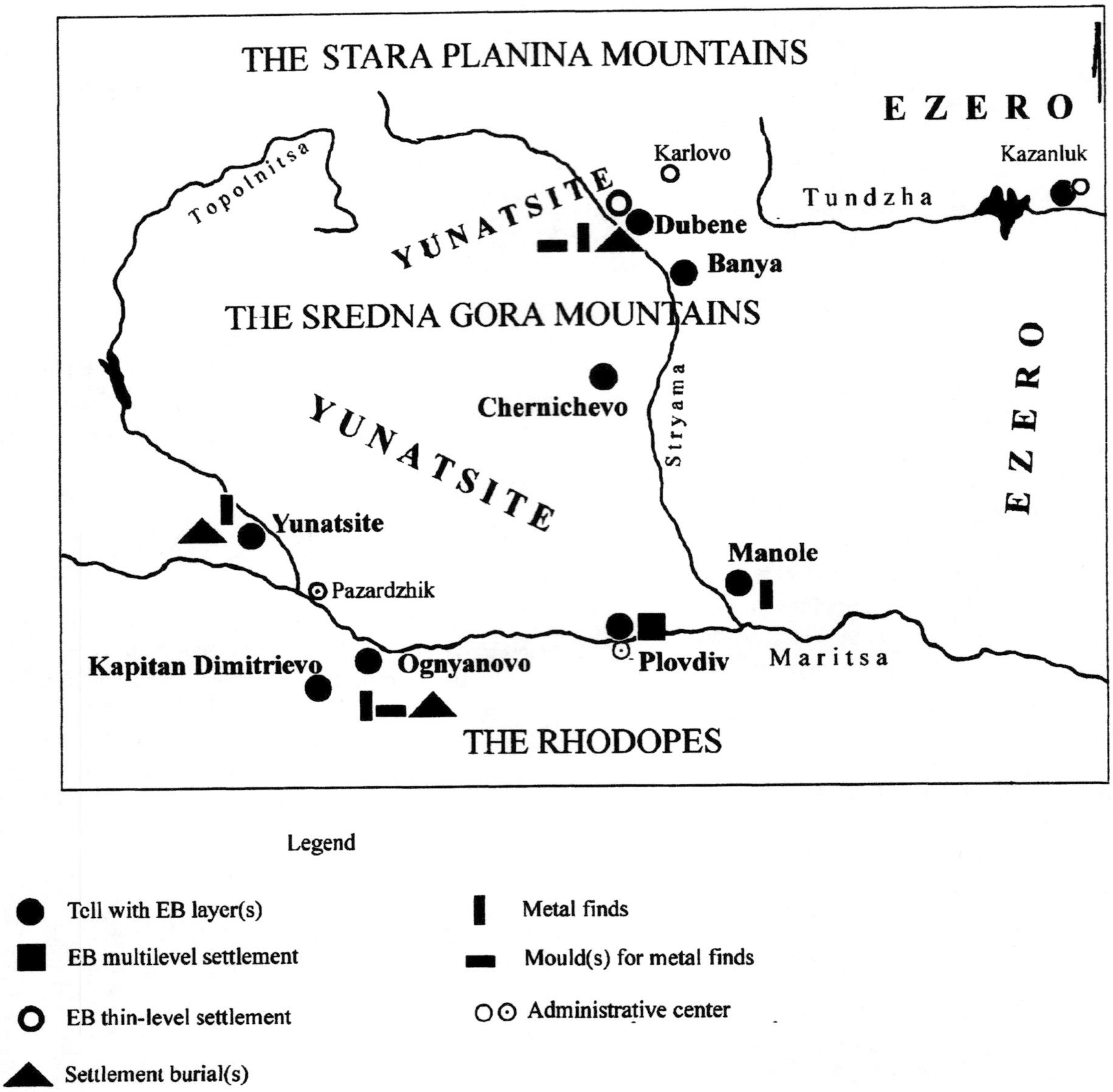

Figure 15.2. Yunatsite culture sites in Western Upper Thrace.

with a predominance of a plain pottery (the third stage) - ***Figures 15.3-15.29***. The small finds include common for southern Balkans (***Figure 15.30: 2-3, 5, 7-8***) and some peculiar ones (***Figure 15.30: 1, 4, 6***).

The development of the ceramic style is best represented by the Yunatsite 17-1 (Early Bronze I-III; Katincharov et al. 1995; Nikolova 1999) and Dubene-Sarovka IIA-B (Early Bronze I-II) sequence (Nikolova 2000a; ***Figure 15.1***). For Early Bronze III, of special importance are the data from Manole (Detev 1981) and Ognyanovo- Maltepe (Detev and Matsanova 1977; Nikolova 1999; see chapter 16).

More than 30 ^{14}C dates originate from Yunatsite 15-3, Dubene-Sarovka IIB1-2, Plovdiv-Nebet Tepe I and Manole I (Nikolova 1998b; 1999; 2000a), which date the Yunatsite culture from later fourth millennium and the third millennium BC (see in details in Nikolova 2000a).

The Yunatsite culture was in close interrelations with the neighbor cultures especially from Eastern Upper Thrace (Ezero culture), the Middle and Upper Struma valley (Pernik group) and Sitagroi IV-V / Dikili Tash group (The Northern Aegean). A considerable number of sites from those and other close or distant cultures include contact direct or indirect data, like Ezero tell 13-1, Karanovo VII and Gulubovo in eastern Bulgarian Thrace (Southeastern Bulgaria); Sitagroi IV-V and Dikili Tash IIIA-B, Skala Sotiros in the Northern Aegean (Northeastern Greece); Kovachevo IIIA-B, Pernik-Krepostta and Radomir-Vakhovo II-III in the middle and upper Struma valley (Southwestern Bulgaria); Bubanj Ib-III, Jelenac, Ostrikovac Ia-d and Novačka Ćuprija in the Velika Morava valley (Eastern Serbia); Kastanas in Northwestern Greece, etc. (Georgiev et al. 1979; Raduncheva 1981; Renfrew et al. 1985; Koukouli-Chrysanthaki 1987; 1989; Demoule and Lichardus 1994; Leshtakov 1996; Stojić 1996; Séfériadès 1996; Hiller and Nikolov 1997; Maran 1998; Nikolova 1999, etc.).

Some distant parallels recently published, such as from Pevkakia Magula (Christmann 1996), give a possibility for a vaster special determination of some common elements in Southeastern Europe. The connections with Northwest Anatolia will be discussed elsewhere.

Dubene-Sarovka

The site of Dubene-Sarovka was documented in 1992 as a low prehistoric tell from Late Copper Karanovo VI culture and from Early Bronze Yunatsite culture. The ceramic style of that site has been a main argument for re-definition of the Early Bronze culture in Western Upper Thrace (***Figure 15.2***) known till 1990s as Ezero culture or Early Bronze Ezero culture and the so-called Middle Bronze Yunatsite culture. A variety of pricked, dot, incised, false-corded, stamped ornamentation (***Figure 15.3***) initiated the search of close interrelation of the Yunatsite culture with the Northern Aegean, Southwest Bulgaria and the Velika Morava valley (Nikolova 1996a; 1996b).

The tell of Sarovka is c. 100 m in diameter, with c. 2-2.50 m thick culture layers preserved. It was localized in 1992 in the course of my surface observation in the upper Stryama valley. P. Detev (1981) mentioned it as a Bronze Age site. Two encrusted vessels from Dubene (Nikolova 1995: fig.1) have been preserved at Historical Museum in Karlovo that have an identical ornamentation with the ceramics from the Sarovka locality. It made possible they to be attributed to that site. The tell is located on the first non-flooded left terrace of the Stryama River (for the paleomorphology of the locality see chapter 28). The uppermost levels were completely destroyed by the recent agricultural activity.

The control drillings in 1992 initiated the systematic investigation of the tell. The excavations in 1993-1996 – when the area of investigation reached more than 3,000 sq. m have given more detail stratigraphic sequence. After a break in 1997–1998, the excavations continued in 1999 - see in detail in Nikolova 2000a: 42-47.

Already from 1992 it was known that the thickness of the cultural layer on the top of the tell is c. 2 m. So, the discovering of equipment at a depth of 1.80 m in H18-4 was not a surprise in 1999. Having in mind the ceramic characteristics at a depth of 1.55 m there, the lowest levels discovered in H18-4 can be chronologically correlated with the levels at depths of 1.37 m and 1.55 m in the south. That correlation is based mainly on the peculiarities of the stratigraphic picture but not on the ceramics that is missing from the earliest levels discovered.

The ceramic contact data (channel pottery, zigzag ornamentation, dot ornamentation) synchronize Dubene-Sarovka-Sarovka IIA with Yunatsite 17-15. But archaeomagnetic samples from later Dubene-Sarovka IIA house fireplace differs from the values of Yunatsite 17-15 curve (Kovacheva et al. 1995, table 5, fig. 5). This fact assumes that the level was chronologically close to Dubene-Sarovka IIB2 phase or was considerably earlier (Kovacheva et al. 1995: 42). The archaeological observations reject the first possibility. After the excavations in 1999, on the terrace top the thickness of the Early Bronze I layer at Dubene-Sarovka documented is c. 0.60-0.70 m. It includes five house levels of possible three building levels, that can be correlated with Yunatsite 17-15 despite the total thickness of the Early Bronze I layer differs.

There are at least two other possible theoretical models of stratigraphic reconstruction: the fifth level along with the houses on the slope could precede Yunatsite 17, or the latter structures chronologically has preceded the earliest H18-4 level, as well as Yunatsite 17. In this case the earliest Early Bronze H18-4 level can be synchronized with Yunatsite 17.

There could also been culture layer from the first stage of the Yunatsite culture at Ognyanovo. The published askoi have close parallels with finds from the earliest Yunatsite II tell levels, as well as the waved-like shaped handles parallel with Boleráz (Baden I), as H. Parzinger (1993:121) has mentioned.

So, the recent archaeological investigation faces the archaeomagnetic investigations from 1993 in different way. From archeological point of view, it is not possible the Early Bronze I house P15-93 to be compared with Early Bronze IIB Apses House No. 1 on the top of the tell as it was mainly concluded. But the archaeomagnetic samples aroused another hypothesis, according to which house P15-93 preceded Yunatsite 17 (Kovacheva et al. 1995). That aspect of the archaeomagnetic analysis is still not possible to be confirmed directly by the pottery from the control trench in 1993 – almost identical with the earliest Early Bronze Yunatsite, as well as by the stratigraphy.

In the context of the 1999 excavations, it seems there is not represented the level corresponding to Yunatsite 15 on the southern periphery. So, both the documented houses (P15-93 and P16-99) can be compared with Yunatsite 16-17. Another possibility occurs the levels below a depth of 1.55 m in trench H18-4 to have been synchronized with Yunatsite 16-17 while one can assume houses P15-93 and P16-99 completely or partially preceded Yunatsite 17. In this case we should assumed the settlement was primary grounded on the terrace slope and then developed in the north.

Recently, the Early Bronze stratigraphic-chronological sequence in Western Upper Thrace represent a system of interrelated sites with similar ceramic style united under the Yunatsite culture definition. The longest stratigraphic column of that culture and of all the Early Bronze Balkans belongs to the Early Bronze Yunatsite tell where 17 Early Bronze horizons have been documented. Dubene-Sarovka tell is the newly investigated Early Bronze site with a long stratigraphic columns in the southern Balkans.

The correlation of the recent data from both sites infers the following synchronization:

Dubene-Sarovka IIA is synchronous with Yunatsite 17-15. For the time being, there are no certain data that the former site was grounded before Early Bronze Yunatsite;

Dubene-Sarovka IIB1-3 corresponds to Yunatsite 14-9 (Early Bronze IIA-B). Plovdiv-Nebet Tepe 11-8 can be attributed to that phase;

Yunatsite 8-1 is synchronous with Manole 1-7 and later Early Bronze Plovdiv-Nebet Tepe (7-4), as well as with Dubene-Sarovka IIC – a phase distinguished mainly basing on the pottery from destroyed levels on the top of the tell (***Figure 15.29***);

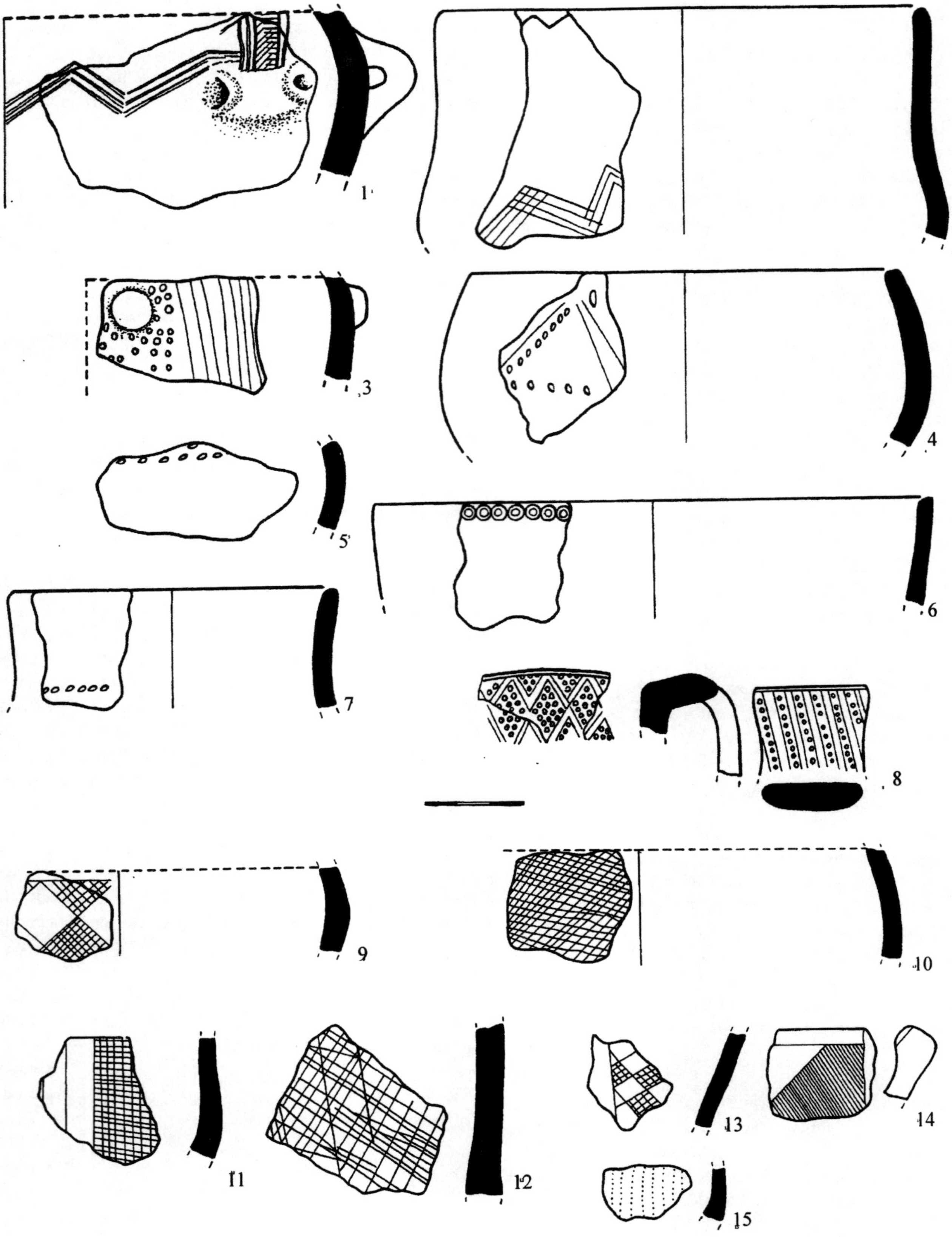

Figure 15.3. Dubene-Sarovka. Pottery from the excavations in 1992 (after L. Nikolova). Early Bronze I (13?, 14-15) and Early Bronze II (1-12).

According to the recent ^{14}C dates, the radiocarbon chronology define only the secondand the third stage of the Yunatsite culture (without its end) - c. 3000 BC – 2200 BC. In indirect chronology of the first stage has been argued elsewhere (Nikolova 2000a: 52-55).

Yunatsite I culture ceramic style

Yunatsite 17-15

The fluted and groove ornamented ceramics are emblematic of the first stage of the Yunatsite culture, which characterize the 17th – 15th horizons on the eponymous tell and Dubene-Sarovka IIA. Hatches on the mouth rims of the pots and on the amphora-like vessels typify the coarse ceramics at Yunatsite. A large amphora-like vessel with brown-black burnished surface and band handles were used as urns at Yunatsite 17-16 (Katincharov and Matsanova 1993: fig. 23: 1-3, 5). An encrusted pottery specifies the Early Bronze I ceramic style on this tell. According to the preliminary information published, in the coarse of the excavations the channel pottery occurred at 15th of Yunatsite and it was the 16th horizon when by the category of the fine pottery the quantity of the channel pottery considerably increased (Nikolova 2000a with the references cited there). It is not clear if the 17th level can be distinguished as an earlier phase of the Early Bronze I occupation of the tell, so the 16th and 15th levels represent the second phase. There is another possibility, 17th and 16th to have represented the earlier Early Bronze I phase and the 15th level the second (later) phase as a transition to the Early Bronze II. By the preliminary information of the pottery discovered, the second division looks more likely.

Dubene-Sarovka IIA ceramic style

The fine black, gray and brown burnished pottery characterizes Dubene-Sarovka IIA.

Storage and technomic smoothed pottery

Those categories are represented by rim-incised and relief ornamented pottery belonging to brown and exception gray and beige slip ceramics (subgroup 1), as well as by brown rusticated (subgroup 2) and brown smoothed pottery (subgroup 3).

Of subgroup 1 pot and urns are the most distributed ordinary ware distinguished by the typical of them relief ornamentation and incisions (***Figure 15.4***). They are with an equal cut or out-turned rims, rounded upper part and a lower part shaped as a truncated cone (***Figures 15.5; 15.6: 1-3***). The obliquely cut rim is a signifying element of that category in EB I (***Figure 15.5: 6, 7***). The relief bands are attached just to the outer wall of the rim and below it. They are also double relief bands preserved under the rim. Small rim sherds mostly exemplify the pots. Based on the upper parts of the vessels preserved, some urns with a shape of truncated cone can be recognized (***Figure 15.5: 8***) being with slightly inverted rim as well. But there is S-profiled pot from pit No. 2, which represents an emblematic EB I shape (***Figure 15.7***).

The fabrics of the pots are characterized by clay with fine and coarse sand. As an exception, a light brown (beige) smooth pot occurs made of paste mixed with sand and organic matter having charcoal like biscuit with white inclusions.

The ***Chart 15.1*** represents the ratio between the color, fabrics and surface treatment of the sherds in ***Figure 15.5***. The predominance of the brown earthenware in the category of the technomic vessels can be inferred as a conclusion for all EB I pots. However, the beige (***Figure 15.5: 6***) and gray pots (***Figure 15.5: 7***) as exception ceramics differ also by they shape and especially by profiling of the rim. The pots belongs to thick-walled vessels (0.8-1.2 cm) but the clay paste is abundant of fine sand and charcoal inclusions (real charcoal or burnt organic ones). However, there is no direct relation between the paste mixture and the thickness of the wall. Other thick-walled pot (***Figure 15.5: 2***) is characterized by abundance of fine and coarse sand. But there are a thin-wall pot (***Figure 15.5: 3***) or moderate (***Figure 15.5: 1, 4-5***) ones with 0.4-0.6 cm thick walls, which are made of clay with an abundance of coarse sand. The sand grains are up to 3 mm in diameter. In the case of sherd of a pot in ***Figure 15.5: 1*** there is a lot limestone inclusions that could belong to crashed bigger pieces. However, it should be stressed the clear documented in the biscuit charcoal inclusion of possible burnt organic.

In compare to the surface walls, the biscuit is of same color (brown or gray [***Figures 15.5: 3, 7 and 8***]) or it is darker in compare to the surface color. It looks that by the firing there was a temperature and duration required for homogenous fabrics that should be reached. There is an instance in ***Figure 15.5*** (No. 1) when the walls are brown but the biscuit is dark grayish. That is the sherd with the most abundant coarse sand limestone and with presence of charcoal.

The surface is smoothed, in some cases the inner wall is better smoothed. There are traces of slip of fine clay but it may document the final process when the pots was washed with fine clay but not a special process of slipping of the technomic earthenware. In ***Figure 15.5*** there is only one sherd with roughly smoothed outer surface (No. 3). It is interesting to note it is the thinnest one and with very high quantity of sand. The sherd is small but some incised lines may document a special treatment with grass.

There is a special element in the fabrics – golden mica. Its quantity is bigger when the sand is more abundant.

The pottery from subgroup 2 is brown (***Figure 15.6: 4-7***). It is containers or wide opened deep bowls with slightly rounded body. The clay is plentifully mixed with fine and coarse sand. The biscuit varies from bright brown to dark brown and dark gray and black. The surface color is bright brown (***Figure 15.6: 1-3***) or ochre brown (***Figure 15.6: 4***).

Slip and burnish technomic pottery (deep containers)

The first subgroup includes brown plain slip pottery – deep containers with equal cut rounded rims (***Figure 15.6: 8-13***). The clay is plentifully mixed with fine and coarse sand, well burnt. The surface color is bright brown (***Figure 15.6: 8-11***) or reddish brown (***Figure 15.6: 12-13***). In some cases (***Figure 15.6: 8***) the slip fell down.

Black slip and burnished pots, jars and deep bowls belong to subgroup 2 (***Figure 15.6: 14-22***). Some of the pottery is gray and black or brown and black. The rims are equal cut or little turn out. As exception, a container with cut from the inner side rim occurs. The profiles by the rims are straight or they are slightly S-profiled.

The possible shapes are necked round-bodied deep containers or wide opened deep containers with slightly rounded body. It is a plain thin to moderate walled earthenware of two fabric types – clay plentifully mixed with fine and coarse sand (***Figure 15.6: 18-19***) with single crushed stone (***Figure 15.6: 14, 21***), as well as fine clay with some powdery-like sand (***Figure 15.6: 15-17,22***). All fabrics include golden mica. The thickness of the wall varies from 0.6 to 0.9 cm.

Slip and burnished technomic and fine pottery

The bowls are one of the most distributed groups of pottery in Early Bronze Age. They are two main morphological varieties, according to the profiling of the upper part of the vessels – the first includes closed shapes with equally cut or inverted rims, while the second includes open shapes with out-turned rims and the biggest diameter is by the rim. Technologically, it is brown, black and gray slip and burnished ceramics. The clay is mixed with fine and coarse sand. The slip of fine clay is thin, 1-2 mm thick.

The typological variety is big. The first number by the type definition defines the category of the bowls.

Type 2.1. Conical bowls with equal cut rounded rims (***Figure 15.9: 1-2***).
Type 2.2.1. Conical bowls with thicken rim from the inner side, sharply cut (***Figure 15.11: 4***).
Type 2.2.2. Conical bowls with incurved and twisted rim, obliquely cut with smooth transition towards the body (***Figure 15.11: 6***).
Type 2.3. Conical bowls with sinuous profile (***Figure 15.11: 5***).
Type 2.4. Bowl woth rounded body and with equally cut rim (***Figure 15.10: 1***).
Type 2.5. Conical bowl with inverted round rim (***Figure 15.10: 3-4***).
Type 2.6. Necked bowl with a lower body part of truncated cone (***Figure 15.8***).
Type 2.7. Round bodied-bowl with slightly inverted rim (***Figure 15.10: 3***).
Type 2.8. Biconical bowl with shorter upper part (***Figures 15.10: 2, 15.10: 4***).

The cups represent a peculiar category among the EB I slip and burnished gray and black pottery ceramics characterized by high handles.

As exception, channeled cups occur with small stringholes lugs (***Figure 15.11: 1-3***).

Channeled black and gray burnished pottery

The channel and grooved black and gray ceramics are the diagnostic group of Dubene-Sarovka IIA and especially of its earlier phase when the channel and grooving and the typical Early Bronze I ornamentation (***Figures 15.9: 5; 15.10: 1-2, 4; 15.11: 2, 5-6, 8; 15.12; 15.13***). That channel pottery includes the type of big high-necked jug (***Figure 15.11: 8***), as well as of askos with a black slip surface, broad band handle and fluted ornamentation of parallel flutes (***Figure 15.12***). The same type is documented at earliest Bronze Age Yunatsite (Katincharov and Matsanova 1993: fig. 15: 1), as well as on the Bereket flat cemetery (Kalchev 1996: fig. 16), but no channels had ornamented the funeral ceramics of the Ezero I culture.

The askos from Dubene-Sarovka has black slip fabric. At Dubene-Sarovka, channels occur as a body ornament of sinuous bowls (***Figure 15.11: 5***), as well as of a handled bowl with rounded body (***Figure 15.11: 2***).

Encrusted pottery

The encrusted pottery seems to become popular already towards the end of the Early Bronze I at Dubene-Sarovka. In the South trench from 1993 it was documented by a cup sherd (***Figure 15.11: 3***). From later EB I pottery it should be mentioned the conical bowls with zigzag incrustation of type 3.1 (***Figure 15.14***) from trench H18-4 (see the discussion in Nikolova 2000a: 44). There is a big high-handled bowl-cup with encrusted ornamentation, which belongs to latest Early Bronze I or to the earliest Early Bronze II (***Figure 15.15***).

Spatial relations of the Yunatsite I culture

Recent data from Western Upper Thrace are limited for a complex definition of the comparative chronology of the Early Bronze I culture there. But some indirect evidence can be used as well. The recent data and especially the ^{14}C dates of the Early Bronze Balkans were discussed elsewhere (Nikolova 1998b). In the context of the Yunatsite culture two problems have been defined:

(1) The correlation of the beginning of Sitagroi IV with the earliest Early Bronze I in Western Bulgarian Thrace;

(2) The relation of the Cernavoda III – Boleráz horizon to the southern Balkans, and particularly to the Yunatsite culture.

(3) The relation of later Baden culture to the Yunatsite I culture.

Those questions aroused in my research in the light of the newly discovered Dubene-Sarovka site. Firstly, it was argued that according to the Dubene data, the Balkan Baden culture completely preceded Sitagroi Va – Yunatsite 14 – Dubene-Sarovka IIB – Ezero 10 (Nikolova 1996a; 1996b; 1999; see also Maran 1998). Also, a hypothesis occurs that it is possible Early Bronze I in the southern Balkans to have been synchronized with later Cernavoda III – Boleráz horizon. In the light of 1999 excavations at Dubene-Sarovka, the chronological position of the pit No. 2 (with the Cernavoda III-Boleráz type bowl, ***Figure 15.8***) can be certainly defined as synchronous with later classical Cernavoda III [Durankulak] – later Boleráz horizon (Nikolova 1999: 250). The *terminus post quem* of the pit corresponds to the house P15-93, so it is possible even two levels to be synchronized with the discussed horizon.

There are different structural elements of the Early Bronze I Balkan chronological system. But the published stratigraphic sequence data are mainly for the later Early Bronze I. So, contact data and comparative stratigraphic observations make it possible a synchronization of the Yunatsite I culture mainly with later Early Bronze I neighbor and distant culture, on the one hand. On the other hand, the absence of the ^{14}C dates from Yunatsite 17-16 and from Dubene-Sarovka IIA requires an indirect absolute chronology of the beginning of the Yunatsite culture in the light of the recent evidence, especially from earlier Sitagroi IV tell and from Ezero 13-12.

Yunatsite culture I – Ezero I culture

The synchronism between Yunatsite 17-15 / Dubene-Sarovka IIA on the one hand, and Karanovo VIIA / Ezero 13-11 / Bereket cemetery / earlier Golyama Detelina on the other hand, is confirmed by the groove pottery as common element, in particular in combination with pricked or stamped small dots (Georgiev et al. 1979: fig. 138c), as well as by askoi (Bereket cemetery; Ezero [Georgiev et al. 1979: fig. 136b, handles; fig. 174g-h]; ***Figure 15.12***). Some other common elements include encrusted ornamentation on the inner side of plate of Dubene-Sarovka (***Figure 15.3: 13***) and earliest Ezero (Georgiev et al. 1979: fig. 138; table 169), as well as jugs with holes under the rim – common element for the Early Bronze I pottery in Bulgarian Thrace (Georgiev et al. 1979: fig. 134c; Nikolova 1998b: plate 5: 1).

For the time being, it is not clear if Yunatsite 17 chronologically precedes Ezero 13, because – from stratigraphic point of view – they seem to have one and the same rhythm of a settlement reconstruction. In other words, three horizons from Yunatsite tell correspond to three horizons of Ezero tell. The similarity between Golyama Detelina pottery (Nikolova 1999: fig. 8.7.1) and Yunatsite 17-16 (amphora-like vessels, cp. Katincharov and Matsanova 1993: fig. 23: 1-3, 5) also makes it possible a synchronization between Ezero 13-11 / Golyama Detelina and Yunatsite 17-15.

In the course of the Yunatsite excavations, it was stressed on the difference in the ceramic style between the earliest Yunatsite (17-16) and earliest Ezero. The absence of the channel pottery at the latter site, as well as the decreasing of such a typical of Ezero 13-11 elements and holes under rim ornamentation was stated (Katincharov et al. 1986; 1987; 1988; 1989), on the one hand.

On the other hand, V. Němejcová-Pavúková (1991: 71 with ref.) had reminded the discovered by her channel fragments from Ezero 11. Recently, the Karanovo excavations confirmed the channel pottery specifies Ezero I culture ceramic style as well.

However, ethnographic peculiarities and different directions of contacts can explain the ceramic difference between Western and Eastern Upper Thrace. On its hand, the pottery from Karanovo VIIA from the recent excavations of S. Hiller and V. Nikolov (1997) parallels with Ezero 13 as well. In the pit B, as well as partially in pits H and D a typical of later Early Bronze I pottery was discovered, with comprehensive analogies in Dubene-Sarovka IIA and Yunatsite 17-15, including channel and groove pottery (Hiller and Nikolov 1997: plate 148: 2,7,16,21; plates 154: 24-27; 155: 2-12; 164: 5); askoi (1997: plate 145: 2,8); relief ornamented jar, pithoi and other ceramic vessels (1997: plate 144: 21-25, 27; 145: 1, etc.); zigzag incised ornamentation over the flat rim of bowl (1997: plate 153: 18), etc.

The sinuous bowls are missing at Karanovo VIIA confirming peculiar common elements in the Early Bronze I Balkan ceramic style from Hotnitsa-Vodopada and Mirovtsi. At the same time, the typical of Ezero 13 ornamentation with small dots also is not represented among the published pottery from Karanovo VIIA pits, as well as encrusted dot ornamentation, which specifies Ezero 12-11. This fact can be explained by a limited pottery discovered during the Bulgarian-Austrian excavations or by possible earlier (pre-Ezero 13) occupation of the Karanovo VII, or by the fact that only Ezero 13 phase exists at Karanovo VIIA. The Karanovo tell and the Ezero tell are in distance less than 20 km, which fact assumes close similarity in the ceramic style between synchronous sites.

Going back to Ezero, the published from 13th level sherds with edges (Georgiev et al. 1979: fig. 134: d) are comparable with the saucier sherds from Dubene-Sarovka IIA (Nikolova 2000a) and Ognyanovo (Leshtakov, chapter 16, ***Figure 16.5: 4***) and Yunatsite (unpublished). Common elements are also the storage vessels with holes under the rim known from Dubene-Sarovka (unstratified, Nikolova 1998b: plate 5: 1), earlier Yunatsite (unpublished) and Ognyanovo (Leshtakov, chapter 16, ***Figure 16.3: 6***). However, the Dubene-Sarovka find is an exception among the numerous known up to now, as well as that from Ognyanovo. One explanation is the Early Bronze I levels are still uninvestigated over vast territory on the former site, but it maybe the fashion of those vessels was not so vast distributed as in Easter Upper Thrace.

Other common element is the relief ornamentation (Georgiev et al. 1979: fig. 134). That is a common Balkan element in the pottery style characterizing Cernavoda III, especially Cernavoda II and Baden culture complex (Nikolova 1999 with the references cited there).

It should be especially stressed on the type of bowls 2.2.1 (***Figure 15.11: 4***). They are known from Dubene-Sarovka IIA, Ognyanovo (Nikolova 1998b: fig. 4:1, Leshtakov chapter 16, ***Figure 16.10: 2***), with parallels at Ezero 13. Despite the parallels between Ezero 13-11 and Yunatsite I culture can be increased, on the whole in the Early Bronze I there are already considerable differences, as the Ezero A1 ceramic style is closer to the eastern Lower Danube, while that of the Yunatsite I culture – to the other ceramic styles in the Central Balkans.

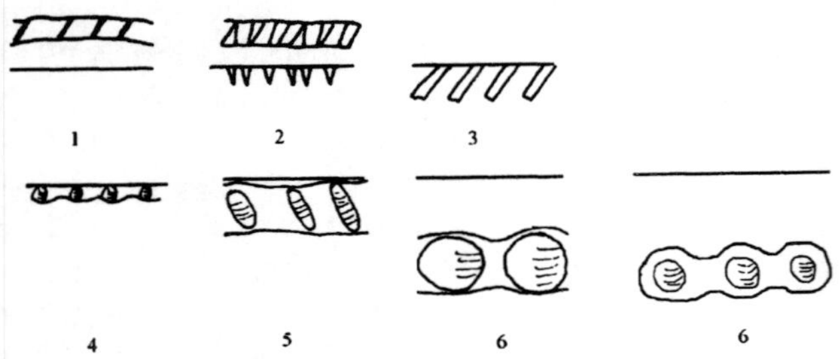

Figure 15.4. Incised and relief ornamentation from Dubene-Sarovka II.

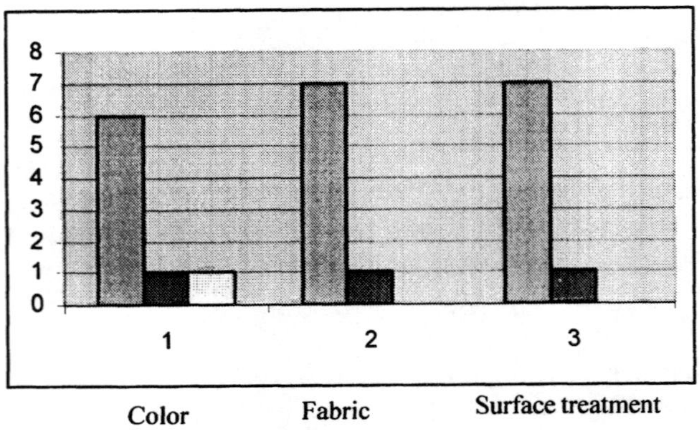

Chart 15.1. Data to Figure 15.5. Color: moderate and dark brown (6), beige (1) and gray ceramics (1). Fabric: clay paste plentfully mixed with fine and rarely coarse (7) and fine and plentfully coarse sand (1). Organic inclusions occur as exception. Smoothed outer surface (7) and rusticated(1) one.

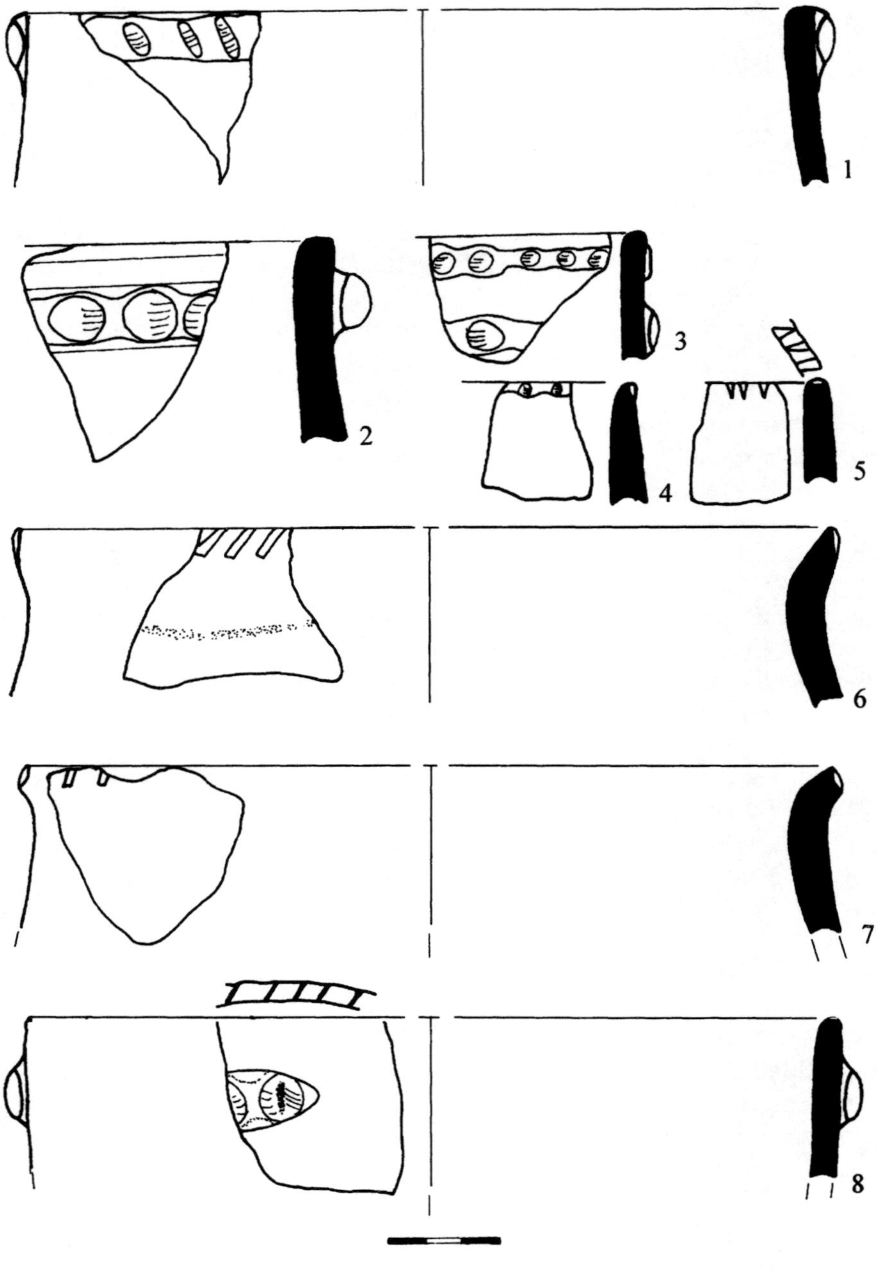

Figure 15.5. Dubene-Sarovka II. A pottery from South sounding (1993) discovered at a depth of 0.30 - 0.70 m below the surface. The cultural layer can be attributed to IIA stage (Early Bronze I).

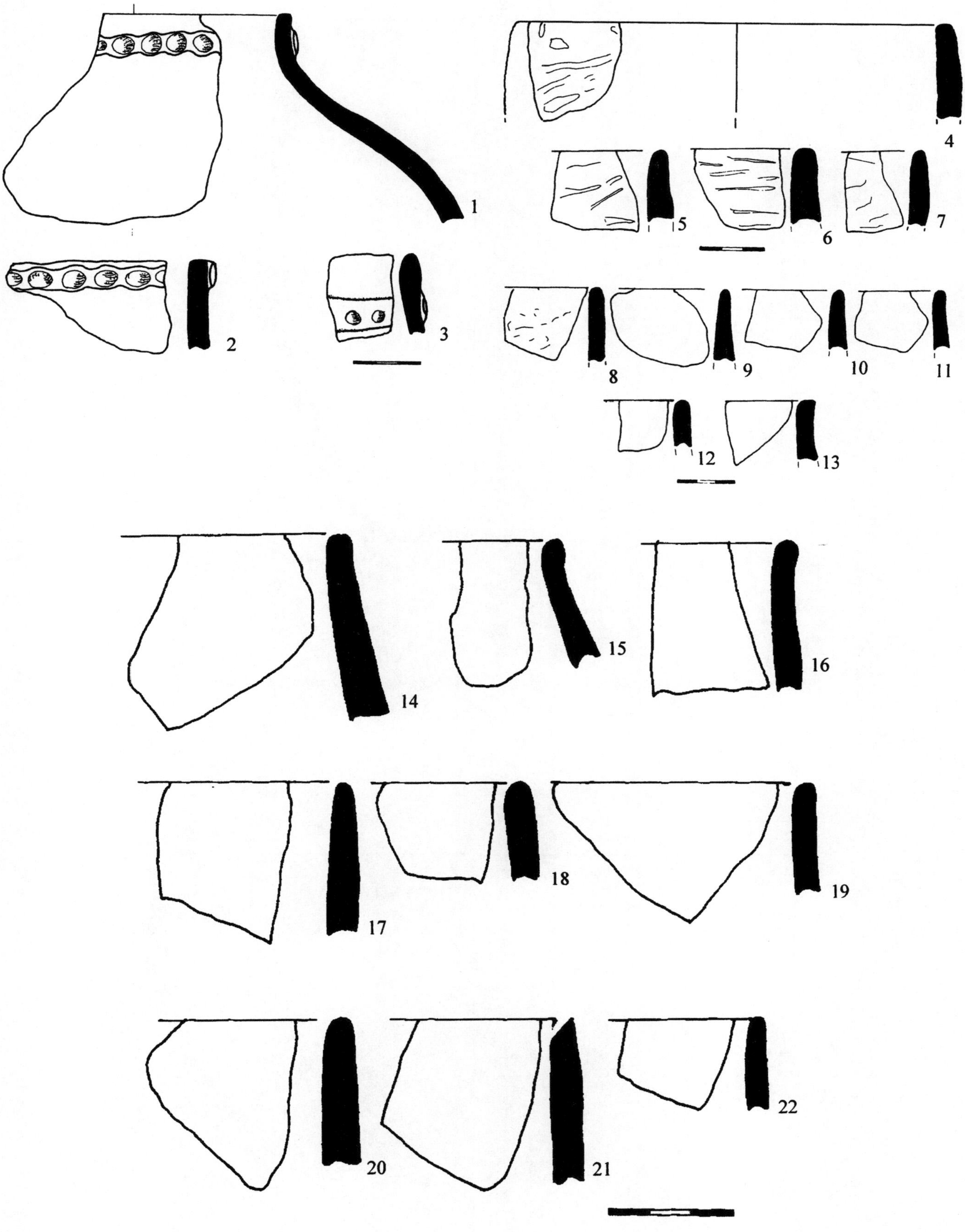

Figure 15.6. Dubene-Sarovka IIA. Early Bronze I pottery.

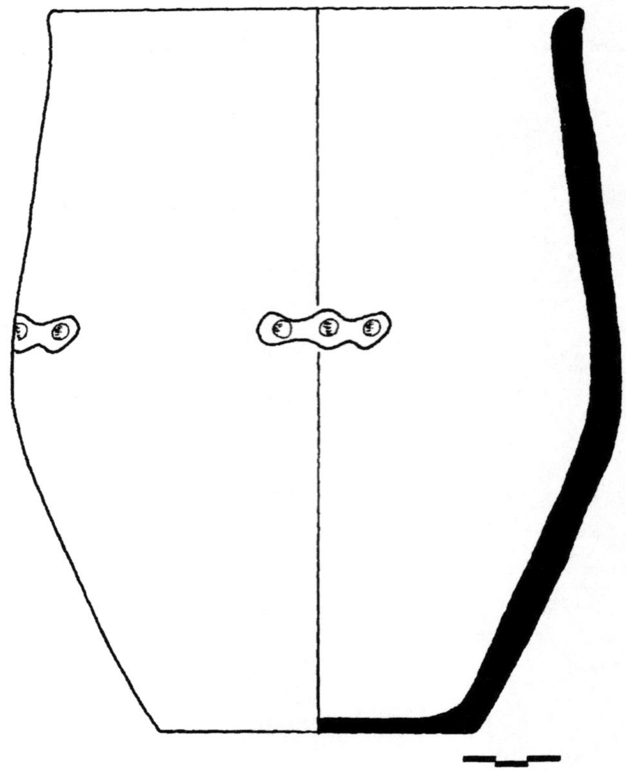

Figure 15.7. Dubene-Sarovka IIA. Brown slip bowl from pit No. 2. Early Bronze I.

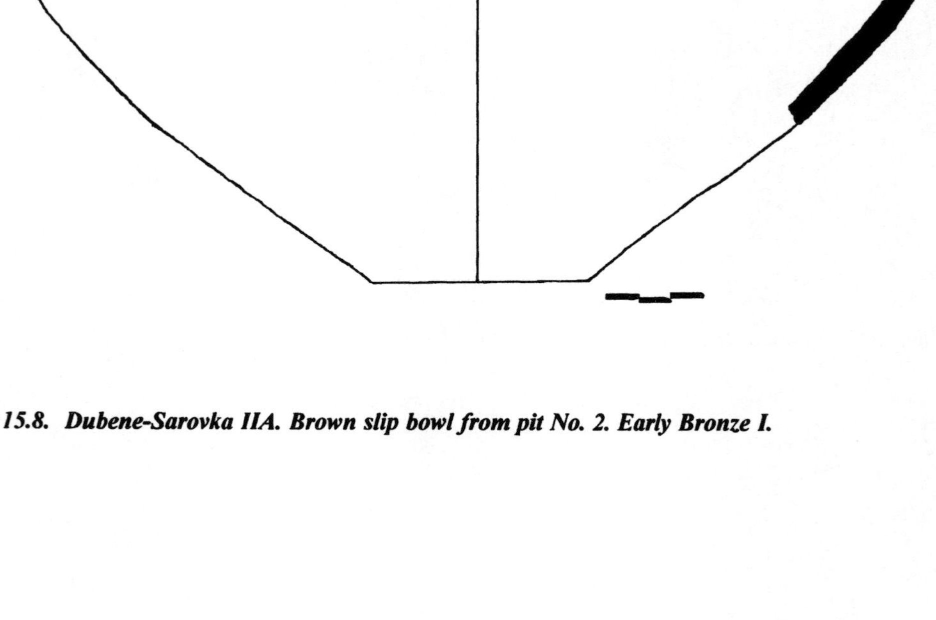

Figure 15.8. Dubene-Sarovka IIA. Brown slip bowl from pit No. 2. Early Bronze I.

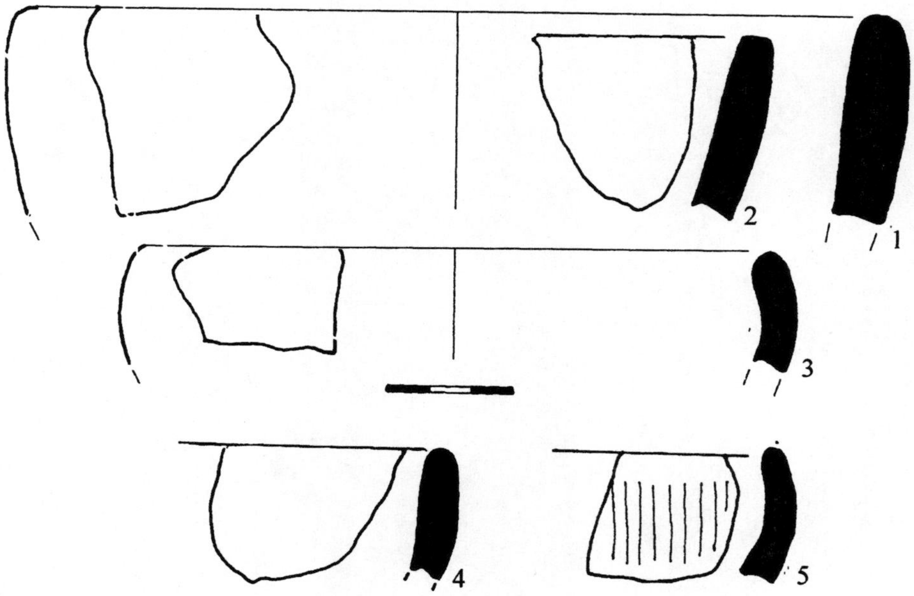

Figure 15.9. Dubene-Sarovka IIA. Brown slip (1-5) and black burnished (6) pottery. Clay mixed with fine and coarse san, well burnt. Over sherd No. 6 there are vertical burnished lines. Early Bronze I.

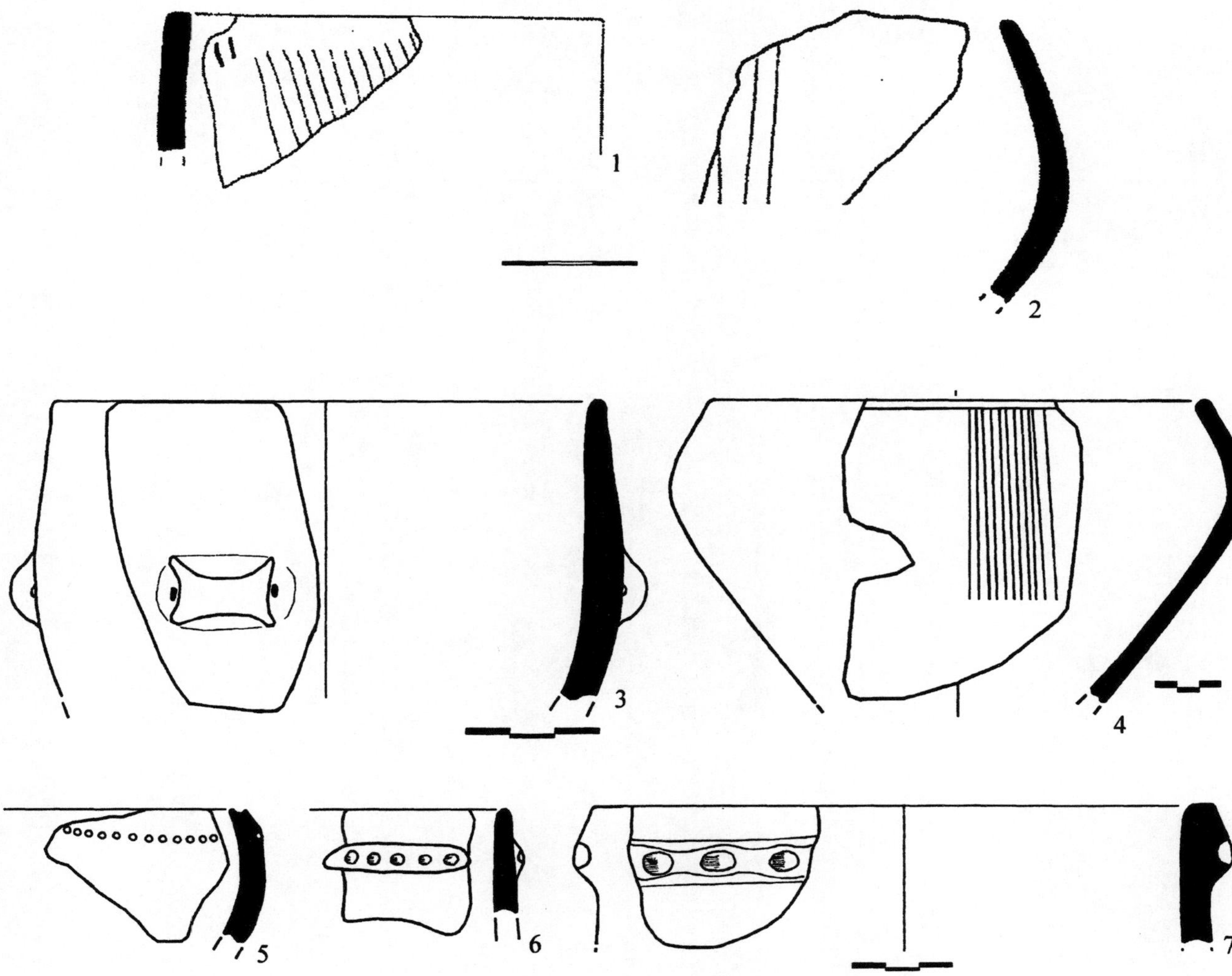

Figure 15.10. Dubene-Sarovka IIA. Bowls . Early Bronze I.

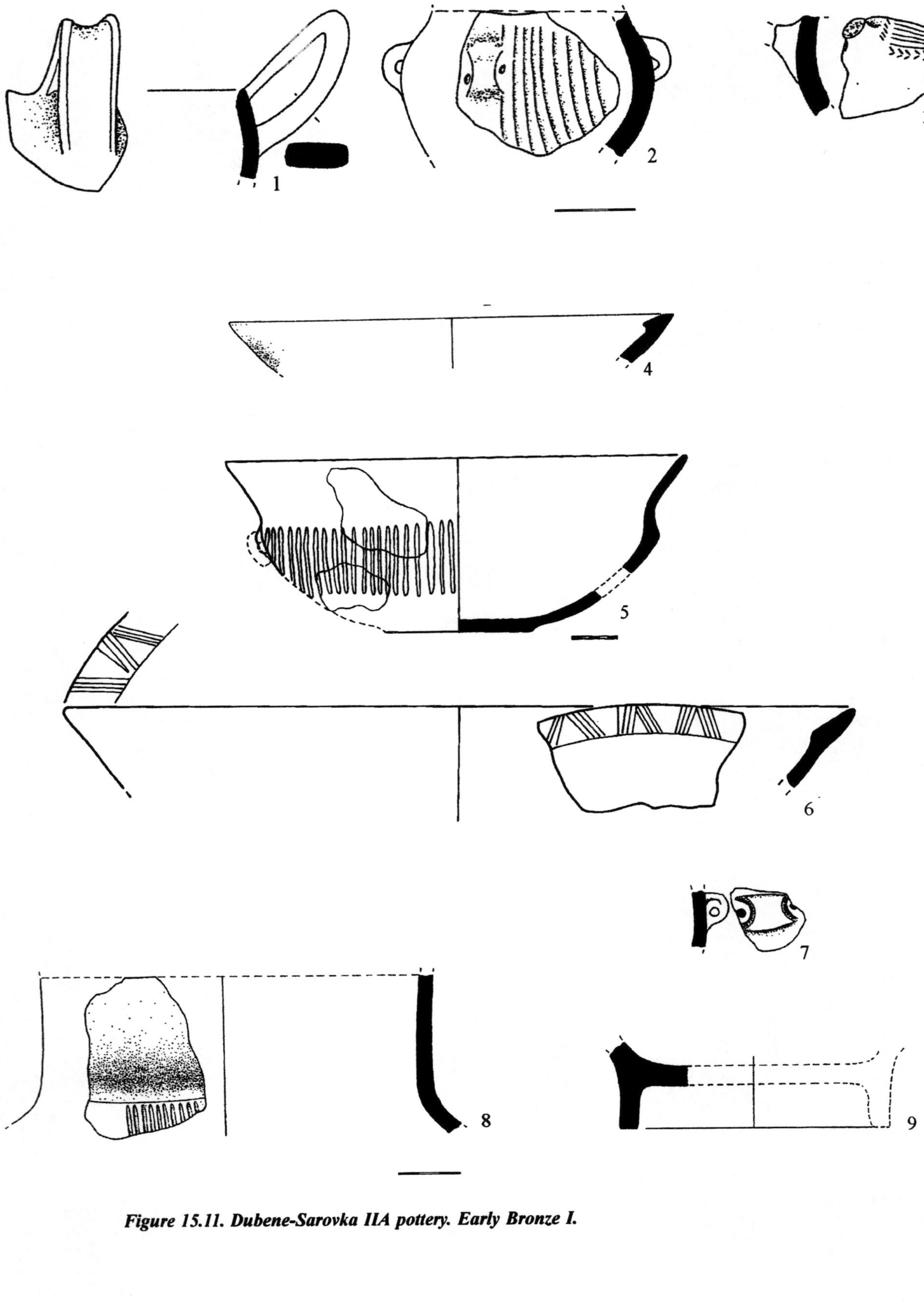

Figure 15.11. Dubene-Sarovka IIA pottery. Early Bronze I.

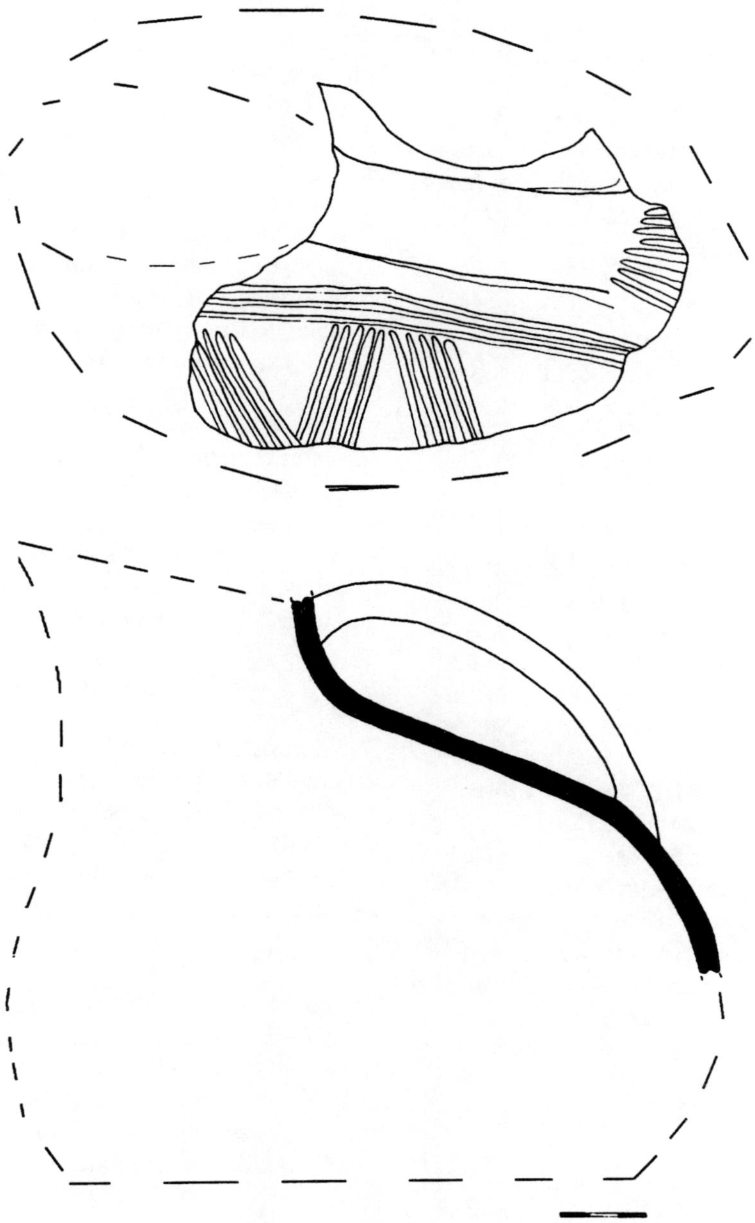

Figure 15.12. Dubene-Sarovka IIA. A black slip fabric. Clay plentifully mixed with fine and coarse sand and single gravel. Squares R12 -S12. A depth of 1.80-2.00 m. Early Bronze and Late Copper Age sherds mixed.Early Bronze I.

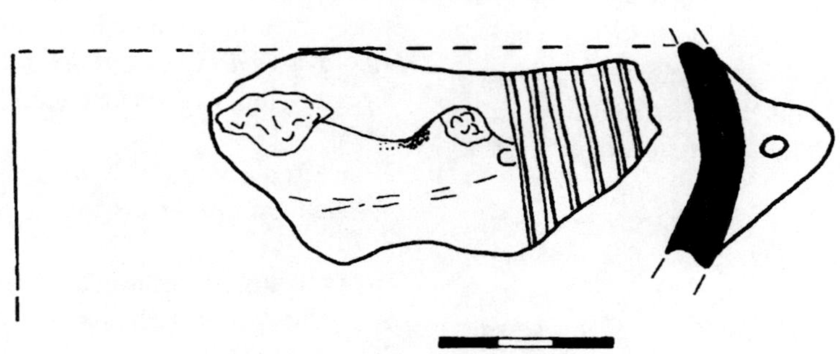

Figure 15.13. Dubene-SArovka IIA. Black fine channeled pottery. Early Bronze I.

Recently, mostly fragmentary pottery have been published from Rupkite-Kaleto (Karasura [Nikolova 2000a: pp. 55, 57]). But it is not accurately dated in the initial publication. As the illustrated evidence indicates, the so-called stages I to III completely belong to Early Bronze I (Rupkite-Kaleto IIA). Then, according to those data, there is a big chronological hiatus in EB II, and again in Early Bronze III the occupation of that site was renewed (the so-called stages IV and V, or Rupkite-Kaleto IIB according to my division).

Of interest is one of the dates from Rupkite-Kaleto IIA (the so-called stages II and III) – Bln-3429: 4790±60 bp, and Bln-3773: 4250±150 bp. The Sum probability of both dates is in the span between 3700 and 3300 BC. Those dates, despite some parallels in the pottery at Ezero 13-11, allow one to include as a hypothesis, the settlement was ground before Ezero 13.

For the time being, we can accept the horizon Yunatsite 17-15 – Dubene-Sarovka IIA – Ezero 13-11 – Karanovo VIIA – Bereket cemetery – Golyama Detelina – Early Bronze I Vesselinovo (Nikolova 1999) – Rupkite-Kaleto (Karasura) IIA, as approximately contemporary, with possible diachronic interrelations of the beginning of different sites included there.

Yunatsite I culture - Sitagroi IV / Dikili Tash IIIA group

To the south, Sitagroi IV and Dikili Tash IIIA from later Early Bronze I give expressive contact data with Yunatsite 17-15 and Dubene-Sarovka IIA. Along with the channel pottery, a cup with a high handle from Dubene-Sarovka parallels to the popular cups from the northern Aegean, but it is ornamented with typical of Bulgarian Thrace ornament consisted of incised triangles (Nikolova 1999: fig.19.2). The common elements of grooving should be stressed despite the ornament from Sitagroi includes zigzag, similar to Cernavoda III – Boleráz pottery.

Sitagroi IV and Dikili Tash IIIA consist of several levels. The evolution of the ceramic style on the former site is commonly described by 'appearance of channel pottery, flourishing of channel pottery and decreasing of this fashion replaced by punched and *pointillé* ornamentation' (Sherratt in Renfrew et al. 1986: 434). C. Renfrew also mentioned three phases distinguished by pottery (1986). But unfortunately, there are not correlation between the ceramic evidence and the stratigraphic sequence (*see* the discussion in Nikolova 1999: 187-91).

Based on the stratigraphy and ^{14}C dates published from Sitagroi IV, two phases of development can be distinguished there. There are demonstrated by ZA32-29 and ZA28-21 levels with at least six floors documented in this sounding (Renfrew et al. 1986: fig. 8.2) and more than a meter thick culture layer. From stratigraphic point of view, the Sitagroi IV is similar to Yunatsite 17-15 and Dubene-Sarovka IIA.

On its side, the similarity of ^{14}C dates from Ezero 13-11 and Sitagroi IV arouses a hypothesis of synchronism between Early Bronze I in Bulgarian Thrace and northern Aegean dated from c. 3300 BC. The earlier values of the sum-probability of the radiocarbon dates from Sitagroi IV (3310-2910 BC by 68.2% confidence) argue that chronology (Nikolova 1998b).

However, the data from Sitagroi IV are incomplete. So, it is possible the beginning of Sitagroi IV even to precede Ezero 13 and Yunatsite 17. One argument is the fact that the earliest pottery is close to that of Final Copper II in lower Danube basin and earliest Early Bronze I Cernavoda III culture, especially that of ZE80 level (Renfrew et al. 1986: fig. 13.6:1). So, the dating of the beginning of Sitagroi c. 3400 BC can be assumed as well. That dating corresponds to the summarized absolute chronology of the beginning of Early Bronze in the Balkans excluding the earlier dates from Hotnitsa-Vodopada 1-2.

The dates from the earlier Sitagroi IV and from Ezero 13 can be used for definition of the beginning of the Yunatsite culture represented at Yunatsite 17, having in mind that the last horizon is synchronous to it immediately precedes Ezero 13. These indirect ^{14}C dates date the beginning of the Yunatsite culture in light of the recent evidence c. 3300 BC.

Yunatsite I culture – Pernik I group

To the west, the evidence from Kovachevo IIIA and Radomir-Vakhovo II corresponds to Yunatsite I culture (Nikolova 1999: 41, 47). Synchronism between Kovachevo IIIA and later Sitagroi IV (ceramic contact data – urns) and contact data – sinuous bowls – between Radomir-Vakhovo II and Dubene-Sarovka IIA confirm this synchronization (Nikolova 1999: 250).

The discovered at Radomir-Vakhovo lid (bowl?) with an incised ornamentation (Maran 1998: plate 2) parallels with Gladnice (Maran 1998: plate 1: 7), Brza Vrba (Maran 1998: plate 3), Jevišovice (Maran 1998: plate 4: 1), late Boleráz at Bratislava (Baxa and Kaminská 1984: plate 11: 1; Němejcová-Pavúková 1992: fig. 6: 1), etc. In my study I assumed Radomir-Vakhovo II (1, or 5 if we accept the sequence numbered from the earliest Early Bronze upwards) chronologically preceded Yunatsite 17 (Nikolova 1999: 250). But it cannot be excluded Radomir-Vakhovo 1 to have been synchronous with Yunatsite 17, respectively with Dubene-Sarovka IIA. At that site, there is one or even two more levels from later Early Bronze I (see Nikolova 1999: 215), but the information is scanty and without stratigraphic documentation published.

Yunatsite I culture – Cernavoda III culture – Orlea type – Baden culture – Coțofeni I culture – Pit Grave Culture

The problem of synchronism between Yunatsite I culture and late Cernavoda III occurs recently. Some contact data between Cernavoda III and Ezero 13 can be explained by direct contacts rather than by diachronic interrelations, but the pottery from the earliest graves of the Golyama Detelina Tumulus is close to earliest Early Bronze Ezero as well (Nikolova 1999). Having in mind the possible connection of population buried there with Pit Grave Culture one, the hypothesis occurs that the last replaced Cernavoda III culture in some parts of the eastern Lower Danube at the time of Ezero 13. There are no arguments for earlier dating of the beginning of Pit Grave

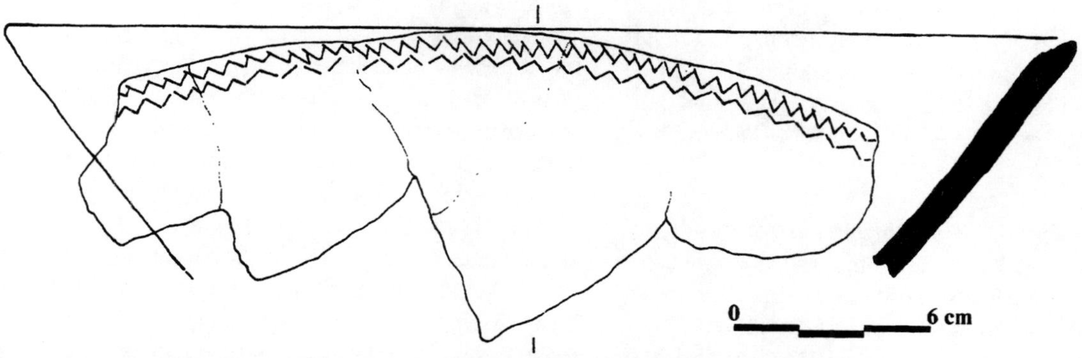

Figure 15.14. Dubene-Sarovka IIA. Encrusted brown slip bowl. Early Bronze I.

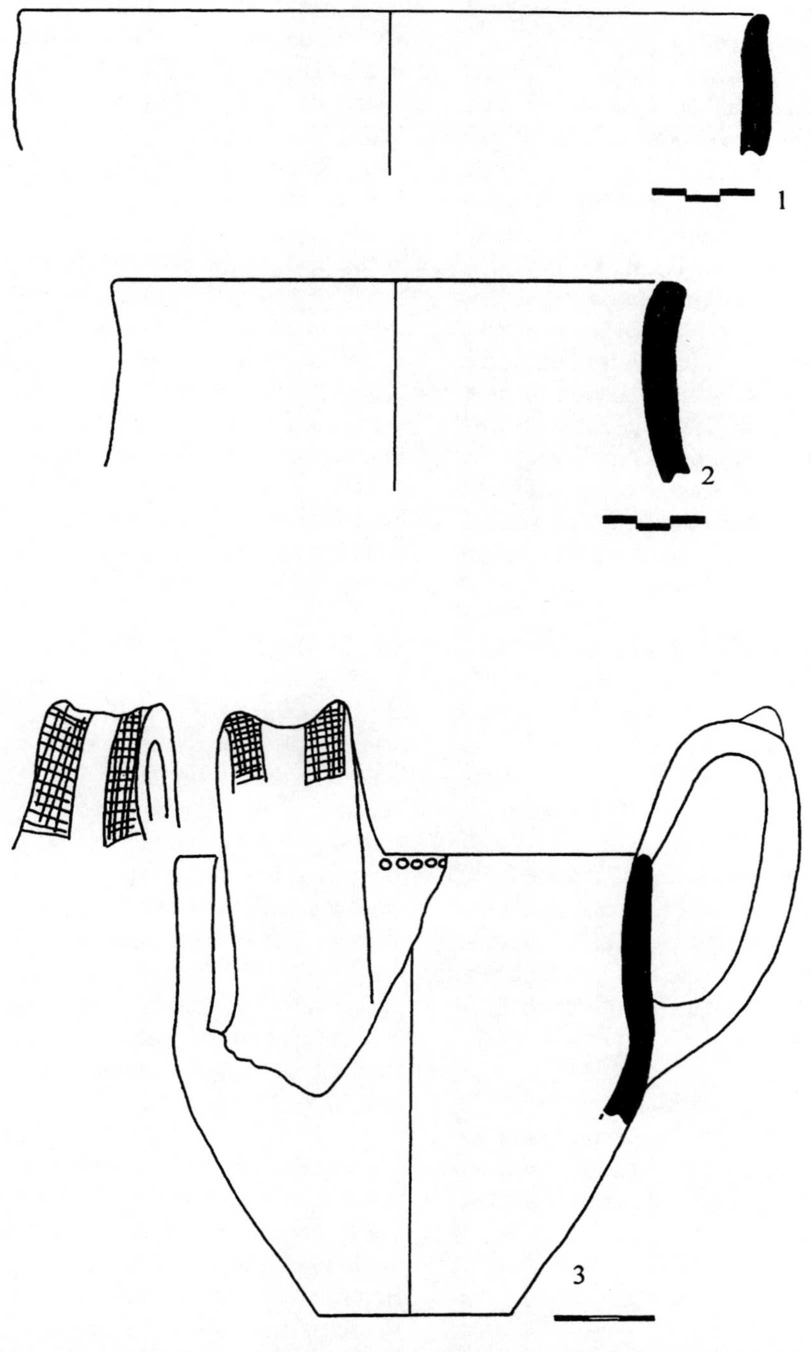

Figure 15.15. Dubene-Sarovka II. Earlier Early Bronze from the western periphery of the settlement.

Culture penetrations in the Balkans, having in mind the late dates from Usatovo corresponding to Early Bronze I. On its side, the triangle-incised ornament at Durankulak (Nikolova 1999: fig. 8.3: 9) indicates the Cernavoda III settlement can be synchronized with Ezero 13 – Yunatsite 17 – Dubene-Sarovka IIA1 as well.

The channel pottery and the later Boleráz – Baden IIa type (channeled handles) from Yunatsite 17 indicate that synchronization. Alder shaped channels known from Yunatsite occur in Baden culture at Ib-IIa stages (Němejcová-Pavúková 1991: fig. 15). The recent data from Karanovo VIIA also indicate the existence of the channel phase in the south, so that the beginning of the Ezero culture (including Ezero 13) also can be attributed to the horizon of later Cernavoda III – Boleráz in the southern Balkans.

Contact data between Radomir-Vakhovo II2 and Orlea type (Orlea-Sadovets group) in eastern parts of the western Lower Danube indirectly synchronize Yunatsite I culture (Yunatsite 16-15, later Dubene-Sarovka IIA) with Orlea–Sadovets group, as well as with the Coţofeni I culture (Nikolova 1999: 250).

The closest sites of Baden culture are from Velika Morava valley – Ostrikovac Ia and Novačka Ćuprija, but they belong to later Baden and the pottery published is limited. Some distant parallels with late Boleráz should be mentioned confirming a possibility Yunatsite 17, respectively Dubene-Sarovka IIA1 to be synchronized with former culture. From Boglarlelle (Hungary) originates a handled bowl woth pedestal bottom (Němejcová-Pavúková 1992: fig. 2: 4). The last element parallels with Dubene-Sarovka IIA (***Figure 15.11: 9***). From the same phase there are sherds with rhomb-incised ornament encrusted (unpublished), that is comparable with the so-called lids from later Boleráz (Němejcová-Pavúková 1992: fig. 6: 1b, 4b).

At Blatne (Slovakia), small plastic applications are attached at ordinary pottery (Němejcová-Pavúková 1984: fig. 4: 18, fig. 5: 9) which parallels with the pot from the Dubene-Sarovka IIA1 pit No. 2 (***Figure 15.7***). Pseudo-handles occur at the same site similar to represented in fig. 2.1: 13, but the bulges from that site parallels with Sitagroi IV (Němejcová-Pavúková1984: fig. 7: 15; fig. 7: 17; Nikolova 1999: fig. 8.5: 4). The vertical and oblique channeling from Yunatsite I culture is comparable with the conception of the ornamentation at Zlkovce (Němejcová-Pavúková 1984: figs. 13-15) despite that other motifs predominate. From the last site the abundance of relief ornamentation and popularity of the big storage vessels should be stressed.

The last, along with incisions, band handles, hang-handles - edges etc. can be used as an argument of synchronization between Yunatsite 17 / Dubene-Sarovka IIA1 and that site, respectively between the beginning of the Yunatsite culture and Boleráz Ic.

Yunatsite I culture – Pevkakia Magula

The published by E. Christmann (1996) pottery from phase 1 at Pevkakia Magula in Thessaly parallels with Early Bronze I Balkans and arouses a possibility of synchronization between Yunatsite I culture and Pevkakia Magula I. The similarity in the pottery includes conical bowls, barrel-shaped containers, S-profiled pots, cups with high handles, ring-shaped bases, pricked ornament; triangle incised motif; rusticated surface with incisions over the rim of some storage vessels, motifs of holes under the rims of vessels, etc. (Christmann 1996: tables 1: 1, 6, 17, 16, 18; 2: 4; 3: 4; cp. ***Figures 15.6; 15: 9; 15.11: 1, 9***). It is under question if the popularity of the saucier in the earlier Yunatsite culture has some relations with the spout-bowls from Thessaly, but according to the stratigraphic data from Pevkakia, they occur already in the Early Bronze I (Christmann 1996: table 3: 1,3).

Considerations

Early Bronze I is divided into two stages Early Bronze IA and Early Bronze IB and with two phases each (Nikolova 1999: 175-198, 250-60). The classical Cernavoda III (with the alder ornament) belongs to Early Bronze IA2, as well as to Early Bronze IB1 and partially corresponds to the earliest Bronze Age phases at Sitagroi IV – Yunatsite (17) – Dubene-Sarovka IIA1 – Radomir-Vakhovo II – Karanovo VIIA – Ezero (13) – Bereket cemetery.

In the light of the recent evidence, the first stage of the Yunatsite culture belongs mainly to the Early Bronze IB (Nikolova 1999: 250; Nikolova 2000a: 61). The ceramic similarity synchronizes Yunatsite 17 with Dubene-Sarovka IIA1, but the question of the contemporary grounding of both sites remains open. Recently, the published by K. Leshtakov pottery from Ognyanovo (chapter 16), confirms the site was grounded in Early Bronze I (Nikolova 1998b). Also, it is possible the Early Bronze I in Western Upper Thrace to have been represented by Plovdiv-Nebet Tepe and Manole-Razkopanitsa (Nikolova 1998b). However, the last data are under discussion.

Yunatsite I culture is synchronous with Ezero I culture (Ezero 13-11, Karanovo VIIA, earlier Golyama Detelina, Bereket cemetery, Rupkite-Kaleto IIA, etc.). In the light of the recent data, the ^{14}C dates from Ezero 13 can be used as one of the indirect records for definition of the beginning of the Yunatsite I culture, despite typologically Ezero 13 and Yunatsite 17 belong to different ceramic styles. To the south, Yunatsite I culture is contemporary to Sitagroi IV, but the comparative-stratigraphic data indicate that it is possible Sitagroi IV to precede Yunatsite I culture, in particular Yunatsite 17.

To the west, the culture under discussion was synchronous with Pernik I group (Radomir-Vakhovo II, Kovachevo IIIA), but it is possible the beginning of Radomir-Vakhovo II to precede Yunatsite 17.

It can be defined diachronous-synchronous interrelations with the northern Balkan Early Bronze I cultures. The beginning and earlier Cernavoda III culture have no synchronous sites in Upper Thrace, according to the recent evidence from Hotnitsa-Vodopada (Nikolova 1999; see also chapter 27), Koprivets, etc. However, it is possible later Cernavoda III (with alder ornamentation) to have been at least partially contemporary with the beginning of the Yunatsite I culture, as well as with earlier Sitagroi IV. On its side, Orlea-Sadovets – Coțofeni I – later Baden horizon is completely contemporary with later Yunatsite I culture (Yunatsite 16-15 and later Dubene-Sarovka IIA).

Newly published data from Thessaly indicate that the synchronization of the Yunatsite I culture can be extended to the south, with Pevkakia Magula 1. On its side, Yunatsite I culture corresponds completely to for-Troy I stage of the Early Anatolian culture (Nikolova 1999: 250). In the Cyclades, later Grota-Pelos (Pelos-Lakkoudes) culture and earlier Kampos (after Maran 1998: table 82) developed parallel with Yunatsite I culture.

Based on the comparative radiocarbon chronology, the beginning of the Yunatsite I culture can be dated from c. 3300 BC, and its end – to the end of the Fourth Millennium BC.

Yunatsite II culture and Early Bronze II cultural interactions in the Balkans

The ceramic style of the Yunatsite II culture

The second stage of the development of the Yunatsite culture comprises the 14th – 9th horizon of the Yunatsite tell, Dubene-Sarovka IIB, earlier Plovdiv-Nebet Tepe and Ognyanovo, surface finds from Banya-Ploskata Mogila, as well as from Dubene-Popovka and Kliment-Stenithi. Encrusted ceramics are diagnostic of that stage, with stamped dots or incised encrusted lines. Various other types of stamps also occurred, as well as false-corded ware (in particular at Dubene-Sarovka IIB). The variety of the encrusted pottery have been already published in detail (Nikolova 1996a; 1996b). Below some additional stratified data from Dubene-Sarovka demonstrate the idiosyncratic elements of the different phases of the Yunatsite II culture.

At the Dubene-Sarovka tell, the second stage of the Yunatsite culture comprises the layer IIB with three phases: Dubene-Sarovka IIB1-3. The false corded ornamentation appeared during this phase, but it predominates during the second and the third phase. The last are well distinguished by stratigraphic sequence. Each of them consists of one to two building levels with reconstruction phases. That sub-division of Dubene-Sarovka IIB1-3 can be accepted to define three phases of the second stage of the Yunatsite culture corresponding to Yunatsite 14-13, 12-11 and 10-9.

Two building levels represent Dubene-Sarovka IIB1 phase, respectively Yunatsite IIA culture. Already in this phase a typical of Early Bronze II Yunatsite culture net-encrusted pottery occurs, in combination with zigzag-incised lines and negative rhomb (***Figure 15.16: 2-3***). A bowl fragmented occurs along with a big shallow cup with pedestal bottom and incised ornamentation and fragment of round-bottom cup (***Figure 15.25: 7-9***). From the same phase an urn discovered in a small ash-pit beside a fragment of floor from the fourth horizon (Dubene-Sarovka IIB1 [***Figure 15.25: 10***]). The phase under discussion includes also a storage vessel with tongue-handle and a bowl with pseudo-handle and bulges (***Figure 15.25: 11-12***). The pottery in ***Figures 15.16-15.18*** represent the first phase of the second stage of the Yunatsite culture as well, but their contents is not connected with architectural remains on the site.

Dubene-Sarovka IIB2 phase represents the Yunatsite IIB2 culture. One building level with reconstructions belongs to that phase including Apses-house No. 1 and Apses-house No. 2. Most of the encrusted pottery discovered at Dubene-Sarovka from the cultivated layer belongs to that and to the following phase (Nikolova 1996a: fig. 10-21; 1996b; 1999: 202-207). Also, a cup with a false-corded ornamentation in combination with net-encrustation represents the second phase of the Yunatsite II culture (***Figure 15.25: 6***). A handled bowl from Yunatsite (***Figure 15.25: 1***), which is close to the Early Bronze III ceramic style originates from the 11th horizon according to the excavators.

At Dubene-Sarovka, the first building horizon from Dubene-Sarovka IIB3 (Yunatsite IIB3 culture) is partially preserved only. At least two house levels characterize that horizon. Plain pottery occurred over two house floors that typify the ceramic style of that kind of earthenware of the third phase of the second stage of the Yunatsite culture in the upper Stryama valley. The round-bottom cups continue from the previous phases (***Figure 15.24: 1***).

Along with a round-bottom cup, a small jug and a small bowl were discovered over the same floor of house F13-94 (***Figure 15.24: 2-3***). At the same depth, over the floor of house K20-96 bowls with discovered with brown to grayish-greenish surface, included a light brown bowl with five bulges (***Figure 15.24: 4-6***). Part of the encrusted pottery from the cultivated layer belongs to that phase as well.

An *askos* with ornamentation executed in the false corded technique, in combination with incised lines was published from Plovdiv – Nebet Tepe (Nikolova 1996a: fig. 22). The composition of chess-like ornament on this vessel characterizes the ceramic style of the second stage of the Yunatsite culture there. On the basis of the limited comparative-stratigraphic data, the askos is dated to the third phase of the second stage of the Yunatsite culture, but it is typologically very similar to the second phase presented at Dubene-Sarovka IIB2.

According to the published encrusted pottery from the Ognyanovo tell, the second stage of the Yunatsite culture is represented there by net like incised lines as an ornament on the inner wall of plates, below the mouth rim (Detev and Mazanova 1977, fig. 8: 12, 13; chapter 16).

The surface find of a bowl fragment with tongue-shaped handle with small holes and encrusted ornamentation from the Banya – Ploskata Mogila also documents the Early Bronze II occupation of this tell. Encrusted circles characterized the pottery from Dubene-Popovka, with parallels at Dubene-Sarovka IIB.

The surface find of a bowl fragment with horizontal stringhole lug and encrusted ornamentation from the Banya – Ploskata Mogila also documents the Early Bronze II occupation of this tell. Encrusted circles characterized the pottery from Dubene-Popovka, with parallels at Dubene-Sarovka IIB. From that site originates the bowl in ***Figure 15.27***, according to the information of N. Subev from Dubene. The bowl belongs to his private collection.

The contacts of the Yunatsite II culture

For the time being, Ezero 10, Sitagroi Va and Gomolava IIIb date the beginning of the Early Bronze II in the Balkans - 3000 BC (Nikolova 1999). Many ceramic data indicate dynamic interrelations of the Yunatsite culture population with the neighbor and distant cultures – direct and indirect contacts, as well as following the fashion and standards distributed in the Early Bronze II Balkans.

Yunatsite II culture – Ezero II culture and Pit Grave Culture

The synchronism between the third phase of the Yunatsite II culture and Ezero 5-4 is confirmed by the ^{14}C dates from Ezero 4. The *terminus post quem* for the end of Ezero 4 is c. 2540 BC-2460 BC corresponding to comparative chronology of this horizon dated the end of the second stage of the Ezero culture c. 2500 BC.

From the latest Early Bronze I – the earliest Early Bronze II can be dated the earlier Pit Grave Culture graves from Northeastern Bulgaria - from Poruchik Geshanovo and Plachidol Tumulus cemeteries, as well as from Eastern Hungary. They correspond to the earliest phase of the second stage of the Yunatsite culture (Nikolova 1999: 220, 255). But contact-data between Dubene-Sarovka (***Figure 15.25: 6***) and Dolno Sakhrane (Nikolova 1999: fig. 9.6: 12) – a round bottom cup with false corded ornamentation dated the tumulus in Early Bronze IIB corresponding to second phase of the Yunatsite II culture. Mednikarovo Tumulus with Ezero B1 encrusted pottery also belongs to the same phase, as well as Turnava I Tumulus with Coţofeni III pottery (Nikolova 1999: 64, 68-69). So, Yunatsite IIA-B culture can be synchronize with classical stage of the Pit Grave Culture in the Balkans.

Yunatsite II culture - Coţofeni II-III culture

The excavations at Dubene-Sarovka in 1992 already initiated the search of the contact and typological data with Coţofeni culture (Nikolova 1995a). In ***Figure 15.32*** some emblematic pottery from Coţofeni II-III in the Northwestern Bulgaria is represented that is comparable with Yunatsite II culture. The bowls in ***Figure 15.32: 1-6*** are common Balkan type. The incisions over the bowl rims characterize mainly earlier Yunatsite culture as well as the bowl with a pseudo-handle parallels with Dubene-Sarovka IIB1 pottery.

The oblique cut small jug (***Figure 15.32: 8***) is close to that from Dubene-Sarovka IIB3 (***Figure 15.24: 2***), but a ring-like pedestal characterizes Dubene-Sarovka IIA (***Figure 15.11: 19***). The round-flatten and rounded bottom cups (***Figure 15.32: 11-12***) are very similar to Dubene-Sarovka IIB cups (*see* for instance the cup in ***Figure 15.24: 1***).

Of special interesting is the bowl in ***Figure 15.32: 10*** from Kozloduj-Saltinov Vodopoj (probably from Coţofeni III culture), which is characterized by typical of the Early Bronze II encrusted ceramic style in the Central Balkans (Nikolova 1996a; 1996b; ***Figures 15.3: 10-11; 15.16: 3; 15.20; 15.23; 15.25: 7***). The analogies with Ostrikovac Ib-d should be stressed where the net incised ornamentation was also very popular (Stojić 1996).

So, the Yunatsite II developed parallel with the Coţofeni II-III culture. It should be stressed there is a typological similarity but it rare concerns the emblematic features of both the ceramic styles. Among the common pottery elements there are the conical bowls with equal cut rims.

The round-bottom cups should be outlined against the peculiarity in the encrustation style of both the cultures. The rims from thicken from their inner side (type Ia3 after P. Roman) is very popular in the Early Bronze II in the Balkans, but that type occurs there in the Early Bronze I, according to the data from Dubene-Sarovka IIA.

So, it is not clear whether the common rim type for the Yunatsite II and Coţofeni II is a result of the synchronous interrelations between the south and north or it already have existed in the later Early Bronze I Coţofeni I culture. Other common elements are the fashion of the ornamentation with incisions of the rims, as well as the applications with pseudo-handles of big spherical bowls. But both the elements also belong to the common Balkan pottery features. However, the oblique cut rimed jug is very similar to the jugs from Dubene-Sarovka IIB3. But the hollowed bottom already parallels with the Early Bronze I pottery. We could assume a southern influence of that type of small jug that we probably used as wine-drink individual supply. Unfortunately, the find from Kozlodui is a surface one. According to the other pottery from that place with parallels at Gradets and Basarabi, it can be attributed to the Coţofeni III culture, but it is not clear whether there was thin-level or multi-level settlement. The recent possible chronology corresponds to the Dubene-Sarovka IIB data as they belong to later Early Bronze II.

The Yunatsite II culture – the Central Balkans

To the west of the Yunatsite culture, expressive contact data synchronize Yunatsite IIA-C with Pernik II group – Radomir-Vakhovo III and Early Bronze II Pernik (the encrusted pottery without stratified data). The similar ornamentation consists of false corded pottery, incised and stamped encrusted ceramics with comparable ornamentation zones (Nikolova 1996a; 1999). The sherds in ***Figure 15.22: 5-6*** indicate that distribution of the similar pottery in the upper Struma valley was connected with an influence from the area of the Yunatsite culture in the Upper Stryama valley (cp. Alexandrov 1995: fig. 7: 104, in Bailey and Panayotov 1995). The contact data between Pernik II group and Bubanj–Ostrikovac II group indirectly synchronize Yunatsite II culture with Bubanj II and Ostrikovac Ib-d, as well as with Jelenac in the Velika Morava valley. Some common incised motifs of the upper Stryama valley and the Velika Morava valley also should be mentioned like net ornamented deep plates with (***Figures 15.3: 10-11; 15.16: 3; 15.23: 4-5***). Very similar to the sherd in ***Figure 15.16: 3*** originates from Gulubnik tell (unpublished). For the

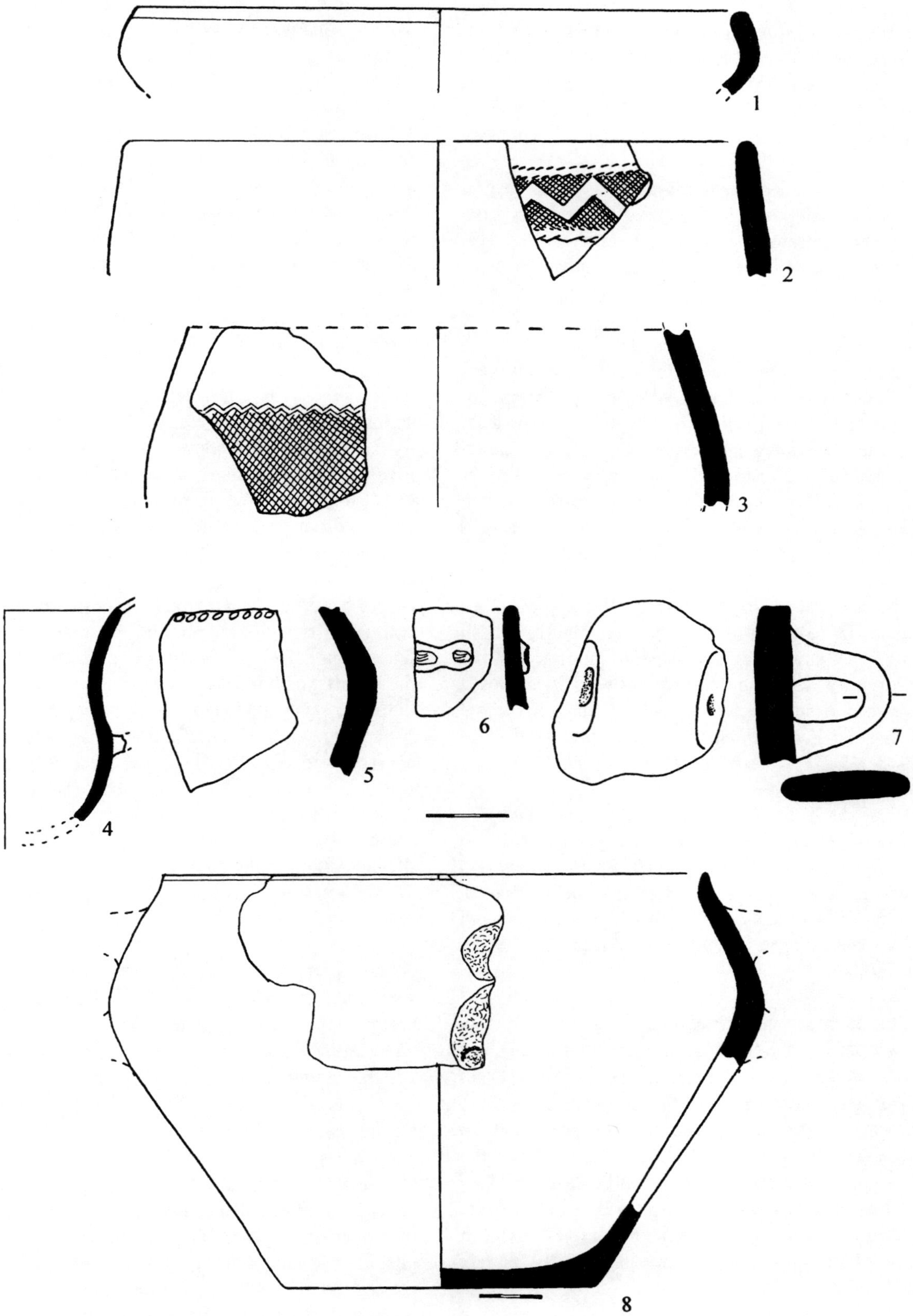

Figure 15.16 . Dubene-Sarovka IIB1. Squares M12-N12, depth - 1.40-1.50 m. That level occurred on the tell below a cultivated layer and there is not a satisfied stratigraphy there. According to the depth, the level belongs to later IIA or the beginning of the IIB stage of the tell but the pottery is characteristic of IIB stage at EB site at Sarovka. Few Copper Age sherds have been discovered with that fragmented bowl. Coarse earthenware (6 and 7) and fine one (1-5 and 8). Black or black and gray burnished pottery (1-5). Despite the concentration of sherds, no architectural feature has been discovered in this sector excavated. Early Bronze II.

time being, it is the earliest stratified of that type. The zigzag element relates the bowl to that in ***Figure 15.14*** dated from the latest Early Bronze I with close parallels at Yunatsite 15. In fact they are discovered at very similar absolute depths, but the context of the bowl in ***Figure 15.16: 3*** differs, as well as there is a natural sloping of the terrain to the west of the site.

The genesis of the motifs under discussion was in the Central Balkans. The increasing of the quantity of such pottery from Dubene-Sarovka indicates that micro-region should be also included in the area of emergence of that innovative ornament vastly distributed in the Central Balkans in very early Early Bronze II.

But there are no ceramics indicated the duration of all these mentioned sites till the end of the Early Bronze II, corresponding to later Vučedol. The unpublished pottery from Negovantsi in the upper Strouma valley parallels to later Vučedol characterized by deep incised encrusted ornamentation and probably defines the end of the Early Bronze II in the western Central Balkans. To that local ceramic style only two sherds from Pernik are comparable (Nikolova 1999: fig. 9.6: 11,15).

Indirect contact data and comparative stratigraphy synchronize Yunatsite IIA culture with the Kostolac culture, as well as Yunatsite II culture - with the Vučedol culture. Popular motifs with zigzag demonstrate share between Yunatsite II culture, Pernik II group, Ostrikovac Ic-d (Stojić 1996) on the one hand, and the Vučedol culture, on the other hand (***Figures 15.16: 3; 15.22: 5-6;*** Nikolova 1999: fig. 9.5).

Some identical sherds between Pernik II group and Dubene-Sarovka (***Figure 15.22: 5-6***) indicate strong influence from the Upper Stryama valley into the upper Struma valley and even possible exchange of pottery. As an instance of opposite direction, it is the bowl in ***Figure 15.21: 1***, which is close to the Kostolac ceramic style, recently analyzed in detail by D. Nikolić (2000).

Moreover, the approximately synchronism between the Apsis House No. 1 from Dubene-Sarovka IIB2 and the Burnt House from Sitagroi Va – respectively between Yunatsite 13-11 and later Sitagroi Va is confirmed by the ^{14}C dates (Nikolova 2000a). Among the parallel between the Yunatsite II culture ceramic style and the Northern Aegean, the peculiar of the former false corded ornamentation should be stressed (***Figure 15.21: 2***; cp. Renfrew et al. 1986: plate B: 3). In the light of this evidence a chronological hiatus between Sitagroi Va and Sitagroi Vb corresponding to Early Bronze IIB2 infers the comparative analysis as well.

Yunatsite II culture – Pevkakia Magula

From phase 3, S-profiled bowl with a plastic pseudo-handle and a flat hole handle (Christmann 1996: plate 23: 9, 14) parallels with Central Balkans Early Bronze II, in particular with Dubene-Sarovka IIB and Banya-Ploskata Mogila. The incisions and especially triangular shaped are common elements as well (Christmann 1996: plate 29: 11, 13). It should be stressed on the bulges attached to walls of bowls, with stratified parallels from Dubene-Sarovka IIB3 (Christmann 1996: plate 32: 5; ***Figure 15.24: 4***).

Considerations

The development of the Early Bronze II ceramic style in Western Upper Thrace indicate a graduate increasing of the fashion of the encrusted pottery. That fashion was common for the Central Balkan areas such as the Northern Aegean, Southwest Bulgaria and Velika Morava valley. That ceramic style of one of the incised, stamped, or pricked geometric motifs with many close parallels including also some common ceramic shapes and manners, like ornamentation of the inner walls of the rims.

If the similarity indicates active interactions and following of popular fashion, the differences, along with the different life style (sedentary in Western Upper Thrace and the Northern Aegean, as well as semi-mobile or semi-sedentary in Southwest Bulgaria and Velika Morava valley), different architectural traditions and settlement patter in peculiar, etc., indicate there were different communities with active interactions.

On its hand, the geographic closeness, the absence of geographic barriers and the presence of such a connective-population factor like the Maritsa valley explain the similarity in the material culture between the Yunatsite and the Ezero cultures. However, the Ezero culture belonged to another culture system, that of the eastern Balkan corded ware, being in close contacts with Pit Grave Culture, so it cannot be identified with the Yunatsite culture. The line of similarity can be extended up to Ezerovo, where similar to Ezero culture jugs occur in Early Bronze II-III (***Figure 15.32***). In other words, the innovation of the popularity of one-handled jugs (***Figure 15.22: 2-9***) was vast spread all over the Balkans, by different types distributed.

As far, as the Lower Danube concerns, some finds like the round-bottom cups and net-encrusted pottery directly or indirectly connect the Yunatsite culture with the Coţofeni II-III cultures. The question, however, is more complicated, as at site like Govora Sat-Runcuri, a typical of Early Bronze II encrusted pottery was distributed, that is connected with the Central Balkan ceramic style (***Figure 15.33***).

In contrast to the Early Bronze I, the ceramic style in the Early Bronze II Balkans developed with peculiarities distributed over smaller territories. In other words, the ethnographic peculiarities considerably increased. On the other hand, the general directions of interaction of the Yunatsite culture remained – not only with Eastern Upper Thrace, but especially with Southwestern Bulgaria and even with the Northern Aegean.

It should be also stressed, despite the Early Bronze ceramic style is characterized by dynamic evolution and vast spread of common motifs (zigzag, triangles, rhombs, or metope ornaments), there are some elements like the zigzag motif, which had a long history exemplified by Dubene-Sarovka IIA-B ceramic style (***Figures 15.3: 2; 15.14; 15.22; 15.25***).

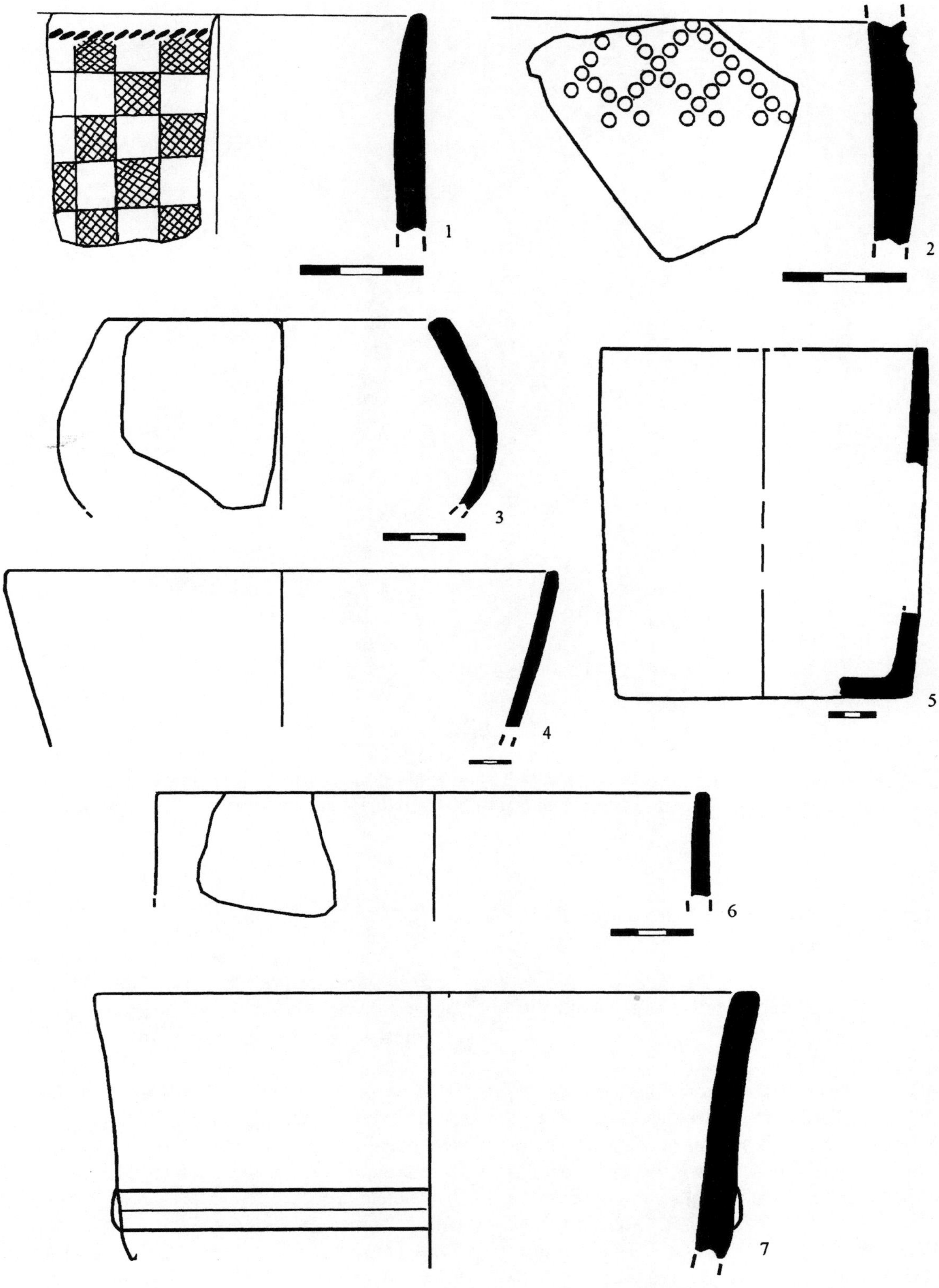

Figure 15.17. Dubene-Sarovka IIB1. Black burnished (3) and beige (7) earthenware. Square N12, depth - 1.34-1.38 m (2); square M12-N12, depth - 1.40-1.50 m (3 and 7). Some Copper Age sherds have been documented in the level. Early Bronze II.

Figure 15.18. Dubene-Sarovka IIB1. Square I12 (N-S - 0,10 m; E-W - 0.70 m). Depth - 1.27 m. Brown (3 and 4) made of clay paste with fine and course sand, as well as brown and black (2) with roughly smoothed surface. Slip outer surface (3). Early Bronze II.

Early Bronze III Balkans and the Yunatsite culture

Emblematic ceramics and periodization of the Yunatsite III culture

Cups with pointed bottoms and with round bottoms are emblematic of Yunatsite III culture, as well as the small jugs with high handles, high holed bottom bowls and tunnel-handled and pseudo-tunnel handled bowls, jugs with buttons attached to the handles, bowls with inverted rim and disk-like high handle (Nikolova 1996a: fig. 23: 2-3; Katincharov and Matsanova 1993: fig. 17; Nikolova 1999: 227-230; ***Figure 15.29***). Stratigraphic sequence from the eighth to first horizons at Yunatsite corresponds to this stage, along with those from the Manole tell (without the uppermost ones). Destroyed levels from same stage are documented by pottery from the fallow land and from pits at Dubene-Sarovka. P. Detev also excavated Early Bronze III levels at Chernichevo tell (Nikolova 1996a: 23:3) and Plovdiv – Philipovo tell (Nikolova 1999: 31). Later Early Bronze deposits from Plovdiv-Nebet Tepe consist of pottery typical of Yunatsite III culture. This is the stage when a settlement on the slope of the hill was also grounded (Koleva 1991; Nikolova 1999: 46). Ognyanovo tell levels belong mainly to Early Bronze III (Leshtakov, chapter 16).

In the Early Bronze III the encrusted pottery style decreased and plain brown and grayish-black burnish ceramics predominated. As exceptions, prick-dot ornamentation and bird-eye-encrusted ornament occur. For the time being, Ognyanovo represents most vastly the ceramic style of the third stage of the Yunatsite culture (see chapter 16).

Because of the absence of published evidence from the Yunatsite tell and the non-complete information from Manole-Razkopanitsa – it is difficult to argue a periodization of the third stage of the Yunatsite culture. Despite mixed artifacts the Manole sequence represents a long duration of pointed bottom cups there. Six Early Bronze III levels can be recognized there divided into three phases by P. Detev. The fourth layer – despite of presence of some Early Bronze III finds - belongs to later periods. Seventeen emblematic types of vessels have been distinguished based on the publications of P. Detev (Nikolova 1999: 228) characterizing the Early Bronze III ceramic style in Western Upper Thrace. Type 12.1 (handled bowl) parallels with Early Bronze IIB1 Yunatsite 11 bowl (***Figure 15.25: 1***) and according to P. Detev, that type characterizes all levels at Manole-Razkopanitsa (1981: 166, fig. 28: 3-4).

Some of the conical bowls from Yunatsite III culture are with pedestal base – common shape for Manole 3 (Detev 1981: fig. 34: 4) and Yunatsite BIII (the third horizon [Katincharov and

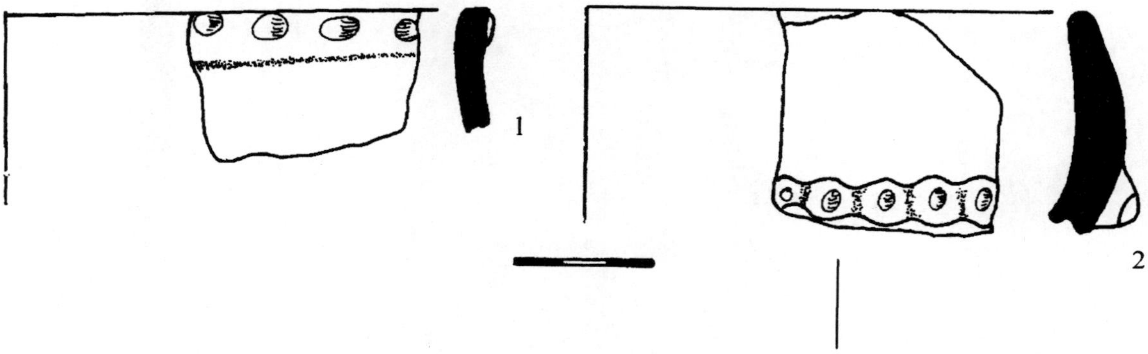

Figure 15.19. Dubene-Sarovka IIB. Square L15. Depth - 1.00-1.10 m. There are no features documented together with the sherds. Earlier Early Bronze II.

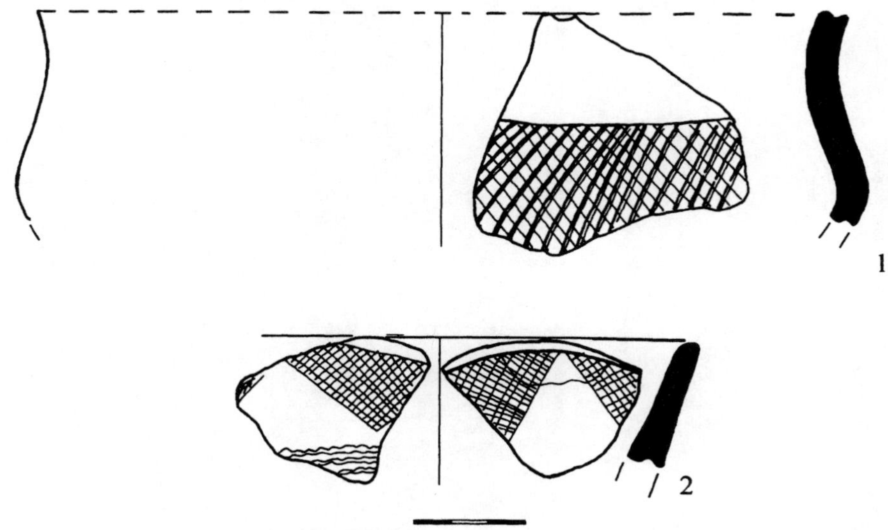

Figure 15.20. Dubene-Sarovka IIB1. Square J13, depth - 1.10-1.20 m.

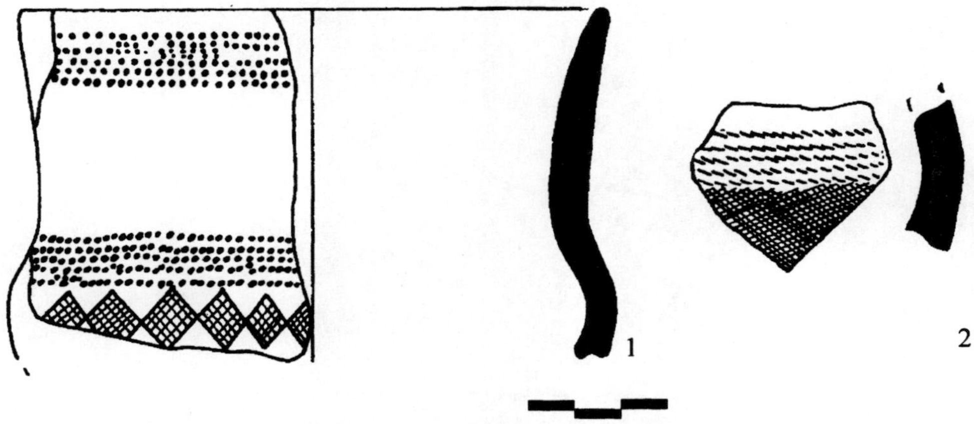

Figure 15.21. Dubene-Sarovka IIB2. 1. Square F18, north. Depth - 0.86-0.95 m. A stratum of brown soil, possible floor level has been documented in the northwest corner of the square by a calcified mat. That area is situated between two burnt houses, well dated from IIB2 phase.

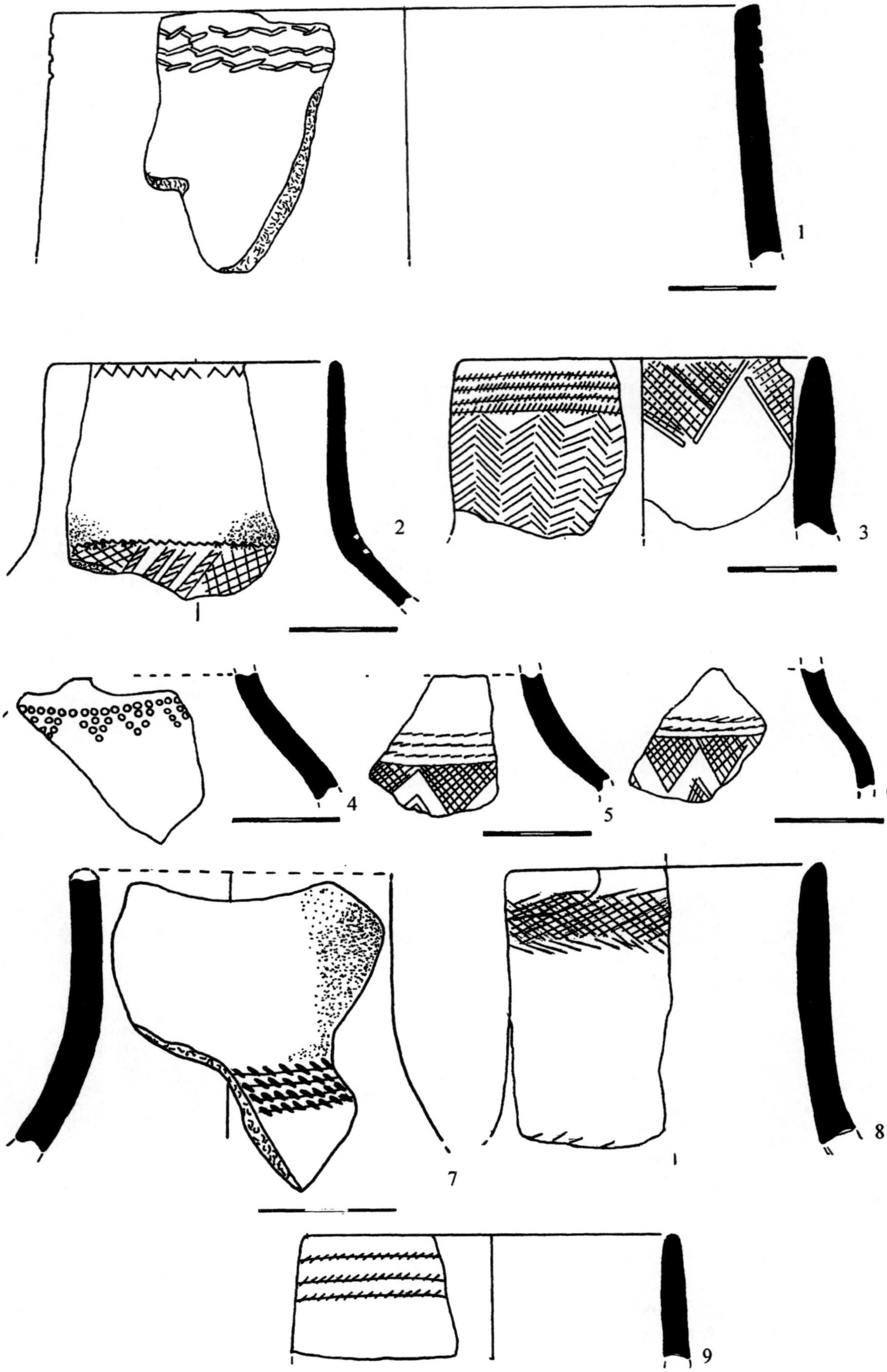

Figure 15.22. Dubene-Sarovka IIB. Early Bronze II pottery (after L. Nikolova).

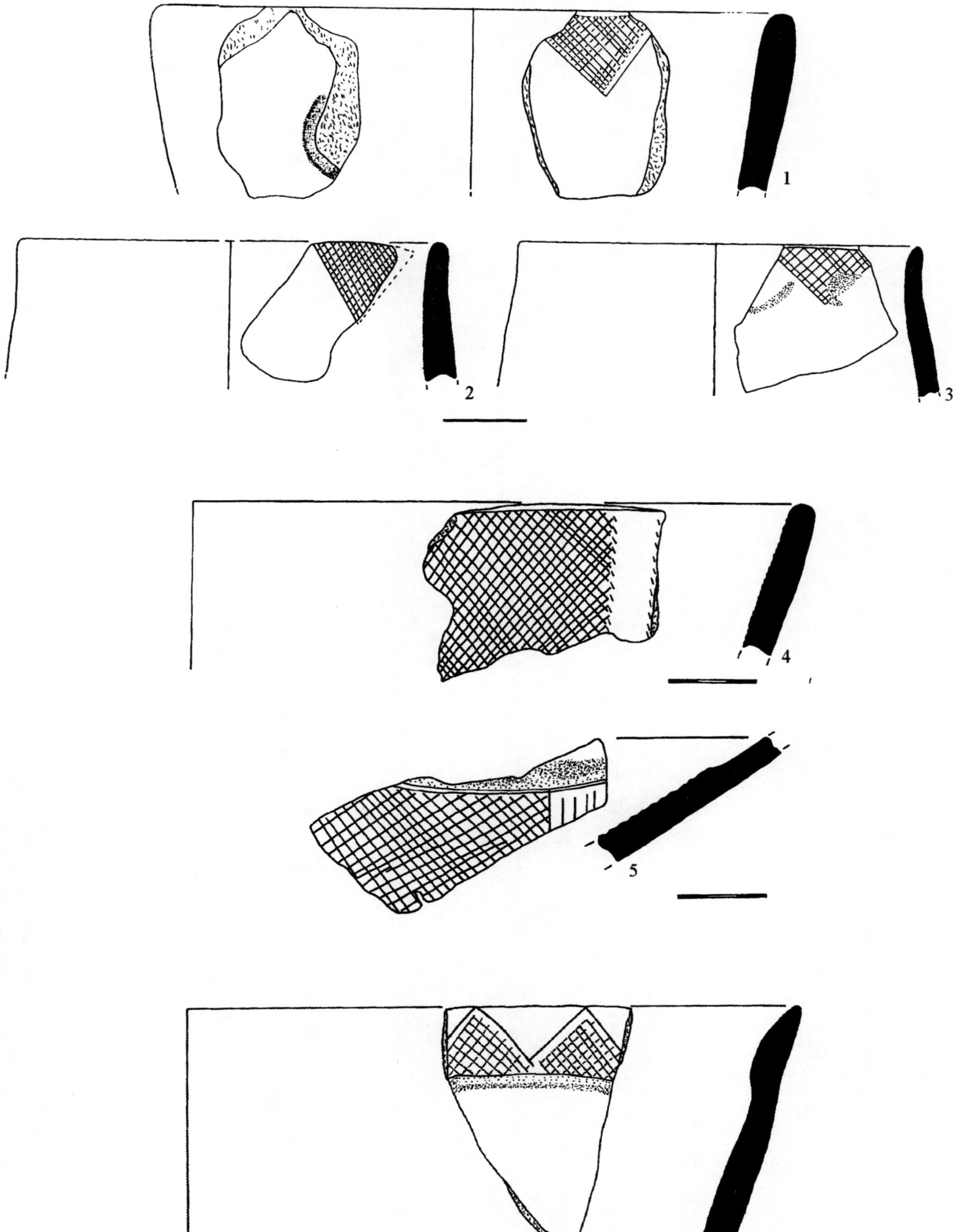

Figure 15.23. Dubene-Sarovka IIB. Early Bronze II pottery (after L. Nikolova).

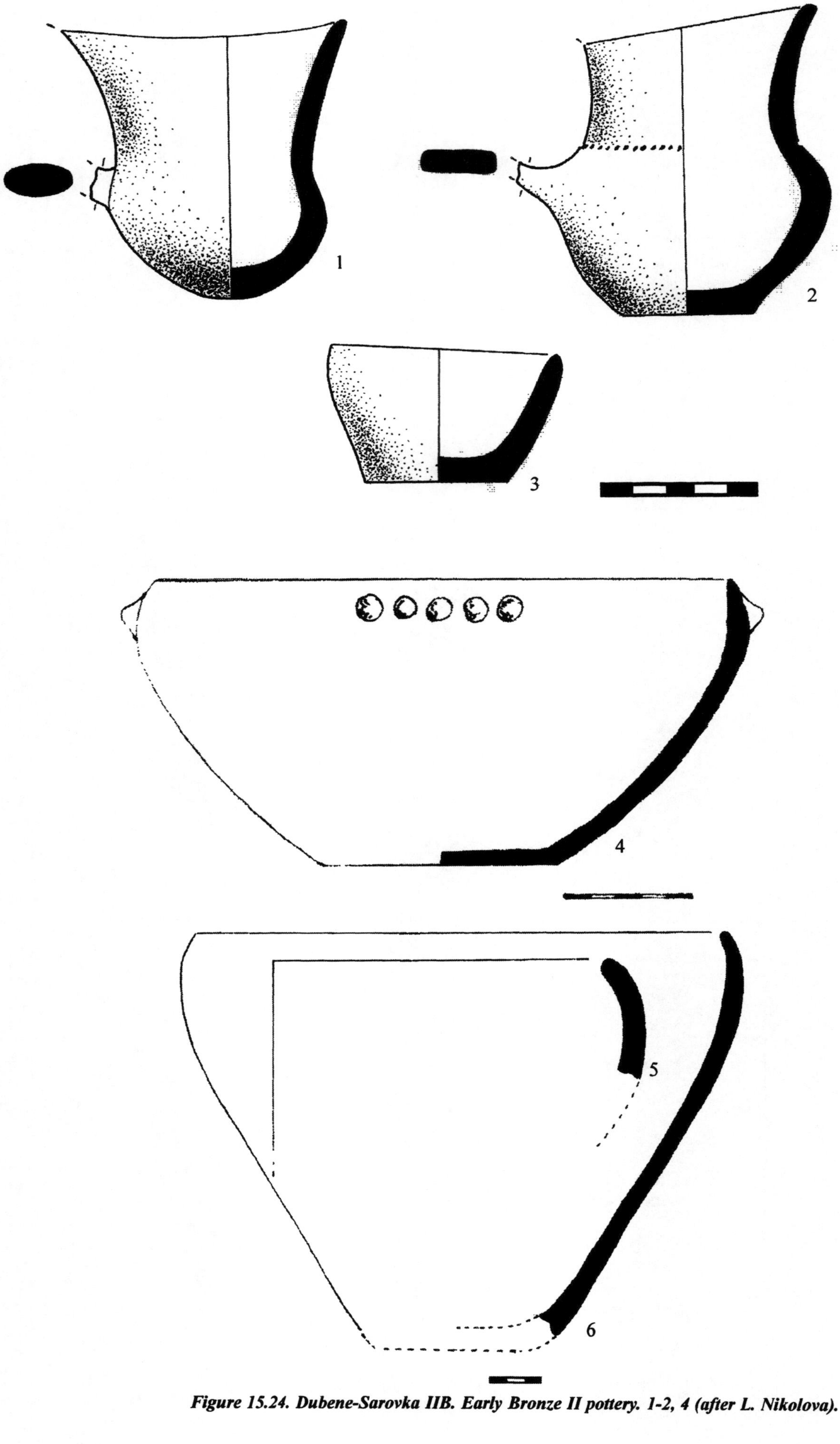

Figure 15.24. Dubene-Sarovka IIB. Early Bronze II pottery. 1-2, 4 (after L. Nikolova).

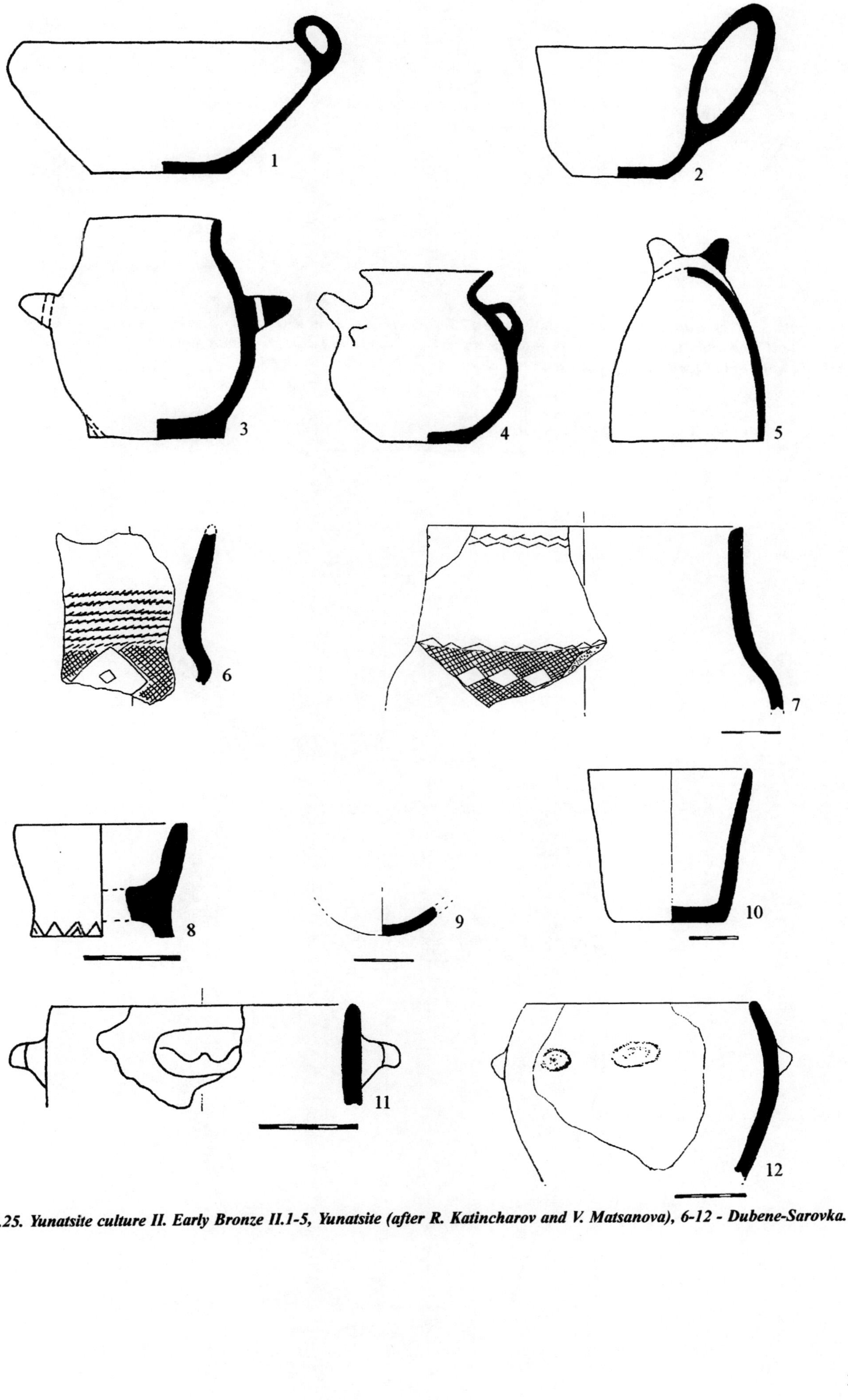

Figure 15.25. Yunatsite culture II. Early Bronze II.1-5, Yunatsite (after R. Katincharov and V. Matsanova), 6-12 - Dubene-Sarovka.

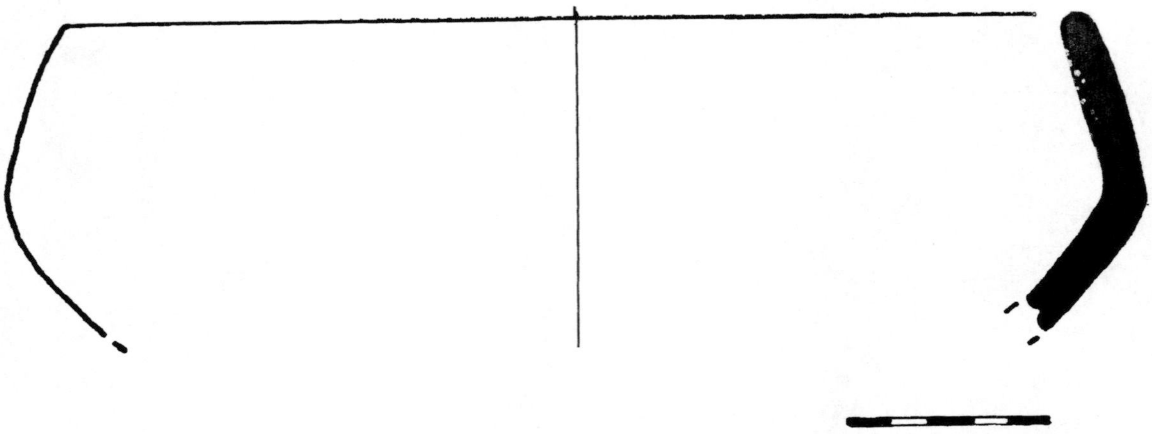

Figure 15.26. Dubene-Sarovka IIB(3). Square F18. Depth 0.50-0.60 m. Ochre brown with gray spots, well-smoothed sherd. The inner surface is gray, roughly smoothed. Clay paste with sand, gravel and golden mica. The sherd belongs to the uppermost documented level at the site.

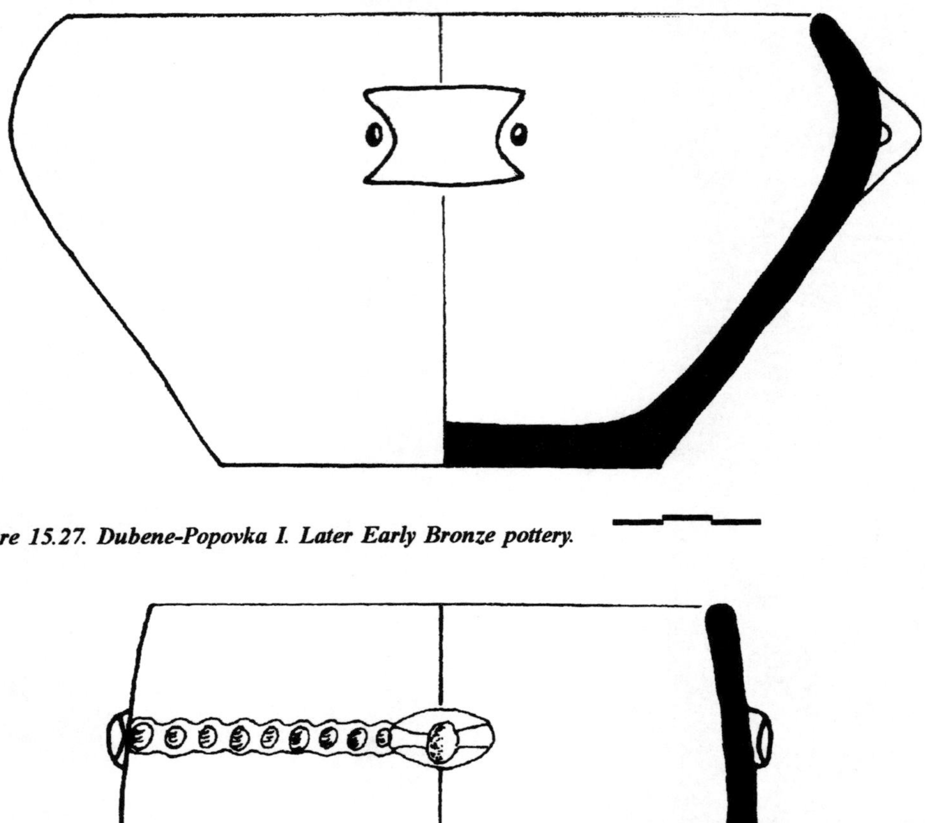

Figure 15.27. Dubene-Popovka I. Later Early Bronze pottery.

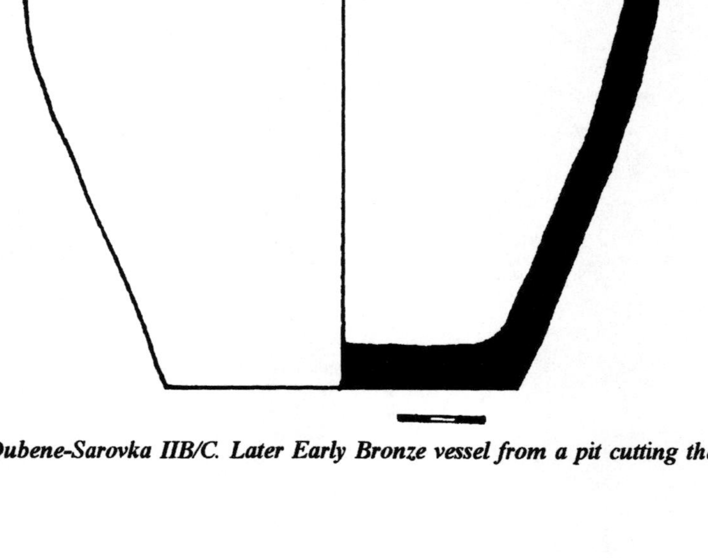

Figure 15.28. Dubene-Sarovka IIB/C. Later Early Bronze vessel from a pit cutting the second level of IIB phase.

Matsanova 1993: fig. 17: 4). So, they specify at least two phases of the Yunatsite III culture.

The high handled jug from Manole 5 parallels with Dubene-Sarovka jug (***Figure 15.29: 4***), but there are no reasons to believe that the last site was occupied for a very long period in Early Bronze III. The jug from Dubene-Sarovka was discovered in the layer of the first horizon, obviously belonging to small pit, in which also the round-bottom cup in ***Figure 15.29: 1*** was deposited.

The tunnel-handled bowl with inverted rim is a typical Early Bronze III Balkan shape, with stratigraphic data from Yunatsite 5. According to the Manole and Yunatsite data, the pointed and round bottom cups characterize all the third stages of the Yunatsite culture (Detev 1981: fig. 37). In ***Figure 15.29: 1-2*** the cups belong to Dubene-Sarovka IIC.

In the light of the recent evidence, it is impossible a periodization of the Yunatsite III culture to be argued based on the ceramic sequence. Therefore, that periodization was elaborated on the based of the ^{14}C dates elsewhere (Nikolova 1999: 227). Three phases were distinguished, to which other one (the last, D) should be edited representing the end of the Yunatsite culture. The ceramic sequence from Yunatsite is still limited published, while the Manole-Razkopanitsa pottery (Nikolova 1999: 227-29; Detev 1981) can be attributed to Yunatsite IIIA-C phases.

A reconstruction of comparative stratigraphy of the Yunatsite III culture is based on the presumption of similar rhythm of the evolution of the settlement pattern, but publishing of the pottery from Yunatsite 8-1 will clarify in details the evolution of the ceramic style.

Spatial relation of the Yunatsite III culture

Yunatsite III – Ezero III culture

The Early Bronze III Ezero culture (Leshtakov 1992; 1993; 1994; 1996; Nikolova 1999; 2000a) is represented by Ezero 3-1 (cp. Nikolova 1999: 256 [Ezero 4, but the ^{14}C dates indicate EB II occupation of that settlement]), Nova Zagora 6-1, later Early Bronze Dyadovo, later Gulubovo, Karanovo VIIB, later Early Bronze Vesselinovo (Nikolova 1999: table 10.4), Early Bronze Kirilovo, Rupkite-Kaleto IIB, etc. The combine probability for radiocarbon dates from Nova Zagora 6-5 – 2570 BC/2560 BC– 2390 BC/2290 BC is correlated to earliest phase of Early Bronze III in the Balkans – respectively to the Yunatsite IIIA culture.

The synchronism of Nova Zagora 6-5 and Yunatsite 8-7 is confirmed by different methods of computer analysis of available ^{14}C dates – combine and sum probabilities for Nova Zagora and sum probability – for Yunatsite 8-7 (Nikolova 2000a: 72).

Some ornamentation similarity between later Nova Zagora (unpublished) and Gulubovo (Leshtakov 1996: fig. 10:9) can be preliminary mentioned confirming the diachronic connection between the third and the fourth phase of the Ezero III culture. Morphological similarity between Manole-Razkopanitsa and Gulubovo excludes the typical of the Yunatsite III culture pointed bottom cups confirming the peculiarity in the culture development between Western and Eastern Upper Thrace. The spout spherical pots share both the sites (Detev 1981: figs. 27:1, 29:1; Leshtakov 1996: fig. 9: 3). The high necked jugs with two handled at the upper part of the neck are almost identical (Detev 1981: figs. 27:4; Leshtakov 1996: fig. 12: 10).

K. Leshtakov publishes some contact data with Ezero III culture from Ognyanovo (chapter 16, ***Figure 16.16: 2***), including the so-called Kirilovo type of cups. The border between both cultures seems to be localized in the Chirpan area, so one of the closest contact-site for the distribution of those types of cups in Western Thrace was Rupkite-Kaleto.

Yunatsite III culture – The Northern Aegean

At Sitagroi Vb, a pointed bottom cup – import or influenced by the Yunatsite III culture, originates from QO 6 (Renfrew et al. 1986: 13.27: 3). That layer is associated with a floor above the Bin complex dating in fact the latest preserved Early Bronze structures on the tell (Vb2).

QO 7-9 belongs in fact to the Bin complex, along with another cup with slightly pointed bottom (Renfrew et al. 1986: 13.20: 3 [QO8]). Both finds in combination with a fragment of possible close to the so-called St. Kirilovo type of cups (Leshtakov, chapter 16, ***Figure 16.16: 2***, cp. Renfrew et al. 1986: 13.20: 7) clearly indicate influence from the Plovdiv region in the Drama valley in the earlier Early Bronze III. In both regions were distributed round-bottom cups of different variants well documented at Plovdiv-Philipopol Restorant (Koleva, chapter 17, ***Figure 17.4: 1-2***), Ognyanovo (Leshtakov, chapter 16, ***Figures 16.13; 16.14: 1,3,5-6***).

Based on the stratigraphic evidence, round bottom cups from Vb1 phase and ^{14}C dates, the Sitagroi Vb is synchronous with the earlier Yunatsite III culture, probably with its first and the second phases.

Another common elements with the Northern Aegean are the spouted bowls (Leshtakov, chapter 16, ***Figure 16.18: 4***) known for instance from Skala Soteros (Papaefthymiou-Papanthimou and Pilali-Papasteriou 1997: fig. 49), but they are also common shape with the Ezero culture – *see* also Nikolova 1999: 225-26 and chapter 16 for Ognyanovo.

In Western Greek Macedonia the basic sites investigated – Kastanas. It is a peculiar ceramic style that follows one of the tendencies in the Early Bronze III towards plain pottery (Aslanis 1985). Along with the emblematic Early Bronze III spouted bowls and disk-handled bowls, as well as common Early Bronze urns and relief ornamentation at the technomic pottery, there are some peculiar elements such as knobs with pits (Aslanis 1985: fig. 25: 1), and high handled cups with possible round base (Aslanis 1985: fig. 21: 9).

Yunatsite III culture – the Horizon of two handled beakers

The small Early Bronze III kantaroi from the upper Strouma

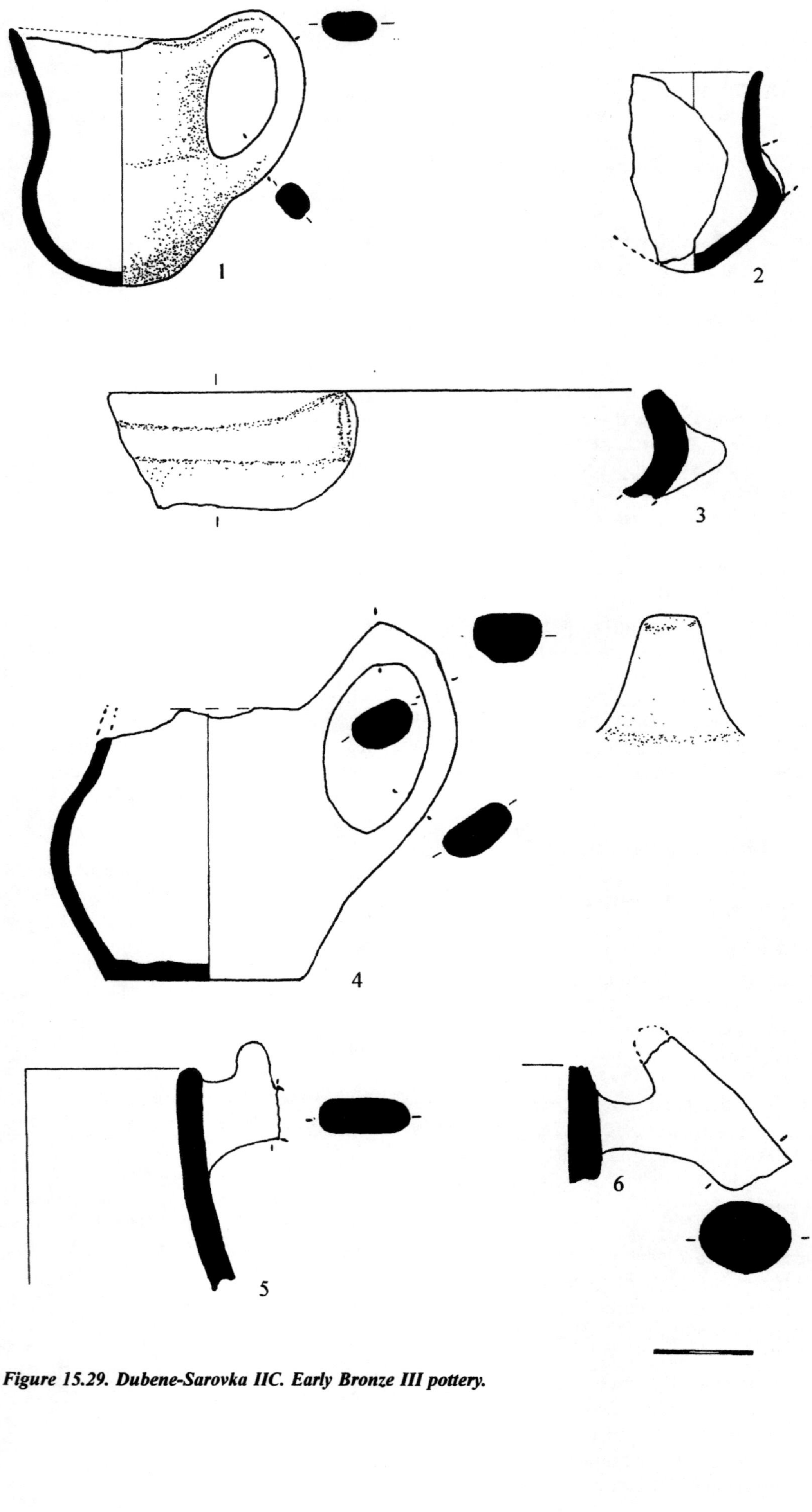

Figure 15.29. Dubene-Sarovka IIC. Early Bronze III pottery.

valley, as well as the two-handled beakers from Vecina Mala (Nikolova 1999: 233-34) have no analogies in the Yunatsite III culture, for the time being. So, mainly the common Balkan comparative chronology synchronizes the third stage of the Yunatsite culture with Pernik III group and Bubanj III culture including also the pointed-bottom cup from Velika Humska Čuka, which is without stratigraphic data.

The common kantaroi for Pernik III and Bubanj III indirect confirm the synchronization between the Yunatsite III culture and the Pernik III group.

Yunatsite III culture – Vinkovci culture and Maros culture

The Early Bronze III Vinkovci culture from the southern middle Danube followed Vučedol culture. The transition between both cultures is well dates in the middle of the third millennium BC. The earlier dates from the Vinkovci culture correspond to earlier Yunatsite III culture as well (Nikolova 1999: 235, 256-57).

There are no expressive contact data between the Yunatsite III culture and the Maros culture. But ^{14}C dates from Mokrin cemetery correspond to later Yunatsite III culture, while some ^{14}C dates from Maros settlements can be correlated with the earlier Yunatsite III culture (Nikolova 1999: table A).

In the northern Balkans the Early Bronze III cultures are better investigated in the Romanian Banat and Transylvania. Gornea-Orleşti, earliest Hatvan, as well as earlier Monteoru from the foot of the southeastern Carpathians and can be synchronized with later Yunatsite III culture (Nikolova 1999: 256-57), and possibly the earliest Vatin.

Considerations

As far as it knows, the ceramic style of the Yunatsite III culture is closer to the Ezero culture, on the one hand, and to the Northern Aegean, on the other hand. Pointed bottom cups occurred in the south as an exception – in the northern Aegean (Sitagroi Vb, as well as in Macedonia (Pelince) and the Velika Morava valley (Velika Humska Čuka [Nikolova 1999: 51). Their connection with Western Upper Thrace seems indisputable. But their distribution in Northwestern Bulgaria, for instance at Bagachina cult place (Bonev and Alexandrov 1996: fig. 46) poses the question of the existence of a Trans-Stara Planina road.

In the interrelation of the Yunatsite culture with the Pernik III group for the time being in not very clear, as in Western Upper Thrace there were not distributed the two-handled beakers. But certainly we can speak of direct and intensive contacts with the area of the Sitagroi Vb in the Northern Aegean.

With regard to the problem of the distribution of the innovations, the Yunatsite culture population was one of the creators of the new style, whom belong the innovation of the pointed bottom cups. Also, the general decreasing of the encrusted pottery style indicates the following of some common tendencies in the distribution of the fashion of the plain pottery. So, we can speak on the one hand, of distribution of materialized innovation, but on the other hand, of distribution of fashion ideas among communities or specialized or semi-specialized potters.

Conclusions

The relations between Baden and Poliochni blue (*azzuro*) phase have been explained by possible movement of population from south to the north (Bouzek 1973: 173). In the light of the recent evidence Baden can be related only with Poliochni black (1). Earlier such a movement was assumed by N. Kalicz based on the parallels in the material evidence between Troy and Baden culture in Hungary (1963). Again - according to the recent record - Baden in Panonia completely preceded Troy I. In the light of the Early Bronze channel pottery, the connections between Thrace and Baden culture look like intensive interaction and distribution of the common fashion, but not as a result of migration of population.

Later, M. Ya. Merpert (1979) assumed the corded ware of the Ezero culture had Central European origin. The Dubene-Sarovka excavations, as well as Yunatsite, clearly broke eventual connection of Western Upper Thrace with Central European corded ware, as only the false was distributed in the Stryama valley, as it was very unpopular at Yunatsite tell.

At the same time, M. Garašanin (1973) has described an archaeological-ethnographic model, according to which in the Early Bronze Age III (according to my periodization system) there developed a big complex in the lower Danube, the Central Balkans and even in Panonia. The differences are explained by 'the different basic autochthonous on which were superimposed new foreign elements which certainly had ultimate connections with the Indo-Europeans stockbreeders of the steppes', as well as because of the invaders the autochthonous people were Indo-Europeanized (1973: 119). That author maintains it is not possible to associate Bronze Age cultural groups in complexes with differentiated ethnic groups while that process of ethnical differentiation took place during the period of transition from the Bronze Age to the Iron Age (1973: 124).

The cultural model argued in this study, has some peculiarities. First of all, based on the case study of Yunatsite culture, it was argued that the population was in active interaction with neighbor cultures, especially with the Central Balkan cultures. In fact the whole Central Balkans were completely aside from the penetration of the Pit Grave culture, the only one with certain 'steppe origin' in Early Bronze Age. In the light of the recent evidence the Indo-Europeanization of the Balkans seems to be more complicated process the genesis of which is deeply in the ancient Antiquity (Nikolova 1998a).

Secondly, the contact-data was applied to all the system of the ceramic style, so it is clear that the ceramic changes in the Early Bronze in the Balkans, exemplified by the Central Balkans cannot be connected with any drastic migration of the population, despite the inner migration was one of the factor of distribution of the innovations, especially in the ceramic style.

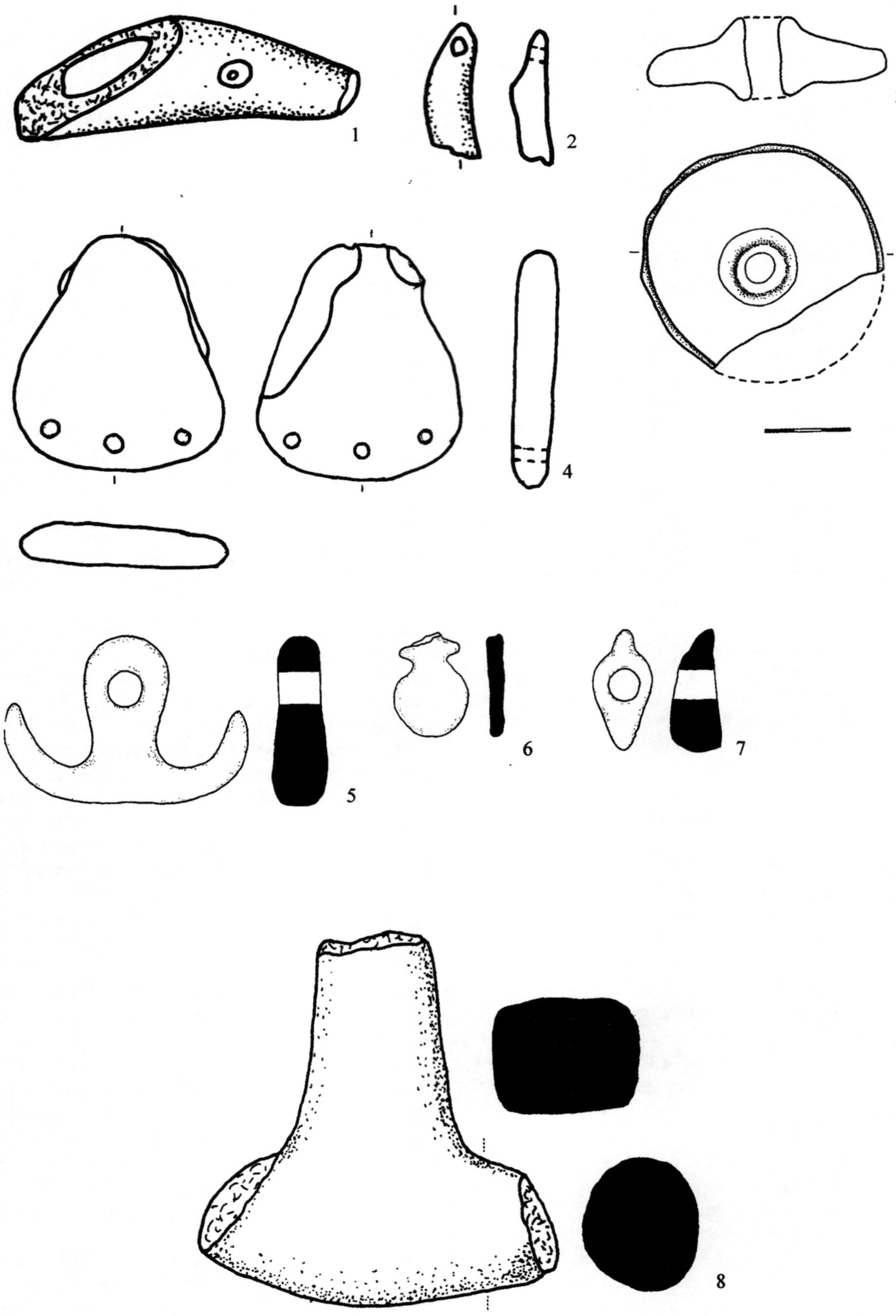

Figure 15.30. Yunatsite culture. Small finds. Early Bronze Age. Dubene-Sarovka (1-5, 9), Yunatsite (6-8, after R. Katincharov and V. Matsanova).

Thirdly, based on the instance distribution of the pointed bottom cups, the application to the ethnical processes can be defined as follows. In the Balkans in Early Bronze Age, especially in the Central Balkans there were different tribes, which high political organization (of chiefdom types, Nikolova 2000b) was one of the reasons for the cultural integration, but also for the preservation of some ethno-cultural peculiarities of the ceramic style. Whether the pointed bottom cups had some religious or other meaning, it is obvious that in region like Ezero culture they were rarely distributed (Konstantia-Asara) but influenced the cultures to south, west and to the north (***Figure 15.36***). The rejection of that innovation from the closest neighbors remains an open question. It is possible to believe the Early Bronze ceramic style is also a record of tribal relations corresponding to the higher levels of the political organization. On its hand, the architecture and even the small finds (***Figure 15.30***; cp. Georgiev et al. 1979; Stefanova 2000: fig.5: 3,5) give number of similarities between both neighbor cultures.

However, the fact that in different but similar way has been seen the forming of the Central Balkan cultural system in Early Bronze II - as Bubanj II culture, after M. Garašanin, or as Central Balkan encrusted pottery style (in my works) - indicates that the peculiarities should be not overviewed. In my opinion, the similarity, which is with different degree between close and distant cultures concerns mostly the techno-economic and techno-cultural characteristic of the Early Bronze population in the Balkans, especially the ceramic styles (***Figures 33.34-33.35***). The main reason was probably the similar social structure. But after all, that is a problem for other discussion.

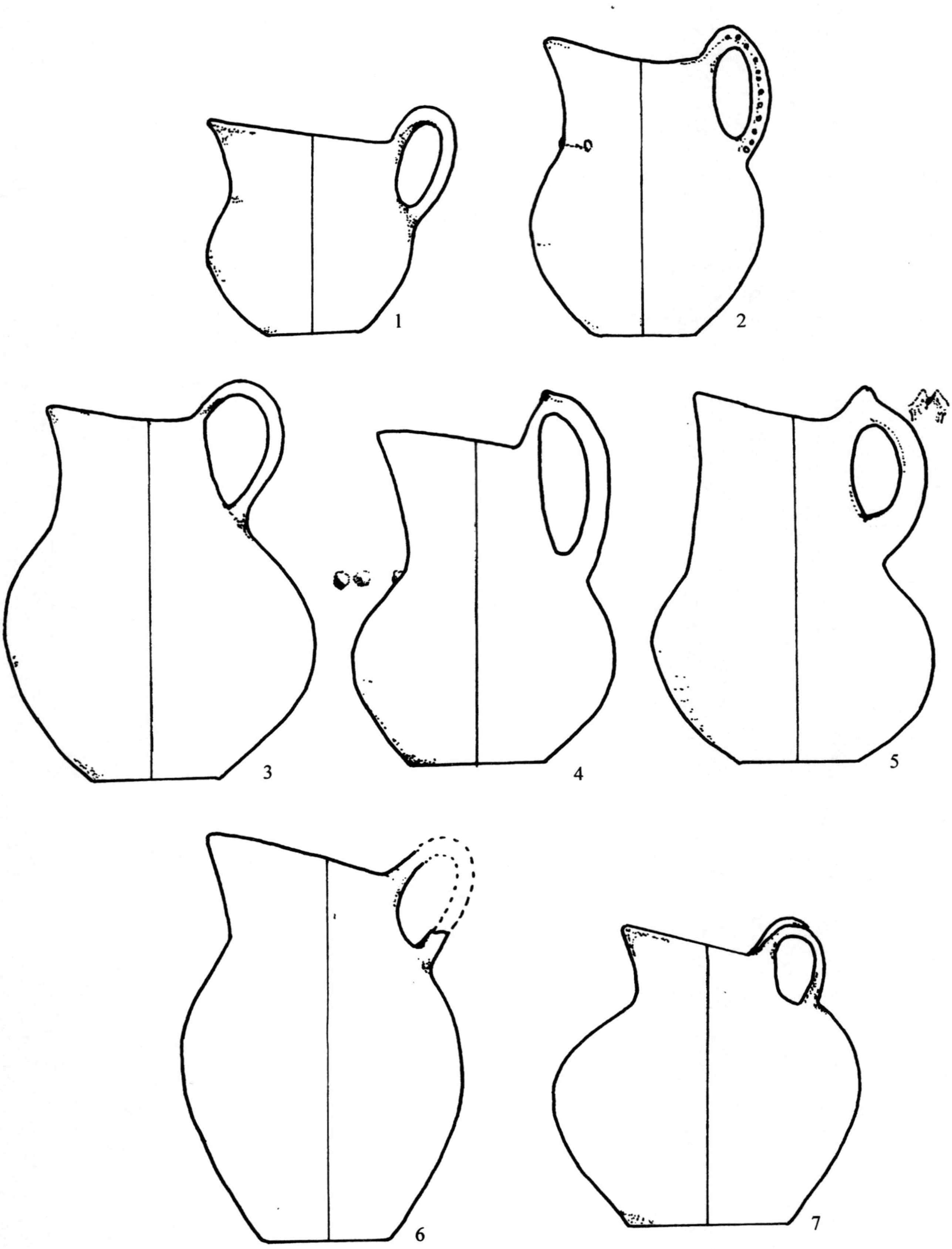

Figure 15.31. Ezerovo. Early Bronze Age pottery (after P. Roman).

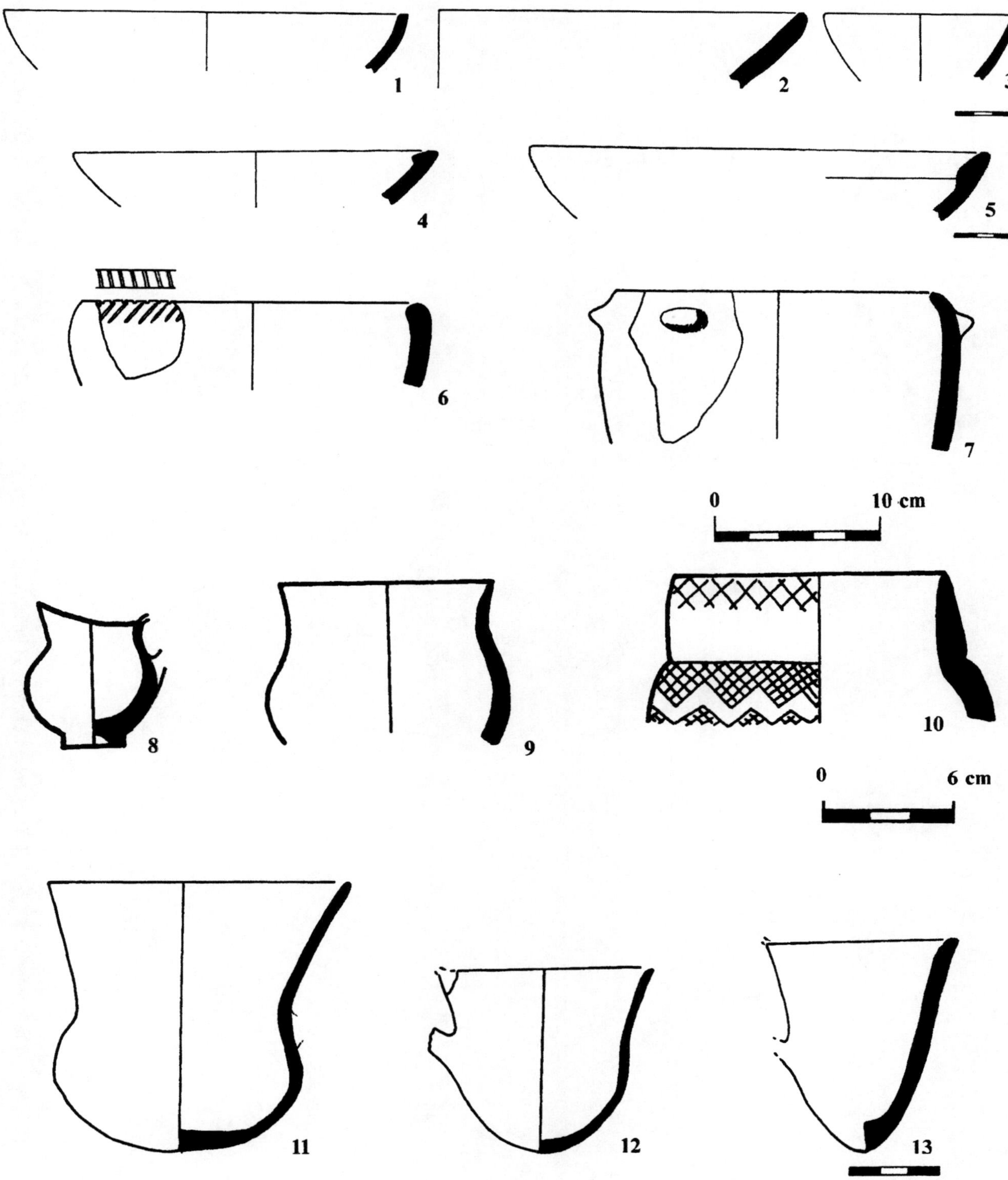

Figure 15.32. Cotofeni II-III culture. Northwestern Bulgaria. Early Bronze II. 6-10, after I. Panayotov and S. Alexandrov.

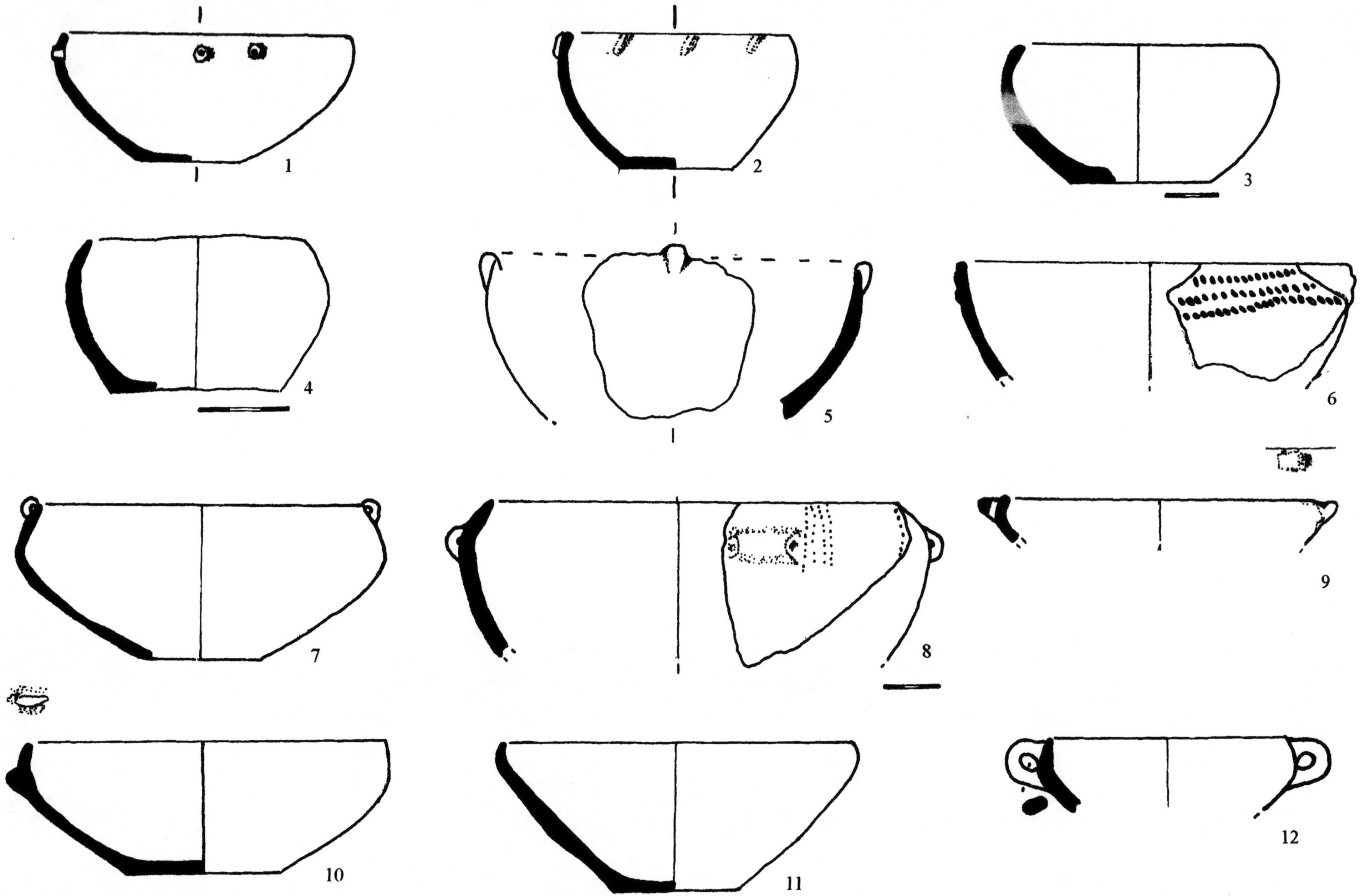

Figure 15.33. Common Balkan Early Bronze ceramic shapes. Bowls from Sitagroi-Dikili Tash group, Ezerovo II group and Glina culture (after C. Renfrew, M. Gimbutas and E. Elster, P. Roman and C. Schuster).

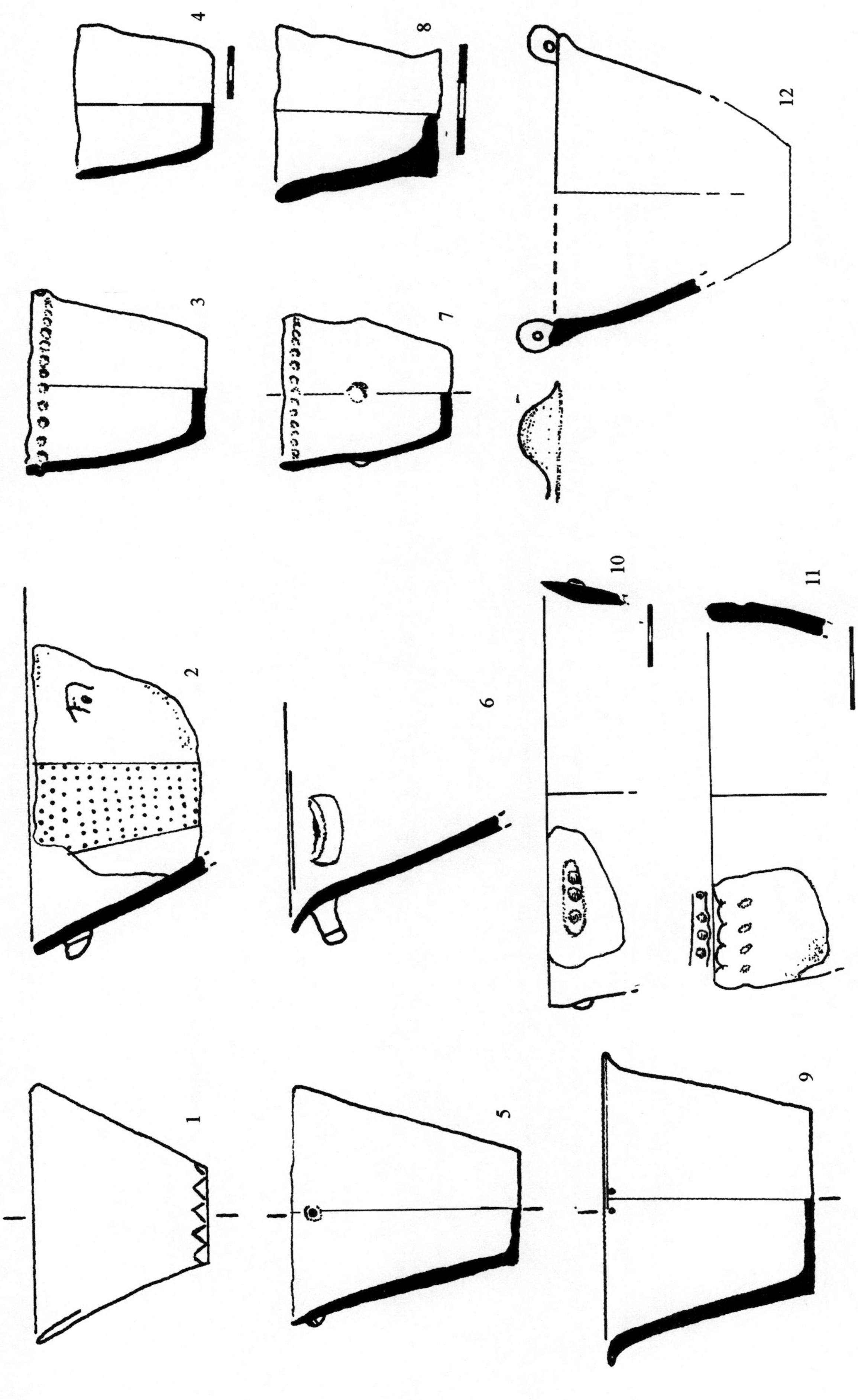

Figure 15.34. Common Balkan Early Bronze ceramic shapes. Urn-like vessels (bowls and pots) from Ezerovo II group and Glina culture (after P. Roman and C. Schuster).

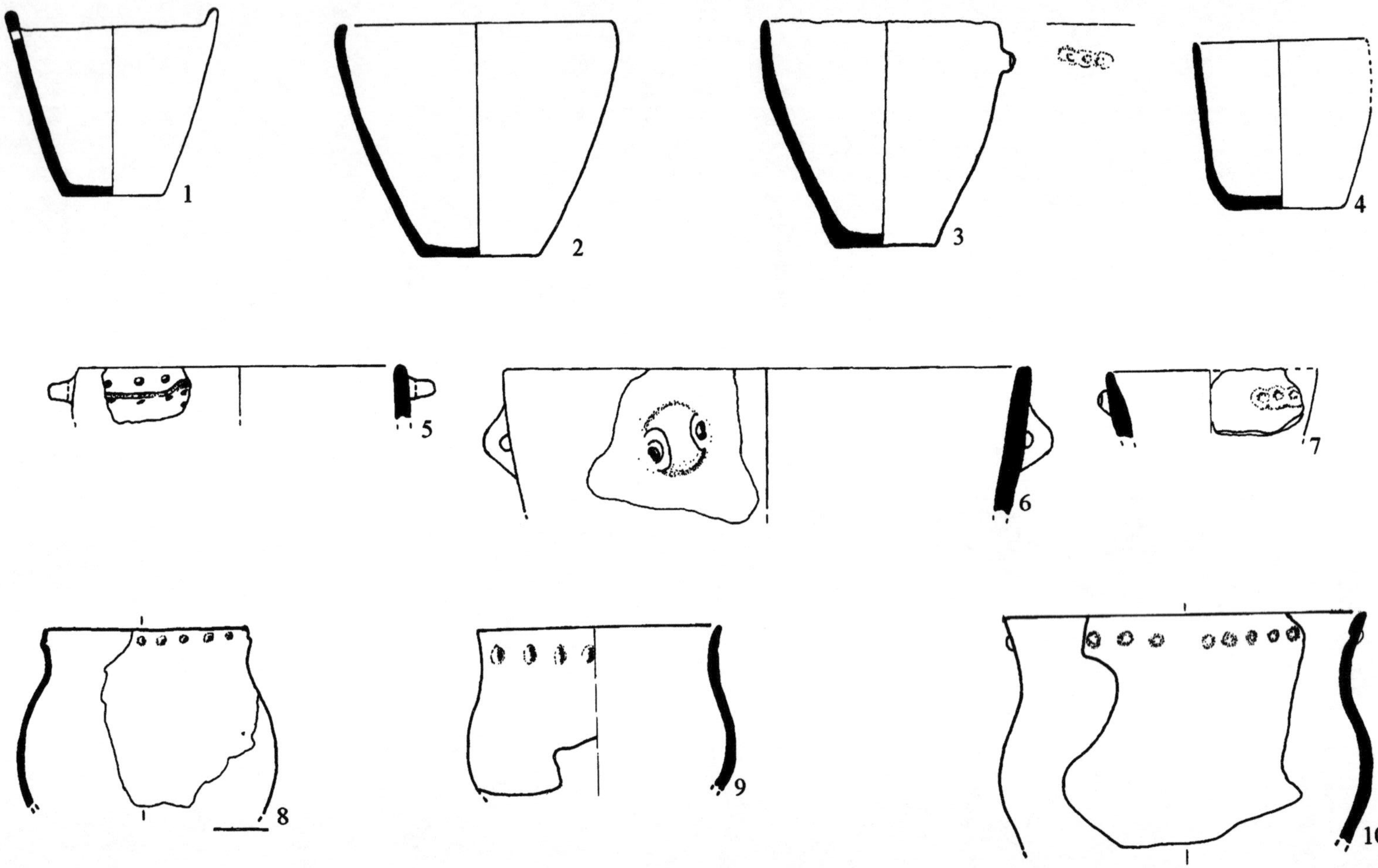

Figure 15.35. Common Balkan Early Bronze ceramic shapes. Urn-like vessels and sinuous pots from Sitagroi-Dikili Tash group (1-4, 7), Yunatsite culture (5-6) and Glina culture (8-10), after C. Renfrew, M. Gimbutas and E. Elster (1-4, 7) and C. Schuster (8-10).

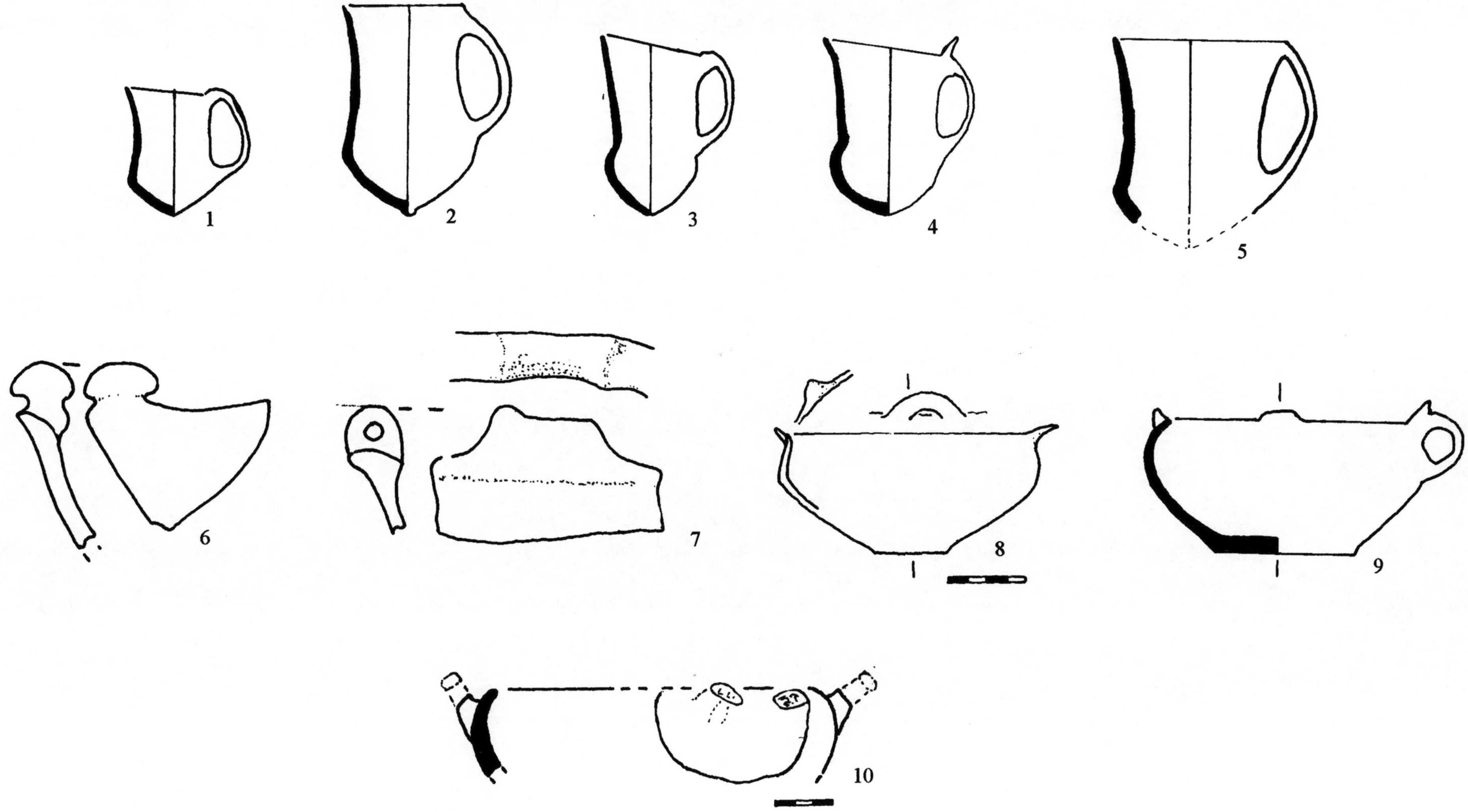

Figure 15.36. Balkan Early Bronze III ceramics. Cups and bowls from the Balkans (after C. Renfrew, M. Gimbutas and E. Elster, P. Roman, C. Schuster, S. Alexandrov, etc.).

CHAPTER 16

THE POTTERY FROM MALETEPE TELL NEAR OGNYANOVO

(CHARACTERISTICS, CHRONOLOGY AND INTERRELATIONS)

Krassimir Leshtakov

Introduction

The prehistoric investigation in Bulgaria has a tradition more than a hundred years long. It is hardly likely to find another country in Southeast Europe with such a big amount of completely investigated or excavated over vast area sites. Despite that fact, exclusively during the recent years the investigations – as methods, documentation and a system of publication – approach the necessary high level. Most of the old excavations are unpublished or preliminary reported in small communications. Even the more detail publication of Maltepe tell excavations near Ognyanovo (Detev and Matsanova 1977) does not respond to the contemporary level of the later Prehistoric research in this region.

In this prelude I would like to focus the attention on some important problems that should be profoundly discussed before starting to create any new models of social development in the Thracian Bronze Age. I mean the recent stage of the empirical source-base.

The investigations of the Bronze Age in Thrace in the last years reached according to some authors prominent results. The data from the excavations are enough for discussing of some especially important of the Thracian prehistory problems. These are the periodization of the Bronze Age, elaborating of the chronology of the distinct stages and their cultural peculiarities, synchronization of the distinct cultures with those from the neighbor regions in the Aegean and Anatolian circle, investigation of the character of the Middle Bronze etc. In this way the value of the big excavations as Ezero, Kazanluk, Nova Zagora, Yunatsite etc. was defended (Katinčarov 1982:117-140). However, there are questions posed at the first time for serious discussion. Along with the question of the formations of the Early Bronze I, under discussion is especially that for the end of the Early Bronze in Thrace, for its content and synchronization with the cultures from Early Bronze 3 with the related events from Anatolia and the Aegean (Leshtakov 1993:556). One of the factors, which makes difficult the decision of those problems is missing of detail publications of sites with long stratigraphic sequence like Nova Zagora tell, the tell near Dyadovo and on the first place Yunatsite tell. The last site has been not published in details, so there are missing enough data to correlate precisely the long stratigraphic sequence with the ceramic style. Therefore, the attempts in this direction have been not exhausted, as they are based on limited background (Garašanin 1996, 100-102). For instance, the author mentioned stressed among the vessels and sherds published from Yunatsite there are missing such comparable with Baden-Kostolac horizon (Garašanin 1996, p.101, comment 16).

There are attempts in spite of the background mentioned to synthesize the data from bigger regions and even in the frames of Southeast Europe and Anatolia (Parzinger1993; Maran 1998; Nikolova 1999; 2000a). This is my personal impression if we continue in that direction, Yunatsite tell will take an unhappy place like that of Ezero and each investigator will concern the re-interpretation of the contents of the ceramic phases at the site and the 'culture' on the whole ... The alternative is one to rely on strong intuition.

One detailed historiographic review of the native and foreign literature immediately brings to light at advanced position the different values of the state of research of the Bronze Age in Thrace. Of the newer summarized works I will allow myself to cite two. According to R. Treuil, Southern Bulgaria consists of much less investigations in compare to Northeastern and Northwestern Bulgaria. The author has considered all the prehistory but not only Bronze Age. That is a reason for the difficulties, which occur by comparing the materials with neighbor countries – Greece, Romania and former Yugoslavia. That author thinks the best published is about "Gumelniţa civilization". As far as the Early Bronze Age concerns, the first successful attempt dates from 1979. Since that date, the problem of the common periodization of the Early Bronze Age, which didn't look resolvable before, had begun a bit to be clarified (Treuil 1983:100).

H. Parzinger has not shared even that only positive estimation and has questioned not without absence of grounds the division of the stratigraphic sequence into ceramic phases (1993: 115-118). In other words the background base of the periodization of the Bronze Age is strongly questioned. But even the different directors of the Ezero excavations have had different opinions for horizon division, their cultural attribution and synchronization with the neighbor cultural occurrence in the Southeastern Europe and Anatolia (opinions of N. Ya. Merpert, G. Il. Georgiev and R. Katincharov published in numerous articles cited elsewhere [Leshtakov 1993]).

Accordingly, the optimistic opinion of R. Katincharov held by others during 1980s needs considerable corrections. Despite that fact in Bulgarian historiography there is no questioned the reliability of the record base that may allow the decision of the main problems. Personally, I think the statement is very exaggerated as the record is very abundant but the information is not published at the level required. Then, if we turn back to the few foreign authors that had courage to investigate the Bulgarian material the optimism transforms almost completely in its antithesis. In these works the fullness of the common publications, possibility of verification of the program statements and even the preciseness of the excavations have been questioned. The publication of close complex of most of the sites excavated was only a dream. The absence of serious argumentation base looks the main weakness of sometime pretentious works, which at the latest intend dues to the selection publication of the representative materials and statistics that have multi variant interpretations. The weaknesses circumstantially mentioned in the foreign investigations seem to have real foundation, on the one hand. On the other hand, they open a vast area for realization of not only one generation of cabinet scholars.

Gradually in the new Bulgarian prehistoric research the reestimation of the data on different sites have started from distant and intimately past. The new view includes whole investigating process – from critical analysis of the stratigraphy and initial publications to new interpretation of the data. That work is without doubt useful but not very beneficial. The archives are often in very bad condition, the sherds have no precise signature, in the field diaries consist of contradiction information etc. Some archaeological sites are excavated completely or are destroyed by later industrial activities, so that it is impossible new control diggings to be elaborated.

All those written above with all strength concerns Maltepe tell new the village of Ognyanovo, Pazardzhik District (***Figure 16.1***).

Maltepe tell. Location, excavations and stratigraphic problems

The site of Maltepe (known also as settlement mound of Ognyanovo) is located in the western part of the Upper Thracian valley, on the right-bank of the Maritsa River. It was c. 3 km southeast of the contemporary village of Ognyanovo, Pazardzhik District. Typologically, it can be defined as a low settlement mound (a tell) as the whole culture layer had been c. 3 m thick. Near the tell there was a limestone spring with high capacity. The first settlement was grounded in the Bronze Age there, as no earlier materials were found there. Also, there are no other contemporary sites or cemeteries discovered around. The closest synchronous sites are Banyata tell near the village of Kapitan Dimitrievo (in c. 10 km distance) and Ploskata Mogila tell near the village of Yunatsite (in c. 12 km), Nebet Tepe in Plovdiv city and Razkopanitsa tell near Manole (***Figure 16.1***).

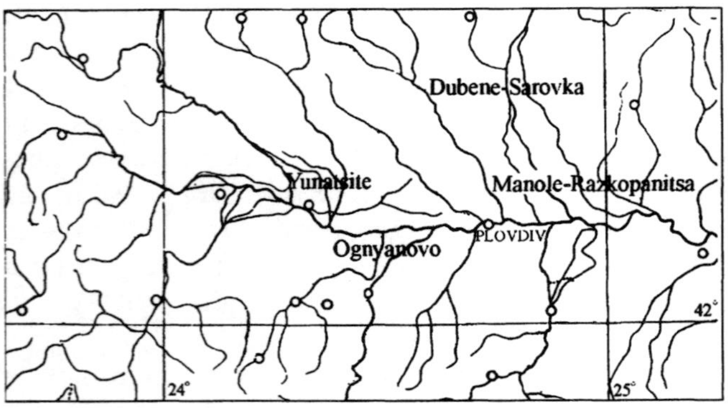

Figure 16.1. Map of the region with man sites mentioned in the text.

It is difficult the initial shape and the sizes of Ognyanovo tell to be defined because of big destruction in the first millennium BC and in the 1st – 4th century AD. A big damage also coursed the construction of the railway line Sofia-Plovdiv, as well as the serving of the stone career located near there. The north end of the tell is carried way by the big seasonal flooding of the Maritsa river and of gradually moving of its bed. Similar is the destiny of Vesselinovo tell, Yambol District, highly damaged by the moving of the Tundzha river bed.

In 1972 P. Detev and V. Matsanova conducted rescue excavations on the tell. Later they published the main results of those investigations (Detev and Matsanova 1977). The cultural layer is almost completely investigated and recently there are insufficient remains only. But in fact the excavated area did not exceed 300 sq. m.

It is interesting the suggestion that after the Bronze Age the site was used for a long time as a holy place. In the period of the Classical Antiquity there was a sanctuary of Hera identified by the devoting reliefs of the Thracian Rider-god (Detev and Matsanova 1977: figs. 18-19). Probably it was connected with the limestone spring located near there.

In the Bronze Age layer there was a pit with 'gray Thracian ceramics', from which a jug is published, wheel-made and completely preserved (Detev and Matsanova 1977: fig. 11, p. 81). The existence of a sanctuary from the classical Antiquity allows the pit to be defined as a cult one. Recently, sanctuaries consisted of the so-called pit-fields were documented in the whole first millennium BC in Thrace. It is possible the pit not to be alone but the rest to have been destroyed by the mentioned above building activities or the excavators to have not registered them. Thracian sanctuaries over settlement mounds have been localized at tells near the village of Dyadovo and Gulubovo, from where devoting reliefs of the Thracian Ridergod originate. But this interesting problem is beyond the goal of this study and will be discussed elsewhere.

Six levels have been distinguished by the excavators in the c. 2,5 m thick Bronze Age layer at Ognyanovo. They were numbered from down upwards, the first to sixth one. Vise versa, in the text 'building horizons' are mentioned in the opposite direction, from up towards the earliest level. The published profiles complicate the stratigraphic reconstruction additionally because of their summary character and very distant connection with the real situation on the terrain. It is very difficult to testify them profoundly. In the context of the commonly documented thickness of one building level in the EBA Upper Thrace – c. 30-35 cm – the total thickness of 2,5 m at Ognyanovo-Maltepe seems to consist of eight or even nine levels. The assumption could be: Either the excavators did not recognize all the levels because of the complicated stratigraphic picture of the site and the numerous later destruction, or the leveling of the registered levels compensate the thickness of the levels comparatively accounted. However, the building levels have been grouped into two cultural layers.

A ditch and rampart surrounded the settlement, but there is a very scanty information about those structures (Detev and Matsanova 1977: 47).

In the course of the rescue excavations, more complete information was collected about the houses from two building levels at the base of the tell. The planes of three small apses-houses have been published, as well as parts of another three. The highest levels were almost completely destroyed; the third one was highly destroyed by later interventions and at the fourth one no houses were registered. The pottery and the tools found in them were not documented and published in complexes.

Very important information comes from a cremation (?) child-grave in urn, which should be connected with the fourth/third or even second (?) building level. It was wrongly dated by the excavators to the second half of the first millennium BC because of the surrounding sherds from "gray Thracian" wheel-made pottery (Detev and Matsanova 1977: 81)[1].

Opinions on the dating of the Ognyanovo tell and some related problems of the periodization of the Early Bronze in Western Upper Thrace

According to the excavators, the earlier levels should be dated to the Late Bronze Age while the upper ones – to the transition to the Iron Age (Detev and Matsanova 1977: 45-46). But all the ceramic parallels given by P. Detev are considerably earlier; i.e. the comparative information was not used of full value (Detev and Matsanova 1977: 81-82). Obviously, the dating followed the model of Manole – Razkopanitsa excavated also by P. Detev and which was also dated to Late Bronze and Early Iron Age, without any grounds (Detev 1950; 1968; 1981). Later, R. Katincharov re-dated the site in the Middle Bronze Age and only some sherds out of context had to be placed in the Late Bronze Age and Early Iron Age (Katinčarov 1981). Recently, L. Nikolova dates all the layer to EB 1-3 (1999: 31, 192). With regard to the beginning of the settlement activity at the place, more precise chronology is to be argued below.

H. Parzinger, who personally knows the materials from Ognyanovo and Plovdiv-Nebet Tepe, has another opinion. He makes the following comments: "The lower situated the fourth and the fifth levels consists of pottery, which is comparable with that from Yunatsite 13-10 building levels ... Levels 1-3 date to Middle Bronze Age ... with materials ... which is the same as those from Yunatsite 1-7" (Parzinger 1993: 121). Accordingly, he also places Ognyanovo in Early Bronze and Middle Bronze Age, but at the same time it is stressed the materials hardly be later than the Early Bronze Age and his horizon 14a-b (Parzinger 1993: 272). The grounds of that statement have been already searched in the newer results of the Yunatsite excavations. But that very unclear dating cannot be argued because of two considerations: it looks incorrectly six building levels from Ognyanovo to be paralleled with at least 9-10 levels at Yunatsite on condition that there are no hiatuses documented at both the sites. On the other hand, the dating of Yunatsite 1-8 in the Middle Bronze Age on the whole is untenable, as it is to be argued in more details.

In general, Yunatsite tell is a main prehistoric site in the northwestern part of Upper Thrace, a standard like Ezero in Southeast Bulgaria. H. Parzinger writes in no uncertain terms that the culture at Yunatsite III (the depth of 2.10 – 1 m) belongs to the early stage of the Middle Bronze Age. In my opinion, that author accepts trustfully, with little corrections, the dating of the uppermost horizons by V. Milojčić (1949: 55-56), as well as of Ezero 3-1 and Nova Zagora 6-1 ("Nova Zagora culture" after R. Katincharov) and thence the later materials from Manole-Razkopanitsa and Ognyanovo to Middle Bronze Age (Mikov 1939: figs. 24-26; 1971: 56; Milojčić 1949: 50; Katinčarov 1972b: 243; 1973: 73-74; 1975b: 4-11; 1981: 120-23; Panayotov 1975: 45; Katinčarov and Matsanova 1993: 155-73; Katinčarov et al. 1995). The question is important, because it opens doors to the very debatable theme of periodization and chronology of the BA in South Bulgaria (Katinčarov 1975a; 1975b; 1982; Panajotov 1975; 1988; 1989; 1995; Leshtakov 1993; 1993a; 1996). That problem is synthesized in the theses of two- and three-partial division of the EBA which influences the identification of the Middle Bronze Age, and on the whole this is the impossible stretching of the problem between the Aegean and the Central European periodization system (Panajotov 1988; 1989; Thomas 1967: 98-112; 1992: 504-515; Erich and Bankoff 1992: 375-79; Parzinger 1993). In Southeast Bulgaria the debate is concentrated on the upper levels at Ezero tell, Nova Zagora tell 6-1, the tell near Dyadovo village and the dating of the Depa amphikypella from Baa Dere and Constantia (Leshtakov 1993). Problems of the western part of Upper Thrace are discussed only on common plan, without confirmation with concrete material (Leštakov 1997 in print). I will summarize a part of already published discussion below, without which the dating of the 1st – the 4th level from Ognyanovo – Maltepe can not be well argued.

In Bulgarian historiography the dating of the Early Bronze – Middle Bronze at Yunatsite is based on the assumed ceramic parallels with sites from Thessaly and South Greece, dated from the end of the third – the beginning of the second millennium BC. Those parallels however, are indirect, as well as there are no true imports. Of importance for the chronology of whole "Yunatsite culture" is the chronology of the enigmatic for its time "Bubanj Hum "in Nišava valley. M Garašanin dates the phase of Bubanj Hum Ib a little before 2 000 BC. That date is conformable to with emergence of the "Pseudo-Minyan" ceramics in Greece from Early Helladic III (Garašanin 1973: 206). As early as 1940s, V. Milojčić asserted the idea of dating of the Velika Humska Čuka from the Middle Helladic period. According to his opinion, of basic importance is the fact that at the last site typical pointed bottom

[1] On the ritual cf. Nikolova 1999: 31, who questions the cremation as it is possible to have been an inhumation with secondary placed ashes in the urn. The same author assumes the burial may date from the end of the Early Bronze I.

cups have been discovered as imports from Yunatsite III (Milojčić 1949: fig. 2: 6-7). The context consists of materials, which are comparable with terminating Macedonian Early Bronze Age (Milojčić 1949: fig. 2: 1-5, 8-10). According to that author, there are sherds of "real Minyan pottery" discovered at the same site (Milojčić1949: 56, fig. 2). This is the way in which the period of the Yunatsite III is synchronized with Middle Helladic II.

Later, the finds from Velika Humska Cuka (excavations in 1934) are also discussed in the Serbian literature (Garašanin 1954: 19-24). But it turned out, that the documentation is missing, because the diaries of the excavations were lost in the time of the Second World War. Despite that fact, Bubanj is synchronized with Early Helladic III, and Bubanj III with an early phase of Middle Helladic. At the root of that assumption lies again those fragments, which "strongly resembled the Minyan ceramic types". The popularity of both the sherds gradually increases in the course of the time and they participate in the core of quite complicated constructions. According to R. Erich, the materials from Lerna date the Minoan influence to the time of Troy II-III (=Early Helladic III) and in this way, the synchronization of Bubanj I with Early Helladic II and of Bubanj II with Early Helladic III becomes possible (Erich 1965: 429-30). In this case Bubanj Hum III gets completely into Middle Bronze Age and it is synchronized with Middle Helladic. Sometimes, by such a dating the newcomers from northeast course the ceramic changes and even the Indo-European invasion in Pelloponess between 2300 and 2200 BC can take part (Weinberg 1965: 304-305).

Recently, M. Garašanin partially revises the synchronization of the upper Bronze Age levels of Yunatsite in relation to "Bubanj-Hum culture", and by way of it – with Early Helladic culture, i.e. the correlation again crosses the already known zones of Struma, Nišava and Vardar valleys. In the opinion of that author Yunatsite 8-7 are contemporary with Bubanj-Hum III, i.e. there is already a considerable revision of the old statement. The stratigraphy of Pevkakia Magula has yet a main value for their relation to Early Helladic. The cups with two handles, symptomatic shape of Bubanj Hum III, Armenochori and Western Bulgaria, which are only assumed for Yunatsite basing on unclear fragments, occur at Pevkakia VII and especially in the "Transitional Period". In the phase of Pevkakia VII there are vessels of the time of Hagia Marina, and the Minoan pottery (known also from Early Helladic III, according to M. Garašanin) occurs in the 'Transition Period'. Accordingly, that author synchronizes Bubanj-Hum III – Yunatsite 8-6 levels with the developed phase of the Early Helladic III and dates them to c. 2200 BC (Garašanin 1996: 104 and note 40). There is not a word for the "Minyan fragments" and their dating to Middle Helladic II, but it follows from the context – if Yunatsite 8-7 levels date to c. 2200 BC, the remained six levels can get up into developed phase of the Middle Helladic period in Greece. However, the opinion of Weinberg from 1965 is repeated, but it is deprived of its "migration" nuances. That consideration does not correspond to the ^{14}C dates from Yunatsite tell (see a last commentary in Nikolova 1999: 256-60 who argues dating of Yunatsite 2-1 levels to 2100-2000 BC).

So, in this curious and unconvincing way there was an attempt the Middle Bronze nature of the "Yunatsite culture" to be argued. Those observations are not original, because they have been struck an impression earlier. For instance, according to V.S. Titov, the thesis under discussion is based on parallels with uncertain stratigraphy and in the final reckoning they cannot be used with a confidence (Titov 1995: 14-17 with ref.).

The elimination of one or several arguments of definite thesis does not mean it is not true anymore. The mathematical problems can be asserted with the help of reduction ad absurdum, but that mechanism is not valid with the same rate for the humanities. For that reason, I do not belief that the uppermost Bronze Age levels at Yunatsite do not get into the beginning of the second millennium BC. I only modestly remind that the thesis should be proved. It is understandable the effort of the authochtonists to fill in one chronological period in the western part of the Upper Thracian valley with cultural contents, but it cannot be made at any cost. It seems for that reason "the critically analyzing authochtonist" C. Renfrew regards by compromise the horizon Yunatsite, Manole-Razkopanitsa and Bubanj Hum II a transition to the Middle Bronze Age (1979: 145, 150).

But going back to the point of the matter, after drop out of the "Yunatsite type cups" as a symptomatic of the Middle Bronze ceramic item, there is almost no one argument which can confirm the idea that Yunatsite 8th-1st levels date after 2000 BC. No one ^{14}C date confirms such an opinion (Nikolova 1999: 227-29). However, there is one find at Yunatsite, which according to the conductors of the excavations is an argument of a high proving value. At the beginning of the renewed excavations at Yunatsite in the course of the cleaning the old V. Mikov's trench, a shaft-hole axe is discovered (Katinčarov and Matsanova 1993: 170, fig. 18: 7)[1]. According to the excavators, the shape and the manner of its casting are arguments, which allow the shaft-hole axe to be dated to the Middle Bronze Age. They belief it is the only find of that type discovered by regular archaeological investigations and without doubt it confirms the assumption that the tell was occupied during the Middle Bronze Age. The significance of the axe for a more precise definition of the absolute and the relative chronology of the Bronze Age in Upper Thrace increases because of the fact that at the same building level there are pointed bottom cups of "Yunatsite type". One of the closest parallels of the find from Yunatsite originates from Poliochni red, from the cache of Vano 829 (Nakau 1997: 637-38, fig. 2 and fig. 3 on p. 640). It is of type Vesselinovo after A. Vulpe (1970) or type 14 after E. N. Chernykh (1978: 150). The axe is published by L. Bernabo-Brea (1964: 661-62, Plate CLXIII) being a rare stratified find of that type in Greece. The type of Vesselinovo is distributed in Bulgaria and Romania, but it is unknown in Anatolia and the southern Black Sea, as it is stressed by G. Nakau. J. Maran (1989, 132-133) dates that type to Early Helladic II-III and also points out its close northern parallels. Accordingly, the axe from Yunatsite 6 can be dated to Early

[1]The terrain data are: Terrain Number 3758 (August 10th, 1976). The sixth building level. Square H_9, western profile. Length - 12.5 cm.

Bronze 3, i.e. its existence in the Middle Bronze Age is possible, but it is not a categorical proof in a chronological aspect.

Therefore, the review of the synchronizations at the beginning of this section is based on the parallels in one general direction: the western part of the Upper Thrace – the Sofia plain – the Nišava valley, and from there to the south towards Macedonia and Thessaly. Those directions partially correspond to Roman Via Diagonalis – a road that is reasonably assumed to have doubled the direction of the prehistoric link connected Anatolia and the Balkans (Mellaart 1971: 129-31). From Niš area to the south the direction is covered with the prehistoric road to Vardar valley that have connected Pelloponess and Northern Greece with Central Europe defined as a "strong axis North-South" in the period of Sitagroi Vb and parallel to the road along the Struma river (Maran 1997: 176-77). This indirect link of the cultural contacts (no matter migrations of population groups or a transport of raw materials or goods are thinkable) is the derivative links of the model of "Kettendatierung". The last consists of a peculiar logic, but it is hardly to believe that is the most rational for a correlation of Upper Thrace with Northern Aegean. From theoretical point of view, this manipulation provokes doubt because of the indirect connection between the compared regions and the intervention of numerous secondary zones with unclear cultural attribution – "variants" of Baden, Kostolac, Vucedol, Cotofeni etc. In west direction the borders of "Yunatsite culture" are not defined and it hardly will be done soon, with all following reasons. Accordingly, the interrelation offered for this direction are indirect and cross many mediators. Most of them represent a weak links in the chain – for instance Sofia lowland and Nišava valley, to the west, and the valley of the middle and upper Struma, to the south. Then the line of the otherwise orderly logical constructions is short. In fact, the most precise parallels of the northwestern Thracian material can be discovered through the Rhodopes in the settlement mounds in Northern Greece. There the stratigraphy is clearer, the chronologies are more certain, so they require a construction of Trans-Rhodope communication-bridge.

But there are two circumstances, which again render difficult the correlation:

1. There are no comparable sites with certain stratigraphy in the Central and Western Rhodopes. Two caves – Haramijska Cave and Yagodina Cave, give a good information about the end of the Chalkolithic and Early Bronze Age, but you cannot rely on their data because the materials of the former sites are unpublished or they are only shortly reported.

2. The problems of the prehistoric roads through the Rhodopes are not investigated. That question was posed just recently on the occasion of trade contacts of the Greek emporium Pistiros (near the town of Septemvri) with the island of Thasos (Domaradzki 1994), but the premature death of the director of the project M. Domaradski made an end of the investigations at their beginning.

If a special elaborated methods will be apply it can be suggested a reconstruction of the Trans-Rhodopes roads in principal. So, a question can be posed if the Rhodopes were a barrier in prehistory or they did not make difficult the contacts and connections of different character. However, it should be stressed there are possible more than one model of the mechanism of the exchange of information from both sides of the mountain.

Analysis of the pottery from Ognyanovo

After above long discussion let return back to the concrete material from Ognyanovo. It should be stressed, the chronology already given by P. Detev, R. Katincharov, H. Parzinger and L. Nikolova has faced some problems. According to the data published, as well as according to the notes in the diary, the distinguishing of distinct levels and especially of complexes is very uncertain. The main role for the chronology of cultural layer of Ognyanovo has the pottery. The small finds are atypical and with some exceptions they can be dated from whole Early Bronze Age. Unfortunately, a bigger part of the ceramics can be defined typologically only. That is because of the later destruction of the EBA layers, as well as because of some mistakes by the excavation, documentation and publishing of the material. Most of the finds have uncertain context. Nevertheless, with a caution required, it can be made attempts of more precise stratigraphy of the site and the position of the vessels.

Equipment, houses and stratigraphic position of the pottery

The sixth building level

It is situated at the base of the cultural layer. Several floors of houses are recovered from it (Detev and Matsanova 1977: fig. 1) but the inventory is unpublished. There are data on fragments collected around the oven of one apses-house only:

- fragments of big vessels – pithoi and big amphorae vessels (***Figures 16.2; 16.3.6; 16.4.2***);
- jug handle with deep oblique incisions and a handle of oblique ridges of 'Mikhalich' style and incised decoration of "western style" (Cf. ***Figure 16.11: 8***);
- lugs of big vessels (***Figure 16.4.5***);
- vertical wide handles with flat section attached at big deep vessels (Detev and Matsanova 1977: figs. 9: 3-4; 11-13; 15-16).

At that level a base of pit was discovered with a big pot in it (Inv. No. A 2324, Cat. No. 27, ***Figure 16.21***). It is remarkable the fact that the burial belonged to cremated child, according to the excavators. The pit was dug from higher building level, probably the third or the second. But it cannot be suggested there was a closed complex as along with the urn a "gray Thracian pottery" was discovered there. Both vessels excavated were clearly diachronous, i.e. it is very possible the early pit to have been cut by a second one, much later (Detev and Matsanova 1977, figs. 10; 11: 2-3). Having in mind the considerable size of the urn-pot, as well as only the bottom of the pit was documented, it can be assumed with certain cogency the niveau of the pit was at least two or three levels above the sixth horizon. In other words, it does not correspond to the 6th or 5th horizon.

The fifth building level

It is the second horizon upwards. Two completely preserved bases of apses-houses have been documented. Outside one of them a round-bottom cup was found (Detev and Matsanova 1977: fig. 17: 1). Among the fragments the conical deep plates with incised ornamentation (***Figure 16.10.1-7***) and pots from the inner side of the orifice are make an impression (***Figure 16.3.1-4***). The handles are wide, with incisions and hollows with arising upwards ends typologically comparable with the fifth horizon. In the course of excavation of their bases, perhaps materials from the upper horizons were intruded.

The fourth building level

There are missing completely preserved plans of houses as only parts of house features were recovered. Some intact vessels cannot be accepted with finds with certain context because of destruction from the Classical Antiquity mentioned above (Detev and Matsanova 1977: figs. 17: 2; 24: 1, 3, 5). The ceramics published as belonging to that building level visibly changes its style. Some shapes and decoration occurs which are typical of a late stage of the Early Bronze Age. Most important are the cups, amphorae and the spout bowls. According to the ceramics, which was processed later the following vessels can be identified:

? A cup with oblique orifice and round bottom with parallels in the 'Sveti Kirilovo' stage in Southeastern Bulgaria, of which three are discovered together, but without clear context (Detev and Matsanova 1977: figs. 15; 24: 1, 3, 5).

? Bowls and plates with spouts and small vertical handle (Cat. No. 5, ***Figure 16.18.4***, Inv. No. 2147). One similar fragment is found as discovered in a house (Detev and Matsanova 1977: fig. 23: 1).

? An intact funnel (unpublished) has been also documented.

? A cup of 'Sveti Kirilovo' type, the only authentic instance of that type, most probably import from Southeast Bulgaria (from the fourth horizon, Inv. No. A 2155, ***Figure 16.16.2***).

? Amphorae like vessels are documented by fragments from the fourth level, some of which are with incised ornament. It is zigzag motif in two rows under the rim of the orifice and two rows over the upper part of the shoulders (***Figure 16.12.6***). Quite a lot sherds with rounded and profiled outside rim, as well as one fragment of amphora-like vessel or of a big jug (***Figure 16.12.5***) originate from the fourth level. The amphorae like vessels had two vertical handles at the widest part of the body. The decoration consists of flat (disk-like) knobs. A part of them are a quite big and may belong to storage vessels.

The third building level

Whole plans of houses are not recovered. The levels is in fact the uppermost for the biggest part of the tell (Detev and Matsanova 1977: 63). At this level more intact vessels have been found, among which some bowls were mentioned as well a bowls on high pedestal decorated with incised standing triangles (***Figure 16.8.6***). The other vessels are mainly cups, but an askoi has been discovered as well (Detev and Matsanova 1977: figs. 24: 1; 29: 1-2; 30: 1; 38: 2).

Among the other fragments they should be mentioned:

? Spouted bowls and plates. Some of the bowls are decorated with stamped circles and rectangular fields, filled with perpendicularly incised horizontal and vertical lines. At the third horizon there are fragments of the same bowls, with equally and obliquely cut rim of the orifice decorated with incisions and two rows of stamped circles. The last are with white encrustation. Some bowls with inverted rim have under the orifice relief bulges, smoothed or with hollows.

At that level the following vessels are represented:

- Bowls with strongly inverted rim (***Figure 16.7.4-6***);.

- One almost intact askos with horizontally shaped body (***Figure 16.17.1***, Inv. No. A 2191).

- Cups with round bottoms close to Sveti Kirilovo type. But they differ in some details and can be named "Sveti Kirilovo type cups, western variant" (cf. ***Figures 16.13*** and ***16.14:1-3, 5-6***).

The second building level

That building level had been preserved over very small area. Parts of floors of some houses have been preserved only (Detev and Matsanova 1977: 70). The following vessels are attributed to that horizon:

- Cups with handle of "Mikhalich" style, which is obviously without stratigraphy and fell in the level accidentally (cf. ***Figure 16.11.7***);.

- A big cup with pointed bottom and rich pointillée decoration over the whole body (***Figure 16.15.4***).

- A spouted vessel with one horizontal handle, which is called in the original publication 'Priam's vessel' (Detev and Matsanova 1977: 67). In fact that is a late replica of the popular in earlier Early Bronze "saucier". Similar spouted vessels (cups or bowls) are popular in the east, along the Maritsa valley, and characterized the Early Bronze Age 3 (Leshtakov 1992: figs. 17e; 32h).

- Fragments of incised and stamped decoration of the 'Yunatsite type' (***Figure 16.12.1-3***).

- Biconical big vessels with cylindrical neck (***Figure 16.12.4***). It has precise analogies at Manole-Razkopanitsa.

The first building level

It is not registered over whole area of the site (Detev and

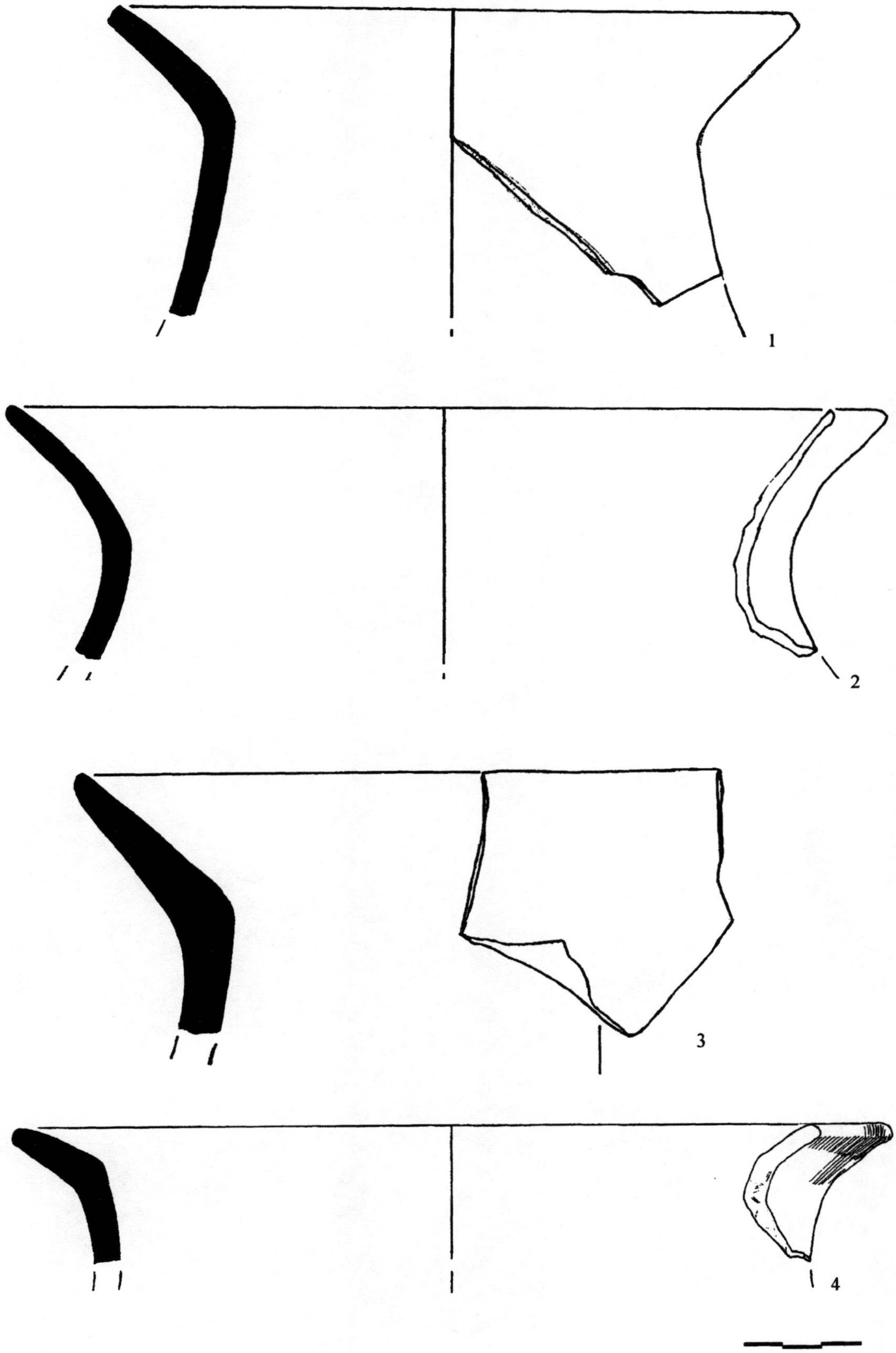

Figure 16.2. Ognyanovo-Maltepe. Early Bronze pottery.

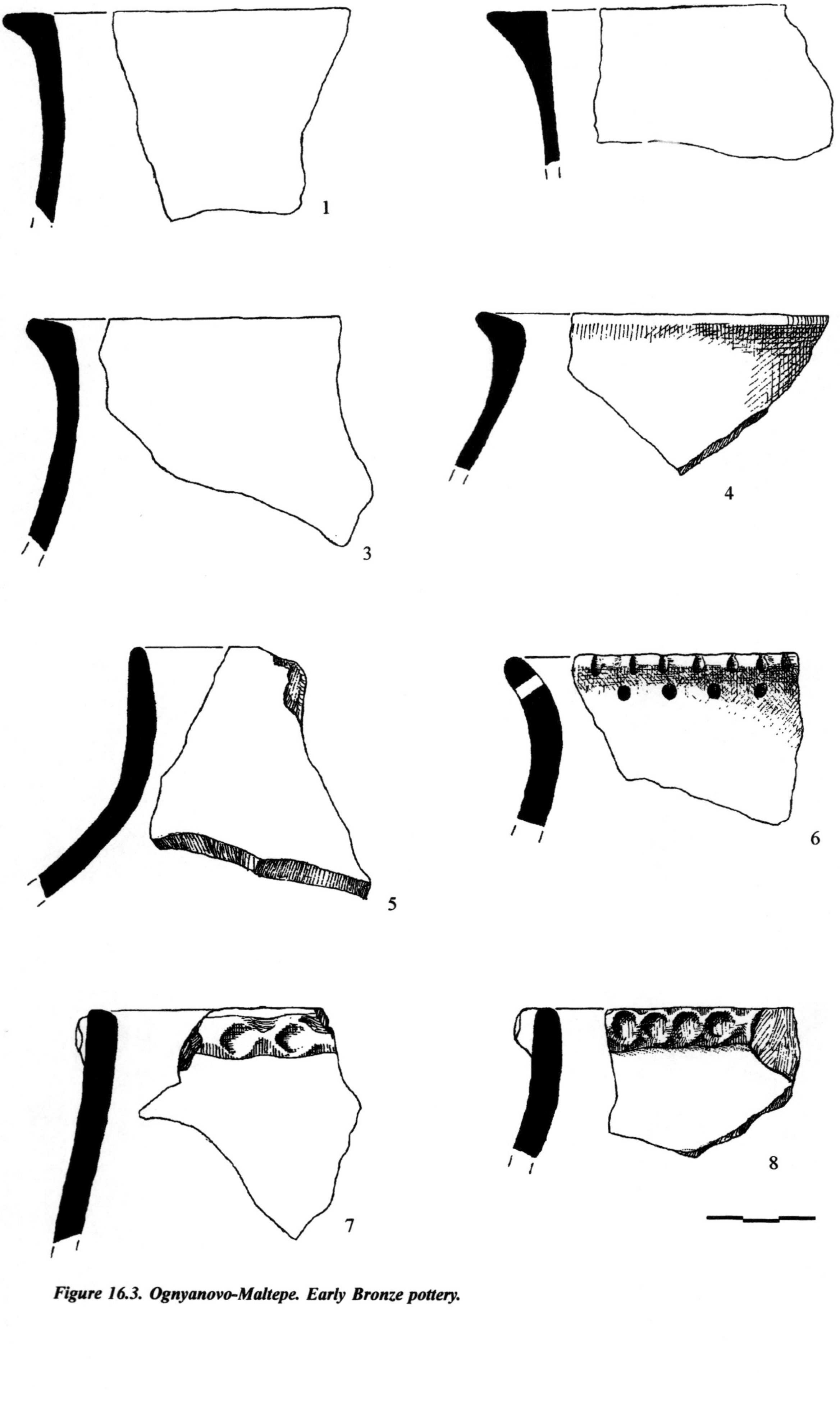

Figure 16.3. Ognyanovo-Maltepe. Early Bronze pottery.

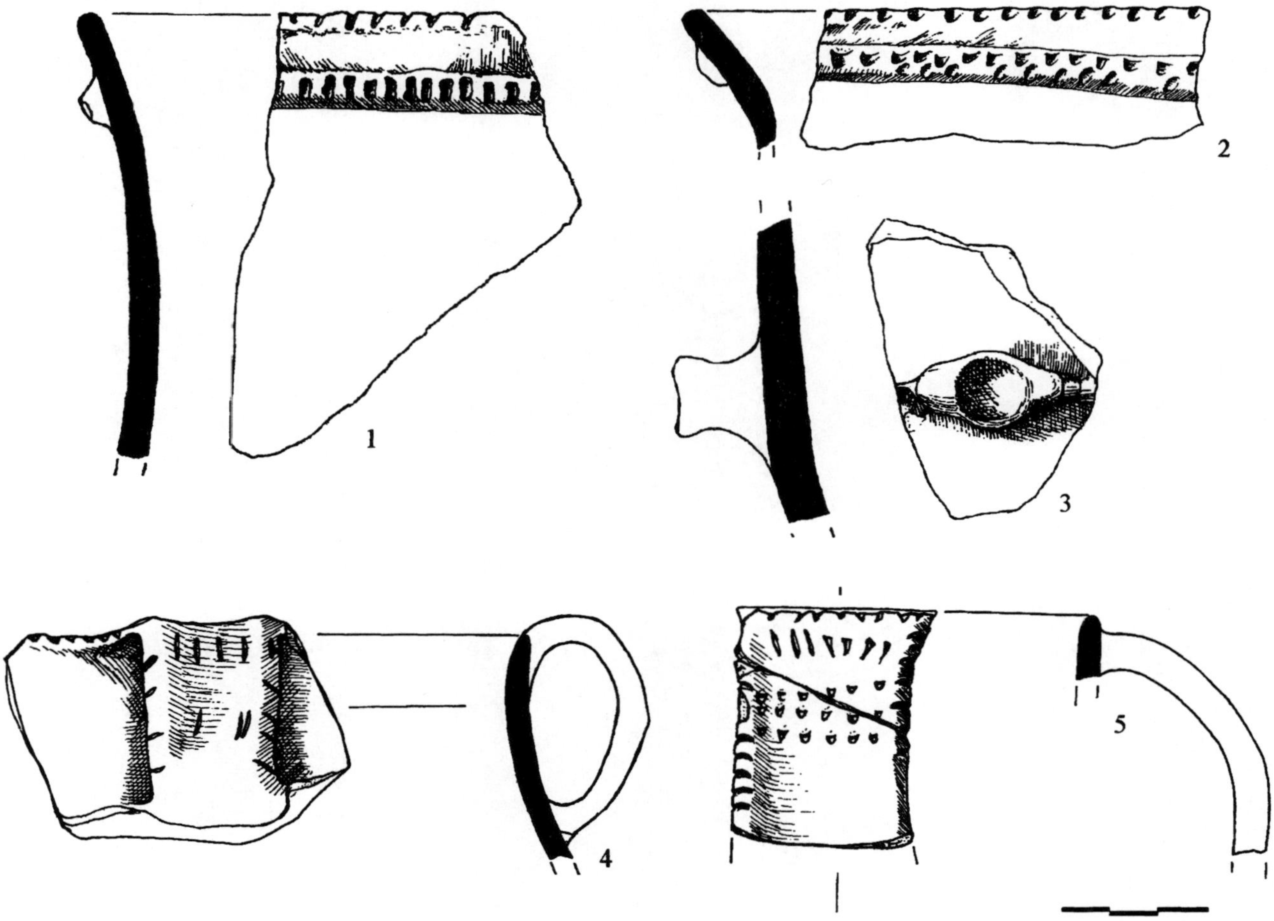

Figure 16.4. Ognyanovo-Maltepe. Early Bronze pottery.

Matsanova 1977: 75-76). Layers from Classical Antiquity and contemporary ditches had destroyed almost completely the layer before the excavations. Therefore, it is hardly to believe the vessels have had in a certain context. To that level the following vessels are attributed:

- Cups of "Yunatsite type" with uncertain stratigraphy (Detev and Matsanova 1977: fig. 30: 2). The second cup published here was not illustrated in the primary publication (***Figure 16.16: 3***).

- A cup with round bottoms, "Sv. Kirilovo type, western variant" (Detev and Matsanova 1977: fig. 24: 4; cf. ***Figure 16.13***).

- A fragment of biconical vessel with incised decoration, probably of a big cup (Detev and Matsanova 1977: fig. 8: 7; identical to this in ***Figure 16.12: 4*** but smaller in size).

- A fragment of jug, non-illustrated in the initial publication (***Figure 16.3: 5***).

Beside the vessels accounted above there are Ognyanovo vessels and sherds without stratigraphic data, only partially published, as follows:

- Bowls with conical shape. They belong to different types (***Figures 16.6: 2, 16.7: 2, 16.8: 1***).

- Bowls with semispherical shape, also of different types (***Figure 16.6: 1, 3***).

- Bowls with moderately and strongly inverted orifice. Some of them have a trumpet handles (or handles ?), and others are with flat vertical and low attached horizontal handles. The decoration consists of pits over the handles and two relief bulges from both sides of the base of the handle (***Figures 16. 6: 6-10, 16.9: 1-9***).

- Two cups with pointed bottoms, but in view of fact it has almost spherical body and possible cylindrical but not funnel-shaped neck, differing from 'Yunatsite type' (***Figure 16.15.1-2***). One of them is richly decorated with incised and pricked ornament (No. A 2159), while the others from the same type (No. A 2162 for instance) are without ornament.

- A big cup or a jug (?) with three incised lines above the bottom and pricked dot of a zigzag technique (cf. ***Figure 16.12: 7***).

- Big shallow cups (also used as bowls?) with vertical handle attached at the rim.

- A cup with a peculiar shape, which cannot be defined typologically because of its fragmented condition (***Figure 16.18: 1***, Cat. No. 11).

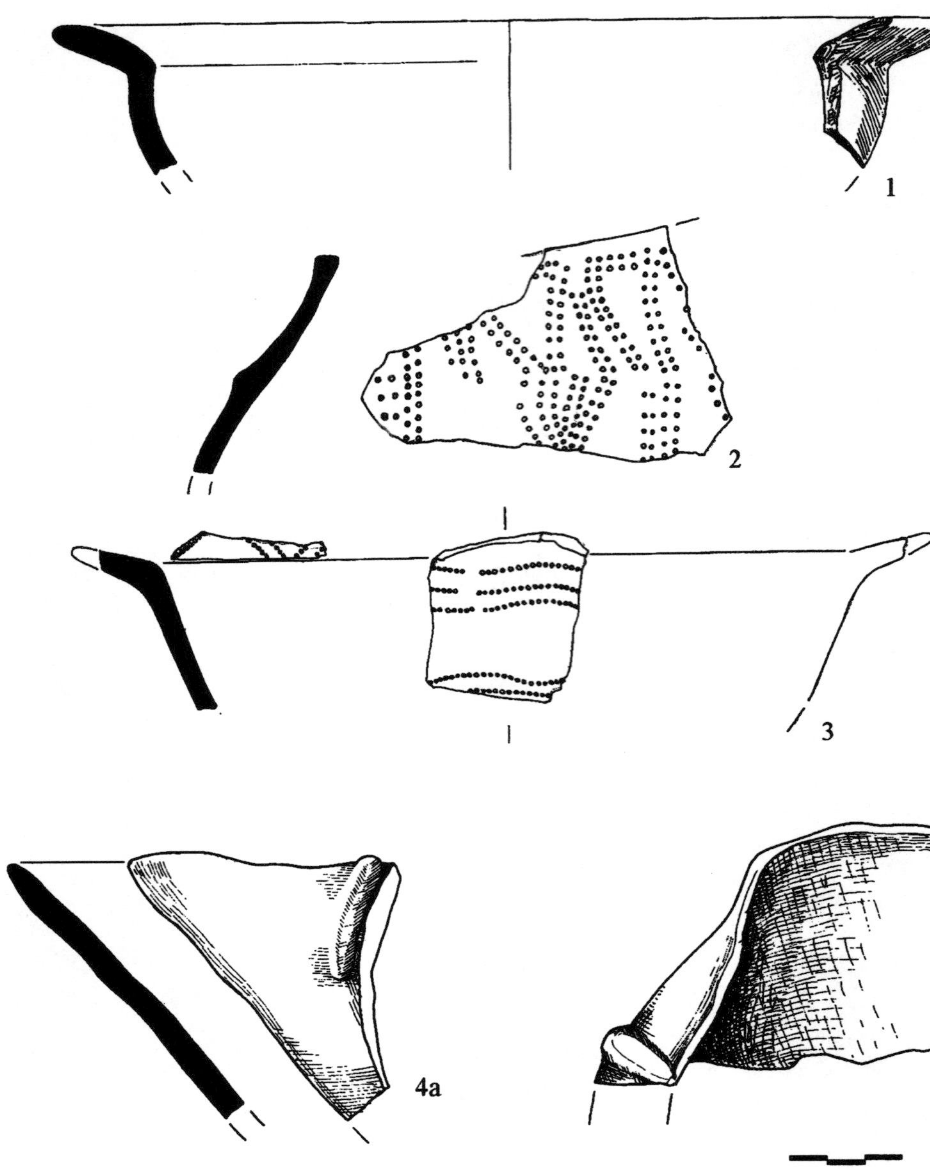

Figure 16.5. Ognyanovo-Maltepe. Early Bronze pottery.

A comparative analysis and problems of chronology (Ognyanovo – Yunatsite – Manole – Plovdiv-Nebet Tepe – Dubene-Sarovka)

The main characteristics of Ognyanovo belong to those known for the Early Bronze in Thrace: surrounding ditch and embankment (with parallels at the tells of Manole, Yunatsite Dyadovo, Nova Zagora, Rupkite etc.), apses-houses, a standard inventory, small number of finds etc. The ceramic vessels and the fragments published, along with the unpublished information added give no argument of possible dating of the site to the Late Bronze and less to Early Iron Age. The technological characteristics of the pottery harmonize with the known from Yunatsite, Manole-Razkopanitsa, Plovdiv-Nebet Tepe, and Dubene-Sarovka in western Upper Thrace. At the same time, at Ognyanovo there is no pottery comparable with Gulubovo 1-4 (Leshtakov 1993a; 1997), so there is no possibility the upper chronological borderline to be defined after the beginning of the second millennium BC.

The pottery from the sixth and fifth levels

The earliest two levels of Ognyanovo can be attributed to the Early Bronze Age II . In this way the assumption there were two groups of levels (VI-V and IV-I) mentioned in the primary publication is confirmed but the dating is not Late Bronze and Early Iron but Early Bronze 2 and Early Bronze 3.

The earlier pottery has parallels at Yunatsite 14-9, the lower building horizons at Plovdiv-Nebet Tepe, at Dubene-Sarovka II (Nikolova 1995, 15-23; 1999), and in Southeast Bulgaria – at the fortified settlement of Mikhalich and at Ezero 8-4 (Georgiev et al. 1979, 502-504). At Ognyanovo 6-5 there are plates with conical shape, sometimes with slightly thickened rim of the orifice from the inner side, and incised decoration. They are especially typical at Yunatsite after the 14th horizon, i.e. in Early Bronze II. To that stage can be dated the jugs and cups with jagged handles, probably influenced by 'Mikhalich style' (Leshtakov 1992: figs. 35-36). The wide handles of big vessels with black polished surface look early (Detev and Matsanova 1977: fig. 9: 3-4, 6), as well as the handles deco-

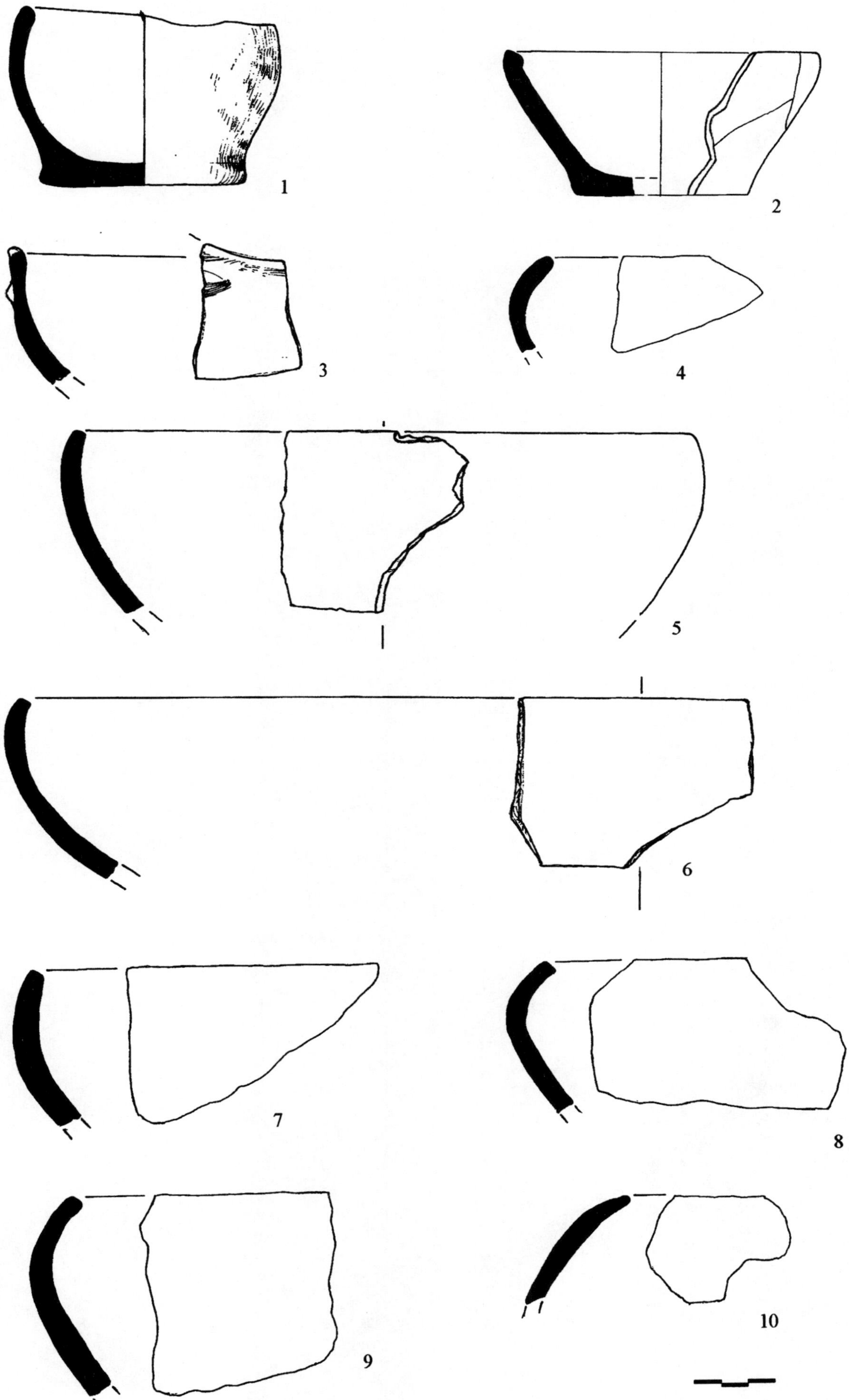

Figure 16.6. Ognyanovo-Maltepe. Early Bronze pottery.

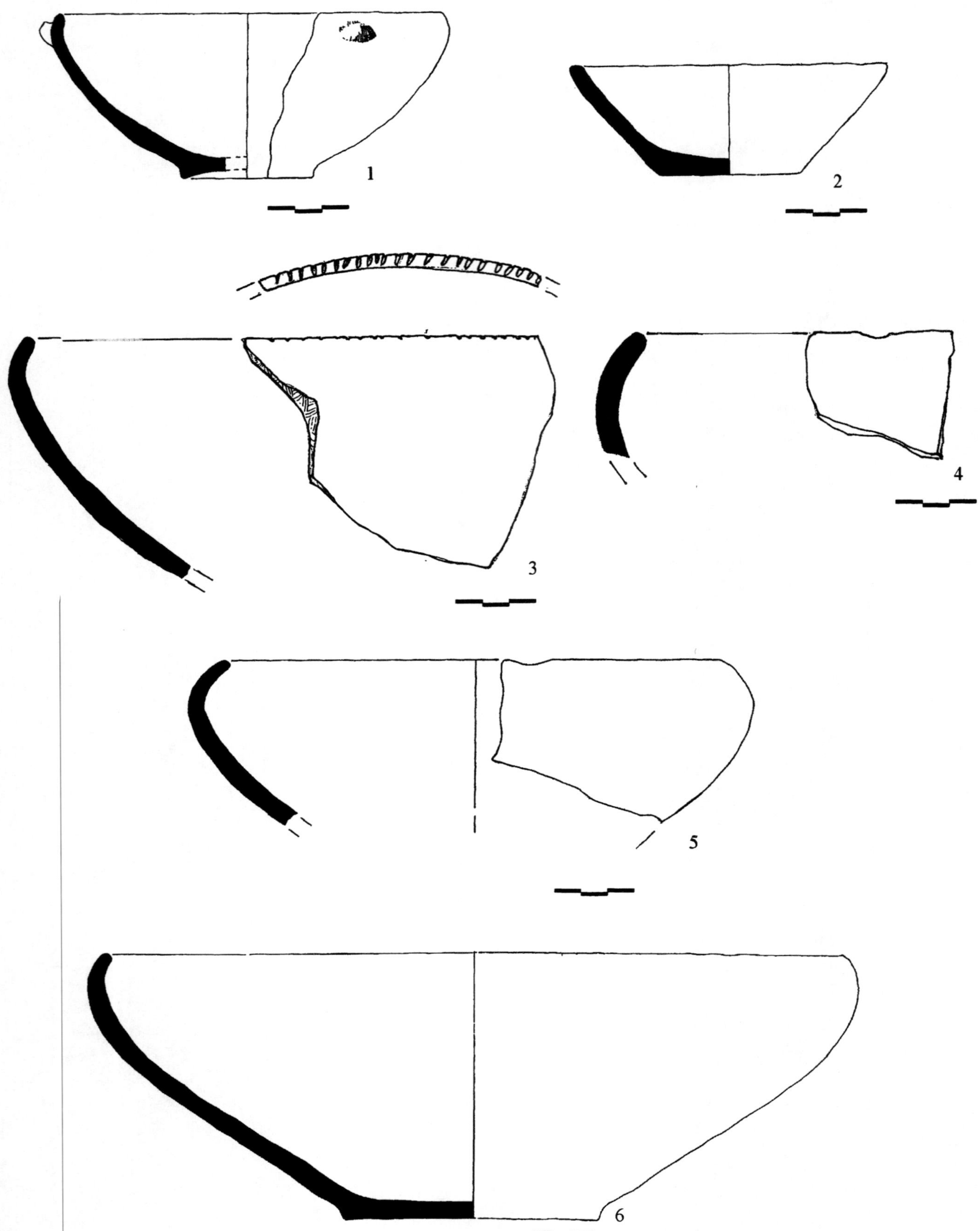

Figure 16.7. Ognyanovo-Maltepe. Early Bronze pottery.

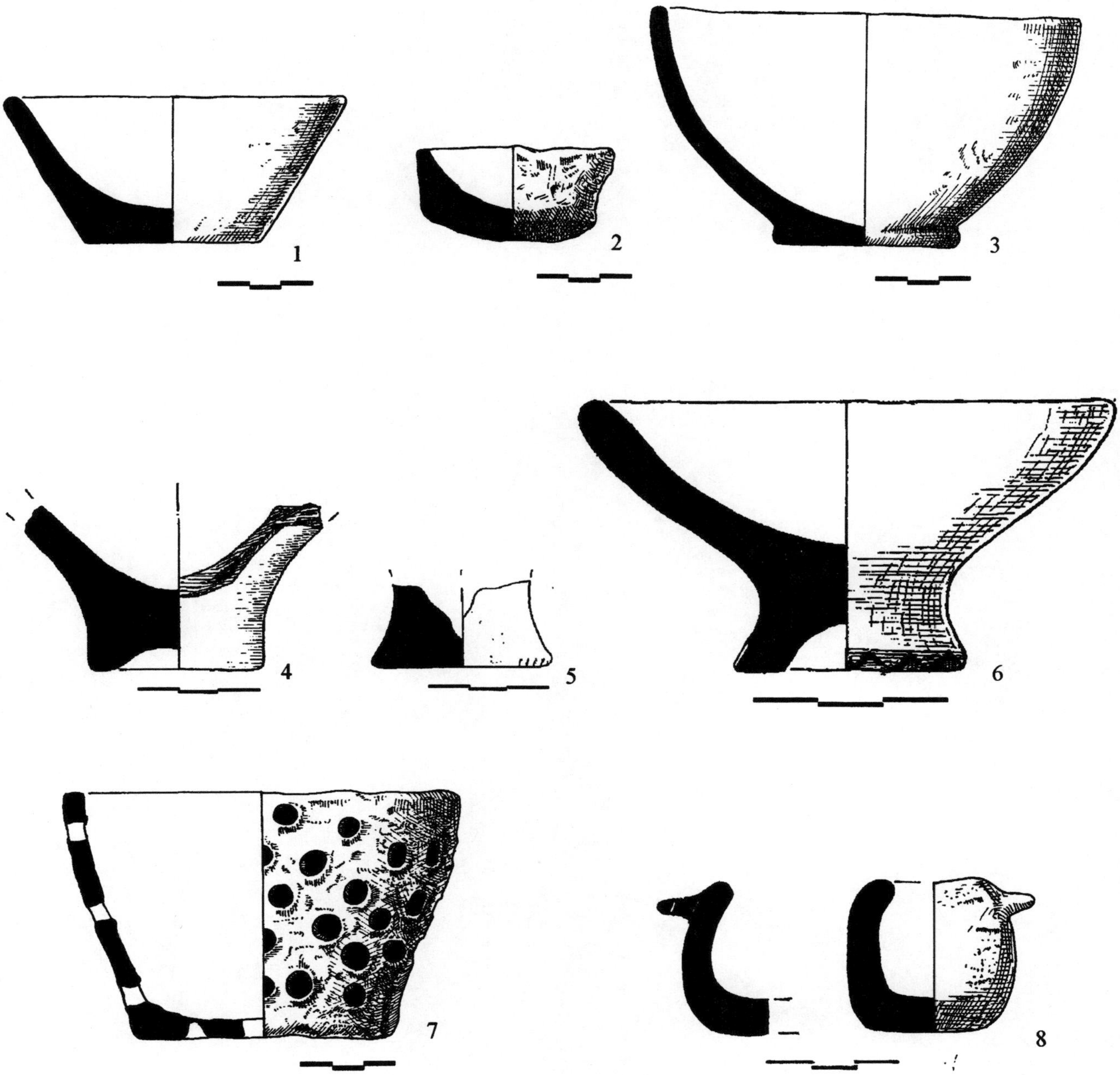

Figure 16.8. Ognyanovo-Maltepe. Early Bronze pottery.

rated with incisions attached at the pots (Detev and Matsanova 1977: fig. 9: 1-2). They all have good parallels at Yunatsite 14-13. But it cannot be excluded a possibility a synchronization with Yunatsite 15, but these are details, which will be discussed later.

The pottery from the fourth to the first horizon

The ceramic shapes and their decoration have many parallels in western Upper Thrace: Yunatsite 8-3/2 (Katincharov and Matsanova 1993), 6th-5th (4th?) level at Plovdiv-Nebet Tepe, Manole-Razkopanitsa (Detev 1950a; 1981) and at Dubene-Sarovka IIC (Nikolova 1995: 15-23). In Southeastern Bulgaria the sites with parallels are Ezero 3-1, Dyadovo 4-2 (Leshtakov 1994, 95-103), Nova Zagora 6-3, Kirilovo etc., dated from 'Sveti Kirilovo stage' (Leshtakov 1992: 38-67 with ref.). The analogies point to a very probable dating to Early Bronze 3. Both the askoi have precise analogies at Yunatsite 8-7.

The bowls with moderately and strongly inverted rims have parallels at the upper levels of Plovdiv-Nebet Tepe. At Ognyanovo, the decoration of pits over the handles and of two relief bulges on the both sides of the handle base, are distributed at the base of the vertical handles of bowls as well. It seems they characterize all the period of the Early Bronze 3. The round-bottom cups, characteristic of the upper horizons, have analogies at Manole-Razkopanitsa, house L. They differ from the typical of 'Sveti Kirilovo' and are closer to those from Sitagroi Vb.

The only cup of Sveti Kirilovo type (from the 4th horizon, No. A 2155) probably was an import from Southeast Bulagria, because it is absolutely identical morphologically and technologically to the samples from Kirilovo, Karanovo, Dyadovo and other tells (Leshtakov 1992: figs. 22; 26). The cups of 'Yunatsite type' belong to the 3rd and the 1st level. The cup

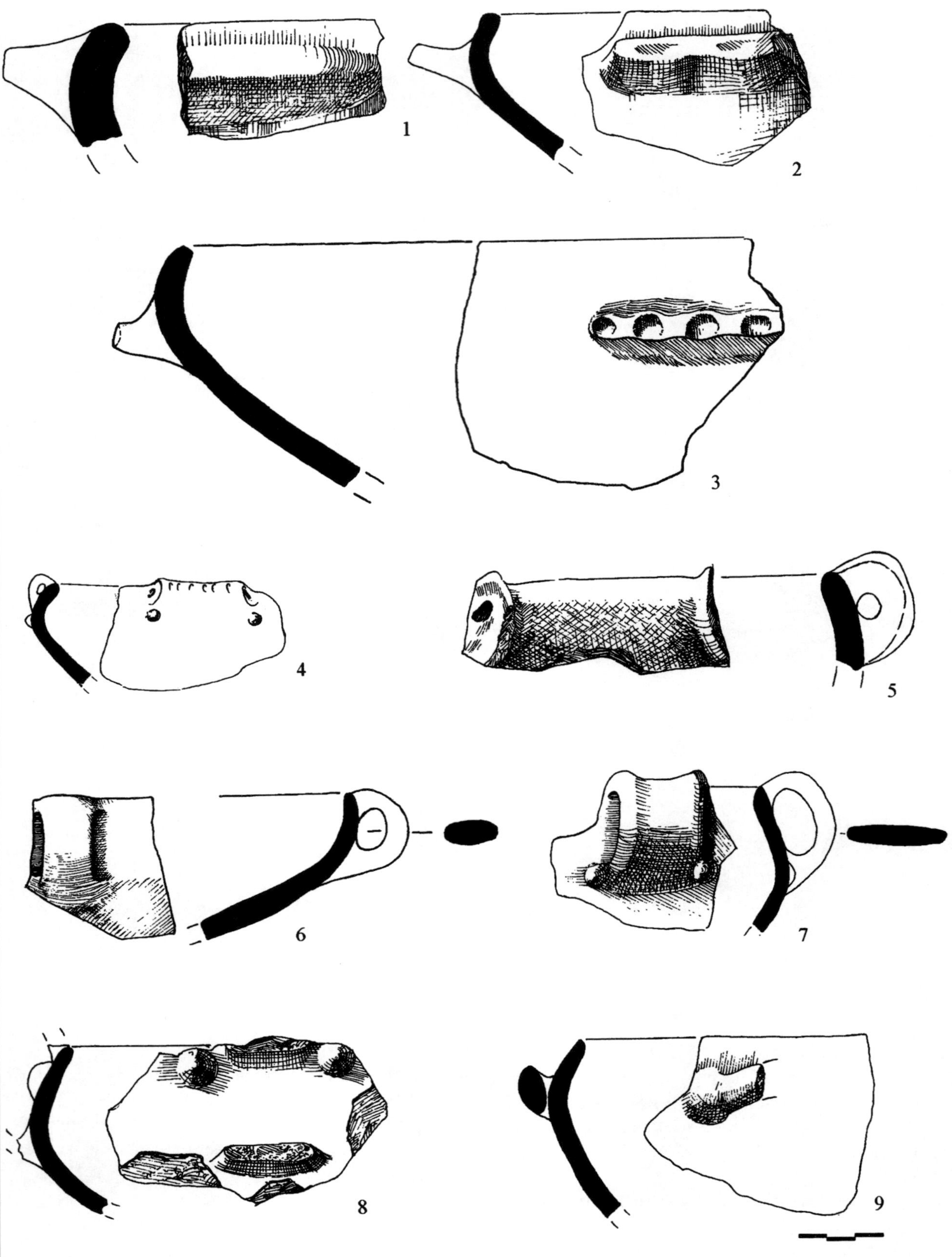

Figure 16.9. Ognyanovo-Maltepe. Early Bronze pottery.

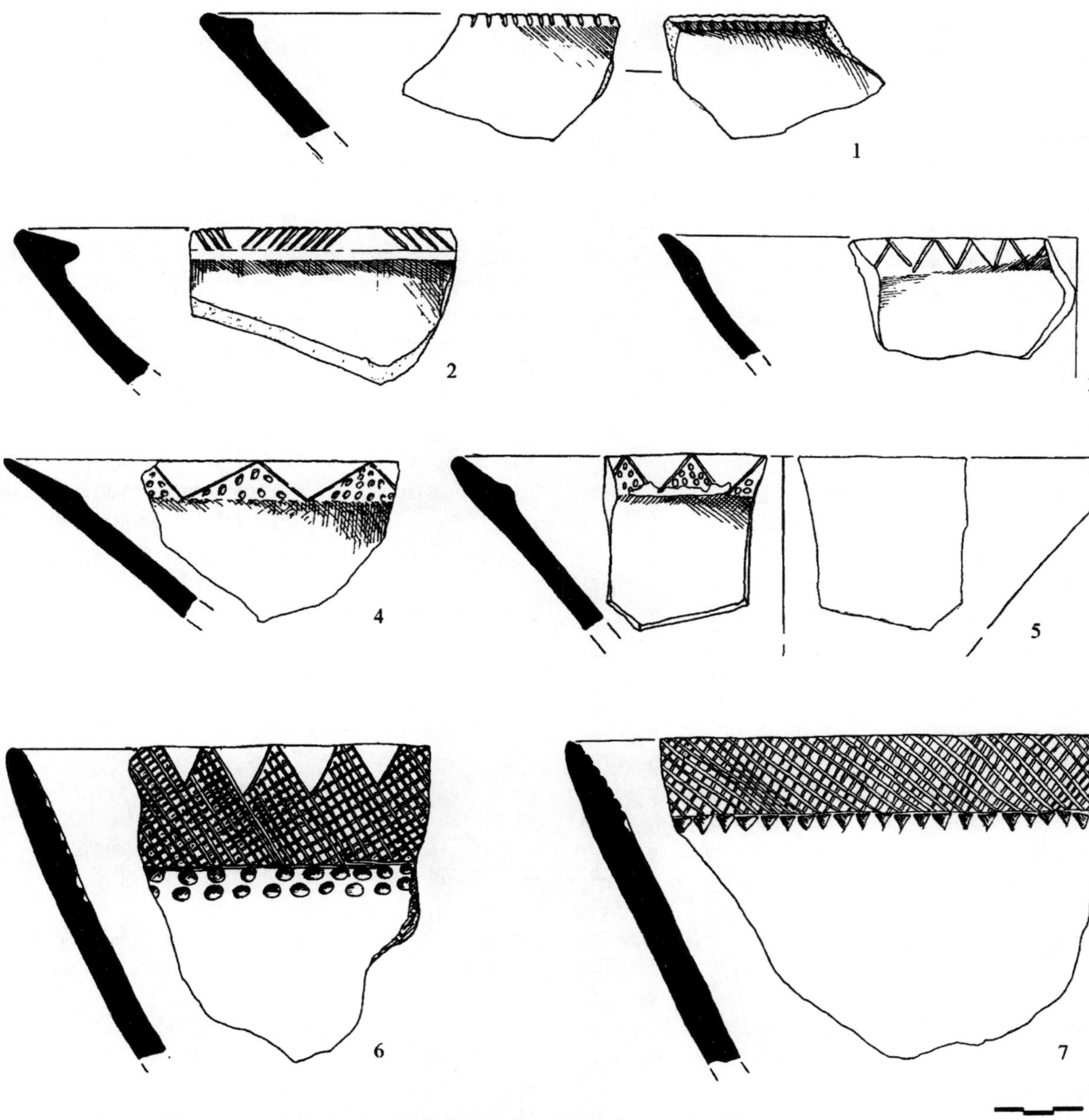

Figure 16.10. Ognyanovo-Maltepe. Early Bronze pottery.

from the 1st horizon has parallels at Yunatsite 7-2, at Sadovets, Northern Bulagria and Sitagroi Vb. More special body shape has one other cup (No. A 2157). Both the cups with pointed bottoms similar to those of 'Yunatsite type' (No. A 2159 and No. A 2162) have parallels at Chernichevo in the Stryama basin but there is missing a stratigraphy (Nikolova 1996a: fig. 23: 3).

One of the askoi at Ognyanovo has vertical structuring of the body (No. A 2192, ***Figure 16.17.2***), which makes it close to the finds from the upper levels at Yunatsite. Biconical vessels with cylindrical necks are registered only by one fragment from the second level. However, it has a very precise parallel at Manole-Razkopanitsa (Detev 1968b: fig. 10:4).

One of the bowls (Detev and Matsanova 1977: fig. 8: 10) has good parallels of the ornament registered over cups after the 8th level at Yunatsite.

Therefore, the existence of two groups of pottery – from Early Bronze 2 (eventually from the end of the Early Bronze 1) and from Early Bronze 3 can be discussed, based on the parallels argued. At the same time, it should be stressed that the pottery did not reach the end of the third millennium BC. Two assumptions occur: The life at the settlement finished before the end of the upper levels of Yunatsite, or the building activity from the classical Antiquity had destroyed the uppermost levels at Ognyanovo.

Parallels of the ceramics from Ognyanovo in Northern Greece and the problem of the Trans-Rhodopes contacts

The pottery from the end of the EBA1 or the very beginning of the EBA2 will be not discussed at least, because of two reasons. Firstly, its context is not certain. Secondly, the nest parallels should be searched at Yunatsite. For the time being, the negative conclusion can be argued that at Ognyanovo is missing channel pottery, registered at Yunatsite 17-15. The other elements illustrated in the figures, could continue longer on such small settlements as Ognyanovo.

The pottery with richly encrusted ornamentation from EBA 2 at Ognyanovo has parallels at all sites attributed to the so-called horizon of the encrusted pottery (Nikolova 1996b, 81-93 with the references cited there). The problem of the distribution of specific elements is well analyzed, so I can only state that the pottery from EBA 2 at Ognyanovo does not differ from the common tendencies. Even such specific elements of the ceramic repertoire as horizontal lugs with tree perforated holes are identical (cp. Renfrew et al. 1986: Fig. 13.13. 5). The existence of few sherds influenced or even imported from the area of the Mikhalich ceramic repertoire is also of importance (cp. ***Figure 16.11: 7-8***). That fact allows the synchronization of both sites to be elaborated on certain base, even on possible real imports.

It was suggested above the basic part of the pottery from Ognyanovo to be dated to EBA 3. Very precise parallels from Northern Greece confirm that chronology.

The synchronization of the Northern Greek sites with the other to the south is well argued in the literature, so that the pottery from Ognyanovo can be discussed in broader EBA Aegean-Anatolian context. At Kritsana VI there is an imported sherd "dark on white" with possible Early Helladic III origin. That allows a synchronization of the Macedonian EBA III with Early Helladic III. In Western Macedonia the contacts with Early Helladic II, III and IIIA can be followed through specific pointed-bottom cups, jugs with low body turning point and shallow bowls with very turned rim and tongue lugs attached (Warren and Hankey 1989: 43). J. Maran (1998) added the parallels and again argued the parallels without considerable change in the opinion, so their detailed analysis here is unnecessary.

The region under discussion has good parallels with the Eastern Aegean areas and especially with Troada, despite the disputed concrete parallels from Sitagroi IV-Vb and Dikili Tash III. According to C. Renfrew (1971: 280) , the beginning of Dikili Tash and Sitagroi IV and was parallel with Troy I, while J. Deshayes (1970; 1973) states, the former preceded the grounding of Troy I. J. Coleman prefers the opinion of C. Renfrew, as the ceramic shaped used in the discussion were distributed at Poliochni II (blue), which was synchronized with Troy I. The Sitagroi IV-V transition is smooth, without hiatus. Sitagroi Va is synchronous with Dikili Tash IIIB and with Troy I. However, Sitagroi Vb should be dated from the time of Troy II-IV and probably was contemporary with Early Thessalian III, after the periodization of E. Hanschmann. In other works, Sitagroi Va with the "Burnt House" is partially contemporary with the very beginning of Troy I, or even preceded the latter. But Sitagroi Vb is contemporary with Troy I / early Troy II (Yakar 1985: 88). Dikili Tash 2-13 on the whole area dated from Early Bronze Age II: levels 13-7 from Early Bronze Age IIa being contemporary with Troy I and Baden/ Pecel; levels 7-5 – synchronous with middle Troy I; and levels 5-2 – with late Troy I. In other words, there are missing EBA III or later Troy II materials (Yakar 1985: 88).

At Pentapolis located near Seres Town, there is a pottery from phase II with direct parallels at Sitagroi Vb (Renfrew et al. 1986: 483). Good parallels with Early Helladic III, as well as a long ^{14}C dates series, which coincides with Sitagroi and the sites from Early Helladic III, confirm that opinion (Parzinger 1993: 130-32, 140-41).

In the region of North Greece there are missing certain finds from Middle Bronze Age (Kull 1988: 50), which is probably due to the gap in the investigations, mentioned in earlier works as well (Theocharis 1972, 28-29). Therefore, the picture is similar to that from Western Upper Thrace. New information represents the finds from Samothrace, which B. Kull did not discuss. But the last author mentioned that island should have taken place of mediator between Anatolia and the northern coast of the Aegean, because of its peculiar geographical position. The pottery from Samothrace dates from the beginning of the Middle Bronze Age, because of Crete imports (Matsas 1991: figs. 10-12; 1: 2-8; 13). But it can be synchronized with Gulubovo, but not with Ognyanovo (Leshtakov 1996, 248-52), which fact confirms additionally "terminus post quem" suggested above.

Let continue with the particular comparisons:

- Amphorae-like vessels, registered by sherds at Ognyanovo, the fourth horizon, have analogies at Sitagroi Vb (Renfrew et al. 1986 Fig. 13.21.1-3, 6-7).

- The fragment of an amphora-like vessel or from a big jug (4th horizon, ***Figure 16.12: 5***) has a precise analogy of decoration (a technique of pricked dots) at Sitagroi Vb (Renfrew et al. 1986: fig. 13.21:7-8).

- A jug with low body turning point (***Figure 16.19:1***) has a good parallel at Sitagroi (Renfrew et al. 1986: fig. 13. 25: 12).

- Analogies of the big cup or jug (?) with three incised lines above the base and ornament of zigzag pricked dots (***Figure 16.12: 7***), can be found again at Yunatsite and Sitagroi Vb (Renfrew et al. 1986: fig.13.24).

- The round-bottom cups have good analogies in Northern Greece (for instance Inv. No. 2183 cp. with Renfrew et al. 1986: figs. 13. 20: 23; 13.20: 12). There is a precise analogy of the cup in Figure 16.13: 4 from the first horizon (Inv. No. A 2187) with a find from Sitagroi Vb as well.

- Relatively big dots pricked with an instrument with a round point decorated big vessels (***Figure 16.12: 5***) are distributed at Sitagroi (Renfrew et al. 1986: Fig. 13. 21: 7).

- The decoration of pits over the handles and of two bulges from both sides of the handle base (Figure 16.9: 7-8) also has parallels in North Greece, again at Sitagroi Vb. But the relief ornamentation is by the handle of amphora-like vessel there. The bowls with incurved rim with relief bands, smooth or with pits, are very similar to those from Argissa and generally in Thessaly (***Figures 16.9: 1-3; 16.20: 1***).

- Some tunnel lugs at bowls also have precise parallels at Sitagroi Vb (Renfrew et al. 1986: fig. 13.26: 14-20 and especially 15, cp. with ***Figure 16.9: 4-5***).

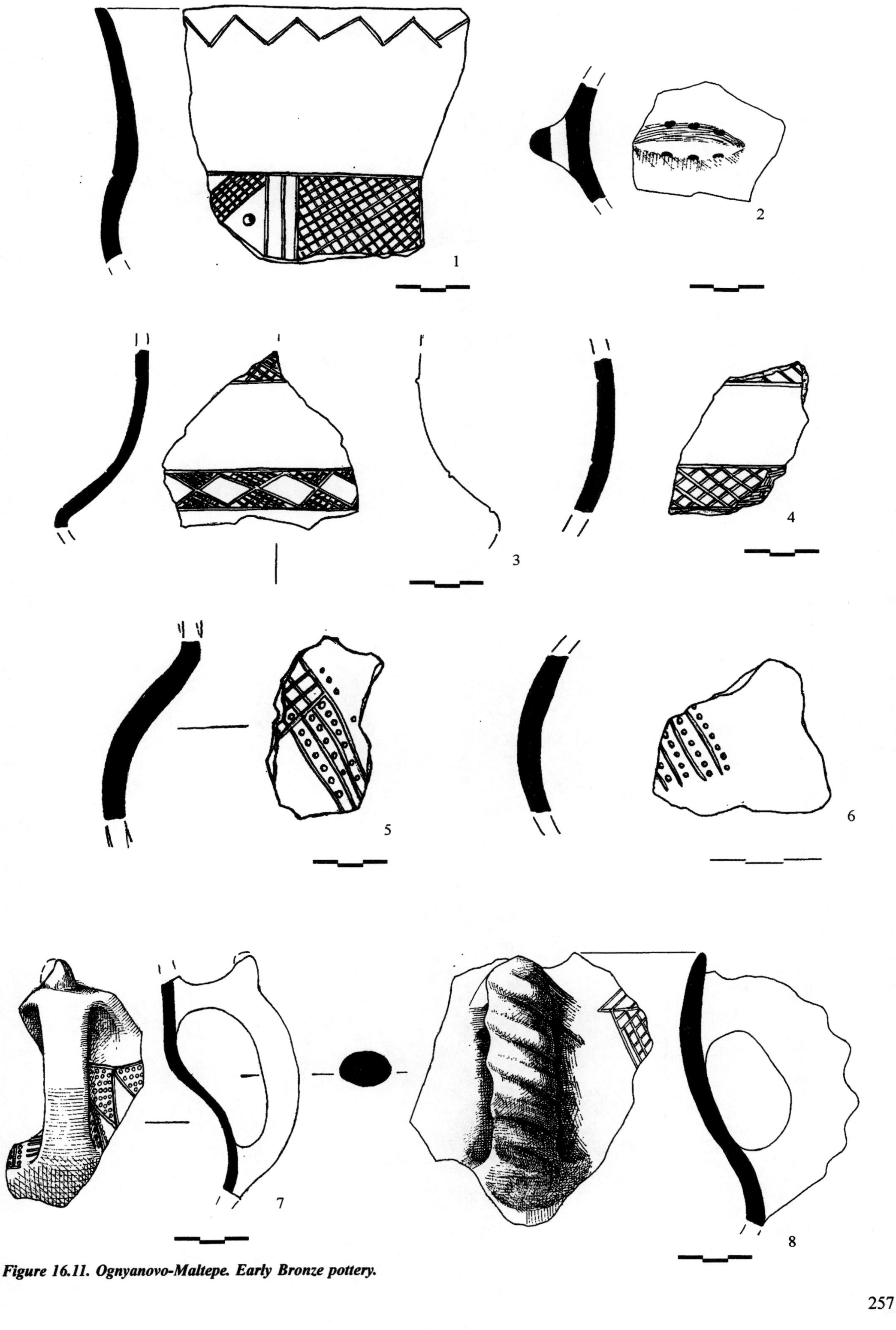

Figure 16.11. Ognyanovo-Maltepe. Early Bronze pottery.

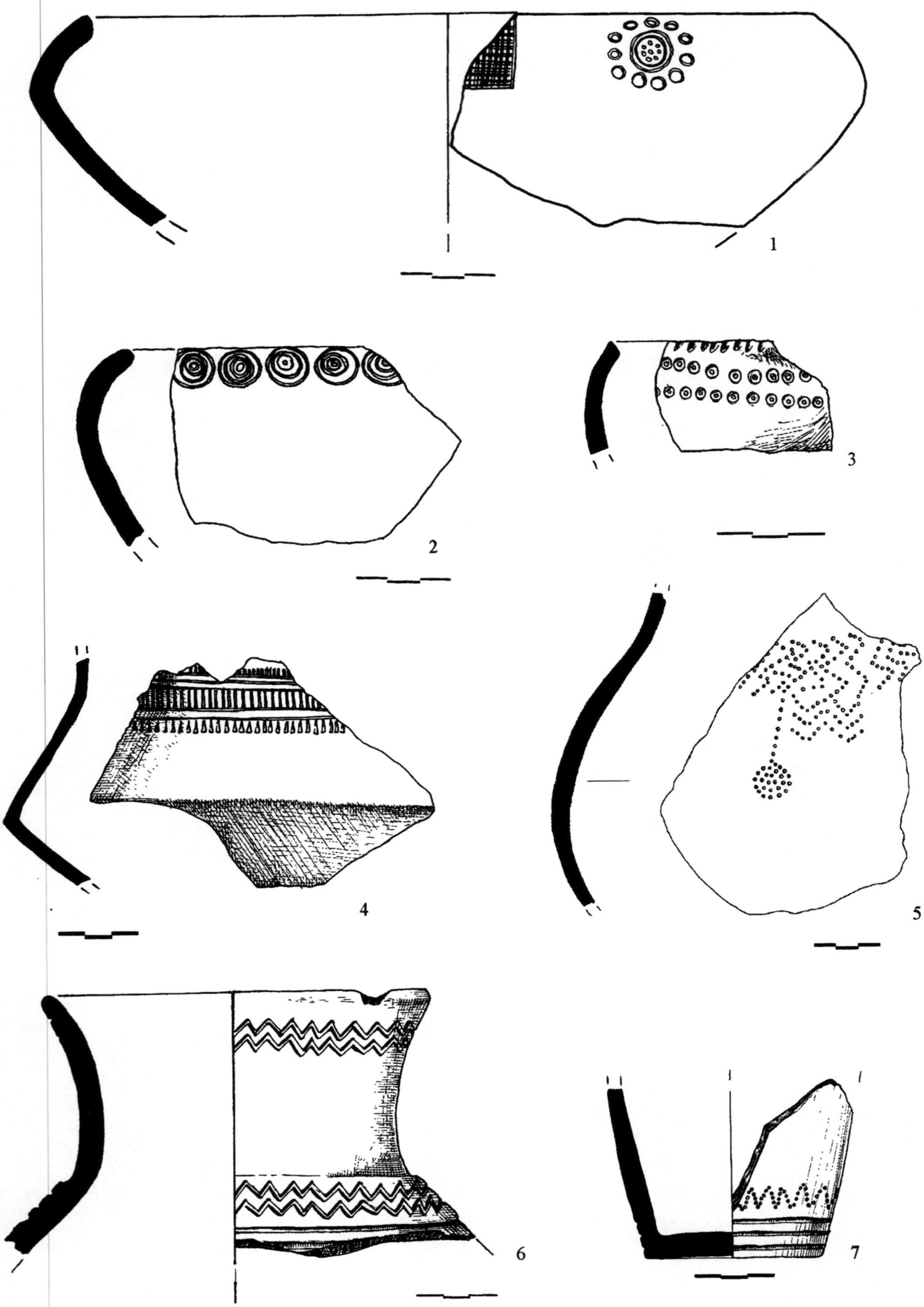

Figure 16.12. Ognyanovo-Maltepe. Early Bronze pottery.

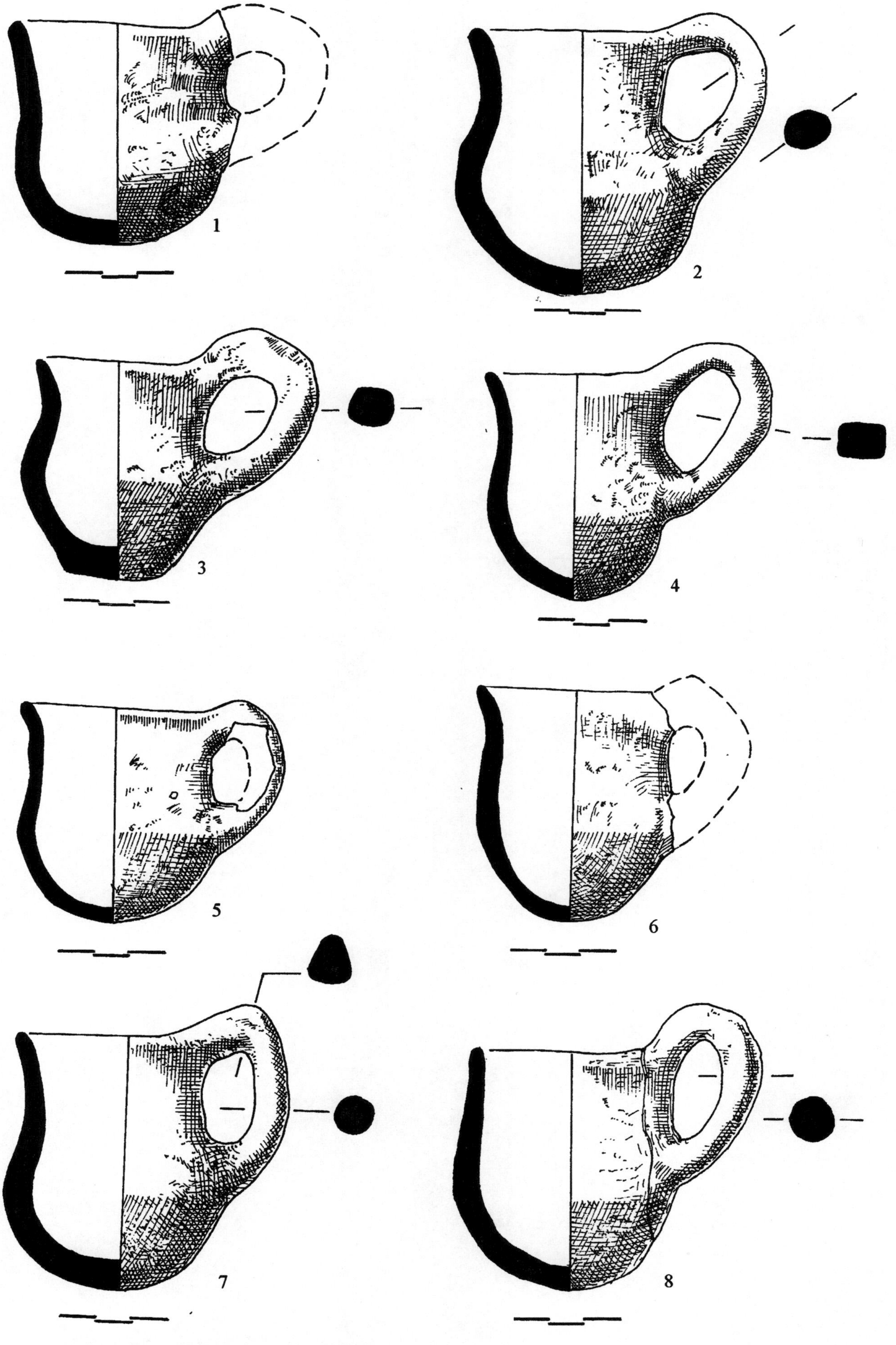

Figure 16.13. Ognyanovo-Maltepe. Early Bronze pottery.

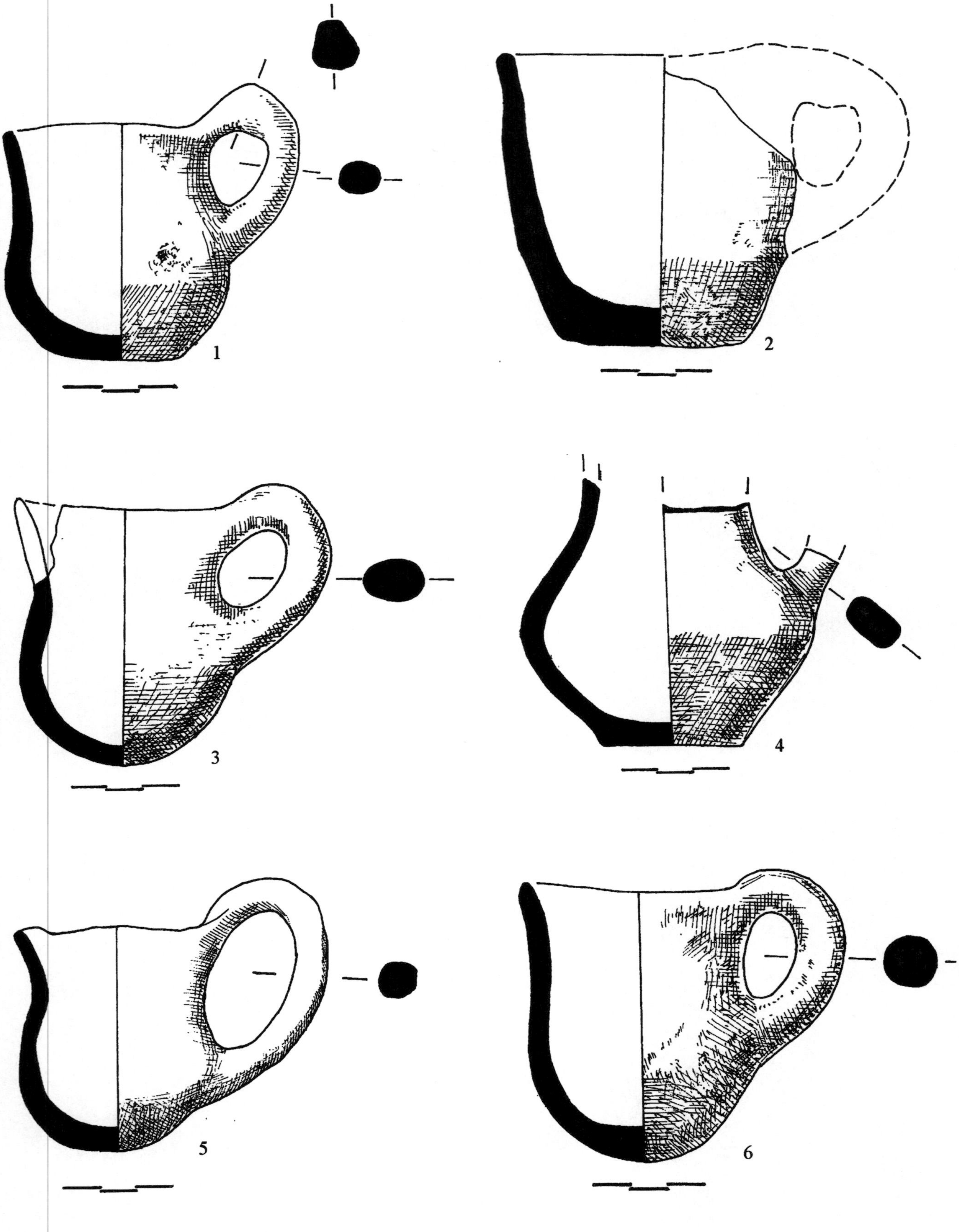

Figure 16.14. Ognyanovo-Maltepe. Early Bronze pottery.

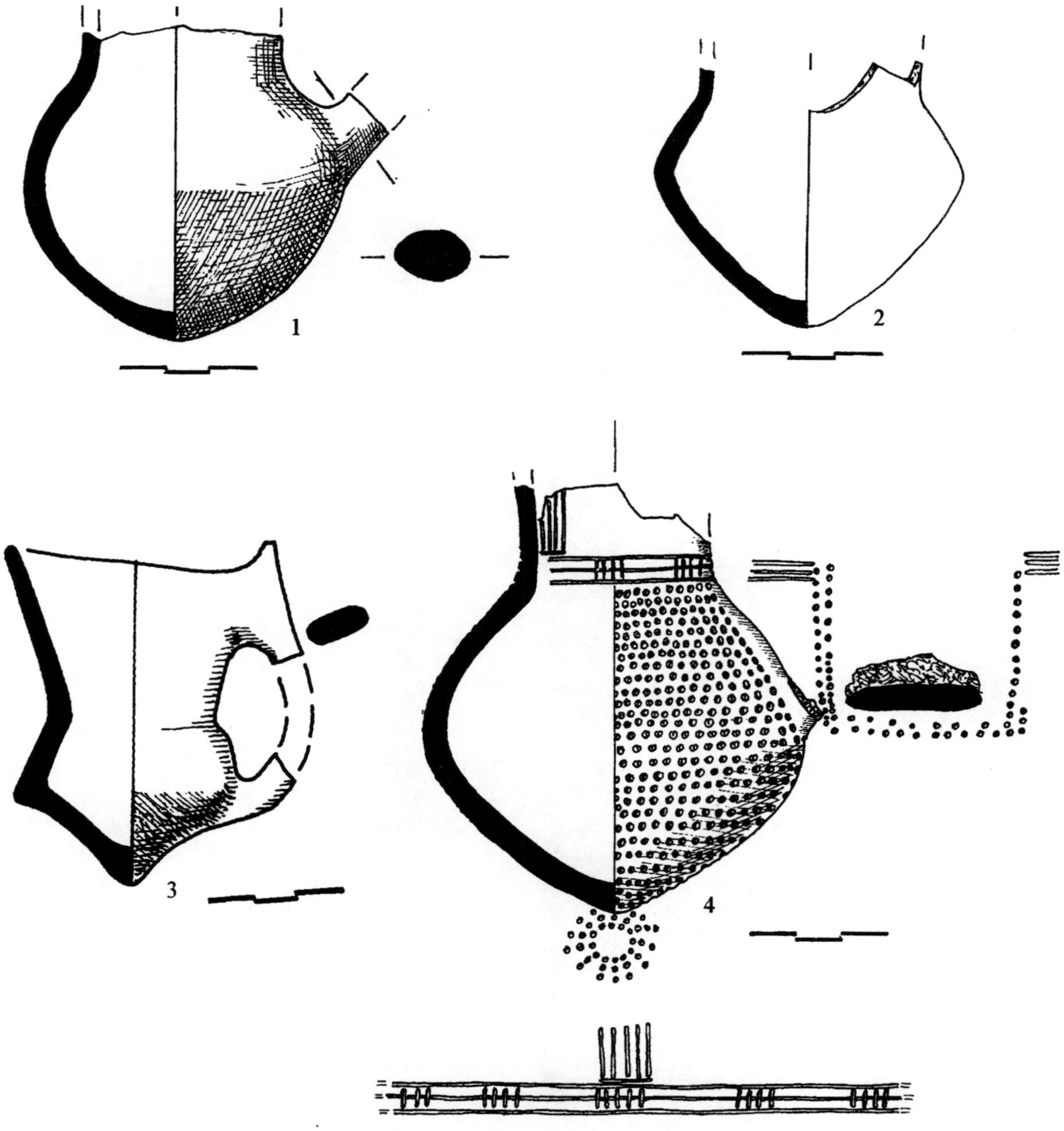

Figure 16.15. Ognyanovo-Maltepe. Early Bronze pottery.

- The peculiar pricked decoration of pricked triangles at Sitagroi Vb (Renfrew et al. 1986: fig. 13.23: 8) and at Ognyanovo is one and the same.

- One of both intact askoi at Ognyanovo (***Figure 16.7: 1***, Cat. No. 29, Inv. No. A 2191, the third building level) belongs to a shape, which was popular in Southeast Europe. The closest analogies exist from Manole-Razkopanitsa (Detev 1971: fig. 1), as well as at Argissa III (Hanschmann and Milojčić 1976: fig. 27: 3). The significance of that vessel for the synchronization between the materials from Ognyanovo, as well as generally of western Upper Thrace is high (Nikolov 1983: 17-18). On its side, Argissa III has well documented contacts with Central and South Greece in Early Helladic III (early Lerna IV) moreover by that askos (Warren and Hankey 1989: 44). In this way it is possible a synchronization between later pottery from Ognyanovo and Early Tessalian III to be found.

Discussion

As the comparative analysis has shown, the main parallels employed originate from Sitagroi. It may look somebody strange that the whole Rhodopes have been crossed in this case, but there is a real background for such a procedure. At that place the extremely complicated and even non-formulated problems of Trans-Rhodope roads have been entered. Belo I will mark the main questions only, which occur by the comparative analysis of the culture in Late Prehistory from both side of the mountains.

It was already G. Il. Georgiev who promoted the idea that the north coast of the Aegean and Upper Thrace was one cultural unit, but he did not give any arguments confirming that thesis (1972b). Later, similar problems occur with regard to the Paradimi phenomenon and the origin of Vinca-Vesselinovo I. If importance in this case is not the origin of the phenomenon

Figure 16.16. Ognyanovo-Maltepe. Early Bronze pottery.

– Anatolia, Thrace, the Aegean or other place, but the fact that the pottery from both sides of the Rhodopes in the time of the Middle and Late Neolithic are amazingly similar. Later, the materials from Dikili Tash and Sitagroi confirmed those observations. The Dikili Tash – Slatino culture even was formulated for the Early Chalkolithic, but it developed in the Struma valley. The Early Chalkolithic in the Rhodopes is completely uninvestigated, so it is unclear if the mountains regions and the Mesta valley were not included in that region as well.

There were not comparative analyses elaborated for Late Chalkolithic, so I will use some instances from the Ph.D. thesis of N. Todorova (2000, forthcoming). The best-investigated Late Chalkolithic site in western Upper Thrace is Yunatsite.

According to N. Todorova, most numerous and precise parallels of the pottery from Yunatsite originate from Sitagroi and Dikili Tash in Drama valley. At the Chalkolithic levels of Sitagroi III there are no certain closed complex documented, but parts of walls, hearths and pits only (Renfrew et al. 1986: 180-81; Andreou et al. 1996: 589). The conical shallow bowls with straight walls and twisted on the inner wall of worn thin rim have parallels (Evans 1986: figs. 12.13: 1; 12.12: 1,4). Similar bowls with graphite ornamentation as those from Yunatsite and Sitagroi have been documented among the pottery from Dikili Tash (Séfériadès 1983: figs. 26-28). The conical bowls with straight walls, vertical upper part and differently shaped rim are identical at both sites (cp. Evans 1986: fig.12.2: 1,3 with ***Figure 16.22: 1***). The same shapes are known from Dikili Tash (French 1961: fig. 9: 5). The biconical bowls with thickened middle point of the body and channels increase the parallels of the bowls from the House 1 at Late Chalkolithic Yunatsite (cp. Evans1986: figs. 12.11: 3; 12.12: 3 with ***Figure 16.22: 2-3***; for Dikili Tash see parallels in French 1961: fig. 13: 4, 5). The list of the analogies can be continued mentioning the following shapes:

- Bowls with spherical and conical lower part and upper part of the body shaped as a truncated cone (cp. Evans 1986: fig. 12.2: 5 with ***Figure 16.22: 4***; Evans 1986: fig.12.10: 5 with ***Figure 16.22: 5-6***). It should be stressed that the bowl from Sitagroi in fig. 12.10: 5 has not only identical shape with a bowl from Yunatsite – House 12, but also an identical ornamentation, as well as there is a similar ornament over a bowl from the first horizon.
- High biconical cups with two handles (cp. the profiles in Evans 1986: fig.12.5: 2 with ***Figure 16.23.1***). The motifs of the graphite ornamentation and the existence of impression roundly shaped, at the base of the handle repeat in details the known from Yunatsite (cp. Evans 1986: fig.12.2: 7 with ***Figure 16.23: 2***).
- The formal characteristics of amphorae-like vessel from Sitagroi is identical with a pot from the first building level at Yunatsite (cp. Evans 1986: fig.12.14: 3 with ***Figure 16.23: 4***). Morphologically, it is similar the richly decorated amphorae-like vessel with a spout (Evans 1986: fig.12.5: 5). Amphorae-like pots also characterize Dikili Tash (Séfériadès 1983: fig. 37).
- Low lids with semi-spheroid shape. Both examples – from Yunatsite and Sitagroi are covered with graphite ornament (cp. Evans 1986: fig. 12.7: 2 with ***Figure 16.23: 7***).
- The closest parallels of the asymmetric vessels from Yunatsite originate from Sitagroi (cp. Evans 1986: fig.12.5: 1; 12.9: 3 with ***Figure 16.23.5***). They have identical profiles and similar decorations.
- There are also parallels among the common material: a fragment of graphite-ornamented pot with prolonged spherical shape and two handles attached at the widest part, from Sitagroi, has exact parallel with a vessel from House 1 at Yunatsite (cp. Evans 1986: fig.12.6:1 with ***Figure 16.23: 6***).

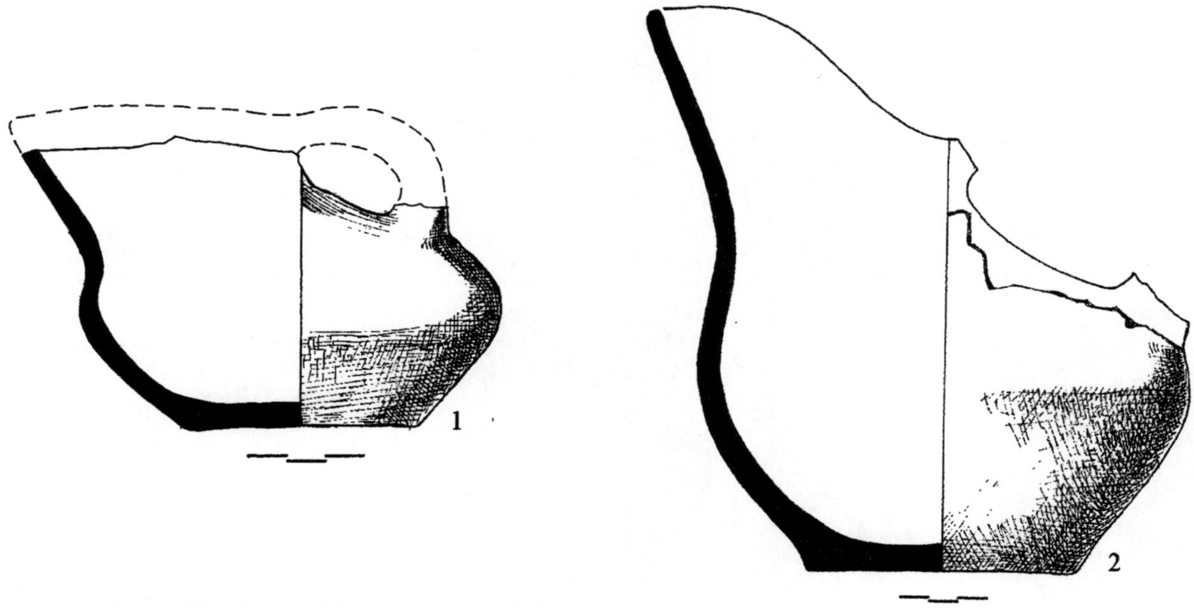

Figure 16.17. Ognyanovo-Maltepe. Early Bronze pottery.

Along with the morphological parallels represented, the pottery from Yunatsite and Sitagroi are very close with regard to the motifs and ornamental schemes used. By the graphite ornament, all motifs, shown as characteristic of Sitagroi, have been documented among the pottery at Yunatsite (Evans 1986: fig.12.1a). The simple linear motifs, wide open spirals etc. are characteristic of the graphite decoration of the vessels at Dikili Tash IIb, Paradeisos (Klisi Tepe) etc. (French 1961: 106-108; Séfériadès 1983:653).

The similarity between both complexes includes the technological characteristics of the vessels (Evans 1986: 393, 397; see chapter 27). Fragments of vessels painted with graphite and defined as similar to those from Sitagroi and Dikili Tash are registered at sites like Dimitra and Statmos Angistas as well (Andreou et al. 1996:589).

All these data show that since the Middle Neolithic there were probably regular contacts through the Rhodopes. One precondition is the fact that the roads through the mountains were passable during the bigger part of the year and they can be passed for less than two weeks. It should be not forgot that despite not so intensively as Thrace and the Aegean, the Rhodopes were also occupied in prehistory. Perhaps the local population initiated some of the contacts for food supply and exchange of resources or animal products. It is even possible more of less permenant directions of contacts' movement to be localized, and their models to be argued. But that question will be elaborated elsewhere. I would like to finish here only stressing that the direct correlation of the culture from western Upper Thrace and that from the Aegean coast of Greece is a perspective direction of investigations. Even more, in my opinion, the attempts will be more fruitful that the attempts of establishment of net dating through Sofia valley and the Struma valley, or to the east – along the Maritsa valley to the Aegean region.

Catalogue

In the catalogue are to be included intact and some fragmented vessels.

The technological characteristics of the pottery include the following data summarized into four groups:

1. Fine paste. The paste is mixed with moderately fine quartz sand of high concentration, as well as with fine husky mica of moderate concentration and a bit organic mixture.

2. Moderately fine paste. The clay paste is mixed with quartz sand of moderate or high concentration, fine mica of moderate concentration and organic matter of moderate to small concentration.

3. Moderate coarse paste. The clay paste is mixed with quartz rock in medium size and of low concentration) and quartz sand (medium coarse, of high and moderate concentration), medium in size organic fractions, of low to moderate concentration; mica of medium concentration.

4. Coarse paste. The clay is mixed with coarse quartz sand of high concentration, comparatively small quartz rock of small concentration and organic matter (straw and grain of high concentration). A black biscuit with two thin brown strips aside.

All sizes of the vessels are given in cm. Abbreviations: H. - height; O. - orifice; B. - base; D. - diameter. HMP – Historical Museum in Pazardzhik Town.

1. ***A plate***. Semispherical body shape. High pedestal, with conical shape. An intact find. Black color surface, with beige spots around the base. Black biscuit. Thick slip. Uneven, well-smoothed surface.

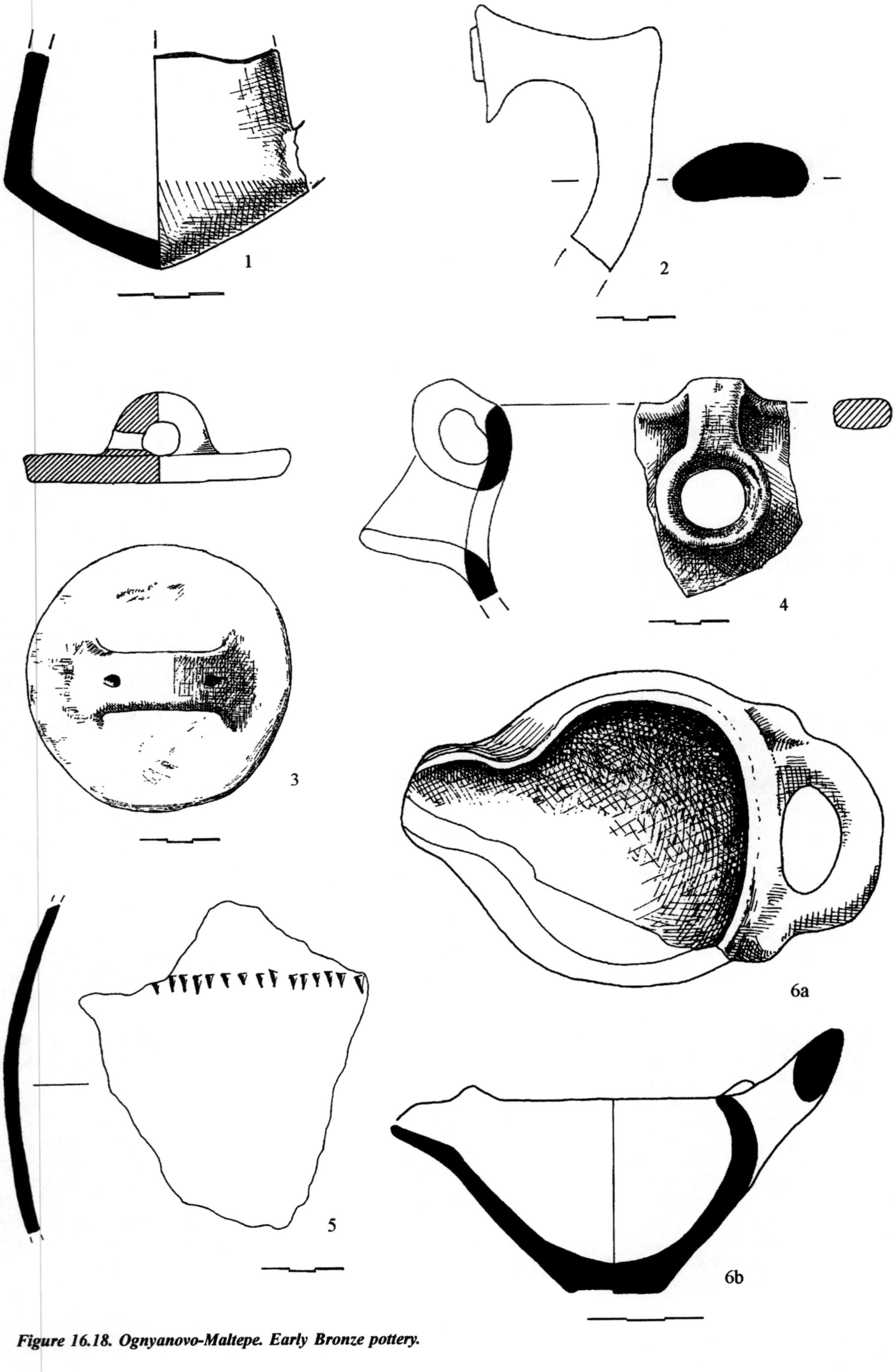

Figure 16.18. Ognyanovo-Maltepe. Early Bronze pottery.

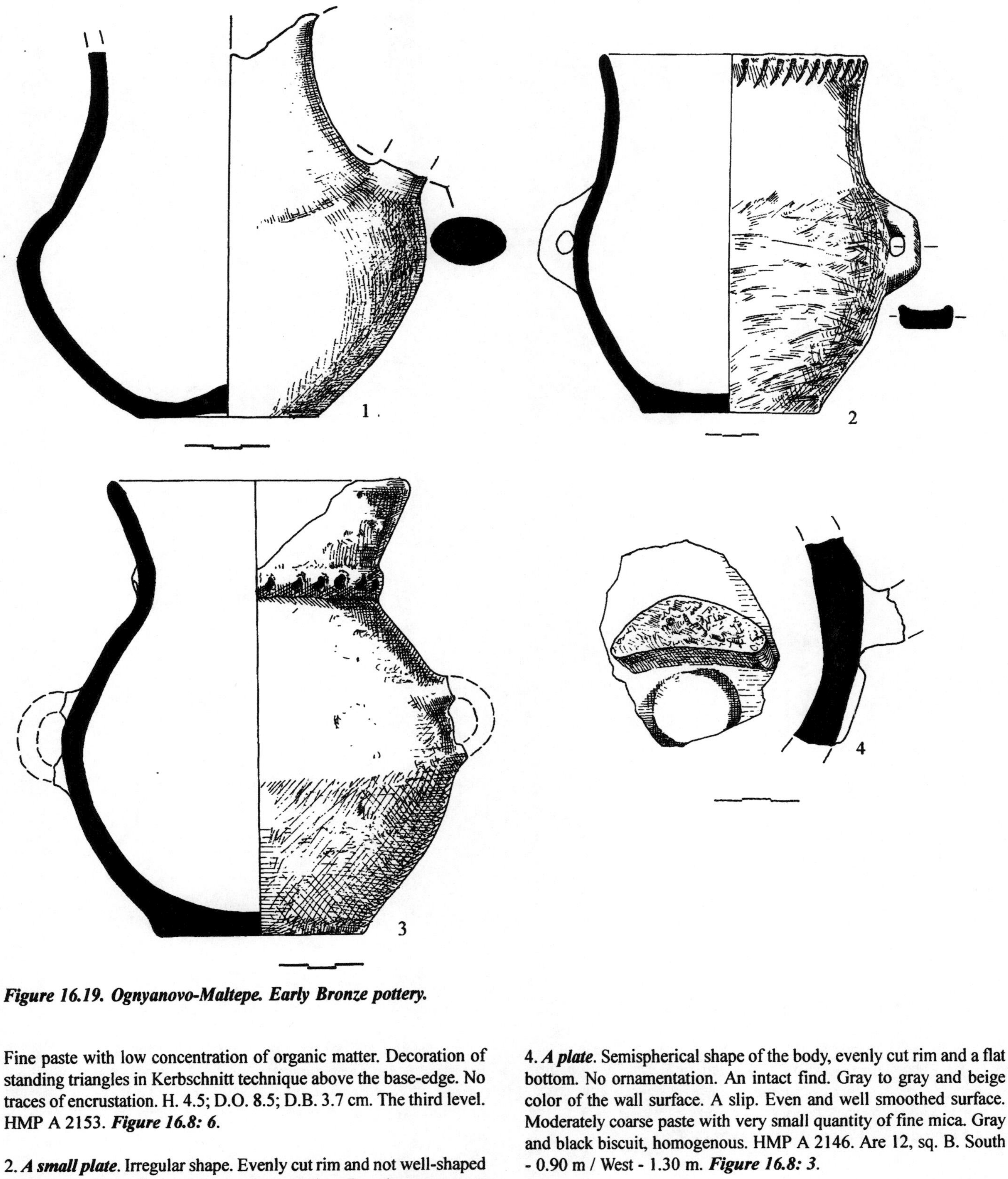

Figure 16.19. Ognyanovo-Maltepe. Early Bronze pottery.

Fine paste with low concentration of organic matter. Decoration of standing triangles in Kerbschnitt technique above the base-edge. No traces of encrustation. H. 4.5; D.O. 8.5; D.B. 3.7 cm. The third level. HMP A 2153. ***Figure 16.8: 6.***

2. ***A small plate.*** Irregular shape. Evenly cut rim and not well-shaped bottom. An intact find, without ornamentation. Rough uneven surface with pores. No slip. Gray and brown color of the wall surface, with gray and light brown spots. Black biscuit with two thin brown strips aside. H. 3; D.O. 6. No stratigraphic data. Historical Museum Pazardzhik, Inv. No. A 2154. ***Figure 16.8: 2.***

3. ***A plate.*** Conical body shape, evenly cut rim and flat bottom. No ornamentation. An intact find. Light gray surface color with light brown spots. No slip. Equal, carefully smoothed wall surface. Moderately fine paste. Light brown biscuit with black heart. H. 4.5; D.O. 10.5; D.B. 5.4. HMP A 2149. Are 13, sq. B. North - 0.30 m / West - 8.90 m. ***Figure 16.8: 1.***

4. ***A plate.*** Semispherical shape of the body, evenly cut rim and a flat bottom. No ornamentation. An intact find. Gray to gray and beige color of the wall surface. A slip. Even and well smoothed surface. Moderately coarse paste with very small quantity of fine mica. Gray and black biscuit, homogenous. HMP A 2146. Are 12, sq. B. South - 0.90 m / West - 1.30 m. ***Figure 16.8: 3.***

5. ***A fragment of tee pot.*** Evenly cut rim, perhaps the body was inflated close to spherical. Brown surface, with dark gray spots. No slip. Carefully smoothed even wall surface. Moderately coarse paste with small quantity of mica. Gray and black biscuit, homogenous. HMP A 2147. Are 11, sq. B. Noth - 2.70 m / South - 0.80 m. ***Figure 16.18: 4.***

6. ***A jug.*** Asymmetrically flattened body with high neck. The rim and the handle are missing. The bottom is small, concave. No orna-

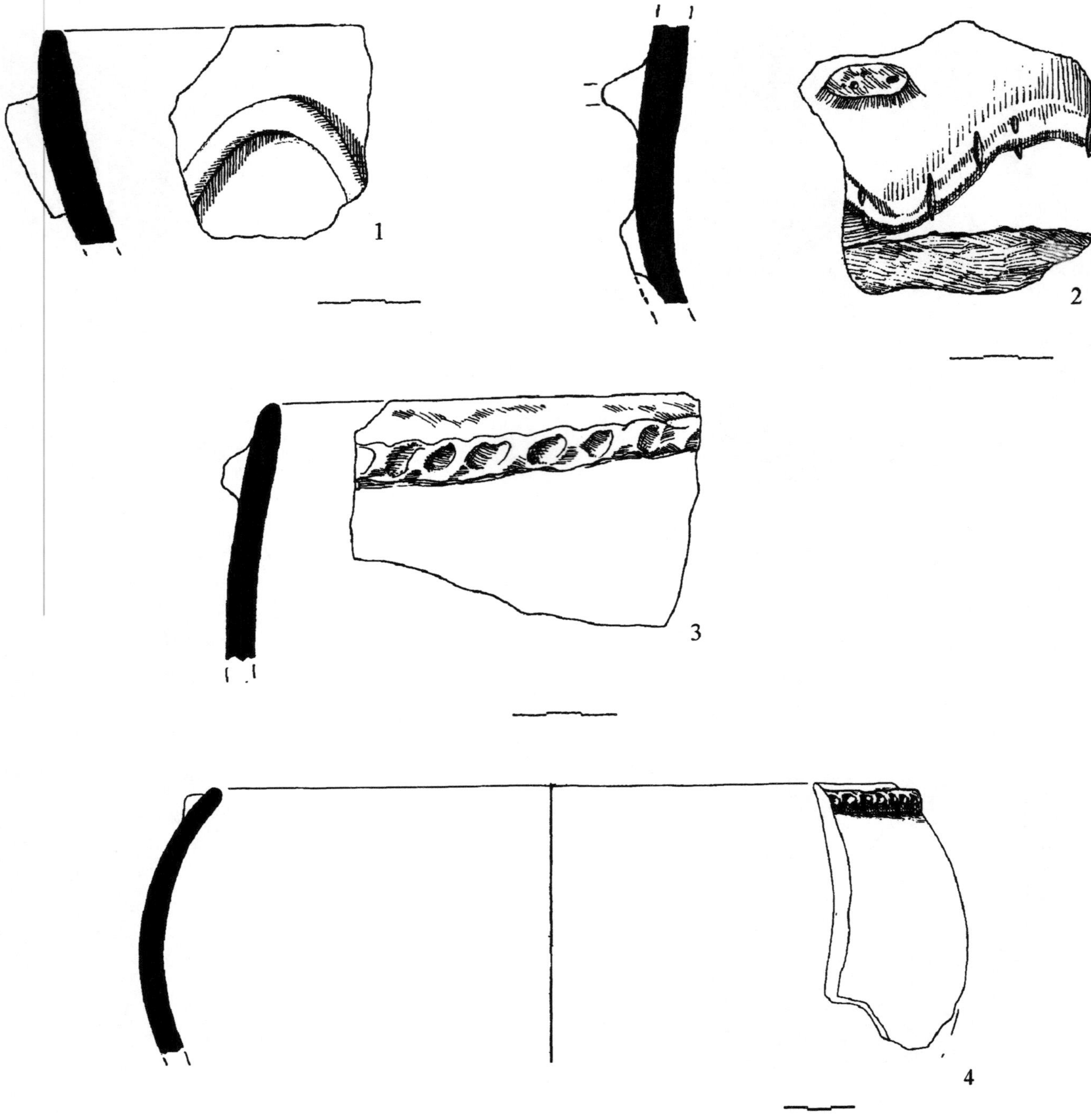

Figure 16.20. Ognyanovo-Maltepe. Early Bronze pottery.

mentation (?). Fragmented. Dark brown wall surface with gray spots. Very thin slip, almost washing only. Even surface with burnished strips made of flat narrow implement. Moderately coarse paste. Brown biscuit with black heart. Preserved sizes: H. 14.7, D. of the body 14.7, D.B. 6. HMP A 2163. ***Figure 16.19: 1.***

7. ***A jug***. Inflated body, almost biconical with smooth transition point. The bottom is flat, with stressed circle edge. The neck is broken at its upper part, and the handle is missing. No ornamentation (?). Fragmented. Gray and black wall surface, homogenous. A slip coated by brush after attaching the handle as there is a space under it uncoated. Even, burnished to brilliancy wall surface. Moderately fine paste. Gray and black even biscuit. Preserved sizes: H. 7.3; D.B. 3.8. HMP. ***Figure 16.14: 4.***

8. ***A cup of Yunatsite type***. Biconical body with sharp edge and shorted lower part, high conical neck and oblique orifice. The bottom is pointed. There is a vertical arch-shape handle with a flat bulge near the orifice, now fragmented. No ornamentation. Brown and black wall surface, homogenous. Thin slip. Even and smoothed surface. Moderately fine paste. H. 8; D.O. 5.5. HMP A 2161. Depth – 0.50 m. Are 1, sq. B. North - 0.50 m / West - 1 m. HMP A 2161. ***Figure 16.15: 3.***

9. ***A cup of Yunatsite type***. Biconical shaped body with a sharp transition. The upper part is slightly inflated body, and the lower one with slightly concave walls. Both body parts are with equal size. High turned down conical neck with evenly cut rim. There is a vertical arch-shaped handle starting from the orifice and reaching the widest body part. No ornamentation. In intact find, with an exception of the broken pointed part of the bottom. Black wall surface with big light beige spots. Dark slip. Evenly burnished in horizontal lines surface. Moderately fine paste. Gray and black biscuit, homogenous. Preserved H. 9.7; D.O. 8; D.B. 7.8. HMP A 2182. ***Figure 16.16: 1.***

10. ***A cup of Yunatsite type***. Flattened and spherical body with smoothed transition, narrow cylindrical neck, pointed bottom. There was a vertical handle with flat section, broken. It is missing the upper part of the neck and the orifice. A decoration of pricked dots over almost whole body with an exception of the space near the handle, incised ornament over the transition from the neck towards the body and at the front part of the neck. No traces of encrustation. Brown wall surface with black spots. Thin slip burnished brilliantly. Fine paste. Brown to black biscuit, homogenous. Preserved H. 9.5. HMP A 2159. ***Figure 16.15: 4***.

11. ***A cup of Yunatsite type***. Biconical body with flattened lower part, sharp transition and pointed bottom. There was a handle, broken. The upper part of the vessel is missing as well. Brown with brown and black spots wall surface. No slip. Even, carefully smoothed surface. Moderately coarse paste. Gary and brown biscuit, homogenous. Preserved sizes: H. 6.2; D. of the body 8. HMP A 2157. ***Figure 16.18: 1***.

12. ***A transition Yunatsite type***. Almost spherical body, cylindrical neck and pointed bottom. There was a vertical handle with oval section. The neck and the handle are broken. No ornamentation. Black and gray wall surface, homogenous. Thin slip. Evenly burnished wall surface. Moderately fine paste. Gray biscuit, homogenous. Preserved sizes: H. 7; D. of the body 7. HMP A 2162. ***Figure 16.15: 1***.

13. ***A cup of western variant of Sveti Kirilovo type***. Prolonged body, almost unshaped body, slightly oblique orifice. Round bottom. There is a vertical high handle with round cross-section, which starts from the rim, raising slightly above and finishes at the widest part of the body attached. Brown with brown and gray spots wall surface. No slip. Even, carefully smoothed wall surface. Moderately coarse paste. H. 7.5. H. with the handle 8; D.O. 6. HMP A 2158. ***Figure 16.14: 6***.

14. ***A cup of western variant of Sveti Kirilovo type***. Prolonged body

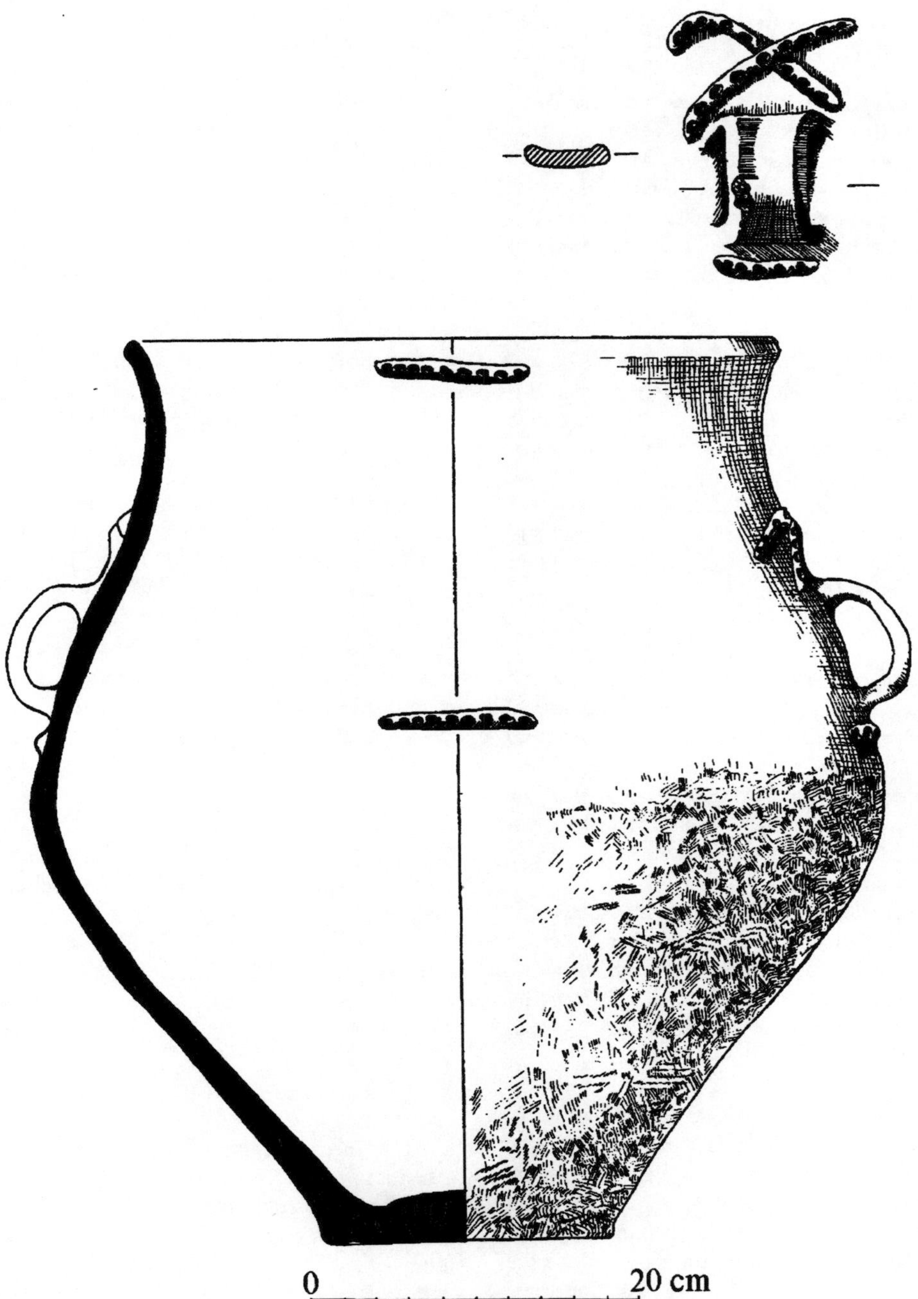

Figure 16.21. Ognyanovo-Maltepe. Early Bronze pottery.

with moderately high neck. The rim is probably slightly oblique, broken against the handle. Rounded bottom. There is a handle with oval section, starting from the orifice and slightly raising above it, as well as finishing attached at the widest body part. No ornament. The front part of the orifice is restored. Uneven wall surface color, from brown and red by the rim to dark brown with black spots under the handle. No slip. Even and smooth surface. Moderately coarse paste. H. 7.3 (?); H. with the handle 7.9 (?); D.O. 5.7 (?). HMP A 2186. ***Figure 16.14: 3.***

15. ***A cup of western variant of Sveti Kirilovo type.*** Prolong body, short widened in the upper part neck, evenly cut rim. The bottom is rounded up to sharp pointed. There is a vertical handle, starting from the rim of the orifice raising up and finishing attached at the widest part of the body. No ornament. An intact find. Brown wall surface, unevenly colored. No slip. Unevenly smoothed surface. Moderately coarse to coarse paste mixed with coarse quartz rock of moderate concentration, coarse quartz sand of moderate concentration and a bit mica. H. 6.3; H. with the handle 7.3; D.O. 5. HMP A 2187. ***Figure 16.13: 4.***

16. ***A cup of western variant of Sveti Kirilovo type.*** Prolonged body, short neck and evenly cut orifice, rounded bottom. There is a vertical handle with round section, which starts from the rim and raises above it finishing attached at the widest part of the body. No ornament. An intact find. Brown and gray wall surface, up to gray towards the handle. No slip. Comparatively even, coarsely smoothed surface. Medium coarse paste. H. 6.6; H. with the handle 8; D.O. 5. HMP A 2188. ***Figure 16.13: 8.***

17. ***A cup of western variant of Sveti Kirilovo type.*** A small inflated body, high neck with slightly inclined orifice. The bottom is leveled with a smooth transition towards walls. There is a vertical handle starting from the rim, raising up and finishing attached to the widest part of the body. No ornament. An intact find. Dark brown to black wall surface, with brown spots. No slip. Uneven, partially smoothed surface. Coarse paste. H. 6.6; H. with handle 7.8; D.O. 6; D.B. 3.2. HMP A 2190. ***Figure 16.14: 1.***

18. ***A cup of transition type of Sveti Kirilovo.*** Inflated body and medium high almost cylindrical neck with slightly oblique orifice. The bottom is rounded. There is a vertical handle, which starts from the rim, raises above it and finishes attached at the widest part of the body. No ornament. An intact find. Light brown with dark gray spots towards the handle. No slip. Roughly smoothed surface. Medium coarse paste. H. 6.2; D.O. 5.7. HMP A 2189. ***Figure 16.13: 5.***

19. ***A cup of western variant of Sveti Kirilovo type.*** Prolonged body, the neck is almost undivided, with evenly cut orifice. The bottom is rounded. There is a vertical handle, broken. No ornament. Restored. Gray and brown wall surface with light brown spots. No slip. Roughly smoothed, uneven surface. Medium coarse paste. H. 6.5; D.O. 5. HMP A 2156. ***Figure 16.13: 6.***

20. ***A cup of western variant of Sveti Kirilovo type.*** Slightly inflated body, short neck, evenly cut orifice with turned outside rim, rounded bottom. There is a vertical handle with oval section, which starts from the rim and slightly raises above it, finishing attached at the widest part of the body. No ornament. An intact find. Gray wall surface with light brown handle and light brown spots around the orifice. No slip. Carefully smoothed even surface. Medium fine paste. Dark gray biscuit with two brown strips aside. H. 7; H. with the handle 8; D.O. 5. HMP A 2184. ***Figure 16.13: 7.***

21. ***A cup of western variant of Sveti Kirilovo type.*** Almost spherical body, short slightly widened towards the orifice neck, evenly cut rim and rounded bottom. There is a vertical handle with oval section, which starts from the rim and slightly raises above it, finishing attached at the widest part of the body. No ornament. An intact find. Gray and brown wall surface, with a big black spot towards the handle. No slip. Uneven surface, well smoothed in the upper part of the body. Medium course paste. H. 7.4; H. with the handle 8; D.O. 5.6. HMP A 2185. ***Figure 16.14: 5.***

22. ***A cup of western variant of Sveti Kirilovo type.*** A little not well shaped body with smooth transition to the neck which is widened towards the orifice. Rounded bottom. There was a handle, broken. No ornament. Gray and brown color. No slip. Uneven, roughly smoothed surface. Medium coarse paste. H. 6; D.O. 5.5. HMP A 2181. ***Figure 16.13: 1.***

23. ***A cup.*** Inflated with short neck narrower in the lower part, S-profiled. Evenly cut rim. Flattened bottom with smooth transition towards the walls. There is a a vertical handle with oval section, which starts from the rim and slightly raises above it, finishing attached at the widest part of the body. No ornament. An intact find. Gray and black wall surface, unequal. No slip. Uneven, roughly smoothed surface. Coarse paste. H. 6; H. with the handle 7; D.O. 4.8; D.B. 2.7. HMP A 2144. Are 13, sq. D. South - 4.70 m / East - 4.10 m. ***Figure 16.13: 3.***

24. ***A cup.*** A truncated cone shape of the body, evenly cut orifice with slightly turned out rim. Flat bottom with smooth transition towards the walls. There was a handle, broken. No ornament. Restored. Light brown wall surface, with gray spots. Thick slip. Comparatively even, burnished without brilliancy surface, carefully smoothed from the inner side. Moderate fine paste. H. 8; D.O. 7.5 (?); D.B. 5. HMP A 2165. ***Figure 16.14: 2.***

25. ***A pot.*** Almost spherical body with truncated cony neck. A flat bottom with a stressed 'heel'. There are two small vertical handles attached at the widest part of the body, broken at their bases. An ornament of horizontal relief band, with hollows in internals, attached over the transition from the body towards the neck. Restored. Brown and gray color of the surface walls. Coarse paste. Black biscuit, homogenous. The bottoms consists of impressions of coarse sand. H. 16.3; D.O. 10.5; D.B. 7.5. HMP A 2151. Are 11, sq. B. West - 3.60 m / South - 0.30 m. The third building level. ***Figure 16.19: 3.***

26. ***A pot ('urn-like vessel').*** Prolonged body, smooth S-profiled. Flat bottom and evenly cut orifice. Two vertical handles attached at the widest part of the body. A decoration of a row of incisions under the rim of the orifice. An intact find. Brown with dark brown spots wall surface. No slip. The body under the neck is covered with barbotine in wide bands. The neck consists of carefully smoothed surface. H. 20; D.O. 14.5; D.B. 10 cm. HMP A 2160. ***Figure 16.19: 2.***

27. ***Urn (a child grave).*** Prolonged biconical shape with smooth transition, conical neck with turned out rim. Flat bottom and flat rim. There two vertical handles at the base of the neck. Relief ornament of bands with pits in intervals under the rim and in the space around the handles. An intact find. Unevenly brown wall surface with dark brown to gray and brown spots. No slip. Smoothed surface with flat implement (1.5 cm wide working part), carefully in the upper part and roughly over the lower part. Coarse paste. The bottom is flat and roughly smoothed. H. 55.2; D.O. 42; D.B. 17.7. HMP

A 2324. ***Figure 16.21.***

28. ***Askos***. Oval body, short straight up neck, strongly oblique orifice, oval flat bottom. Fragmented. The front part of the vessel is preserved, the handle is broken, as well as the upper part of the vessel. No ornament. Dark brown to brown and black wall surface, smoothly changing into black spots. Burnished even surface. Moderate fine paste. Homogenous biscuit with brown to black color, corresponding to the surface one. H. 19.6; D.O. 11.5; D.B. 9.2. HMP A 2324. ***Figure 16.17: 2.***

29. ***Askos***. Ellipsoidal body, orifice and bottom. There is a handle, broken nowadays, which lower end was attached at the widest part of the body. The upper part of the orifice and the handles are missing. Restored. Beige and gray wall surface, homogenous. Thin slip. Burnished up to brilliancy surface of the outer wall. Fine paste. The bottom is carefully smoothed. H. preserved 7.5. HMP A 2191. ***Figure 16.17: 1.***

30. ***Strainer*** (?). Truncated cone turned down with evenly cut orifice and flat bottom. The walls are perforated. The bottom is with a big hole. No ornament. An intact find. Dark brown to gray wall surface, with gray and black spots. No slip. Coarse, unsmoothed surface. Coarse paste. Brown biscuit, homogenous. H. 7.5; D.O. 11.5; D.B. 7.5. HMP A 2324. A 2143. Are 12, sq. B. West - 4.50 m / South - 0.40 m. ***Figure 16.8: 7.***

31. ***A small pot***. Irregular spherical body with turned out rim and flat bottom. There is one tongue handle, holed. No ornament. An intact find. Gray and brown surface, homogenous. No slip. Smoothed, even surface. H. 3.5; D.O. 2.5; D.B. 2.5. HMP A 2166. The third level. Are 5, Sq. D, a house, near the base of oven. ***Figure 16.8: 8.***

32. ***A lid***. Disc-like shaped, with a arch-shaped handle. The last has two holed at its bases, perhaps for cord. No decoration. Intact find. Brown surface with a big black spot. No slip. Relatively evenly, carefully smoothed, in some places burnished surface. Semi-fine paste. H. with the handle 3.4; D. 10; HMP A 2152. Are 7, sq. B, North - 6.10 m / West 0.50 m. ***Figure 16.18: 3.***

33. ***A cup***. Almost spherical body, relatively evenly cut rim, and one handle, raising a little bit above the rim. Round bottom. No ornamentation. Intact find. Gray and brown surface, with a black spot under the handle. No slip. Carefully smoothed surface from the rim to the widest body part. Semi-coarse paste. H. 7.5; D. rim 5.5; HMP no inventory number. Are 7, sq. B, North - 6.10 m / West 0.50. ***Figure 16.13: 2.***

34. ***Saucier***. Incurved evenly cut rim, and one horizontal handle raising a little bit above the rim. Flat base. Decoration. Restored. Gray and black surface, homogenous. No slip. Carefully smoothed. Semi-fine paste. H. 4.9; D. rim 9 x 7.5; D. B. 2.4; HMP 2180. Are 7, sq. A, level 5 (the second building level?). August 16, 1973. ***Figure 16.18: 6.***

35. ***A cup of "Sv. Kirilovo type"***. Almost spherical body with slightly pointed bottom, asymmetrical neck and one high handle with a relief bulge at its highest point. Incised line at the turning point between the neck and the body. Intact find. Brown color of the surface with black spots at the base of the handle and under the rim. Slip and burnished surface. Semi-fine paste. H. with the handle 12.7. HMP A 2155. ***Figure 16.16: 2.***

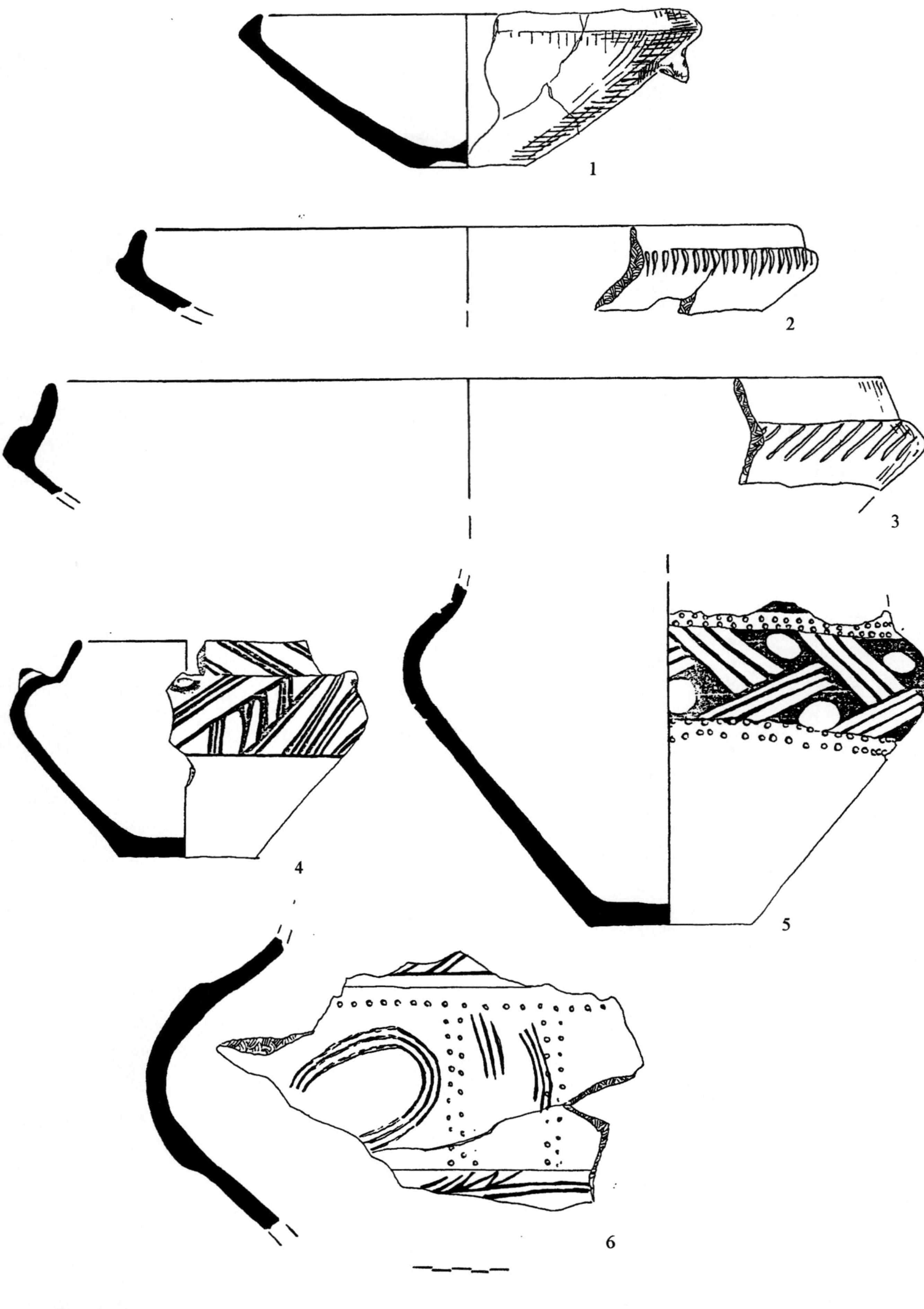

Figure 16.22. Yunatsite tell. Pottery from Late Chalkolithic.

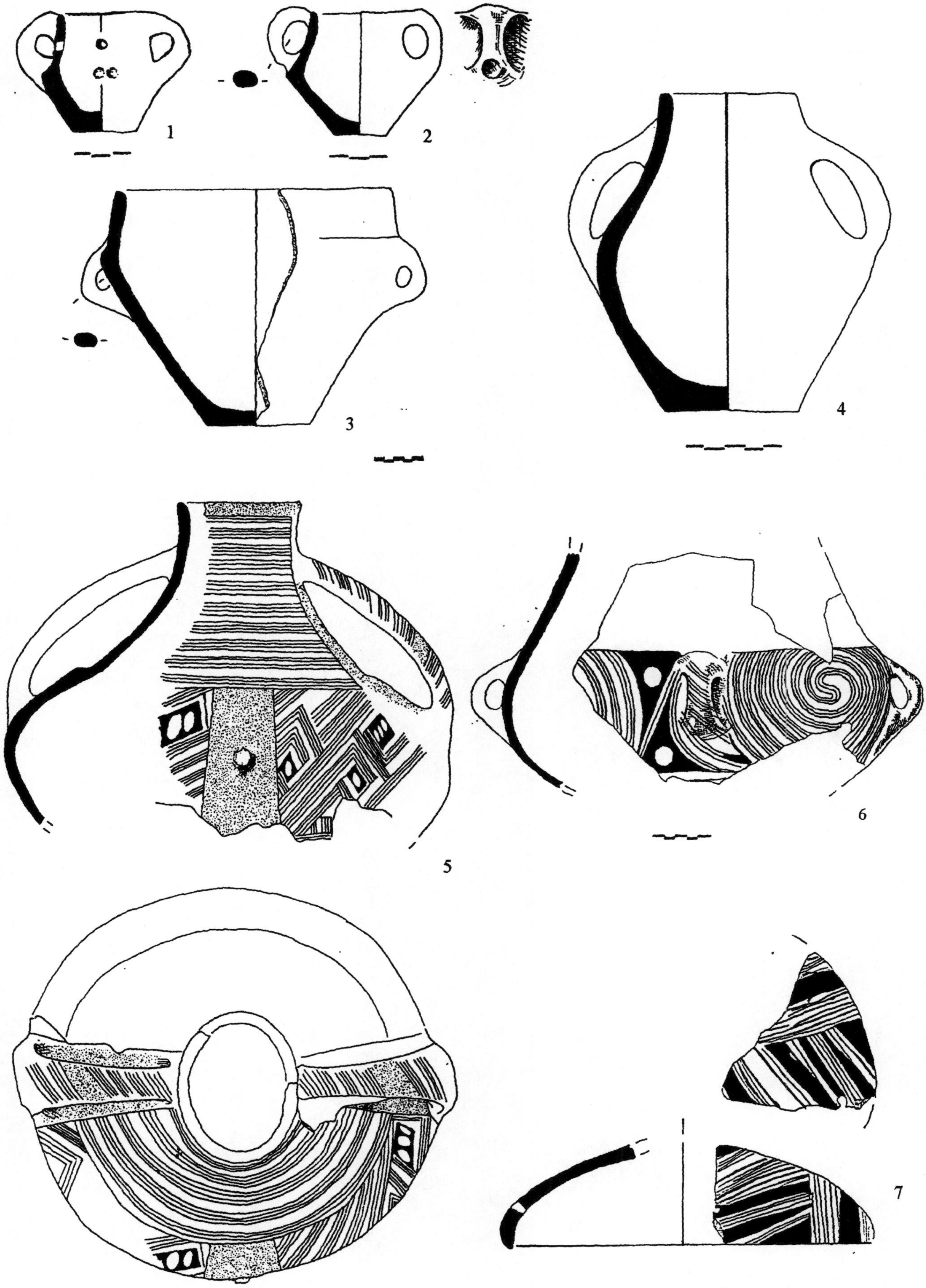

Figure 16.23. Yunatsite tell. Pottery from Late Chalkolithic.

CHAPTER 17

AN EARLY BRONZE SETTLEMENT ON THE WESTERN SLOPE OF NEBET TEPE IN PLOVDIV

Bistra Koleva

Introduction

The hewn into the rock record on the north slope of Nebet Tepe was considered of D. Tsonchev who suggested the Trihulmie hills was that part of the contemporary town milieu where the earliest piling up of Plovdiv Town (Tsonchev 1938). From this moment till the end of 1990s, with short breaks, parallel with investigation of the Roman town, in the lowland, Trihulmie hills have been intensively investigated. Because of different reasons, the attention was on the small sector over the highest part of the three hills (Botusharova 1963; Detev 1963; Pejkov 1995). As an exception, Memlivi Hane site was investigated, at Dzhambaz Tepe, attributed to the Bronze Age. Unfortunately, the results of the excavations are unpublished. In the course of the last 10 years, the intensive building that covers all the territory of the Trihulmie hills give me a unique possibility to investigate an archaeological site up to the continent (***Figure 17.1***). Despite the excavations are over small terrain and have mainly sounding character the results obtained are extremely important marking a new stage of the investigation of Trihulmie hills.

The goal of this communication is to publish a stratigraphic picture obtained in course of the rescue excavations in 1993-1995 in two neighbor sectors of Philipopol Restaurant and T. Hrulev Street at the western slope of Nebet Tepe.

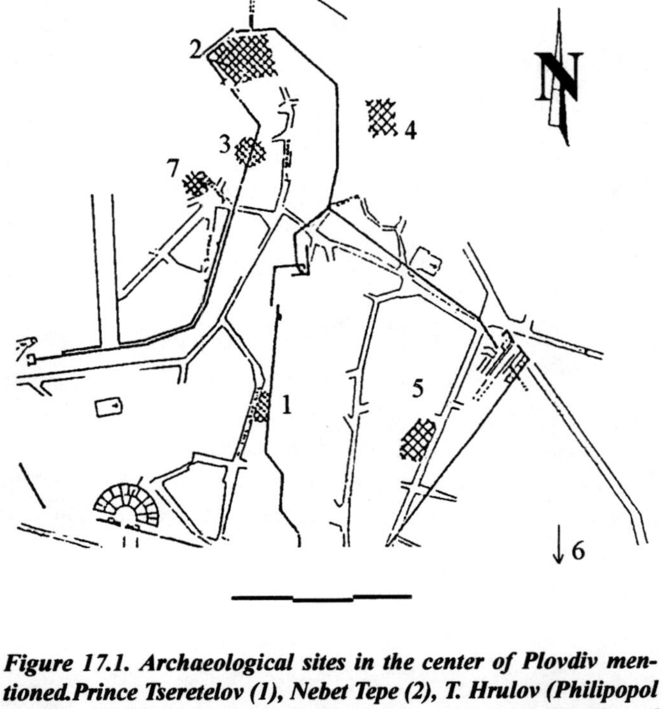

Figure 17.1. Archaeological sites in the center of Plovdiv mentioned.Prince Tseretelov (1), Nebet Tepe (2), T. Hrulov (Philipopol Restaurant [3]), Kamchiya (4), M. Panaret (5), M. Luiza (6) and 4th January (7).

The site is situated on the first terrace of the western slope of Nebet Tepe and in c. 300 m from Nebet Tepe site. The investigations were provoked from erection of two small buildings, that defined the limited size of the trench. The area of the excavations was additionally limited by the grounds of the tholos from the end of the Bronze Age and the beginning of the Iron Age, as well as by the contemporary destruction of the second building sector. Along the eastern borders of both the building sectors passes the wall from Antiquity (***Figure 17.2***).

Stratigraphic observations

In both the soundings the excavations reach the continent. The following structure of the cultural layer was documented

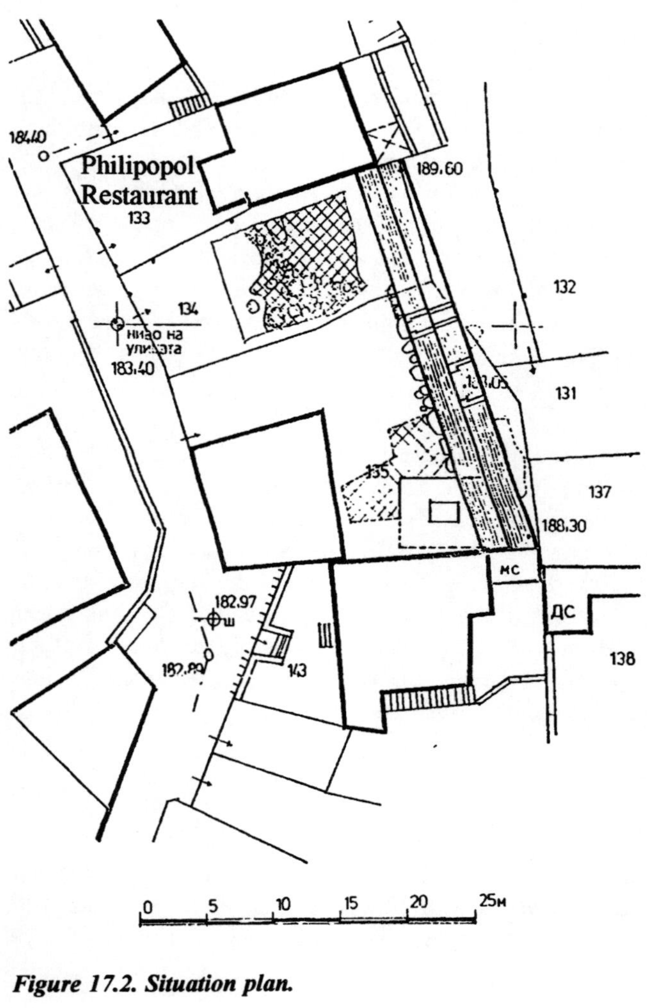

Figure 17.2. Situation plan.

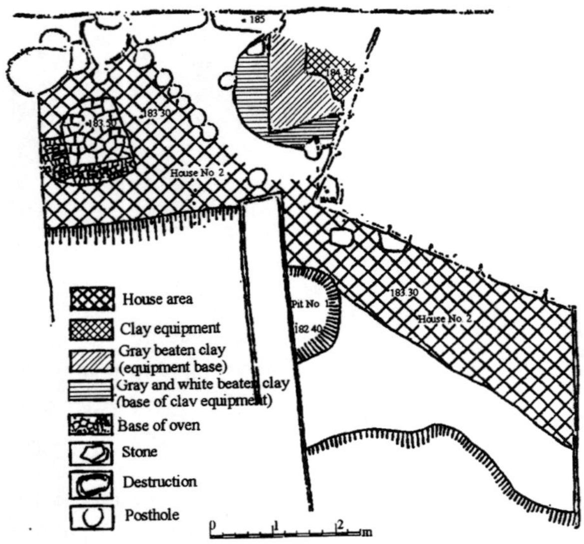

Figure 17.3. The first buiding level. South section.

in this part of the Trihulmie hills:

Layer A. It belongs to Early Iron Age. In the south it is completely destroyed by the grounds of the Late Antiquity tower.

Layer B. It has resinous black color with yellow strata in some places, ashes and charcoal. Five building levels have been distinguished by floor levels and grounds of equipment belonging to the Early Bronze Age.

The first building level

It was investigated in both the sections.

The north section. After excavating of the floor construction of the tholos, the grounds of two ovens have been documented (***Figure 17.3***). Both ovens are rectangular with rounded corners. The oven No. 1 was cut by the grounds of the Late Antiquity wall. In the north of the ovens a clay beaten wall was traced with width of 0.5 cm. Wall of the house or other traces of its construction wasn't been discovered. In the same sector two rows of holes with a diameter of 0.3 m have been excavated that cross radially the ground of the tholos from the upper layer and belong to its carrying construction of its roof.

The south section. To the north of the tower a base of an oven horseshoe shaped was excavated. There is a wide floor before its opening that reaches the middle point of its long walls. The floor and the base of the oven are precisely covered with clay. The floor of the house is beaten clay with gray color. In 0.80 m south of the oven a row of seven holes with diameter of 0.20 m was excavated. There is no other evidence of the house. Beside the oven, over the floor, there was a one-handled cup discovered (***Figure 17.4: 1***).

The second building horizon

The north section. That building level is distinguished by the floors of two partially recovered houses. The floor of the first house is covered with clay, smoothed. From the north and from the east two rows of holes have been discovered. The second floor in completely under the north profile. It should be stressed the compactness of the house building which is similar to that on the tells.

The south section. No equipment was discovered. In the space north and west of the tower it was excavated a small strip of house floor limited from a row of post-holes. Over the floor there was well-preserved cup with one handle discovered.

The third building level

The north section. Over whole area of the trench there was a floor of one house discovered. It is rectangular oriented with its long wall east-west. From the east it is cut by the stronghold wall, and from the west – part of it continues in the unexcavated area. The oven discovered in the east half was destroyed by the stronghold wall as well. It is probably horseshoe shaped, at the opening of which there is a ledge 0.12 m thick. Along its long walls there was hipped shatters.

The south section. In this trench the forehead of the continental rock was recovered. There were traces of treatment. There were no floors and equipment discovered. In west direction the filling with amorphous with resinous color.

The fourth building level

Of this horizon areas of two houses floors have been investigated. The House No. 5 is rectangular oriented with its long walls north-south but its long east wall is destroyed by the stronghold wall. The house is stood out by its massive floor construction. Before heaping of the main construction material – the clay with bright yellow color 0.40 m thick – a grating of wood beams with round section 0.25 m in diameter. The grating consists of three rows of prolong-beans and four across the short side. Two ovens have been discovered over the floor – one almost in the middle, very destroyed, and another with horseshoe shape, in 0.8 m from the north short wall. There were traces of the arch. The second type equipment in that house is a pit for ashes and charcoal. It is dug with a cylindrical shape and diameter of 0.45 m. In front of the pit and the oven with horseshoe shape there were parts of two clay axes discovered.

Of the second house there was only a small section excavated, in the northwest corner of the sounding. Its distinguishing was possible basing on a small fragment of a oven that continued in the north profile.

The fifth building level

The lowest horizon was grounded immidiately over the rock continent. For building of the houses in some places the rock was leveled. The floors were built without additional isolation, in contrast to the house documented from the following horizon. Bases of two very destroyed ovens were documented from the fifth horizon.

Pottery and chronology

The pottery discovered is very limited. Shallow and deep bowls, pots, cups and jugs are the mostly documented ceramics. Some of them are prick decorated or incised and white encrusted. Some of the deep bowls consist of pseudo-tunnel handles. But there is no pointed bottom cups documented.

The round-bottom cup from the second horizon parallels with Nebet Tepe 5 (Pejkov 1995: fig. 9) and Dubene IIC (Nikolova 1996a: fig. 23: 2), Ognyanovo (Leshtakov, chapter 16, ***Figures 16.13, 16.14: 1, 3, 5-6***). The encrusted bowl fragment has analogies at Ezero 12-3 (Georgiev 1979: Table 178; Leshtakov 1993: Figure 1), as well as the models of axes of types 3-11 (Georgiev et al. 1979: Table 229). The parallels allow the settlement to be dated from the later Early Bronze II and earlier Early Bronze III, according to the recent data. The analysis of the stratigraphic situation registered on the western slope of the Early Bronze settlement show a small sector has been excavated of the settlement registered for the first time on the hill (Pejkov 1995). Unfortunately, for the time being, it is not possible its territory and especially its borderlines to be defined more precisely.

Catalogue

1. A cup. Biconical, small bottom and high orifice slightly turned out. A high handle starts from the orifice. Burnished surface with gray and black color. Height – 4.9 cm, Diameter of the orifice – 4.7 cm. South sector. The first building level. Inv. No. 1150. ***Figure 17.4: 1.***

2. A cup. Biconical, very small bottom, with high orifice and slightly turned out rim. High handle started from the rim. Gray and black color, burnished surface. Height – 4.6 cm, diameter of the orifice – 4.4. cm. South sector. The second building level. Inv. No. No. I-1151.***Figure 17.4: 2.***

3. A bowl. Body with truncated cone shape, inverted rim, and ellipsoidal bottom. Two tongue-like handles with vertical edges aside are attached by the orifice. Gray and black color, burnished surface. A small part is preserved. Reconstructed. Height – 9.8 cm; diameter of the orifice – 17.6 / 14.7 cm. North sector. The second building level.***Figure 17.4: 3.***

4. A model of axe, clay. The working part is preserved only. Incised ornament over both sides incisions in vertical rows. Gray and brown color, smooth surface. Length – 5.9 cm; width – 4.9 cm. North sector. The fourth building level. A house.***Figure 17.5: 5.***

5. A model of axe, clay. The working part is preserved only. Trapezoidal shape with small round opening for a handle. There is a row of horizontal cuts over the peripheral edges. Gary and brown color, smooth surface. The shape is identical with the vast distributed stone axes and belongs to type 3 of the clay models of axes from Ezero. Length – 5.1 cm; width – 3.6 cm. North sector. The fourth building level. A house.***Figure 17.5: 6.***

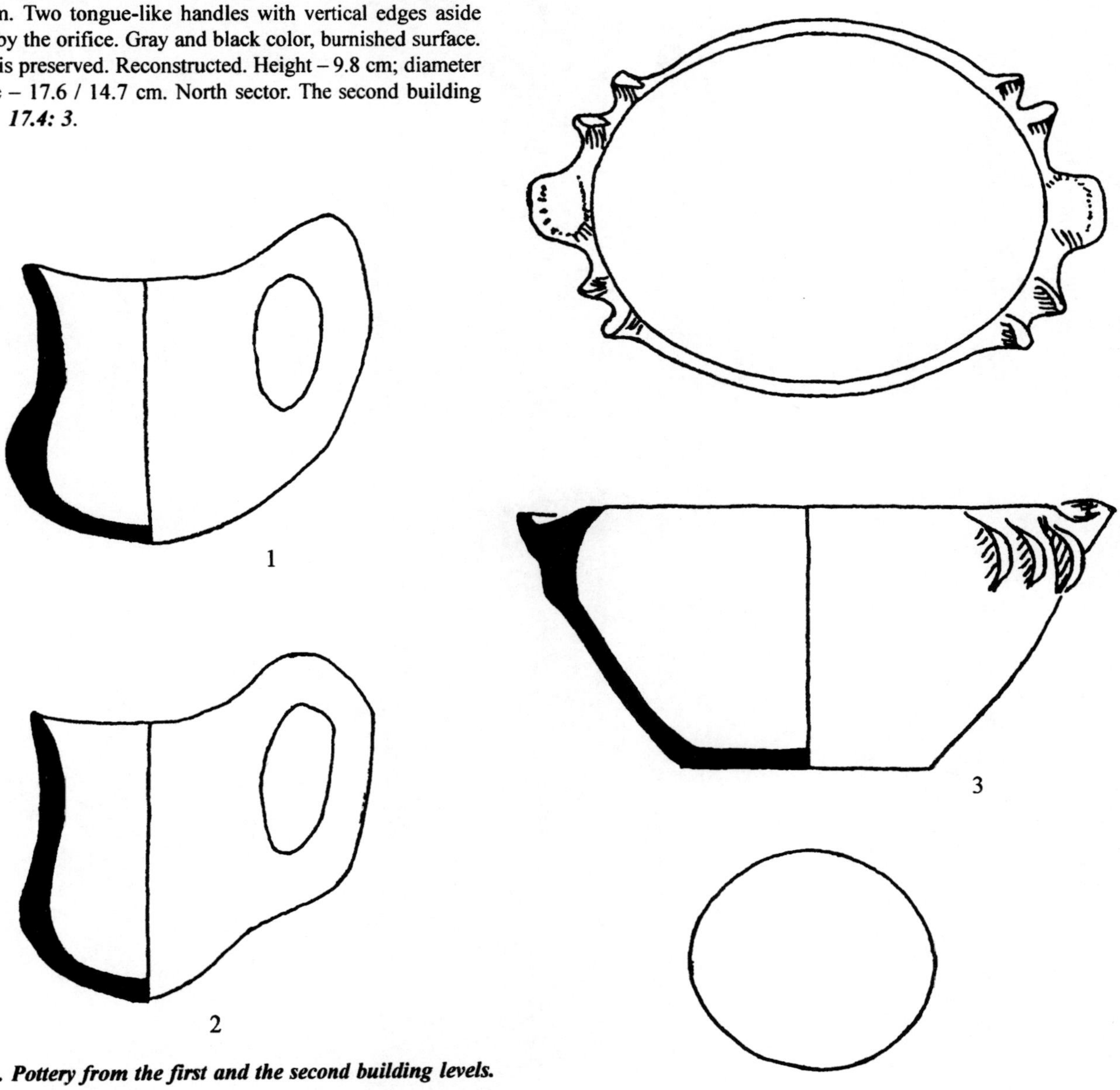

Figure 17.4. Pottery from the first and the second building levels.

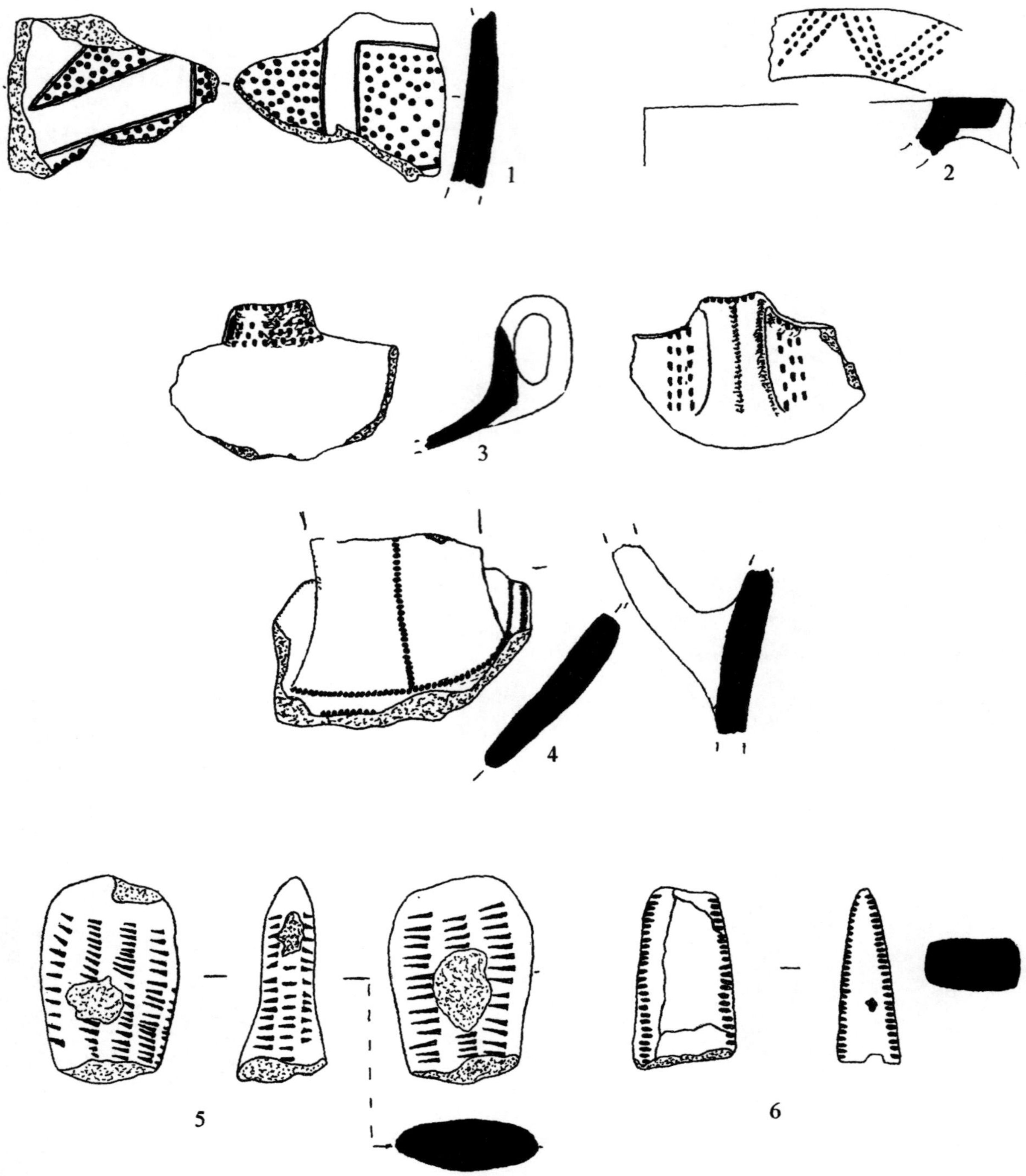

Figure 17.5. Pottery and models of axes fragmented.

CHAPTER 18

BRONZE AGE GOLD AND ELECTRON ADORNMENTS FROM TABASHKA CAVE IN THE OSUM RIVER DRAINAGE

(NORTHERN CENTRAL BULGARIA)

Martin Hristov

Introduction

In 1987 several finds were registered in the Depot of Historic Museum of Lovech Town. They originated from Tabashka Cave located at Bash-Bunar site near town that is crossed by Osum River. The finds were preliminary mentioned in the literature as originating from Vasil Levski Cave (Kitov et al. 1991: 49).

V. Mikov carried out the first investigations in Tabashka Cave (1928-1929: 316-317). The excavations of N. Dzhambazov from 1960 defined a new stage of investigations (1963: 195-237). The stratigraphic sequence has be divided into two groups: Quaternary dated to Late Palaeolithic, and Holocene to Neolithic, Chalkolithic, Bronze Age and classical Antiquity. The Relative chronology of the Holocene strata is as follows: Early Neolithic, Late Chalkolithic, two horizons of Bronze Age, and classical Antiquity (Dzhambazov 1963: 223-235; Todorova 1986: 131; Zmejkova 1990: 106; Leahu 1978).

The gold finds were discovered accidentally by the visiting of the cave by spelaeologists and the stratigraphic context of the finds is not clear. IT consists of six gold spiral pendants and a piece of arc-shaped object, probably a fragmented breastplate.

Gold spiral pendants

The are shaped by a round in section wire with a diameter varying – the middle part is flatten and the ends are pointed but without a final sharp edge. All of them consist of 1.5 turns. The sizes vary (***Table 18.1; Figure 18.1***).

These pendants can be dated generally to the Early Bronze Age having numerous parallels in the EB Balkans. The Bulgarian parallels consist of:

The grave from Belitsa (Northeast Bulgaria);
Madara – Tumulu I, grave 3 (Mikov 1934: 431 and fig. 289);
Kalugeritsa – Tumulus III, grave 4,
Plachidol I – Tumulus III, grave 6 (Panayotov 1989: 75, 112, 122);
Goran-Slatina - Tumulu II, grave 4; Tumulus II, grave 18; Tumulus III, grave 3; Tumulus III, grave 7; Tumulus III, grave 8; Tumulus V, grave 1; Tumulus IX, grave 2 (Kitov et al. 1991: 56 and fig. 33; 60 and fig. 35c; 60 and fig. 41f; 73 and fig. 46g; 73 and fig. 47d; 80 and fig. 54f; 100 and fig. 72e);
Lovech (Kitov and Pavlov 1973: 6-7 and fig. 54f);
Turnava – Tumulus I, grave 7; Tumulus I, grave 8; Tumulus I, grave 9; Tumulus III, grave 1 (Nikolov 1976: 43 and fig. 10a; 43 and fig. 12b; 43 and fig. 10b,c; 46 and fig. 12a,b);
Mihajlovo (Nikolov 1976: 46 and fig. 12d), published later as belonging to a grave (Nikolov 1982: 195).
Turnak (Torbov 1994: 17 and fig. 7);
Troyanovo – Kamenna Mogila Tumulus, graves 1 and 2 (Buyukliev 1964: 64 and fig. 6);
Pet Mogili – Tumulus II, grave 3 (Kuncheva-Rousseva 1994: 92 and fig. 5);
Mednikarovo- Iskritsa – Mogila III, grave 2 (Panayotov and Alexandrov 1995: 90, 94 and fig. 10);
Yunatsite (Mikov 1940: 71 and fig. 14);

In the neighbor regions similar gold spiral pendants have

Pendant No.	Material	Wire turns	Ends	Middle part	Diameter cm	Thicknes of wire, cm	Weight g	Carats cm	Inventory No. (HM – Lovech)	Figure No.
1	gold	1.5	sharp	flatten	1.38	0.16-0.3	5.3	23	1607	18.1: 1
2	gold	1.5	sharp	flatten	1.35-1.46	0.18-0.36	5.3	23	1608	18.1: 2
3	gold	1.5	sharp	flatten	1.3-1.4	0.15-0.3	4.7	23	1609	18.1: 3
4	gold	1.5	sharp	flatten	1.26-1.37	0.15-0.3	4.9	23	1610	18.1: 4
5	gold	1.5	sharp	flatten	1-1.1	0.07-1.1	1.1	23	1611	18.1: 5
6	gold	1.5	sharp	flatten	0.098-0.15	0.09-0.16	1.1	23	1612	18.1: 6

Table 18.1. Gold spiral pendants from Tabashka cave.

been discovered in Romania, as well as in Yugoslavia like at:

> Glăvăneştii Vechi – Tumulus I, grave 10 (Comşa 1987b: 372 and fig.s 11-2);
> Verbiţa – Tumulus II (Berciu and Roman 1984: 17 and figs. 1-3);
> Poiana (Dunâreanu-Vulpe 1938: 162 and fig. 14: 2,3);
> Zimnicea (Alexandrescu 1974: plate 8: 1-7, 10-14; plate 9: 1-2);
> Sucidava-Celej (Nica 1982: fig. 6: 2,3);
> Transylvania (Popescu 1956: 203 and fig. 119);
> Vojlovica (Jovanović 1976: 17 and plate 4: 2) etc.

In the Northwest Black Sea along with Nerushaj – Tumulus I, grave 14 and Glubokoe – Tumulus I, grave 26 (Shmaglij and Chernyakov 1970: 12 and fig. 64: 21; 45 and fig. 64: 15), it should be stressed the finds from the Basarabi variant of the Cernavoda I culture, and Usatovo culture (Manzura 1993: 45, 47, 50 and figs. 4,6,9), as well as many other finds along the eastern lower Danube, in the Dniester basin, Moldova and Romania (Yarovoj 1985: 80; Dergachov 1986: 56; Zirra 1960: 112; Alekseeva 1976: 185).

Most of the finds in the Balkans and the Northwestern Black Sea belongs to Pit Grave culture graves (*see* also Panayotov 1989: 47; Kitov et al. 1991: 11; Nikolova 1999: 306-307). However, there are some finds, which belong to settlements (Sucidava-Celej and Yunatsite) or flat cemeteries (Zimnicea) and are not connected with the PGC cultural pattern. According to the archaeological context, the gold pendants from Tabashka cave belong to that second group of finds in the Balkans.

According to Kitov et al. (1991: 27), because of the shape and the drawing apart of the wire the former pendants 'seem preceded with several centuries the cemetery of Goran-Slatina'. The last Tumulus cemetery – located not far from Lovech Town – consists of close analogies at Turnava, and has been connected with the initial penetration of the distribution of the PGC in Bulgarian lands (Kitov et al. 1991: 49; cp. Alexandrov 1994: 88; Nikolova 1999: 381, 383). If we accept the spiral pendants were already spread in the Balkans before the PGC, the *terminus ante quem* of the Tabashka Cave gold spiral finds is EBA I. They probably belong to a Bronze Age level in the cave which was recorded by Dzhambazov (1963: 233).

The upper Bronze level is dated to Middle Bronze Age – Tei II culture (Leahu 1978: 61; Zmejkova 1990: 106) but there are no published material from the earlier level. It is possible the finds to belong to that phase of occupation of the cave.

The breastplate from electron

The breastplate is a moon-shaped made of electron (***Figure 18.2***). It is broken into two parts. The plate is widened toward the middle part. There are double openings by the ends. One of the pairs consists of hole, which are broken. Length – 18.8 cm, max width – 1.6 cm, thickness – 0.06 cm, weight – 24.6 g. Inventory No. 1613, Historical Museum – Lovech. An existence of double openings, one broken, probably indicates the breastplate was reused.

Two breastplates, similar to the Lovech find, were documented in Emenska Cave – layer B after Ya. Nikolova and N. Angelov (1961: fig. 16). The excavator dated the layer to Early Bronze Age. But there is no evidence of the context of the discovered metal finds. Later, V. Leahu working with the material from that cave date the Bronze layer there to Tei III phase (1978: 54). To the end of Middle Bronze Age they are dated by I. Panayotov (1984: 14, comment 40), as well as by I. Ilcheva (1982: 21). The Middle Bronze Age is their relative chronology according to E. Chernykh (1978: 126, 162-163), but there are different criteria for a definition of the last period.

Having in mind the close typology between the breastplates from Emenska Cave and Tabashka and the fact that there is a Tei II material in the former cave, it can be assumed the electron breastplate from Tabashka Cave dates also to the Middle Bronze Age.

Conclusion

Generally speaking, the published finds – six spiral pendants by gold and a breastplate by electron belong to Bronze Age. In this communication it is assumed the pendants date from Early Bronze Age, with *terminus ante quem* Early Bronze I, while the electron breastplate belongs to Middle Bronze Age based on the parallels at Emenska and the possible similar context (earlier Tei culture). However, the question of precise dating of those finds remains because of the absence of clear context of their discovering. However, the newly published pendants are the richest collective find of that type (4.1 after Nikolova) in northern Bulgaria, while the electron breastplate indicates that this type of jewelry – already known from Emenska Cave – became popular in later stage of the Bronze Age. The spiral pendants from Tabashka Cave also increase the instances of not – PGC cultural context of their distribution to the south of the Danube River.

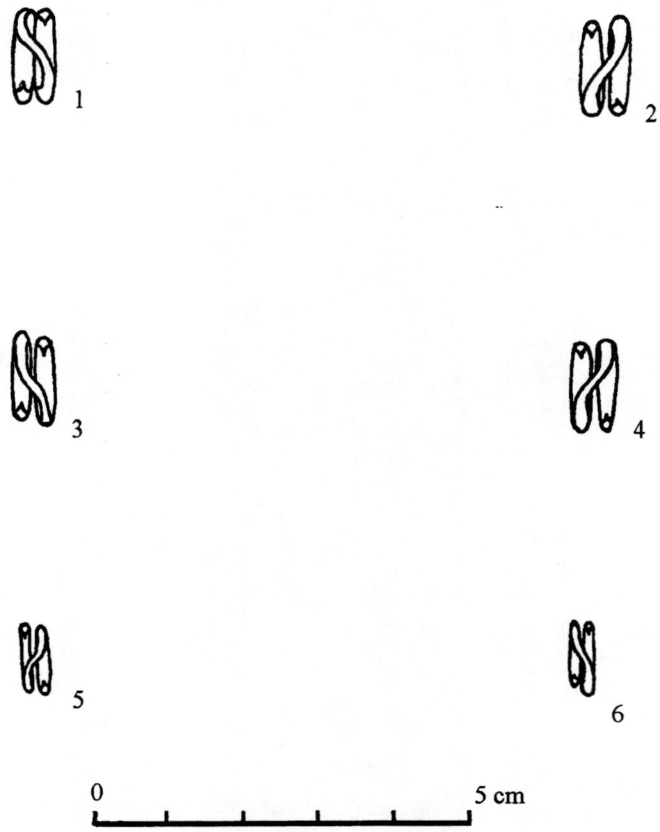

Figure 18.1. Gold pendants from Tabashka cave, Lovech District (Bulgaria).

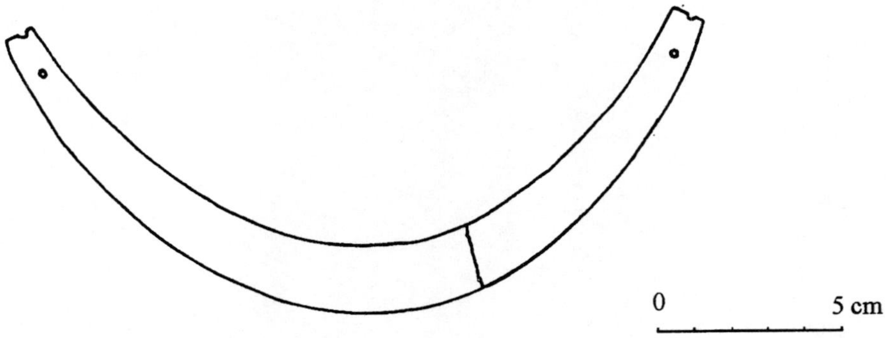

Figure 18.2. An breastplate from electron. Tabashka cave, Lovech District (Bulgaria).

CHAPTER 19

MIDDLE AND LATE BRONZE AGE OF SOUTHEASTERN EUROPE (ROMANIA AND BULGARIA)

Cristian Schuster

Introduction

According to the opinion of most Romanian specialists, namely the ones who adopted the chronological system conceived by P. Roman (1986), the Early Bronze Age (EBA) from the region under study, particularly the extreme east of Oltenia, Muntenia and Dobrudja, begins with the funeral appearances of Zimnicea-Olteniţa-Călăraşi-Mlăjet type. They have correspondents south of the Danube, in Bulgaria - at Batin - and north of the Meridional Carpathians, in South-eastern Transylvania, in the Bârsei Land - at Zăbala (Roman 1986; Schuster 1997; Schuster 1998a; Székely 1997). We have to mention that we find hard to be documented by now the assumption of some archaeologists, who consider that the EBA started with the Coţofeni Culture (Vulpe 1995; 1997a).

If we would follow the distribution of EBA in South and South-Eastern Romania, we could see that the time span between the transitional period from Eneolithic and Middle Bronze Age (MBA) is "filled" in the Western and Central Muntenia. Subsequently, during the Middle and Late Bronze Age (LBA) it would strongly be developed there the Tei culture. This would occur after the up mentioned horizon, namely Zimnicea-Zimnicea-Olteniţa-Călăraşi-Mlăjet, with the aspects specific to the EBA, like the Glina Culture (Schuster 1997, 29-117), the Odaia Turcului (Tudor 1982, 59-75; Tudor 1983, 108-111; Vulpe 1990, 105-111; Băjenaru 1996, 82-83; Schuster 1997, 119-125) and Năeni-Schneckenberg groups (Vulpe, Drâmbocianu 1981, 171-193; Vulpe 1983, 71-75; Vulpe 1990, 105-111), as well as the horizon of stone cist burials (Schuster 1997, 126-131).

Tei culture

The analyses of the Năeni-Schneckenberg ceramics, of the sherds belonging to the stone cist burials horizon from Northern Muntenia or of those coming from the recent finds of Bucureşti-Militari Câmpul Boja type, located in Central Muntenia, show no connection to the Tei Culture. Regarding the Odaia Turcului, the author of the first investigations has stated that the up mentioned cultural aspect had played no role in the development of the Tei Culture (Tudor 1982, 74). In his turn, V. Leahu remarked that Odaia Turcului has exerted an influence "in a less degree" upon the formation of the Tei Culture (Leahu 1995, 154). The only cultures involved would have been the Glina Culture and a possible horizon with scratched ceramics and textile impressions, shaped as a honeycomb (Roman 1986, 31). As concerns the latter aspect (Lazarovici 1998), which is specific to the Romanian Banat (Gogâltan 1993, 51-64; Gogâltan 1995, 55-79; Gogâltan 1996, 43-67; Gumă 1996, 21), Oltenia (Boroneanţ 1971, 5-12; Roman 1981, 166; Roman 1986, 31; Petre-Govora 1988, 137-146; Petre-Govora 1995, 41; Nica 1996, 18-20) Tranylvania (Roman 1984, 269; Roman 1988, 219-222; Roman, Németi 1990, 36-37; Ciugudean 1996, 110-112; Cavruc, Cavruc 1997, 157-172; Cavruc 1997a, 17; Cavruc 1997b) and to the Yugoslavian Vojvodina (Falkenstein 1998, pl. 12/8-9), as I already stated on another occasion (Schuster 1997, 161), I am reluctant when considering its presence in Muntenia, due to the lack of certain discoveries of this kind.

In contradiction to my opinion, when discussing the formation of the Tei Culture, V. Leahu had pointed out that to the various elements of influence "should also be added, for its genetic role, the ceramics with grooved decoration, that is extremely close to the phase Tei I materials, yet, without being identical". With no other details, he would note that "such a ceramics has been identified in the South central zone of Muntenia but, in uncertain, not-checked conditions"(Leahu 1995, 254).

If we think about the uncertainties that are still connected with the genesis of the Tei Culture, it should not be excluded (even if we could not see for the moment their direct descent in Muntenia) that, the population from which subsequently would emerge the Tei communities, on its way towards the territory under study, has been "contaminated" or even came as direct descender of the communities belonging to the horizon with scretched ceramics and textile impressions shaped as honeycomb (Schuster 1998b, 143).

This latter manifestation, coming from the west, probably under the pressure of the Glockenbecher cultural phenomenon, by a diffusion along the Danube and its tributaries Mureş and Tisa, has covered rather fast the western parts of Romania. Its movement eastwards, namely towards Muntenia, could have been slowed down or even stopped by eastern penetrations, by those coming from the north-Pontic regions, or from south of the Danube.

Today, it is clear that the only phase of the Glina Culture which could be considered as contributing to the formation of the Tei Culture is the classic one (Schuster 1997, 86, 161), more exactly the Govora Sat-Runcuri phase (= Glina III, by Roman) (Roman 1985, 279-297). Its presence is well documented in the Glina settlements from Muntenia, all located on the banks of the Argeş River or on its direct tributaries (Tudor 1982; Tudor 1983; Schuster 1992a, 35-41; Schuster 1992b, 7-10; Schuster 1995, 53-63; Schuster, Popa 1995a, 20-26). This would be the exact location where later would exist the strongest and most constant core of the Tei Culture (Schuster 1994). Excepting some vessel shapes, especially those belonging to coarse wares (Schuster 1997 annex 4, T. 1-6) and some types of handles (tube-shaped) (Schuster 1997, 66), one could establish certain ornamental analogies between those two cultures. We refer to those conceived with the "little broom" (Schuster 1997, 71), but also to the technique of filling the impressions with white matter (Schuster 1997, 70-71).

During the EBA, in the northeastern area of Muntenia and Southern Moldova, as well as in the curvature region of the Carpathians Mountains, appeared the Monteoru Culture that lasted up to the LBA. This is one of the most important cultures of the Bronze Age in Romania. Roman (1986, 45) considered that its evolution has started the earliest in the phase EBA IIb, namely in parallel with Glina IV in Western Romania (in Oltenia), with Makó, also in Western Romania (in Banat), Hungary, Yugoslavia, with Nyirség, Vinkovci, Somogyvár, from the same regions, Odaia Turcului, the find of Bucureşti-Militari Câmpul Boja type and the stone cist burials horizon from Northern and Central Muntenia.

According to E. Zaharia (1987; 1995), this culture appeared with the phase Monteoru IC4 in the EBA, having its basis on the local communities and receiving southern influences, like Cernavodă II, Folteşti II, Bucureşti-Căţelu Nou, Coşereni, Zimnicea, or even those from the Ezero milieu from Bulgaria. In her view, „in the beginning phase, the culture shows the general characteristics of the Romanian EBA, related to the Glina-Schneckenberg cultures“ (Zaharia 1995, 190-1999). She questions the conclusions of Alexandru Vulpe regarding the Schneckenberg-Năeni group and its presence at Sărata Monteoru (Zaharia 1987, 39 sq.).

Even after the issue of the article written by E. Zaharia (1987) on "The beginning phase of the Monteoru Culture in the light of the researches from Sarata Monteoru", Vulpe maintained his opinion that the finds in the mentioned Monteoru IC4 1 settlement were in fact the traces of an Odaia Turcului community type (Vulpe 1990, 111 and fig. 2/1-8), while Monteoru IC4 2 would belong to the Năeni-Schneckenberg group (Vulpe 1990, 111 and fig. 2/9-12). According to this scholar, only what is known as Monteoru IC4 3 could be considered as a first clear expression of the Monteoru Culture (Vulpe 1990, 111), which would still belong to the MBA (Vulpe 1997, 46).

In Dobroudja, the EBA is relatively poor in vestiges. By its geographic position, between the Black Sea and the Danube, this territory, and especially its northern part, meant for most of the time a region into which nomadic populations have penetrated from the steppe north of the Black Sea. As concerns cultural phenomena, Southern Dobroudja is mostly connected with Bulgaria.

Today it is clear that the range of the Tei Culture is concentrated in the Central Muntenia but, in some evolution stages this area is surpassed and thus, Tei groups penetrate both northwards, beyond the Meridional Carpathians, in the Bârsei Land and southwards, across the Danube, in Bulgaria.

Some investigations undertaken in the past few years in sites with Tei Culture vestiges, like those from Bucureşti-Chitila Fermă (Boroneanţ 1984, 157-166; Boroneanţ 1985, 22-29; Boroneanţ 1992, 109-117), Mogoşeşti (Schuster 1990, 103-105; Schuster 1996, 102-103; Schuster, Popa 1995a, 33-36; Schuster, Popa 1995b, 147-156; Schuster, Popa 1997, 63-77), Popeşti-Nucet (Palincaş 1996, 239-288; Palincaş 1997, 173-192; Fischer 1996, 289-295; Vulpe 1997, 163-171), Bila-La Fântână (Schuster, Popa 1995a, 41), Căscioarele (Sîrbu, Damian 1992, 11-16), Cătunu (Stoica 1989, 259), are important for their contribution to a better understanding of the mentioned culture development. The Tei Culture was split into five phases (I-V), with their sub-phases (regarding this topics see recently: Leahu 1992). Following some older investigations at Bungetu-Brăteştii de Sus and Costeştii din Vale, Ion Chicideanu (Chicideanu 1973, 27-43; Chicideanu 1977, 225-238; Chicideanu 1980, 127-128; Chicideanu 1982, 101-116) has defined a new supplementary phase at the beginning of the culture, called the Bungetu phase. But, a closer analysis of the ceramics from the mentioned sites and also from Căscioarele, leads to the conclusion that the respective phase was, in fact, a first sequence (Ia) of the culture and not an individual, well defined phase.

If the distribution in the field of those three spots with Tei vestiges Ia is to be taken into consideration, one could easily notice that Bungetu and Costeştii din Vale are located somewhere in the central part of the Dâmboviţa County, close to the Argeş and Dâmboviţa Rivers, whereas Căscioarele is not far from the Danube. Could it mark, prior to the Tei settlements, some penetrations of communities with grooved ceramics and textile as well as honeycomb decorations of Gornea-Orleşti type? The banks of the Danube have always been routes of swinging for populations who came both from east and west. It is true that, during the EBA, these paths have mostly been used by the nomads coming from the north-pontic steppe. But, it is possible that, this flow could have turned less intense towards the end of the EBA, thus allowing a reverse penetration.

The Tei I ceramic fragments discovered in South-Eastern Transylvania, at Sfântu Gheorghe-Örkö (Székely 1973, 129-134; Leahu 1997, 123), as well as south of the Danube at Devetaki and Kadăkioi (Leahu 1983), prove the penetration of some groups belonging to this culture, that probably occurred even in its initial stage. But, on the other hand, those material traces could simply be some Tei imports in foreign milieus.

It is possible that, subsequently, after a time of development, the Ib phase Tei communities (Leahu 1992, 69) could have continued their habitation in the initial territories (the second level Bungetu-Brăteştii de Sus), but, at the same time, they conquered new ones, especially in Central Muntenia, reaching up to the Mostiştea River.

The ceramic analysis allows us to consider the Tei I phase, with its two different moments (Ia and Ib) in which some scholars saw local and partly contemporaneous phenomena

(Boroneanţ 1981, 214), as a beginning stage of the culture and, in a wider context, as the first period of the MBA from the Central and Western Muntenia. It is possible that this time span could have also been preceded by other cultural aspects, which are not known yet, additional to the mentioned ones, but which ensured the connection between the EBA III (by Roman) and MBA. This passing stage we have called sometime 'Übergangsperiode' (Schuster 1996, 343).

Inner factors, especially at the beginning, and outer ones towards the end, would lead to the crystallisation of the second stage of the mentioned culture, namely the classical one. This time span has to be considered the second sequence of the MBA in the zones under study (MBA II). The classical phase comprises the Tei II phase (Leahu 1992, 63, 69), divided, in its turn, in two other moments: Tei IIa (Leahu 1992, 69-70) and Tei IIb (Leahu 1992, 70), namely MBA IIa1 and MBA IIa2 (Schuster 1998, 145), and the Tei III phase (Leahu 1992, 63-64, 70), also with two subdivisions: Tei IIIa (Leahu 1992, 70-71) and Tei IIIb (Leahu 1992, 71), respectively MBA IIb1 and MBA IIb2 (Schuster 1998, 145). The harmony of this periodization underwent some disturbances. Thus, V. Boroneanţ (Boroneanţ 1981, 214) believes that the levels I and II in the Tei site from Bucureşti-Chitila Cărămidărie correspond "to the end of the phase I and II established by V. Leahu. The human community from Chitila Cărămidărie takes over the stock of the phase I and continues the evolution towards the subsequent stages" thus, to a moment that could hardly be included into one of the phases defined by V. Leahu. This conclusion would confirm the opinion of the last mentioned scholar, according to whom it "appears more than certain the fact that between the phases I and II has existed another cultural aspect that could not be detected and defined by the archaeological research" (Leahu 1966, 154).

Also hard to be assigned to a certain phase are the vestiges from Mogoşeşti, where the fine and the semi-fine ceramics bear decoration elements which are specific to the Tei II and Tei III phase, which fact gave the possibility to some of the excavations' authors to consider them as "Mogoşeşti type materials" (Schuster, Popa 1997, 66). This would prove that, on the mentioned spot, could be identified "a community that reflects the moment of transition from Tei II to Tei III" (Schuster and Popa 1995b, 149).

The vestiges from Bucureşti-Chitila Cărămidărie and Mogoşeşti could eventually be considered as local finds. But, the region where they were located could not provide, by its relief (at least in the manner it looks now), the necessary conditions for developing strictly local occurrence, isolated by the rest of the range. Still, if we take into account the environment of the time in which the Tei Culture has evolved - forests, places which could easily be flooded or, on the contrary, very arid ones - we have to give another, nuance importance to the forwarded hypothesis.

Together with the transition to the second phase, the number of Tei settlements grows, especially in Central Muntenia, in the region of Bucharest and its surroundings. In turn, it is surprising the lack of settlements in the eastern part, in the region of Mostiştea River. Is it possible on the respective territory to have still existed the Tei I settlements? Or could the Tei II phase be a local expression that evolved from Tei just in the perimeter of Bucharest, as well as north and south from it, as seem to indicate some materials? What if the Mostiştea Valley was simply abandoned by the Tei communities right after the phase I? This latter assumption would be in a total contradiction with the hypothesis that is generally accepted by specialists, that Mostiştea was deserted just at the end of the phase II and the beginning of the phase III.

The second expression of the classical period of the Tei Culture, respectively of the MBA, from the mentioned region, as I already established, is the phase III of the culture. Now, except some changes that could be detected in the ceramics, is notable the increase number of settlements and the occurrence of the first proves regarding the funeral rite, unknown for the beginning phases of the culture. Settlements Tei III have been unearthed in Central Muntenia, in Transylvania,in the Bârsei Land (Leahu 1966, 17; Leahu 1997, 124-127; Schuster 1997b, 131-135) and south of the Danube, in Bulgaria (Leahu 1983). The Tei III ceramics has been also identified at Troy, in Turkey (Leahu 1983). What determined that increase in the settlements' number and, at the same time, the extension of the cultural range was recently reanalyzed by V. Leahu (1997, 123-129) and myself (Schuster 1997b). In this research stage, we consider appropriate to be accepted the idea that, the outer factor, namely the penetration of some nomadic elements from the east - which in the opinion of Leahu were part of a wider phenomenon, a hypothesis which is not accepted by A. Vulpe (1995, 20) - has pressed on the eastern limit of the territory inhabited by the Tei communities (Leahu 1997, 125; Schuster 1997b, 131-132). This phenomenon joint with the inner factor, probably a consistent demographic growth, has determined the movement of some Tei groups northwards, beyond the Meridional Carpathians and southwards, across the Danube. As I already stated above, this process could have been initiated earlier, in the Tei II phase (Schuster 1998, 148).

Even if the archaeological discoveries are not always too eloquent, we could admit the physical existence of the Tei Culture bearers both in Transylvania and in Bulgaria and not only the presence of some ceramics imports. We should not forget that in South-Eastern Transylvania have been found Tei vestiges in caves (Schroller 1933, 75, nr. 68; Nicolăescu-Plopşor, Păunescu, Pop 1962, 118; Leahu 1966, 52-53; Boroffka 1994, 54). They were real refuges, that point to the temporary use of those natural shelters by groups of individuals in movement at that time. The Tei penetration was also accompanied by some Monteoru elements (Székely 1971, 329; Zaharia 1990, 37). It should be not excluded the idea that the Petrişoru Racoviţeni group could have played a significant role both in South-Eastern Transylvania and probably in North-Eastern Muntenia (Oancea 1982, 54, 57; Leahu 1997, 127).

For the lands south of the Danube, the new finds confirm the statement that here have strongly manifested themselves the Tei communities, fact which is in deep accordance with the older hypothesis of a South Danubian variant of the Tei Culture (Alexandrescu 1973, 83).

It is evident that, the Tei ceramics from MBA II, namely from the phases II and III, as well as the Mogoşeşti type materials, could hardly allow us to presume a penetrating evolution. In turn, subsequently, during the LBA, as also noted by Valeriu Leahu "it must be taken into account the too well outlined occurrence of the Tei IV phase, its striking differentiation in comparison with the preceding phase. It is thus imperative the hypothesis of a transitional aspect, from the Tei III sites and manifestations" (Leahu 1992, 68,72).

Significant and somehow surprising is the fact that, starting with the phase III it is notable the first funeral vestiges of the Tei Culture. We mention here the burials from Chirnogi (Morintz, Ionescu 1968, 103-105; Leahu 1969, 27; Morintz 1978, 48; Leahu 1987, 49), Căscioarele (Şerbănescu, Bălteanu 1998) and probably Sitaru (Comşa and Schuster 1997, 137-145). Therefore, as I stated some other time before, during the phase Tei III "it is not excluded that we could talk not only of changes in the material culture, but also in the one of the spiritual life" (Schuster 1997b, 133), fact which could have its roots in the preceding period. We take into consideration here the contact between the Tei population with the one that came from the north-Pontic milieu, as pointed out by the discoveries from Baldovineşti.

Of course, for the MBA in Muntenia, the Dobroudja, Oltenia and other Romanian territories like Moldova and Transylvania, we have to analyze also the relations with the neighboring lands - recent Bulgaria, Yugoslavia, Hungary, Republic of Moldova - together with the manifestations specific to the Balkans, Anatolia and to the north-Pontic space. A strong opinion trend connects cultures like Tei, Monteoru, but also Wietenberg and Otomani from Transylvania, with the Mycenean civilization. Even if there are some elements in common as concerns the material culture of the Southern Balkans and north-Danubian lands, as result of some exchanges of goods but also of cultural influences, there must not be neglected the connections with the MBA and LBA from the north-Pontic zone and with the central European ones. Evidences in this sense are goods like the cheek pieces made of bone, discovered in the Tei, Monteoru and Wietenberg milieus, which have analogies both in the west and east (Boroffka 1998). In the Monteoru Culture, for example, I. Chicideanu has noticed the penetration of some eastern goods (Chicideanu 1995, 224-225) like: adornments, harness pieces or weapons (buckles, hatchets, cheek pieces, daggers, chisels, axes, arrow points), coming from the range of Sabatinovka and Noua Cultures. From the north and North-West have intruded into the Monteoru Culture bone pieces from the Nitra milieu and bronze pins from the Lausitz Culture (Chicideanu 1995, 225 sq.). From the southern space, Egyptian-Mycenean one, through the Mures Culture from Western Romania have come the glass pearls (Chicideanu 1995, 238).

These prestige goods, most of which were found in burials, but also in settlements, are results of some exchanges. The bearers of the Monteoru Culture, rich in salt but poor in ores, had economic relations with their neighbors from Transylvania, with the populations from south and by these, with other remote communities.

The Tei IV and V communities, that in our opinion correspond to the LBA I, have strongly manifested themselves in Central Muntenia and south of the Danube, as pointed out by the latest archaeological investigations from Bulgaria.

Recently, Nona Palincaş has taken over an older opinion trend, that was abandoned in the meantime, according to which, the Tei IV-V and Verbicioara IV-V phases were a single cultural phenomenon that included, besides Oltenia, Western and Central Muntenia and territories south of the Danube, in Bulgaria. In her attempt of unifying the Tei IV phase with the Tei V one and their "separation from the Tei Culture and unification with the Govora group (of the Verbicioara Culture) under the denomination of Govora-Fundeni group" (Palincaş 1996), she analyses the kantharos type vessels and the horizontal handles of other receptacles. The simple or angled horizontal handles preceded the Tei IV and V phases. They were discovered in the Tei II and III milieus (Leahu 1975, 111; Leahu 1992, pl. XIV/2; Schuster and Popa 1995a, 38), both in Romania and Bulgaria - Devetaki, Emen, Vişegrad, Pchelarovo-Grădişte (Leahu 1992, pl. XIV/4-5; Nekhrizov 1995, 315 and fig. 3/27; Palincaş 1996, fig. 15/a-b; Schuster and Popa 1997, fig. 3/1) and we must look for their origin south from the Danube (Leahu 1975, 111).

Regarding the three kantharoi type vessels- the big one with a narrow, truncated-conical neck and strong shoulder, the second one small, with a short, cylindrical or truncated-conical neck and globular body, the last one with a wide, cylindrical neck - have been identified in the late Tei milieus (Leahu 1966, 9 with footnote no. 1; Sandu 1992, pl. IX/3; Palincaş 1996, fig. 13/a-c) and late Verbicioara (Crăciunescu 1992, pl. 4/6, 9; Palincaş 1996, fig. 13/a-c). The distribution of the spots where such vessels have been found, has outlined two clear ranges, separated east of the Olt River by a space where such finds are missing. This fact points out the existence of two distinct centers or cultural diffusion that, according to the fashion, produced the same kind of receptacles. The kantharos it the vessel type specific to the LBA of the Lower Danube, afterwards being present in the whole North-Balkan region and reaching up to Anatolia. Wider analogies for kantharoi from Romania could be established with the Western and eastern zone of the Rhodope Mountains, with Macedonia, Bulgaria etc. (Nekhrizov 1995, 314). This is why, one must be careful when choosing this kind of vessel as a reason of a cultural unit.

As we could see, starting with the phase Tei III, eastwards from the Mostiştea River, were more and more strongly represented nomadic population groups coming from the east. In LBA I, must be also included the phases Verbicioara IV and V together with the final aspects of the Zimnicea-Plovdiv (§erkovna by Bernhard Hänsel) and Noua-Sabatinovka-Coslogeni Cultures (Schuster 1996, 147-148). The latest researches strongly lead to the conclusion that these cultures of the Balkan-Danubian space were connected with Greece and North-Western Anatolia (see more recently: László 1997; Koppenhöfer 1997; László 1998). Without re-discussing things, it is worth to be mentioned that lately there are increasingly greater in number the archaeological sites from the mainland and islands of Greece where „Barbarian Pottery“ has been found. We mention here Athens, Mycene, Lefakandi, Tiryns, Aigeira, Sparta, Kalopodi, Pellana, Delphi, Teichos Dymaeon (Achaia), Perati, Komnos and Chania in the western part of Crete and Troy VIIb1-2 in Turkey (Rutter 1977; Harding 1984; Koppenhöfer 1997).

The occurrence of this ceramic type is not a hazard but, as already stated before, it is a 'Balkanization process' of Greece and of some parts from the Near East, up to Troy (Hänsel 1976, 235), determined by a population movement from the north, maybe due to Thracian communities – Phrygians, Mysi – (László 1997, 125; László 1998, 44). What generated their motion to the south it is hard to tell. László noted that „the penetration of some elements having origins in the Thracian area took place first towards Macedonia and Greece and somewhat later towards North-Western Asia Minor“ (László 1998, 43). As explanations of these movements to south must not be neglected the climate factor, the damages determined maybe by some earthquakes, knowing that Romania, Macedonia and Greece are sensible from this viewpoint, or volcanic eruptions, accompanied by earthquakes and tsunami waves. All these, associated with an economic and military weakening of the East-Mediterranean states, the occurrence of new fighting techniques (in which infantry gained new attributions) and the creation of more efficient weapons, would ease the march to the south of some North-Balkan population groups.

It must be stated that probably, the phases of northern population penetrations are contemporaneous with what we called LBA II in Muntenia, when in the region it is know a cultural mixture, and possibly and ethnic one too, coming from the preceding cultures - Coslogeni, Zimnicea-Plovdiv, Tei (Schuster 1996, 147).

When analyzing some discoveries from Southern Muntenia and Northern Bulgaria, M. Gumă speaks of an ethnic-cultural group, which he called Zimnicea-Novgrad (Gumă 1995, 109-110). In his opinion, this manifestation, which is „specific to the transitional period to the Early Iron Age... represents a totally different ethnic-cultural entity in comparison with the others groups...The genetic background of this group were the entities of the Zimnicea-Plovdiv-Şerkovna type from the Late Bronze Age in the same area, to which some contributions of Late Tei or Coslogeni type and even influences of Late Gârla Mare type must be also added“ (Gumă 1995, 109).

As pointed out by the chart conceived by M. Gumă, which renders the range of the respective manifestation, this group spreads from the flow of the Jiu River and up to the flow of the Argeş River into the Danube, thus, covering just a restrained strip of land north of the Danube (Gumă 1995, pl. XVII). A comparative analysis of the ceramics discovered in the Zimnicea-Novgrad range with the one from the Central Muntenia, discussed above, shows similarities but differences as well, fact which makes us believe that those two phenomena were partly contemporaneous (Schuster 1996, 147).

According to M. Gumă (1995, 110), the Zimnicea-Novgrad group must be placed in Ha.A-Ha.B1, namely in the second half of the 12th century and up to the 10th century B.C. It would be contemporaneous „with the other groups of transitional period of South-Western Romania, Northern Serbia and North-Western Bulgaria, but connected rather with other similar groups of the Lower Danube and the East Carpathians regions. The Zimnicea-Novgrad group is practically situated at the boundary between the eastern cultural complex...and the western one illustrated by the latest „Urn fields“ entities“ (Gumă 1995, 110).

Some conclusions

1. Even if it was not our purpose to create an exhaustive image of the MBA and LBA in the Lower Danube region, it is our impression that we managed to tackle their main issues.

2. It is evident that, both on the northern and southern banks of the Danube, the ethnic-cultural mechanisms which led to the formation of the MBA are still partly veiled by the lack of knowledge, the same like those from the end of the LBA and the beginning of Hallstatt.

3. Yet, the latest archaeological investigations have started to put away that veil, by unearthing new materials, which enhance the information concerning the above mentioned sequences of time;

4. Even if we consider the case of a well-defined civilization, like the Tei Culture, there are still many unknown aspects. This is especially given to the fact that; its periodization is mostly based on ceramics typology and less on stratigraphy. Also, in connection to that culture, there is a need to better correlate the north of the Danube discoveries with the older or more recent ones from the south. Thus, it seems the entire image regarding the range of this cultural expression will certainly achieve new interpretations.

5. The phenomena, which occurred on the Lower Danube during the Middle and Late Bronze Age, are evidently, in close connection with those, taking place in Central and Southern Europe. In this context, should also be considered those developing east of the Prut River. Unfortunately, some manifestations of the MBA and LBA on the Lower Danube had an unilateral expression, namely, their traces could be certified exclusively by settlements or, on the contrary, by burials or necropolises. Just in rare cases could be investigated settlements, as well as cemeteries.

CHAPTER 20

ENCRUSTED POTTERY CULTURE ALONG THE LOWER DANUBE

(ON THE EXAMPLE OF ORSOYA CEMETERY)

Alexander Bonev

Introduction

The second half of the second millennium BC was a notably dynamic process in the development of tribes and populations in the Lower Danube basin, the eastern Balkan Peninsula and the Aegean. It was the time of the flourishing of the bronze metallurgy, when the bronze implements and weapons completely became dominated in the life of society. The accelerating search for copper, tin and precious metals lead to an activation of the mutual contacts to the highest end of the scale in the historical development there. Besides, the development itself in southeast Europe moved along an unequal line. In fact, regions like Thrace and Thessaly, prospering in the previous epoch, left off performing a primary role. Minoan Crete from the periods of Old and New Palaces, along with the archaic Kingdoms following on the Mainland – Tiryns, Pilos, Orchomenos and mostly Mycenae, with which names will be connected whole the epoch, arose as main centers of the incipiently early European civilization.

They undisputedly influenced their neighbors, especially in Late Bronze Age cultures in the North. This influence was particularly strong, but as it could be seen at first sight in a narrow geographical region along both the sides of the Danube River, between the embouchements of the Tisza – Drava and Jiu – Ogosta Rivers. In the last half of the second millennium BC, an extremely rich, distinctive and attractive culture developed there, recently known as ***Encrusted Pottery Culture along the Lower Danube*** (EPCLD) in the specialized literature.

The 'River civilization' in question, EPCLD, was discovered and for the first time investigated in northeast Serbia, at the end of the last and the beginning of recent century. The Serbian archaeologist M. Vasić unearthed the cemetery near Žuto Brdo published at the beginning of 20th century. Then, the considerably intensive investigations continued in northern Serbia. Attractive results were obtained by excavating cemeteries and settlements of that culture (Bonev 2000 with the references cited there).

The investigations of EPCLD on the territory of Romania began notably later, in compare to Serbia. Excavations were carried out mainly after the Second World War. But we should stress they approached significant results. The excavations of the Cârna cemetery with 116 cremations could be pointed out as the greatest achievement of the Romanian archaeology. The results of these investigations were published in time, at very high scholar level. It is worse mentioning the monograph of V. Dumitrescu (1961) has remained a fundamental classical study in the investigation of that culture yet.

In Northwest Bulgaria, there have begun a research on EPCLD at the latest. As a matter of fact, at least for the initial stage, it is difficult one to speak about purposeful investigations. There was rather a collection of data on that culture published by V. Mikov (1970), as no regular excavations were performed.

The state of these affairs generally has changed after the beginning of the excavations of the cremation cemetery near the village of Orsoya, Lom District. The excavations continued 10 years – from 1969 to 1979. For the time being, this is the richest site of Encrusted Pottery Culture along the Lower Danube excavated. T. Filipov published incomplete information about the excavations (1978). Typological systematization of all known finds from Orsoya has been recently elaborated, along with a chronological definition, an evaluation of the cultural context and the role of society to which the burial ground have belonged in the processes of formation of the whole image of the Early Thracian civilization (Bonev 1988; 2000).

There are over 180 finds from Orsoya cemetery preserved (Nos. 1-184 in Bonev 2000. The typology of the ceramic finds includes 28 types of different kinds of objects (***Figure 20.1-3***). Their variety is considerable, as well as other types can be added.

Typological variety

Burials urns (***Figure 20.1: 1-2***) is the most numerous material distinguished into five groups and characterized by rather different types and typological variants. On the one hand, it should be mentioned that a small number represents some of the types from the different groups, even in some cases there are only single items, some of which unique. Some of them are very rich ornamented.

The vessels of kantharoi group have a short neck, shaped as a truncated cone in the upper part, with equal cut orifice. The body is shaped as a truncated sphere. The two handles are attached to the orifice and the upper part of the body slightly protuberant. The kantharoi are richly ornamented with an incised, pricked and stamped ornament performed in *Furchenstich* technique. All typical of Encrusted Pottery Culture along the Lower Danube motifs are included. The encrustation is ubiquitous (***Figure 20.1: 3-7***).

Figure 20.1. Encrusted pottery from Orsoya cemetery.

Of the burial-goods found at the cemetery of Orsoya, the cups deserve a consideration (***Figure 20.1: 8***). The utilitarian function distinguishes them from the urns and the kantharoi. The decoration is typical of Encrusted Pottery Culture along the Lower Danube. There are also other encrusted vessels as bowls, kiathoses, lids etc. (Bonev 2000).

The zoomorphic vessels are a remarkably attractive type of vessels (***Figure 20.2***). In compare to the other sites of EPCLD where only single instances are documented that type is well represented at Orsoya. The vessels exemplify birds (water-ones or doves?), with a train stressed, a short body with a shape of truncated cone, rounded turning point and a body

Figure 20.2. Encrusted zoomorphic vessel from Orsoya cemetery.

with a shape of truncated sphere and short cylindrical base holed. Some of them have well shaped bulges (Buckel), but others consist of small band handle as well. Without exception, all the zoomorphic vessels consist of lids of the so-called 'Combrero' type. The ornamentation includes almost all the motifs and technique characteristic of EPCLD and is very rich.

The anthropomorphic figurines (the so-called bell idols) represent very stylized human figurines (***Figure 20.3***). A massive cylindrical bulge is shaped into a head. The breast (the upper part of the body) is cylindrical. A pinch shapes the waist while the lower part of the body consists of hollowed cylinder. The figurines have gray, gray-and-black and reddish brown color, as well as well smoothed surface. They are ornamented with an extremely rich geometric decoration. In fact, they represent all typical technique and ornamental motifs of EPCLD. It should be mentioned an interesting fact. The ornamentation of all the mentioned above objects is based on the principle of symmetry. In other words, the motifs follow in order and there is a precise sequence in their decoration, irrespective of the kind of the vessel – urns, kantharoi or zoomorphic vessel. The principle is broken by the bell idols – the front, back and the side parts of the figurines are ornamented in different way, despite the motifs are united from a compositional point of view. I cannot answer the question what was the reason, but I think one should not ignore the opinions of some researchers that the ornamentation represents a female costume richly decorated with typical of that time motifs. That is another question what is the origin of those motifs. The Orsoya collection of anthropomorphic figurines is the richest known from EPCLD. Among the other finds there are models of axes, boat, throne, etc.

Some of the problems of the morphological characteristics of the pottery from Orsoya cremation graveyard were discussed in the context of their typological definition. Despite of the similarity with the other pottery finds from Encrusted Pottery Culture along the Lower Danube, there is peculiarities in morphology and the variety of the number representing the different types, as well as exception for that culture finds.

Most of the finds from the Orsoya Graveyards are richly ornamented with geometric motifs executed by a false corded technique, in combination with an encrustation with à white matter. The secondary ornamental techniques are those of the incision and stamping. The main ornaments can be divided into two groups. The linear ornament belongs to the first group including horizontal strips with incisions between, zigzag lines, single and interconnected triangles hatched, simple and complex meander, rhombs, and the sign of 'boomerang'. The curve-linear ornament characterizes the second group – garland, spiral, the sign if 'lily', S-sings, concentric circles and solar signs in different variants.

It should be stressed the familiar way of treating of the main motifs by choosing of different combinations. In other words, at least on the topic of the published finds yet, I could not find any doubling of the combinations used. There are no identical objects with one and the same ornamentation, even from one and the same type. Despite the similarity in the main motifs, there are also no identical parallels with the ornamentation from the other known sites of EPCLD – Žuto Brdo, Korbovo, Ostrovul Mare, Gărla Mare, Balta Verde, etc. (see also Shalganova 1993; 1995).

It was stressed elsewhere (Bonev 1988: 49-54), the similarity between the ornamentation of Orsoya and Cârna infer that both the cemeteries were synchronous, as a whole. Therefore, the chronology of V. Dumitrescu on the Cârna cemetery graveyard (1961) is acceptable for Orsoya as well. The fact that the rich collection of the zoomorphic figurines and the models of axes with a rhomb shape from Orsoya have only a few analogies at Cârna does not contradict such a conclusion. However, there are some peculiarities with chronological values. For instance, the absence of the anthropomorphic idols with mobile heads at Orsoya infers a later dating of the last cemetery. So, the beginning of the Orsoya cemetery can be dated to later 15th century BC. I could not recognize any phases in the development of the graveyard. With some reserves, I suggest the end of the Orsoya graveyard dates to c. 1100 BC.

Discussion and conclusions

From cultural and chronological point of view, especially the ornamental system at Orsoya reflects a strong south influence. On its side, the population of EPCLD remarkably contributed to the distribution of that influence (respectively of the so-called Mycenaean *Koine* of the ornamental style) to the west, along the Danube. Clearly, it had close contacts with the Aegean cultural circle. The southern influence resulted in a distribution of motifs not only unknown before in the area of Danube, but also unknown for its social environment – lotus, palm, papyrus and plants. It should be stressed also the wide distribution of the spiral motifs and of S-sign, combination of rhombs and spirals, etc. Those elements in fact also influenced the neighbor cultures like Otomani – Füzesabony, Wietenberg, etc. (see for instance Boroffka 1994 with ref.). The metal finds (for instance the so-called developed type of Vesselinovo) characterized by a rich ornamentation also reflect the Mycenaean ornamental style. So, the emergence of the Orsoya cemetery corresponds to the emergence of the so-called Mycenaean *Koine* to the north, in the Lower and Middle Danube basin, on the one hand.

On the other hand, the end of the cemetery also reflects the southern cultural processes, because it corresponds to the final phases of the Great Aegean population movement. It other words, to the graduate decreasing of the Mycenaean style.

What was the historical destiny of the EPCLD population,

Figure 20.3. Anthropomorphic figurines with encrusted ornament. Orsoya cemetery.

respectively of the Orsoya inhabitants, it is difficult one to say. The problem is multi-aspect. In historical context, at the end of the second millennium BC, because of the objective reasons – gradually introduction of the iron and the re-allocation of the spheres of influence, the possible demographic 'boom', went to general change of the ethno-demographic map of Southeastern Europe, the Aegean and western Asia Minor. The Hittite was destroyed for a short for that time period and the main Mycenaean centers were in ruins. Different ethnical names occurred in the Hellenic ancient record, among which the Thracian was as a common ethnos-name. In the material culture there are some continuation elements – as kantharoi and kyathoses, as well as a development of the ornamental system known from EPCLD. Main motifs and technique of the beginning of the Iron Age are genetically connected with EPCLD. Because of that fact, having in mind the communication and information system of the ornament in the ancient societies, they were not only preserved but also developed in new context. That is an important fact inferring continuity in the culture on the border between the second and the first millennium BC over vast area of the Balkans. Therefore, in my opinion, the cradle of the earlier Thracian civilization can be searched in the area of the Encrusted Pottery Culture along the Lower Danube, which most representative record is the Orsoya cemetery.

CHAPTER 21

CERAMICS AND CLAY SOURCES: HEAVY MINERAL STUDIES FROM KAMENSKA CHUKA, SOUTHWEST BULGARIA

Mark Stefanovich, George Ajdanlijski, H. Arthur Bankoff, Strashimir Strashimirov, Ilija Kulov and Nathan Meyer

Introduction

Two main north-south routes through the Balkans link the Aegean through the Central Balkans with temperate Europe: the Morava-Vardar (Axios) River valley and the Struma (Strymon) River valley. (Stefanovich and Bankoff 1998:255-58). The Struma River rises southwest of Sofia, Bulgaria and flows south and southeast into the Strymonic Gulf in northern Greece (Aegean Sea). The Blagoevgrad basin, the largest of the five Struma valley basins south of Sofia encompasses approximately 150 square kilometers along the middle Struma and the lower Bistrica Rivers between the Vlahina Planina and the Rila range (essentially from Dupnica to Blagoevgrad) (Stefanovich and Bankoff 1998:257).

The excavation and survey of Kamenska Chuka and its environs comprises the first phase of the Struma Valley Project, a larger program of archaeological survey, excavations, and analyses designed to fill an informational gap about the most important north-south route. Over the long term, the project plans to study human occupation, especially, along the middle Struma River valley through time. Today's rapid growth of urban centers, the expansion of mechanized farming and extensive illicit digging threaten to eradicate existing sites and make survey and excavation in the region particularly urgent. The excavations at Kamenska Chuka, five kilometers south of Blagoevgrad in southwestern Bulgaria, were begun in 1993 and are projected for completion in 2000. The first three seasons (1993-95) of the Kamenska Chuka excavations are described in a preliminary report (Stefanovich and Bankoff 1998:255-338). Initial results from Kamenska Chuka show that the northern periphery of the Aegean world in western Bulgaria in the Later Bronze Age extended at least to the middle Struma River basin.

At the regional scale, the Struma Valley Project has brought together specialists in archaeology, geology, and palaeoecology with the explicit goal of contextualizing the region and exploring its connections with the contemporaneous neighboring cultures. Thus far, at the microregional scale, specific emphasis has been given to understanding the relationship between the site of Kamenska Chuka and its surroundings.

The subject of this report are the results of the preliminary geological survey of clay sources and analysis of the heavy mineral content of selected ceramic sherds conducted by the University of Mining and Geology, Sofia. The objective of the study was to determine whether the pottery found at Kamenska Chuka was produced from locally available clays or from sources farther afield.

Site Location

The site of Kamenska Chuka consists of Eneolithic remains (V-IV millennium BC) and a Later Bronze Age (c.14th-13th centuries BC) structure occupying a commanding height (elevation 404 meters above sea level, location 41° 59"; 23° 06") above the middle course of the Struma River. Geologically, the basin is composed of low-lying alluvium surrounded by colluvial slopes and older eroded metamorphic hills. The neighboring mountains rise to over 2000 meters. The site itself rests on a outcrop of metamorphic micaceous gneiss and schist. The larger settlement area encompasses an area of approximately 200 x 150 meters on the highest point of the hill, and is definable by the extent of heavy sherd, daub, and rock deposit on the surface (Stefanovich and Bankoff 1998:258, 266-69 [***Figure 21.1***]). An abrupt drop of some two meters sharply delineates the "mound" of Kamenska Chuka from the surrounding hill slope on the north and east where the Eneolithic settlement designated Strumsko has been excavated (Ste-fanovich and Bankoff 1998:262-63). The south and west sides of the hill are at a much steeper angle, making approach from those directions nearly impossible. From the top of this hill one can see the entire southern part of the Blagoevgrad basin. To the northwest is the housing development of Strumsko. To the north lies the city of Blagoevgrad, near the probable location of the town of Scaptopara. To the south lies the Kresna defile marking the edge of the basin, through which the Struma River and the ancient and modern roads wind towards the Aegean.

Materials and methods

To determine the possible connection and use of the available raw materials surrounding Kamenska Chuka a program of geological and mineralogical studies has been initiated. One of the first tasks was to identify the source of the pottery found at Kamenska Chuka. In order to do this, all natural clay outcrops around the archeological site, as well as pottery fragments from two Trenches - numbers 11 and 12 were sampled and studied for their heavy mineral content. (See Stefanovich and Bankoff 1998: 264, Fig 11 for trench layout through 1995). Trenches 11 and 12 are located within the walls of the Later Bronze Age structure. All of the samples were collected from

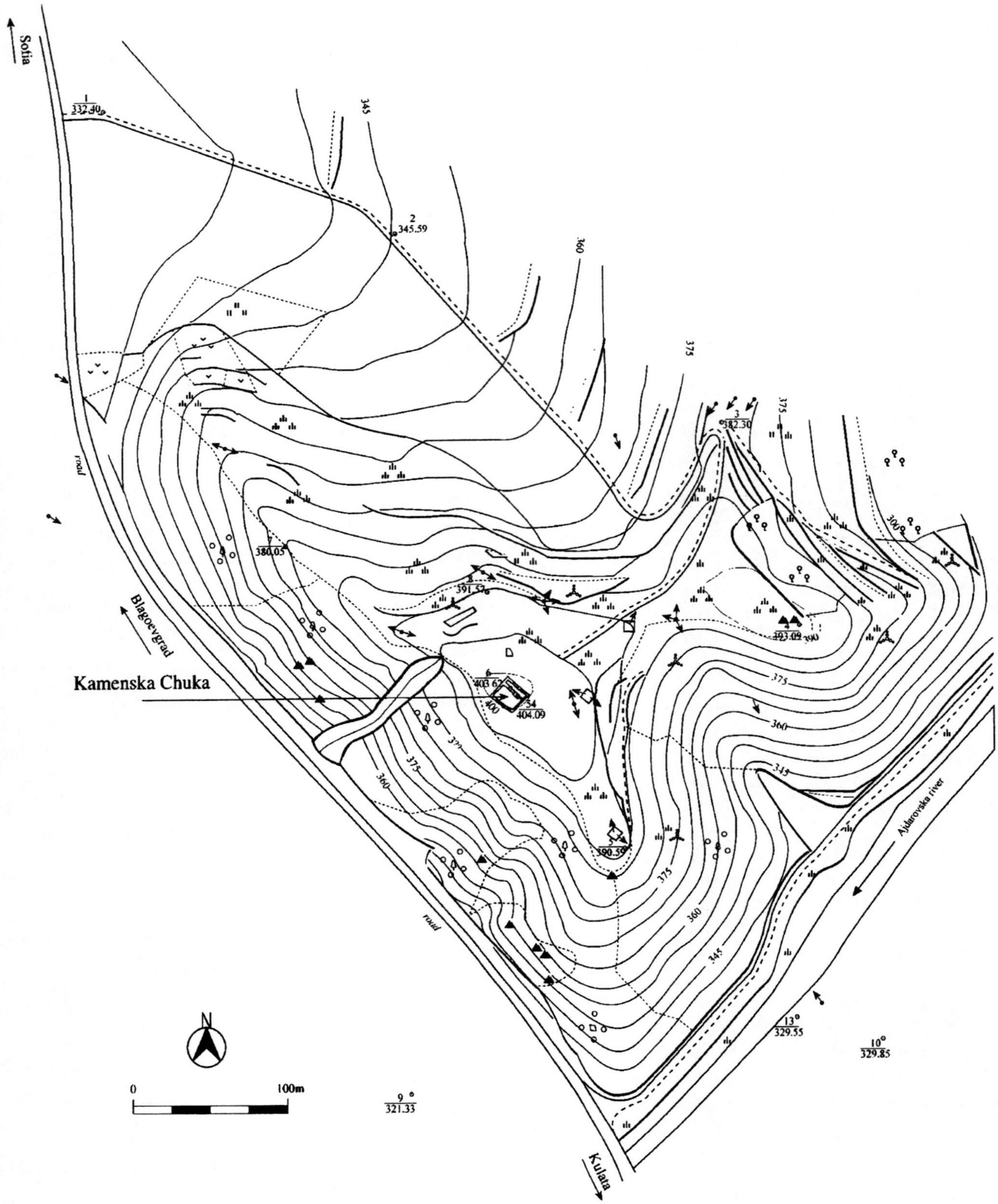

Figure 21.1. Topographic plan of the plateau around Kamenska Chuka mound.

the 1995 excavations.

During the field work four natural clay-containing outcrops (9 samples) around the archeological site were sampled. The distance from Kamenska Chuka to the clay sources are: field point 71 – 200 – 250 meters, field point 74 – c. 1500 meters, field point 75 – c. 2000 meters and field point 76 – c. 3000 meters. All of them are situated northeast of Kamenska Chuka (***Figure 21.2***) and comprise fluvial and proluvial gravel-sandy successions with intercalation and lenses of sandy and aleurolitic clays. Both coarser (laboratory numbers 75, 77, 80 and 82) and finer (laboratory numbers 74, 76, 78, 79 and 81) parts of the outcrops were sampled. The average weight of each of the nine samples is 10 kg. After standard field washing the magnetic fraction was removed from the remaining heavy fraction.

Before beginning the specialized heavy mineral laboratory procedures, the pottery fragments were observed under a binocular stereo microscope and classified according to their internal structure. As criteria for this classification we used the degree of internal homogeneity and the existence and type of the zonality. Representative samples of each of the determined groups (39 samples) were subjected to the standard laboratory treatment for heavy mineral study - crushing, sieving and washing. The heavy mineral fraction obtained from every ceramic sample, together with the probes from the field sampling, went through brommophorm and electromagnetic separation. The remaining electromagnetic and nonmagnetic heavy mineral fractions were the object of the mineralogical, crystalomorphological and statistical study under a binocular stereo microscope. The percentage content of each mineral was evaluated by using the "Composition Chart for Estimating Percentage Composition (AIG Data Sheet 15.1,2) ".

The cluster analysis of the remaining mineralogical data was carried out using the Pearson method. The crystalomorphological investigation of the detrital zircon was based on the methodology, proposed by Pupin (Pupin 1980).

Geological setting

The studied area represents the southern part of the Blagoevgrad graben, which developed in Precambrian rocks and is filled by Neogene sediments (***Figure 21.2***). The graben base in the area of archaeological investigation is represented by metamorphic rocks of the Prarhodopha (Ograzhden) Supergroup (Archaic ages) (Zagorchev1984) represented here by the Troskovo Group (amphibolites, biotite and two-mica gneisses and schists) with the general orientation of the schist's cleavage 37/40 and the Maleshevska Group (biotite and two-mica gneisses, augen-gneisses) with the general orientation of the foliation 50/50-60. The highest area of the Kamenska Chuka mound is of two-mica gneisses. The eastern parts of the mound are of augen-gneisses belonging to the Maleshevska Group. These petrographic types are the main constituents of the Kamenska Chuka mound and were used as the main building materials for the Later Bronze Age structure.

More important for the provenance investigation of the pottery origin are the characteristics of the sedimentary successions filling the graben. This investigation is based on the stratigraphic scheme proposed by Bakalov (1977; 1983) and its paleogeographic reconstructions. The sedimentary complex is subdivided into two parts: upper - represented by Quaternary fluvial and proluvial deposits, and lower - the Blagoevgrad System (according to NACSN 1983, this is an allostratigraphic unit with the rank of an allogroup). On the basis of the intrabasinal interruption of the processes of sedimentation this unit is subdivided in four Intertems (according to NACSN, 1983, it is an allostratigraphic unit with the rank of an alloformation), from bottom to top: the Stob, Belopolci, Elijci, and Slatino Intertems. Based on lithic characteristics and stratigraphic position they are subdivided into Formations and Members. Here we describe just those that have outcrops near the archeological excavations of Kamenska Chuka.

Two of the above-mentioned Intertems, Stob and Belopolci, were sampled. The Elijci Intertem is eroded and the Slatino Intertem is not developed in the studied area. The Stob Intertem here is represented by the fluvial successions of the Eminovtsi Formation, the proluvial-fluvial sediments of the Pokrovnik Formation (Ajdarovo Member) and limnic sediments of the Dzherman Formation (***Figure 21.2***). They cover a significant part of the metamorphic floor to the north and east of the Kamenska Chuka mound. Just limited areas are built by the predominantly fluvial successions of the Barakovo Formation (Chaprashlaka and Chaprashlaka-Belopolci Member). Every one of these units has natural outcrops around Kamenska Chuka. In the sediments of the Chaprashlaka Member there are no raw materials that can be used to make pottery. All Quaternary sediments around Kamenska Chuka, resulting from the recent position of the Struma River, are agriculturally and industrially altered to a significant degree and cannot be used for the purposes of this investigation.

The lowest part of the section is built by the Pokrovnik Formation (Ajdarovo Member - ad N_2), represented by the alternation of grayish-green (5Y 7/2-6/4) massive gravel, sands and sandy clays (***Figure 21.3a***) and outcrops in the northern (***Figure 21.2***, field point 74) and part of the southern slope of the Ajdarovo River. The clay bodies form lenses, enriched by an aleurolitic fraction and compose 1/5 of the studied outcrop. The gravel is polymictic (predominantly metamorphic rocks, dominated by 20% amphibolites). The degree of lithification is low - it is easy to dig by hand. Age - N_2.

To the north of the Ajdarovo River the sediments of the Dzherman Formation (dz N_2) cover the rocks of the Pokrovnik Formation. They form a significant part of the top of the mounds north of Kamenska Chuka. In the sampled outcrop (***Figure 21.2***, field point 75) they are represented by massive aleurolitic and sandy clays with lenses of polymictic sands and gravels (***Figure 21.3b***). The clays are grayish-green (10Y 7/2) and the degree of lithification depends on the carbonate content - there are levels with clay-carbonate concretions and lenses. The average silt and sand content in the clays is 30%. Gravels are mainly from well-rounded leucocrate granites, pegmatites and quartz. The mezocratic granites, amphibolites and lower-triassic sandstones are present in lesser quantities. The age of the rocks is Pontian.

To the northeast the Dzherman Formation is separated by a fault from the brecca-conglomerates of the Eminovo Forma-

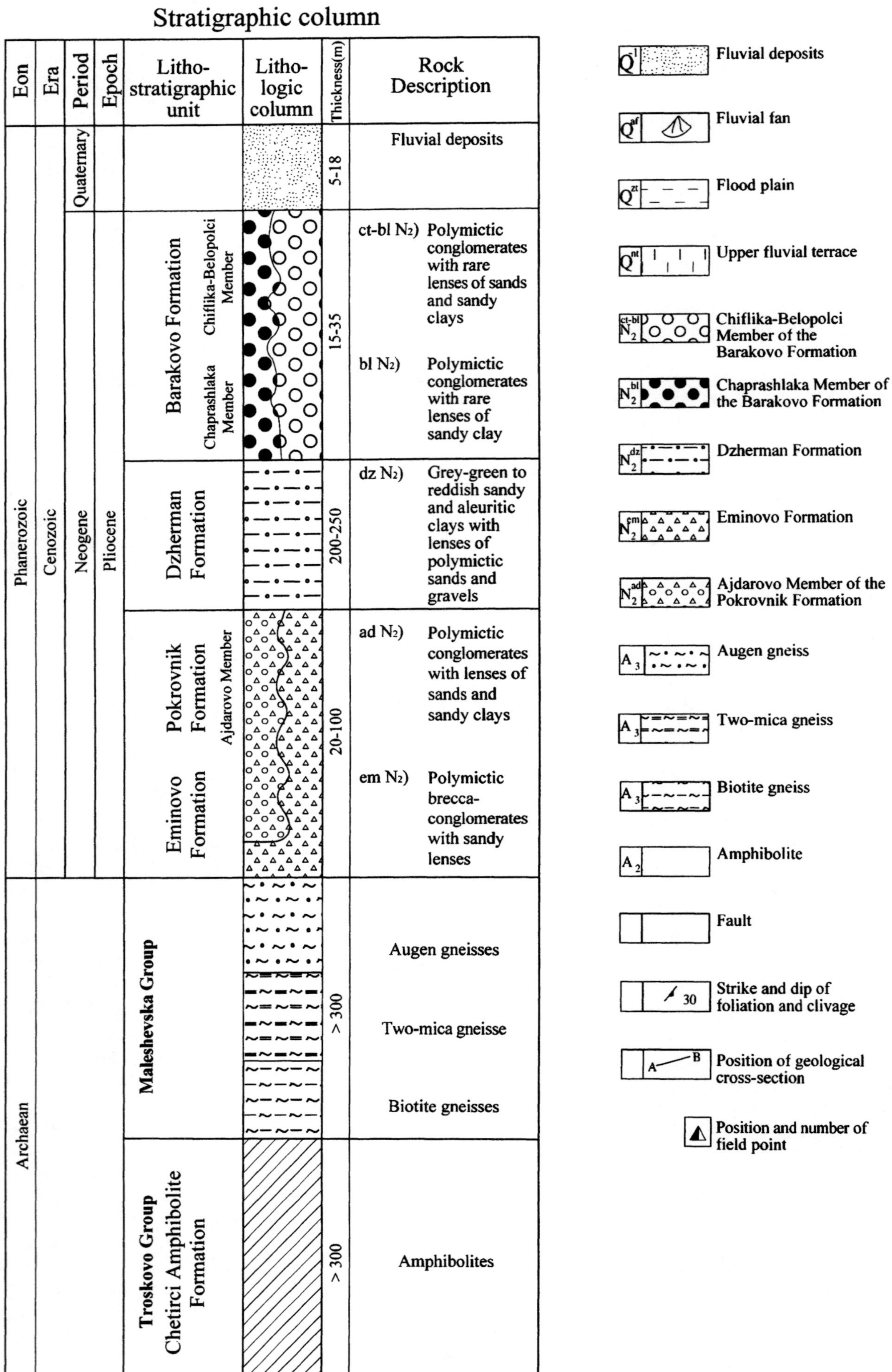

Figure 21.2. Stratigraphic column and schematic geological map of studied area (after P. Balakov 1983) with the samples locations marked.

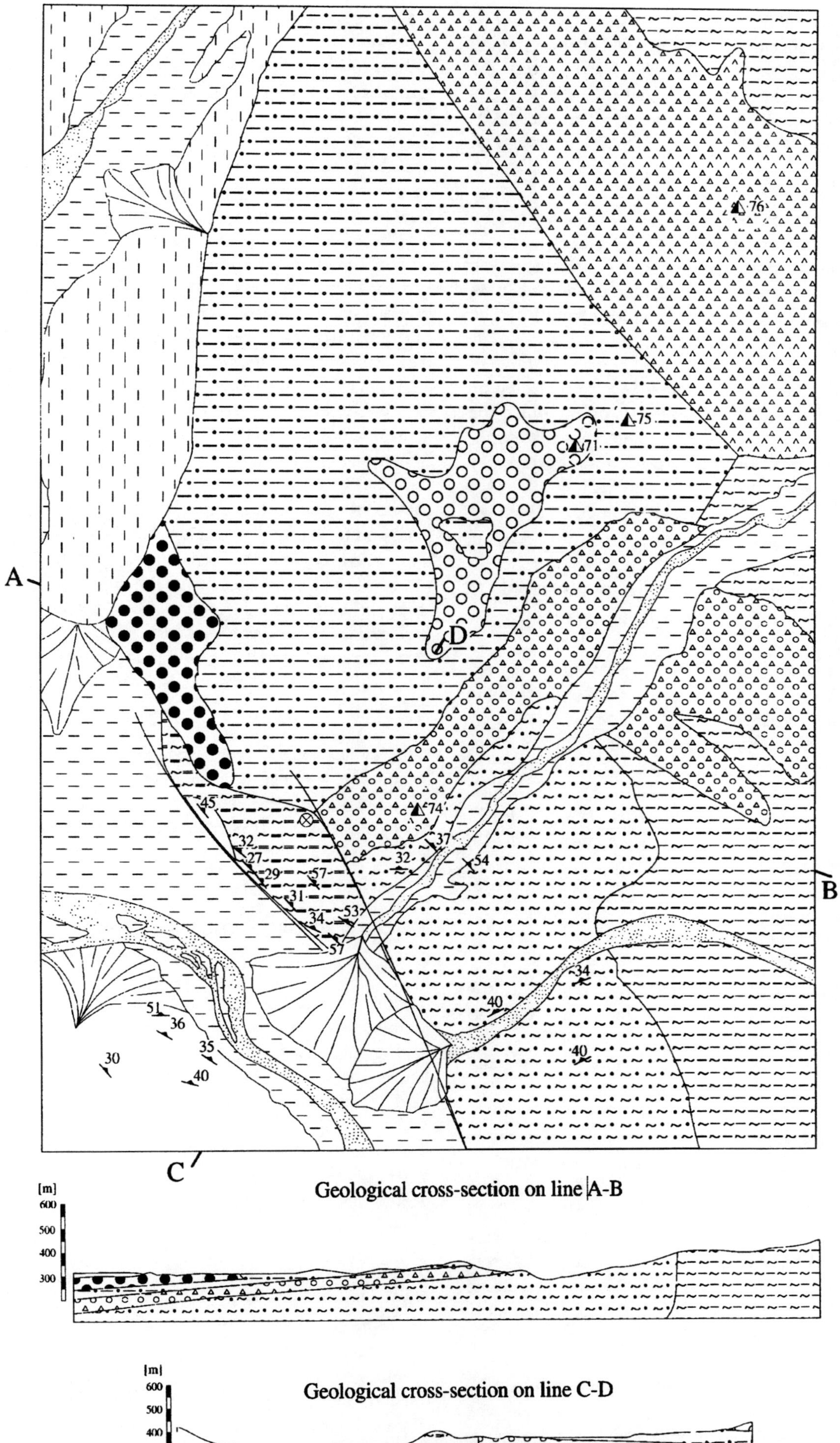

Figure 21.2. Stratigraphic column and schematic geological map (continued).

tion (em N_2), which fill the graben to its eastern edge in this area. Here coarse sediments predominate and clay-containing lenses can be observed only in some parts. The color of the sediments is grayish-green (5Y 6/2). All lithological types studied are poorly sorted and have strongly variable grain-size. In some places graded bedding is visible. The age is Pontian.

To the northeast of the archeological excavations, covering a limited part of the mound, is situated the Barakovo Formation (Chaprashlaka-Belopolci Member - cf N_2). Coarse-grained (gravel) sediments predominate and polymictic sands and grayish-green sandy clays are found in smaller quantities. The age is Pontian.

The borders of the Kamenska Chuka mound are presented by normal faults with a northwest-southeast orientation. The mound, which is a small-scale horst, topographically dominates the area because of the lithological features of the rocks that were used to build it. They represent a high metamorphic complex, which are more or less liable to weathering than the sediments around the horst.

Mineral composition of the heavy mineral fraction

The mineralogical characterization of the electromagnetic fraction is given in ***Table 21.1***. Garnet is the main mineral in the fraction. Red colored garnet predominates the raw clay samples. The studied pottery samples are characterized by either pink garnet or red, or a mixture of both.

Ilmenite is present in just half of the pottery samples. In the clay samples it varies form 3-4 to 25%. Its concentration in the pottery samples is not related to the degree of alteration of the clays in them. Hematite is present in all clay samples (1-10%) and in most of the pottery samples, where it varies significantly from single grains to 15%.

The epidote is better represented in the clay samples (those from the Barakovo and Pokrovnik Formation) and in some of the pottery samples, where it varies from 1-2 to 13%. In two of the sampled field units the epidote content is very insignificant and varies.

The blue-green amphiboles and pyroxene are represented in some of the clay samples. The hornblendes, both green and brown, are present in insignificant quantities (4-5%) in just one of the raw material samples which comes from the sediments of the Barakovo Formation. In almost all of the pottery samples the hornblendes have been found in quantities from single grains to 2-3% and in only some of them are they found to be the dominant electromagnetic mineral (See laboratory numbers 96, 97 and 104 all of which come from Trench 11).

Some of the frequently found minerals in the electromagnetic fraction, such as martite, are more useful for determining the raw clay samples than the pottery samples. The martite is present only in the Barakovo and Pokrovnik Formation where its average content is around 5-6%. In the pottery fragments it is between 2 and 10%, and only in three of the samples was it more than 20%.

Tourmaline and monazite are rare and can be used only for correlation between the raw material samples.

The nonmagnetic heavy fraction is more variable in terms of content and structure (***Table 21.2***). The analysis showed that one of the pottery samples has no nonmagnetic heavy mineral content.

The dominant mineral in the nonmagnetic heavy fraction is zircon. Its content varies between 40 and 70% and is very rarely less than 20% (laboratory numbers 79, 80 - Dzherman and Eminovo Formation, 125 and 127 - sherds from Trench 12). Interesting results, connected with the definition and recognition of each of the examined sources of raw clay materials, were obtained after the statistical processing of the crystalomorphological parameters (***Figure 21.4***). The diagram shows that every sampled unit is recognizable and definable.

Rutile varies from single grains to 25%. This mineral is more common for the sediments from the Pokrovnik and Eminovo Formation (field points 74 and 76) and for the ceramics from Trench 11 in Kamenska Chuka. The content of apatite is variable - in some samples it is very stable (Barakovo Formation - field point 71), in other it varies in a large diapason (Dzherman and Eminovo Formation - field points 75 and 76). The average apatite content in the ceramics is 10%, but in some samples it reaches 40%.

The nonmagnetic sphene (titanite) is a rarer mineral in the samples. It is found in different quantities in only two of the sampled outcrops (field points 71 and 74) and in several pottery samples from Trench 11. In the rest of the samples it is either missing or it appears as single grains.

Barite was found in only one of the sampled outcrops - Dzherman Formation, field point 75. Pyrite comes in several samples as single grains, some of them strongly oxidized.

Leucoxene, andalusite, anatase and kyanite were found predominantly in the sampled raw materials as a single grain and can't be used for correlation. Together with the corundum, they were eliminated during the statistical processing and correlation.

The obtained results show that there are enough criteria for distinguishing all sources of raw clay materials available near Kamenska Chuka. Electromagnetic fraction is more appropriate for this purpose.

Ilmenite is at a maximum in the samples from the Barakovo Formation (***Figure 21.2***, field point 71), while in the samples from the Dzherman Formation (***Figure 21.2***, field point 75) and Eminovo Formation (***Figure 21.2***, field point 76) it is found in smaller quantities. Epidote is abundant in the sediments of the Barakovo and Pokrovnik Formation (***Figure 21.2***, field point 74). All of the sphene in the Barakovo Formation is nonelectromagnetic. Sphene is rarer in the samples from the Dzherman and Eminovo Formation. Brown hornblende appears only in the samples from the Barakovo Formation, blue-green amphiboles are present in the samples from the

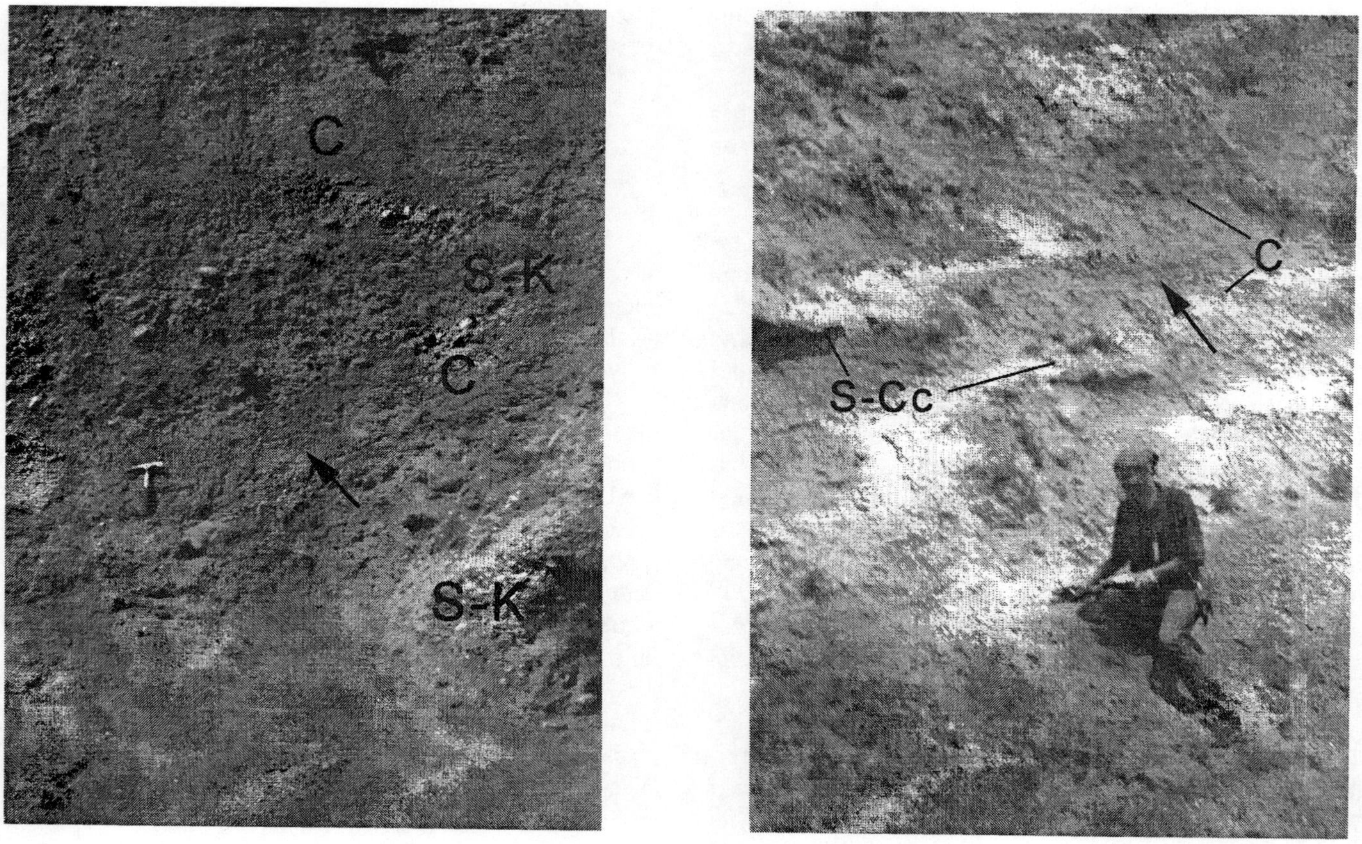

Figure 21.3. Sampled outcrops of raw clay minerals around the Kamenska Chuka mound.

a) Fluvial succession typical for the Pokrovnik Formation, represented by sandy and gravelly dominated channel deposits and clay dominated lenses connected with overbank sedimentation

b) Clay dominated succession, typical for the Dzherman Formation, with visible sandy levels containing carbonate concretions

——→ – Place of clay sampling
C – Clay
S – Sand
K – Conglomerates
Cc – Carbonate concretions

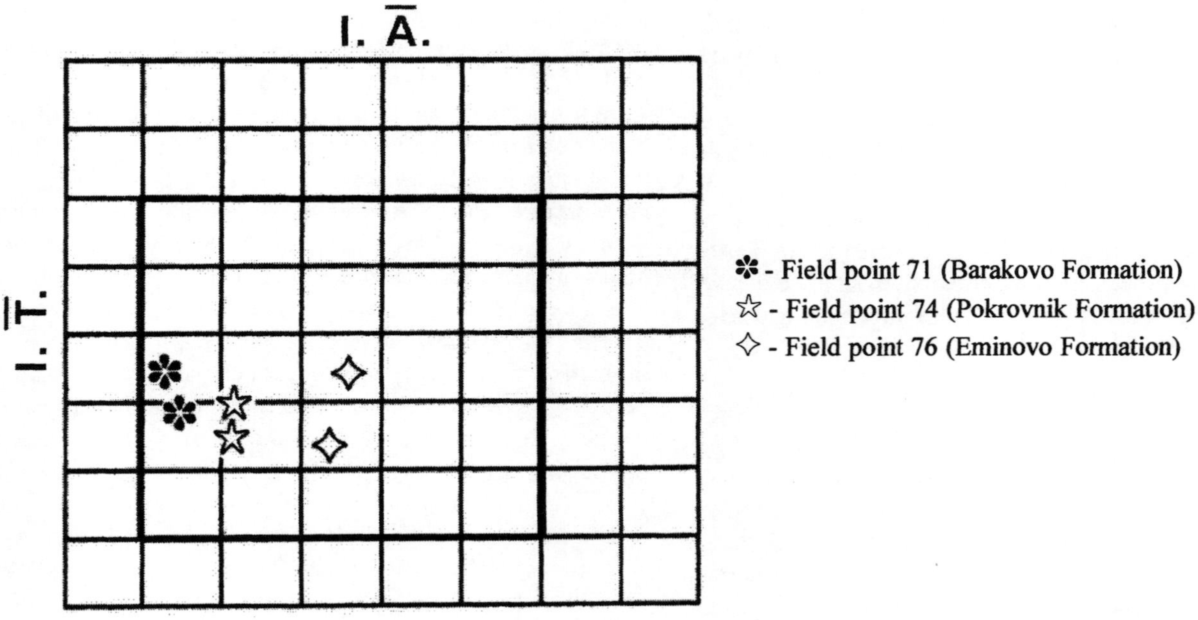

Figure 21.4. Type-crystalomorphological statistical distribution of the studied detrite zincons from the sampled outcrops with raw clay materials.

Pokrovnik and Eminovo Formation. Monazite is present only in the sediments of the Pokrovnik Formation.

The correlation between the raw clay materials and the sampled pottery on the one hand, and between the pottery samples on the other hand, is based on a cluster analysis (***Figure 21.5***). The value of the criterion for the statistical significance of the coefficient of correlation of the mineralogical data from the electromagnetic heavy fraction is 0.95, while the one from the nonmagnetic heavy fraction is 0.75. After analyzing both diagrams we distinguish two groups of samples of high correlation - one that incorporates some of the sampled outcrops and pottery fragments and another which contains only pottery samples. The first group contains the samples from the Barakovo and Pokrovnik Formation (***Figure 21.2***, field points 71 and 74) and three pottery samples from Trench 12 (laboratory numbers 105, 116 and 128). The second cluster group includes pottery fragments from Trench 12 (laboratory numbers 111, 112, 113, 115, 117, 118, 119, 120, 122 and 123). There are several other small groups of samples with a significant degree of correlation, for example samples with laboratory numbers 94 and 95, 96 and 97, 101 and 109.

The results show that from the studied outcrops, the ones from the Barakovo and Pokrovnik Formation (field points 71 and 74) are the closest possible sources of raw clay material for the studied pottery from Kamenska Chuka. The pottery samples with significant correlation with these sediments have a coarse-grained internal structure, high degree of porosity and significant content of mica and rock fragments. Only one of them shows a more significant alteration - asymmetrical zonation, a result of the firing of the vessel.

Discussions

Geological Considerations

In our study we excluded the magnetic fraction (present only in the clay samples), here represented predominantly by magnetite and insignificant quantities of pyrotine and chromite, because of its easy and complete oxidation during a vessel's firing. The hematite and limonite in the pottery, produced after firing, cannot be compared as to quantity and specific features with the magnetite in the field samples too. It is more or less the same with pyroxene (Px), blue-green amphiboles (Am) (ferro- and hydro- containing minerals) and epidote. Depending on the temperature treatment, sphene may also change its magnetic features. Because of its structure, barite is easily distorted even after a short treatment with a high temperature. Therefore, the use of these minerals for cluster analysis is limited.

Crystalomorphology of detrital zircon and garnet is a very powerful tool for heavy mineral comparison because of its stability during erosion, sedimentary transport, accumulation and firing. The small size of the individual pottery samples did not allow the extraction of a good number of chrystalomorphologically well recognizable zircon and garnet grains that could be used for correct statistical processing. On the contrary, single grains of the stable minerals such as andalusite, corundum, kyanite and anataze allowed us to make conclusions about the provenance genesis of the sediments and pottery.

Geological Conclusions

There are enough features of the heavy mineral fraction to recognize every sampled outcrop of possible clay raw materials around Kamenska Chuka, to establish the possible connection between these materials and the sampled pottery fragments, and to group the fragments into groups with possible similar provenance.

Archaeological Considerations

Unbeknownst to the geologists, the sherds sampled in this analysis fall chronologically into two main groups. The first (Samples 105, 116, 128) are coarse wares of Eneolithic or Copper Age date (5th-4th millennium BC). They are found here in secondary context, either included in the fill of the double wall construction of the Later Bronze Age structure or in the fill layer which later intentionally covered the mound (cf. Stefanovich and Bankoff 1998). Their fabric, technique of manufacture, and associations are similar to those found in excavated Eneolithic settlements of southwestern Bulgaria. The other sherds (laboratory numbers 90 – 104, 106 – 115, 117 - 127) are both coarse and fine ceramics whose wall thickness range from 4.67 mm to 25.45 mm. These sherds are in primary context within the ruins of the structure. These are Later Bronze Age in date (14th-13th centuries BC), contemporaneous with the use-life of the Kamenska Chuka structure. Some of the sherds in the ceramic assemblage from which these samples came has formal analogies to those found from this period in northern Greece, Macedonia, Albania, and Serbia (Stefanovich and Bankoff 1998).

These two groups, distinguished as they are culturally, typologically, and chronologically, are also apparently distinguished by choice of clay sources. Although the sample size in this pilot study is small, it seems relatively certain from the results (see ***Figure 21.4***) that the Eneolithic potters utilized outcrops close to home, while the Later Bronze Age pottery seems to be made of clay from as yet unidentified other sources. Several possible explanations may be offered for this distinction. It is possible that the local clays were considered of sufficient quality by the Eneolithic potters. On the other hand the thinner-walled finer wares of the Later Bronze Age may have required clays of a different composition, although much of the earlier Eneolithic pottery is also of a high quality and thin walled. It appears that in the Later Bronze Age there was a preference for clays that might have come from sources outside the survey area, although not necessarily that far away from the site. If, in fact, the choice of clay is based on characteristics important for the manufacture of a particular type of fabric, one should expect that the clay used in later coarse wares (e.g., large storage jars) will vary from that of the contemporaneous fine wares. Analyses and comparison of the composition of more precisely documented Later Bronze Age coarse ware vessels from the 1999 excavation might indicate that these were made from raw material from the same clay sources as the Eneolithic coarse wares. (During the 1999 season in situ vessels and sherds were excavated that are defi-

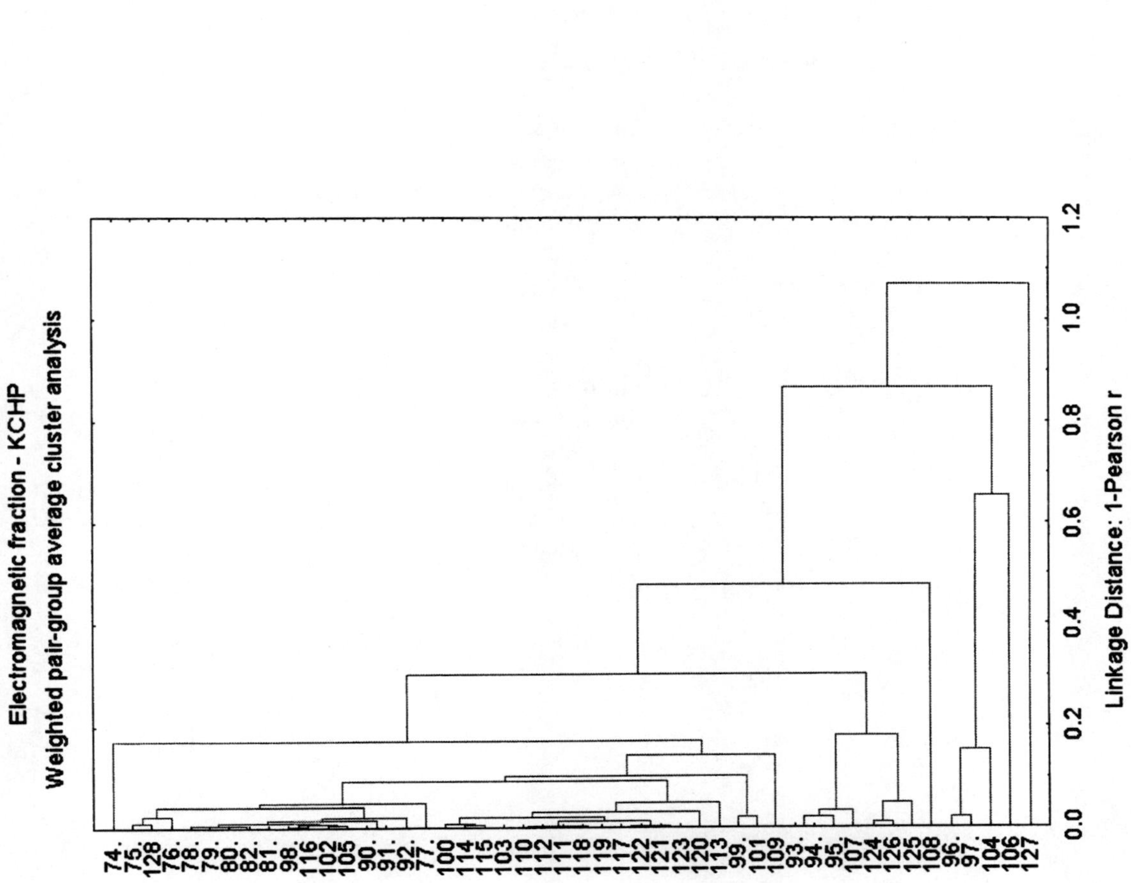

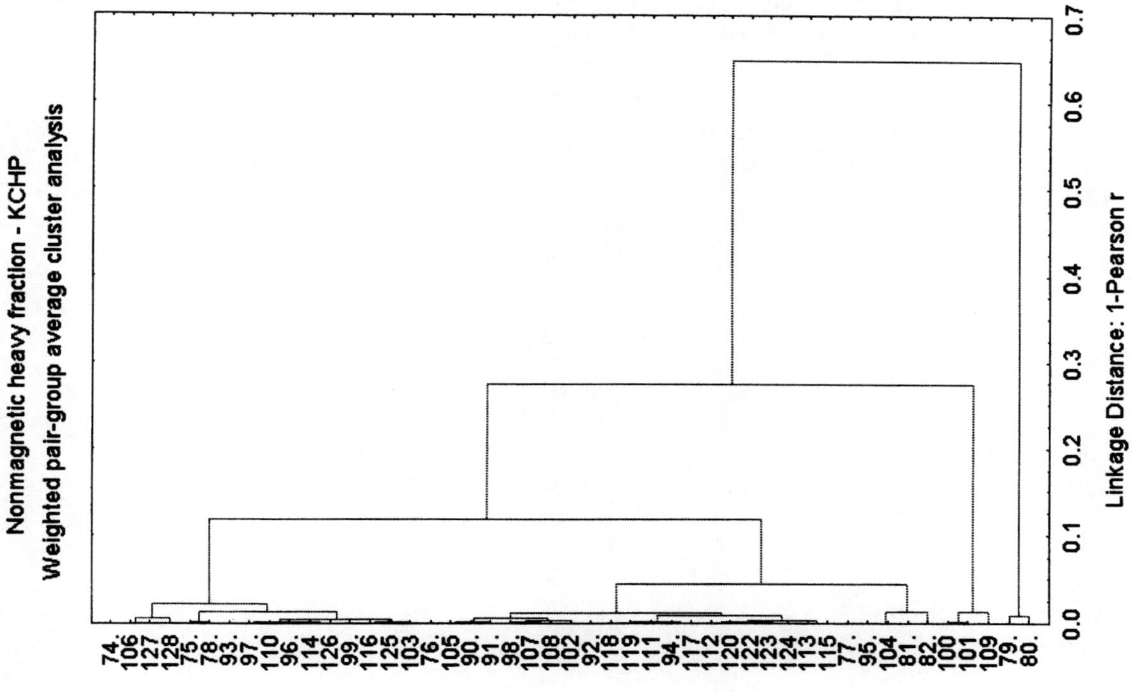

Figure 21. 5. Weighted pair-group average cluster analysis of the mineral content of a) electromagnetic and b) nonmagnetic haevy fraction.

nitely assignable to the Later Bronze Age floor levels, while in earlier seasons most of the pottery rested on the floor level. As a result of the destruction of the Later Bronze Age structure by earthquake and fire and the ultimate mounding of the destroyed remains in the IIIrd century AD many of the sherds are depositionally mixed.) Certainly, it seems likely that the large storage jars found in the Kamenska Chuka structure (Stefanovich and Bankoff 1998: Figs. 31,32,79,80,82) were made locally.

It is also possible that the later ceramic assemblage contained a high proportion of imported finer wares. As mentioned above, this assemblage has typological similarities to that found at sites to the south, west and east. If one is indeed dealing with imports, comparison of the heavy mineral fraction of these samples with those of similar wares from the area of presumed origin would be of interest even if the clay sources were not known. Again, before conclusions are drawn, a much larger sample size would be necessary, and one must eliminate the possibility that the clays in these wares from different sites are similar due to some similarity in clay processing or manufacture of the pottery rather than the source of the raw material.

It is certainly no surprise that two completely different cultural groups inhabiting the same site at two separate times should use different sources for their potting clay, although neither should it surprise us if they had both used the closest available local sources. Economic and social changes, as well as temporal distance, separate the Eneolithic from the Later Bronze Age cultures of southeastern Europe. Increased craft specialization, industrial rather than home production, participation in a wider transactional sphere for both luxuries and necessities, finished goods and raw materials, can all be coaxed out of these results. These conclusions will have to await further data collection and analyses if they are to be dependable and far reaching. The results of this study, intriguing as they may be for the speculative prehistory of the region, are just a beginning.

Acknowledgements

To our colleagues Jordan Gentchev for his support during the field sampling and laboratory studies, Valentin Popov for mineralogical examination of part of heavy mineral fraction samples and Panaiot Bakalov for the interesting discussion on the stratigraphy of the studied area.

Appendix

Aabbreviations used in the text

Am - blue-green amphibole
Ant - anatase
Ap - apatite
Ba - barite
Ep - epidote
Ga - galenite
Gr - garnet (red and pink)
Hb - green and brown hornblende
He - hematite
Ilm - ilmenite
Ko - corundum
Ky - kyanite
L. frac. - light fraction
Lpi - limonitizated pyrite
Lux - leucoxene
Mn - monazite
Mr – martite
NAM – not available to measure
P – sherd number
Pi - pyrite
Px - pyroxene
Rf - rock fragments
Rt - rutile
Sl - mica
Sp - sphene
T - Trench
Tr – tourmaline
U - Unit
Zr – zircon

Field №	Vessel Wall Thickness (mm)	Lab. №	Gr	Ilm	Ep	Sp	Px	Hb	Am	Tr	Mn	Mr	Lpi	Sl	He	Rf
71/1		74	40	25	8	1	*	5		*		10	*		5	5
71/2		75	60	15	5	*	*	4			*	6			4	5
74/1		76	55	15	10	5	2		2			5	*		2	5
74/2		77	50	5	10	10			2		*	2	*		10	10
75/1		78	84	3	7	*	*		3						2	
75/2		79	80	7	*		*	*					*		1	10
75/3		80	85	5	1			*					*		2	5
76/1		81	60	5	*	2			3			2			*	23
76/2		82	75	4	1	*			*			2	*		1	15
T-4/B-80/U-57/P-3782	25.45	90	50	3				3				2			7	35
T-11/B-81/U-57/P-3769	13.5	91	50	5		*		2				3			7	33
T-11/B-81/U-57/P-3774	8.62	92	50	*		5		5				*		*		40
T-11/B-80/U-56/P-1	NAM	93	50			*		3				15			15	17
T-11/B-80/U-56/P-3671	8.85	94	10					*				3			3	84
T-11/B-80/U-56	4.67	95	30		*			3				10			15	42
T-11/B-80/U-56	NAM	96	35	*				60				5			*	
T-11/B-80/U-56	16.41	97	35	10				50				5			*	
T-11/B-74/U-38	7.82	98	55			*		2				5			3	35
T-11/B-73/U-28/B-5819	12.59	99	50	10				15				5			1	20
T-11/B-74/U-17/P-3146	7.68	100	60	10	5		*	5				15			*	10
T-11/B-74/U-43/P-3134	13.42	101	50	*				15				5				30
T-11/B-73/U-23/B-5815	11.24	102	40		2		*	5				5				40
T-11/B-81/U-55/P3735	12.97	103	60	*	1	*		1				10			*	28
T-11/B-81/U-57/P-3777	13.28	104	5					80				1			*	14
T-12/B/U-301/P-3339	17.13	105	30		3	*		2				2			*	63
T-12/B/U/P-3856	15.79	106	60		2			2				2		*	*	34
T-12/B/U-338/P-3861	14.28	107	15		3	*		*				5			5	72
T-12/B/U-338/P-3862	7.4	108	20	*	13	*	*	13				3			3	47
T-12/B/U-338/P-3358	10.97	109	60	20		20		*								
T-12/B/U-338/P-3859	6.43	110	35			*						5			*	60
T-12/B/U-338/P-3852	7.22	111	5				*	*				1		*		94
T-12/B/U-305/P-3348	13.5	112	66	*	3	1		*			1	10			*	20
T-12/B/U-308/P-3366	11.18	113	10	*	*	2		*				3				84
T-12/B/U-336/P-3658	10.8	114	30	5	5	*		1				10				50
T-12/B/U-336/P-3663	20.25	115	40	5	5	*		1				10	*		*	3
T-12/B/U-316/P-3393	16.74	116	30			*		*				2				66
T-12/B/U-316/P-3391	15.57	117	30	*		*		3				10				54
T-12/B/U-327/P-3458	9.11	118	25					*				5			*	70
T-12/B/U-327/P-3461	6.33	119	40		1	*		1			*	10		*		48
T-12/B/U-327/P-3455	13.71	120	40	10				*				15			*	35
T-12/B/U-327/P-3459	15.48	121	30	*		*		*				10				60
T-12/B/U-332/P-3471	11.21	122	30					2				10		*	*	58
T-12/B/U-332/P-3472	11.51	123	30	*		*		*				10			*	68
T-12/B/U-332/P-3478	12.19	124	20		*							10				70
T-12/B/U-332/P-3487	NAM	125	20		*	*						20				60
T-12/B/U-332/P-3481	12.3	126	30		*	*						20			*	50
T-12/B/U-332/P-3477	17.48	127	13	37		*	*	2				37				10
T-12/B/U-332/P-3473	25.1	128	40	5		*	*	2				5				48

25 – weight percentage
* – single grains
NAM – not available to measure
See appendix for used abbreviations

Table 21.1. Mineralogical characteristics of the eletrometric heavy fraction.

Field №	Vessel Wall Thickness (mm)	Lab. №	Zr	Rt	Ap	Sp	Lux	And	Ant	Ky	Pi	. fra	Ba	Ko	Ga
71/1		74	50	10	30	10	*	*		*	*				
71/2		75	62	10	25	1	*	*	*						
74/1		76	72	15	10	3				*					
74/2		77	75	20	3	2									
75/1		78	55	10	20					*		15			
75/2		79	15	*	55	*						25	5		
75/3		80	7	*	50							35	5		
76/1		81	75	20	*	*						5			
76/2		82	60	25	5	*						10			
T-11/B-80/U-57/P-3782	25.45	90	50	10	5	*						35			
T-11/B-81/U-57/P-3769	13.5	91	50	10	2	*						37			
T-11/B-81/U-57/P-3774	8.62	92	50	*	2	*						48			
T-11/B-80/U-56/P-1	NAM	93	50	*	10	*						40			
T-11/B-80/U-56/P-3671	8.85	94	80	7	3	5						5		*	
T-11/B-80/U-56	4.67	95	74	20	3	3									
T-11/B-80/U-56	NAM	96	40	2	10							48			
T-11/B-80/U-56	16.41	97	50	*	10							40			
T-11/B-74/U-38	7.82	98	68	10	5	2					*	15			
T-11/B-73/U-28/B-5819	12.59	99	30	*	10							60			
T-11/B-74/U-17/P-3146	7.68	100	25	2	25							48			
T-11/B-74/U-43/P-3134	13.42	101	25	*	25							50			
T-11/B-73/U-23/B-5815	11.24	102	40	5	5				*		*	50		*	
T-11/B-81/U-55/P-3735	12.97	103	50	2	15	*						34			
T-11/B-81/U-57/P-3777	13.28	104	40	10	*	*						50			
T-12/B/U-301/P-3339	17.13	105	60	10	5	*						25		*	*
T-12/B/U-338/P-3856	15.79	106	40	*	20							40			
T-12/B/U-338/P-3861	14.28	107	60	5	5	*						30			
T-12/B/U-338/P-3862	7.4	108	70	5	3	*		*			*	21			
T-12/B/U-338/P-3358	10.97	109	50	*	40			*				10			
T-12/B/U-338/P-3859	6.43	110	70	2	15						*	13			
T-12/B/U-338/P-3852	7.22	111	98	*	1	1					*				
T-12/B/U-305/P-3348	13.5	112	60	*	5	*					*	34			
T-12/B/U-308/P-3366	11.18	113	33	2	5	*						60			
T-12/B/U-336/P-3658	10.8	114	20	*	5							75			
T-12/B/U-336/P-3663	20.25	115	55	5	10	*					*	30			
T-12/B/U-316/P-3393	16.74	116	30	*	10							60			
T-12/B/U-316/P-3391	15.57	117	80	3	*	1						16			*
T-12/B/U-327/P-3458	9.11	118	60	*	2	*						37			
T-12/B/U-327/P-3461	6.33	119	80	*	2	*						18			
T-12/B/U-327/P-3455	13.71	120	75	*	5							20			
T-12/B/U-327/P-3459	15.48	121													
T-12/B/U-332/P-3471	11.21	122	40	*	5	*						55			
T-12/B/U-332/P-3472	11.51	123	50	*	5							45			
T-12/B/U-332/P-3478	12.19	124	20	*	3							77			
T-12/B/U-332/P-3487	NAM	125	15	*	5	*						80			
T-12/B/U-332/P-3481	12.3	126	20	*	5	*					*	75			
T-12/B/U-332/P-3477	17.48	127	10	*	5	*						85			
T-12/B/U-332/P-3473	25.1	128	25	*	10							65			

25 – weight percentage
* – single grains
NAM – not available to measure
See appendix for used abbreviations

Table 21.2. Mineralogical characteristics of the nonmagnetic heavy fraction.

Chapter 22

Bird-Images on Early Iron Age Pottery from South-Eastern Thrace

Krassimir Nikov

The constantly increasing amount of information about the ceramic production in South-Eastern Thrace from Early Iron Age allows the raising of several new and seemingly non-traditional problems. An interesting and specific aspect in this regard presents the images of birds on the pottery of this age (***Figures 22.1-22.2***). Up to now, the author is aware of 13 different samples: Pshenichevo (Čičikova 1972); Glavan (Nikov 1996); Kapitan Andreevo (unpublished); Lenovo (Detev 1960); Constantia - 3 (Hänsel 1976: Taf. 24: 16; Aladzhov and Balabanyan 1984, as well as unpublished); Rizia (Τριανταφγλλος 1987: fig. 16); Roussa (Triandaphyllos 1983: fig. 16); Malkoto Kale (Fol 1982); Ovcharovo and Gulubovo tell (unpublished).

Geographically, they are spread so as to concentrate considerably in a region with its center the middle course of the Maritsa River, north-west slopes of Sakar Planina Mountains, the lower course of the Sazlijka and Blatnitsa Rivers (***Figure 22.3***). The most of them probably come from settlements but have no certain stratigraphic position. The only one vessel with entirely preserved birds decoration is discovered in a funeral context of a destroyed tumulus, but unfortunately there is no information about its original place (Τριανταφγλλος fig. 16).

The comparison of the different samples shows a considerable difference in the technique and artistic manner of execution (***Figure 22.1***). Some of the images are pressed with a whole stamp (***Figure 22.1: 1-2 and 4***), other are represented by separate pseudo-corded incisions (***Figure 22.1: 6, 8 and 10***) or other incised pattern (***Figure 22.1: 5 and 9***). A special case in the sherd from Constantsia (unpublished) on which the bird image is incised after the firing of the pot. Almost all the representations reveal a certain primitivism and unequal level of the artists' abilities and experience in the techniques applied. The legs in cases when they are preserved are elongated and always in a static position (***Figure 22.1: 8-10***). The bodies are leaf-shaped – a form probably preferred for its easier execution. The long neck is another characteristic feature typical of the long-leg aquatic birds (***Figure 22.1: 5-10***). The stork from Konstantsia is the most elegant and the best executed bird image known (***Figure 22.1: 5***). As the pottery is highly fragmented, certain elements of the images are often missing. Despite this one can say that they are depicted mainly on the upper part of the body and some times just below the rim (Glavan [***Figure 22.1: 2***]).

These images have a central and some times independent role in the overall decorative design. They are isolated images (Glavan [***Figure 22.1: 2***]), grouped in panels in the center of the upper part between the handles (Roussa [***Figure 22.1:8***]), in radial composition on the widest part of the body (Konstantsia [***Figure 22.1: 10***]), or in files (Lenovo [***Figure 22.1: 4***]). In some cases beneath the birds there are elements that could be identified as a land or water surface (***Figure 22.4: 1 and 2***). The birds are always presented in right profile and a tendency towards realism of the depictions is obvious. The characteristic elements of the aquatic birds are stylized, but easily recognizable. One can identify ducks (***Figure 22.1: 1***), storks (***Figure 22.1: 5 and 8***) and cranes (***Figure 22.1: 10***). Interpreting them we should always bare in mind the collective manner of life of these migratory birds.

The bird representation from Malkoto Kale near Sozopol Town is a quite isolated example for its geographical position and the specifics of the artistic manner (***Figure 22.1: 9***).

It is possible only 2 different ceramic shapes to be defined that bare birds' decoration – belly-handled amphorae (Roussa, Glavan) and pithoi with relief bands (Lenovo). This second form often bares rich and quite uncommon decoration (on the specific shape and its decoration see Nikov, in print).

The highly stylized pattern, till recently considered as a separate ornamental group, could be also considered with the bird repertoire. The sherd from Ivantzi supports this suggestion (***Figure 22.2: 1***). It comes from the Thacian fortress Kulata nea the village of Ivantsi, Kurdzhali District. The sherd gives us a good ground for interpreting those motifs as stylized bird images. The region of their distribution is larger but its center coincides with the first group territory (the middle and lower course of the Maritsa River and the lower course of the Suzlijka River [***Figure 22.3***]). Usually the first and the second group decorations are found together, but never on one and the same sherd. The second group pattern decorates belly-handled amphorae, pithoi with relief bands and bowls.

The grate number of variants in decoration is presented here with the most typical examples (***Figure 22.2***). These patterns are pressed with whole stamp. The figures are presented often with right profile (***Figure 22.2: 1-9***), but unlikely the first group, in several cases they are in left profile. Birds in heraldic pose are depicted on only one of the sherds (***Figure 22.2: 15***).

The usage of whole and almost identical stamps does not require special artistic abilities – a fact that could be a probable explanation of the larger area of their distribution and suggests the idea for the existence of a center from which the motifs were spread round. The bird images of the first group depicted by a whole stamp are an exception. Highly stylized

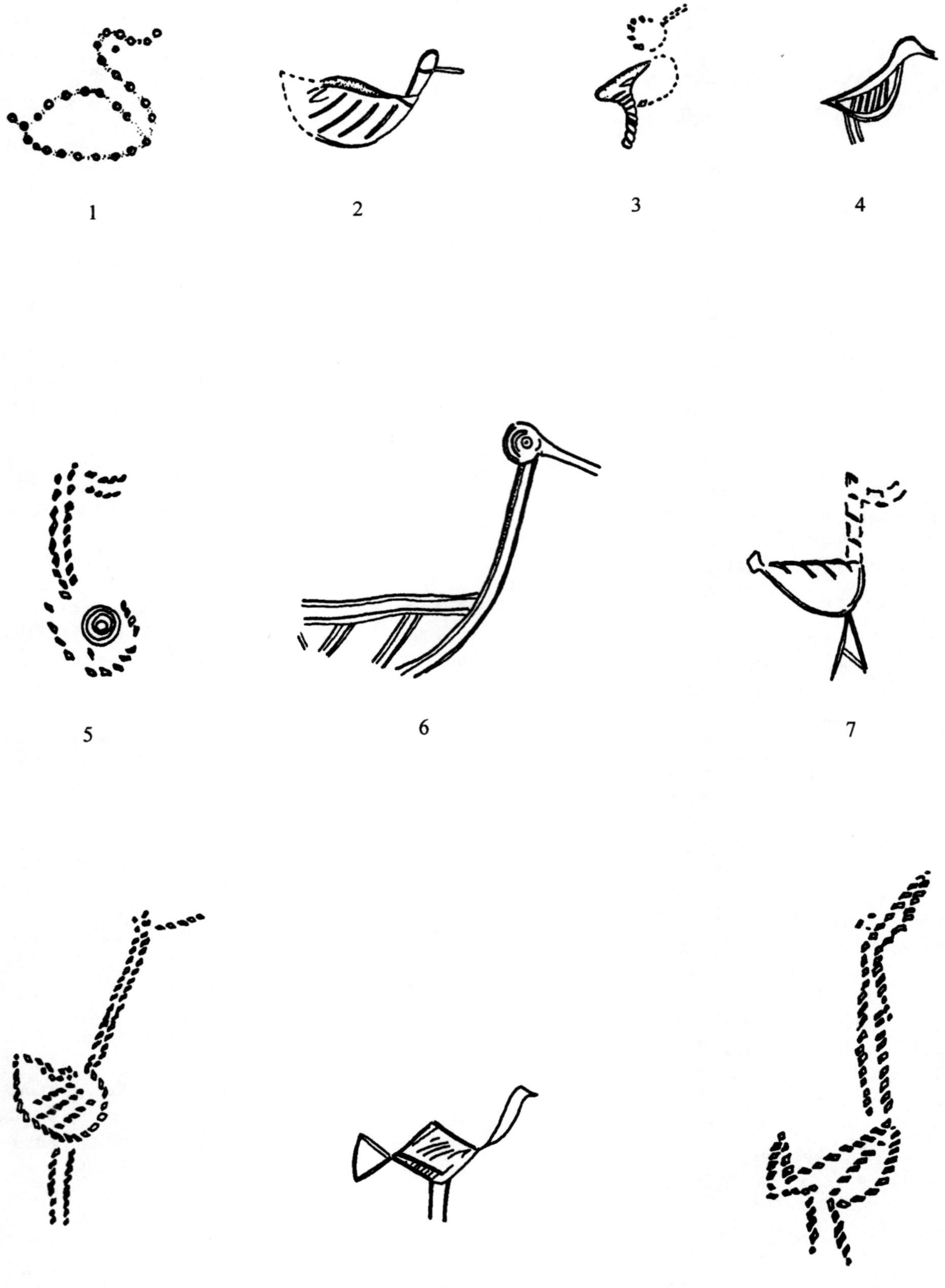

Figure 22.1. The first group of birds images. 1. Pshenichevo. 2. Glavan. 3. Kapitan Andreevo. 4. Lenovo. 5 and 7. Rizia. 6 and 10. Konstanzia. 8. Rousse. 9. Malkoto Kale.

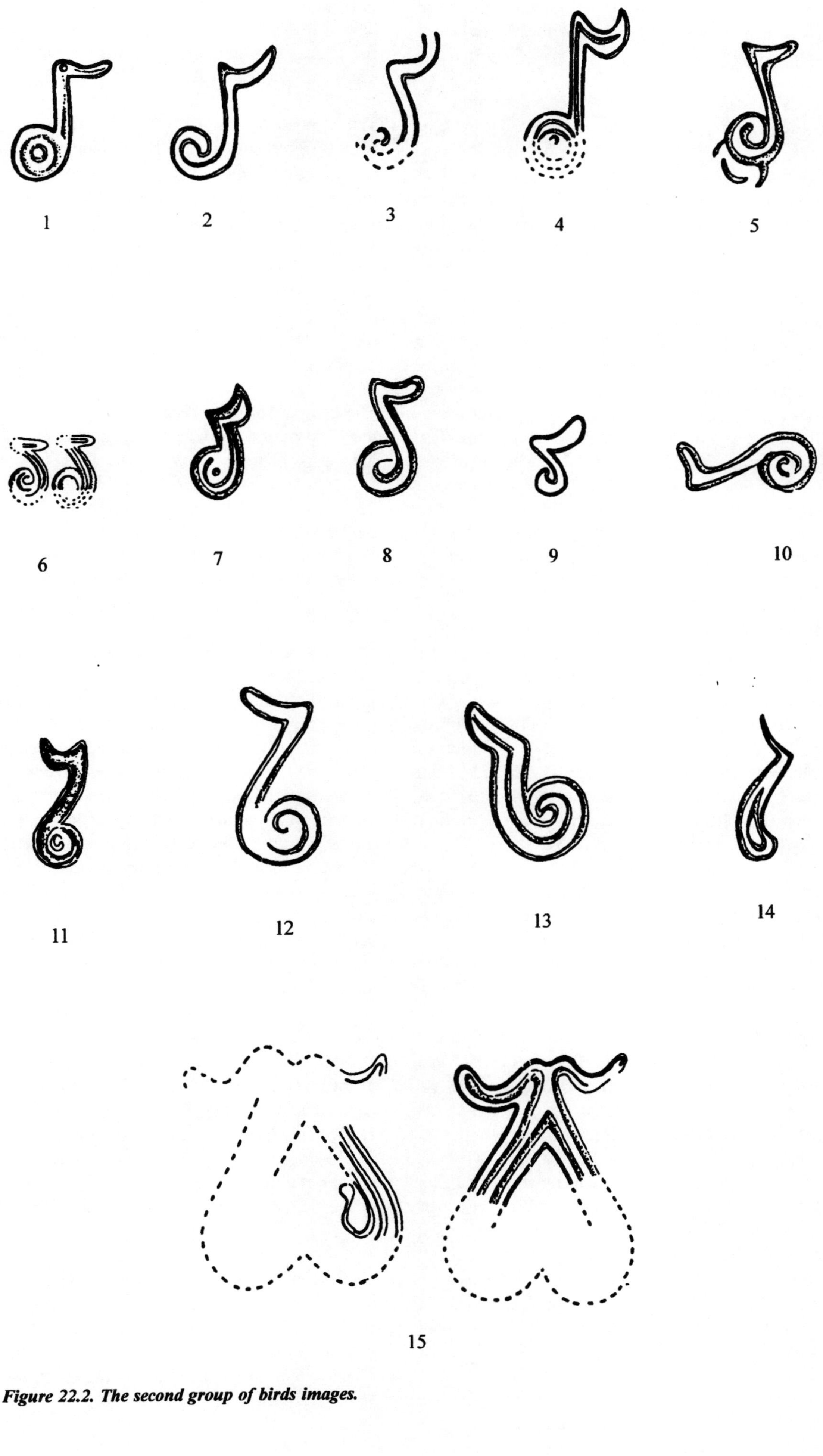

Figure 22.2. The second group of birds images.

images executed with a cord-like pattern are known only from one sherd (Glavan [unpublished]). Usually the whole stamp images are arranged in files.

The considerable variety of the stamps reveals once more the attempt for a realistic representation.

The circumstances of discovering do not allow us neither to accept nor to reject the existence of both the groups in one and the same time. One can do an attepmt to turn to the situation in Helladic in the Dark Ages (Coldstream 1977: 115, 142).

For the time being there are no sherds with the second group pattern from burials. The pottery is found in a mound near the village of Glavan, but without stratigraphic data (on its decoration see Georgieva 1991, fig. 1e). The sherds probably originate from the bird images, but they have lost their previous symbolic value, and are used as a decorative motif. The lost of their original meaning is suggested first by their participation in different combinations with other ornamental elements and by the variability of their position and place in the overall design (***Figure 22.2: 10***). The sherd with file arranged ornaments remind us the Late Geometric decoration in Helladic.

The fact that both the groups are typical just for one compact territory evokes the problem for the cultural specifics of the region. The lack of this type of decoration in the neighboring areas, together with the rare finds of ornithomorphic plastics, so typical of the other parts in the Balkan Peninsula, confirm this suggestion (Georgieva 1977). The only two examples of ornithomorphic plastics from South-Eastern Thrace come from burials and probably should be connected with different cultural traditions (for Glavan see Katincharov and Dimitrov 1978, and for Roussa, Triandaphyllos 1987: fig. 14.1). It's difficult not to judge about the symbolic meaning and connections between both the categories of birds representations. Only the find from Roussa where bird figurines are discovered together with birds decorated ceramics, allow us to connect them with funeral rites. In all other cases some other meaning is possible.

The stylized motifs on ceramics also appear on one of the Razlog stone plates (Toncheva 1984). Despite the fact that the stone plates come from considerably distant region, the manner of depicting is one and the same. The motifs there are combined with bird protomes' bark. There are different opinions on the origin and dating of the Razlog plates (Hänsel 1969), but their peculiar character is indisputable. That fact gives us a ground for a similar explanation at least for a part of the first group of bird images in pottery decoration. This idea is supported by the territories connected with the Hallstatt culture region where the motifs in question are found on a great variety of forms and materials, but they are not found in pottery decoration (except the Bassarabi area [Kossak 1954; Tasić 1983; Briard 1976]).

As a confirmation for the cultural specific of the region we should add the exclusive variety of the decorative pattern and its combination. The validity of this statement is best illustrated by the materials from the middle Maritsa valley. The different sites certain preferences in the kinds of the depicted bird. It is another argument necessitating the identification of the local variants of ceramic decoration in South-Eastern Thrace in Early Iron Age.

The compact territory in which the decoration is spread lead us to the problems for its origins and time of emergence. The contemporary knowledge does not allow us to search its prototypes in the Late Bronze Age. This statement is based on the difference in the composition principles, the techniques and the execution manners. Stylized bird images are depicted on a bowl from the Late Bronze burial newr Ovcharitsa (Maritsa-Iztok). The execution of the birds is similar to those on the belt from Burzitsa. This similarity, the date of the belt and the fact that the bowl is found at a burial reveal the lack of continuity of these images[1].

The comparison with regions where similar depictions from the same period are widespread give opportunity for resolving the problem.

The art styles of the painted pottery in the Aegean world in the Early Iron Age offer a plenty of comparative data for certain conclusions – namely mainland Greece and the islands in the Geometric period and the area of the Phrygian painted pottery in the period of the developed style.

The bird depiction appears for the first time in Attic in the late ninth – early eighth century BC (Coldstream 1977: 78). Inspired by Attic, the motif becomes soon a favorite one on the territory of Hellas. Its emergence in Phrygia is considered as a result of independent trend of development (Sams 1974: 170; 1978). The two regions show difference in the kinds of the birds preferred. During the Geometric period the presentations of birds connected with water dominate in Greece. Thus the birds from Thrace stay closely to the Greek artistic tastes but their realism is a distant feature for the region. The common characteristic of the regions in question is the representation of the birds in right profile (with the exception of the heraldic compositions). In Thrace one can not find the animals common in Attica and Phrygia. The situation here reminds for Corinth where the horse images, so popular in Attica (Benson 1970: 31), could not be found, and for Phrygia, where the favorite animals are the lion and the goat (Sams 1974). Some depictions from Thrace show great similarity in the overall design with the Attic painted pottery from the Late Geometric period (***Figure 22.4***; cp. Borell 1978: figs. 1: 53; 2: 85; 5: 5; 7: 47). Another similarity is found in the manner of execution. Despite the technique differences, the Thracian images remind the open-work technique invented for bird images in Attic (***Figure 22.1: 2, 5, 8 and 10***; cp. Benson 1970: 70-72).

Now we come to the question of the emergence of these motifs in Thrace: Are they a result of cultural interactions or a product of an independent line of cultural development? The first suggestion is supported by the bird decorated pottery from Thasos. Similar depictions are known from Troy VIII and

[1] I would like to thank T. Kuncheva-Russeva for the infomation. See also Kuncheva-Russeva, chapter 25.

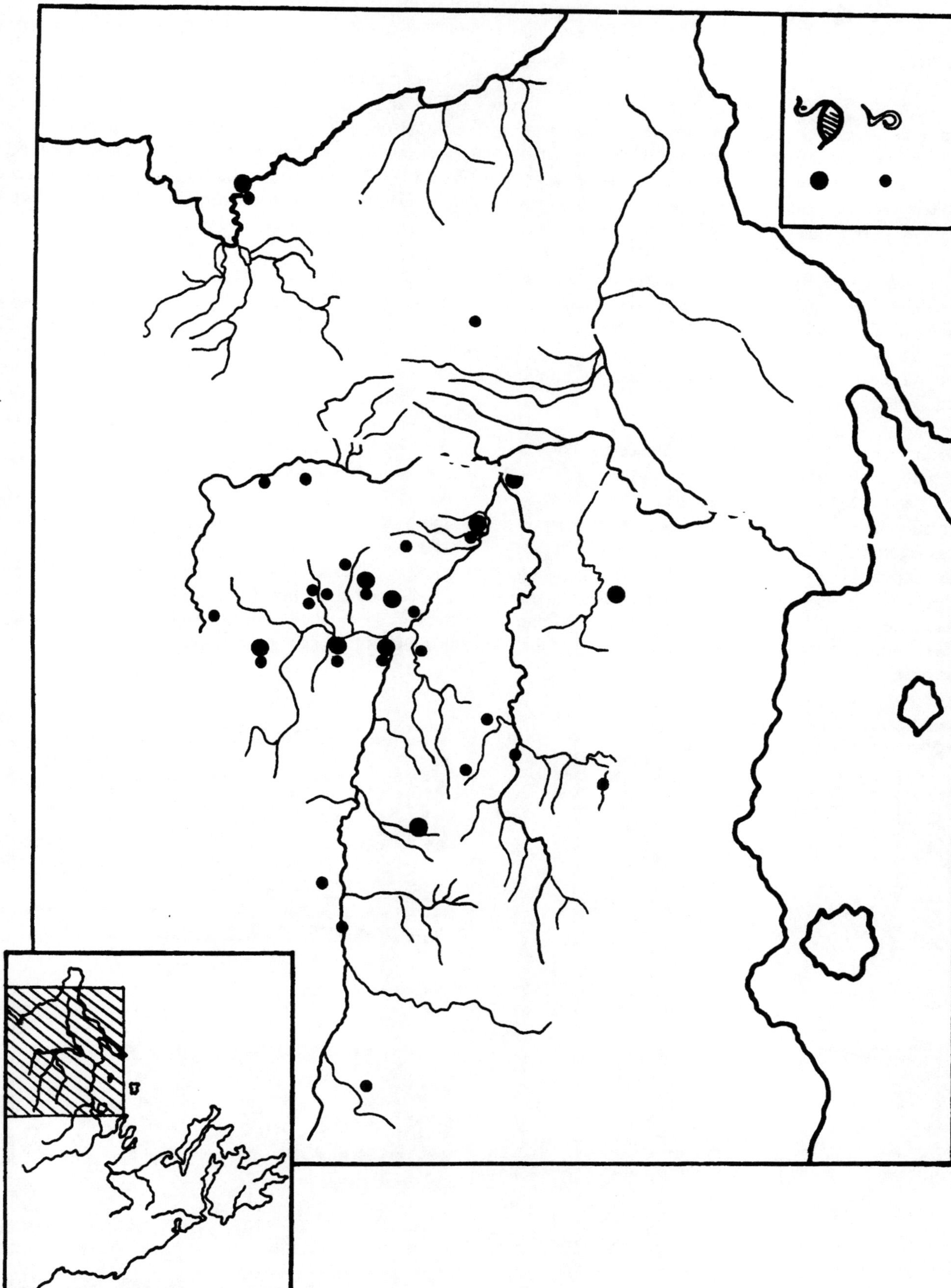

Figure 22.3. The distribution of pottery with birds images from South-east Thrace.

Lesbos and are stylistically connected with the so-called G 2-3 pottery (Bernard 1970: 31). A possible explanation of the diffusion of this pottery are some intensive cultural interactions in the Northern Aegean but the dating of this process in the seventh century BC, especially for Thasos is uncertain. The unclear stratigraphy of Thasos Artemision, as well as the presence there of earlier (?) pottery does not reject the earlier dating of those with bird images (Bernard 1970: 140-42, fig. 52).

Taking into consideration the close connections of Thasos with the mainland, one could expect exostence of early painted pottery in Thrace. In this respect quite important is the report for discovering of early painted Greek pottery in the Viza area (Özdoğan 1985). Arguments for early cultural contacts could be also seen among the materials from the Western Black Sea Coast. The finds from Berezan and Histria stimulate investigations in these directions, but unfortunately they are a subject of a long-term discussion (Graham 1990; Boardman 1991).

Accepting this idea, one should bare in mind that the cultural contacts do not always mean cultural influence.

A model of cultural interactions between Greece and Phrygia could support the idea of existence of a much more independent line of cultural development. There are plenty of arguments for the cultural domination of one or another region in the eighth century BC (Medvedskaya 1986: 113-114). Despite the different opinions, both the regions share a lot of common cultural features. Probably Dumbabin correctly states that the common elements in the Early Iron Age Geometric styles from Italy to Syria were suited in some way to the mentality of that age (Dunbabin 1957; Bouzek 1991). Accordingly, there is no distribution of direct imitation but a movement of artistic principals and specifically applied artistic ideas (Medvedskaya 1986: 115). This thesis is supported by the painted pottery from Iran (Sialk B cemetery [Medvedskaya 1986: 92-106]). Similar explanation is acceptable for the bird decorated pottery from South-Eastern Thrace. There is no answer to the question what is the initial agent for the emergence of the bird decorated pottery in the vast areas from Anatolia to the Middle and Lower Danube. The bird images from the Bassarabi region show considerable differences in the manner of depicting and belong to another cultural group (Tasić 1983: 80). The comparison between Bassarabi and Thrace reveals clearly cultural relationships of the regions with the ideas inspired by the Aegean world.

The emergence if these representations is an indication that the geometrical pattern could not longer satisfy the artistic tastes (Bommelaer 1985). The bird images appear in almost one and the same time in Greece (the end of the ninth – the beginning of the eighth century BC), in Phrygia (in the middle of the eighth century BC), and in the Lower Danube basin (in the eighth century BC). The birds images from Thrace could be most probably dated back in the eighth century BC. The appearance of these representations in the transitional period between the first and the second stage of the Early Iron Age is symptomatic. It is the commonly accepted time when the iron takes a leading position in the ancient economy. Although it is known in Thrace before this date, its wide distribution towards the eighth century BC leads to a formation of a new system of ideas and concepts, as well ax of a new mentality in which the birds images have a certain role.

Figure 22.4. Birds' motifs. 1 and 2. South-east Thrace. 3 and 4. Attica.

Section 3

Archaeology of Ritual

Chapter 23

Pit-Grave Culture Tumulus Graves near Izbul

(Northeastern Bulgaria)

Georgi Kitov

Introduction

The modern investigation of the Pit Grave Culture Tumulus graves in Northeastern Bulgaria has begun with the excavations of tumuli near the village of Belogradets in middle of the 1970s, conducted by G. Toncheva (Tončeva 1978). That Tumulus cemetery is located on the border between the Shumen and Varna Districts. In 1980s the investigations of I. Panayotov followed in Dobrich District (1989). Unexpectedly, in 1990 my expedition unearthed two tumuli from the Early Bronze Age in the area of Roman Tumulus cemetery near the village of Izbul, Novi Pazar District.

The Tumulus cemetery of Izbul has been already known to brothers Shkorpil as early as the end of the last century. They accounted 43 tumuli and divided them into four groups (Shkorpil and Shkorpil 1898: 13 sq.). G. Feher seems completely excavated one tumulus, and another partially but there are no certain data on these excavations. Thirty years ago, S. Damyanov investigated the fourth of the tumuli, which we named Damyanova Mogila (Tumulus) in his memory.

Recently – almost a century after the terrain observation of brothers Shkorpil – only 26 tumuli are preserved. The rest of the 43 documented burial mounds perhaps are destroyed and erased from the surface in course of the intensive agricultural activity in past decades.

The cemetery is located on both sides of the Izbul – Shumen Road, the south and south-west of the village of Izbul. The tumuli heights vary from almost invisible up to 10 m. Most of them are destroyed by treasure-hunters and economic trenches. Half of them are located in areas of vineyards or a cereal cultivation.

In 1989-1990 nine tumuli were investigated[1]. Seven of them – as were expected – belonged to Roman Period (the 1st – 3rd century). Numerous cremations *in situ* were unearthed. The equipment is one and the same – simple pits or pita with one – two offsets, with vertical or sloped walls. The walls of the pits preserved are smooth and firmly brick-like burned. Some of them are very rich, but other graves are poor. There is no relationship between the size of the tumuli and the richness of the graves.

Two of the tumuli documented on the south-eastern periphery of the cemetery were erected in the Early Bronze Age. They are located over the ridge of very low eminence – a position typical of other mounds from this epoch in Bulgaria (Kitov et al. 1991: 5). The tumuli are usually sprawled in result of winds, as well as of the destruction and dispersing of the uppermost levels by cultivation. Perhaps there had existed other tumuli from this period erased from the surface. The discovered tumuli are in distance of c. 10 km from the investigated by R. Popov burial mounds near Endzhe (recent Tsarevbrod [Popov 1930-1931]) and not far from Madara, Kalugeritsa and Kyulevcha Tumulus cemeteries of Pit-Grave Culture (Mikov 1934; 1936).

In tumulus No. 23 (Darina Mogila) 6 graves were investigated, 5 of which belong to EBA and one – to Early Iron Age. In tumulus No. 24 (Vladova Mogila) 7 graves were unearthed, all of which belong to EBA.

Darina Tumulus (No. 23)

The tumulus was located 3 km south of the village of Izbul, on the south-eastern periphery of the Tumulus group and about 200 m north-west from tumulus No. 24. It was erected over the ridge of very low eminence between the villages of Izbul and Vurbyane. The recent shape of irregular segment of sphere is a result of evaporating and sprawled because of strong winds and agricultural activity (***Figure 23.1***). Height – 2-m. Diameters – 33-38 m. The tumulus was named Darina by the name of Darina Vulcheva, at that time a graduate student at the Section of Thracian Archaeology by the Archaeological Institute – Sofia. At the moment of the excavations (21st – 31st of August 1990) a wheat block was spread over and around. No archaeological finds were documented over the surface and in the destroyed uppermost levels of the tumulus. A grid layout (30 x 30 m) of 5 m squares, respectively 36 squares were set. Peripheral parts of the tumulus were not included in the grid layout. The excavations were made with the assistance of bulldozers C 100 and DM 108. The graves were investigated by hands.

The highest point of the tumulus is point 25. The levels of some of the others points are given in ***Table 23.1***.

Six graves were documented – all of them in the central part of the tumulus. The relative chronology of the graves is as follows: Nos. 5 and 6 are primary dug into the ancient humus and continent. Grave No. 4 follows destroyed in the course of the digging of grave No. 1, then Nos. 1 and 2 succeed. Grave

[1]Second director of the excavations of Izbul cemetery in 1990 was G. Atanasov from Museum of History in the town of Shumen. The municipality of Nova Pazar sponsored the investigations represented by Z. Strandzhev. D. Vulcheva, S. Spasov, Z. Zlatev, D. Tsankova, P. Arsov, A. Kitov, T. Ribarova, V. Dimitrov, students from Pedagogical Institute in Shumen Town, workers from N. Pazar etc. participated in the excavations

Point number / m below the zero point	***Point number / m below the zero point***	***Point number / m below the zero point***
Point 3 – 0,84	Point 21 – 1,02	Point 36 – 1,04
Point 4 – 0,84	Point 22 – 1,04	Point 37 – 0,73
Point 5 – 0,94	Point 23 – 0,54	Point 38 – 0,60
Point 6 – 1,12	Point 24 – 0,15	Point 39 – 0,54
Point 9 – 0,77	**Point 25 – 0,00**	Point 40 – 0,74
Point 10 – 0,60	Point 26 – 0,36	Point 41 – 1,00
Point 11 – 0,50	Point 27 – 0,70	Point 42 – 1,14
Point 12 – 0,69	Point 28 – 1,04	Point 44 – 1,08
Point 13 – 0,95	Point 29 – 0,72	Point 45 – 0,94
Point 15 – 0,84	Point 30 – 0,50	Point 46 – 0,99
Point 16 – 0,54	Point 31 – 0,30	Point 47 – 1,02
Point 17 – 0,36	Point 32 – 0,14	Point 48 – 1,42
Point 18 – 0,16	Point 33 – 0,35	Point 49 – 1,40
Point 19 – 0,40	Point 34 – 0,85	
Point 20 – 0,67	Point 35 – 1,07	

Table 23.1. The levels of some points of Tumulus No. 23. Comment: The lowest levels do not correspond to the real height of Darina Mogila given, because its periphery left outside the square net. The measuring were made by the former technician at the Museum of History Atanas Atanasov.

No. 3 was dug into the tumulus later belonging to Early Iron Age. The burials are numbered and will be described bellow according their succeeded discovering.

By the description of the orientation firstly is given the position of the skull.

Grave No. 1. Sq. 15 and sq. 21, in the central part of the tumulus. It destroys the grave pit of burial No. 4. Secondary. At a depth of 1.10-m under the surface. A grave pit with not clear shape because of the partially destroyed walls in course of the machine pressure. Traces of wooden cover of the grave fell into the pit. The shape and the number of the woods is not recognizable. An inhumation. Traces of floor covering. Comparatively well preserved bones. Orientation: South-West – North-East. Position – turned to semi-right with crouched legs. The skull turned to the right, broken. Straight spinal column. Chest turned to semi-right. The arms crouched in the elbows at right angle. The legs fell to the right, close each to other. A weak ochre coloring of the forehead, the right bone of the shoulder and the vertebrae of the spinal column. Inventory:

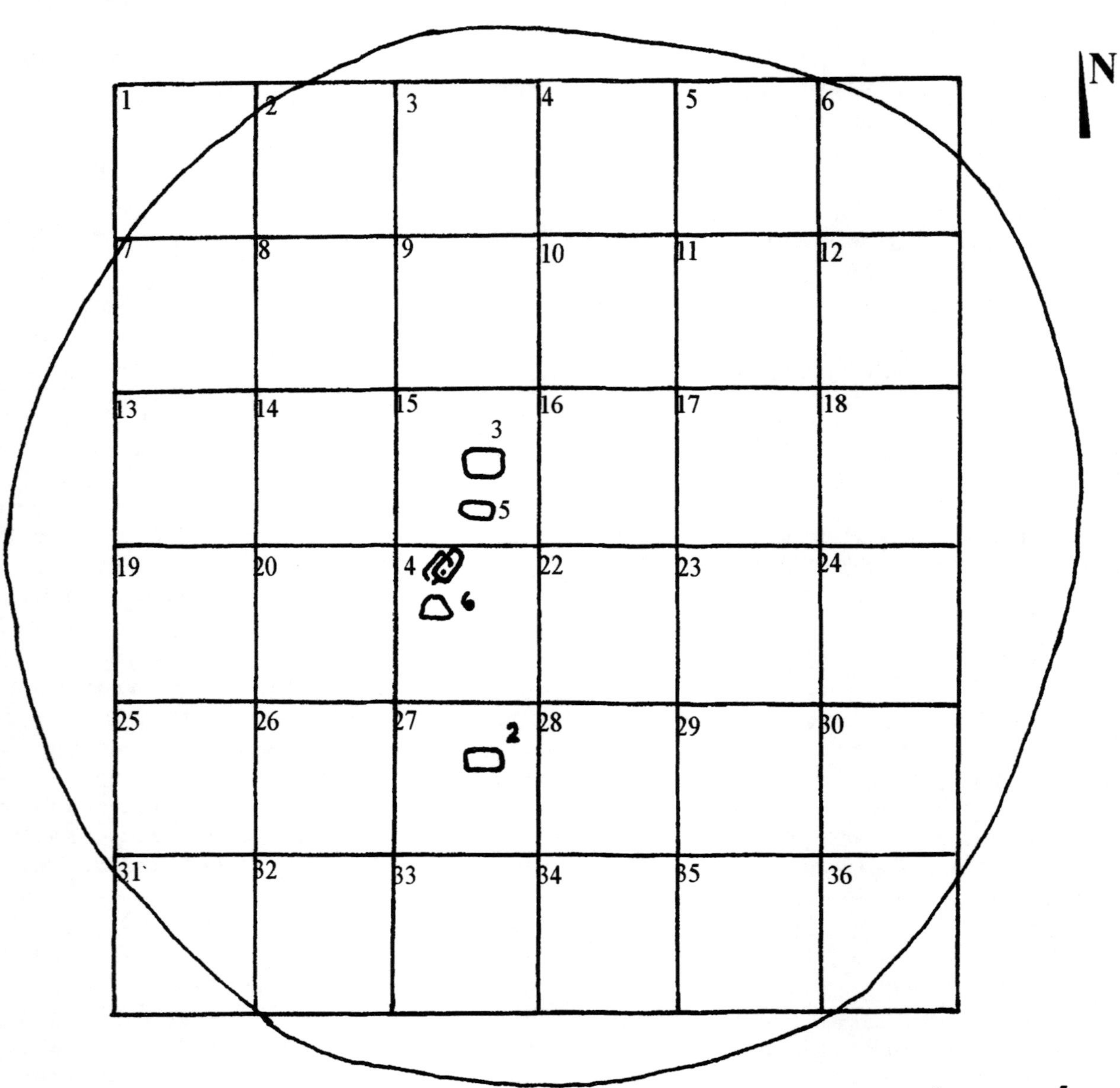

Figure 23.1. Darina Tumulus . Izbul cemetery.Plan of the tumulus with the square net of the excavations and location of the graves.

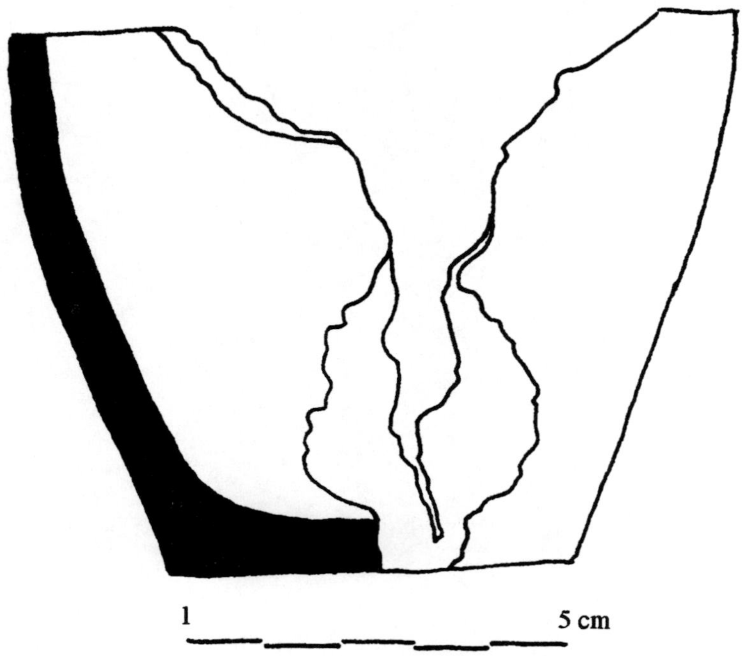

Figure 23.2. Darina Mogila. A ceramic cup from grave No. 1.

in the area of the pelvis – a clay cup horizontally placed. Excavation inventory No. 424. Description: a fragmented cup of two sherds preserved. A truncated conic shape, lightly invert rim and flat bottom. The paste mixed with sands, weak baked. Sizes: diameter of the rim – 9-cm, height – 7.5-cm, thickness – 0.4 cm (***Figure 23.2***).

Grave No. 2. Sq. 27, in the central zone, 7.5-m south – south-west from the geometric center of the tumulus. Secondary. At a depth of 0.6-m from the surface. The borders of the pit were not distinguished. There were no traces from wooden covering or floor covering. Bad preserved and mixed bones. Only the skull and bones from the arms are preserved with preserved length of 0.78-m. Orientation: east – west. The position is not distinguishable. The skull lays to the left, without traces of ochre coloring. No inventory.

Grave No. 3. Sq. 15. In the central part, in 2.5 m north-west from the geometric center of the tumulus. Secondary. Depth of 0.5 m from the surface. Inhumation in a pit without documented shape and size. Well preserved bones of an adult and very poor preserved bones of a small child. Orientation of the adult West (the skull) – East. Extended on back position. Next to the right shoulder a small child with a skull to the west. Inventory: 4 ceramic vessels – a cup, a bowl, a small bowl, and a bowl with two tongue handles. The grave belongs to the Early Iron Age.

Grave No. 4. Squares 15 and 21. It is located in the central part of the tumulus, in 2.5 m to the west – south-west of the geometric center. Secondary. Almost completely destroyed by the digging of the pit for grave No. 1. Depth of 1.1 m from the surface. Inhumation in a pit without documented shape and size. Poor preserved bones, collected in a heap between the skull and the right arm of the skeleton of grave No. 1. Bones preserved of the ribs, lower jaw, pelvis, legs, and a skull with the upper jaw. Fragments of the skull also under the legs of the skeleton of grave No. 1. The position and orientation are unclear. It looks possible the body to have been placed with a head to North-East if we have in mind the bones of the skull that were under the legs of the later grave. Traces of wooden cover. Red coloring over some of the skull's and one of the femurs. No inventory.

Grave No. 5 (Figure 23.3). Sq. 15, the central zone, in 2.5 km north-west of the geometric center of the tumulus. Primary. Depth of 1.5 m from the surface. Inhumation in rectangular pit with rounded corners, 1.1 x 1.55 m (eat-west direction). Depth – 0.95 m. Traces of wooden cover fell into the pit down. Without bottom spread. Very well preserved skeleton. Orientation East (the skull) – West. Position – inhumation in crouched position to the right. Arch-turned spine column, some vertebrae; the right shoulder-blade and the clavicle slightly replaced, probably from the animal intruded in. The left arm straight, in a secondary position under the pelvis after subsiding. No coloring. Length of the skeleton – 1.2 m, straight length – 1.69 m. No inventory.

Grave 6 (Figure 23.4). Sq. 21. The central zone, in 2.7 m north-west from the geometric zone of the tumulus. Primary. Depth – 1.6 m from the surface. Inhumation in step trapezoidal pit, dug into the old humus and up to 0.35 m in the continent. Sizes: 0.75 m (west wall) x 1.6 m (east wall). Depth of 0.72-0.78 m. Up to depth of 0.47 the walls are straight. Then a step follows with a width of 3-9 cm. The bottom of the pit is rectangular in plan. Traces of wooden cover, without bottom spread. Comparatively well preserved skeleton. Orientation east (the skull) – west. Position: inhumation to semi-right with crouched. legs. The right arm is straight, lying under the left knee joint, the left one – crouched at obtuse angle towards the

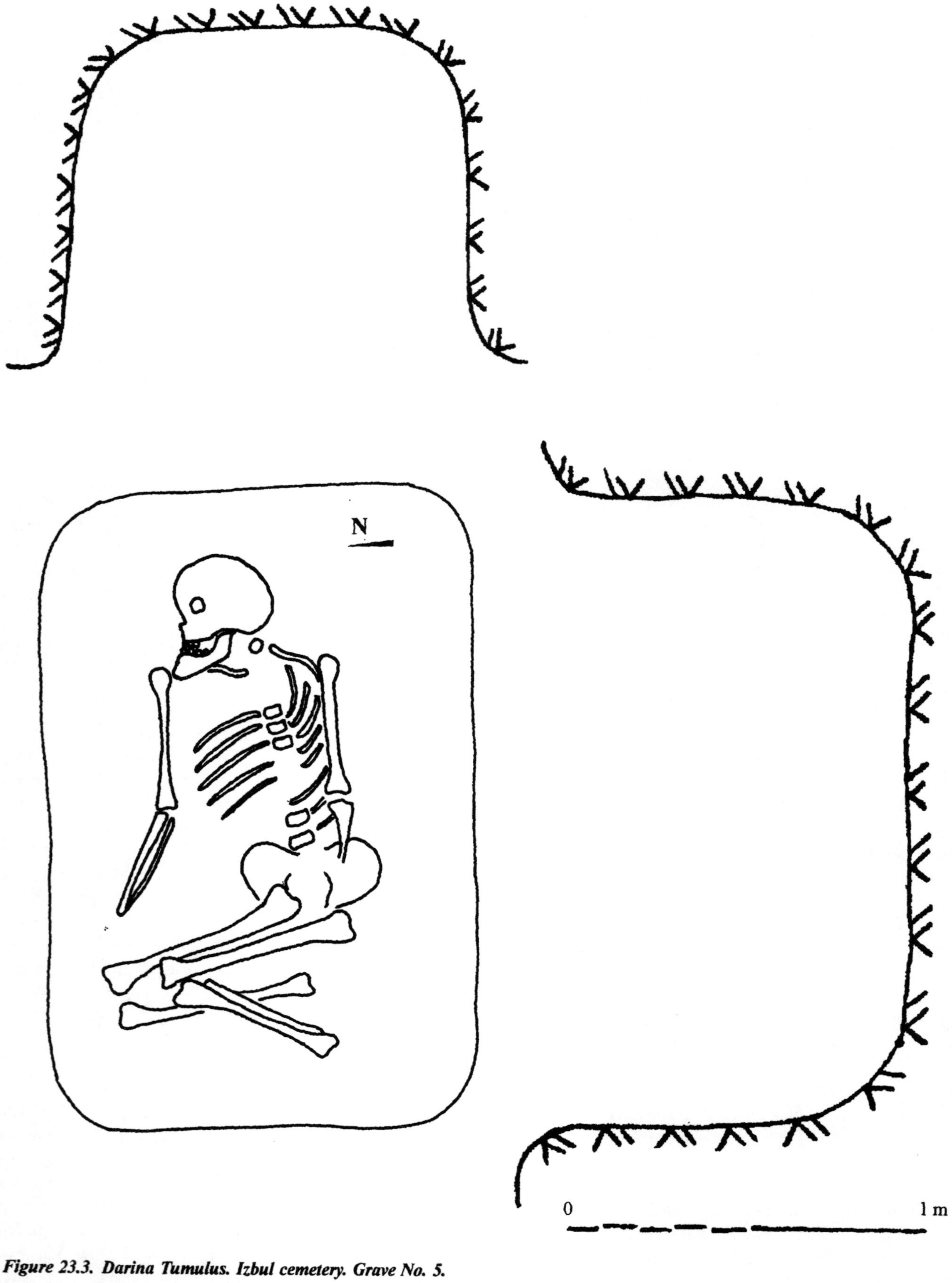

Figure 23.3. Darina Tumulus. Izbul cemetery. Grave No. 5.

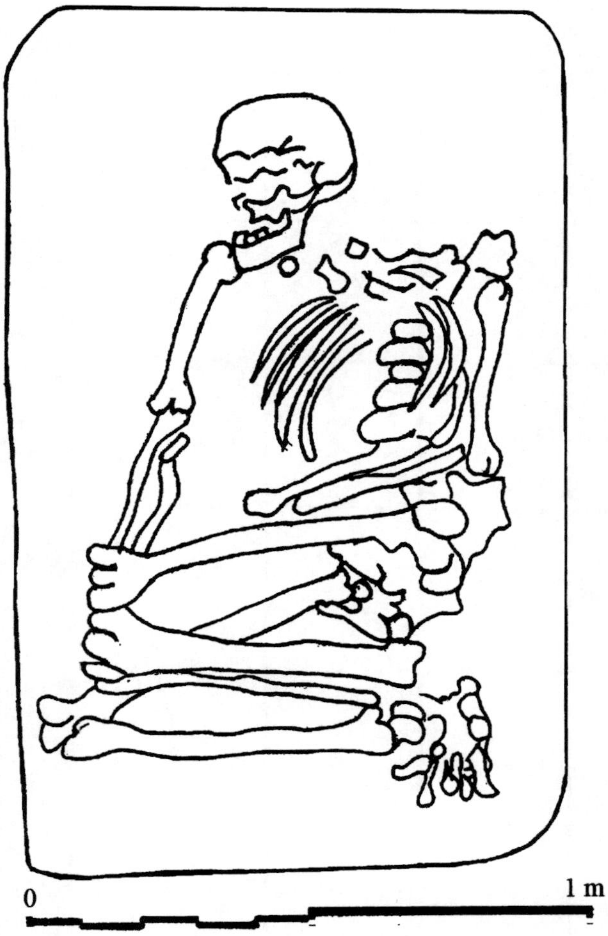

Figure 23.4. Darina Tumulus. Izbul cemetery. Grave No. 6.

stomach area. Red coloring in red over the skull (the back of the head and the forehead), shoulder bones, ribs and the pelvis. Length of the skeleton – 1.1 m, straight 1.89 m.

Vladova mogila (no. 24)

It is 200 m south-east of tumulus No. 23, located on the south-eastern periphery of the tumulus cemetery, over the highest part of the hill between Izbul Village and Vurbyane Village. It is also evaporated of the winds and the agricultural activities, shaped as a segment of a sphere, being before the excavations in a wheat field (***Figure 23.5***). Height – 2.3 m, diameter 35 m (North – South) x 40 m. It is called Vladova tumulus after the name of one of the participants in the expedition – Vladimir Dimitrov, a schoolboy at the Language school in Pravets Town at the time of the excavations. Square net 35 (North-South) x 40 m, 56 squares 5 x 5 m. Excavated as the former tumulus.

Seven graves from the Pit Grave Culture have been excavated in the central zone and to the west of the axis line. Because of the monotype heaping, the stratified layers have been not documented. The relative chronology is also impossible to be defined. It is only clear grave No. 4 was primary. The graves are numbered and are to be reported according to their order of discovering.

The highest point of the tumulus is point 41. The levels of some of the others points are given in ***Table 23.2***.

Grave No. 1. Sq. 36, the central zone, 2.5 m south of the geometrical center of the tumulus. Secondary. Depth – 0.5 m from the surface. Inhumation in obviously ellipsoidal pit, dug in the tumulus, old humus and the continent. Sizes: 1.1 x 1.6 m (east-west), depth – 0.40 m. The body was placed over the rock continent dug. Traces of cover and flooring are missing. Badly preserved bones. Orientation east (the skull) – west. Position – crouched to semi-left with legs strongly bent. The skull is destroyed lying to the left. The arms are crouched in elbows at acute angle. Phalanxes of the hands and of the feet are missing. No ochre coloring. Length of the skeleton preserved – 1.0 m.

Grave No. 2. Sq. 20, north-west section, in 10 m north – north-west from the geometric center of the tumulus. Secondary. Depth – 0.30 – 0.40 m from the surface. The grave is completely destroyed by deep ploughing. The shape and the size of the grave pit is not possible to be established. Remnants of bones displaced give no possibility the orientation, position and the sizes to be defined. The are no traces of coloring over the bones remained. No inventory.

Grave No. 3 (***Figure 23.6: 1***). Sq. 29, the central zone, c. 5 m north-east of the geometrical center of the tumulus. Secondary. Depth – 1.5 m from the surface. Inhumation in a grave pit dug into the tumulus, old humus and the continent including the rock base. The pit is rectangular in plan with rounded corners and sizes 1.10 x 1.80 m (south-west – north-east), depth – 0.85-1.00 m. The bottom is uneven, the transition from the walls toward it is by rounding. Over the uneven walls of the pit there are traces of the implements used it to be dug, especially in the rock continent. Traces of flooring or wooden color have been not registered. Badly preserved bones. Orientation (South-West – North-East). Inhumation in crouched position to the right. The right arm very slightly bent into the

Point number / m below the zero point	*Point number / m below the zero point*	*Point number / m below the zero point*
Point 5 – 1.32	Point 30 – 0.76	Point 51 – 0.43
Point 10 – 1.73	Point 31 – 0.35	Point 52 – 0.92
Point 11 – 1.45	Point 32 – 0.17	Point 53 – 1.47
Point 12 – 1.19	Point 33 – 0.45	Point 54 – 1.98
Point 13 – 0.97	Point 34 – 0.84	Point 55 – 1.86
Point 14 – 0.99	Point 35 – 1.48	Point 56 – 1.48
Point 15 – 1.22	Point 36 – 2.01	Point 57 – 1.14
Point 16 – 1.56	Point 37 – 1.67	Point 58 – 0.88
Point 17 – 1.96	Point 38 – 1.25	Point 59 – 0.80
Point 18 – 2.36	Point 39 – 0.71	Point 60 – 1.00
Point 19 – 1.83	Point 40 – 0.20	Point 61 – 1.35
Point 20 – 1.36	**Point 41 – 0.00**	Point 62 – 1.79
Point 21 – 0.88	Point 42 – 0.31	Point 65 – 1.74
Point 22 – 0.56	Point 43 – 0.79	Point 66 – 1.47
Point 23 – 0.49	Point 44 – 1.40	Point 67 – 1.25
Point 24 – 0.74	Point 45 – 1.96	Point 68 – 1.16
Point 25 – 1.16	Point 46 – 1.92	Point 69 – 1.37
Point 26 – 1.65	Point 47 – 1.35	Point 70 – 1.62
Point 27 – 2.15	Point 48 – 0.88	Point 71 – 2.02
Point 28 – 1.62	Point 49 – 0.46	
Point 29 – 1.24	Point 50 – 0.32	

Table 23.2. The Levels of some points of Tumulus No. 24. Comment: The measuring were made by the former technician at the Museum of History Atanas Atanasov.

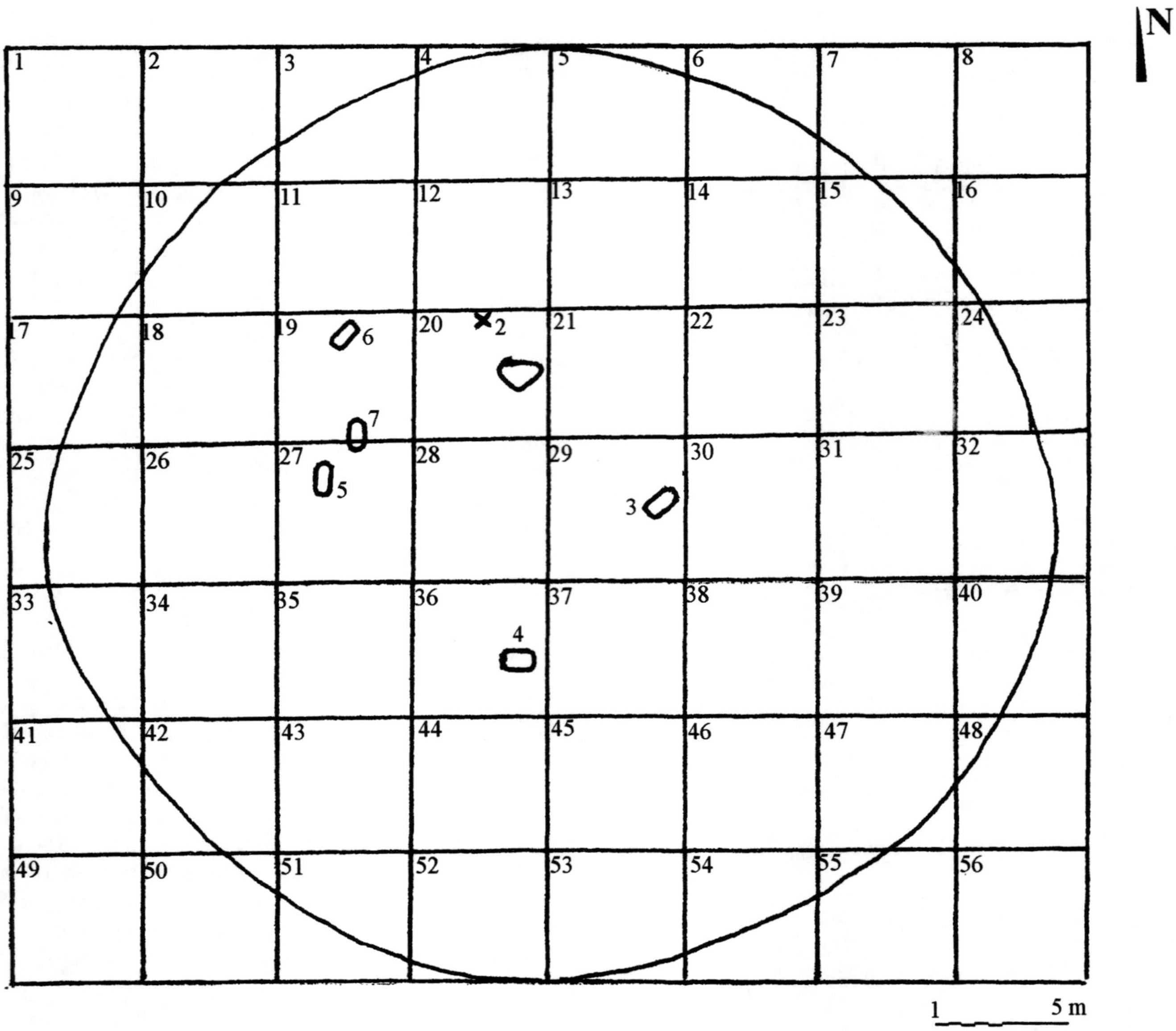

Figure 23.5. Vladova Tumulus. Izbul cemetery.Plan of the tumulus with the square net of the excavations and location of the graves.

elbow, straight phalanxes. The left arm is bent at right angle. The left leg is stronger crouched in the knee. The phalanxes of feet are missing. There are traces of red ochre over the skull and the extremities. The length of the skeleton in crouched position – 1.0 m.

Grave No. 4. Squares 20 and 21, in the north section, in 7.5 m north from the geometric center of the tumulus. Primary. Depth – 1.6 m from the surface. Inhumation in irregular pit changing its shape in the depth, but closest to triangular with rounded corners. Dug into the old humus, continent and the basic rock. Sizes 1.95 (east-west) x 2.1 m, depth up to 1.1 m. No traces of flooring and cover. At different levels in the pit there were discovered spread around human bones in very bad condition, without traces of red ochre. Probably the grave was dug into the continent, perhaps, plundered.

Grave No. 5 (Figure 23.6: 2). Sq. 27, west section, in c. 8 m west – north-west from the geometric center of the tumulus. Depth – 0.50 m from the surface. Secondary. Inhumation in a grave pit dug into the tumulus, old humus, the continent and the continent rock, irregular ellipsoidal shape in plan. Sizes – 0.90 x 1.35 m (North – South), depth 0.80 m including 0.05 - 0.10 m into the rock. The walls from the east and west are inclined, vertical from the south, unclear shaped from the North. No traces of flooring and cover. The skeleton is preserved in very bad condition, only parts from the extremities are preserved. The position of the last indicate the body probably was placed in crouched position to the right with a head to the south. Preserved length of the skeleton – 0.65 m. No inventory.

Grave No. 6 (Figure 23.6: 3). Sq. 19, north-west section, c. 11 north-west of the geometric center of the tumulus. Depth – 0.50 m from the surface. Secondary. Inhumation in a grave pit dug into the tumulus, old humus, the continent and the continent rock with rectangular shape with rounded corners in plan. Sizes – 1.20 x 1.90 m in north east – south-west direction, depth – 0.40 m. The walls are reclined, so that the bottom is oval in plan. Uneven walls and bottom. No traces of flooring and wooden cover. The skeleton preserved in bad condition. Orientation: North-East (the skull) – South-West.

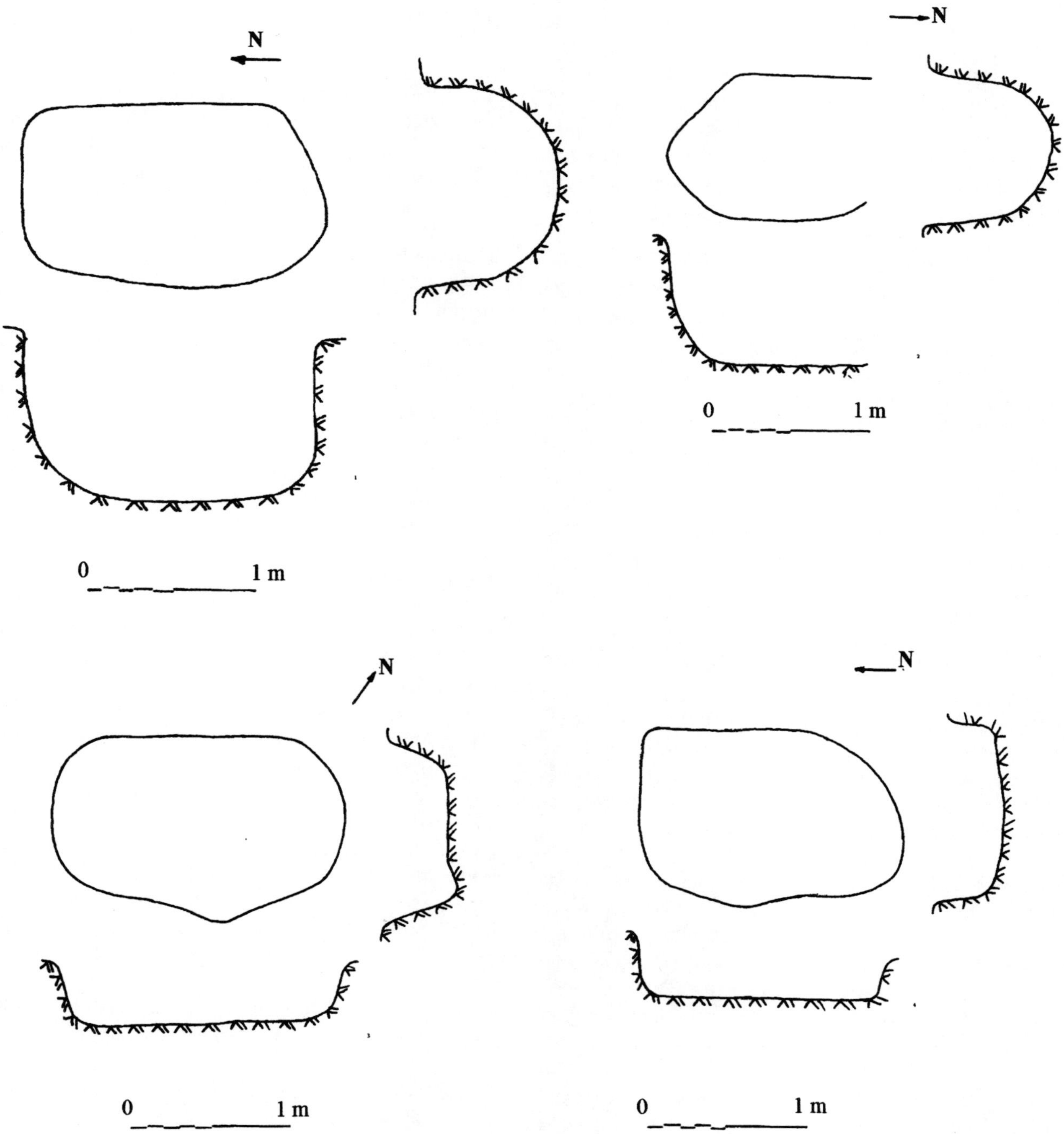

Figure 23.6. Vladova Tumulus. Izbul cemetery. Plan and sections of the burial pits of graves No. 3 (1), No. 5 (2), No. 6 (3) and No. 7 (4).

Position – on back with extended extremities. The precise position of the last is unclear because of their bad condition. No ochre coloring. Length of the skeleton – 1.30 m. No inventory.

Grave No. 7 (Figure 23.6: 4). Sq. 19 and 27, north-west section, c. 9 m north-west of the geometric center of the tumulus, next to grave No. 6. Depth – 0.50 m from the surface. Secondary. Inhumation in a pit dug into the tumulus, old humus, the continent and the continent rock rectangular shape with rounded corners in plan. Sizes – 1.20 x 1.50 m in direction North – South, a depth of 0.40 m. No traces of flooring and wooden cover. The skeleton preserved in bad condition. Orientation North (the skull) – South. Position – crouched to semi-right with very strongly bent legs, straight right arm, and the left arm bent at obtuse angle in the elbow. Traces of red ochre over the bones of the feet, pelvis and the spine vertebrae. Length of the skeleton – 1.15 m. No inventory.

After the investigation of the graves the better preserved skeletons were taken while those in bade condition remained unmoving. Most of the pits were preserved and refilled again, then in each a glass bottle was placed with a short text and contemporary coins.

Discussion and conclusions

All twelve graves from Izbul cemetery consist of most of the characteristics of the PGC burial ritual (Kitov et al. 1991; Nikolova 1999: 369-389). All graves are unearthed in the continent or in the grave mound. Four of the graves from the upper levels of the tumuli are partially destroyed and the borders of the pits were not distinguished. At the same time, one of the graves was plundered already in Antiquity and the record for this grave is limited.

According to the number of discovered graves – 5 and 7 respectively, both tumuli are close to the mound cemeteries with the biggest number of placed graves. With the exception of the Nova Zagora Tumulus phenomenon (34 graves), only in several PGC mounds in Bulgaria are discovered more than 5 graves – Kalugeritsa V and Belogradets by 6 in each, Goran – Slatina III (9), Goran – Slatina V, Turnava and Plachidol (by 10 in each [Kitov et al. 1991: 28]).

The tumuli from the Izbul cemetery were in distance 200 m each from other. They were with similar diameter (35 – 40 m) but with different heights - respectively 2 and 3.5 m. The second tumulus was not only higher but it also erected in a locality dominated the surround terrain.

Grave pits characteristic of the PGC were excavated - rectangular with round edges (four from eight preserved – one pit in tumulus No. 23 and three pits in tumulus No. 24) or oval (two in the tumulus No. 24) with parallels at Goran-Slatina (Kitov et al. 1991: 34 sq.). At the same time, new types were documented – a step-like trapezoid (one grave pit at Darina Mogila) and triangular (one grave pit at Vladova Mogila).

In the first tumulus the primary graves might be two – one rectangular pit grave and the trapezoid pit grave. The pit of grave No. 1 from the Darina Mogila tumulus partially destroyed the grave No. 4. The bone of the earlier burial were collected and placed in one of the corners of the new burial. But this act was not done carefully, so some of the bones from the destroyed grave were to be found in the original position. In the Vladova Mogila tumulus the primary is the grave with triangular pit. It was destroyed and probably destroyed and probably plundered in the course of the equipment of other graves.

There is no evidence for additional heaping of the tumuli. Floor covers were documented only in two graves. It might be the missing of floor covers to have been specific characteristics of PGC burial rituals in Shumen District. In this case the doubt that the earlier excavators did not pay attention on this element of the PGC burials will die out. The existence of wooden covering of the pits was documented only by four graves in Darina Mogila tumulus – by both primary and two secondary graves.

All burials are single inhumations. The orientations of the inhumations are documented certainly in 10 graves (***Figure 23.7***). It is highly varied. However, most of the skeletons are with head in the east – 3 cases. By two inhumations are oriented in the north-east and south-east, as well as by one burial – in east – north-east, south and north. Both primary graves in Darina Mogila tumulus have east orientation – in the east and east – north-east. The orientation of the inhumation from the primary grave in Vladova Mogila tumulus cannot be defined. The new excavations confirm the geographical determination of the orientations of the PGC inhumations – from North-East to North-West. The orientation is smoothly changed from east toward west and the last predominates in Northern central Bulgaria (Kitov et al. 1991: 42). At the same time, the south and north orientations are often documented in Southern Bulgaria (Buyukliev 1964; Batsova and Kunchev 1974), which are exception at Izbul characterizing two graves.

The position of the inhumations in Izbul graves is documented by 7 cases. The crouched inhumations to the semi-right predominated – 3 cases. Two single inhumations are in crouched position to the right and there are by one instance of crouched to semi-left and extended on back inhumations. It cannot be distinguished and interconnection of the positions with the primary and secondary graves, with the shapes of the pits or between the position of the inhumations in both tumuli.

There is no instance for using of the ochre as a burialgood. But six skeletons are partially colored in red. By 5 of them the sculls are also colored, along with legs, pelvis, feet and rare the vertebrae of the spinal column. The ochre coloring makes an impression with its limited using in compare to Goran – Slatina cemetery, for example (Kitov et al. 1991: 46).

All graves are extremely poor. Only in grave No. 1 from the Darina Mogila tumulus was found a small clay cup fragmented.

The graves from Izbul tumuli demonstrate an insufficient diversion from the documented up to now about ten peculiarities of the PGC burial practices, then they should be attributed to this culture. At the same time it is an important new record for the richness of its characteristics in the Balkans. On the other hand they confirm the observation that one of the centers of distribution of the PGC was located in the Shumen District. It should be expected that future investigations in Izbul area would document new tumuli of this culture.

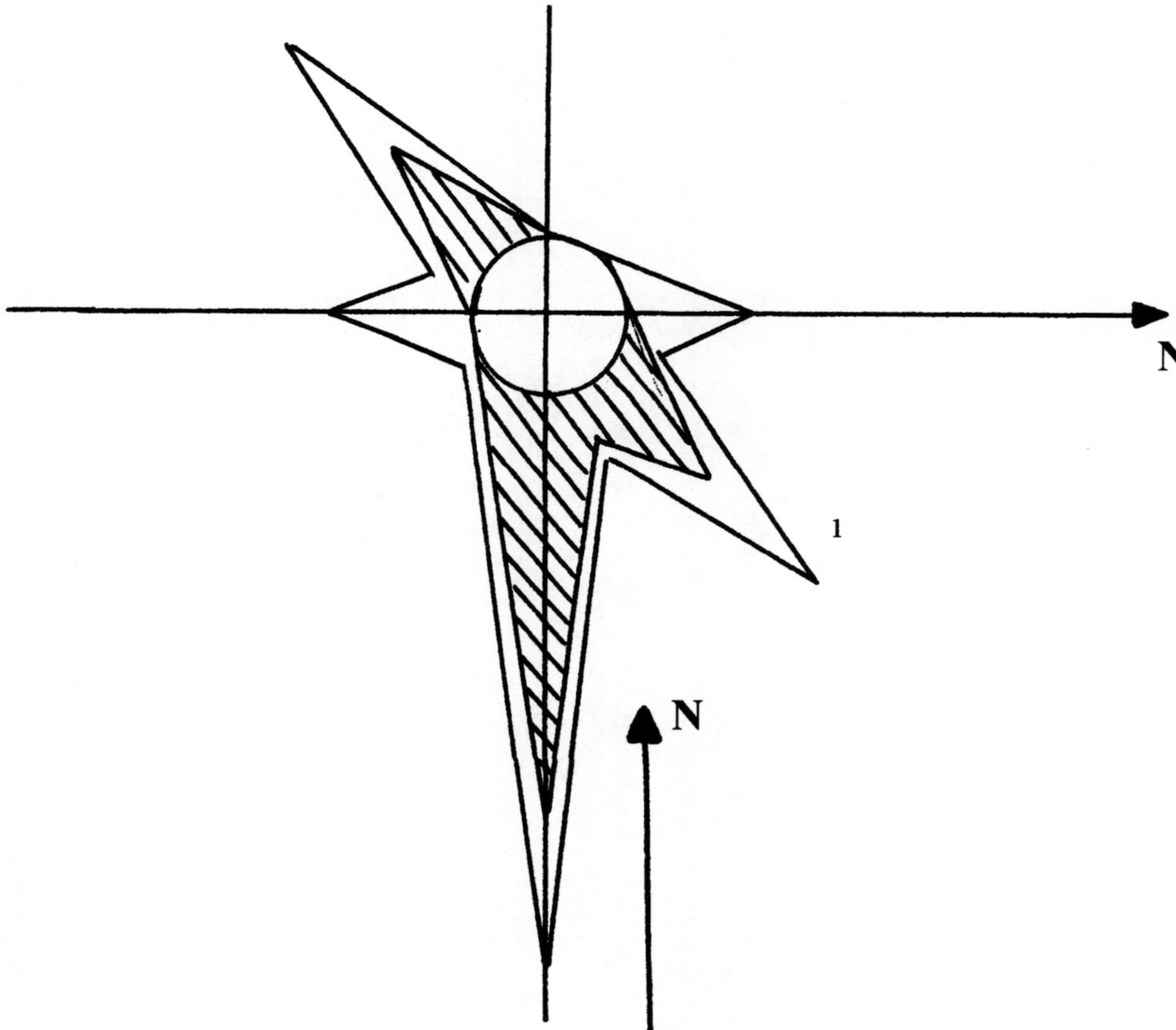

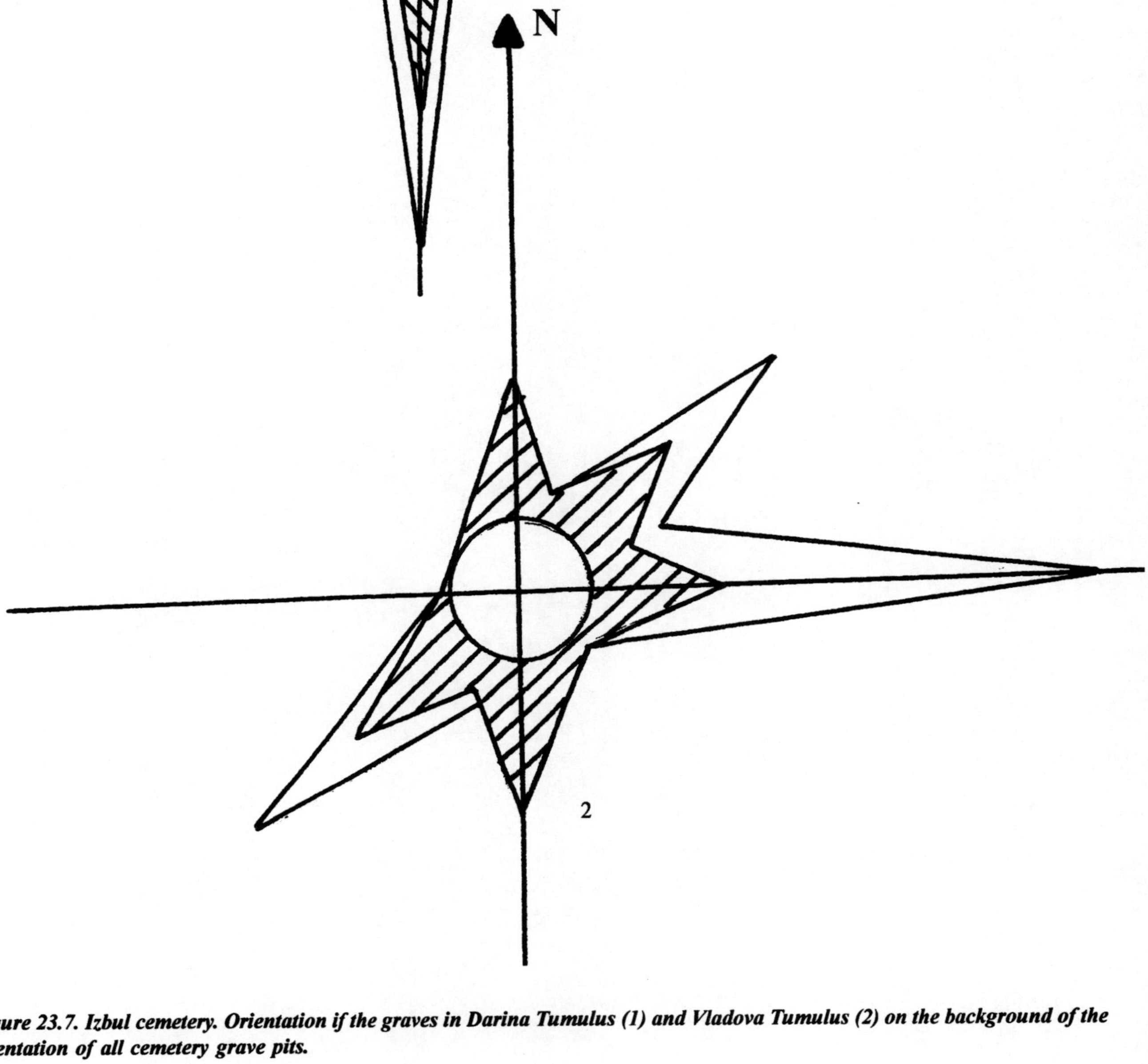

Figure 23.7. Izbul cemetery. Orientation if the graves in Darina Tumulus (1) and Vladova Tumulus (2) on the background of the orientation of all cemetery grave pits.

CHAPTER 24

A CROMLECH NEAR DOLNI GLAVANAK IN THE EASTERN RHODOPES

(PRELIMINARY COMMUNICATION)

Georgi Nekhrizov

Location of the monument and archaeological characteristic of the region

The Megalithic monument is unearthed in Bunar Altu locality, in 1.2 km west from the village of Dolni Glavanak, Madzharovo Community, Haskovo District. It is situated over slightly manifested ridge, 305 m above the see level, which is a part of the east slope of a hill arising at the left bank of the Arda River. Orographically, it is south bifurcation of the big east Rhodope ridge of Gorata.

As a result of the archaeological excavations in the coarse of the Middle Arda drainage it was documented a high concentration of monuments of the Thracian culture. A considerable compactness of sites from the first millennium BC is documented in the micro region, where is located the equipment - subject of this publication. There is a direct visual connected from it to a cult place from the Early Iron Age situated in 1 km northwest on the ridge of the Taushan Tepe Hill. Over the ridge of the hill located on the left bank of the Arda River, in 2.6 km southwest is established the Papaz Kaya fortification where pottery from Late Iron Age is documented. Over the same ridge in Petrovo Kale locality, in 2.4 km south, there are two burial mounds. Burial mounds are registered northwest, north and northeast from the site on the land of Dolno Sudievo, Ruzhenovo, Zlatoustovo and D. Glavanak Villages. Settlements with material from Iron Age are documented near Selska Polyana (6 km northeast of the site), Malko Bryagovo Village (8 km east-northeast) and Gorno Pole (5.4 km southeast). On the land of Gorno Pole there are also known groups of rock niches in Hambar Kaya locality (3.5 km southeast), two rock sepulchres in Kosharite locality (4 km southeast) and in Furenzhika locality (5 km southeast), as well as a Thracian sanctuary with rock hewing on Sivri Dikme Mount (6.8 km east-southeast). The Thracian sites registered – 4 sites, 3 cult places, 5 Tumulus cemeteries, 2 rock sepulchres and a group of rock niches indicate the existence of developed settlement system in the neighborhood of the megalithic monument and intensive life in the first millennium BC.

Description of the monument

The borders of the equipment are fixed by vertical placed big, roughly pyramidally shaped blocks, cut from local hard volcanic rock. The bases of the blocks lay directly over the continent rock without places shaped for them or cutting hollows for impaling. Their shape and the heaping of smaller stone around insured the stability of the equipment. The height of the blocks is 1.50-1.70 m each, with medium size of the base 1.20 x 0.90 m. The vertical blocks are placed in approximately equal distance from each other, as smaller are placed horizontally between. They describe circle with 10 m inner diameter, which is open from southeast. There are 9 vertical blocks preserved in place. There fell down near their initial place, but two or three are missing. A lot of broken stone was heaped around the vertical blocks along the slopes of the slightly mounting ridge.

To the south of the big equipment were registered two smaller ones. Equipment No. 2 is located 15.30 m southeast of the first one. It is a fence with ellipsoidal shape and outer diameters east-west 3.4 m and north-south 2.8 m shaped of medium in size broken stone placed in one or two rows. Equipment No. 3 is 18.70 m south-southeast from the basic one. It is erected in the same way as the second one and also has ellipsoidal shape. Its diameter are east-west – 6.40 m, and north-south – 5.40 m.

The archaeological investigation

The megalithic equipment was registered in May 1998 in the course of terrain investigations on the territory of the Middle Arda River drainage on the land of Madzharovo community. The archaeological investigations was performed in July 1999 thanks to sponsoring of Mobikom.

The whole equipment was investigated archaeologically and a strip 1.5 m wide outside the row of stone blocks (***Figures 24.1-24.2***). The cultural layering inside the equipment was excavated up to 0.60-0.70 m depth as in some places the continent rock was recovered. Outside the equipment the layer 0.50 m thick was excavated. The cultural layer consists of crumbly black soil, abundantly mixed with stone. It has homogenous structure and no stratigraphic levels have been distinguished. There was a concentration of archaeological material (pottery, animal bones, plaster, flint artefacts), as the finds were concentrated mainly around the stone blocks.

The two smaller equipment were investigated as well. Traces of burial were documented in equipment No. 2, cremation from the Late Hellenistic epoch (middle of the third – the first century BC). A glass bead was placed as a burialgood. In equipment 3, from which only the southwest sector was investi-

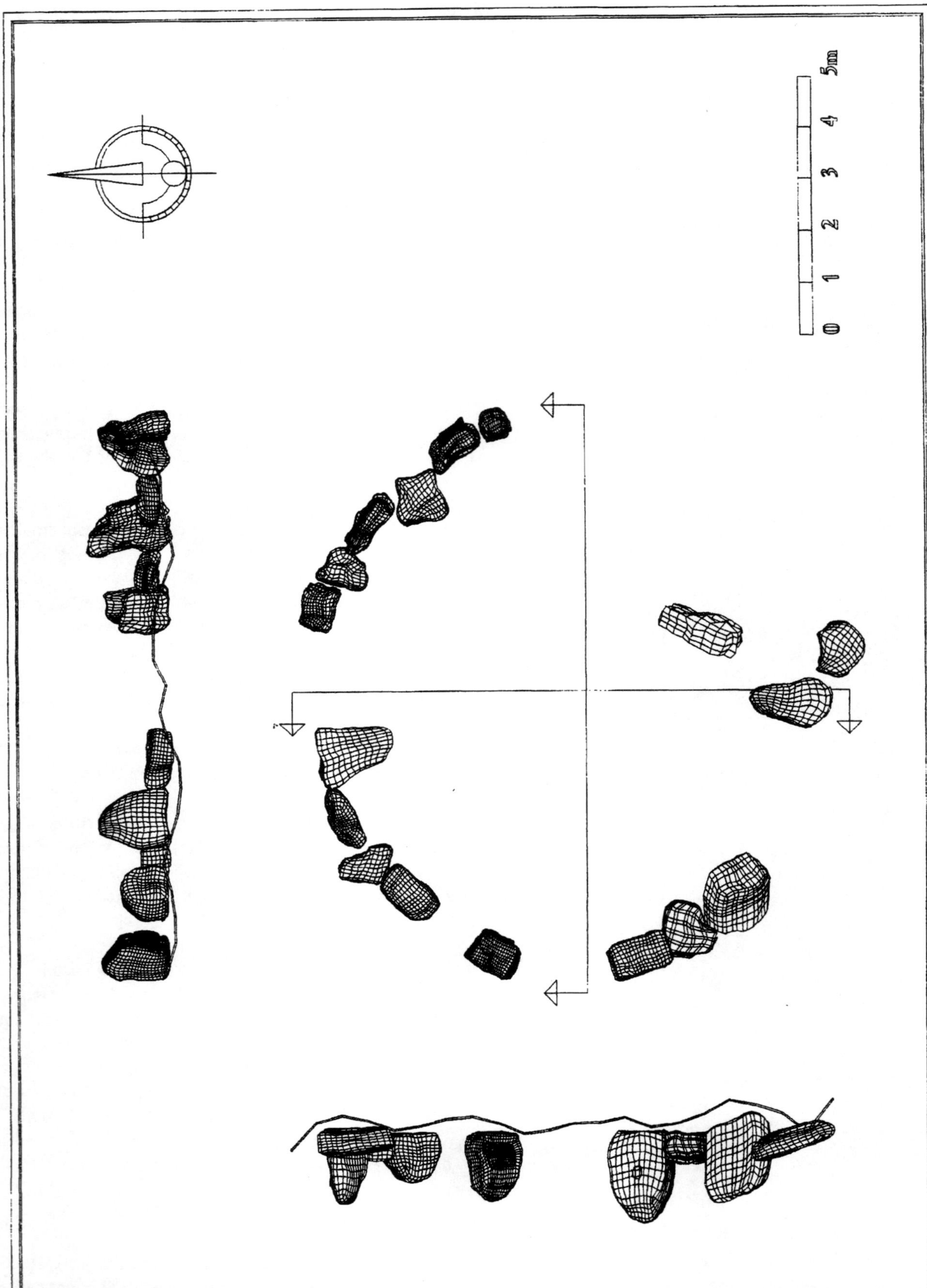

Figure 24.1. Plan and section of the megalithic monument from Dolen Glavanak. Author: Arch. Yana Topalova.

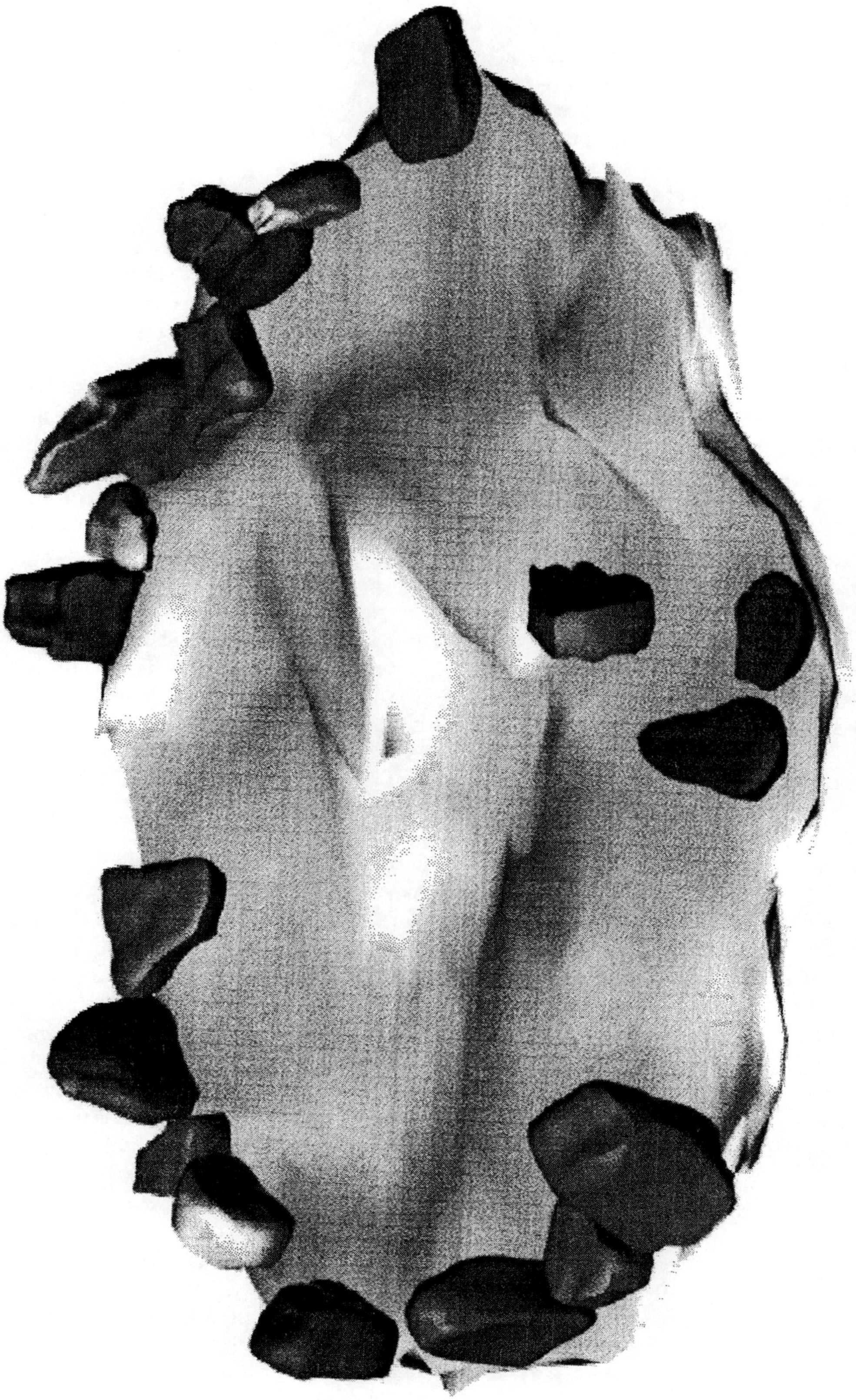

Figure 24.2. 3D model of the cromplech from Dolen Glavanak. Author: Arch. Yana Topalova.

gated, a big quantity of Early Iron pottery have been investigated. A considerable part of ceramic fragments were found from both sides of the stone row marking the borders of the equipment.

The finds

In all mechanical levels the artefacts from Early Iron Age predominate. The fragments of hand made pottery found can be divided into two technological groups, with several subgroups each. In the group of the fine pottery with burnished and smoothed surface, the plates and the storage vessels are most often distributed. The cups are rarely documented. Single finds represent the categories of jugs, kantharoi-like and amphora-like vessels. The statistics of the pottery indicate that most distributed technique of ornamentation is the incised one. Most of half of the ornamented fragments are decorated with incised motifs.

The stamp pottery is on the second place, 40.5 % of the ornamented fragments. Among the last, the number of the different compositions including the stamped S-element is three times over the motifs consisted of stamped circles. The channels are distributed considerably rarer, only among some types of cups and bowls.

Different types of pots represent the group of the technomic pottery. The are decorated with plastic band, by incision or prick ornament.

The fragments if hand made and wheel pottery from Late Iron Age are of considerably less quantity. The technomic pottery dominates among the hand made pottery, as pot is again the most represented shape. The wheel sherds belong to vessel of the so-called Gray Thracian earthenware.

It is interesting to note the existence of Middle Age fragments in the uppermost levels, dated from 11th-12th century.

Several metal objects have been also discovered (***Figure 24.3: 1-3***). Most important is one well-preserved bronze fibula. It consists of single spiral, with arched shaped bow and triangle plate of the pin-holder. A spiral of iron fibula was discovered so far, without clear type, as well as a bronze pin. Very corrode iron objects – a horse-shoe, a knife, a plate and forged nails.

Chronology

The erection and the beginning of megalithic equipment's exploitation near Dolni Glavanak can be dated from the second phase of the Early Iron Age (8th-6th centuries BC). The ceramic material discovered definitely defines that chronology (***Figure 24.3: 4-5***) and it is confirmed by the bronze fibula. The last belongs to type A I 2, variant 'γ' after D. Gergova. The distribution of that type of fibulae is dated from 8th-6th century BC and they are distributed mainly to the south of the Stara Planina Mountains (Gergova 1987: 22 sq., tables 1: 14-17; 2: 18).

The intensity of the finds and the thickness of the cultural layer indicate a long period of active functioning of the monument. Its exploitation includes chronologically the second stage of the Early Iron Age and continues in the next for some time, as the finds from Late Iron Age (5th-1st century BC). Towards the end of that epoch the cremation in equipment 2 was performed. Equipment 3 existed synchronous with the first as in and around similar acts were elaborated traces by the materials discovered.

After a long break, in the Middle Bronze Age that place again was episodically visited and perhaps there was a ritual activity (possible pledge).

Interpretation

The location, plan scheme, the way of erection and the finds of the site near Dolni Glavanak village infer it was a cult place. Conditionally, it can be attributed to the most distributed type of cult places in the Eastern Rhodope – located over ridge and surrounded by stone fence territory. Those sites occupy ridge of hills of mounts, and the sacral place is surrounded by not high stone wall. Following the rule, the wall is erected on the periphery of the ridge, built of broken stone, without soldering. It is not deeply grounded, up to 1 m thick and not very high. It most often surrounds oval and the inner area is rarely more than 1 Decare.

At all sites of that group a high concentration of archaeological material is documented: ceramic fragments, plasters, animal bones, flint artefacts, spindle whorls, anthropomorphic figurines, metal weapons, adornments, fibulae, and coins. In some cases a concentration of finds near the stone equipment is documented. It is not rare the cases when traces of ritual activities from different epochs are registered, i.e. the tradition of in course of thousands these places are a secret places (Nekhrizov 1999b).

The cult places located over ridges are most distributed type sanctuaries in the mountain areas of Thrace in the first thousand BC (Domaradzki 1994: 70). The site of Dolni Glavanak contains rather functional characteristics, which allow it to be attributed to the most popular places of ritual activities in the mountain areas of Thrace. At the same time, the construction manners used by its erection and the monumental affect contrast to the rest known. The big stone blocks vertical placed in an equal distance one from other shaping a circle, are close to the ideas of the megalithic constructions. By analogy with similar monument of the megalithic culture from Western Europe, the site of Dolni Glavanak can be defined as a megalithic monument of cromlech type.

Grounded in the land stones defining an oval are known from the areas of distribution of the megalithic culture in Thrace. Brothers Shkorpil wrote about several dozens of such monuments – circles of grounded stone in the area of Hlyabovo Village in the Sakar Mountains. Those circles were in diameter of 6 m, and the stone, which surrounded them had height of 1 m (H. and K. Shkorpil 1888 : 24-27, illustration 1). Later, G. Bonchev precisely localized those monuments and actualized the information about them (Bonchev 1901: 695 sq., fig. 6). Brothers Shkorpil indicated similar monuments in North-

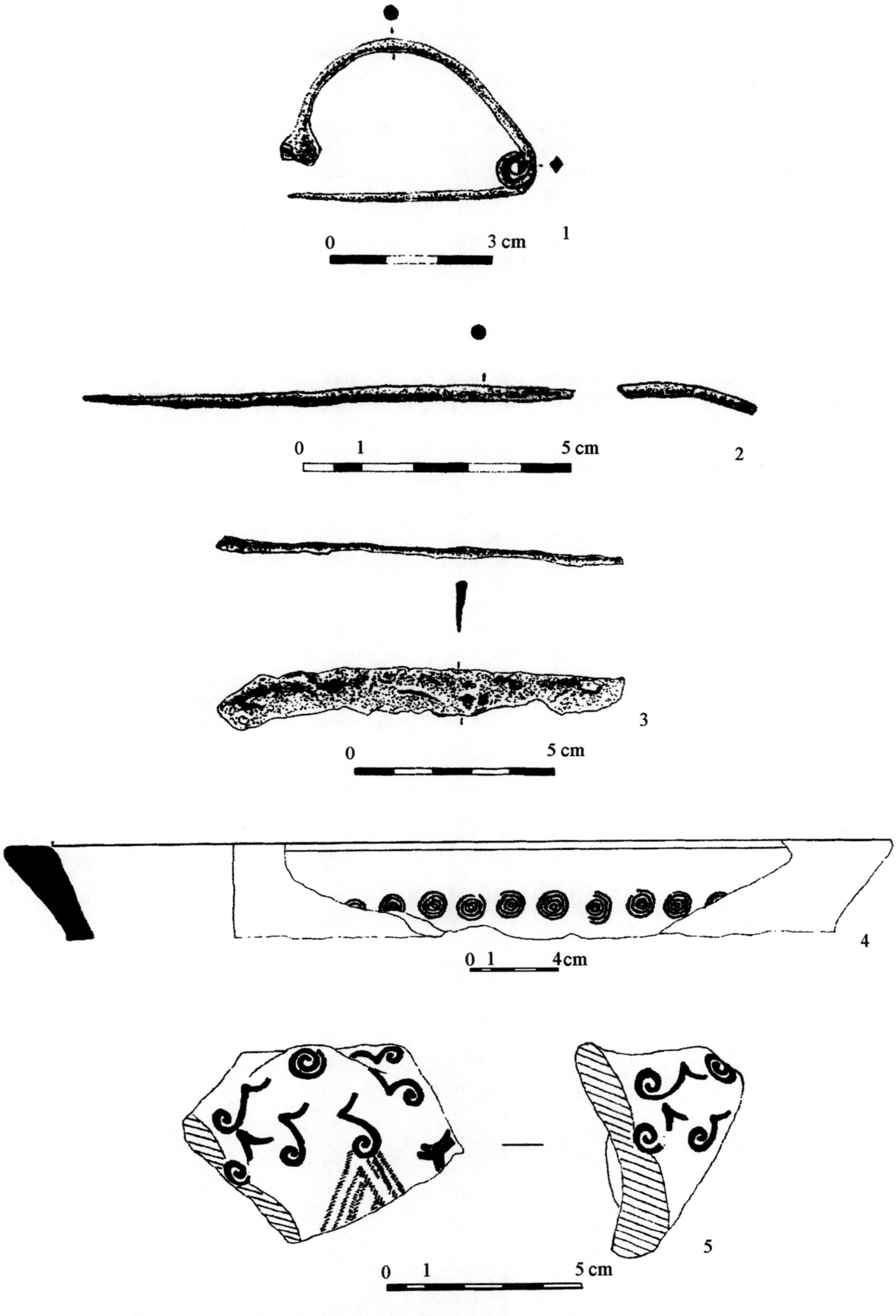

Figure 24.3. Archaeological finds from the area of Dolen Glavanak monument.

east Bulgaria near the villages of Kamen Bryag and Ttyulenovo (H. and K. Shkorpil (1892: 68-70). D. Tsonchev in his study on Haskovo mineral springs mentioned several 'stone circles' with a diameter of 4 to 15 m (Tsonchev 1940: 86-89). Monuments with similar character are discovered on other places in the Thracian land, but its function is still not categorically defined (Delev 1982: 418 sq.). According to brothers Shkorpil, it is places for meeting of the local population in the period of religion feasts (H. and K. Shkorpil 1888: 27; 1896: 14). G. Bonchev stressed in the Sakar Mountains those monuments are always near the dolmen cemeteries and defined them as surrounding destroyed or enclosures of unbuilt dolmens (Bonchev 1901: 696). It is important to note most of the monuments were to a big extent destroyed and grounded at the moment of discovering, so none is preserved. This is the reason no excavations to have been elaborated of such monuments.

In Eastern (Turkish) Thrace there are well-preserved complexes of grounded stone of big size. But they are not investigated and published in details (Özdoğan 1982: 47).

Outside the zone of distribution of the Thracian Megalithic monuments, there is a group registered in the Western Rhodope, of different constructions oval or circle in plan, corridor, lines and heapings of medium stone. One stone and earth equipment is investigated archaeologically, in which materials from the Roman period have been discovered (the second half of the third century BC) and single fragments of Late Bronze and Late Iron Age. According to the investigators of those equipment located near the big tumulus cemetery Kochan – Satovcha, there were rituals connected with the burial practices (Domaradski 1999: 74 sq., 98, figs. 47-52, 64-68).

Conclusion

The site of Dolmen Glavanak belong to the zone of distribution of rock sepulchres, graves and niches (Nekhrizov 1996: 9, fig. 1; 1999a: 25 sq.), which are accepted as peculiar characteristics of the megalithic culture, chronologically synchronous with the dolmens (Delev 1984; 1993: 173). There are no dolmens on the territories of the monuments discussed. Then, despite typological similarity the monument of Dolen Glavanak considerably differs of the stone enclosures from Sakar, Srandzha and Eastern Thrace.

In most cases they the last are grounded in a some distance from each other stone plates of gneiss or granite, with close to circle shape. That is the way in which the surrounding dolmens stone circles are erected, perhaps defending the tumulus heaping. In those areas megalith sites of both kinds are distributed, often very near, which gives a ground of G. Bonchev to assume they had similar or identical function.

It is possible the reason the rock grave equipment to be preferred in the area of Dolen Glavanak monument to be due to the existence in this region of soft sedimentary or volcanic rock, advantageous of cutting and the absence of rock for making stone plates. Perhaps because of this reason the cromlech from Dolen Glavanak is made of volcanic rock massive blocks, which in contrast to all other sites of that type were not grounded in the terrain and placed over the ground.

Despite there are no direct parallels of the newly discovered megalithic monument, there is no question about its belonging to the Thracian culture. It is outstanding instance of the locally artistic use of the principals of erection of the megalithic monuments that appeared in certain regions in the Thracian land in the first millennium BC.

Chapter 25

Late Prehistoric Burials in the Area of Ovcharitsa II

Tatyana Kuncheva-Russeva

The site of Ovcharitsa is located on high terrace on the left side of Ovcharitsa River, Radnevo District (Southeast Bulgaria), in the area of the Maritsa-Iztok Mining complex. In 1984-1989 a defended site was excavated there (Ovcharitsa II). A culture layer from Early Bronze Age and destroyed Iron Age layer were documented at the site. The settlement from the Early Bronze Age was defended by a ditch from the north and with a wall that surrounded it from all the sites (Kănčev and Kănčeva-Russeva 1996).

Eleven inhumation graves were excavated on the northernmost periphery of the settlement, to the north and south of the ditch and partially in dwelling No. 3 located next to the ditch (***Figure 25.1***). Around the skeletons, by their recovering, there were discovered single animal bones, but there are no certain arguments they were interrelated with the burials. It is possible they to belong to the settlement, that is located nearby.

The grave equipment consists of pits with rectangular or irregular shape. The pits of graves Nos. 7-9 are rectangular in plan with round corners. Their long sides are oriented eat-west but No. 7 with slightly deviation to the north. The grave pits are dug into the continent at a depth from 1 m to 1.20 m. Grave No. 4 is destroyed by the digging of grave pit No. 3 that may indicate a long using of the cemetery.

All the burials are inhumations in crouched position to the left, but there are two burials, which are very destroyed and the body position in is unclear (***Table 25.1***). The arms are bent in the elbow, placed in front of the chest, the legs are bent under the sharp angle. The orientation varies, the head is oriented to the West, South-West, East, North-East and South respectively. There is no a special arrangement of the graves.

The closest parallels originate from the flat cemetery from Late Bronze Age in the town of Nova Zagora (Kunchev and Kuncheva 1990: 8-11, figs. 1-7).

At the cemetery of Ovcharitsa II the burials recovered belong to adults and one child. The bones of most skeletons have been well preserved - see the Annex. Two of them are anthropologically investigated. They belong to a juvenile and male adult.

Of the 11 graves recovered 5 (45%) consist of burialgoods (Nos. 1,2,6,8-9). The grave inventory is monotype. It consists of pottery only (***Cat. Nos. 1-4; Figures 25.2***). The burial gifts are arranged in different places – near the head, legs or the arms of the dead. Without being a rule, the pottery placed by the arms dominates (60% of the graves with an inventory).

The grave inventory represents different shapes, which are in almost identical quantity ratio. They are kantharoi, bowls and cups. The cup from grave No. 2 is strongly fragmented, so it is not clear its exact shape and size. Common feature is the bad burning and the existence of numerous mixtures in the clay.

Of interest is the ornamentation over the bowls from grave No. 8, placed near the arms of the adult (***Cat. No. 3; Figure 25.2: 4***). The incised ornament, which covers all the body and the vessel bottom consists of a combination of geometric figurines and strongly stylized flying birds. The ornamentation follows the traditions of "the Balkan geometric style" characteristic of the Culture with encrusted pottery in Northwest Bulgaria, Southwest Romania and Northeast Yugoslavia in the Late Bronze Age (Gergova 1982: 14). The decoration of the bowl from Ovcharitsa II cemetery has exact parallels in the silver belts from Burzitsa treasure, Varna District (Gergova 1982: figs. 5a, 6, 9b, 10a with the reference cited there). Similar is the bowl ornamentation from grave No. 4 from Manchova Tumulus near the village of Malka Detelina in Maritsa-Iztok complex (Kunchev 1991: 43, fig. 4d) and the ceramics from Tei IV-V culture (Leahu 1966: fig. 33: 6, 9; 40: 7). The treasure of Burzitsa e dated from after the 12th century BC, according to D. Gergova (1982: 10), and the grave from Manchova Tumulus, as well as Tei culture – from the Late Bronze Age (Kunchev 1991: 43; Leahu 1966).

The symbolism of the Late Bronze Age distributed in the following Early Iron Age. Most often its main motifs were distributed – cross and water bird, as well as some ceramic shapes, which fact once more confirms the continuity and the difficulty to mark border line between both epochs (Georgieva 1977: 64; Chichikova 1968: 16).

The kantharoi-like vessels from Ovcharitsa II (***Cat. Nos. 1, 4; Figure 25.2: 1,2***) have the closest parallels at Bukyovtsi, Vratsa District (Nikolov B. 1964: figs. 8b, 9b). They are a characteristic ceramic style distributed in the Late Bronze and Early Iron Age.

The grave inventory and the parallels cited allow us to attribute the graves to the transition from the Late Bronze Age to Early Iron Age and to the beginning Iron Age. We think, the investigated graves are a part of a cemetery belonging to a destroyed settlement of that period (materials from Early Iron Age are documented over the territory of Ovcharitsa II).

Catalogue of the vessels

1. Kantharos-like vessel. Equal cut, straight rim, biconical body

No.	Grave equipment	Depth	Body position	Orientation (head)	Size of the skeleton crouched		State	Inventory	Age and gender
					Length	Width			
1	Irregular shaped pit	1.20 m	Hocker aside	South	0.53 m	0.40 m	Destroyed	Kantharos	Child
2	Irregular shaped pit	1 m	Hocker to the left	East	0.75 m	0.40 m	Destroyed	Bowl	Adult
3	Irregular shaped pit	1 m	Hocker to the left	North-East	0.80 m	0.35 m	Destroyed from grave No. 4	No	Adult
4	Irregular shaped pit	1.20 m	Hocker to the left	North-East	0.90 m	0.40 m	Good	No	Adult
5	Irregular shaped pit	1.20 m	Hocker to the left	South	1 m	0.70 m	Good	No	Adult
6	Irregular shaped pit	1.10 m	Hocker to the left	West	0.90 m	0.50 m	Good	Cup	Adult
7	Rectangular, East-West with deviation to the north, 1.20 x 0.60 m	1.20 m	Hocker to the left	East with deviation to North	0.80 m	0.40 m	Good	No	Juvenilis, 16-17
8	Rectangular, East-West 1.50 x 0.60 m	1.30 m	Hocker to the left	East	1.20 m	0.40 m	Good	Bowl	Male, 40-45
9	Rectangular, East-West 1.20 x 0.45 m	1 m	Hocker to the left	North-East	1.06 m	0.40 m	Good	Kantharos	Adult
10	Irregular shaped pit	1.20 m	Hocker to the left	West	1.05 m	0.40 m	Destroyed	No	Adult
11	Irregular shaped pit	1.20 m	Hocker to the left	South-West	0.50 m	0.30 m	Destroyed	No	Adult

Table 25.1. Ovcharitsa II burials.

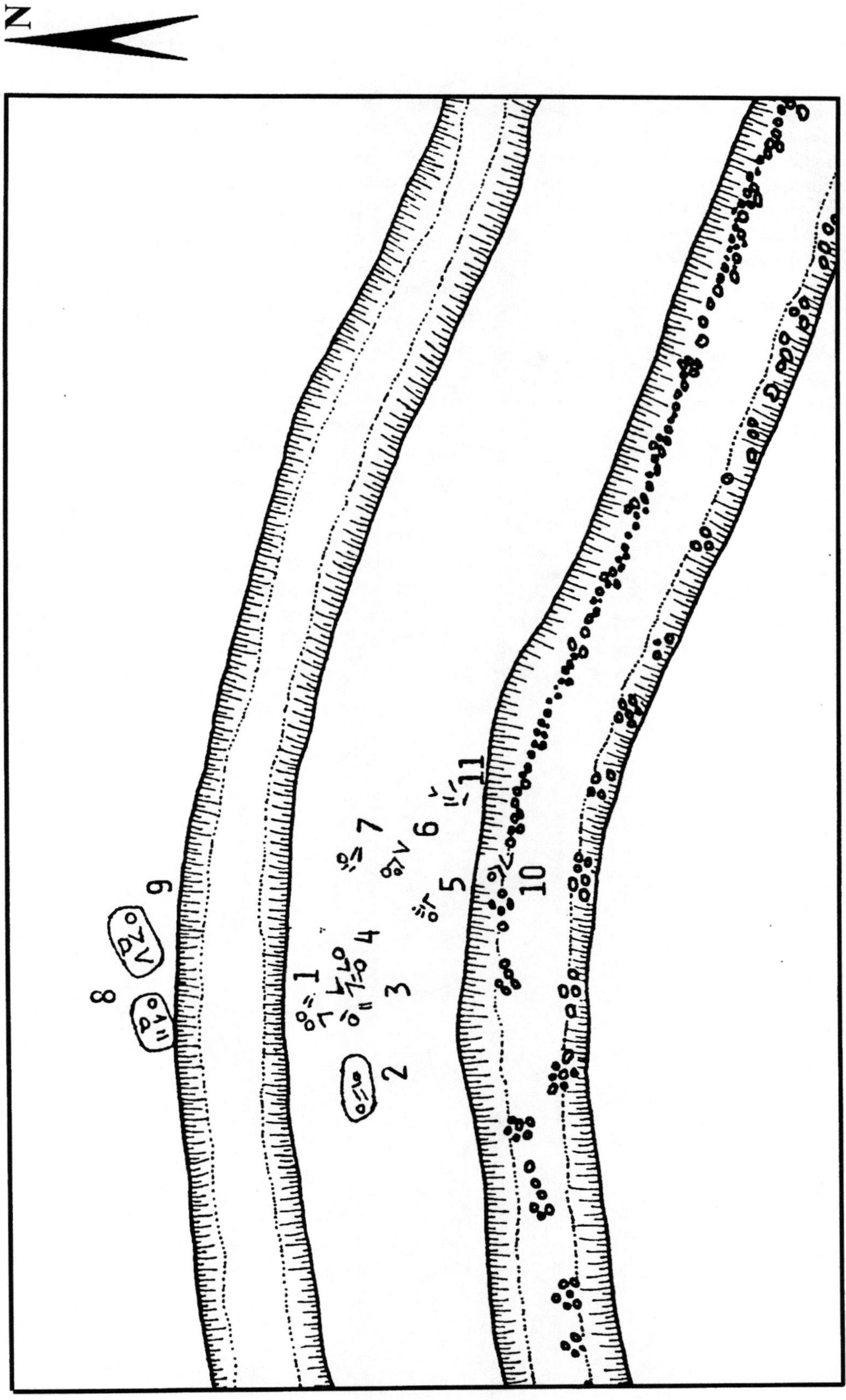

Figure 25.1. *Plan of the Ovcharitsa II cemetery located in the erea of the Early Bronze defend system.*

and low hollowed base. Two arch-shaped high handles are attached to the orifice. The biscuit is of fine clay mixed with sand and stone. Gray colored, with brown flacks. Not very well smoothed surface and not very well burnt. Height with the handles – 12 cm, height without handles – 9.5 cm, diameter of the orifice – 10 cm, diameter of the body – 12 cm, diameter of the bottom – 4.7 cm. Restored. Grave 1. ***Figure 25.2: 1.***

2. A cup. Oblique cut from the inner side rim, truncated-conical body and flat bottom. Arch-shaped handle vertical handle attached at the orifice and the lower part of the body. Biscuit of coarse clay, not well purified, mixed with sand, stone and mica. Dark gray colored with light flacks. Not very well smoothed surface. Badly burnt. Height – 11.5 cm, diameter of the orifice – 11.2 cm, diameter of the base – 6 cm. Restored. Grave 6. ***Figure 25.2: 3.***

3. A bowl. Turned out rounded rim, biconical body and low hollowed base restored. The ornamentation over all the surface of the body and the bottom consists of incised ornament – angles, triangles, rhombs, net, stylized birds. It is made of not well purified clay, mixed with sand, stone and mica. Dark gray colored. Well-smoothed surface. Badly burnt. Height – 10.5 cm, diameter orifice – 12 cm, diameter body – 14.2 cm. Restored. Grave No. 8. ***Figure 25.2: 4.***

4. Kantharos-like vessel. Round and slightly turned out rim, high almost cylindrical upper part and biconical lower part, high hollowed base. Two very high handles attached to the orifice. The handles are decorated in the upper part with zoomorphic ornament. Well purified clay mixed with sand. Lightly gray colored, with brown flacks. Well smoothed surface. Well burned. Height with handles – 17.3 cm, height without handles – 14 cm, diameter orifice – 11.2 cm, diameter body – 13 cm, diameter base – 6.3 cm. Restored. Grave No. 9. ***Figure 25.2: 2.***

Annex
Anthropological data for skeletons from Ovcharitsa II

Slavcho Cholakov

Grave No. *7.* A skeleton of Juvenilis, 16-17 years old. The bones are badly preserved. Single fragments of the skull are saved, thorax, and the long bones of the extremities. The bones are moderately massive, with stressed relief of the joins of the muscles, and without complete growth, impossible to be measured.

Grave No. 8. A skeleton of male adult, 40-45 years old. Of the skull single fragments are preserved, from the crown and the scruff of the head, as well as forehead bone and parts of the jaws. The skull bones are massive, with well expressed relief. Of the skeleton are preserved a part of the extremities, spinal column, pelvis and the thorax. The long bones are very massive, with strongly expressed relief (showing well developed muscles, and robustness stressed. The height, according to the size of the bones – 181 cm. Pathological (diseased) changes have been observed at the spondyles, a result of chronic spondylitis, which strongest damaged the necked parts of the spine. Arthritis modifications have been discovered on the surface of the small joints. Among the skeleton's bones there were numerous fragments of animal bones.

Figure 25.2. Burialgoods from Ovcharitsa II cemetery. Grave No. 1 (1), grave No. 6 (2), grave No. 8 (3) and grave No. 9 (4).

Section 4

Case Studies

CHAPTER 26

LATE CHALKOLITHIC CERAMIC STYLE AT YUNATSITE TELL (APPROACH TO THE SYSTEMATIZATION OF THE CERAMICS FROM THE NEWLY EXCAVATED LEVELS)

Nadezhda Todorova and Velichka Matsanova

Introduction

The Late Chalkolithic in Thrace is well investigated and well known period at first view only. There is a rich ceramic material collected from the excavated sites, but most of it is unpublished. Many sites excavated are poorly published, as Bikovo, Gabarevo, Lenovo, Mechkyur, Kostievo, Kolena, Devebargan, Zavet, known in the scholar literature by single short publications only (Detev 1954; 1962b; Mikov 1926-1931; 1961; Georgiev 1948; Popov 1922-1925; Seure and Degrand 1906). Materials from some sites like Dolnoslav, Mudrets, Stara Zagora Mineralni Bani, Stara Zagora – Okruzhna Bolnitsa, Chatalka (Delimonevo Kale) etc. are completely or almost unpublished. The only information comes from the short reports to the Archaeological Institute with scanty data. In many cases the authors spare unfoundedly small place for the pottery, the descriptions are common, and serious attempts for systematization are missing. As a rule, the pottery is published selectively. The more peculiar fragments or whole shapes are represented there, which do not completely describe the complex. The quantity and especially the quality of the illustrations in most publications make it difficult or impossible complete using of the information.

In the last years the publications concerning the Late Chalkolithic did not considerably increased. Better is represented the material from Dubene-Sarovka (Nikolova and Zakharieva 1994, chapter 6) as in the article the material is systematized, but unfortunately it is fragmented, as well as the results of drilling excavations at Gulubovo tell (Stefanova 1991). A part of the materials from Yunatsite (Mazanova 1992, Matsanova 1994), single fragments from Perushtitsa (Koleva 1988), and recently materials from Kapitan Dimitrievo (Matsanova in Nikolov et al. 1999: 77-94) and Rupkite-Kaleto tell has been published as well.

To the publications above contrasts the recently published Late Chalkolithic pottery from Karanovo, where despite the considerable fragmentary character of the material there is a detailed technological characteristics and an attempt of typology and classification of the pottery (Schrol 1997).

On the whole, as a matter of fact, there is no detail classification and typology of the Late Chalkolithic pottery from the Upper Thracian valley, as well as there are missing significant investigations on the technology of the Chalkolithic pottery. One explanation is the fact that the excavations usually reveal a big quantity of ceramic and its primary systematization is highly labor consuming.

In this approach to the systematization of the Late Chalkolithic pottery in Upper Thrace the basic material from the recent excavations at Yunatsite tell, the Upper Maritsa valley are to be analyzed.

Yunatsite tell. Common characteristics

Location and culture environment

The site is located 1.5 km of Yunatsite village, Pazardzhik District. The settlement is situated over a low terrace on the right side of the old bed of Topolnitsa River (Batakliev 1969: 490). Not far, to the south, are the fist unflooded terraces of the Maritsa River. The north slopes of the Rhodopes raise ca. 15 km south of the site. The south slopes of the Sushtinksa Sredna Gora border the region from the north. The Ihtiman Heights are located to the west. On the whole, the country in the neighborhood is flat, with slight sloping to southeast.

The tell of Yunatsite had visible height over the surrounding terrain c. 9 m. The shape is oval, slightly prolong in east-west direction with diameters of the base 110 x 100 m. Military trenches and water channel have destroyed the western periphery of the site in recent times.

The closest prehistoric site are Tekera tell, near Vetren and Banyata Mogila near the village of Kapitan Dimitrievo, located 9-10 km southwest of Yunatsite tell. C. 12 km east is situated Ognyanovo tell.

It should be especially stressed the key geographical position of the site. The Yunatsite tell is situated in the region between Eastern and Western parts of the Balkan Peninsula where from most ancient Antiquity times the big Eurasian road from southeast to northwest passed, for transmission of people, ideas, constant trade and culture exchange. The western parts of Upper Thracian valley offer direct access to Sofia Field, respectively to contemporary Northwestern Bulgaria, Eastern Serbia and Romania. In southeastern direction, the Maritsa River provides reliable water road to Turkish Thrace, the Aegean and Northwestern Asia Minor. In the region are located the most accessible passes through Western Rhodopes massif – direct connection with the regions of Western Thrace, Thessaly, and Eastern Macedonia. That favorable location of the settlement on the cross-road of the main Balkan arteries makes possible the existence of a complex where elements from different synchronous cultures could be traced. The existence of such complexes allow precise relative dating of the distinct archaeological cultures, their phases and synchronization, as well as analysis of the ways and manners of realization of the interactions between the population with neighbor regions (for instance, existence of exchange trade).

Excavations

The first description of the tell belongs to brothers Shorpil (Shkorpil H. and Shkorpil K. 1898: 81). Evidence of the site published V. Mikov as well (1933: 84). The last author excavated for the first time the tell in the autumn of 1939. A trench, 25x10 m, was excavated up to 9 m depth, without reaching the sterile layer (Mikov 1937-1939: 55-84; Katincharov et al. 1995:13).

In 1976 significant investigations began at the site conducted by R. Katincharov and V. Matsanova. In 1981 the expedition became international with participation of N. Ya. Merpert, V.S. Titov, etc. Since that time to 1995 the whole layer of Early Bronze Age has been investigated in the east one third of the tell, as well as partially the first and the second building level of the Chalkolithic. The common stratigraphy of the tell was elaborated precisely (Katincharov et al. 1995: 24-33).

Stratigraphy

In the course of the excavations in 1976-1985 the following stratigraphic sequence was documented (Katincharov et al. 1995: 15):

Layer A – a depth from 0.30 to 1.00 m.
Middle Age cemetery, 13th – 14th century.
Settlement and castle – 2nd-4th century.
Thracian settlement from Late Iron Age – 6th-5th BC.

Layer B – a depth from 1.00 to 6.20 m.
17 building levels from Bronze Age.
Layer C – depth from 6.20 to 9.00 m.
Chalkolithic.

At the time of the investigations materials from Late Bronze Age and destroyed structures from Early Iron Age have been discovered as well.

In the last years (1989-1998) the archaeological excavations were concentrated over investigation of the Late Chalkolithic building levels of the site. The bases of ŕ part of the houses from the 17th building level from the Early Bronze Age are situated over the destruction of the first building horizon of the Chalkolithic. Moreover, there is a stratigraphic hiatus at the central profile of the site.

The partially investigated Chalkolithic building levels have recovered an extremely complicated stratigraphic picture. The problems of planigraphy and the stratigraphy of the site are beyond the subject of that paper, then, only basic data will be shortly represented here. The data from the recent excavations of the Chalkolithic layer in the eastern part of the site show the existence of 2 building levels belonging to the Copper Age without fixed hiatus between. Both settlements were fired. Since 1986 on 13 houses from the first and second horizon have been registered and partially investigated. As the pottery of the both horizons does not show any differences, the analysis of the material from both the horizons will be commonly treated.

Technological data on the pottery

For the technological description of the intact vessels and the fragments from the first and the second horizons partially the scheme of G. Schneider is applied (Schneider et al. 1989). The following main criteria are used: Contents of the paste, slip, color and treatment of the surface.

Based on the correlation of the characteristic mentioned above (see Annex 1) 9 technological groups are distinguished:

Fine black burnished ware. The number of the vessels from that group is considerably small. Three categories of vessels are represented – shallow and deep bowls, cups and pots. The pottery of that technological group has compact and homogenous black color. It is made of extremely fine clay paste with insignificant quantity of mixture of very fine organic matter and fine sand. Almost in all cases the inner and the outer surface is covered with slip (fine, semi-fine or compact), very rare there is self slip. As a rule, the vessels are burnished inside and outside. The burnishing is up to glance or in strips. Rarely there are vessels with smoothed surface, as the smoothing is always homogenous. That manner of surface treatment is characteristic mainly of the inner surface of the vessels from the category of the bowls and the cups and pots (B.I.2.1; D.V.1.2; E.VII.2.2). a considerable part of the shapes under question are decorated with graphite ornament (A.II.2.1; A.IV.3.1; B.V.2; D.V.1.2; E.VII.2.2).

Fine gray and gray and black ware. That group is also not numerous. The category of the plates (A.II and A.VII and rarely others) mainly represents that group. The vessels have light gray and dark gray and black color of the surface. The coloring is homogenous. The contents of the clay does do differ from that of the black fine pottery. In all cases the surface is covered inside and outside with fine and semi-fine slip and is evenly (A.II.2; A.VII.2) or in strips (A.IV.3.1) burnished. Th inner surface of some plates are decorated with graphite (A.II.2; A.IV.3.1).

Black and gray and black burnished ware. A considerable part of the ceramics belongs to that group. It is mainly distributed in the category of the shallow bowls – almost all basic shapes are represented (A.I; A.II.2; A.IV.1, A.IV.2; A.V.2; A.V.3; A.V.4), as well as distinct types of bowls, cups, pots and lids (B.III.2.1; B.VII.4; D.V.3; E.VI.2; E.VI.3; E.VII.1; F.I.1). The surface color is compact black and gray and black, homogenous. The vessels were made of semi-fine or semi-coarse clay. The clay mixture consists of one and the same elements – fine quartz, small and medium big quartz stone, fine organic matter. The only difference is in their concentration. Most of the pottery has slip surface, which is fine, semi-fine and compact. Self-slip surface and washing surface are rarely distributed. All variants of the burnishing of the surface are documented – evenly, in strips, burnishing to shine surface. Most often the evenly burnishing of the vessels inside and outside is documented. The decoration with graphite is vast distributed among the vessels of that technological group. The ornament is placed over the inner surface of shallow bowls of types A.II.2; A.V.2; A.V.3 and over the outside surface of cups of types D.V.1; D.V.2; D.V.3; D.V.4, as well as at the upper part of distinct types of pots and amphorae-like vessels (Ĺ.VII.1; F.I.1).

Light brown burnished ware. The vessels from the categories of shallow bowls, bowls, cups, pots and miniature earthenware are approximately equally distributed in that technological group (Ŕ.I.1; A.V.1.1; B.VII.1; D.I; D.II; K.IV; M.I.1.; M.II.2.1; M.III.3). The pottery has homogenous (evenly) light brown color. The clay used is semi-fine. The mixture consists mainly of fine quartz sand and quartz stone in small and medium size. The slip of the surface is semi-fine or self-slip. It should be mentioned the fact that the analyzed pottery of that technological group does not include compact or semi-fine slip. The only exception is a lid (shape K.IV) covered outside by compact slip. The outer surface of the vessels is burnished – evenly or in strips. There is one instance of evenly smoothed surface. The ceramic biscuit has uniform light brown color. The vessels mostly belong to the group of the unornamented pottery.

Brown and dark brown burnished ware. The number of the vessels of that technological group is not big. Bowls and some types of cups are represented in (Ŕ.IV.1.2; A.VII.1.2; D.V.2.2). The pottery has compact and homogenous brown and dark brown color. The clay used is semi-fine, rarely semi-course, mixed with fine sand, quartz stone and fine organic matter. The vessels' surface is covered with fine and semi-fine slip. Self slip is rarely distributed. The surface is evenly burnished, by the bowls – inside and outside. The vessels of that technological category normally have no ornamentation, with exception of cups of type D.V.2.2, which has a graphite ornament in its upper part.

Brown, black and brown and gray and brown burnished ware with irregular coloring of the surface. Most categories of vessels represent that group. The biggest is the number of the shallow bowls. The vessels of the rest of the categories – cups, amphorae-like vessels, pots and etc. – are approximately proportionately distributed (Ŕ.II.2; A.IV.2; A.V.2; A.V.3; A.V.4.1; A.VIII.3; B.I.2; B.III.2.2; B.V.1; D.IV.1; D.IV.2; D.V.2.2; E.IV.2; F.III etc.). The pottery has irregular color of the surface, which is due to the way of burning, but probably in some cases to the secondary burning. There are spot of irregular shape on the inner and the outer surface. The color of the spots varies from light beige to dark brown and from light gray to black. The clay used is semi-fine and semi-coarse, as the mixture is identical with those described by the group of the brown burnished pottery. The outer surface, and by the bowls the outer and the inner surface, is covered with fine and semi-fine slip, rarely self-slip is documented. The vessels are burnished, as at the closed shapes in the categories of the cups, the inner surface is always evenly smoothed. A considerable number of the vessels are decorated with graphite (Ŕ.II.1.1; A.II.2; D.IV.2; E.IV.2; F.III).

Brown and dark brown coarse ware. The vessels of that technological group are not numerous. The categories of bowls, pots and storage vessels are mainly distributed B.I.1; B.II.1; B.V; B.VII.3; E.VIII.1; J.III.1.2; J.IV.1; J.V.2.2). The pottery has homogenous brown color, sometimes the inner and the outer surface of the vessel is differently colored. The clay used is semi-coarse, rarely coarse. The mixtures consist of quartz and stone in different sizes and high concentration. Organic matters are registered as well. Fine, semi-fine and self slip characterize most of the vessels, which cover only the inner surface. In some cases the upper outer parts of some types of bowls and storage – depositories are slipped as well as rarely the upper parts of some types of bowls and storage vessels. The slip parts are homogeneously smoothed, rarely burnishing is distributed. The outer surface of the vessels normally is roughed by organized barbotine, uneven and very rarely evenly smoothed. The decoration is mainly relief.

Brown, black and brown and gray and brown rough ware with irregular coloring of the surface. To that technological group a big quantity of vessels belong, mainly of the category of bowls, pots and storage - depositories (B.I.; B.II.1; B.V.1; B.VI.1; B.VI.2; B.VI.3; E.II.1; E.IV.1.2; E.VI.1; E.VIII.1; J.III.1.2; J.V.1.2; J.V.2.1; J.V). The pottery has irregular color of the surface, identical to the described for group No. 6. The clay used is semi-coarse and coarse. The mixture of quartz stone, sand and organic is of high concentration. As by the vessels of the previous group, the inner surface or the upper outer parts of the closed shapes are slipped. The surface treatment is identical to that of group No. 7. The inner surface is evenly smoothed rarely irregular smoothing is distributed. The outer surface is consciously roughed (dragging, barbotine). The decoration of the vessels is mainly relief.

Red and brown rough ware. Storage vessels – depositories of types (J.I.2; J.III1.1; J.III.2; J.IV.2; K.II) represent that technological group. The pottery has red and brown homogenous and irregular color of the surface. The clay is rough with considerable quantity of mixtures of relatively coarse organic matters, medium and big quartz stone and highly concentrated sand. The outer surface is unslipped and consciously roughed. The inner surface is covered with fine slip or self slip and evenly smoothed. The only used technique of decoration is relief one.

The analysis of the technological groups shows that in the complex vessels with dark coloring dominate with different nuances of black, gray and black and dark brown. The group of the light brown pottery is not numerous. The vessels with uneven coloring of the surface are often distributed. But it is important to stress on the fact the pottery analyzed originates from burnt horizons and because of that reason the color of the vessels should be not leading in the distinguishing of the certain technological groups. The importance of that criterion should be not overvalued as the ceramic complexes at other Late Chalkolithic sites indicate very different picture in relation to the color variation documented at Yunatsite. Among the pottery at Telish and Sudievo clearly dominate lighter colors – light brown, beige, light beige etc. and it cannot be excluded the fact that those sites exemplify a kind of another regularity.

As far as the clay mixtures concern, at Yunatsite dominate the quartz sand and stone in different sizes. The organic matter is relatively often used, but mixtures of mica is rarely distributed and the limestone inclusions, as well as 'shamot' are exceptions (see the Annex No. 1).

It is interesting to note the small quantity of vessels made of fine clay – mainly cups, shallow bowls and small bowls. Main part of the pottery belongs to the group of the gray and black and black burnished and dark brown and black and brown burnished ceramics, so that almost all categories are distributed (see above). Traditionally, the 'kitchen' earthenware, as well as the storage vessels - depositories, is made of coarse paste with roughly treated or consciously roughed surface.

The biggest part of the pottery from the first seventh technological groups has an additional slip documented – over the whole surface or over distinct part of the vessels, related to the shape and the function of the vessel. Slip and self-slip vessels are with carefully treated burnished surface.

On the whole, the technological characteristics defined do not show considerable differences among the known from the Late Chalkolithic sites in the region of the parts of the Upper Thracian valley – Karanovo (Schrol 1997: 287), Gulubovo (Stefanova 1991: 145), Gudzhova Mogila, Iskritsa, etc. Unfortunately, there are no considerable data on the pottery from sites like Dolnoslav, Stara Zagora-Mineralni Bani, Chatalka, Sveti Kirilovo, Bikovo, etc. Because of that reason, the detail technological comparison cannot be elaborated.

Classification

The classification of the pottery is specified on the base of intact ceramic shapes originated from Chalkolithic levels and fragmented pottery of closed complexes existed.

At the base of the classification are the formal and functional characteristics of the vessels. The distinguished categories of earthenware are shallow bowls, bowls, jugs, cups, pots, amphorae-like vessels, 'vases', askoi, storage vessels - depositories, lids, pot-stands and miniature vessels.

Some of the basic methods used in accomplishing classification of the ceramic materials from Yunatsite are accepted from the Vossen's study, which are used in the contemporary American archaeology (Vossen, 1970:63-66). The structure and terminology proposed by K. Leshtakov for the pottery from Early and Middle Bronze Age (Leshtakov 1988:1) is also partially applied. Functional and formal criteria define the elaboration of the distinct elements of the scheme presented (category, shape, type, sub-type, and variant).

According to functional criteria three groups of pottery can be defined – fine table, coarse kitchen, and depositories - storage. Those groups characterize the whole prehistoric pottery (Todorova 1982) – cp. chapter 6. Those three basic categories have along with functional, a technological characteristics as well. They can be related to the technological groups defined, as the distinct technological groups can be connected with one or more functions.

A determination of the category is the next niveau in the classification of the ceramic information. The formal and the technological criteria are generally important for the dividing of the distinct categories of vessels. For definition of the categories the contemporary terms for the vessels with similar shape are used, as well as some names are accepted from the Ancient Greek language (askoi and amphora).

The basic shapes are defined based on the common structure of the vessels (spherical, prolong spherical, conical, biconical, etc.). The stable correlation of formal criteria, especially profiling of the vessels defines the distinct types. The subtypes are distinguished on the base of the peculiarities of the shaping of the distinct elements – rims, handles and necks. Varient is rarely used and marks existed details of shaping of the rims, handles, etc.

For definition of the distinct elements of the classification elaborated the following terms are used:

A. Category.
I. Main form
1. Type.
1. Sub-type.
a. Variant.

The classification of the pottery from Yunatsite is given in Annex No. 2 and in ***Tables 26.1-26.9***.

The Annex No. 2 represents a considerable variety of forms, types and sub-types and it is not an easy task to define one or more leading shapes and types. Here we will show some of the traditional, mostly distributed in the complex shapes and will briefly mention the analogies of the more interesting, specific shapes.

Of the categories of the shallow bowls open conical forms are wide distributed, which are with straight walls and thickened differently shaped rim (A.II; ***Figure 26.2: 1,9-12***), as well as the conical plates with short vertical and inclined upper part (A.IV, A.V; ***Figures 26.1:3,4; 26.2: 2-4***). As a rule, the first type vessels are richly ornamented with graphite, while the other two decorations are distributed rarely and poorly. That 'popular shape' has extremely numerous parallels, which originate from rather vast geographical area.

Special attention should be paid on some variants of form A.II (***Figure 26.2: 11***), which have their close analogies among the pottery at Kolarovo (1995:133, fig.13: 453). The shallow bowls with conical shape and incurved rim (A.VI) and with semispherical shape, and thickened rim (A.III; ***Figures 26.1: 1,2; 26.2: 6-8)*** of type 'Kum Tepe Ib' (French 1961: 111; Maran 1998: 40) are relatively often distributed among the pottery from Yunatsite. They have many parallels in the pottery not only from the Aegean and from Northwestern Anatolia, but also in Thrace and Northern Bulgaria –

Figure 26.1. Yunatsite tell. Late Chalkolithic pottery.

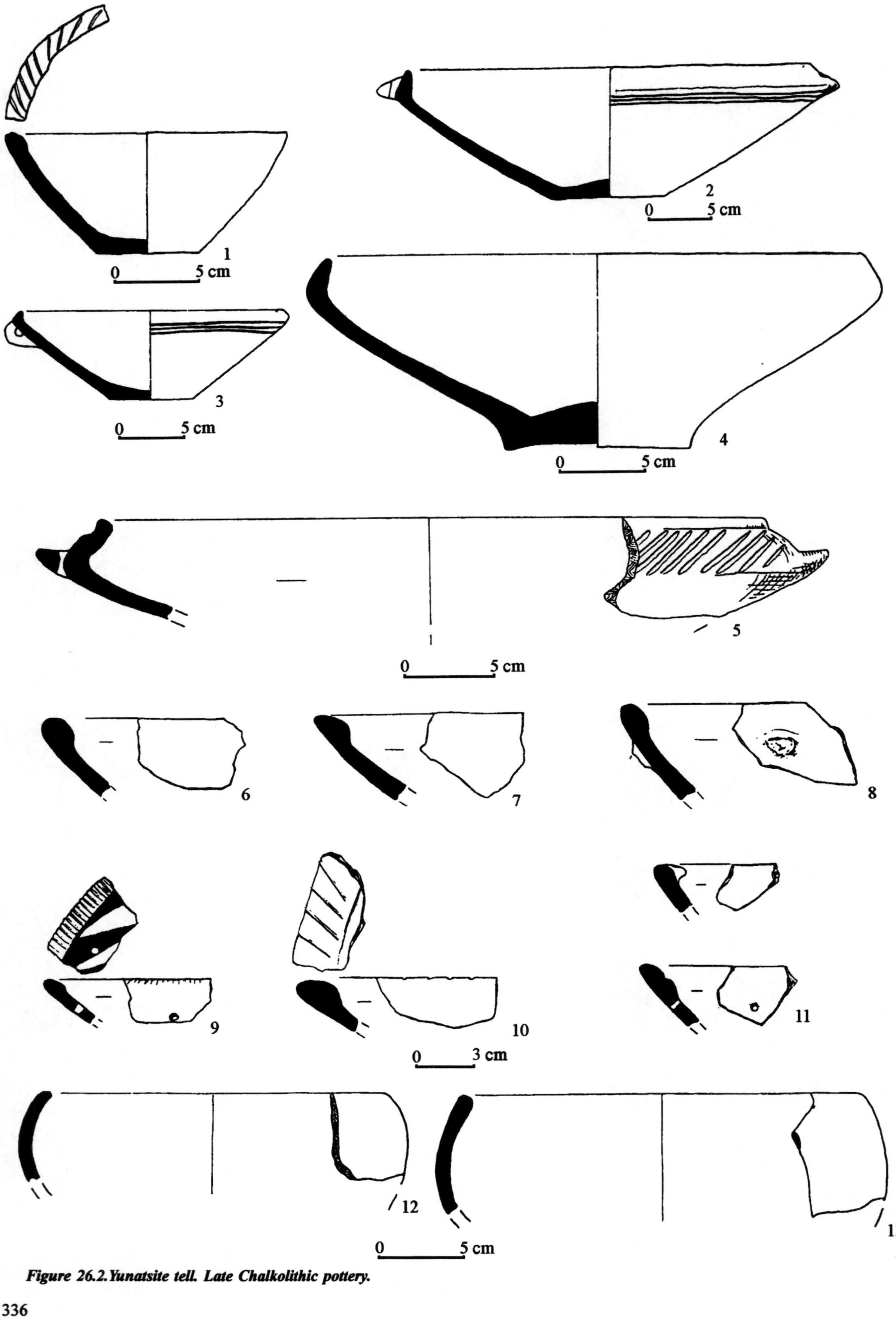

Figure 26.2. Yunatsite tell. Late Chalkolithic pottery.

Karanovo (Schlor 1997: fig. 131: 4-6), Iskritsa – pit 4, Sadovets (Simeonova 1968: fig. 3.5; fig. 5.5) and Okol Glava.

The stage of investigation of the Late Chalkolithic pottery in Upper Thrace as well as in the region of Turkish Thrace (lower stream of the Maritsa River and its tributaries) does not allow straightforward explanation of that similarity and the temptation to be searched in the synchronous occurrence looks hasty.

Besides the shaped mentioned above, it should be stressed on the distribution of ceramic complex of shallow bowls with carinated profile and out-curved rim (A.VII) and of many fragments of biconical shallow bowls with thickened and differently shaped turning point (A.VIII, ***Figure 26.2: 5***). A bigger part has exact parallels among the Late Chalkolithic materials from the area of Krivodol-Sălcuța (Petkov 1950: fig. 98; Nikolov 1984; Berciu 1961: figs. 85: 6, 11; 101: 9, 10 etc.), but single fragments (A.VIII) are registered among materials from Bikovo (Detev 1954: fig. 45).

A considerable variety of forms and types characterizes the category of the bowls. The bowls with simple semispherical shape are numerous (B.I; ***Figures 26.2: 12, 13; 26.3: 1, 2, 4***), as well as those with short neck differently shaped (B.II, B.III; ***Figure 26.3: 3,5***). The spouted bowls are a charcateristic peculiarity (B.VI, ***Figure 26.3: 6, 7***). Close analogies of that shape is known from Krivodol (Mikov 1948: fig. 31b), Sălcuța I (Berciu, D. 1961: fig. 85: 15), Karanovo (Schlor, I. 1997: fig. 137:18), Peklyuk (Petkov 1964: fig. 17), Okol Glava (Petkov 1950: figs. 98, 100, 103), Kapitan Dimitrievo (Petkov 1950: fig. 8: 15l, 16l), Sudievo, Mudrets-Gudzhova Mogila and Telish.

The biconical bowls with thickened turning point and two vertical handles attached at the upper part represent other typical of Yunatsite shape (B.VII, ***Figures 26.4: 5-7, 26.5: 7***). A graphite geometric ornament is often painted over the upper part of the vessel, sometimes in combination with decoration of red distemper. Bowls with spherical and conical lower parts are characteristic of the complex as well and the upper one shaped as truncated cone (B.V), as some of them consist of rich graphite ornamentation (***Figure 26.4: 3***) or in combination with red paint (***Figure 26.4: 2***).

In the category of the cups biconical shapes with two vertical handles at the upper part dominate (D.IV, D.V). According to the manner of shaping of the turning point and the rim, several types and subtypes of the basic shapes are distinguished (see Annex No. 2). As a rule, the upper part of the vessels is decorated with relatively simple graphite ornament (***Figure 26.5: 1-6***), sometimes in combination with distemper painted ornament located around the handles. In some cases the rim and the cross-section of the cups is oval. At most of the vessels are turning points are sharp, stressed and on the whole they strongly resemble the cups with identical profile but without handles, which are characteristic of the pottery from Late Chalkolithic in eastern parts of Upper Thrace and in Northeastern Bulgaria (Stefanova 1991: fig. 13e, d; Schrol 1997: fig. 132: 13-15, 21; Popov 1916-1918: fig. 105b; 124a, 125a, b; Georgiev and Angelov 1957: fig. 33: 2; 45: 2,3; Todorova 1978: fig. 48a; Todorova et al. 1983: fig. 87: 1, 2, 5, 9, 15). Close analogies of the shape from Yunatsite is known from the Chalkolithic layers at Bikovo tell (Detev 1954: fig. 74; Georgiev 1961: fig. 33: 2), Kapitan Dimitrievo-Banyata Mogila (Detev 1950: fig. 8: 15k) and Dolnoslav (Raduncheva 1991: 110). Precise parallels of the shapes and motifs originate from Sitagroi (Evans 1986: fig. 12.2: 7; 12.5: 2; see also the discussion in chapter 16 in this book).

The cups with two handles are one of the most characteristic shapes in the Late Chalkolithic pottery in the regions of Western Bulgaria, the area of Krivodol-Sălcuța. In contrast to known from Yunatsite and other sites in Upper Thrace, the two-handled cups from the region mentioned are the profiles are smooth, rounded, the walls +are unflatted at the medium point and there are instances of maximum widened part located in the lower part of the body, as well as with almost round bottom (Mikov 1948: fig. 30b,c,n,o; Petkov 1964; personal observation of cups from Telish).

Of special interest is a small fragmented cup from the first building level on the tell (***Figure 26.7: 6***; Cat. No. 41). The vessel has spherical and conical shape and small flat bottom. The outer surface is covered with very fine red slip, burnished to glance. The inner surface is treated relatively careless, with some traces of partially smoothing. The biscuit of the vessel has homogenous light brown color, and good burning. The character of the slip, the shape of the vessel, as well as some peculiarities in the technology has no analogies among the pottery at Yunatsite and the complexes in Thrace. It can be assumed that the cup is an import. One relatively exact parallel originates from the 'middle' level at Pevkakia Magula (Weisshaar 1989: fig. 42: 13).

Only one whole vessel (***Figure 26.7: 5***) represents the category of the jugs. The shape is identical to materials from Telish 3 (personal observation). Probably the vessels from that category were numerous but it is difficult to define their real number by the shreds preserved only.

The vessels of pot category show considerable variety of that shape. Basic shape and most typical is that of pots with spherical and biconical, gently profiled body, medium high and clearly distinguished cylindrical or conical neck and two horizontal handles at the upper part (E.II, E.IV). In most of the cases front parts of the vessels are ornamented with horizontal rows of nail incisions (***Figure 26.6: 5-7***). Most precise analogies originate from Late Chalkolithic complexes on the territory of Northwestern and central Northern Bulgaria – Sadovets (Simeonova 1968: fig. 12: 19), Devetaki cave (Mikov and Dzhambazov 1960: 65: fig. 47), Zaminets A-C (Nikolov 1975: figs. 23, 35, 41, 75, 88, 98), Krivodol 3 (Nikolov 1984: 57, 60), Telish, as well as in eastern Serbia like from Bubanj (Tasić 1995: plate 2), etc.

Other one characteristic shape at Yunatsite is the biconical pots with thickened and gently shaped turning point (E.VII.2). Normally, the upper part of the vessel is decorated with graphite ornament, but over the turning point there are rows of shell incised (***Figure 26.4: 1***). In contrast to the shapes represented above, a considerable part of the analogies has 'eastern' origin. Close parallels are documented among the material from Karanovo (Schlor 1997: figs. 132: 17, 18; 137: 4,5,7-9), Gulubovo (Stefanova 1991: figs. 13d, 15a), Russe (Georgiev and Angelov 1957: figs. 33: 1, 42: 4), Hotnitsa (Todorova 1978: fig. 35), Devetaki cave (Stoyanov 1992b: 15) etc.

Of special interest are some fragments of closed shapes, probably pots thickened outside and differently shaped rim (***Figure 26.10: 1-5***). The vessels with profiles shaped in this way are not typical of the ceramic complex from Yunatsite and for the time being, there are single instances documented. One of the sherds originates from a house and contrasts to the rest of the pottery in the closed complex. Along with the unusual for Late Chalkolithic shape in Thrace, the fragment shows peculiarities regarding its technological characteristics as well. The sherd has an intensive black slip, the surface is burnished to glance and its decoration is not typical of Thrace. (***Figure 26.10: 5***). The only relatively exact parallel originates from Late Chalkolithic site of Kolarovo (Pernicheva, L. 1995: fig.13: 483). The rest few fragments (***Figure 26.10: 1-4***) do not differ technologically from the common material at the site, and some are decorated with graphite. The only peculiarities the turned outside, thickened and rounded rim. Those 'untypical' shapes have only analogies among the published pottery from the Chalkolithic layers of Kaleto tell near the village of Rupkite.

The amphorae-like vessels from Yunatsite (F.I-IV) show rather big quantity of variants in the manner of structuring of the body and position of the handles. The shape is typical of the complex, as many intact vessels are registered, as well as sherds. Some of the vessels are decorated in their upper part with graphite painted geometric ornament (***Figure 26.6: 3,4***) and incised ornament (***Figure 26.6: 1***), but there are undecorated ones (***Figure 26.6: 2***). One of the shapes of the amphorae-like vessels (F.II.2, ***Figure 26.6: 1***) has many and precise parallels among the materials in Upper Thrace – Bikovo (Detev 1954: fig. 480, Karanovo (Schrol 1997: fig. 137: 1-3), Stara Zagora – Miniralni Bani (Fol et al. 1982: 124, 141c). Another region with identical Late Chalkolithic finds is Western Bulgaria – Peklyuk (Petkov 1964: fig. 7a), Krivodol (Mikov 1948: fig. 31d), Telish, etc.

Of interest are two vessels (G.I, G.II) with high narrow cylindrical necks and flatted spherical body. One of the vessels has a small vertical handle attached at the middle point of the body and is decorated with complicated graphite ornament (***Figure 26.7: 2***). Precise analogies of that shape are missing. The second vessel has two big vertical handles at the upper part. Very rich graphite painted decoration characterize that vessel in combination with red distemper paint (see Leshtakov, ***Figure 26.23: 5*** in this book). The closest parallels of the shape and the style of decoration of that vessel originate from Late Chalkolithic levels at Sitagroi (Evans 1986: figs.12.5: 1, 12.9: 3). In both cases the vessels are asymmetrical, with oval in plan orifice and slightly flatted body. They have identical profiling and shape of the handles and represent not only the most precise, but in fact the only analogies of the 'amphora' from Yunatsite (more detail comparison between the pottery from Yunatsite and Sitagroi see in chapter 16).

There are no numerous askoi among the intact vessels and fragment pottery at Yunatsite (***Figure 26.7: 7***).

Of special interest is one unique shape, without analogies among the published Late Chalkolithic materials in the Balkans (***Figure 26.8***). The orifice is obliquely cut and rounded, and the handle is attached at the rim. The whole surface is decorated with complicated composition painted with graphite. The askos does not differ technologically from the other pottery at the site. Characteristic is the extremely well performed and the compact, burnished to shining slip. The rather 'untraditional' shape indicates more peculiar function of the vessel.

The category of the lids includes few basic shapes only (K.I-IV). The number of the vessels and sherds is not considerable. One of the relatively often distributed shapes in the complex are the so-called pot-stands (***Figure 26.7: 9, 8***), which have close analogies at Krivodol (Mikov 1948: fig. 39d), Okol Glava (Petkov 1950: fig. 105c) and Telish.

Some shapes of depositories (pithoi) are reconstructed only, so the distinguishing of certain types is problematic. On the whole, the storage vessels with prolong spherical or spherical shape characterize the complex, with differently shaped, wide or narrowed upper part (neck) – J.I, J.III, J.IV. As a rule, the surface of the vessels is consciously roughed, with exception of the upper parts, which are slipped and carefully treated. In most cases the barbotine is organized, as with fingers groups of vertical, horizontal and oblique lines are shaped (***Figure 26.9: 1, 3-5***), in combination with plastic bands, finger impressions and differently shaped knobs. The definition of the analogies of the depositories from Yunatsite to a big extend is difficult because of the limited number of published and known shapes of that type. Generally speaking, the vessels have prolonged spherical shape, low narrow conical neck and two (or four) vertical handles attached at the middle body point (***Figure 26.9: 2***). They have analogies among materials in the area of Northwestern Bulgaria – Krivodol (Mikov 1948: fig. 31f) and Telish (personal observation). In some cases big, but more precisely made vessels represent the same shape, with slip and burnished surface and rich graphite decoration. There are only few sherds known from Yunatsite (***Figure 26.23: 6***), which have good parallels at Sadovets, as complete analogies characterize the ornamental system as well (Simeonova 1968: fig. 6:10; Todorova 1992: fig. 6; 1986: fig.18).

Discussion and conclusions

The analysis of the pottery from Yunatsite performed above clarify in more details the ceramic complex has close correlations with Telish-Redutite 3. The pottery from Telish has certain stratigraphic context and originates from closed complexes. Most of the shapes distinguished are represented at the ceramic complex at Telish. The similarity between both sites includes not only the formal-typological closeness of the material, but also Telish II-III and Yunatsite have nu-

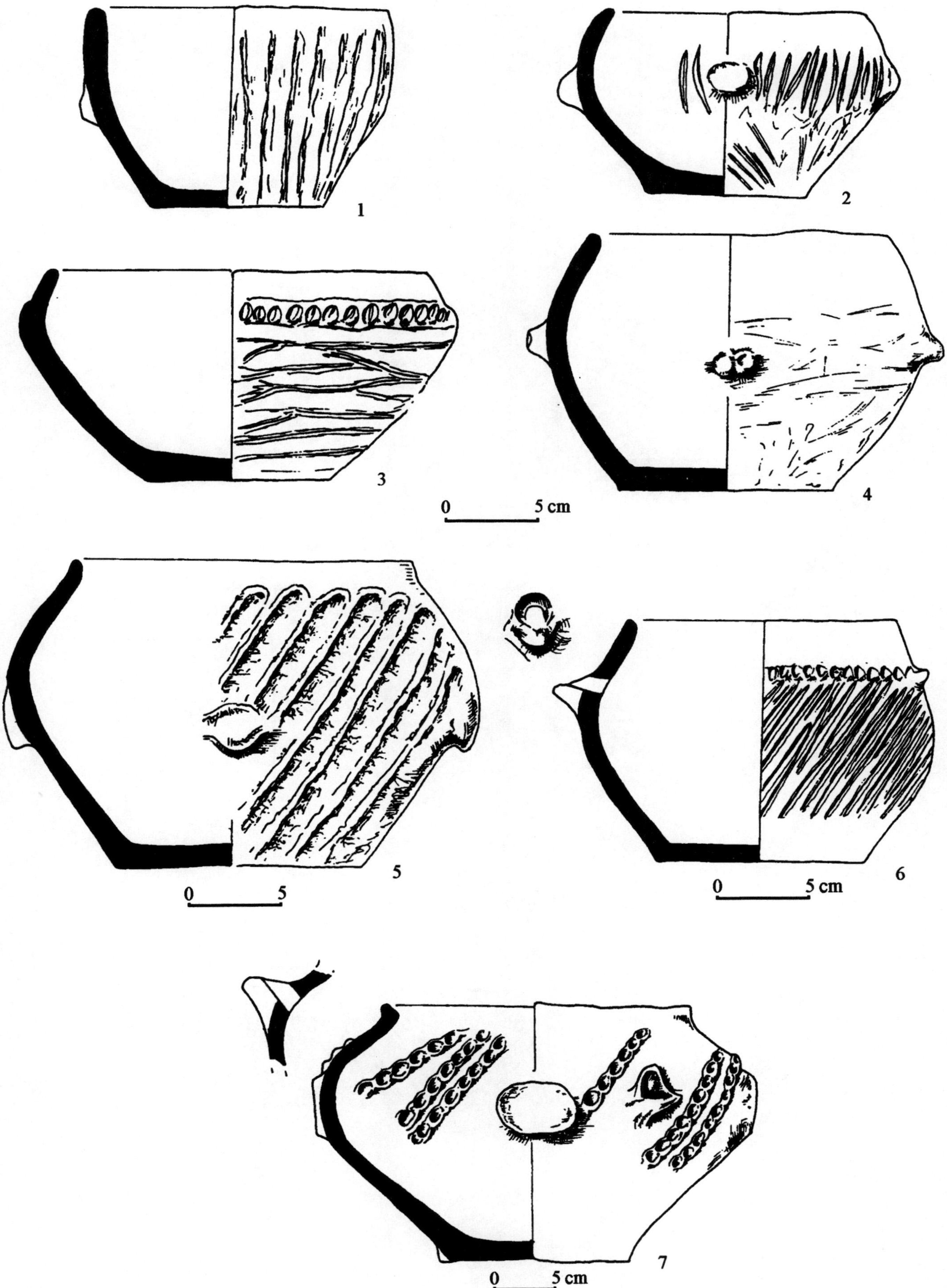

Figure 26.3. Yunatsite tell. Late Chalkolithic pottery.

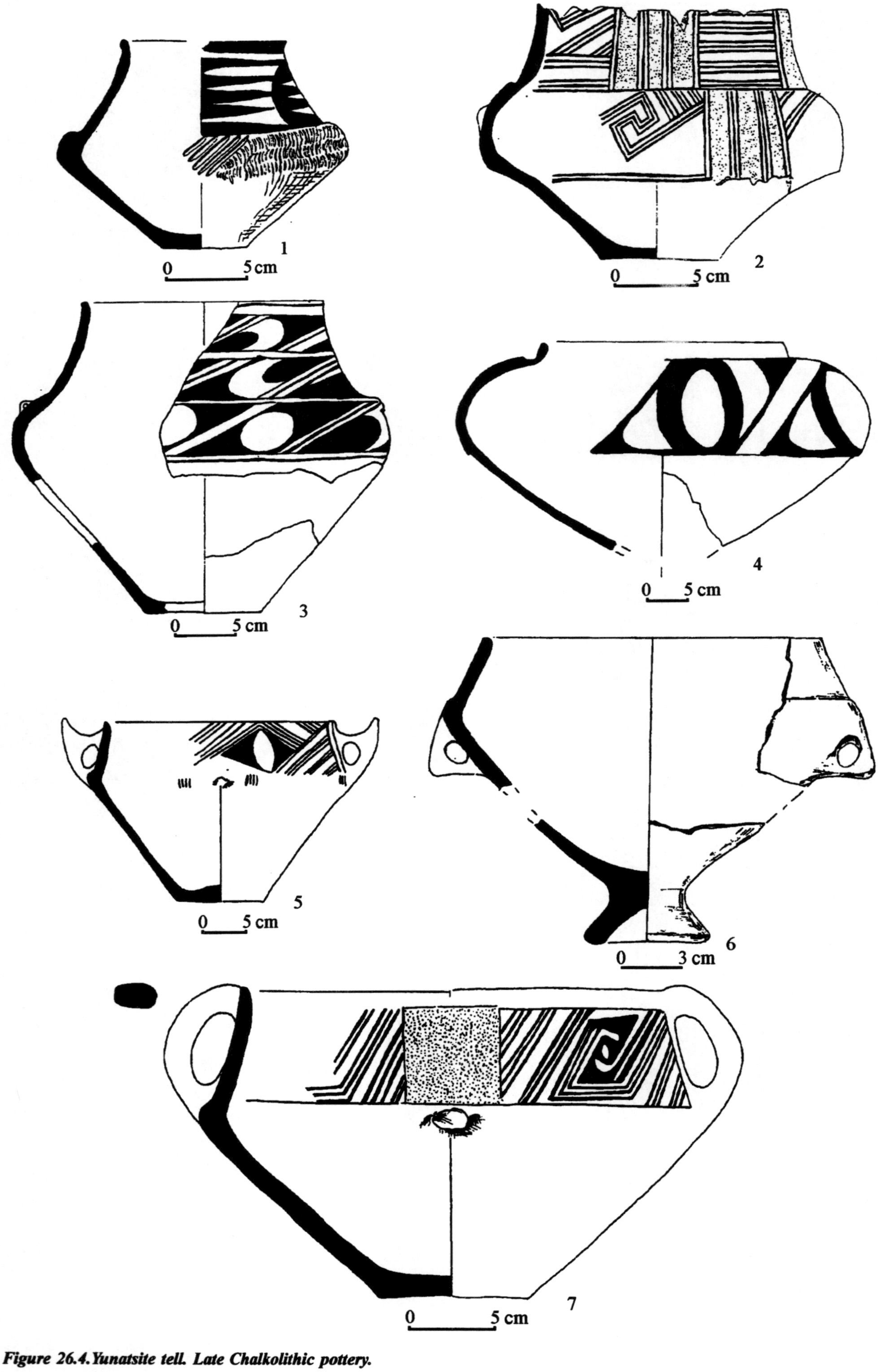

Figure 26.4. Yunatsite tell. Late Chalkolithic pottery.

merous correlations regarding the technique of decoration. Of interest is the existence of similar, often identical ornamental motifs and schemes. At the same time, there are differences, which will be discussed elsewhere. They concern the technological characteristics of the vessels, the paste used, the mixtures and the manner of burning. The analysis of the material indicates they were different at the distinct sites, on the one hand. On the other hand, there are some differences with respect to formal peculiarities. Not all known from Telish shapes and types have direct correlations with Yunatsite, as well as by some categories there are considerable differences. Those conclusions are based on the considerable part of the whole vessels and mass material from Telish II-III (unpublished).

There are numerous and precise parallels of the vessels at Yunatsite with materials from Sitagroi IIIb, Sălcuţa I-III, Kolarovo I-II, Sadovets, Kapitan Dimitrievo-Banyata Mogila, Bikovo, Krivodol, etc. Important are the analogies, which originate from sites located in the eastern parts of Upper Thrace. There are exact analogies of the material from Yunatsite at the last Chalkolithic levels at Karanovo, Sudievo, Gulubovo – Asara, Mudrets – Gudzhova Mogila, etc.[1] Of interest is the fact that the pottery from Yunatsite has good analogies among the materials in the western Balkan regions – Krivodol-Sălcuţa complex, as well as among the Late Chalkolithic sites from eastern Upper Thrace. Accordingly, the problems of the existence of the so-called 'contact' or 'transition' zone occur of primary importance, as the mixture of distinct elements and existence of cultural elements, characteristic of different geographical regions. In connection with that problem is the question of the manner and ways of realization of the contacts between the distinct regions (Renfrew 1979). Unfortunately, there are no numerous sites to be used for comparison of distinct complexes and horizons with the methods of the 'chain dating' in Upper Thrace, which could allow elaboration of stable, based on certain facts relative chronology of the Late Chalkolithic period there. Recently, L. Nikolova has elaborated a compendious investigation in this direction where the questions of the relative and the absolute chronology have been analyzed in details, and a number of conclusions concerning the synchronization of the distinct occurrences have been argued (Nikolova 1999: 71-89).

The key geographic position of the Yunatsite tell is one of the reasons the material from that site and especially the pottery to take a considerable place in the future elaboration of precise periodization and chronology of the Late Chalkolithic in the Upper Thracian valley and Western Bulgaria. The fact that a considerable part of the vessels analyzed originate from closed complexes and have certain stratigraphic position, as well as the precise stratigraphic observation allow tell Yunatsite to be defined as a guide site for periodization and chronology of the Chalkolithic and Bronze Age cultures documented there.

Annex 1

Technology

Contents of the paste

Fine paste

The mixtures used consist of fine organic substances and of quartz of low or very low concentration. Small quartz rock of very low concentration was sometimes used, as well as rarely mica of very low or low concentration. The clay was mostly used for shallow bowls of type II.2, two handled cups of type V.1.2 and very rarely that fine clay was used for production of bowls and some of the miniature vessels.

Semi fine paste

The main mixtures used were fine quartz sand of small and high concentration, small quartz rock with different concentration for the different vessels (from very low to high) and semi-big quartz rock of very low concentration. Fine organic mixtures were often used of low to high concentration. A small quantity of semi-fine organic matters of very low concentration was rarely added. That clay characterizes mainly the vessels of category of shallow bowls and deep bows (different types), two handled cups of type V.2-3, considerably rarely the pots and other categories (pot-stands, miniature vessels).

Semi-coarse paste

The mixtures used were small quartz rock of high and low concentration, semi-big rock of low, medium or high concentration. There is often organic mixture of fine and not very fine inclusions. Mica of low concentration was rarely added. The main mixture used is quartz stone – small and semi-big of high and very high concentration for the distinct vessels. Very fine and fine organic matters were very often added. Mica of very low concentration was very rarely distributed.

Of all categories vessels most distributed are those with semi-coarse clay (c. 50%). The clay is used for producing of vessels mainly of the categories of bowls, storage vessels, amphorae-like vessels and rarer shallow bowls.

Coarse paste

The main admixtures are quartz rock in medium size and rarely small of high or very high concentration, big rock of high and medium concentration, medium fine sand of high concentration. There were semi-fine and coarse organic matters of medium and high concentra-

[1]We are grateful to Dr. K. Leshtakov, V. Gergov, T. Kuncheva-Rousseva and G. Shejleva who allowed us to work with unpublished ceramic materials from Mudrets-Goudzhova Mogila, Telish-Redutite, Sudievo and Iskrista, which gave a possibility for more detailed spatial determination of the newly excavated Late Chalkolithic pottery from Yunatsite.

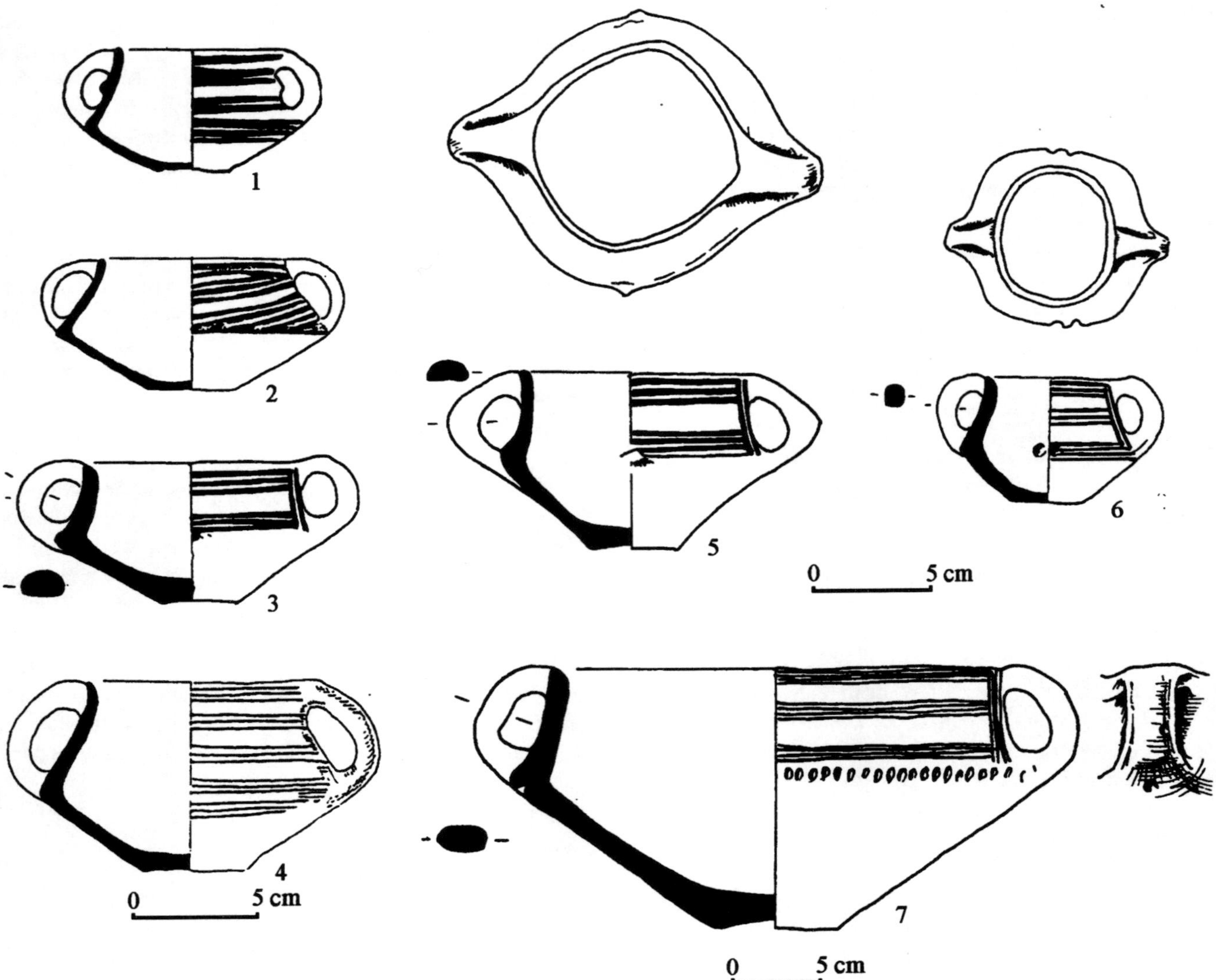

Figure 26.5. Yunatsite tell. Late Chalkolithic pottery.

tion as well. In some cases 'shamot' was added. The coarse clay was used mainly for producing of depositories, some types of bowls, lids and single miniature vessels.

Slipped and wash surface

Slipped surface

The analysis of the ceramic material from the first and second Chalkolithic building levels at Yunatsite reveals the following types of slip: fine slip; semi-fine slip, compact slip, colored slip, self-slip and thin slip – wash.

Fine slip. Thin layer of fine clay covers the surface of the vessels. The slip is approximately 0.2-0.3 mm thick and it sometimes evenly differs at the biscuit. As a rule, the slip is burnished homogeneously or in strips, rarer there is evenly smoothing of the surface. The fine slip is documented almost by all categories of vessels.

Semi-fine slip. It is mostly sitributed. The thinkness of the slip is 0.5 mm. Most of the semi-fine slip is evenly burnished, rarely there is burnishing in strips and evenly burnished surface.

Compact slip. It is thicker, up to 1 mm. It is clearly visible in the biscuit of the vessels. The slip is almost always evenly burnished, in some cases there it is burnished to shine. Smoothing and burnishing in strips is very rarely distributed, mainly over the inner walls of the vessels.

Colored slip. There is only one instance of that manner, a cup of type D.VI. The outer surface is covered with thin bright red slip, burnished up to shine.

Self slip

There is a thin layer, which occurred by burnishing of unslipped vessels. It is sometimes very difficult to distinguish fine slip from self-slip. The vessels with self-slip usually have burnished surface, often in strips. Some types of bowls and distinct closed shapes have evenly smoothed inner surface. There are no instances of burnishing to shine.

Wash

The wash is very thin layer of fine clay. It occurred by the burnishing or smoothing of the vessels, which were only 'dipped' into or 'wash' with very diluted fine clay. It is rarely distributed and it is difficult to distinguish the wash from self-slip. In most cases the wash covers only some parts of the vessels – necks of the pots, upper parts of some types of pithoi (E.II.2.2, E.IV.1.2, J.III.1.1). The wash can be evenly burnished or in strips, as well as homogeneously smoothed.

Unslipped surface

It is not small the amount of the pottery without additional slip, self-slip or wash. Those are the vessels of different categories of semi-coarse, and coarse clay. The outer surface of those vessels is often consciously made roughed through dragging or barbotine, unevenly smooth or through smoothing. As a rule, the inner walls have traces of evenly smoothing. It should be stressed the absence of unslipped vessels from the category of the cups.

The following definitions have been used for descriptions of the manners of treatment of the surface: Unsmoothed surface; irregularly smoothed surface; evenly smoothed surface; burnished in strips surface; evenly burnished (without shine) surface; shine burnished surface; roughed surface.

Unsmoothed surface

The walls have no traces of additional treatment. In most cases the outer surface of the vessel is smoothed only. As the irregular smoothing, unsmoothed surface can be related to the unslipped pottery made of semi-coarse and coarse clay. The vessels with unsmoothed surface are not distributed in most categories of vessels. Several manners were distinguished.

Irregularly smoothed surface

That manner is relatively rarely distributed. There are clear traces of careless smoothing over the outer and inner surface of the vessels. The suing of instrument is often documented, most probably putty knife of wood or bone, hard 'broom' etc. The irregular smoothing is always connected with unslipped surface and is not documented by the categories of bowls and amphorae-like vessels.

Evenly smoothed surface

Along with the evenly burnishing it is the widest manner of surface treatment. It is registered by all categories of vessels, but weakest by the bowls where the burnished surface dominates. There is no clear connection with the over-layering of the vessel walls. The evenly smoothed surface characterizes the slip, as well as unslipped pottery. There are very often vessels and fragments with evenly smoothed inner surface in combination with burnished outer surface. The last is especially characteristic of cups, pots, amphorae-like vessels and some types of bowls.

Burnished in strips surface

The burnishing is irregular. There are distinct groups of strips, which mark the direction and manner of burnishing. In almost all cases registered the strips are horizontal. Most probably a bone or stone smoother was used for the surface treatment. A considerable part of the vessels before the burnishing were slipped, as fine slip and self-slip was most distributed, rarer semi-fine slip. There are no instances registered of combination of compact slip with strips burnished.

Evenly burnished surface

The surface is evenly burnished, lusterless. In most cases the burnished in this manner vessels have fine, semi-fine or compact slip, rarely there is self-slip or wash. There is no interrelation between the color of the vessel and the existence of burnishing, as by the rest ways of surface treatment. The evenly burnishing is especially characteristic of the vessels of the categories of shallow bowls where it is distributed among all basic forms. It should be stressed in the categories of pots, amphorae-like vessels, pithoi and some types of bowls the burnishing is documented over the outer surface only.

Shine burnished surface

The surface is evenly burnished, polished and has strong shine. The vessel surfaces with shine burnished are covered with compact slip. In on of the cases there is bright red coloring (D.VI). The other vessels have evenly compact black color, as the burnishing is registered over the outer surface only.

Roughed surface

The vessels and the fragments with consciously roughed surface are numerous. The widest technique distributed is barbotine, in rare cases there are chaotic dragging. That manner, along with the last two

Annex 2

Classification of the pottery from Yunatsite

A. A CATEGORY OF SHALLOW BOWLS

A.I. Deep shallow bowls with conical shape.
A.I.1. Straight-sided.
A.I.1.1. Straight-sided, straight orifice.
A.I.1.1.a. The rim is rounded.
A.I.1.1.b. The rim is evenly cut.
A.I.1.2. Straight walls, the orifice is slightly out curved.
A.I.1.3. Straight walls, the orifice is with slightly thickened inner wall and rounded rim.
A.I.2. Walls in lowly constructed at the orifice.

A.II. Flaring conical shallow bowls with straight-sided.
A.II.1. Unthickened orifice.
A.II.1.1. Round rim.
A.II.1.2. Straight rim.
A.II.2. Thickened inturned rim.
A.II.2.1. Rounded rim.
A.II.2.1.a. A ledge attached at the rounded part of the rim.
A.II.2.1.b. Rounded rim with a groove at the middle point.
A.II.2.2. Evenly cut on the top of the orifice rim.
A.II.2.3. The rim is obliquely cut on the inner side of the orifice.
A.II.3. The inturned rim, highly thickened and twisted on the inner side of the orifice.
A.II.4. The rim is slightly thickened on the inner side.
A.II.4.1. The rim is rounded.
A.II.4.2. The rim is worn thin.
A.III. Open shallow bowls with semi-globular or conical shape and lowly squat walls.
A.III.1. Unthickened orifice.
A.III.1.1. The rim is worn thin and rounded.
A.III.1.2. Equally cut rim.
A.III.2. The rim is thickened and rounded.
A.III.3. Lowly inclined orifice with lowly incurved thickened rim.
A.III.4. The orifice is lowly outcurved, rim is slightly thickened and rounded.
A.IV. Shallow bowls with conical body and short and straight upper part
A.IV.1. The orifice is straightly cut on the inner side and squat on the outer side. – without "of the walls"
A.IV.1.1. The carination is gently shaped on the inner side.

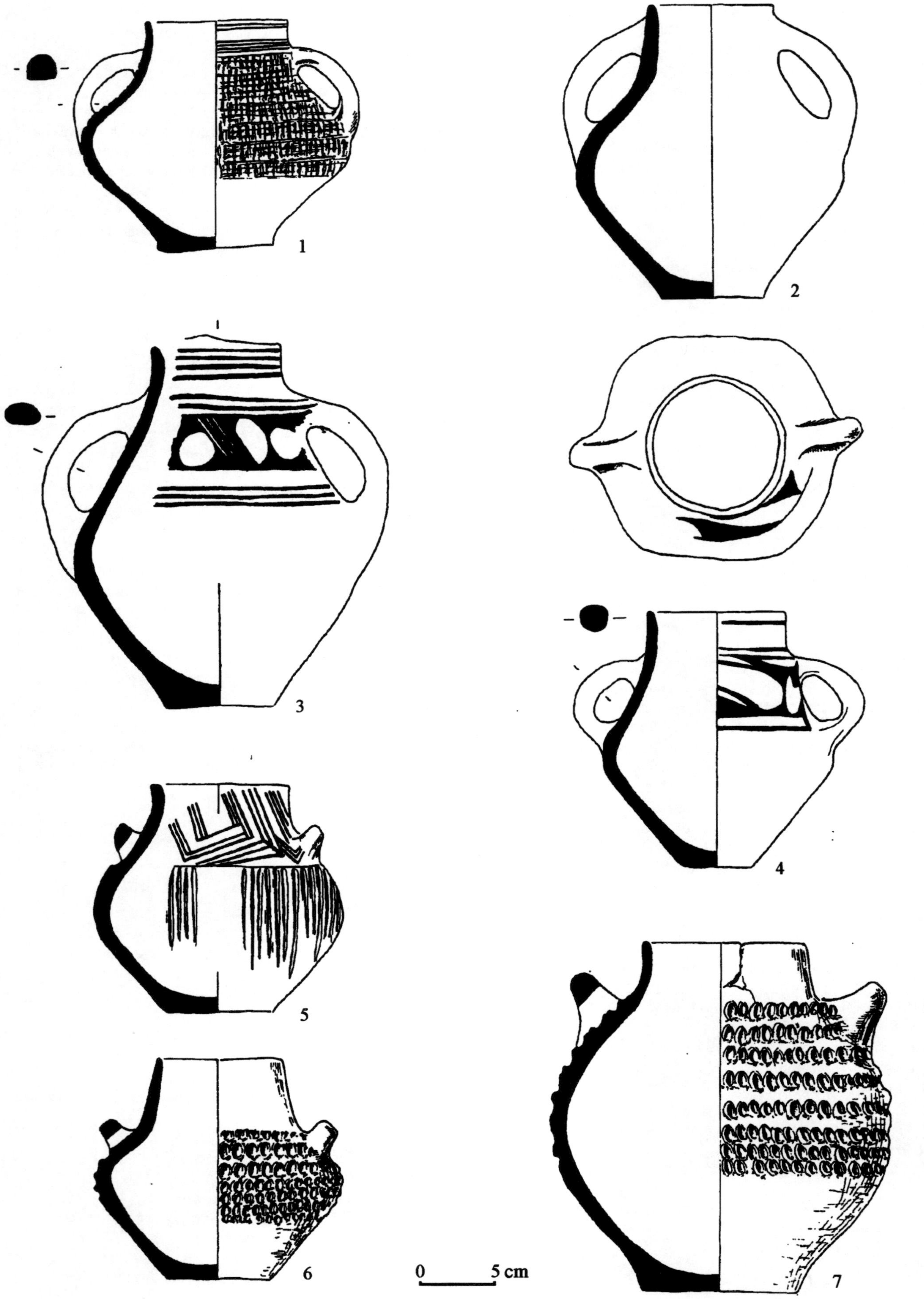

Figure 26.6. Yunatsite tell. Late Chalkolithic pottery.

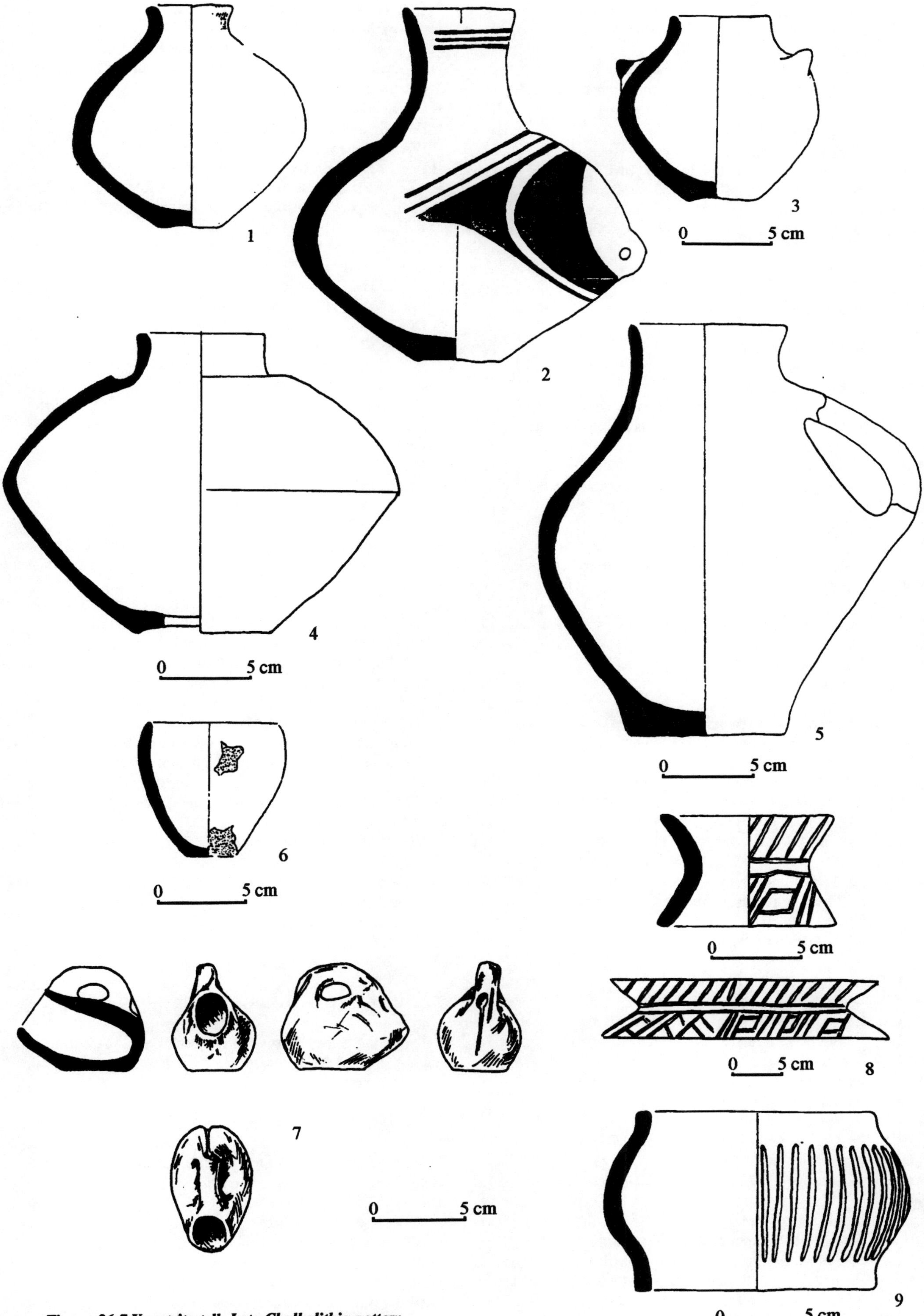

Figure 26.7. Yunatsite tell. Late Chalkolithic pottery.

A.IV.1.2. Inclined orifice with a sharp carination.
A.IV.2. Straight orifice the rim worn thin on the inner side.
A.IV.3. Straight orifice with sharp carination at the both sides
A.IV.3.1. Rounded rim.
A.IV.3.2. The rim is worn thin and rounded.
A.V. Shallow bowls with conical body, straight walls and short inclined upper part.
A.V.1. The upper part is very short.
A.V.1.1.Thickened and rounded rim.
A.V.1.2. Thickened rim with straight inner wall.
A.V.2. Short upper part. Highly incurved and thickened rim.
A.V.3. Short upper part. Highly incurved rim, worn thin.
A.V.4. Short upper part. Highly incurved rim, without thickening.
A.V.4.1. Rounded rim.
A.V.4.2. The rim is evenly cut.
A.V.4.3. The rim is worn thin.
A.VI. Conical shallow bowls with straight walls and incurved orifice.
A.VI.1. Lowly incurved rim.
A.VI.1.2. Evenly cut rim.
A.VI.2. The rim is highly incurved.
A.VII. Shallow bowls with sharp carination and out curved rim.
A.VII.1 The upper part is straight the rim is slightly curved.
A.VII.1.1. The rim is rounded.
A.VII.1.2. The rim is worn thin.
A.VII.1.3. The rim is highly thickened on the top and rounded.
A.VII.2. Carinated bowls with outcurved orifice.
A.VII.2.1. The carination is gently shaped inside.
A.VII.2.2. The carination is sharply shaped inside
A.VII.2.3. Highly everted prolong rim, gently shaped carination.
A.VII.2.4. The rim is thickened and obliquely cut on the inner side..
A.VIII. Biconical shallow bowls with thickened and differently shaped carination.
A.VIII.1. The carination is sharply shaped inside.
A.VIII.1.1. Round rim.
A.VIII.1.2. The rim is worn thin.
A.VIII.2. The carination is rounded outside the wall.
A.VIII.2.1. The rim is rounded.
A.VIII.2.2. The rim is equally cut.
A.VIII.2.3. The rim is slightly profiled outside and rounded.
A.VIII.3. The carination is thickened.
A.VIII.3.1. The upper part is very short.
A.VIII.3.1.a. There is small horizontal handle attached at the carination.
A.VIII.3.1.b. Without handle.
A.VIII.3.2. The upper part is higher, slightly inclined in.

B. A CATEGORY OF DEEP BOWLS.

B.I.1. Open semi-spheroid deep bowls.
B.I.1.1. The rim is rounded.
B.I.1.2. The rim is equally cut.
B.I.1.3. The rim is worn thin.
B.I.2. Semi-spheroid bowls with incurved upper part
B.I.2.1. The rim is rounded.
B.I.2.2. The rim is obliquely cut.
B.I.2.3. The rim is worn thin.
B.I.2.4. The rim is slightly thickened and equally cut.
B.II. Necked semi-spheroid bowls.
B.II.1. The carination is gently shaped. The orifice is slightly out curved.
B.II.1.1. The rim is rounded.
B.II.1.2. The rim is worn thin inside and rounded.
B.II.2. Well-shaped short neck and inverted orifice.
B.II.3. Well-shaped neck. The orifice is straight and thickened inside.
B.III. Bowls with flatted spheroid shape and short neck.
B.III.1. The neck is very short, gently shaped.
B.III.2. Short neck, divided from the body by a groove. Thickened inside rim and rounded.
B.III.2.1. Straight walls, without thickening.
B.III.3. Moderately high neck, divided from the body by a groove. Thickened rim, rounded.
B.IV. Biconical deep bowls.
B.IV.1. Biconical deep bowls with straight walls.
B.IV.1.1. The carination is gently shaped.
B.IV.1.2. The carination is sharply shaped.
B.IV.2. Biconical deep bowls with straight lower part and slightly quad upper part. The orifice is weakly everted, without thickening, rounded.
B.IV.3. Biconical bowls, with short upper part and thickened outside carination.
B.IV.3.1. The rim is rounded.
B.IV.3.2. The rim is thickened outside and inside.
B.IV.3.3. The rim is worn thin and rounded.
B.IV.3.4. The rim is flat cut.
B.V. Deep bowls with spheroid and conical lower part and upper part shaped as truncated cone.
B.V.1. Lower straight walls. The carination is gently shaped.
B.V.2. The walls in the lower part are slightly inverted. The carination is stressed.
B.VI. Deep spouted bowls.
B.VI.1. Semi-spheroid body.
B.VI.1.1. The lower parts of the body walls are squat.
B.VI.1.2. The lower parts of the body walls are slightly inclined
B.VI.2. Flatted spheroid body with differently shaped upper part.
B.VI.2.1. Very short neck. The rim is slightly everted and thickened on the inner side.
B.VI.2.2. Medium high, clearly separated from the body neck. Rounded rim.
B.VI.3. Biconical body with thickened carination.
B.VII. Biconical deep bowls with two vertical handles attached at the upper part of the body and thickened carination.
B.VII.1. Biconical bowls. The walls in the lower body part are slightly squat. Weakly everted rim.
B.VII.2. Biconical bowls. The walls are squat in their lower part. The handles are attached at the carination. Small conical pedestal base.
B.VII.3. Biconical bowl. Straight walls. The carination is stressed, thickened and rounded.
B.VII.3.1. Vertical strap-handles.
B.VII.3.2. Vertical handles with horn-like projection.
B.VII.4. Biconical, shallow bowls. Straight walls. The carination is thickened and rounded.
B.VIII. Biconical bowls with very wide flat bottom – “pixis” .
B.VIII.1. Spheroid “pixis”
B.VIII.2. The “pixis” shaped as truncated cone.

C. CATEGORY OF JUGS

C.1. Prolong body. High cylindrical neck gently shaped. Rounded, slightly profiled out-curved orifice. There is a handle under the rim of the vessel.

D. CATEGORY OF CUPS

D.I. Handless cup. Prolong S-profile. The widest part is thickened. Thinned rim inside and outside.
D.II. Handless cup. Almost biconical body with short cylindrical neck. Pedestal conical base.
D.III. One handled cups. Biconical shape and short wide neck.
D.III.1. Straight orifice. The rim is rounded.
D.III.2. Slightly everted and thinned rim.
D.IV. Two-handled, biconical high cups.
D.IV.1. Without thickening at the carination point. The handles

raise slightly higher above the orifice.
D.IV.1.1. The rim is straight and rounded.
D.IV.1.2. The rim is slightly out curved, rounded.
D.IV.2. Thickened carination.
D.V. Two-handled, biconical cups.
D.V.1. Sharply shaped carination without thickening and slightly out curved rim.
D.V.1.1. The rim of the orifice is equally cut.
D.V.1.2. The rim is rounded.
D.V.1.3. The rim is worn thin and rounded.
D.V.2. Sharply shaped carination with thickened, and slightly profiled out curved rim.
D.V.2.1. The rim is equally cut (with rhomb cross section).
D.V.2.2. The rim is worn thin and rounded.
D.V.3. Sharply shaped carination and straight rim.
D.V.3.1. The rim is rounded.
D.V.3.2. The rim is worn thin and rounded.
D.V.4. Gently shaped carination. The rim is rounded.
D.V.4.1. Round cross-section.
D.V.4.2. Oval cross-section.
D.VI. Handless biconical small cups with sharp shaped carination
D.VI.1. High.
D.VI.2. Short
D.VII. Cups with spheroid and conical shape. The rim is worn thin outside, and equally cut
Import (?).

E. CATEGORY OF THE POTS

E.I. Pots with prolong spheroid body, without neck.
E.I.1. The walls are squat.
E.I.2. The walls are slightly unflatted.
E.II. Necked with prolonged spheroid body.
E.II.1. Medium high conical neck.
E.II.2. Medium high cylindrical neck.
E.II.2.1. The upper part of the body is unflatted.
E.II.2.2. The lower part of the body is slightly inclined toward the flat base.
E.III. Pots with flatted spheroid body.
E.III.1. Pots with flatted spheroid body and slightly out curved rim.
E.III.2. Pots with flatted spheroid body and short neck.
E.IV. Pots with flatted spheroid body and high conical neck.
E.IV.1. Pots with flatted spheroid body. Two horizontal handles attached under the neck.
E.IV.1.1. The rim is rounded.
E.IV.1.2. The rim is worn thin and rounded.
E.IV.2. Pots with flatted spheroid body. The transition towards the neck is shaped with a deviation.
Slightly out curved rim.
E.V. Pots with S-profile.
E.VI. Biconical pots with slightly shaped carination without thickening.
E.VI.1. The walls are slightly unflatted. No neck. Small horizontal handles under the orifice.
E.VI.2. Straight walls. Short, gently shaped neck.
E.VI.3. The walls are slightly unflatted. Short cylindrical neck shaped with a deviation.
E.VII. Biconical pots with thickened and gently shaped carination.
E.VII.1. The thickened carination is gently shaped.
E.VII.2. The thickened carination is shaped with deviation.
E.VII.2.1. The rim is straight and equally cut.
E.VII.2.2. The rim is slightly out-curved, worn thin and rounded.
E.VIII. Biconical pots with sharply shaped carination.
E.VIII.1. The rim is straight and rounded.
E.VIII.2. The rim is slightly out-curved and worn thin.
F. Category of amphorae-like vessels.
F.I. Prolong spheroid body with gently shaped cylindrical neck.
F.I.1. Cross-section close to square. Wide conical neck. The vertical handles started from the middle point of the neck and are highly widened aside.
F.I.2. Narrow conical neck. The handles are attached low under the orifice.
F.I.2.1. Oval cross-section of the neck.
F.II. Flatted spheroid body with slightly shaped cylindrical neck.
F.II.1. The handles start just under the orifice.
F.II.2. The handles starts from the middle point of the neck.
F.III. Biconical amphorae-like vessels with gently shaped carination.
F.IV. Biconical amphorae-like with sharply shaped carination.

G. 'VASE'

G.I. Flatted spheroid body with high cylindrical neck. The middle body part is thickened. Two vertical handles started from the medium point of the neck. Their lower ends are attached at the widest part of the body.
G.II. Flatted spheroid body and high narrow cylindrical neck. Small vertical handle attached at the widest body part.

H. CATEGORY OF ASKOI

H.I. Oval body with irregularly shaped slightly flatted body. The orifice is obliquely cut raising up. The rim is worn thin and rounded. Oval cross-section of the orifice. Handle attached to the wall of the upper body part.
H.II. Unique shape.

J. CATEGORY OF DEPOSITORIES (PITHOI)

J.I. Prolonged rounded body, short vertical neck. No handles.
J.I.1. The rim is equally cut.
J.I.2. The rim is rounded.
J.II. Prolonged , rounded body and high gently shaped neck.
J.III. Prolonged, spheroid body and short narrow neck.
J.III.1. Straight walls at the lower part. Clearly distinguished cylindrical neck.
J.III.1.1. The rim is obliquely cut. Two (or 1) big handles are attached to the upper body part.
J.III.1.2. The rim is rounded. Two handles at the medium part of the body.
J.III.2. The wall of the lower body part are slightly unflatted. The turning point towards the neck is gently shaped. Two vertical handles attached to the widest part of the body.
J.IV. Depositories with flatted spheroid body.
J.IV.1. Without neck.
J.IV.2. Gently shaped, short and wide neck.
J.V. Biconical depositories with thickened carination and two vertical handles.
J.V.1. Biconical depositories with sharply shaped carination.
J.V.1.1. The rim is rounded.
J.V.1.2. The rim is worn thin on the inner side of the wall.
J.V.2. Biconical depositories with rounded thickened carination.
J.V.2.1. The rim is obliquely cut and rounded.
J.V.2.2. Th rim is rounded.
J.VI. Open, deep depositories with conical shape and slightly inverted upper part.

K. CATEGORY OF THE LIDS

K.I. Oven lid. Conical shape. High handle.
K.II. Flat lids with weakly inverted profile. A big handle attached to the upper wall.
K.III. High lid with semi-spheroid shape.
K.III.1. Equally cut base and opening at the middle part.
K.III.2. Rounded base, everted and pouring rim.
K.IV. Lids with conical shape.

K.IV.1. High conical lid. Horizontal opening at the upper part.
K.IV.2. Low conical lids.
K.IV.2.1. The base is slightly inverted, gently shaped.
K.IV.2.2. The base is refracted.
K.IV.2.3. The lower part is carinated, so that the rim is slightly everted.

L. CATEGORY OF THE POT -STANDS

L.I. Pot-stands with unflatted middle body part and equally cut cylindrical bases.
L.II Pot-Stands with inclined at the middle part walls.
L.II.1. Short stands. The walls are slightly inclined. The bases are rounded.
L.II.2. High stands. The walls are slightly inclined. The bases are rounded.
L.II.3. High stands. The walls are highly inclined. The bases are rounded.

M. MINIATURE VESSELS

M.I. Shallow bowls.
M.I.1. Conical shape with short vertical upper part. The rim is worn thin.
M.I.2. Biconical shallow bowls.
M.I.2.1. The rim is straight and rounded. Four horizontal handles.
M.II. Deep bowls.
M.II.1. Bowls with semi-spheroid shape. The rim is rounded.
M.II.1.1. Flat base.
M.II.1.2. Round bottom and four vertical handles.
M.III. Pots.
M.III.1. Pots with prolonged, rounded body. No neck.
M.III.2. Pots with prolonged rounded body, short conical neck and everted rim.
M.III.3. Pots with spheroid shape.
M.III.3.1. Very short, gently shaped neck. Two horizontal handled under the orifice.
M.III.3.2. Short cylindrical neck clearly shaped. Two horizontal handled under the orifice.
M.III.3.3. Short, very narrow neck. The rim of the orifice is highly everted. Unique shape.
M.IV. Model of a cup with two handles.

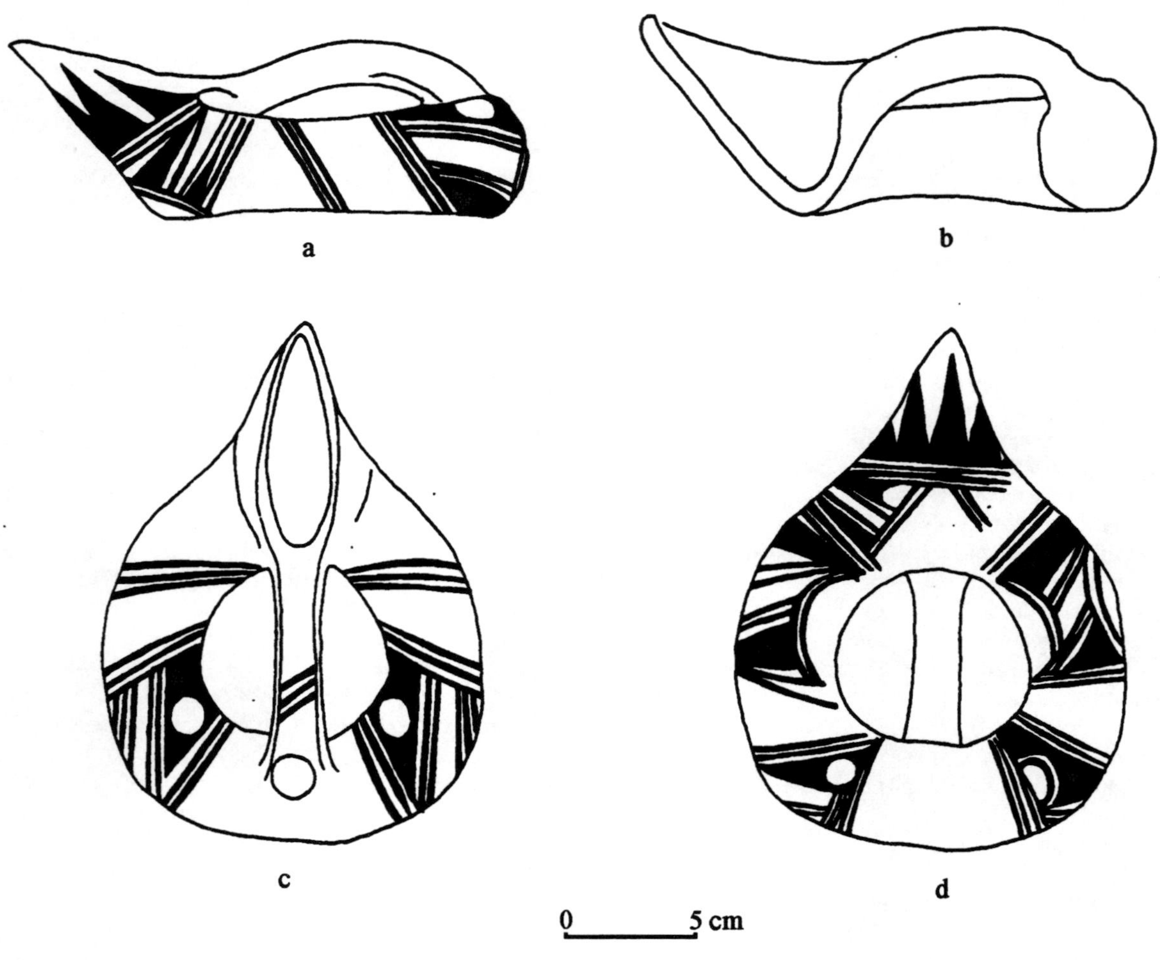

Figure 26.8. Yunatsite tell. Late Chalkolithic pottery.

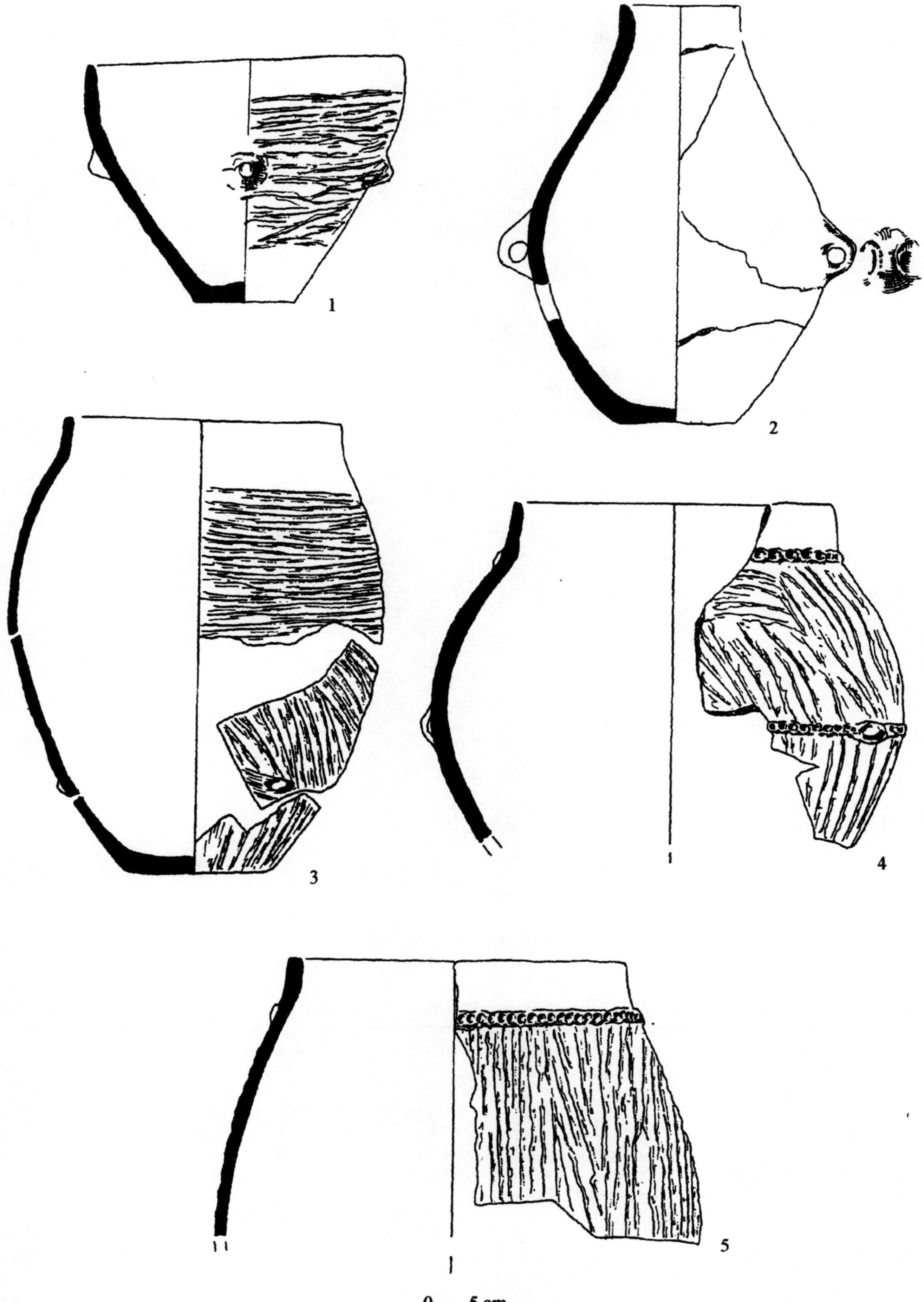

Figure 26.9. Yunatsite tell. Late Chalkolithic pottery.

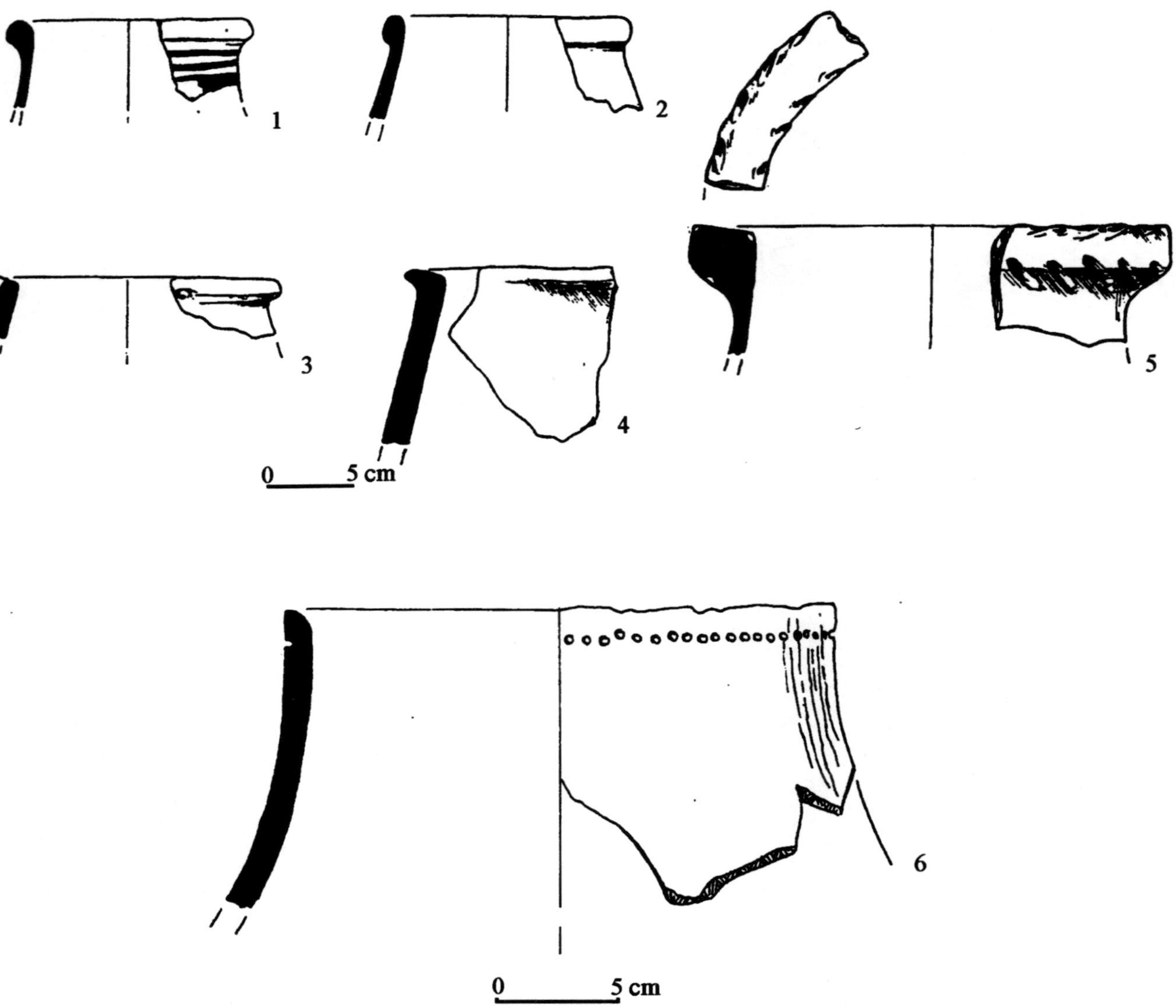

Figure 26.10. ***Yunatsite tell. Late Chalkolithic pottery.***

Annex No. 3

Catalogue of the intact vessels

1. ***A shallow Bowl.*** Light brown, uneven color inside and outside. The surface is slipped and burnished in strips outside and inside. Semi coarse paste. Diameter orifice – 39 cm; diameter bottom – 8 cm; height – 12 cm. House 12. No. 2047. Historical Museum Pazardzhik No. 4849. ***Figure 26.1: 1.***

2. ***A shallow Bowl.*** Dark brown color outside and inside. Slipped surface, burnished inside, self-slip outside, burnished in strips. The paste is semi-fine. Brown biscuit. Diameter orifice – 44 cm; diameter bottom – 9.8 cm; height – 14 cm. House 4. No. 2046. ***Figure 26.1: 2.***

3. ***A shallow Bowl.*** Black color outside and inside, single light brown spots. Slipped and burnished surface. The paste is semi-fine. Diameter orifice – 42 cm; height preserved – 12 cm. House 3. ***Figure 26.1: 3.***

4. ***A shallow Bowl.*** Light brown color with a black spot outside and homogenous inside. Slipped surface and burnished in strips inside and outside. Semi-fine to fine paste. Diameter orifice – 31 cm; diameter bottom – 8 cm; height – 13 cm. House 5. No. 1923. Historical Museum Pazardzhik No. 4750. ***Figure 26.1: 4.***

5. ***A shallow Bowl.***. Dark brown to black color outside and inside. Self slip surface burnished in strips. The paste is semi-coarse. Diameter orifice – 16 cm; diameter bottom – 5.2 cm; height – 7.1 cm. House 4. No. 1753. Historical Museum Pazardzhik No. 4296. ***Figure 26.2: 1.***

6. ***A shallow Bowl.***. Brown, irregular outside, with black and light brown spots; brown homogenous inside. Very fine slip (wash) surface, burnished in strips outside, unslipped smoothed inside. Coarse paste. Bottom – medium sized and big stone, big organic matters. Diameter orifice – 33 cm; diameter bottom – 8.5 cm; maximum diameter – 35; height – 11 cm. House 5. No. 1921. Historical Museum Pazardzhik No. 4762. ***Figure 26.2: 2.***

7. ***A shallow Bowl.*** Gray and black color with single brown spots

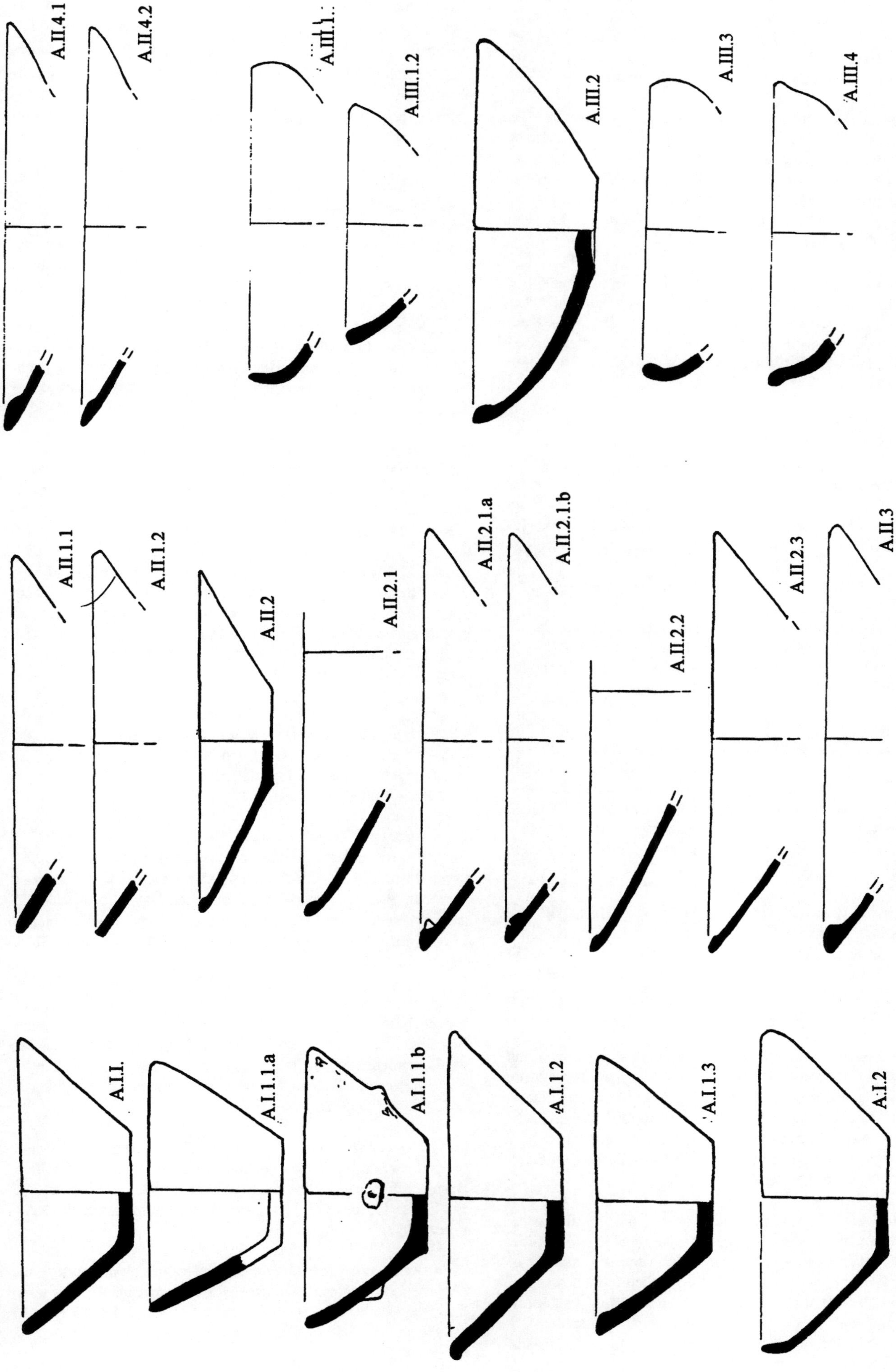

Table 26.1. Yunatsite tell. Classification of pottery. Late Chalkolithic.

outside, black inside. Slipped surface, burnished in strips outside, evenly burnished inside. The paste is semi coarse. Bottom – sand, organic matter. Diameter orifice – 15 cm; diameter bottom – 4.5 cm; maximum diameter – 15.7; height – 5.3 cm. House 1. No. 1816. Historical Museum Pazardzhik No. 4787. ***Figure 26.2: 3.***

8. ***A shallow Bowl.*** Light brown, irregular inside and outside. Slip surface, burnished inside and outside. The paste is semi-fine. Black biscuit with two dark brown strips. Bottom – at the lower part of the vessel, by the bottom, there is additionally attached band of paste. Diameter orifice – 30 cm; diameter bottom – 10 cm; maximum diameter – 32; height – 11 cm. House 5. No. 2106. Historical Museum Pazardzhik (1996). ***Figure 26.2: 4.***

9. ***A deep bowl.*** Dark brown to black color, black inside. Smoothed surface in the upper part inside and outside, without slip; the lower part is consciously roughed. Paste is semi-coarse. Bottom – sand. Diameter orifice – 16.4 cm; diameter bottom – 9 cm; height – 10.4 cm. House 4. No. 1760. Historical Museum Pazardzhik No. 4430. ***Figure 26.3: 1.***

10. ***A deep bowl.*** Brown, irregular color, with darker and lighter spots outside and dark brown to black outside. Smoothed surface outside and inside, without slip. The paste is semi-coarse. Bottom – sand and fine organic matter. Diameter orifice – 14 cm; diameter bottom – 8.4 cm; maximum diameter – 16.5 cm; height – 9.7 cm. House 12. No. 1965. Historical Museum Pazardzhik No. 4793. ***Figure 26.3: 2.***

11. ***A deep bowl.*** Beige and brown irregular color, with black spots. The surface is without slip, in the upper part burnished outside and inside. Paste is coarse - semi-coarse. Black homogenous biscuit. Bottom –sand and fine organic matter. Diameter orifice – 20 cm; diameter bottom – 9.5 cm; maximum diameter – 23 cm; height – 11.2 cm. House 12. No. 1963. Historical Museum Pazardzhik No. 4792. ***Figure 26.3: 3.***

12. ***A deep bowl.*** Gray and brown, irregular color, dark brown with black spots outside and inside. The surface is without slip, roughly smoothed (perhaps with a hard broom). Paste is coarse. Bottom – with sand and clear traces of fine cloths. Diameter orifice – 16 cm; diameter bottom – 11 cm; maximum diameter – 20.5 cm; height – 14 cm. House 1. No. 1394. Historical Museum Pazardzhik No. 3890. ***Figure 26.3: 4.***

13. ***A deep bowl.*** Gray and black irregular color, reddish spots outside and inside. The surface is without slip, burnished outside under the rim, the lower part of the body is with barbotine organized in oblique vertical bands. Paste is medium coarse. Diameter orifice – 18 cm; diameter bottom – 13 cm; height – 16 cm. House 1. No. 1390. Historical Museum Pazardzhik No. 3887. ***Figure 26.3: 5.***

14. ***A deep spouted bowl.*** Brown irregular color, single darker and lighter spots outside and inside. Slipped surface, smoothed outside and inside in the upper part, but the lower part towards the bottom is coarse, unsmoothed. Paste is semi-coarse. Bottom – sand. Diameter orifice – 14 cm; diameter bottom – 11 cm; maximum diameter – 18.5 cm; height – 13.3 cm. House 3. No. 1912. Historical Museum Pazardzhik No. 4733. ***Figure 26.3: 6.***

15. ***A deep spouted bowl.*** Light brown, reddish outside and inside. The surface is without slip, roughly smoothed inside and outside. Paste is coarse. Bottom – very coarse, a considerable amount of stone up to 6-7 mm big. A addition band attached to the lower part of the vessel, perhaps for stabling of the bottom (?). Diameter orifice – 21 cm; diameter bottom – 14.5 cm; maximum diameter – 32 cm; height – 19 cm. House 3. No. 2048. Historical Museum Pazardzhik No. 4851. ***Figure 26.3: 7.***

16. ***A pot.*** Black color outside and inside. Fine slipped surface, smoothed in the upper part outside and inside, the lower part outside is without slip, smoothed and sandy. Fine paste. Diameter orifice – 9.5 cm; diameter bottom – 4.8 cm; maximum diameter – 16.5 cm; height – 12 cm. House 12. Depth – 5.20 m, No. 2030. Historical Museum Pazardzhik No. 4833. ***Figure 26.4: 1.***

17. ***A deep bowl.*** Gray and brown, irregular color, reddish burnt spots. Slipped surface, burnished outside, smoothed inside. Fine paste. Evenly brown biscuit. Diameter orifice – 14.5-15 (?) cm; diameter bottom – 7 cm; maximum diameter – 20.6 cm; height preserved – 15 cm. House 12. No. 2048. Historical Museum Pazardzhik No. 4851. ***Figure 26.4: 2.***

18. ***A deep bowl.*** Brown color, irregular outside, gray and brown inside. Slipped surface outside, in strips inside. Paste is semi-fine. Black biscuit with brown strips. Diameter orifice – 21.4 cm; diameter bottom – 10 cm; maximum diameter – 33 cm; height reconstructed – 25 cm. House 1. ***Figure 26.4: 3.***

19. ***A deep bowl.*** Brown color outside and inside. Fine slip surface, burnished in its upper part, the lower part is smoothed, sandy. Paste is semi-fine. Dark brown biscuit with two brown strips. House 1. Diameter orifice – 28 cm; maximum diameter – 48 cm; height preserved – 23 cm. House 1. ***Figure 26.4: 4.***

20. ***A deep bowl.*** Gray and black to black outside and inside. Slipped surface in the upper part and burnished. The lower part is without slip, smoothed, sandy. Paste is semi coarse. Diameter orifice – 28 cm; diameter bottom – 10 cm; maximum diameter – 31 cm; height – 21 cm. House 1. No. 1497. Historical Museum Pazardzhik No. 4023. ***Figure 26.4: 5.***

21. ***A deep bowl.*** Brick red outside, secondary burnt, brown inside. The surface is with slip, evenly burnished outside, in strips inside. Paste is semi-fine, sandy. Brick red biscuit. Diameter orifice – 20 cm; diameter bottom – 7.4 cm; maximum diameter – 24 cm; height – 17.3 cm. House 2. ***Figure 26.4: 6.***

22. ***A deep bowl.*** Very dark irregular color, spots outside and inside. Slipped surface, burnished, in fields with red ochre – unsmoothed, covered with fine slip or self-slip, smoothed inside. Semi-fine paste. Restored. Bottom – coarse organic matter. Diameter orifice – 25 cm; diameter bottom – 9 cm; maximum diameter – 30 cm; height – 18 cm. House 1. No. 1393. Historical Museum Pazardzhik No. 3888. ***Figure 26.4: 7.***

23. ***A two handled cup.*** Black color outside and inside. Slipped surface, burnished to shining outside, smoothed inside. Paste is fine. Black biscuit. Diameter orifice – 6 cm; diameter bottom – 2.8 cm; maximum diameter – 8.6 cm; height – 5 cm. House 3. Sq. K8, depth – 5.63 m; vessel 10a. No. 2042. Historical Museum Pazardzhik No. 4843. ***Figure 26.5: 1.***

24. ***A two handled cup.*** Gray and black color outside and inside. Slipped surface, burnished outside and inside. Paste is semi-fine. Diameter orifice – 8.5 cm; diameter bottom – 3.5 cm; height – 5.5 cm. House 1. No. 1373. Historical Museum Pazardzhik No. 3898. ***Figure 26.5: 3.***

25. ***A two handled cup.*** Gray and black to black color outside and inside. Slipped surface, burnished outside and inside. Paste is semi-fine. Diameter orifice – 9 cm; diameter bottom – 3.5 cm; height – 7 cm. House 1. No. 1374. Historical Museum Pazardzhik No. 3897. ***Figure 26.5: 4.***

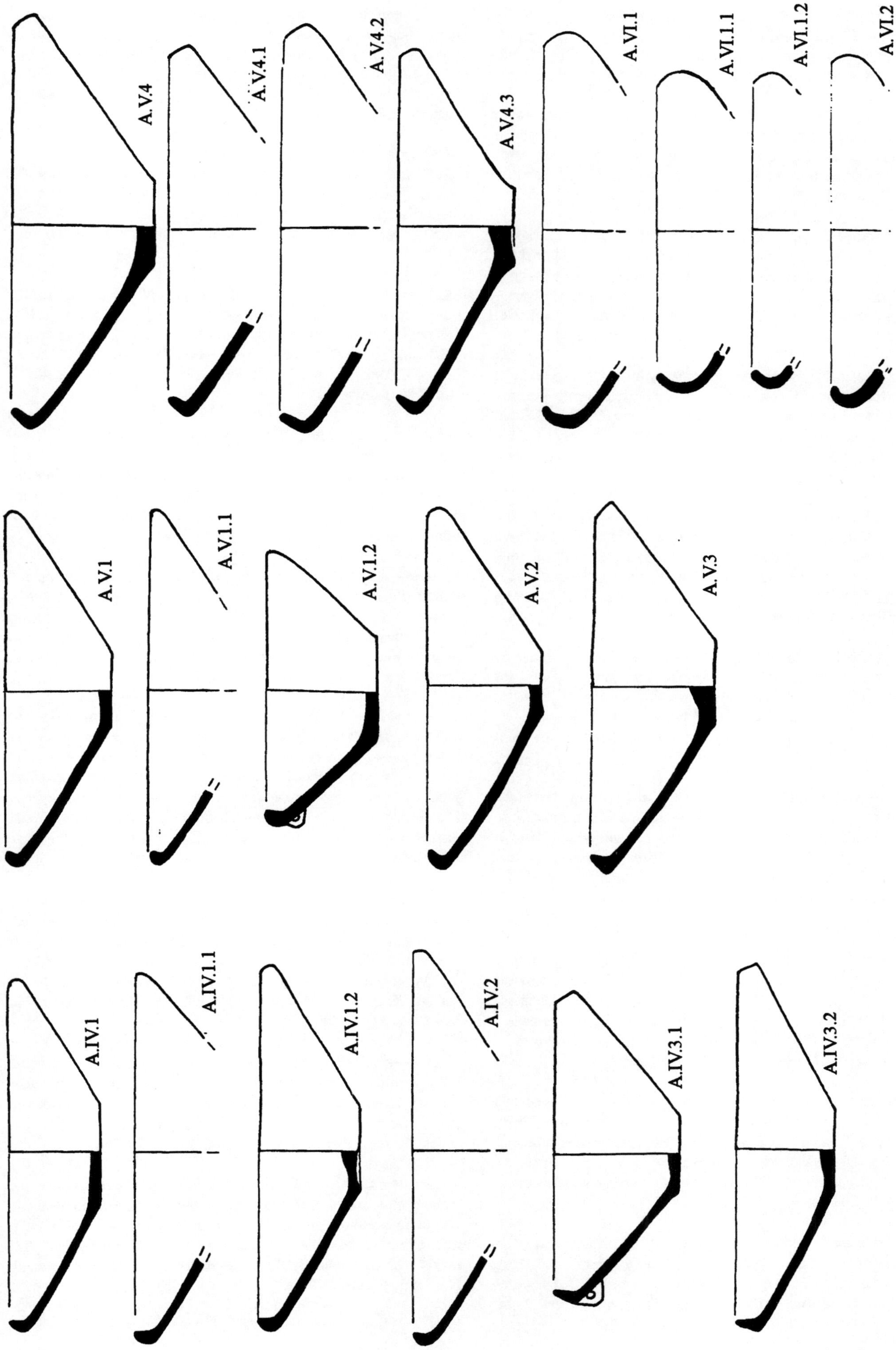

Table 26.2. Yunatsite tell. Classification of pottery. Late Chalkolithic.

26. ***A two handled cup***. Black color outside and inside. Slipped surface, burnished outside and inside. Paste is semi fine. Diameter orifice – 5 cm; diameter bottom – 2.8 cm; height – 5 cm. House 1. No. 1378. Historical Museum Pazardzhik No. 3896. ***Figure 26.5: 5.***

27. ***A two handled cup***. Light gray, gray and beige outside, brown spots inside. Fine slip surface outside, burnished; inside – smoothed. Paste is semi-coarse. Black biscuit with two brown strips. Diameter orifice – 8.4 cm; diameter bottom – 3.8 cm; height – 7.5 cm. The first horizon. ***Figure 26.5: 6.***

28. ***A two handled bowl***. Black irregular color, single dark brown spots. Surface with slip in trips, smoothed with traces of burnishing outside, evenly smoothed outside. Paste is semi coarse. Bottom – sandy, organic matter. Diameter orifice – 26.4 cm; diameter bottom – 7 cm; maximum diameter – 30 cm; height – 15.2 cm. House 4. No. 1758. Historical Museum Pazardzhik No. 4429. ***Figure 26.5: 7.***

29. ***An amphora-like pot***. Brown, irregular color outside and inside. Surface with self slip outside in the upper part, in the rest – evenly smoothed. Paste is semi-coarse. Restored. Diameter orifice – 8.2 cm; diameter bottom – 6.7 cm; maximum diameter – 16 cm; height – 12.8 cm. House 3. No. 1918. Historical Museum Pazardzhik No. 4747. ***Figure 26.6: 1.***

30. ***An amphora-like pot***. Brown irregular color, single darker and reddish spots outside and inside. Slipped surface, smoothed outside. Paste is semi-coarse. Bottom – stone of medium size. Diameter orifice – 7.6 cm; diameter bottom – 6.4 cm; maximum diameter – 16.3 cm; height – 17.4 cm. House 4. No. 1761. Historical Museum Pazardzhik No. 340. ***Figure 26.6: 2.***

31. ***An amphora-like pot***. Light brown irregular color, with darker spots outside and inside. Slipped surface, burnished outside and smoothed evenly inside. Paste is semi-coarse. Restored. Diameter orifice – 7.5 cm; diameter bottom – 6.4 cm; maximum diameter – 17 cm; height – 21 cm. House 1 (?). No. 1392. Historical Museum Pazardzhik No. 3885. ***Figure 26.6: 3.***

32. ***An amphora-like pot***. Dark brown, irregular color, gray and black spots outside and inside. Fine slip (washing) surface, burnished in the upper part outside, the rest of the vessel is smoothed. Paste is semi-coarse. Bottom – sandy. Diameter orifice – 8.4 cm; diameter bottom – 4.2 cm; maximum diameter – 14 cm; height – 15 cm. House 4. No. 1752. Historical Museum Pazardzhik No. 4292. ***Figure 26.6: 4.***

33. ***A pot***. Dark brown to black color, spots outside and inside. Slipped surface, burnished in the upper part outside and inside, the lower part is smoothed. Paste is semi-coarse. Bottom – sandy. Diameter orifice – 8.5 cm; diameter bottom – 6.4 cm; maximum diameter – 15.2 cm; height – 13.5 cm. House 4. No. 1757. Historical Museum Pazardzhik No. 4431. ***Figure 26.6: 5.***

34. ***A pot***. Gray and brown, irregular color, gray spots. Slipped surface, in the upper part burnished, in the lower part smoothed. Paste is semi-coarse. Bottom – 'shamot', sandy. Diameter orifice – 7.5 cm; diameter bottom – 6 cm; maximum diameter – 15 cm; height – 13 cm. Sq. T6, heaping. No. 1789. Historical Museum Pazardzhik No. 4670. ***Figure 26.6: 6.***

35. ***A pot***. Black color outside and inside. Slipped surface, burnished in the upper part outside, without slip in the rest part, smoothed. Paste is semi-coarse. Brown biscuit with two thin black strips outside and inside. Diameter orifice – 10 cm; diameter bottom – 9 cm; maximum diameter – 20.5 cm; height – 20.3 cm. Sq. M9, heaping. No. 2065. ***Figure 26.6: 7.***

36. ***A spherical pot***. Light brown, little brick reddish outside and inside. Self-slip surface burnished outside and inside. Paste is semi-fine, very sandy. Homogenous biscuit, light brown. Diameter orifice – 4.3 cm; diameter bottom – 3.1 cm; maximum diameter – 12.4 cm; height – 11.7 cm. House 4. No. 1979. Historical Museum Pazardzhik No. 4783. ***Figure 26.7: 1.***

37. ***'A vase'***. Gray and black, irregular color, single gray and brown spots. Slipped surface, burnished outside and inside. Paste is semi-coarse. Diameter orifice – 6 cm; diameter bottom – 4.5 cm; maximum diameter – 17 cm; height – 19 cm. House 12, next to grave No. 76. No. 1982. Historical Museum Pazardzhik No. 4797. ***Figure 26.6: 2.***

38. ***A pot***. Light brown, little brick-reddish color. Self slip surface burnished outside and inside. Paste is semi-fine, very sandy. Homogenous biscuit, light brown. Smoothed bottom. Diameter orifice – 5.3 cm; diameter bottom – 3.5 cm; maximum diameter – 10.6 cm; height – 10 cm. House 2. No. 1684. Historical Museum Pazardzhik No. 4286. ***Figure 26.7: 3.***

39. ***A pot***. Compact red outside, gray and beige spots inside. Compact slip surface, burnished to shining outside, self slip evenly smoothed inside. Paste is semi-fine. Homogenous black biscuit. Diameter orifice – 7 cm; diameter bottom – 8 cm; maximum diameter – 21 cm; height – 16 cm. House 3. No. 2050. Historical Museum Pazardzhik No. A780. ***Figure 26.7: 4.***

40. ***A jug***. Light brown, irregular color, grayish spots outside and inside. Surface with fine slip (self slip ?), burnished outside, smoothed inside. Paste is semi-fine. Gray biscuit with two brown strips. Diameter orifice – 8.5 cm; diameter bottom – 8.5 cm; maximum diameter – 18.5 cm; height – 22 cm. Sq. M9, heaping. No. 2062. ***Figure 26.7: 5.***

41. ***A cup***. Import (?). Bright red color outside, brown inside. The surface is with very fine slip and burnished to shining. There are two irregular brown spots. The inner surface is roughly treated and partially smoothed. Paste is semi-fine with a lot of coarse quartz sand. Red and brown biscuit, in the middle thin dark brown strip. Diameter orifice – 7.5 cm; diameter bottom – 2.5 cm; maximum diameter – 8.5 cm; height reconstructed – c. 7.5-8 cm. Sq. P. ***Figure 26.7: 6.***

42. ***An askos***. Miniature vessel. Dark brown to black, irregular color, single spots outside and inside. The surface is without slip, smoothed, with traces of washing. Paste is semi-coarse. Bottom – smoothed. Diameter orifice – 2.4-1.7 cm; diameter bottom – 6.8 cm; width – 4.5 cm; height of the body – 4.5 cm; height with the handle – 5.7 cm. Sq. C6, depth – 6.92 m. No. 1774. Historical Museum Pazardzhik No. 4664. ***Figure 26.7: 7.***

43. ***A pot-stand***. Reddish and brown, irregular surface color, beige and gray and beige spots outside and inside. Smoothed surface without slip. Paste is semi-coarse. Diameter bases – 9.2 cm; diameter in the middle part – 6.5 cm; height – 6 cm. House 1. No. 1437. Historical Museum Pazardzhik No. 4044. ***Figure 26.7: 8.***

44. ***A pot-stand***. Brown up to black in some places color of the surface, spots outside and inside. The surface is slipped, in the middle part there are traces of burnishing, smoothed outside. Paste is semi fine. Diameter bases – 12.8 cm; maximum diameter – 16.5 cm; height – 10 cm. House 1. No. 1381. Historical Museum Pazardzhik No. 3893. ***Figure 26.7: 9.***

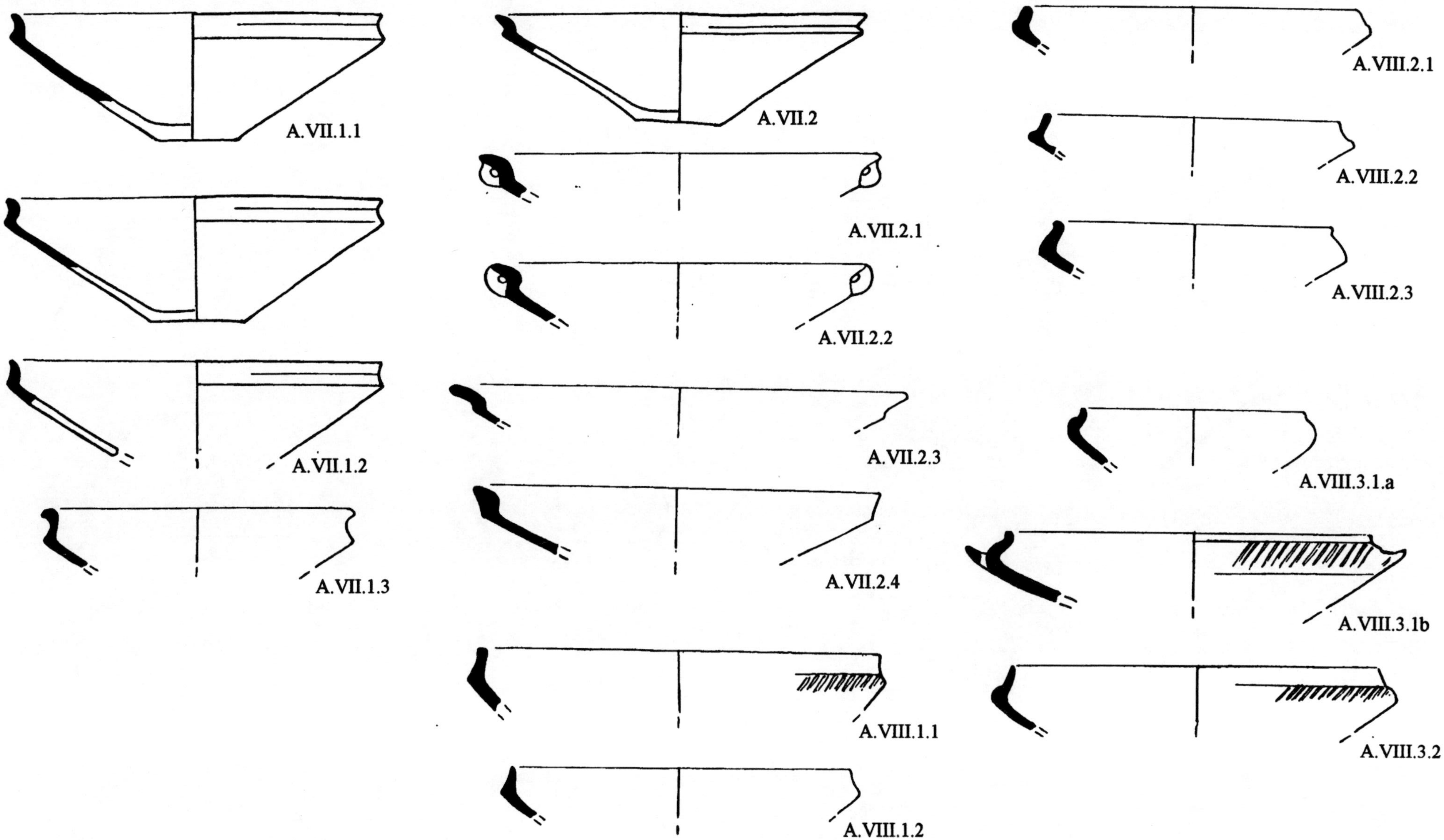

Table 26.3. Yunatsite tell. Classification of pottery. Late Chalkolithic.

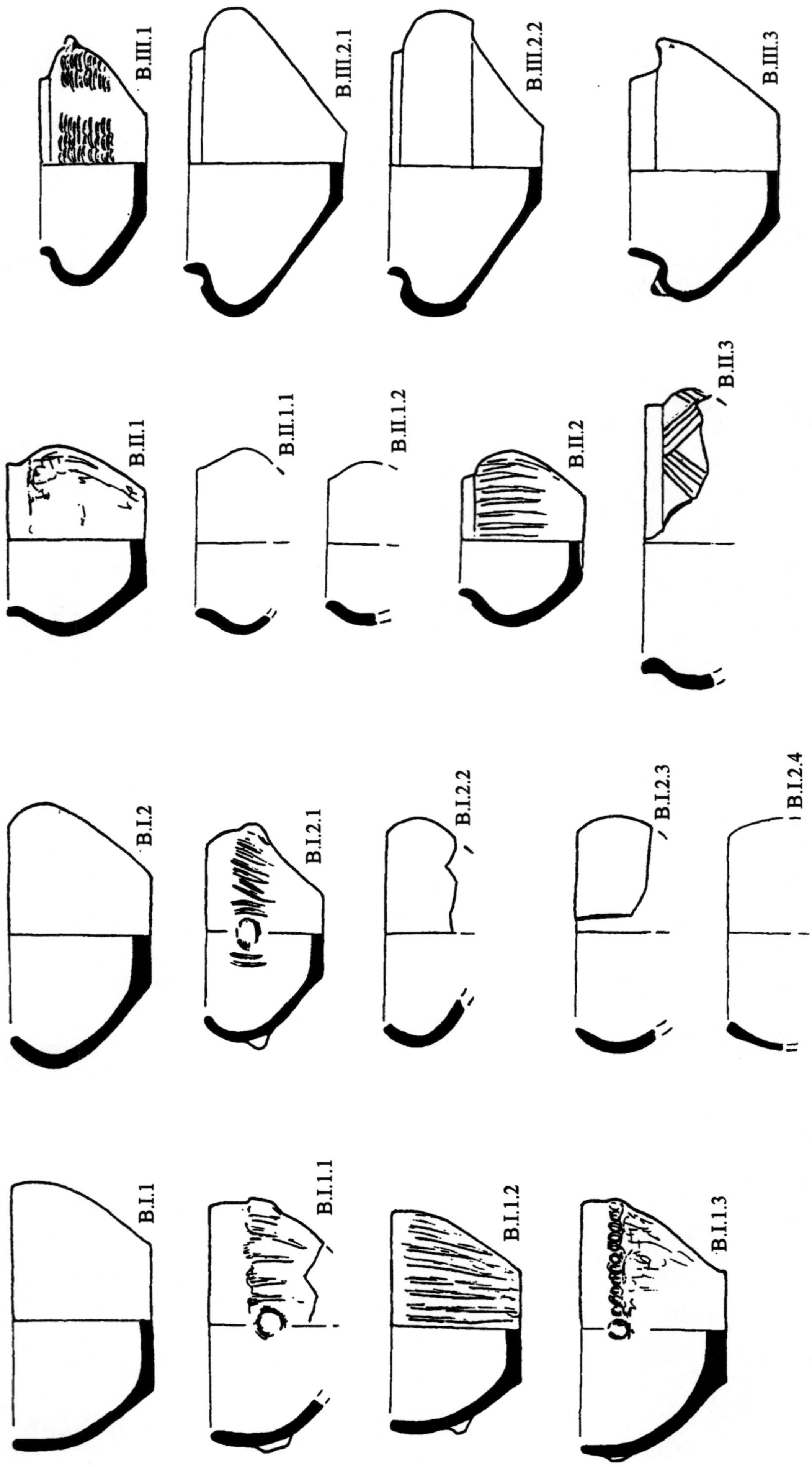

Table 26.4. Yunatsite tell. Classification of pottery. Late Chalkolithic.

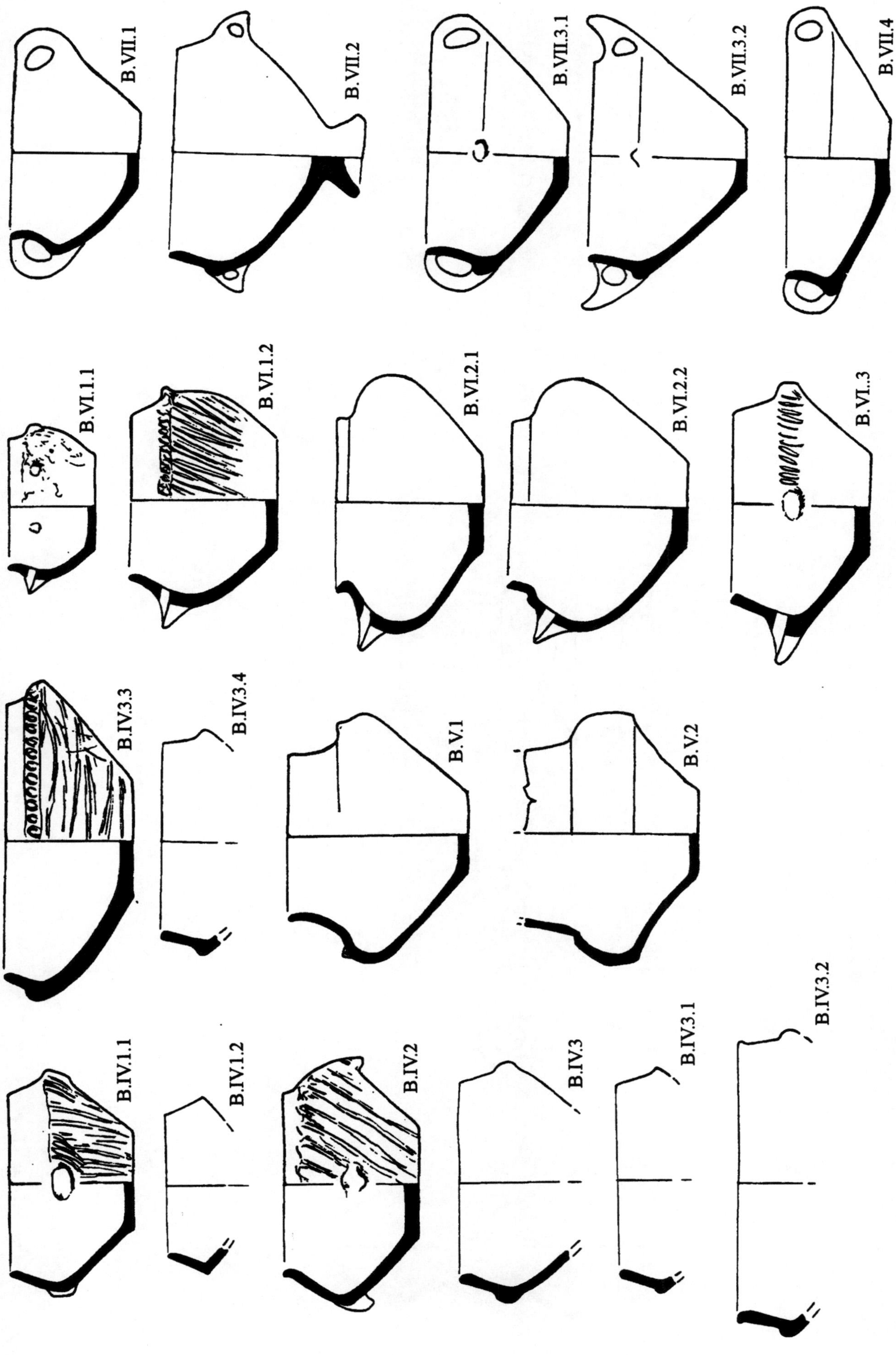

Table 26.5. Yunatsite tell. Classification of pottery. Late Chalkolithic.

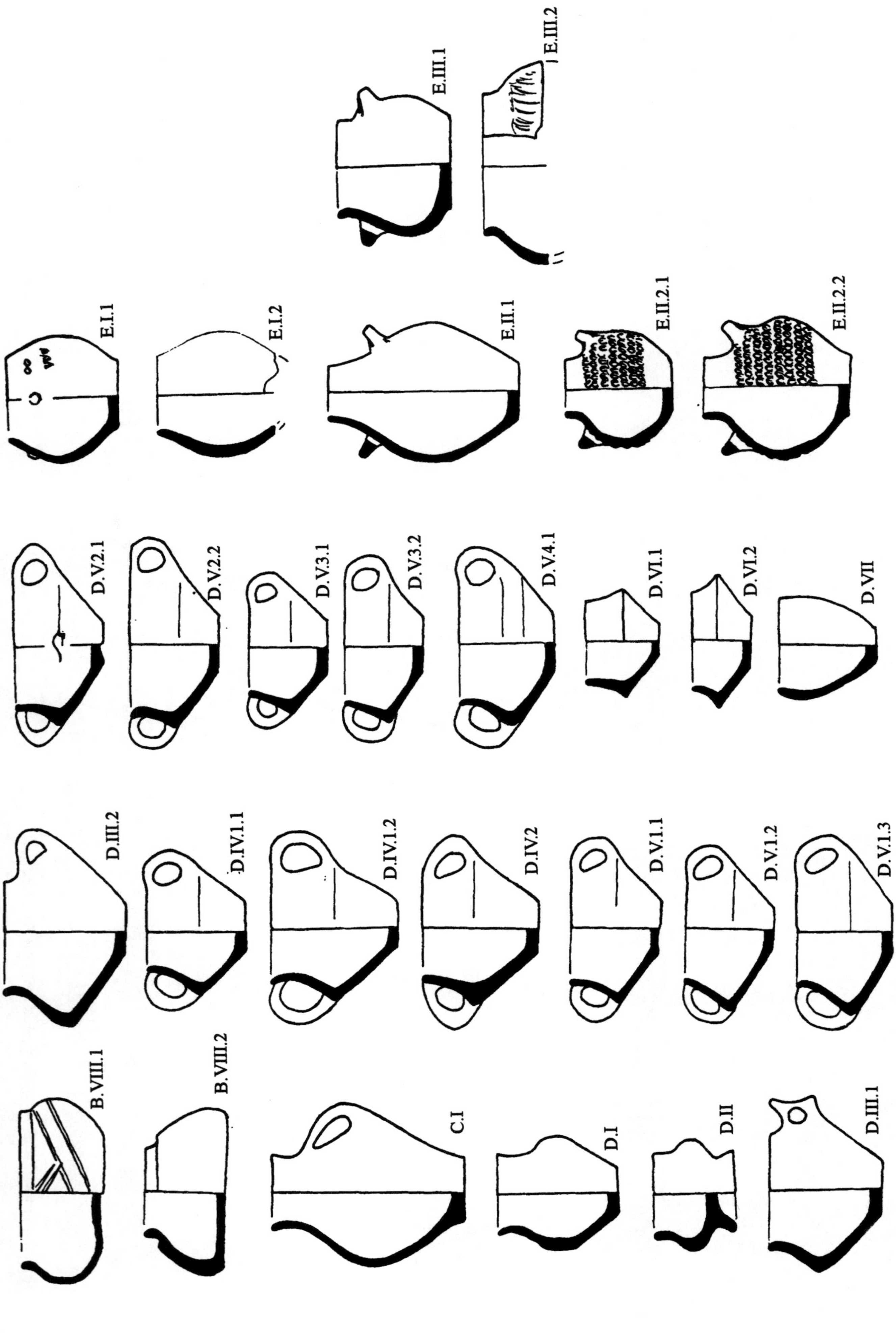

Table 26.6. Yunatsite tell. Classification of pottery. Late Chalkolithic.

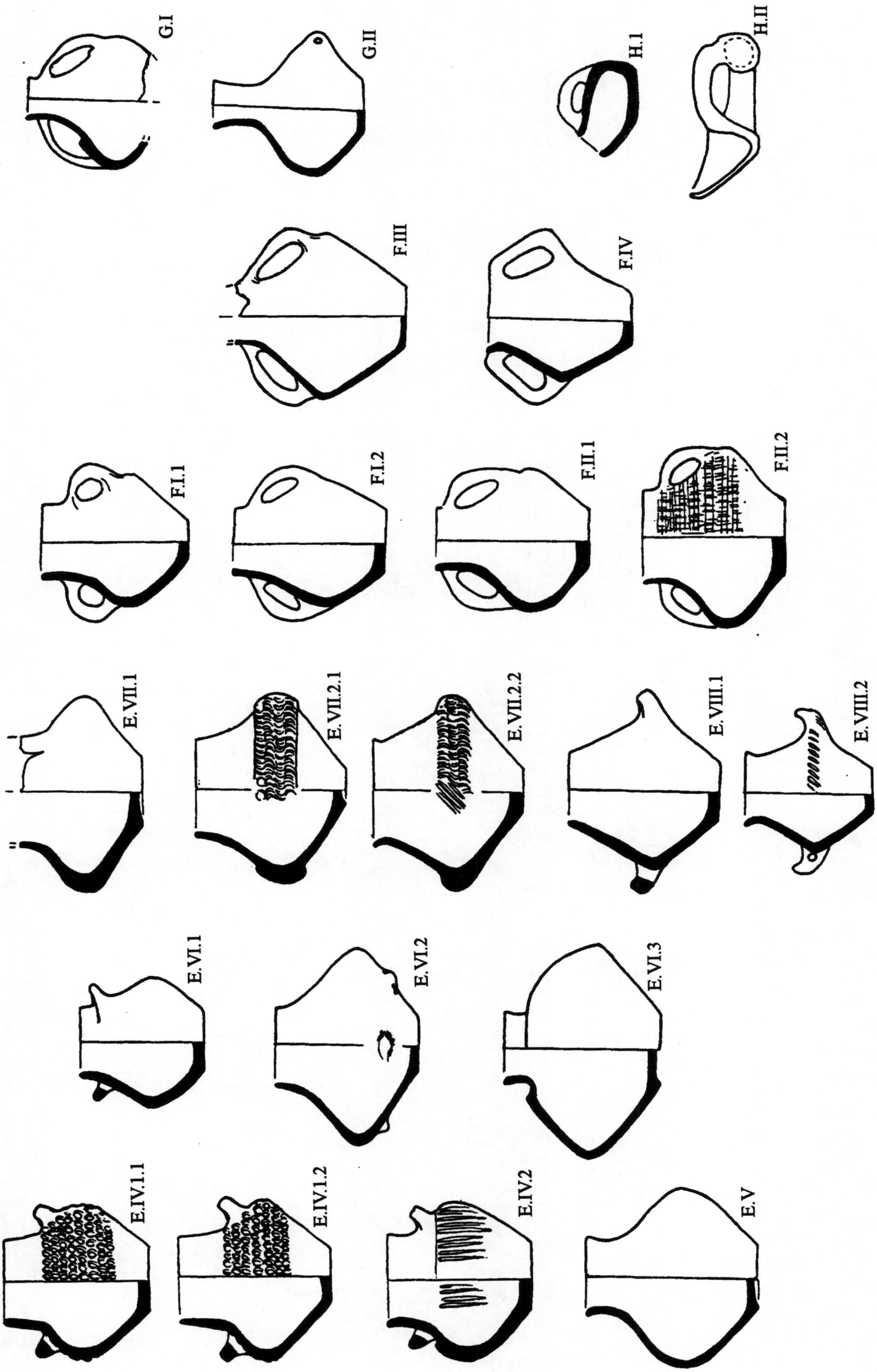

Table 26.7.Yunatsite tell. Classification of pottery. Late Chalkolithic.

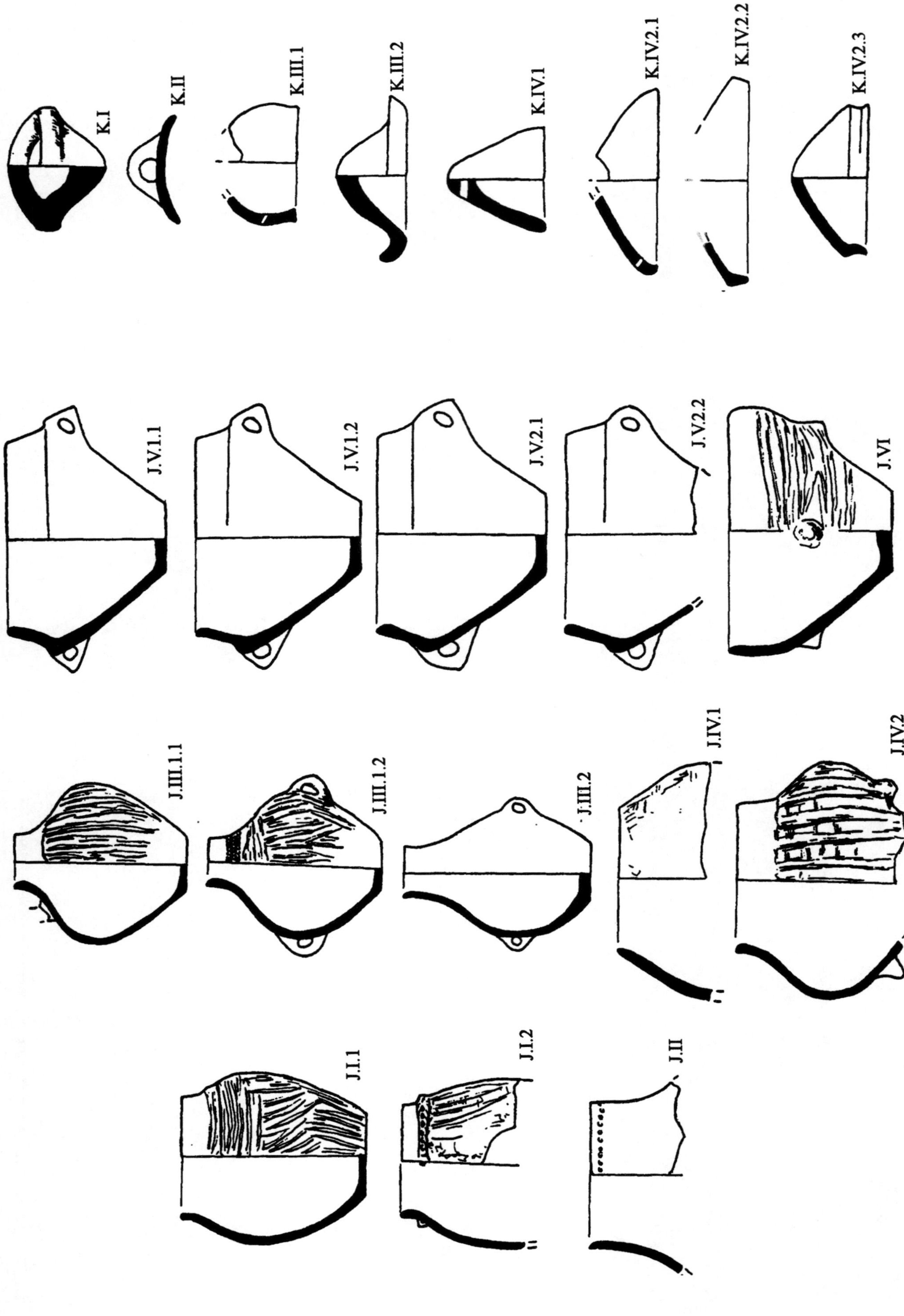

Table 26.8. Yunatsite tell. Classification of pottery. Late Chalkolithic.

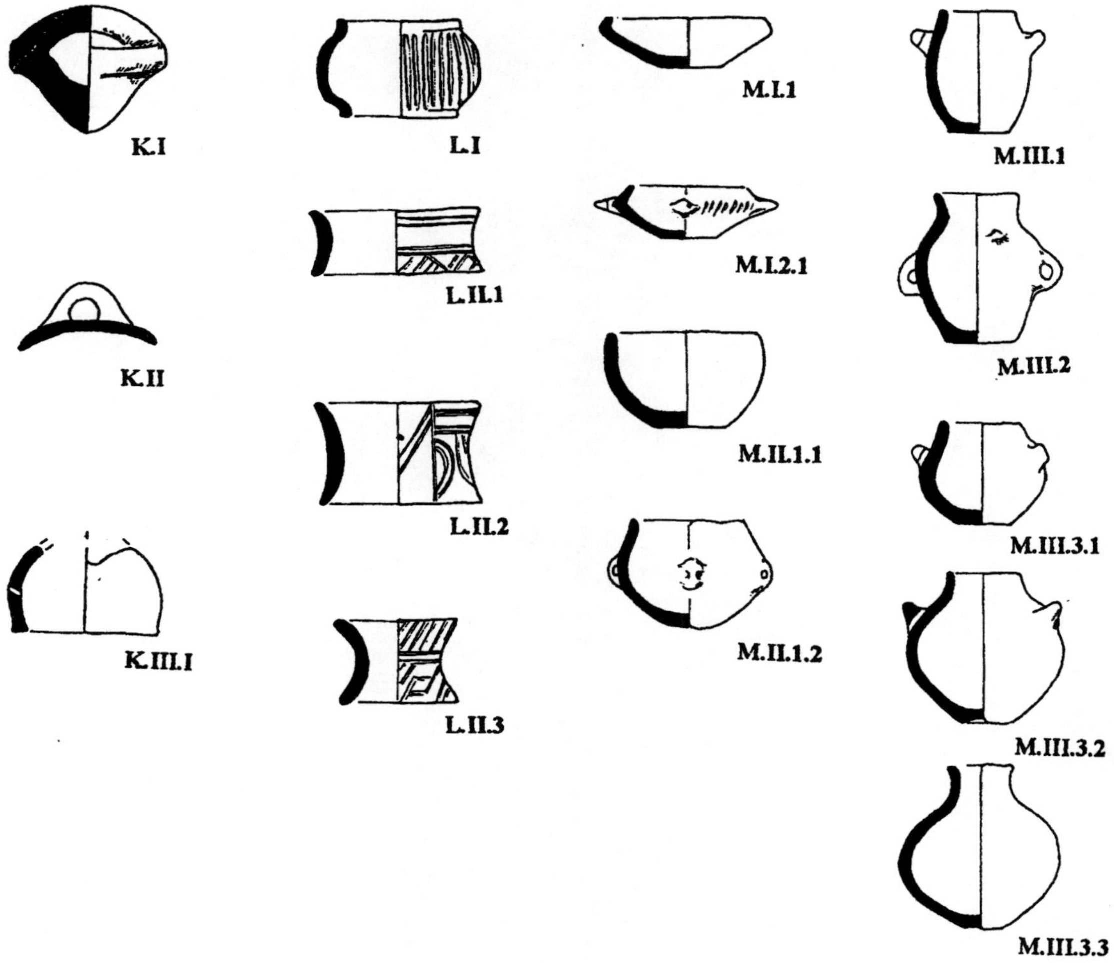

Table 26.9. Yunatsite tell. Classification of pottery. Late Chalkolithic.

45. ***An askos***. Unique shape. Black color outside, in the lower part single lighter, brown spots, brown inside. The surface is compactly slipped, burnished to shining, only the back part of the vessel is without slip, where it is unornamented, very coarse, unsmoothed. There are traces of cutting of the paste inside. Paste is semi-fine. Brown biscuit, thin black strip. Length – 17 cm; width – 14 cm; height – 4 cm; maximum height – 7 cm. House 4. No. 1664. Historical Museum Pazardzhik No. 4305. ***Figure 26.8: 1.***

46. ***A depository***. Brown, irregular color, dark brown and light brown spots outside and inside. Surface without slip, evenly smoothed in the upper part outside and inside, the middle part is roughed (barbotine). Paste is semi-coarse to coarse. Additionally attached band at the bottom. Diameter orifice – 39 cm; diameter bottom – 11.8 cm; height – 28 cm. House 5. No. 1920. Historical Museum Pazardzhik No. 4749. ***Figure 26.9:1.***

47. ***A depository***. Reddish and brown color outside, compact black inside. The surface is with compact brown slip outside, evenly smoothed, roughly treated, smoothed with a broom (?) inside. Paste is semi-coarse (almost coarse). Black biscuit, brown strip towards the outer wall. Diameter orifice – 14 cm; diameter bottom – 14 cm; maximum diameter – 35 cm; height– 49 cm. House No. 1.***Figure 26.9: 2.***

48. ***A depository***. Red and brown color outside and inside. The surface is with slip and carefully smoothed in the upper part, smoothed inside. The lower part is consciously roughed (barbotine). Coarse paste. Red and brown biscuit. House 3. ***Figure 26.9: 3.***

49. ***A depository***. Brown, irregular color, single reddish and black spots outside and inside. The surface is with slip and smoothed in the upper part outside and inside, the rest part is roughed (barbotine). Paste is coarse. Homogenous brown biscuit. Diameter orifice – 28 cm; maximum diameter – 40 cm. House 12. ***Figure 26.9: 4.***

CHAPTER 27

TECHNOLOGICAL ASPECTS OF THE POTTERY FROM HOTNITSA-VODOPADA

Vulka Ilcheva

The archaeological investigations of Hotnitsa-Vodopada were carried out between 1978 and 1990. Two settlement levels have been investigated. The sites is attributed to latest Cernavoda I (Ilcheva 1996 with the references cited there) or to the earliest Cernavoda III as the fist phase of the Early Bronze I (Nikolova 1999). According to the 14C dates, it dates from the second quarter of the fourth millennium BC (Nikolova 1999).

This chapter provides the volume with the results of the analysis of the technological aspects of the pottery from Hotnitsa-Vodopada. The statistics included is based on the analysis of all fragmented pottery discovered in three pits (Nos. 5, 6 and 11), a total of 1,659 fragments. The typological variety includes different types of cups, jugs, plates, bowls, pots, deep big vessels and unique vessels – ***Figures 27.1-22.10*** (Ilcheva 1996). The pottery from Hotnitsa-Vodopada is mostly unornamented. The ornamentation is an exception. The ornamentation distributed relief, incised, stamped, corded, painted, channel, etc. (see in details Ilcheva 1996).

Technological groups

The pottery discovered at the site under discussion belongs to two technological groups: fine and coarse. The latter predominates and the ration is 1:4. In both cases the clay paste is mixed with sand, organic matters and powdered shell. In some cases those mixture components have been discovered in combinations. Six sorts of fine pottery has been distinguished according to the clay mixtures (Ilcheva 1996):

Group FP [Fine pottery]1. Brown smoothed earthenware made of fine clay paste. It is most numerous among the groups of fine pottery. Mostly small vessels with 4-6 mm thick walls belong to that group. It is the most numerous group of fine pottery (30%).

Group FP2. Gray slip pottery, homogenous or with some reddish spots. The clay paste was mixed with fine sand. Gray biscuit, 6-8 mm thick, with reddish strips from both sides of the kernel strip. The gray color may do to the reducing final firing (700°). 26% of the fine pottery belongs to that technological group.

Group FP3. Unslipped smoothed to burnished brown pottery. Organic clay mixture of dung. Brown biscuit, 6-8 mm thick. The organic matters had burnt by the firing, living pores over the surface walls. 9% of FP.

Group FP4. Gray slip pottery. The clay paste was mixed with fine sand and some fine organic matter. Slip of fine covers the outer walls while the inner walls of the vessels are unslipped, smoothed, and with pores of the burnt organic matter. Gray and brown biscuit, 7-9 mm. Reducing atmosphere of firing. 3% of FP.

Group FP5. Gray slip pottery. The clay paste was mixed with powered shell and organic matters. The outer walls have smoothed surface, the inner ones porous. Gray and brown biscuit. 16% of FP.

Group FP6. Beige and brown or red and brown slip pottery. Clay paste consisted of powered shell. Gray to black biscuit. 15% of FP.

The coarse earthenware is divided into 7 kinds mainly based on the character of the clay paste mixture.

Group CP [coarse pottery]1. Beige to light red smoothed earthenware. Clay paste richly mixed with organic matter, without sand. Smoothed outer and inner walls. 26% of CP.

Group CP2. Light beige made of clay paste mixed with powered shell and chaff. The biscuit is crumbly. 13% of CP.

Group CP3. Gray and black smoothed earthenware. Black biscuit, crumbly. 12% of CP. There are small fractions of the shell over the surface.

Group CP4. Light gray slip earthenware. Clay mixed with coarse quartz sand. A slip of fine clay covers the outer and inner walls, often fell. Gray biscuit. 30% of CP.

Group CP5. Beige earthenware. Clay paste made of fine organic matters and sand. Gray biscuit. 9% of CP.

Group CP6. Beige and brown or gray and brown slip earthenware. Brown biscuit. 13% of CP.

Group CP7. Red earthenware. Black biscuit with 'shamot' (ceramic fractions). The outer walls have red color, the inner walls are gray to beige. Big thick-walled pithoi belong to that group. 1% of CP.

Surface treatment and color

After vessels' shaping, the surface of the walls was additionally treated. The vessels with slipped inner and outer walls have the highest percent of distribution (39.7%). The slip consists of fine clay. In some cases only outer or inner wall of the vessels is covered with a slip, but the another one is smoothed or rough. 20.58% of the vessels have smoothed surface, followed by those with rough (14.92%) and porous surface. The percent of the burnished pottery is low. It includes plates, jugs and bowls of fine painted pottery with burnished outer walls.

The highest percent belong to gray and black pottery vessels (21%) followed by brown and black pottery.

Conclusions

Technologically and typologically the pottery from Hotnitsa-Vodopada cannot be connected with the Late Chalkolithic traditions. Its characteristics are closer to the Bronze Age. In other words they show a genesis of ceramic technologies and typologies, which are developing later in the Early Bronze Age. In compare to the quality of the pottery is worse. It is close to pottery from Shemshevo-Klise Bair and Kachitsa D sites (see in details in Ilcheva 1997), but there are some differences among the ceramic complexes mentioned. For instance, at Shemshevo-Klise Bair the fine pottery predominates, at Kachitsa the coarse prevails over the fine a bit only, while at Hotnitsa-Vodopada the ration between the fine and the coarse is 1:4.

At Hotnitsa-Vodopada there the variety of clay pastes is higher, while at Shemshevo-Klise Bair and Kachitsa D there are only three – four kinds, as well as the shell clay mixture is missing. The typological variety is also lower, as at Hotnitsa-Vodopada the cups with very high handles occur for the first time in the Middle Yantra region. In my opinion there are chronological differences and Hotnitsa-Vodopada follows Shemshevo-Klise Bair and Kachitsa D.

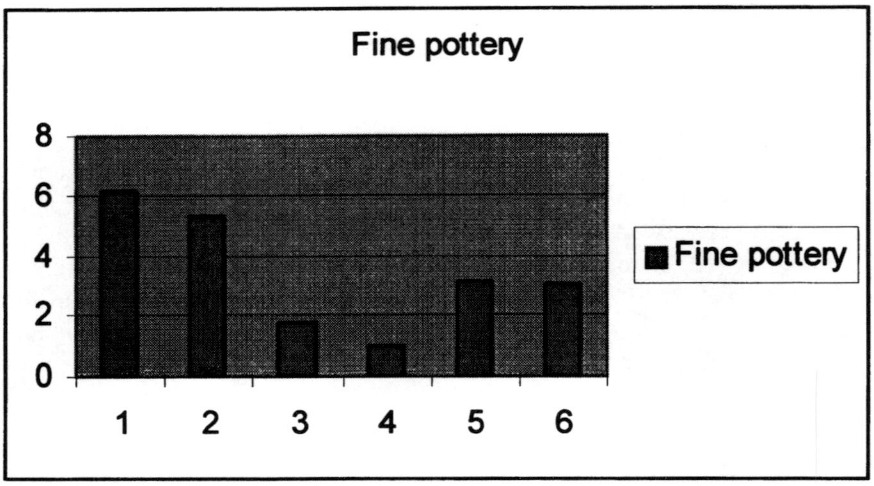

Chart 27.1 Fine pottery ratio of the different groups to the total ceramics statistically examined.

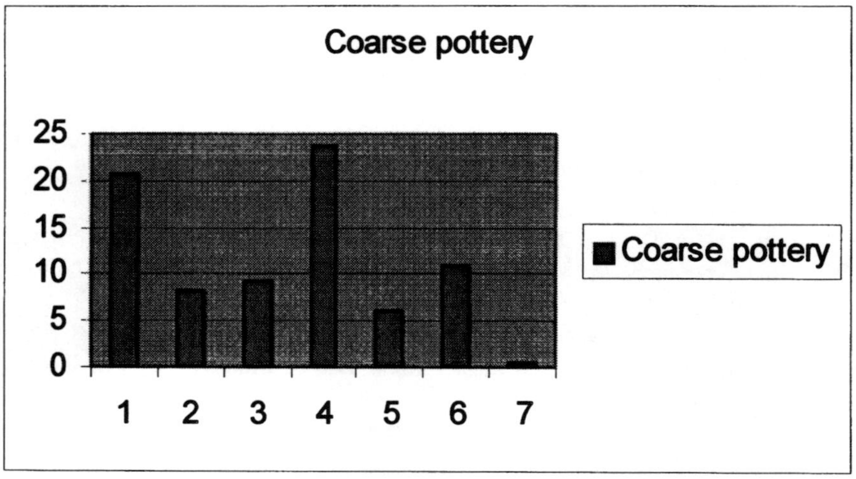

Chart 27.2. Coarse pottery ratio of the different groups to the total ceramics statistically examined.

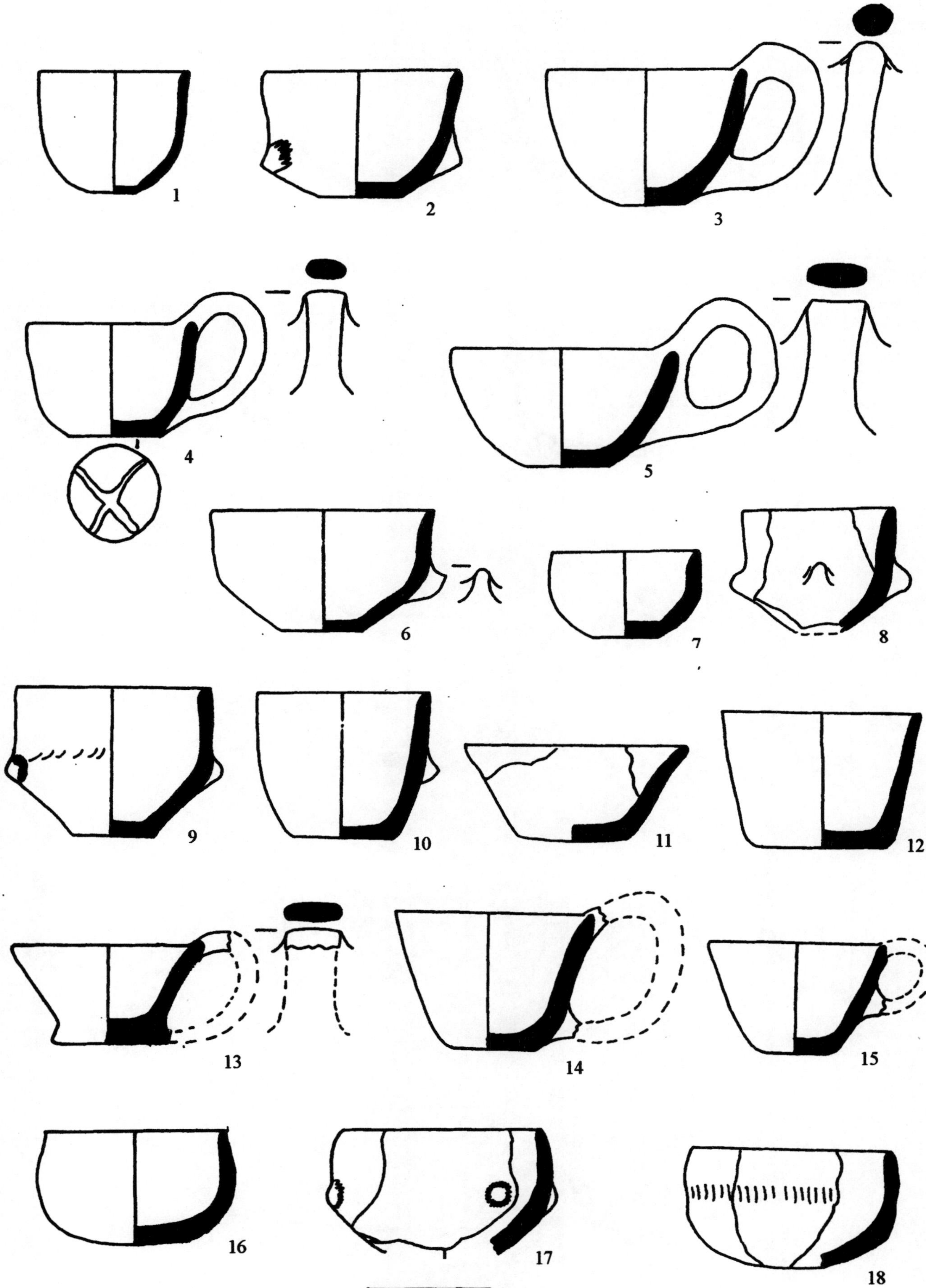

Figure 27.1. Hotnitsa-Vodopada. Cups. FP1 (1, 2, 9), CP1 (3, 4-8, 10, 12-13, 17-18), CP2 (11) and CP4 (14-16).

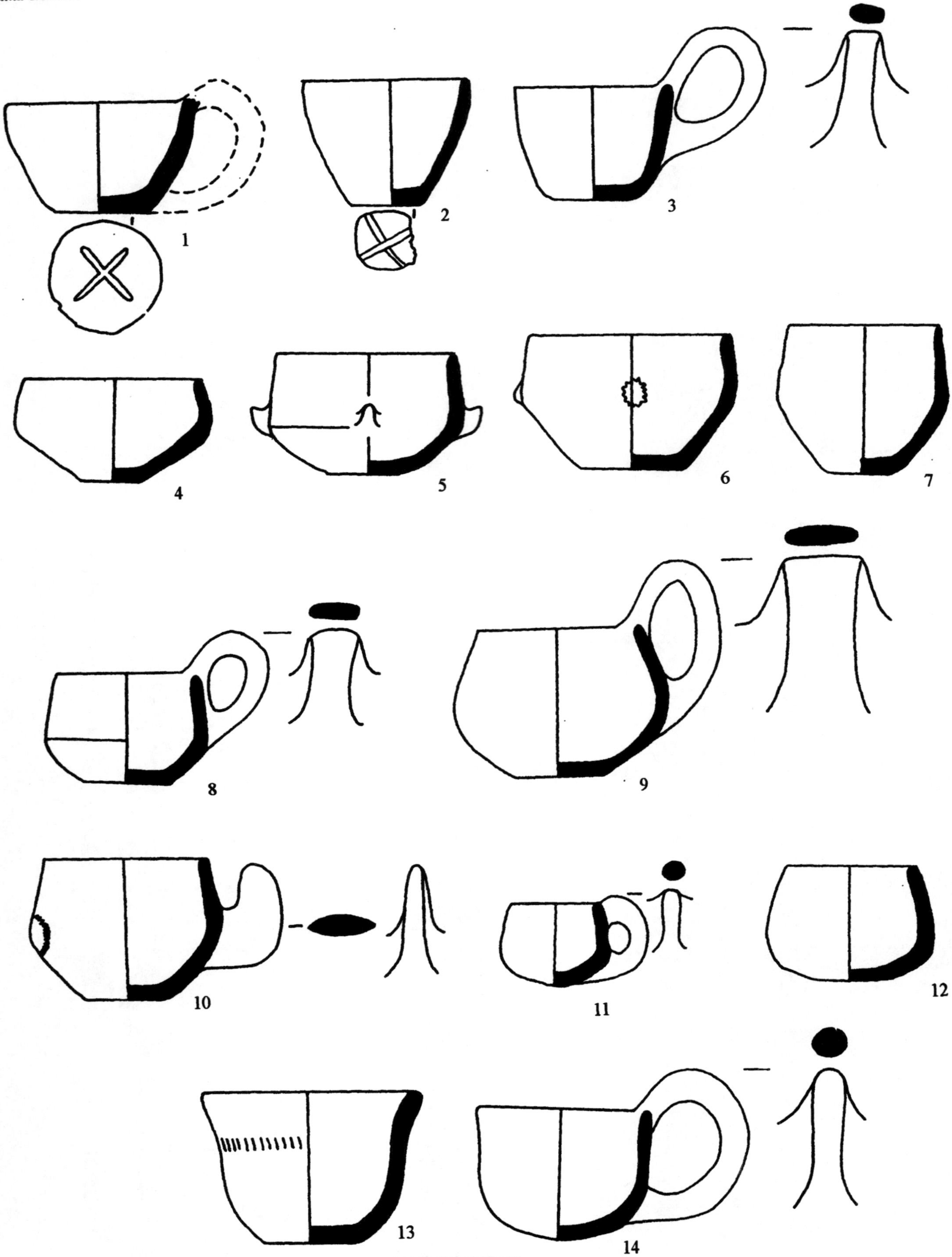

Figure 27.2. Hotnitsa-Vodopada. Cups. FP1 (2, 14) and CP1 (1, 3-10, 12).

Figure 27.3. Hotnitsa-Vodopada. Cups. FP1 (2, 4, 10-11) and CP1 (1, 3, 5-9).

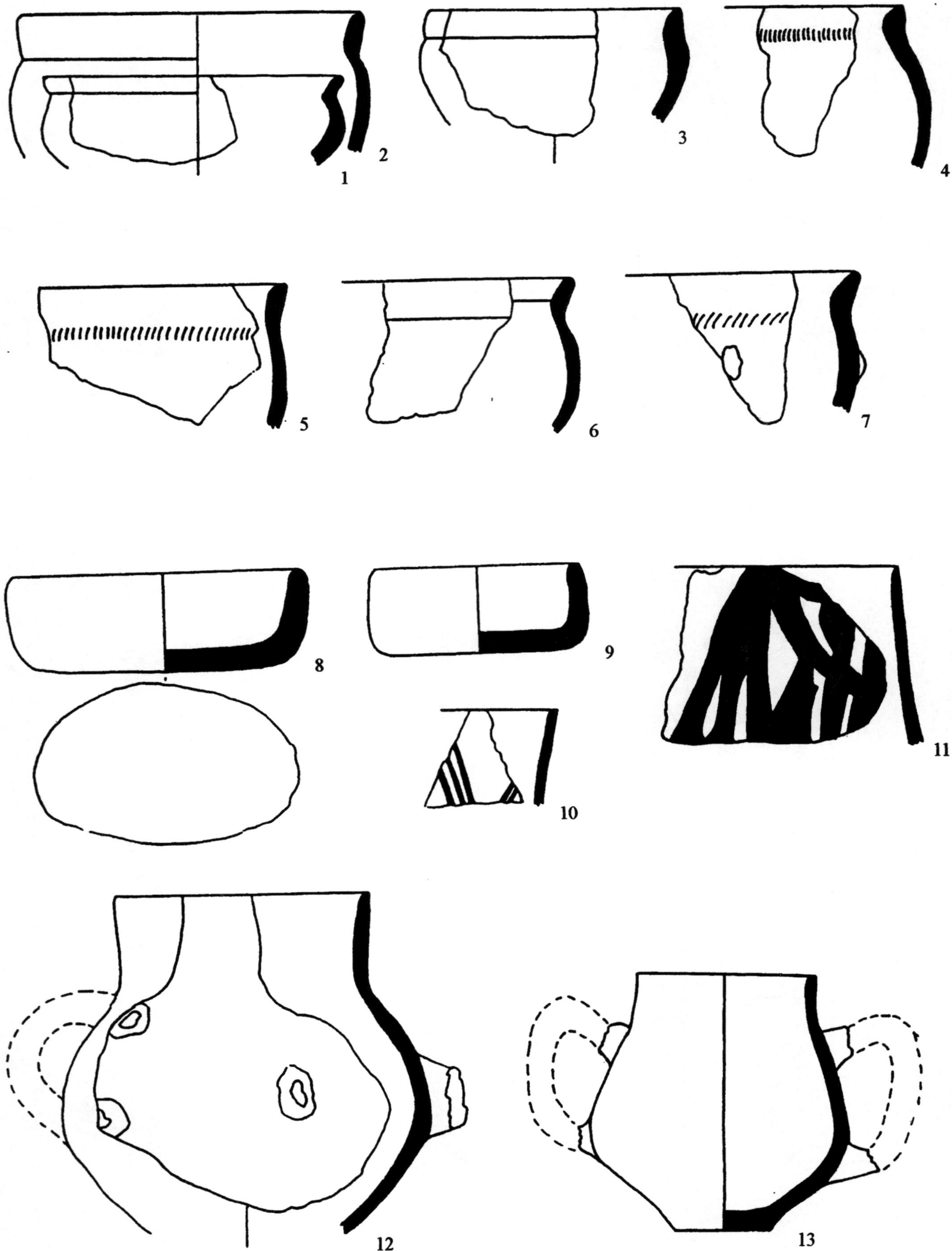

Figure 27.4. Hotnitsa-Vodopada. Plates, bowls and jugs.FP1 (10, 11-12), FP2 (2, 3, 6, 13), CP4 (4, 5, 7-9) and CP6 (1).

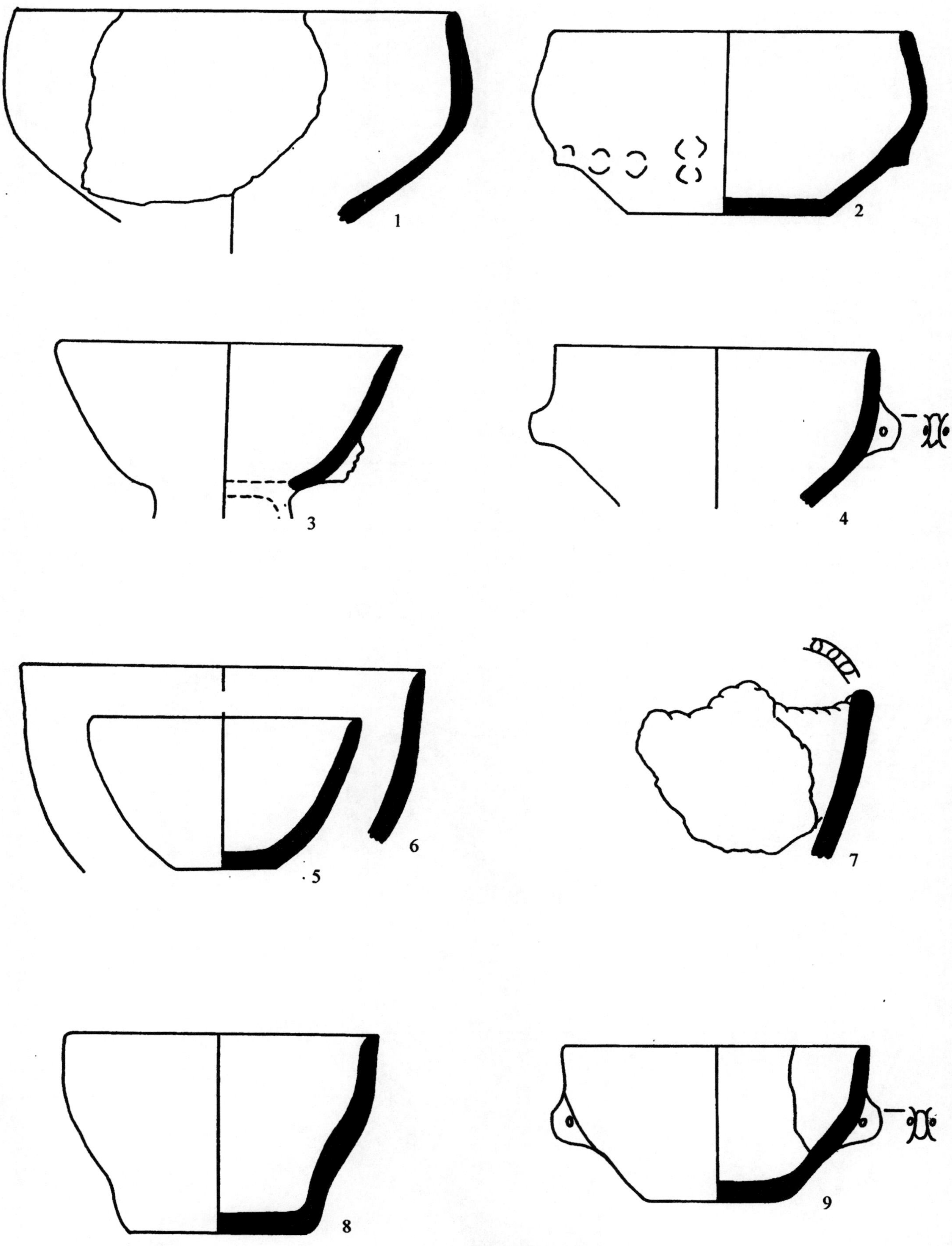

Figure 27.5. Hotnitsa-Vodopada. Plates and bowls. FP1 (9), FP2 (3-4), CP1 (1-2, 5-6), CP4 (8) and CP5 (7).

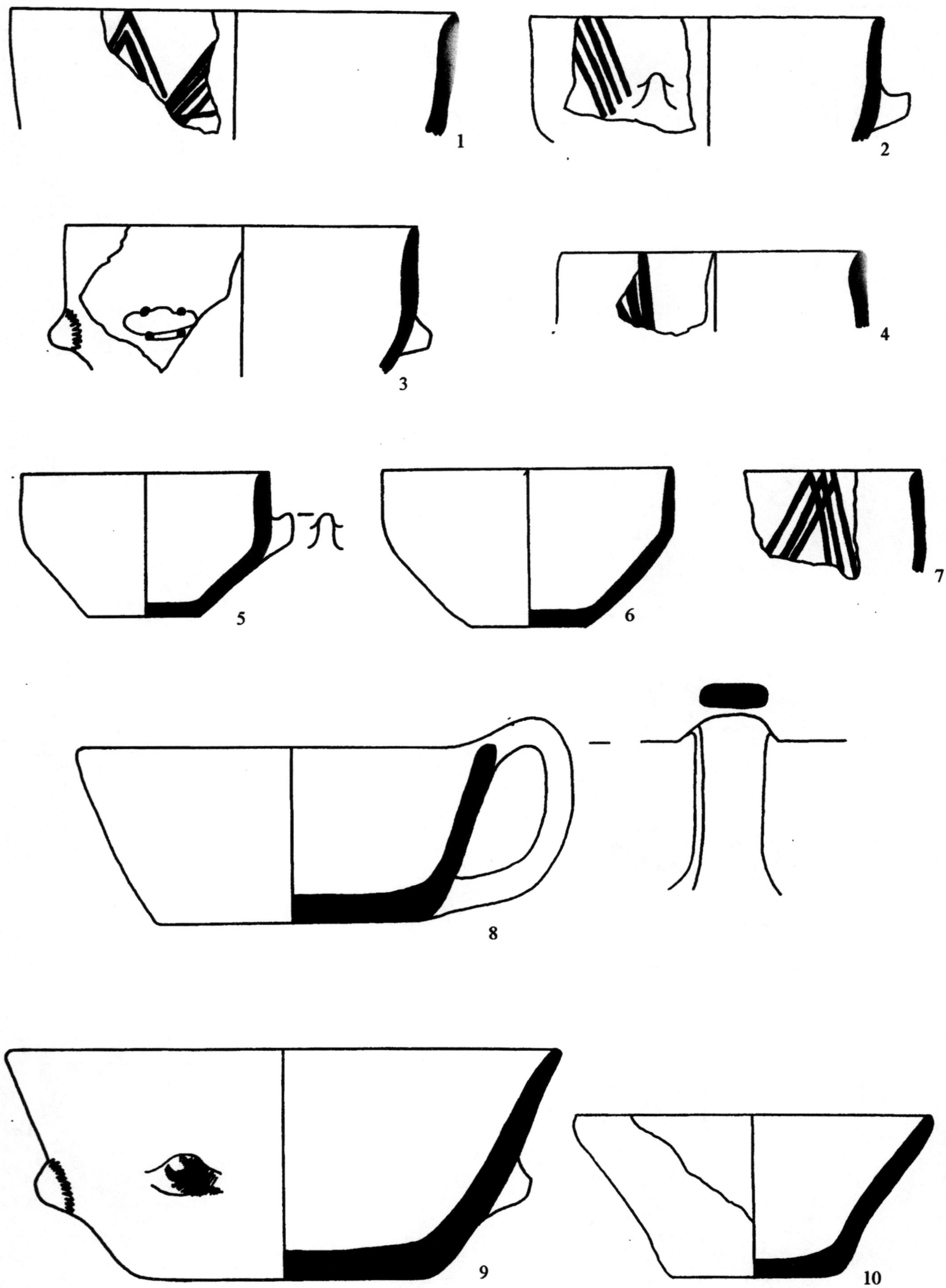

Figure 27.6. Hotnitsa-Vodopada. Plates and bowls. FP1 (1-2, 4, 7), CP1 (5-6, 10), CP4 (8) and CP5 (9).

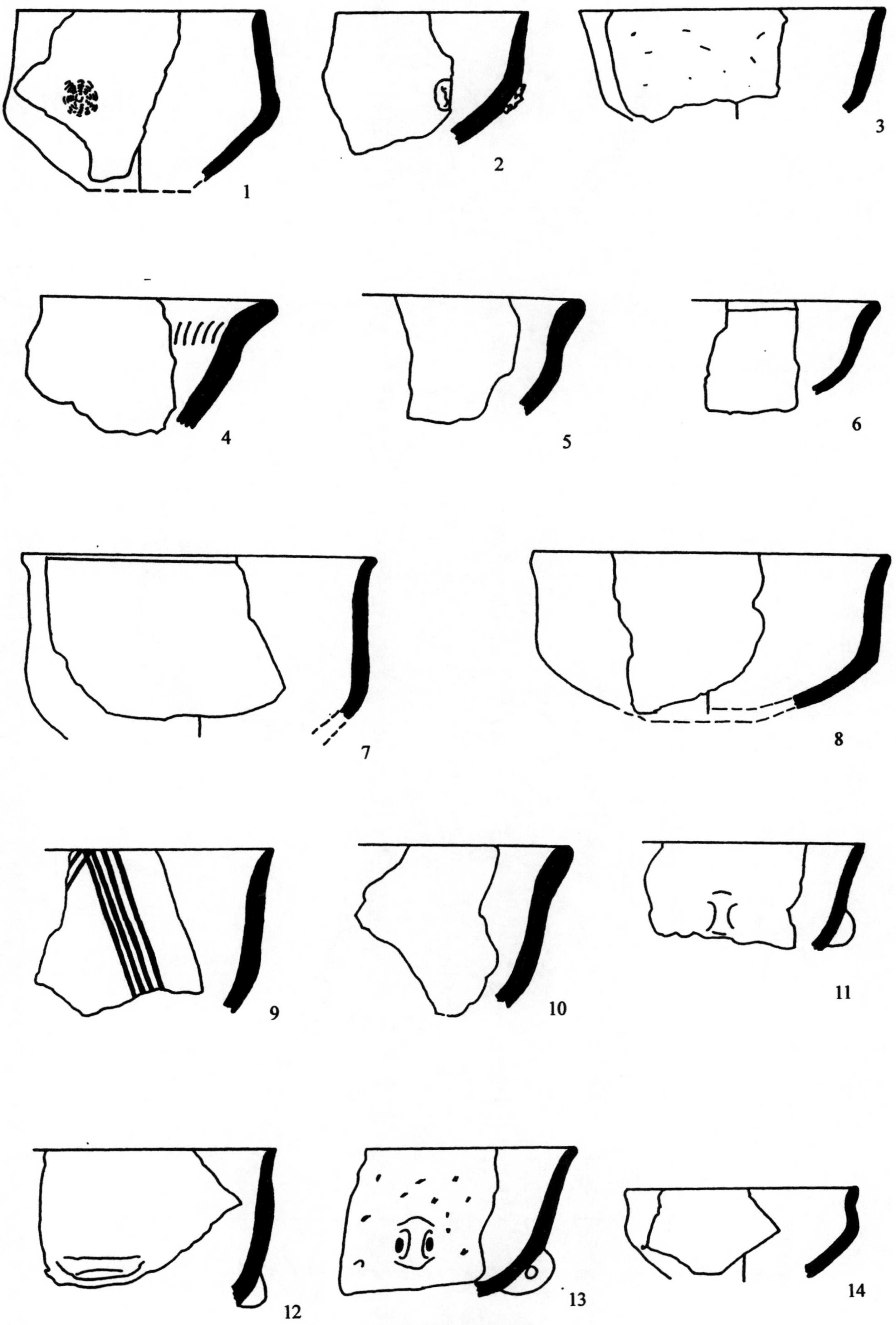

Figure 27.7. Hotnitsa-Vodopada. Plates and bowls. FP1 (3, 7, 9), FP2 (6), FP3 (2, 14), CP1 (8, 13) and CP4 (1, 4-5, 10, 12).

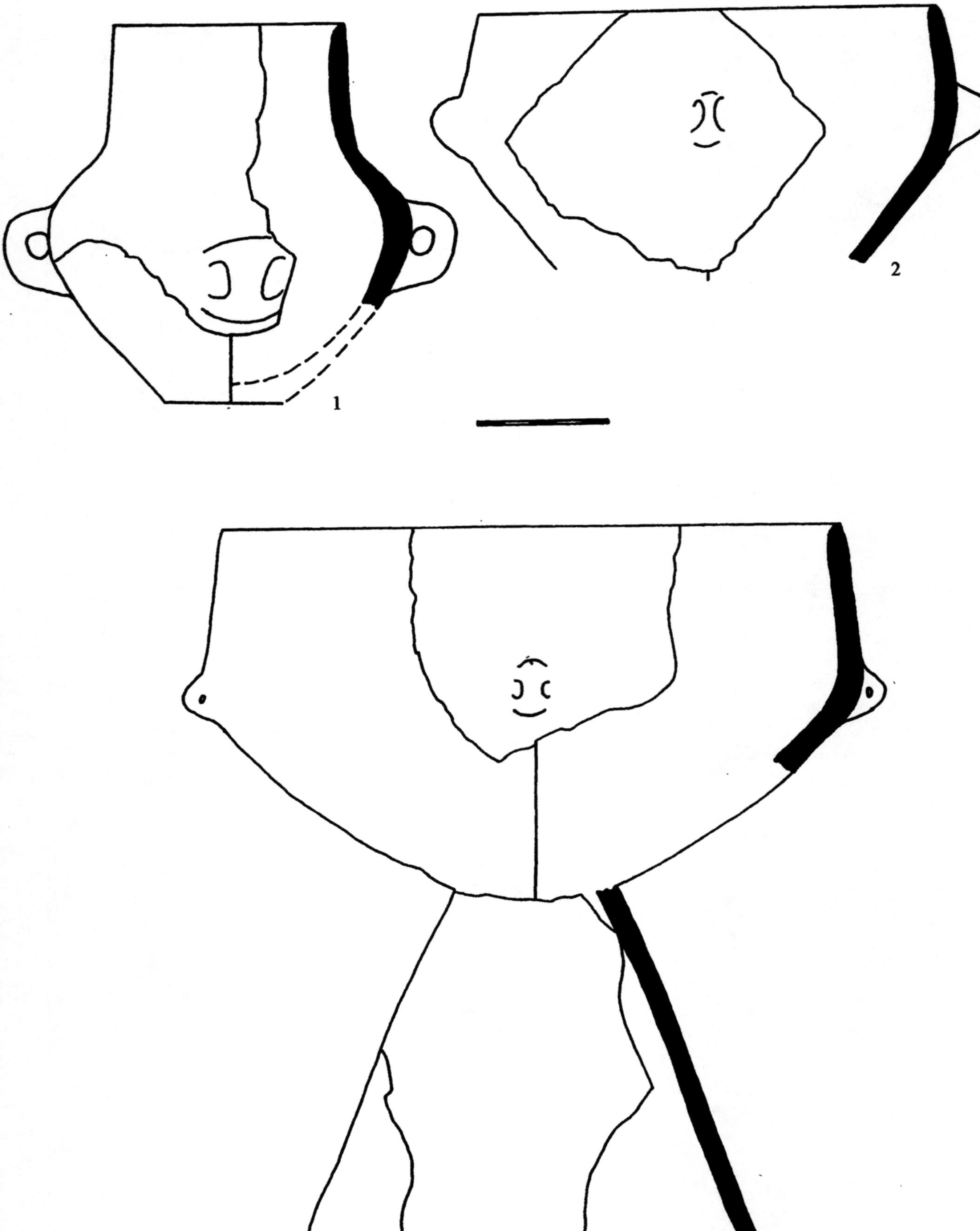

Figure 27.8. Hotnitsa-Vodopada. A jug and bowls. FP2 (3), FP4 (2) and CP1 (1).

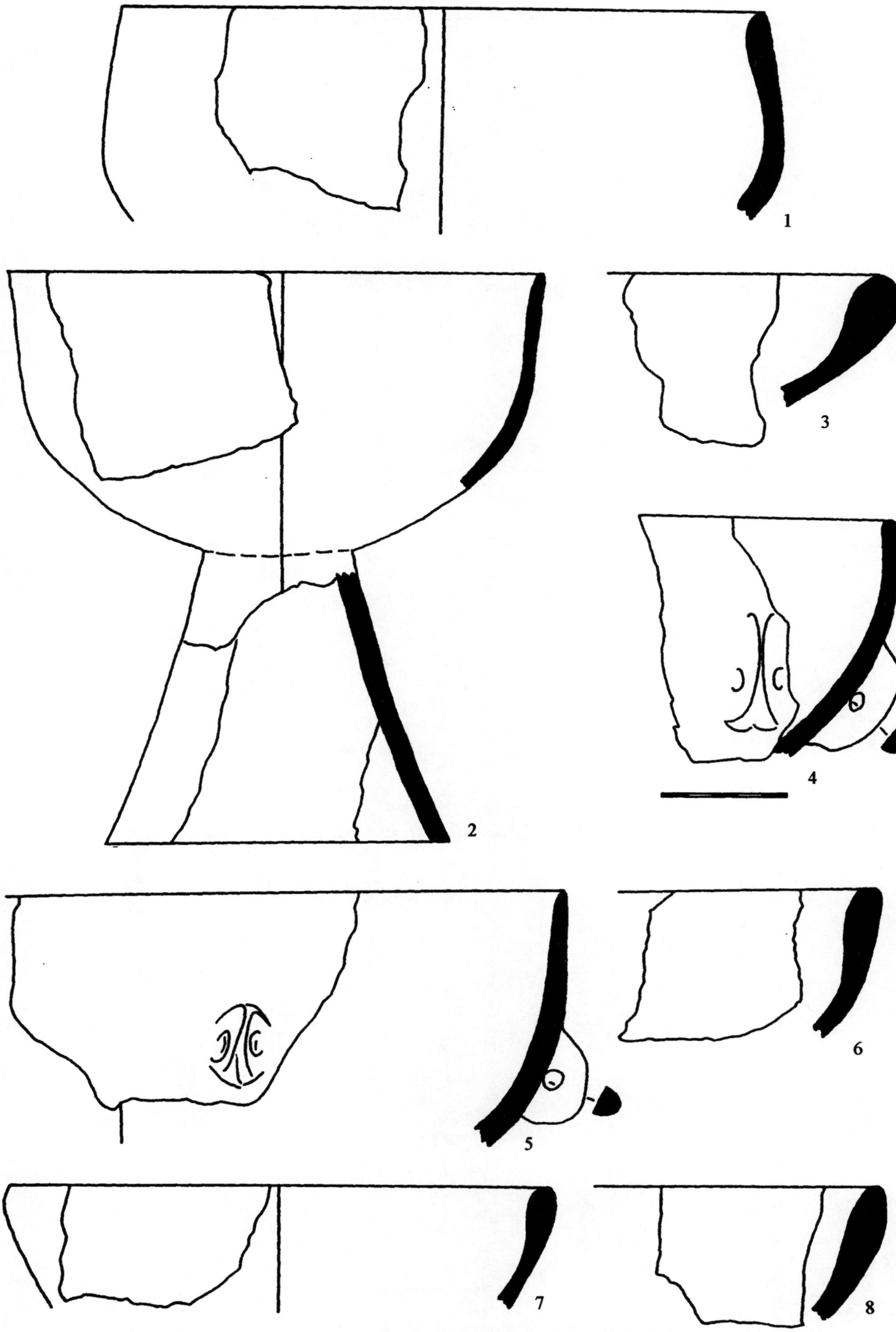

Figure 27.9. Hotnitsa-Vodopada. Bowls. FP1 (1), FP4 (2), CP1 (4, 5), CP4 (3, 7, 8) and CP5 (6).

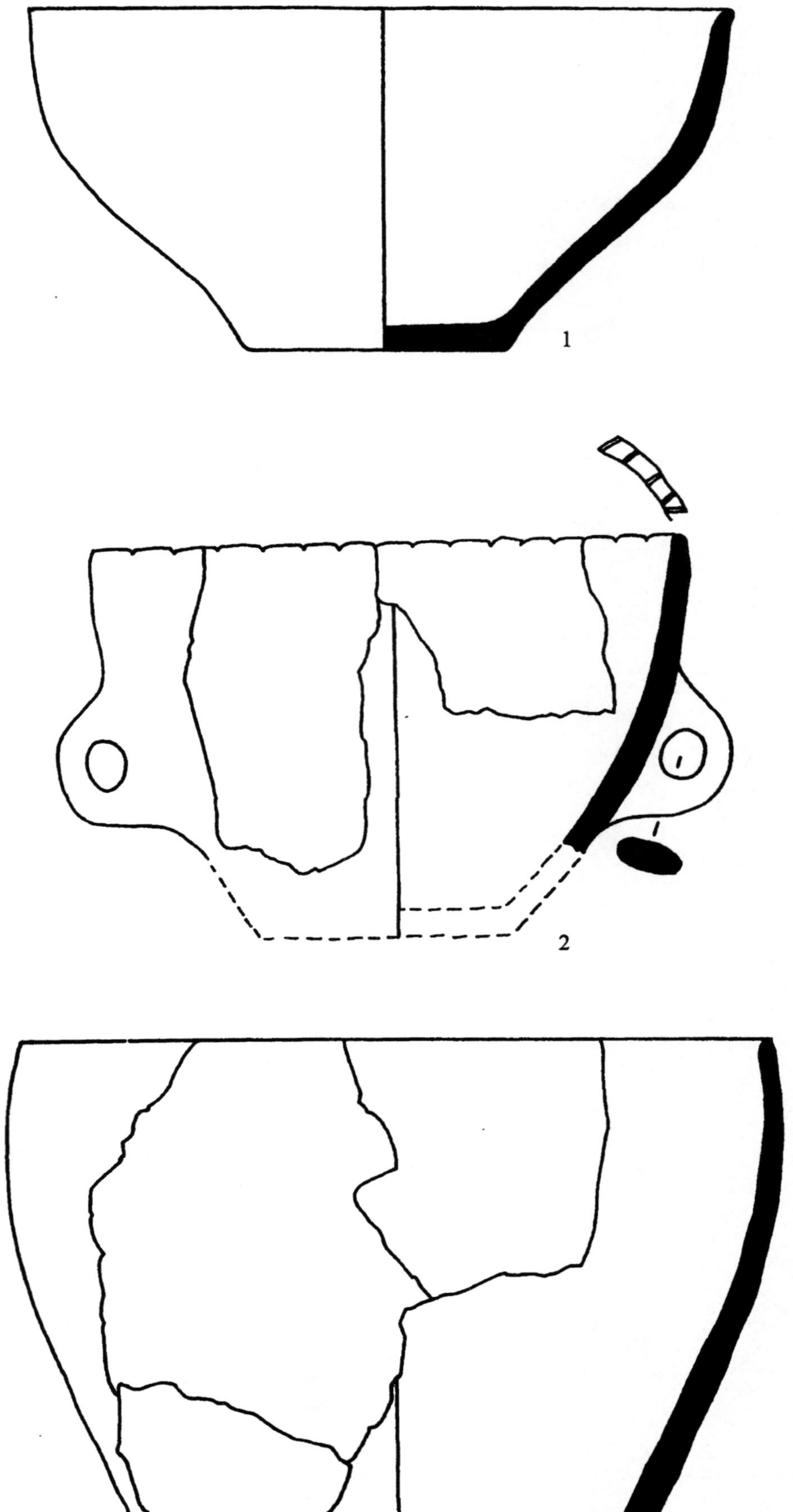

Figure 27.10. Hotnitsa-Vodopada. Bowls.FP2 (1), CP4 (2) and CP5 (3).

Chapter 28

Geomorphological Investigations of Dubene-Sarovka Area and the Sanctuary near the Village of Dositeevo

Rositsa Kenderova

This study represents the results of geomorphological investigation of two newly excavated prehistoric sites in Southern Bulgaria – Dubene-Sarovka in northwestern Upper Thrace and Dositeevo Sanctuary in the eastern Rhodope Mountains. The purpose is characteristics and origin of the different levels in the sectioned performed to the defined at both sites, with a possibility some applications concerning the palaeoclimate and natural environment developed in the southern Balkans to be argued.

Dubene-Sarovka

Description

The geomorphologically investigated site is situated on the excavated area of the Dubene-Sarovka tell (excavations of L. Nikolova). The site is located south-west of Dubene-Village, Sarovka Locality. The layers studied belongs to Early Bronze I-II and the layers below those levels.

Geomorphologically, that locality is a part of the hollow bottom of Karlovo morphostructure (D. Kanev) and of left catchment area of the Stryama River. In this area the river flows south of the inner hollow hills of Balyovitsa (375.6 m), 1 km of the excavations respectively. There is a channel between, which flows between Balyovitsa and another inner hollow hill of Konnika (394.2 m).

The tell is situated at c. 10 m above the recent bed of the Stryama River. According to different authors (Kanev 1988; Lukanov 1992, etc.), there is a movement of the river bed in the tile of whole Quaternary.

Material and Method

In the area investigated there is a section described (***Figure 28.1***) attached to the eastern profile of the central wall. The geomorphologic section is identical with sounding H18-4.

The main purpose of the investigation is the layering of the different layers to the analyzed, so their origin to be proved. Two horizons were of special interest: that between 172-216 cm, which is a sterile layer, as well as the level at 153-173 cm, a hearth floor and is to be archaeomagnetically investigated as well.

Macro scope, granulemetric, and mineralogical analyses were used as the methods of G.Baltakov and L.P. Seryabnikov was applied (1970).

Results

Five levels were investigated by *granule metric method* (***Figure 28.2***). The results of that analysis show the following more important peculiarities:

- Variety of the fractions at all levels: from coarse broken stone to clay;

- In all the levels (the cultural and the sterile ones) the clay predominates, as its quantity varies from 83.5% in the lowest level to 22.8% in the levels between 152 and 172 cm.

- The sand, as well as the rock and the gravel are relatively homogeneously distributed in the profile – between 12 and 29 % with exception of level between 96 and 101 cm, where the sand quantity is 8.46%.

- The layer in the level between 172 and 216 cm indicates sand and clay filling with a participation of gravel (12%) and single rock (8.9%).

All the rock and a part of the gravel were investigated by morphoscope analysis (***Figure 28.3***). The data of that analysis show the following peculiarities:

- Well rounded pieces predominate in all the levels;

- The petrographic structure analyzed is identical with the province of the Stryama River fed up, that is 'South Bulgarian granite-like fractions', as well gneiss and granite and gneiss (after the Geological map of Bulgaria, list Troyan and Karlovo, Scale 1: 100 0000).

- There are some black spots of Fe^{3+}, which a proof there was difficult water-exchange;

- There are traces of anthropogenic (mostly thermal) intervention over the surface of the rock from all the levels, with exception of the level between 173 and 261 cm.

- It is worth mentioning between 132 and 147 cm: 1. 64% is the maximum coefficient of rounding of the pieces; 2. 2.88% of the pieces have disc-like and flat and prolong shape, then, the were trailed, and another 12% were rolled; there are traces of thermal treatment over the surface of 62% of the pieces – these are cracks, compact overlaying clay, sand and gravel, as well as broken rock; 4. Maxi-

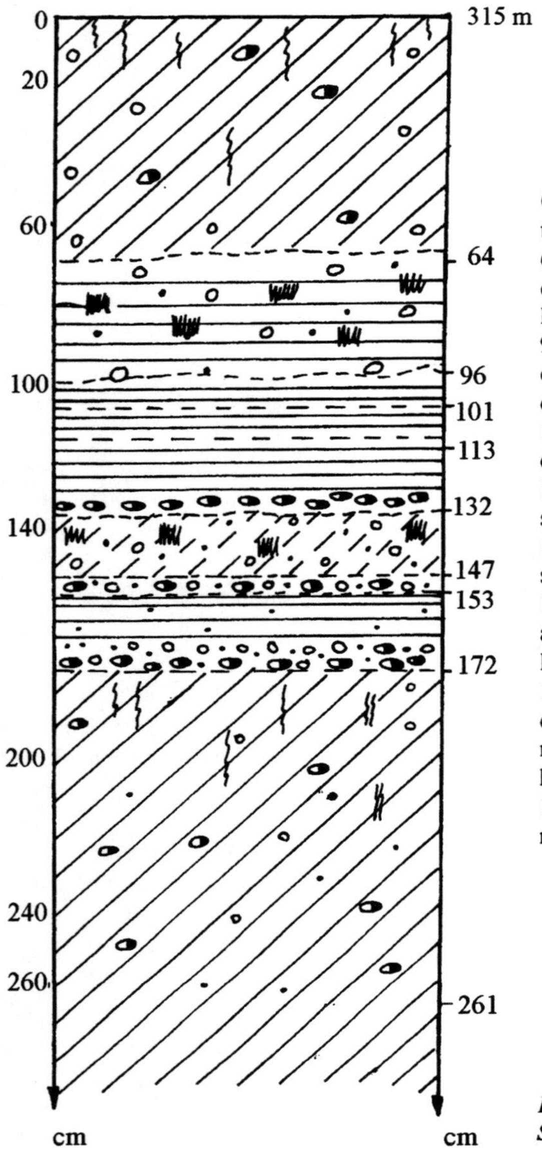

0-64 - soil, dark, clayey with a small amount of sand; organic matters
64-96 – clayey, with inclusions of rounded gravel, sand; charcoal, cultural layer named in the archaeological documentation as Early Bronze IIB.
96-101 – clayey, light yellow (10YR 6/4 after the table of Mounsel); cultural layer from Early Bronze II according to the archaeological documentation (see Nikolova, chapter 15)
101-113 – light yellow clay (10YR 6/4 after the table of Mounsel); cultural layer from earlier Early Bronze II;
113-132 – brick reddish from Early Bronze II ; sample for morphoscope analysis was taken;
132-147 – clayey and sandy; whitish; gravel inclusions; existence of small spots of gray sand; charcoal; a cultural layer; Early Bronze I/II;
147-153 – sandy and clayey or clayey and sandy; inclusions of gravel and sand; organic inclusions of root; charcoal; a cultural layer; Early Bronze I;
153-173 – clayey; floor plaster; at the same depths there is a hearth documented at the western exposition. Its floor was made of rounded rock. A sample was taken from there as archaeo-magnetic samples has been taken from the same structure;Early Bronze I;
173-261 – sandy and clayey; black; organic of root; inclusions of rounded up to the second or third degree rock and gravel.

Figure 28.1. The section of Sarovka. It is located on the left side of the Stryama River in Karlovo Lowland, and has east exposition.

mum sizes of the pieces of the sample are along axe ‘a’ 52 mm, along axe ‘b’ 52 mm, and along axe ’c’ 36 mm.

- Following features characterize the rock from the hearth floor at a depth of 173-216 cm is (***Figure 28.4***): 1. Relatively similar sizes along ‘a’, ‘b’ and ‘c’ axes; 2. All the material is rounded, as the coefficient of rounding (K) in 24% of the cases reaches the highest (4th) degree; 3. 64% of the rock have disc-like and flat and prolong shape, and 37% spheroid and cylindrical shape. The former were trailed, but the latter rolled.

The samples from four levels were investigated by the *mineralogical analysis* (***Figure 28.5***) – at 113-131 cm depth, at 132-147 cm depth, 153-172 cm depth, and 172-261 cm depth. The data indicate:

1. Forming of two levels, named conditionally upper (up to 153 cm) and lower (up to 261 cm). A background for such a division is the existence and the quantity of magnetite and hematite, and well we of amphibole and chlorite.

2. The increasing of the quartz at a depth of 132-172 cm we connect with slightly warming of the climate in relation to the lowest level (172-261 cm) and upper level (up to 132 cm).

3. The presence of the heavy minerals mentioned proves the existence of a feeding-up province consisted of granite-like and gneiss.

Dositeevo site

Description

The sanctuary in the region of Dositeevo Village dates from Late Bronze – Early Iron (the first layer) and fourth – third century BC (the second layer). Geomorphologically, it is situated on isolated single hill, a stronghold. It raises on the left valley slope of the Baluchkata River (left tributary of the Maritsa River) at a height of 115 and 133.40 m. The stronghold consists of several massifs of basic rocks and located between platforms of not big sizes. The Basic rocks in the region are two-mica, medium grain gneiss and granite and gneiss, and the platforms between are made of unsoldered superimpose structures of sand and clay.

Dubene

Level at a depth of 96-101 cm — **Weight – 91.1 gr.**

D mm	> 20.0	20.0 - 15.0	15.0 - 10.0	10.0 - 6.25	6.25 - 5.25	5.25 - 5.00	5.00 - 4.75	4.75 - 4.50	4.50 - 3.50	3.50 - 3.00	3.00 - 2.50	2.50 - 2.00	2.00 - 1.60	1.60 - 1.25	1.25 - 1.00	1.00 - 0.80	0.80 - 0.63	0.63 - 0.50	0.50 - 0.40	0.40 - 0.31	0.31 - 0.25	0.25 - 0.10	0.10 - 0.07	>0.07
Gr.					3.3							1.8			2.7	0.4			0.1		4.5	2.2	0.1	76
%					3.62							1.97			2.96	0.43			0.1		4.93	2.41	0.1	83.4
	Rock – -%			Gravel – 5.59%												Sand – 8.46 %							Clay – 83.53%	

Dubene

Level at a depth of 96-101 cm — **Weight – 91.1 gr.**

D mm	> 20.0	20.0 - 15.0	15.0 - 10.0	10.0 - 6.25	6.25 - 5.25	5.25 - 5.00	5.00 - 4.75	4.75 - 4.50	4.50 - 3.50	3.50 - 3.00	3.00 - 2.50	2.50 - 2.00	2.00 - 1.60	1.60 - 1.25	1.25 - 1.00	1.00 - 0.80	0.80 - 0.63	0.63 - 0.50	0.50 - 0.40	0.40 - 0.31	0.31 - 0.25	0.25 - 0.10	0.10 - 0.07	>0.07
	15.4	20.4	5.0	-	4.3	1.5	1.3	0.7	2.4	1.5	0.8	0.2	1.7	1.1	2.1	0.5	4.0	1.6	1.3	3.0	3.5	40.0	4.6	83.1
%	7.7	10.2	2.5	-	2.15	0.75	0.65	1.2	1.2	0.75	0.4	0.1	0.85	0.55	1.05	0.25	2.0	0.8	0.65	1.5	1.75	20.0	2.3	41.55
	Rock – 20.4%			Gravel – 48.8%												Sand – 26.95%							Clay – 43.85%	

Dubene

Level at a depth of 96-101 cm — **Weight – 91.1 gr.**

D mm	> 20.0	20.0 - 15.0	15.0 - 10.0	10.0 - 6.25	6.25 - 5.25	5.25 - 5.00	5.00 - 4.75	4.75 - 4.50	4.50 - 3.50	3.50 - 3.00	3.00 - 2.50	2.50 - 2.00	2.00 - 1.60	1.60 - 1.25	1.25 - 1.00	1.00 - 0.80	0.80 - 0.63	0.63 - 0.50	0.50 - 0.40	0.40 - 0.31	0.31 - 0.25	0.25 - 0.10	0.10 - 0.07	>0.07
Gr.	42.9	47.5		15.4	3.1	0.6	0.6	0.4	2.1	0.3	0.9	1.0	1.1	1.5	2.6	0.5	2.4	1.7	2.3	3.3	1.4	14.0	2.2	52.3
%	21.45	23.75		7.7	3.1	0.3	0.3	0.2	1.05	0.15	0.45	0.5	0.55	0.75	1.3	0.25	1.2	0.85	1.15	1.65	0.7	7.0	1.1	26.1
	Rock – 45.2%			Gravel – 14.8%												Sand – 12.8%							Clay – 27.2%	

Dubene

Level at a depth of 96-101 cm — **Weight – 91.1 gr.**

D mm	> 20.0	20.0 - 15.0	15.0 - 10.0	10.0 - 6.25	6.25 - 5.25	5.25 - 5.00	5.00 - 4.75	4.75 - 4.50	4.50 - 3.50	3.50 - 3.00	3.00 - 2.50	2.50 - 2.00	2.00 - 1.60	1.60 - 1.25	1.25 - 1.00	1.00 - 0.80	0.80 - 0.63	0.63 - 0.50	0.50 - 0.40	0.40 - 0.31	0.31 - 0.25	0.25 - 0.10	0.10 - 0.07	>0.07
Gr.	38.5			11.2	2.0	0.6	0.6	0.6	3.0	1.8	1.2	0.2	2.3	1.3	1.9	0.4	2.1	0.7	1.0	1.6	1.3	4.9	1.3	21.5
%	38.5			11.2	2.0	0.6	0.6	0.6	3.0	1.8	1.2	0.2	2.3	1.3	1.9	0.4	2.1	0.7	1.0	1.6	1.3	4.9	1.3	21.5
	Rock – 38.5%			Gravel – 26.7%												Sand – 12 %							Clay – 22.8%	

Dubene

Level at a depth of 96-101 cm — **Weight – 91.1 gr.**

D mm	> 20.0	20.0 - 15.0	15.0 - 10.0	10.0 - 6.25	6.25 - 5.25	5.25 - 5.00	5.00 - 4.75	4.75 - 4.50	4.50 - 3.50	3.50 - 3.00	3.00 - 2.50	2.50 - 2.00	2.00 - 1.60	1.60 - 1.25	1.25 - 1.00	1.00 - 0.80	0.80 - 0.63	0.63 - 0.50	0.50 - 0.40	0.40 - 0.31	0.31 - 0.25	0.25 - 0.10	0.10 - 0.07	>0.07
Gr.	-	9.4	7.6	8.8	0.8	0.2	-	0.6	2.8	1.2	0.5	0.2	2.8	1.6	4.5	1.0	4.7	3.3	3.4	7.4	4.1	34.1	8.3	92.7
%	-	4.7	3.8	4.4	0.4	0.1	-	0.3	1.4	0.6	0.25	0.1	1.4	0.8	2.25	0.5	2.35	1.65	1.7	3.7	2.05	17.05	4.15	46.35
	Rock - -%			Gravel – 12%												Sand – 29%							Clay – 50.5%	

Figure 28.2. Granule-metric contents of the layers.

No.	Sizes (mm)			Petrographic sort	Coefficient of rounding	Shape		
172-261	a	b	c			c/b	b/a	Name
1	14	10	9	Granite	II	0.9	0.71	Spherical
2	6	6	2	Amphibole gneiss	III	0.33	1	Discoid
3	12	6	1	Amphibole gneiss	II	0.16	0.5	Flatted and prolonged
4	10	6	3	Granite	I	0.5	0.6	Flatted and prolonged
5	7	6	2	Granite	II	0.33	0.85	Discoid
6	8	7	3	Granite	II	0.42	0.87	Discoid
7	10	8	5	Granite	I	0.62	0.8	Discoid
8	11	8	6	Granite	II	0.75	0.72	Spherical
9	10	9	6	Granite	I	0.66	0.9	Spherical
10	9	8	6	Amphibole gneiss	III	0.75	0.88	Spherical
11	12	8	6	Granite	II	0.75	0.66	Spherical
12	7	6	2	Amphibole gneiss	I	0.33	0.85	Discoid
13	9	5	1	Amphibole gneiss	I	0.2	0.55	Flatted and prolonged
14	8	6	1	Amphibole gneiss	V (III-I)	0.16	0.75	Discoid
15	7	6	5	Granite	II	0.83	0.85	Spherical
16	9	8	6	Granite	III	0.75	0.88	Spherical
17	9	7	2	Amphibole gneiss	V (IV-I)	0.28	0.77	Discoid
18	8	5	2	Amphibole gneiss	IV	0.4	0.62	Flatted and prolonged
19	8	6	2	Granite	II	0.33	0.75	Discoid
20	8	5	2	Amphibole gneiss	V (IV-I)	0.4	0.62	Flatted and prolonged
21	6	6	4	Amphibole gneiss	II	0.67	1	Spherical
22	10	8	6	Granite	III	0.75	0.8	Spherical
23	10	7	7	Granite	III	1	0.7	Spherical
24	10	7	5	Granite	II	0.71	0.7	Spherical
25	7	6	5	Granite	III	0.83	0.85	Spherical
26	7	5	3	Granite	V (IV-I)	0.6	0.71	Spherical
27	7	4	3	Granite	III	0.75	0.57	Cylindrical
28	6	5	2	Amphibole gneiss	III	0.4	0.83	Discoid
29	7	4	3	Granite	II	0.75	0.57	Cylindrical
30	6	4	3	Granite	II	0.75	0.66	Spherical
31	7	5	5	Granite	II	1	0.71	Spherical
32	6	5	4	Granite	IV	0.8	0.83	Spherical
33	5	5	4	Granite	IV	0.8	1	Spherical
34	8	3	3	Granite	III	1	0.37	Cylindrical
35	8	5	4	Granite	III	0.8	0.62	Cylindrical
36	8	4	3	Granite	III	0.75	0.5	Cylindrical
37	6	5	2	Amphibole gneiss	II	0.4	0.83	Discoid
38	7	5	3	Granite	III	0.6	0.71	Discoid
39	7	4	3	Amphibole gneiss	II	0.75	0.57	Cylindrical
40	6	5	2	Amphibole gneiss	I	0.4	0.83	Discoid

Figure 28.3.1. Morphoscope analysis of overlayering at the different levels.

No. 132-147	Sizes (mm)			Petrographic sort	Coefficient of rounding	Shape		
	a	b	c			c/b	b/a	Name
1	41	36	15	Amphibole gneiss	IV	0.41	0.87	Discoid
2	31	21	6	Amphibole gneiss	III	0.28	0.67	Discoid
3	25	20	7	Amphibole gneiss	IV	0.35	0.54	Flatted and prolonged
4	37	25	6	Amphibole gneiss	IV	0.24	0.67	Discoid
5	27	19	7	Amphibole gneiss	III	0.36	0.70	Discoid
6	20	14	11	Amphibole gneiss	II	0.78	0.70	Spherical
7	27	19	9	Pegmatite	I	0.47	0.70	Discoid
8	15	12	6	Pegmatite	IV	0.5	0.8	Discoid
9	19	14	5	Amphibole gneiss	IV	0.35	0.73	Discoid
10	21	14	4	Amphibole gneiss	II	0.28	0.66	Discoid
11	19	11	7	Amphibole gneiss	IV	0.63	0.57	Flatted and prolonged
12	17	11	4	Amphibole gneiss	II	0.36	0.63	Flatted and prolonged
13	23	18	10	Pegmatite	II	0.55	0.78	Discoid
14	19	11	7	Amphibole gneiss	II	0.63	0.57	Flatted and prolonged
15	21	16	4	Amphibole gneiss	III	0.25	0.76	Discoid
16	17	12	5	Amphibole gneiss	III	0.41	0.70	Discoid
17	14	6	5	Amphibole gneiss	IV	0.83	0.42	Cylindrical
18	16	12	5	Amphibole gneiss	IV	0.41	0.75	Discoid
19	13	9	4	Amphibole gneiss	I	0.44	0.69	Discoid
20	20	11	7	Amphibole gneiss	IV	0.63	0.55	Flatted and prolonged
21	21	20	4	Amphibole gneiss	IV	0.4	0.47	Flatted and prolonged
22	17	10	6	Amphibole gneiss	IV	0.6	0.58	Flatted and prolonged
23	16	12	5	Amphibole gneiss	III	0.41	0.75	Discoid
24	17	11	6	Amphibole gneiss	IV	0.54	0.64	Flatted and prolonged
25	16	10	5	Amphibole gneiss	II	0.5	0.62	Flatted and prolonged
26	13	8	3	Amphibole gneiss	IV	0.37	0.61	Flatted and prolonged
27	15	15	5	Pegmatite	IV	0.33	1	Discoid
28	14	7	5	Amphibole gneiss	IV	0.71	0.5	Cylindrical
29	17	5	3	Amphibole gneiss	IV	0.6	0.29	Flatted and prolonged
30	12	9	4	Pegmatite	IV	0.44	0.75	Flatted and prolonged
31	12	9	6	Pegmatite	IV	0.66	0.75	Spherical
32	14	12	3	Amphibole gneiss	IV	0.25	0.85	Discoid
33	11	9	4	Amphibole gneiss	IV	0.44	0.81	Discoid
34	12	7	4	Pegmatite	IV	0.57	0.58	Flatted and prolonged
35	13	8	6	Amphibole gneiss	IV	0.75	0.61	Cylindrical
36	13	8	3	Amphibole gneiss	IV	0.37	0.61	Flatted and prolonged
37	11	8	3	Amphibole gneiss	IV	0.37	0.72	Discoid
38	12	10	3	Amphibole gneiss	IV	0.3	0.83	Discoid
39	14	6	3	Amphibole gneiss	IV	0.5	0.42	Flatted and prolonged
40	11	6	3	Amphibole gneiss	IV	0.5	0.54	Flatted and prolonged
41	12	6	2	Amphibole gneiss	IV	0.33	0.5	Flatted and prolonged
42	10	6	3	Amphibole gneiss	IV	0.5	0.6	Flatted and prolonged
43	12	8	2	Amphibole gneiss	IV	0.33	0.66	Discoid
44	9	7	3	Amphibole gneiss	IV	0.42	0.77	Discoid
45	9	7	4	Amphibole gneiss	IV	0.57	0.77	Discoid
46	13	7	3	Amphibole gneiss	IV	0.42	0.53	Flatted and prolonged

Figure 28.3.2. Morphoscope analysis of overlayering at the different levels.

No. 172-261	Sizes (mm)			Petrographic sort	Coefficient of rounding	Shape
	a	b	c			
1*	73	50	28	Applet	V (III-II)	Discoid
2	60	34	15	Pegmatite	II	Flatted and prolonged
3*	59	33	27	Pegmatite	V (IV-III)	Cylindrical
4	62	39	33	Pegmatite	IV	Cylindrical
5	44	25	22	Pegmatite	III	Cylindrical
6	53	36	12	Pegmatite	IV	Discoid
7*	58	57	35	Pegmatite	V (IV-III)	Discoid
8	58	52	20	Granite	IV	Discoid
9	49	40	13	Granite-gneiss	III	Discoid
10	32	20	15	Pegmatite	III	Cylindrical
11	51	30	19	Granite-gneiss	II	Flatted and prolonged
12	51	41	22	Pegmatite	III	Discoid
13	45	23	17	Applet	III	Cylindrical
14*	53	29	27	Applet	V (III-II)	Cylindrical
15	37	34	12	Pegmatite	III	Discoid
16	41	23	15	Pegmatite	III	Flatted and prolonged
17	27	19	15	Pegmatite	III	Spherical
18	37	35	14	Pegmatite	II	Discoid
19	45	37	19	Pegmatite	II	Discoid
20	40	39	11	Pegmatite	I	Discoid
21	27	24	17	Pegmatite	III	Discoid
22	29	19	14	Granite-gneiss	III	Spherical
23*	31	23	17	Pegmatite	V (IV-I)	Spherical
24*	34	23	13	Pegmatite	V (IV-I)	Discoid
25*	27	21	14	Granite	IV (III-II)	Spherical
26	51	22	17	Granite	IV	Cylindrical
27	36	20	9	Applet	IV	Flatted and prolonged
28*	23	21	12	Pegmatite	V (IV-II)	Discoid
29	24	22	12	Pegmatite	IV	Discoid
30	23	19	14	Pegmatite	IV	Spherical
31	29	24	14	Pegmatite	IV	Discoid
32	25	23	8	Applet	IV	Discoid
33	30	28	9	Applet	IV	Discoid
34*	29	21	19	Pegmatite	V (IV-I)	Spherical
35	34	21	11	Pegmatite	III	Flatted and prolonged
36	28	19	10	Pegmatite	III	Discoid
37	30	22	10	Gneiss	I	Flatted and prolonged
38	45	22	12	Gneiss	IV	Flatted and prolonged
39	30	23	14	Applet	II	Discoid
40	25	21	19	Granite	V (IV-II)	Spherical
41	39	22	9	Gneiss	III	Flatted and prolonged
42	20	18	6	Gneiss	V (IV-I)	Discoid

Figure 28.4. Morphoscope analysis of the rock from a hearth floor. * Rounded pieces with traces of additional destruction as a result of hearth heating, but not of mud-flow process.

Depth (cm)	Heavy fraction %	Magnetite %	Hematite %	Amphibole %	Chlorite %	Biotite %	Quartz %	Muscovite %	Calcite %	Feldspar %
113-132	4.7	0.7	0.5	2.8	0.2	0.5	89.90	0.4	3.8	1.2
132-147	7.2	0.3	1.3	3.0	0.7	0.7	91.40	0.3	4.0	3.7
153-172	4.8	-	0.7	3.2	0.7	0.2	91.40	-	1.5	2.3
172-261	6.2	-	-	6.0	-	0.2	91.10	-	-	2.7

Figure 28.5. Mineral spectrum of overlayering at the different levels.

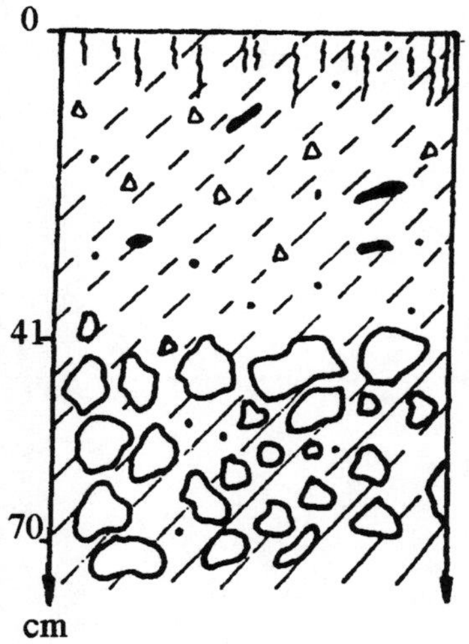

133 m above see level

0-41 cm, a dark and sandy layer, with inclusions of gravel, rock and organic matter; low level of rounding; numerous sherds spread, from Late Bronze – Early Iron Age;
41-70 cm, a dark, clayey, compact layer with a lot of gravel and rock, rounded; organic matters are missing; in the lower level the natural gneiss rock occurred.

Figure 28.6. Dositeevo sanctuary. A section.

The Baluchkata River cut in this region at a depth of c. 30 m and there are no river terraces visible on the slope. Similar strongholds and cutting exist along the neighbor rivers as well. To the west is located the Biyukdere River, and to the east are the Salarska and Yerusalismska Rivers. There are cut in sub-hill surface. According to G. Krumov (1984), it is of Pliocene-Quaternary age and also shape similar strongholds. The sub-hill surface has a character of "Hogback" as a steeply raising slope of a river valley, which reaches up to rounded watershed ridge covered by numerous similar strongholds built of the basic rock.

It was not accidentally chosen that stronghold to have been a sanctuary place. There is a good view over the located settlements in the neighborhood, some of which are along the sides of the contemporary reservoir and other are located towards the mouth of the Golyamata Reka and Biyukdere Rivers.

Material and method

In the region of the stronghold there are descriptions of two points and one is analyzed as well. Basic of the geomorphologic investigation is that which is located in the western part of the stronghold, in the area of the so-called Cave locality. The point situated in the eastern part of the flattened platform is secondary. The reason is the natural structure of the western point and the anthropogenetic activity, carrying through a new structure of the layering in the eastern part of the sanctuary.

Granule-metric, morphoscope and mineralogical analyses, as well as an analysis of the quartz have been performed. For definition of the sizes and the rounding of the pieces tables are used elaborated by G. Baltakov (1988) and L. Serebryannyi (1980). The results of the granule-metric and mineralogical analyses are given numerically and in charts. The following coefficients are also accounted: medium diameter, medium proportion and a coefficient of sorting. The rounding of the pieces is also defined.

A special attention are to be paid on the levels consisted of ceramics. As it consists mostly of clay and quartz sand, they were included formally in the group of the medium stable to denudation and erosion rock. From this position, their rounding give important information of if the levels are formed 'in situ' or they were re-layered.

Results

The main section (***Figure 28.6***) is located in the lower western part between two blocks over a slope under 15-30?. The blocks are independent, but it is visible, they are cut from the located above rock massif. That process of cutting was along a quartz vein c. 20 cm wide. The layers in the section build two clearly distinguishable layers: upper from 0 to 0.41 and lower from 0.41 to 0.70 m thick. Underneath the main rock is visible (***Figure 28.7***).

In the upper layer there are sherds with slightly rounded edges, as well as organic inclusions – bark, leaves and roots. Clay and sand predominate, total of 57.90%. Slightly rounding characterize some of the bigger pieces – gravel and broken stone. The sorting was defined as "very bad" (***Figures 28.8.1 and 28.9.1***).

At the lower level the layering has another structure. The big fractions predominate – broken stone and gravel, a total of 68.26%. All the pieces are compactly each to other; the sorting is 'bad' and there is no rounding, as well as there is no pottery and organic matters (***Figures 27.8.2 and 27.9.2***).

The secondary section is situated close to the channel dug for the flowing water coming from the reservoir located in c. 1 km up the coarse of the river. The field diary reveals in the granule-metric contents of the layering in the whole depth (c. 70 cm) is homogenous and is characterized of predominating sand and clay. The big pieces have a low coefficient of rounding.

Discussion and conclusions

As far as Karlovo hollow concerns, the investigations up to now (A. Lukanov) show the region located c. 2 km west of the archaeological site of Sarovka has been a local area of elevation, then, an erosion rapid of the Stryama River. That is the region between the inner hollow hill of Balyovitsa and such located to the south of it, with a height of 372 m. East of that rapid there are moments of cutting documented, which have been changed with such of raising. Along whole left side of the Stryama River there is a alluvial cover formed (up to 80 cm).

The data obtained allow us to think the layering in the level between 172 and 261 cm characterize peripheral areas of the river channel, which has no high order. The samples round, their sizes and shape verify the water channel character. Most of that rock (57.5%) was rolled, and the rest part (42.5%) trailed. That fact, as well as the sizes of the rock (not more that 55 mm along axe 'a' prove the river has no high order.

However, over that former (or abandoned) bed it was possible a soil layer to be formed, as well as to be cultivated. Evidence of such process is the humus root. The documentation of a settlement over that ground means the area was relatively stable surface and was not any more (or was at least less) flooded by

O=41 cm

Parameters	Broken stone	Gravel				Sand				Clay		
L	100-10	10-9	9-5	5-2	2-1	1-0.8	0.8-0.4	0.4-0.25	0.25-0.1	0.1-0.08	0.08-0.063	<0.063
9	121.3	-	33.7	27	28.5	3.1	64.9	40.3	62.4	18.7	5.6	94.5
%	24.26	-	6.74	5.4	5.7	0.62	12.98	8.06	12.48	3.74	1.12	12.9
Total %	24.26	17.84				34.14				23.76		

41-70 cm

	Broken stone	Gravel				Sand				Clay		
L	100-10	10-9	9-5	5-2	2-1	1-0.8	0.8-0.4	0.4-0.25	0.25-0.1.	0.1-0.08	0.08-0.063	<0.063
9	116.7	-	77.3	99.1	47.7	2.2	20	12.5	12.8	17.7	2.2	91.8
%	23.34	-	15.46	19.82	9.54	0.44	4	2.5	2.56	3.54	0.44	18.36
Total	23.34	44.82				9.5				22.34		

Figure 28.7. Granule contents of overlayering.

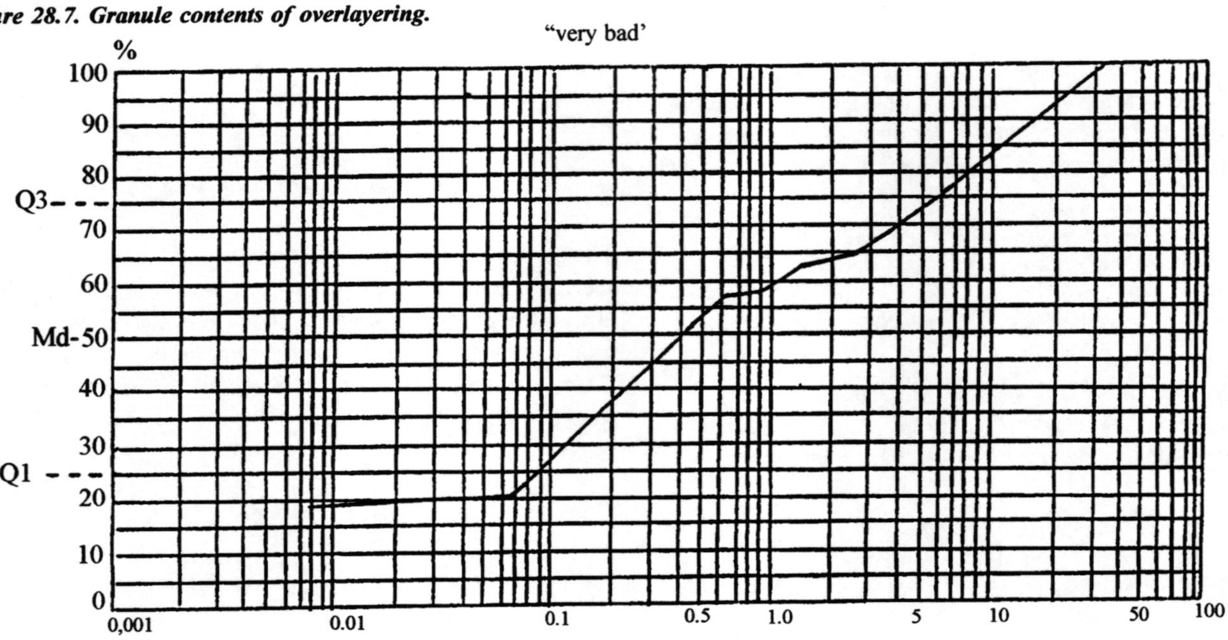

Figure 28.8.1. Cumulative curve. Md=0.4 mm; Mx=5.5 mm; So=2.51 mm. 0-41 cm.

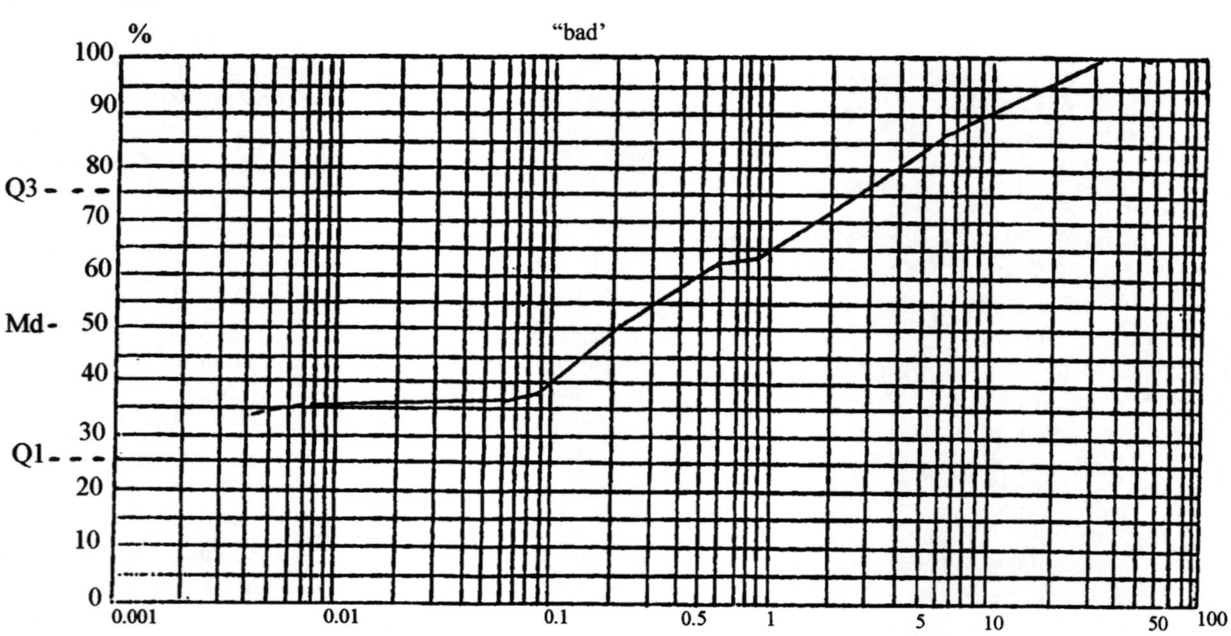

Figure 28.8.2. Cumulative curve. Md=0.2 mm; Mx=2.65 mm; So=1.40 mm. 41-70 cm.

No.	a	b	c	Sort	C_o	Average C_2 of the sort	Shape
1	23	18	14	Quartz	0	0	Spherical
2	10	9	8	Quartz	0	0	Spherical
3	32	23	14	Granite-gneiss	0	0	Discoid
4	24	19	8	Granite-gneiss	0	0	Discoid
5	25	24	10	Granite-gneiss	0	0	Discoid
6	20	19	11	Granite-gneiss	0	0	Discoid
7	26	22	10	Granite-gneiss	0	0	Discoid
8	10	8	5	Granite-gneiss	0	0	Discoid
9	15	12	4	Granite-gneiss	0	0	Discoid
10	13	10	3	Granite-gneiss	0	0	Discoid
11	11	11	5	Granite-gneiss	0	0	Discoid
12	16	11	5	Granite-gneiss	0	0	Discoid
13	12	10	6	Granite-gneiss	0	0	Discoid
14	16	16	4	Granite-gneiss	0	0	Discoid
15	13	13	3	Granite-gneiss	0	0	Discoid
16	15	11	6	Granite-gneiss	0	0	Discoid
17	22	10	4	Granite-gneiss	0	0	Discoid
18	21	14	9	Granite-gneiss	0	0	Discoid
19	20	16	9	Granite-gneiss	0	0	Discoid
20	18	12	1	Granite-gneiss	0	0	Discoid
21	14	11	5	Granite-gneiss	0	0	Discoid
22	15	11	4	Granite-gneiss	0	0	Discoid
23	12	9	4	Granite-gneiss	0	0	Discoid
24	14	11	5	Granite-gneiss	0	0	Discoid
25	11	10	5	Granite-gneiss	0	0	Discoid
26	11	9	4	Granite-gneiss	0	0	Discoid
27	11	8	4	Granite-gneiss	0	0	Discoid
28	12	8	3	Granite-gneiss	0	0	Discoid
29	12	10	9	Granite-gneiss	0	0	Spheroid
30	10	8	5	Granite-gneiss	0	0	Spheroid
31	14	10	8	Granite-gneiss	0	0	Spheroid
32	29	16	4	Granite-gneiss	0	0	Flatted and prolonged
33	17	10	6	Granite-gneiss	0	0	Flatted and prolonged
34	16	10	4	Granite-gneiss	0	0	Flatted and prolonged
35	42	20	19	Granite-gneiss	0	0	Cylindrical
36	18	11	9	Granite-gneiss	0	0	Cylindrical
37	20	8	8	Granite-gneiss	0	0	Cylindrical
38	12	7	6	Granite-gneiss	0	0	Cylindrical
39	15	9	6	Granite-gneiss	0	0	Cylindrical
40	11	7	7	Granite-gneiss	0	0	Cylindrical
41	54	36	30	Granite-gneiss	0	0	Spherical

Figure 28.9.1. Morphoscope analysis of the pieces over 9 mm.0-41 cm.

No.	a	b	c	Sort	C_o	Average C_2 of the sort	Shape
1	30	17	15	Granite-gneiss	0	0	Cylindrical
2	15	17	10	Granite-gneiss	0	0	Spherical
3	25	11	6	Granite-gneiss	0	0	Flatted and prolonged
4	26	15	6	Shale	0	0	Flatted and prolonged
5	21	16	10	Granite-gneiss	0	0	Cylindrical
6	18	10	10	Granite-gneiss	0	0	Spherical
7	20	13	12	Granite-gneiss	0	0	Spherical
8	16	18	7	Granite-gneiss	0	0	Discoid
9	15	12	5	Granite-gneiss	0	0	Discoid
10	10	15	8	Granite-gneiss	0	0	Spherical
11	14	10	5	Granite-gneiss	0	0	Discoid
12	10	13	4	Granite-gneiss	0	0	Discoid
13	13	10	6	Granite-gneiss	0	0	Cylindrical
14	16	8	2	Granite-gneiss	0	0	Discoid
15	10	11	5	Granite-gneiss	0	0	Discoid
16	10	8	6	Granite-gneiss	0	0	Spherical
17	12	9	6	Granite-gneiss	0	0	Discoid
18	14	10	6	Granite-gneiss	0	0	Discoid
19	12	11	4	Granite-gneiss	0	0	Discoid
20	17	8	7	Granite-gneiss	0	0	Cylindrical
21	13	8	6	Granite-gneiss	0	0	Cylindrical
22	9	6	6	Granite-gneiss	0	0	Spherical
23	15	7	4	Granite-gneiss	0	0	Discoid
24	9	10	4	Granite-gneiss	0	0	Spherical
25	10	8	6	Granite-gneiss	0	0	Spherical
26	10	7	5	Granite-gneiss	0	0	Discoid
27	8	7	7	Granite-gneiss	0	0	Spherical
28	9	7	4	Granite-gneiss	0	0	Discoid
29	24	24	9	Feldspar	0	0	Discoid

Figure 28.9.2. Morphoscope analysis of the pieces over 9 mm.41-70 cm.

river water. Accordingly, after forming of level 173-261 cm the Stryama River departed to the south and the territory of the settlement have been raised at some level above the river channel.

The existence of river material (173-261 cm) and 'deviation' of the river to the south have been fixed in the mineral specter of the conditionally defined 'upper' and 'lower' levels. Several peculiarities prove it as:

1. Existence of magnetite and hematite in the upper level at one and the same time, which is a sign of a process of chemical evaporation, and then, of existence of an exposed but not permanently flooded of river water surface. In the lower level there are missing such minerals, probably brought way from the river water.

2. There is relatively smaller quantity of heavy fractions in the 'upper' level and bigger one in the lower level. The quantity of heavy minerals characterize water channel's processes, and the smaller one (when the other preconditions are equal) evaporation processes.

3. The existence of calcite in the 'upper' level can be e plained by agrarian and technical activity (dunging; irrigation, etc.).

The analyzed material from the hearth floor at a depth of 153-173 cm is channel's, but not water's, then, the channel of the Stryama River was closer to the ancient settlements as recently. It is possible following short term flooding and accumulation of the material, which had no pouring or intensive character.

The moving of the Stryama River's channel from north to the south have been expained (D. Kanev, A. Lukanov) by the big alluvial quantity, which the Stara Planina Mountains tributaries drift. They 'prop up' and push out the water channel. In my opinion, that process increased in the first half of the 19th century BC when the slopes of the Stara Planina Mountains deforested as a result of trees' felling and fires. Intensive erosion processes developed over those slopes. But in the region of the lowland the process of accumulation predominated. The channel built, which passes between the inner hollow hills of Balyovitsa (375.6 m) and Konnika (394.2 m) in fact is a old channel of the Stryama River which was strengthened and made deeper. In the course of the time from the forming of the tell till nowadays the moving of the water channel from the north to the south was not more that 1 km and most probably in c. 600-700 m distance.

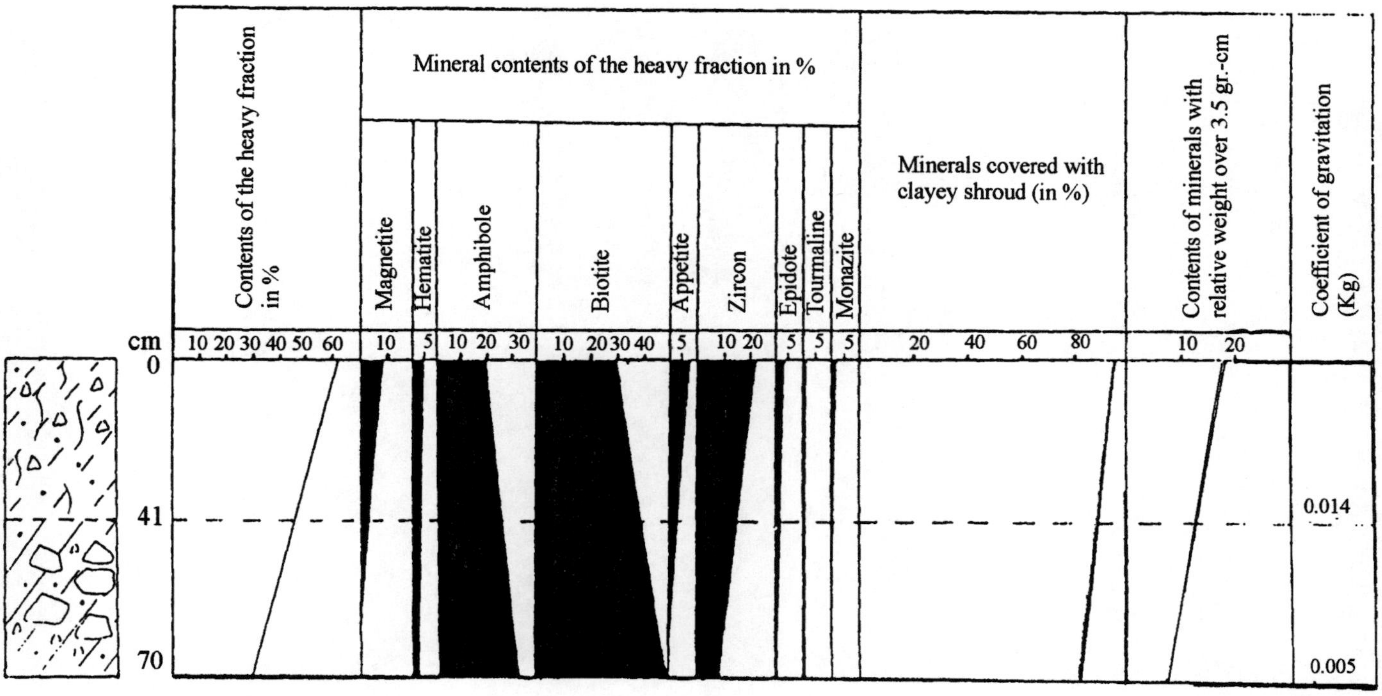

Figure 28.10. Chart of the mineral contents of the layering.

The data above, as well as the analysis of the quartz infers the following conclusions concerning Dositeevo site.

In the western part of the stronghold there is a cone shaped. Its upper part starts between both blocks, then it is fan-like widened. The cone consists of two layers. A stream forms the upper one started from the higher parts of the stronghold. That stream is short, not permanent and of torrential rain character. The torrential rain courses a movement of big pieces of broken stone and sherds. The way of the stream passes between both blocks where, on the highest point, the bigger pieces have re-layered. With decreasing of the sloping of the terrain the smaller pieces re-layered. A proof of that process is the relatively vast participation of clay and sand, sorting, the contents of organic and ceramics, high contents of amphibole and heavy fractions which are relatively unstable minerals etc (***Figure 27.10***).

Ceramics, which are distributed in the upper layer are probably brought from the upper mount part of the sanctuary. The low coefficient of rounding is a proof of that suggestion.

The lower layer contents of broken stone and blocks have another origin. The absence of rounding of the pieces and their compact position, 'the bad' sorting and rounding, the high quantity of unstable minerals, the absence of pottery and organic etc. indicate it is a landslide cone. There is not high probability the structure to have been of scree. In the scraps between the bigger pieces there was layering of sand and clay, a result of the erosion activity of the small stream. Traces of chemical evaporating are visible in the whole cone, over minerals and rocks: the presence of magnetite and hematite at one and the same time, of vermiculite, uneven surface of the mineral apatite, etc.

The sub-stronghold platform on the eastern edge is built of its destroyed material. The way passed is short. Then, the material was layered over the platform and filled the uneven areas continuing the fan process 'in situ'. But it is possible that surface to have been anthropogenetic (for instance, its is not possible to be explained the high coefficient of the sorting) as there are in 20-30 m distance only, a channel coming from the reservoir.

Most probable the structure used for the sanctuary, to have 'endured' catastrophic natural accident in historic time – earthquake. That this the reason of cutting of two big blocks and shaping of the Cave, as well as of shaping of the lower layer in section No. 1. Then, a little change in the climate took place, which was visible by the quantity of the quartz in section No. 1 (***Figure 28.11***). Of them, nowadays, most active are the linear erosion and delluvial washing.

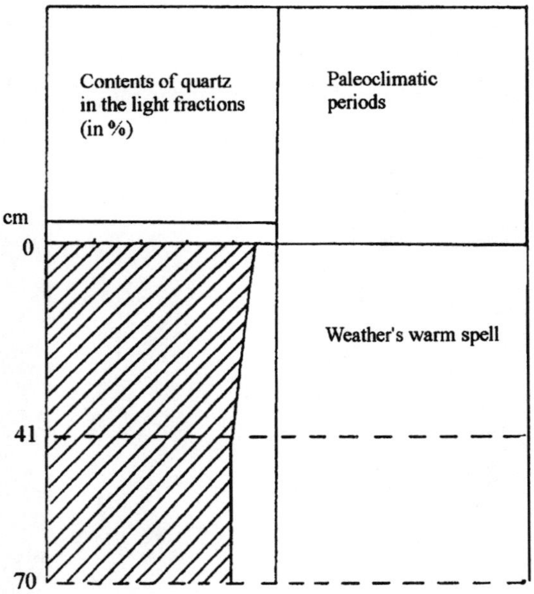

Figure 28.11. Contents of quartz of the light fraction.

A Tender Subject

In the theoretical literature the term *innovation* mainly concerns the distribution of the invention. In my understanding the innovation includes the invention and the distribution of the inventions. In this case we are faced with one of the more difficult problems in the investigating of the Balkan Prehistory – the problem of the origin of many similar components of the prehistoric cultures, as well as the way of adoption and transformation of the innovations. There are three main theories explaining the similarity – culture diffusion (or interaction), migration of population and independent appearance. The interaction as an explanation for distribution of innovation arouses similar problems as for stylistic similarity, as well as the possibility of changes in the process of the adaptation infers a complex of theoretical applications (see for instance Hagestrand 1967; Trinklaus 1987: 8-9 with ref.; Bailey 1998; Stefanovich 1999: 32 sq.).

The individual contributions in this collective study in different way face the problems of the technology, styles and society in prehistory from the point of view of the nature and distribution of the innovations between the Alps and the Black Sea. The territory is traditionally interconnected with regard to the prehistoric cultures, as the Danube River as an active communicative line (Childe 1929). But each research generation finds out new problems and new reasons for definition of such cultural interrelation. Anyway, the main focus in the book is the Balkans.

The flint industry (Gatsov, chapter 1 and Tsonev, chapter 2) argues an ideas of the distribution of knapping technology in the Early Neolithic in the Balkans. According to I. Gatsov, there is a new technological impulse in the period of distribution of the white painted pottery in the Balkans, as well as there is a difference between the monochromic and Mesolithic (Epi-Paleolihic) technological standards. The former suggestions generally correspond to the data on the pottery and the Balkans Anatolian interrelations of the ceramic styles. However, they do not answer the question of the ways of the distribution of the innovations as the migration of population can coexist with trade and other kinds of cultural contacts between distant and closer cultures including the distribution of the innovations.

The forming of the settlement structures in the Balkans in the Early Neolithic is characterized by complex of features, which continue to favor the settlement continuity in the following periods on one and the same or closer localities. N. Tasić (chapter 3) argues the places for habitation were interrelated with the salt resources and the salt distribution was one of the most important innovations in the earlier Balkan prehistory. It remains a question of a future analysis more precise definition of diary of the ancient population, in which without doubt the salt took important place. This is in fact the question of N. Russel (chapter 4) and her analysis is based on spatial dimension of the observed osteological data from Opovo. On his side, H. Greenfield (chapter 7) gives a sort of negative answer to the question if the innovation of the metallurgy influenced the style of life of the population with regard of the food preparing. At the same time, A. Chokhadzhiev represents a hypothesis of a function of ceramic vessel from Copper Age Dyakovo in the Struma valley (chapter 9). Based on typological parallels, that author concludes the vessel was used as a spinning bowl. The functionally of those kinds of vessels reasonably poses the problem, if that innovation was known in the Copper Age Balkans, why it was not wide distributed by the ceramic productions. Were there other possible vessels, for instance of wood that may also have functioned as spinning bowls? M. Hristov considers newly investigated copper finds from the Lower Dnube (chapter 10). For Later Copper Age, A. Raduncheva (chapter 13) argues a complex society based on the patriarchy, which reaches the threshold of the state but as a result of complex of reasons that process was stopped in the Balkans. As far as Northern Italy concerns, E. Kokelj (chapter 14) questions the terminology and the interpretation of the Later Prehistory in the Northern Alps.

There is a big body of newly published data in section 1, which in various way face the problems of the innovations in Neolithic and Copper Age. Prehistoric sequence in the upper Maritsa valley is reported by V. Nikolov (chapter 5) in the light of renewed investigations at Kapitan Dimitrievo tell, which give completely new cultural characteristics of Western Upper Thrace, especially for later Neolithic. Another micro-region in Western Upper Thrace, which is newly investigated, is the Upper Stryama valley (Nikolova, chapter 6). A special attention was paid in that chapter on the expressive data for the contact with Western Bulgarian land indicating a communication line that connected Upper Thrace with Western Balkans through the uppermost drainage of the Stryama River, through Zlatitsa-Pirdop to Sofia Field. Therefore, along with the Topolnitsa River, the Stryama valley was an active contact zone in prehistory argued as a frontier cultural zone.

S. Chokhadzhiev (chapter 8) argues a dynamic development of the Copper Age graphite ornamentation in Western Bulgarian land, with inner rhythm of evolution of the graphite motifs. Reasonably, a special attention in this section is paid on the Final Copper Age. L. Pernicheva (chapter 12) provides the volume with full publication of the Final Copper Kolarovo site from Southwestern Bulgaria. The last is one of the most important records of the analysis of the transformation in the ceramic style in the Balkans in the earlier Fourth Millennium BC. Additionally, for the first time a typology of latest Karanovo VI pottery with many mixed Krivodol-Sălcuța-Bubanj characteristics is elaborated (Todorova and Matsanova, chapter 26). The interaction theory can be applied for the Final Copper Age in the Upper Maritsa valley. L. Leshtakov (chapter 16) included in the interaction zone also the Northern Aegean as there is a similarity between the newly published pottery and some features of the later Sitagroi III ceramic style. The only problem is that there are not investigated the earlier levels from Yunatsite, as well as later Sitagroi III ceramic style is very poorly documented. So, the parallels mentioned confirm the argued with analogies from Western Bulgaria and by ^{14}C dating existence of Final Copper phase at Sitagroi (Nikolova 1999). But it is possible some of the typological parallels, which exist between Yunatsite and Sitagroi to reflect a diachronous interrelation as well.

The second section represents investigations on the Bronze Age and Early Iron Age in the Balkans. Dubene-Sarovka gives a lot of new evidence on the Early Bronze ceramic style, which are interpreted (Nikolova, chapter 15) as an evidence of the cultural integration in the Balkans. The thesis argued in that chapter is the distribution of the innovations in the ceramic style was based on the intensive cultural contacts between close and distant cultures. Of special importance for the population in the Upper Stryama valley was the cultural integration in the Central Balkans marked by similar development in the ceramic style – from channel to encrusted and predominated plain pottery styles. In Early Bronze I the ceramic style was close, however, with Baden complex, so the intensive cultural integration concerned a vast territory between the Black Sea and the Alps (Bodensee). Despite some chronological problems, which are to be discussed elsewhere, the dynamic cultural interaction characterized the mentioned territory in the later fourth millennium BC, and the Central Balkans were an integrative part of those processes. In Early Bronze II the ceramic style from Dubene-Sarovka clearly indicates intensive interactions with the neighbor cultures. The similarity of the Yunatsite II ceramic style is very strong again with the Northern Aegean (Sitagroi Va – Dikili Tash IIIB), as well as with Pernik group from middle and upper Struma valley. That is a period of very peculiar manifestation of the Balkan cultural integration. Western Upper Thrace ceramic style follows the distribution of the fashion of the encrusted pottery, also known from Kostolac and Vučedol cultures, as well as from Coțofeni style, etc. At the same time, there was an increasing of the regionalism in the development of the ceramic styles represented by many local peculiarities. The upper Stryama valley functions again as a frontier between the Eastern Balkan corded ware ceramic style (Ezero A2-B1, Ezerovo, Cernavoda II) and the encrusted stamped and pricked pottery known from the cultures to the west. In such way at Dubene-Sarovka IIB developed a style with wide distribution of the false-corded encrustation style in combination with incised, stamped, pricked and other techniques of encrustation.

K. Leshtakov again for the first time represents a complete publication of the preserved pottery from Ognyanovo, from the excavations of P. Detev and V. Matsanova (chapter 16). By many examples, that author confirms there was Early Bronze I phase at Ognyanovo, as well as Early Bronze II. But it looks that the tell was however, mostly occupied in the Early Bronze III. To Early Bronze II and III belongs the newly excavated levels on the slope of Nebet Tepe Hill in Plovdiv, represented in the volume by B. Koleva (chapter 17). All three last publications mentioned on the pottery from Western Thrace demonstrate a dynamic development of the ceramics style of the Yunatsite culture in the Early Bronze.

The published by M. Hristov gold and electron adornments originate from a region of intensive occupation of the Pit Grave Culture in Early Bronze Age, the Upper Osum basin (chapter 18). But that author suggests that the discovering of the earrings in a cave maybe indicate it is wrong to connect those kind of adornments only with that culture in the Lower Danube. Also, from Bronze Age is dated the innovation of the electron technology of the goldsmith, which may connect the Balkan Bronze Age jewelry with the Aegean-Anatolian traditions. At the same time, C. Schuster (chapter 19) analyses the later Bronze Age development connected with Tei culture including data from both sides of the Danube River. It is interesting the comparison between Late Bronze in the eastern Lower Danube (see also Comşa 2000), which contracts to the western Lower Danube where the Encrusted pottery culture along the Danube developed with the characteristic pottery (Bonev, chapter 20). That ceramic style connects the land of Northwestern Bulgaria with the Middle Danube on the one hand, but as A. Bonev argues in more details elsewhere (2000), some southern influences contribute to the development of that peculiar ceramic style of encrusted pottery. At the same time, the possible southern interrelation of the birds' images from Thrace discusses K. Nikov (chapter 22) facing the tender question of the continuity and discontinuity between the Late Bronze and the Iron Age in the Southern Balkans.

From technological point of view, the close and dynamic interrelation between the ceramic production and the resources in prehistory argue M. Stefanovich with collective (chapter 21). V. Ilcheva (chapter 27) summarizes her observation of the existence of different technological groups among the pottery from Hotnitsa-Vodopada, which is connected with the formation of the Early Bronze ceramic style in the lower Danube. The geomorphologic investigations of R. Kenderova (chapter 25), on their hand, supplies the volume with an analysis of social prehistoric environment in Southern Bulgaria. The question of the interrelation between the geomorphologic and anthropogenetic activities at some points in prehistory occurs, as well as for the possible interrelations between the environmental changes and cultural changes.

The section 3 contributes to the problems of the volume by publishing new tumulus graves from Pit Grave Culture from Northeastern Bulgaria (Kitov, chapter 23), which give some new evidence of the typology of ritual of that cultures in the Lower Danube. The graves from Ovcharitsa (Kuncheva-Russeva, chapter 25) are a very important new record, which documents the tradition of the flat cemeteries in Southeastern Bulagria. Such evidence is very rare from that region. Unfortunately, the absence of evidence from third and second millennium BC makes it difficult a bridge of cultural interconnection to be traced between the Bereket cemetery from Early Bronze I and the later Bronze Age – earlier Iron Age tradition of inhumations in flat cemeteries. However, again in Thrace the inhumations dominate.

Without doubt, G. Nekhrizov (chapter 24) has one of the most important studies for the topic of the innovations in the Balkans. That excavator documents the tradition of cromlechs in the Western Rhodopes region dated from Early Iron Age. The author argues that cult practice was wider distributed, but the other data are scanty and incomplete. The cromlech from Dolni Glavanak exemplifies how similar cult practice can occur in different points of Europe without visible cultural and chronological connection representing possibly similar ideas.

The hypotheses argued in this volume in turn open a discussion for newly formulated problems.

Lolita Nikolova

REFERENCES

ADSHEAD S.A.M. 1992 *Salt and Civilization*. New York.

ALADZHOV D. AND BALABANYAN D. 1984 'Pametnitsi ot starozhelyaznata epokha v Haskovski okrug.' *Thracia* 6, p. 216, fig. 31: 1.

ALEKSEEVA I.L. 1976 'O drevneshej eneoliticheskikh pogrebeniyakh Severo-Zapadnogo Prichernomor'ya.' *Materialy po arkheologii Severnogo Prichernomor'ya* 8, 176-86.

ALEXANDRESCU A. 1974 'La necropole du bronze ancien de Zimnicea (jud. de Teleorman).' *Dacia*, N.S., 18, 79-93.

ALEXANDROV G. 1981 'Novootkriti praistoricheski selishta v Mikhajlovgradski okrug.' *IMSZB* 6, 27-45.

ANATI E. 1962 *Prehistoric trade and the puzzle of Jericho. Bulletin of the American Schools of Oriental Research.*

ANDREOU ST., FOTIADIS M. AND KOTSAKIS K. 1996 'The Neolithic and Bronze Age of Northern Greece.' *AJA* Vol. 100, No. 3, 537–97.

ANGELINI A., BAGOLINI B. AND PASQUALI T. 1980 'Acquaviva di Besenello (Trento).' *Preistoria Alpina* 16, 67-69.

ANGELOV N. 1959 'Zlatnoto sukrovishte ot Hotnitsa.' *Arkheologiya* 1-2, 28-46.

ARANDJELOVIĆ-GARAŠANIN D. 1954 *Starčevačka kultura.* Ljubljana.

ASLANIS I. 1985 *Kastanas*. Volkerspeiss. Berlin. Prähistorische Archäologie in Südosteuropa.Vol. 4.

ASLANIS I. 1992 'Das Chalkolithikum in Nordgriechenland.' *Studia Praehistorica* 11-12, 184-95.

AVRAMOVA M. 1981 'Praistoricheskata keramika na Tsenralnite Rodopi.' *Izkustvo* 9-10, 79-82.

BAGOLINI B. (ED.) 1985 *Archeologia a Spilamberto. Ricerche nel territorio (Spilamberto - S.Cesario)*. Spilamberto.

BAGOLINI B. AND BARFIELD L.H. 1991 'The European Context of Northern Italy during the Third Millennium.' In: J. LICHARDUS (HG.), *Die Kupferzeit als historische Epoche*, 287-97.

BAGOLINI B., BARFIELD L.H. AND CREMONESI G. 1987 'La fine del neolitico.' In: *Il Neolitico in Italia. Atti della XXVI Riunione Scientifica IIPP, Firenze 7-10 novembre 1985*. Vol. 1, 79-88.

BAGOLINI B. AND BIAGI P. 1983 'Il Carso ed il Friuli nell'ambito del Neolitico dell'Italia settentrionale e dell'area balcano-adriatica.' In: *Atti del convegno sulla preistoria del Friuli-Venezia Giulia, 28-29 aprile 1981. Atti della Società per la Preistoria e la Protostoria del Friuli Venezia Giulia* IV (1978-1981), 187-205.

BAGOLINI B. AND BIAGI P. 1988 (1987) 'Distribution, Chronology and Cultural Significance of the "Metopal" Wares of Northern Italy.' *Natura Bresciana* 24, 183-187.

BAGOLINI B., BIAGI P. 1990. 'The Radiocarbon Chronology of the Neolithic and Copper Age of Northern Italy.' *Oxford Journal of Archaeology* Vol. 9 No.1, 1-23.

BAGOLINI B., CARLI R., FERRARI A., MESSORI A., PASQUALI T. AND PESSINA A. 1991 (1989) 'Il sepolcreto eneolitico del Dos de la Forca (Mezzocorona-Trento).' *Preistoria Alpina* 25, 121-64.

BAGOLINI B., CREMONESI G. 1988 'La distribuzione della ceramica a squame.' In: *L'età del Rame in Europa. Congresso Internazionale, Viareggio 15-18 ottobre 1987*. Rassegna di Archeologia 7, p. 634.

BAGOLINI B. AND FASANI L. 1982 'Problemi sulla fine del Neolitico ed il passaggio all'età del Bronzo nel versante meridionale della regione alpina centrale. In Il passaggio dal Neolitico all'età del Bronzo nell'Europa centrale e nella regione alpina. Problemi cronologici e terminologici.' In: *Atti del X Simposio Internazionale sulla fine del Neolitico e gli inizi dell'età del Bronzo in Europa, Lazise-Verona 8-12 aprile 1980*, 343-55.

BAGOLINI B. AND GRIFONI CREMONESI R. 1994 'Il Neolitico italiano: *facies* culturali e manifestazioni funerarie.' *Bullettino di Paletnologia Italiana* 85, 39-170.

BAGOLINI B., LANZIGER M. AND PEDROTTI A.L. 1992 'Rinvenimento di quattro statue-stele ad Arco (Valle del Sarca – Trentino Meridionale).' In: *Atti della XXVIII Riunione Scientifica IIPP*. Firenze, 355-70.

BAGOLINI B. AND PEDROTTI A.L. 1998 *Atlas du Néolithique Européen*. Vol. 2. L'Europe occidentale. L'Italie septentrionelle. ERAUL. Liège.

BAILEY D. 1998 'On Being Famous through Time and across Space.' In: BAILEY D. (ED.), *The Archaeology of Value*, 1-9.

BAILEY D. (ED.) 1998 *The Archaeology of Value. Essays on Prestige and the Processes of Valuation.* BAR. Oxford. BAR International Series 730, 106-130.

BAILEY D. AND PANAYOTOV I. (EDS.) 1995 *Prehistoric Bulgaria.* Prehistory Press. Madison. Monographs in World Archaeology No. 22.

BĂJENARU R. 1996 'Odaia Turcului, Cronica cercetărilor arheologice. Campania 1995.' In: *A 30-a Sesiune anuală de rapoarte arheologice*, Brăila, 82-83.

BAKALOV P. 1977 'Stratigrafski edinitsi v Blagoevgradskiya greben, organicheni ot nesuglasiya.' *Paleontologiya, stratigrafiya i litografiya* 7, 60-74.

BAKALOV P. 1983 *Stratigrafiya, usloviya i obstanovki na obrazuvane na pliotsen-plejstotsenskite naslagi, zapulvashti Blagoevgraskiya greben.* Dissertation (Unpublished).

BALTAKOV G. 1988 *Kvaternerna geomorfologiya i paleogeografiya.* Sofijski Universitet. Sofiya.

BARFIELD L.H. 1981 'Patterns of North Italian Trade 5000-2000 b.c.' In: BARKER G. AND HODGES R. (EDS.), *Archaeology and Italian Society. Prehistoric, Roman and Medieval Studies.* Papers in Italian Archaeology II. BAR International Series 102, 27-51.

BARFIELD L.H. 1983 'The Chalcolitic Cemetery at Manerba del Garda.' *Antiquity* 57, 116-23.

BARFIELD L.H. 1985 'Burials and Boundaries in Chalcolithic Italy.' In: MALONE C. AND STODDART S. (EDS.), *Papers in Italian Archaeology* IV. The Cambridge Conference, Part ii, Prehistory. BAR International Series 244, 152-76.

BARFIELD L.H. 1986 'Chalcolithic Burial Ritual in Northern Italy: Problems of Social Interpretation.' *Dialoghi di Archeologia (terza serie – anno 4)* 2, 241-48.

BARFIELD L.H. 1988 'L'età del rame nell'Italia settentrionale. The Chalcolithic of the Po Plain.' In: *L'età del Rame in Europa. Congresso Internazionale, Viareggio 15-18 ottobre 1987.* Rassegna di Archeologia 7, 411-418.

BARFIELD L.H. 1996 'The Chalcolithic in Italy: Consideration of Metal Typology and Cultural Interaction.' In: *XIII International Congress of Prehistoric and Protohistoric Sciences Forlì – Italia – 8/14 September 1996.* 10. *The Copper Age in the Near East and Europe.* Colloquium XIX. *Metallurgy: Origins and Technology*, 65-74.

BARFIELD L.H. AND BAGOLINI B. 1976 *The Excavation on the Rocca di Rivoli - Verona 1963-68.* Memorie del Museo Civico di Storia Naturale di Verona (II serie). Sezione Scienze dell'Uomo 1.

BARFIELD L.H., BUTEUX S. AND BOCCHIO G. 1995 *Monte Covolo: una montagna e il suo passato. Ricerche archeologiche 1972-1994.* Birmingham University Field Archaeology Unit.

BATALKIEV I. 1969 *Pazardzhik and Pazardzhishko.*Istoriko-Georgafski pregled. Sofia.

BATOVIĆ Š. 1979 'Jadranska zona.' In: *PJZ 2. Neolitsko doba.* Sarajevo. 473-634.

BATSOVA E. AND KUNCHEV M. 1974 'Novootkriti trakijski pogrebeniya v Novozagorsko.' *Arkheologiya* 1, 50-57.

BAXA P. AND KAMINSKÁ L. 1984 'Nové nálezy bolerázskej skupiny z Bratislavy.' *Slovenská Archeológia* Vol. 32, No. 1, 179-94.

BENECKE. N. 1998 'Animal Remains from the Neolithic and Bronze Age Settlements at Kirklareli (Turkish Thrace).' In: BUITENHUIS H., BARTOSIEWICH, L. AND CHYKE A. M. (EDS.), *Archaeology of the Near East.* Groningen. ARC Publications Vol. 18, No. 3, 172-79.

BENSON J. 1970 *Horse, Bird and Man.* Amherst.

BERCIU D. 1961 *Contribuţii la problemele neoliticului din România în lumina noilor cercetări de la Dunărea de Jos.*. Academiei Republica Socialiste Romania. Bucureşti.

BERCIU D. AND ROMAN P. 1984 'Morminetele tumulare de la Verbiţa (jud.Doji).' *Thraco-Dacica* 5, 15-21.

BERNABO-BREA L. 1964. *Poliocni. Citta prihistorca melläisola di Lenmos.* I. Roma.

BERNARD P. 1970 'Céramique de la premiére moitié du VIIe siècle à Thasos.' *BCH* 88, 31, 140-42, fig. 52.

BIANCHIN CITTON E. 1995 'Il complesso funerario di Sovizzo (località S. Daniele, Vicenza).' In: ASPES A. AND FASANI L. (EDS.), *Guide archeologiche. Preistoria e Protostoria in Italia 7. Veneto e Friuli-Venezia Giulia*, 142-49.

BINDER D. AND PERLES C. 1990 'Strategies de gestion des outillages lithiques au neolithique.' *Paleo* 2, 257-83.

BINFORD L. 1981 *Bones: Ancient Men and Modern Myths.* Academic Press. New York.

BINFORD L.R. AND BERTRAM J.B. 1977 'Bone frequencies – and attritional processes.' In: BINFORD L.R. (ED.), *For Theory Building in Archaeology.* Academic Press. New York, 77-153.

BLUSZCZ A., HERCMAN H., PAZDUR A. AND PAZDUR M. F.1992 'Radiometric dating. In: J. K. KOZLOWSKI, LAVILLE H. AND B. GINTER (EDS.), *Temnata Cave. Excavations in Karlukovo Karst Area.* Jagellanian University. Krakow. Vol. I.1, 223-39.

BOARDMAN J. 1991 'Early Greek Pottery on Black Sea Sites?' *Oxford Journal of Archaeology* Vol. 1, No. 3, 387-90.

BOMMELAER J. 1985 'L'apparition de la figure animee sur les vases grec de style geometrique.' *KTEMA* 10, p. 17.

BONCHEV G. 1901 'Megalitni pametnitsi v Sakar Planina.' *Sbornik NUNK* 18, 659-703.

BONCHEV G. 1923 (1922-1923) 'Mineralite v Bulgariya.' *GSU* Vol. 19, No. 1, 1-204.

BONEV A. 1988 *Trakiya i Egejskiyat svyat prez vtorata polovina na II khilyadoletie pr.n.e.* BAN. Sofia. Razkopki i prouchvaniya 20.

BONEV A. 2000 (1999) *The LBA Cremation Graveyard of Orsoya.* Reports of Prehistoric Research Projects 2-3, 1-31.

BONEV A. AND ALEXANDROV G. 1996 *Bagachina. Selishte ot kusnata kamennomedna epoha i trakijski kultov centur.* OOD Polimona. Montana.

BORELL B. 1978 *Attisch geometrische Schalen.* Mainz am

Rhein.

BOROFFKA N. 1994 *Die Wietenberg-Kultur. Ein Beitrag zur Erforschung der Bronzezeit in Sьdosteuropas.* Parts 1-2. Dr. Rudolf Habelt GmbH. Bonn. Universitдtsforschungen zur prдhistorischen Archдologie. Vol. 19.

BOROFFKA N. 1998 'Bronze- und früheisenzeitliche Geweihtrensenknebel aus Rumänien und ihre Beziehungen.' *Eurasia Antiqua* 4, 81-134.

BORONEANŢ V. 1971 'Gornea-Vodneac un nou aspect al epocii bronzului descoperit în zona Porţilor de Fier.' *Revista Muzeelor* Vol. 8, No. 1, 5-12.

BORONEANŢ V. 1981 'Cercetări arheologice privind cultura Tei de la Chitila-Cărămidărie.' *Cercetări arheologice în Bucureşti* 3, 195-225.

BORONEANŢ V. 1984 'Un aspect cultural de la începutul primei epüoci a fierului. Date preliminare.' *Thraco-Dacica* Vol. 5, 156-66.

BORONEANŢ V. 1985 'Contribuţii recente privind începuturile primei epoci a fierului în centrul Cîmpiei Române. Descoperirile de la Chitila-Fermă,' *Izvoare arheologice bucureştene* 2, 22-29.

BORONEANŢ V. 1992 'Date noi privind aspectul cultural Chitila-Fermă de la începutul primei epoci a fierului.' *Cercetări arheologice în Bucureşti* 4, 109-117.

BORRELLO M.A. 1984 *The Lagozza Culture (3rd millennium b.c.) in Northern and Central Italy.* Studi Archeologici III. Istituto Universitario di Bergamo.

BOSSERT H. 1942 *Anatolien. Kunst und Handwerk in Kleinasien von den Anfängen in der Griechischen Kultur. Die ältesten Kulturen des Mittelmeerkreis.* Band 2. Berlin.

BOTUSHAROVA L. 1963 'Stratigrafski prochvaniza na Nebettepe.' *GNAMPl* 5, 61-76.

BOUZEK J. 1991 'Cyprus and Europe: Bronze and Early Iron Ages.' In: *The Civilizations of Aegean and their Diffusion in Cyprus and Eastern Mediterranean 2000-600 BC. International Symposium.* Larnaka, p. 69.

BOYADZIEV Y. 1995. 'Chronology of Prehistoric Cultures in Bulgaria.' In: D. BAILEY AND I. PANAYOTOV (EDS.), *Prehistoric Bulgaria,* 149-91.

BRANIGAN K. 1974 *Aegean Metalwork of the Early and Middle Bronze Age.* Clarendon Press. Oxford.

BRESSAN F. 1982 'Il Ciondar des Paganis (Faedis-Udine).' *Preistoria Alpina* 18, 111-120.

BRIARD J. 1976 *L'âge du Bronze en Europe barbare, des Megalithes aus Celtes.* Paris.

BROGLIO A. AND PERINI R. 1995 'Romagnano III (Romagnano, Trento).' In: BROGLIO A. (ED.), *Guide archeologiche. Preistoria e Protostoria in Italia* 4. *Paleolitico, Mesolitico e Neolitico dell'Italia nord-orientale,* 102-117.

BRUKNER B. 1990 'Typen und Siedlungen Modellen und Wohnobjekte der Vinča-Gruppe in der panonnischen Tiefebene. ' In: SREJOVIĆ D. AND TASIĆ N. (EDS.), *Vinča and Its World,* 79-83.

BRUKNER B. 1991 'Die letzte Vinča-Pločnik (D2) Phase in Jugoslawien.' *Banatica* 2, 201-207.

BUDJA M. 1998 'Clay tokens – accounting before writing in Europe.' *Documenta Praehistorica* 25, 219-35.

BURKERT W. 1987 'Offering in Perspectives: Surrender, Distribution, Exchange.' In: LINDERS T. AND NORDQUIST G. (EDS.), *Gift to the Gods.* Acta Univ. Ups. Uppsala. Boreas 15, 43-50.

BUYUKLIEV H. 1964 'Dve rannotrakijski mogilni pogrebeniya ot Starozagorsko.' *Arkheologiya* 4, 61-65.

CALLEY S. AND GRACE S. 1988 *Technology and Function of Micro-borers from Kumartepe (Turkey).* BAR. Oxford. BAR International Series 411, 69-89.

CAMPANA N., MAGGI R. AND NEGRINO F. 1995 'Valle Lagorara (Maissana, La Spezia). ' In: MAGGI R., MARTINI F. AND SARTI L. (EDS.), *Guide archeologiche. Preistoria e Protostoria in Italia* 6. Toscana e Liguria, 130-37.

CAMPANA N., FRANCESCHI E., MAGGI R. AND STOS GALE Z. 1996 'Grotticella sepolcrale di Val Frascarese (Genova): nuove analisi dei reperti metallici.' In: *L'Antica età del bronzo in Italia.* Atti del Congresso di Viareggio, 9-12 gennaio 1995. Firenze. 556-557.

CAMPANA N., MAGGI R., STOS GALE Z. AND HOUGHTON J. 1996a 'Miniere e Metallurgia in Liguria fra IV millennio e IV secolo B.C. Un Excursus.' In: *La Miniera l'Uomo e l'Ambiente. Fonti e metodi a confronto per la storia delle attività minerarie e metallurgiche in Italia.* Convegno di Studi - Cassino, 2-4 giugno 1994, 15-52.

CARANCINI G.L. 1993 'Primi sviluppi della metallurgia nell'area medio-tirrenica nel quadro della protostoria peninsulare.' In: *Vulcano a Mezzano. Insediamento e produzioni artigianali nella media valle del Fiora nell'età del bronzo.* Comune di Valentano – Museo Civico, 125-50.

CARANCINI G.L. 1996 'La metallurgia dell'antica età del bronzo.' In: *L'Antica età del bronzo in Italia.* Atti del Congresso di Viareggio, 9-12 gennaio 1995. Firenze, 33-56.

CARR CH. 1991 'Left in the dust: Contextual information in model-focused archaeology.' In: KROLL E.M. AND PRICE T.D. (EDS.), *The Interpretation of Archaeological Spatial Patterning.* Plenum. New York, 221-56.

CASTEEL R. W. 1973 *Some Biases in the Recovery of Archaeological Faunal Remains.* Proceedings of the Prehistoric Society 39, 382-88.

CAUVIN M.-C. 1994 'La circulation de l'obsidian au Proche - Orient Neolithique.' In: H. G. GEBEL AND KOZLOWSKI S. K. (EDS.), *Neolithic chipped stone industries of the Fertile Crescent.* Studies in early Near Eastern Production, Subsistence and environment 1, 15-22.

CAVRUC V. 1997a 'Zoltan.' In: CIUGUDEAN H. (ED.), *The Bronze Age in Transilvania.* Alba Iulia.

CAVRUC V. 1997b 'The Final Stage of Early Bronze Age in South-Eastern of Transylvania (in the Light of New Excavations at Zoltan).' *Thraco-Dacica* 18, 97-133.

CAVRUC V. AND CAVRUC G. 1997 'Aşezarea de epoca bronzului de la Zoltan.' *Angustia* 2, 157-72.

CAZZELLA A. 1994 'Dating the 'Copper Age' in the Italian Peninsula and Adjacent Islands.' *Journal of European Archaeology* Vol. 2, No. 1, 1-19.

CAZZELLA A. AND MOSCOLONI M. 1992 'Neolitico ed Eneolitico.' In: *Popoli e civiltà dell'Italia antica.* 11. Biblioteca di Storia Patria, Bologna.

CHAMPION T., GAMBLE C., SHENNAN S. AND WHITTLE A. 1984 *Prehistoric Europe.* Academic Press. Orlando, FL.

CHAPMAN J. 1988 'Ceramic Production and Social Differentiation.' *Journal of Mediterranean Archaeology* No. 2.

CHAPMAN J. 1989 'The Early Balkan Village.' *Varia Arhaeologica Hungarica* 2, 33-53.

CHAPMAN J. 1996 'Enchainment, Commodification, and Gender in the Balkan Copper Age.' *Journal of European Archaeology* 4, 203-242.

CHAPMAN J. 1998 'Objectification, Embodiment and the Value of Places and Things.' In BAILEY D. (ED.), *The Archaeology of Value*, 106-130.

CHAPMAN V. 1974 *Salt Marshes and Salt Deserts of the World.* Cramer.

CHERNYKH E. 1978. *Gornoe delo i metallurgiya v drevnejshej Bolgarii.* BAN. Sofia.

CHICHIKOVA M. 1968 'Keramika ot starata zhelyazna epokha v Trakiya.' *Arkeologiya* 4, 15-27.

CHILDE G. 1929 *The Danube in Prehistory.* London.

CHOKHADZHIEV S. 1989 'Opit za prilagane na nyakoi matematiko-statisticheski metodi pri obrabotkata na rannoeneolitnata keramika ot Zapadna Bulgariya i Severna Gurtsiya. ' *IIMKn* 1, 69-120.

CHOKHADZHIEV S. 1991 'Novo kusnoeneolitno selishte v basejna na reka Struma.' *IIMKn* 3, 23-38.

CHOKHADZHIEV S. 1993 'Kharakteristika na edna nova faya na ranniza eneolit v dolinata na r. Struma.' In: NIKOLOV V. (ED.), *Praistoricheski nahodki*, 103-110.

CHOKHADZHIEV S. 1994 'Zhilishte ot posteneolitniua period v basejna na reka Struma.' In: PANAYOTOV ET AL. (EDS.), *Maritsa-Iztok*, 39-46.

CHOKHADZHIEV S. 1997 *Slatino – Praistoricheski selishta.* Vital. Veliko Turnovo.

CHOKHADZHIEV S. 1999 (IN PRINT), *Periodizatsiya na eneolita ot basejna na reka Struma.*

CHRISTMANN E. 1996 *Die deutschen Ausgrabungen auf der Pevkakia-Magula in Thessalien.* II. *Die frühe Bronzezeit.* Dr. Rudolf Habelt GmbH. Bonn. Beiträge zur ur- und frühgeschichtlichen Archäologie des Mittelmeer-Kulturraumes. Vol. 29.

ČIČIKOVA M. 1972 'Nouvelles données sur la culture thrace de l'époque du Hallstatt en Bulgarie du Sud.' *Thracia* 1, p. 90, fig. 11.

CIUGUDEAN H. 1996 *Epoca timpurie a bronzului în centrul şi sud-vestul Transilvaniei.* Bibliotheca Thracologica 13, Bucureşti.

CLASON A. T., AND PRUMMEL W. 1977 'Collecting, Sieving and Archaeozoological Research.' *Journal of Archaeological Science* 4, 171-75.

COCCHI GENICK D. AND GRIFONI CREMONESI R. 1989 *L'età del Rame in Toscana.* Viareggio.

ČOCHADŽIEV S. 1984 'Ausgrabungen an der prähistorischen Siedlung beim Dorf Djakovo, Kreis Kjustendil.' *Studia Praehistorica* 7, 64-80.

ČOCHADŽIEV S. 1998 'Contribution to the Research of the Earliest Copper Extraction and Processing in the Struma River basin.' *Archaeologia Bulgarica* No. 2, 10-14.

ČOHADŽIEV S. AND GENADIEVA V. 1998 'Contribution to the Study of the Early Neolithic in the Struma River Basin.' In: STEFANOVICH M. ET AL. (EDS.), *In the Steps of James Harvey Gaul.* Vol. 1. James Harvey Gaul Foundation. Sofia, 79-89.

COLDSTREAM J. 1977 *Geometric Greece.* London.

COLES J. M. AND HARDING A. F. 1979 *The Bronze Age in Europe.* St. Martins Press. New York.

COMŞA A. 2000 'Demographic Aspects of the Bronze Age in Romania.' In: NIKOLOVA L (ED.), *Analyzing the Bronze Age*, 43-52.

COMŞA E. 1969 'Données concernant la civilisation Vinča du sud-ouest de la Roumanie.' *Dacia* N.S. 13, 11-44.

COMŞA E. 1976 'Quelques considérations sur la culture Gumelniţa (L'agglomération Măgura Jilauei).' *Dacia* N.S. 20, 105-127.

COMŞA E. 1978 'L'utilisation du curve par les communautés de la culture Gumelniţa du territoire Roumain.' *Studia Praehistorica* 2, 109-120.

COMŞA E. 1987a *Neoliticul pe teritoriul României. Consideraţii.* ARSR. Bucureşti.

COMŞA E. 1987b 'Morminetele cu ocru din movila I de la Glăvăneştii Vechi.' *SCIVA* Vol. 38, No. 4, 367-87.

CONNAH G. 1991 'Bunyoro: Seeking the Origins of an African Kingdom.' *Antiquity 65*, 479-94.

CORNAGGIA CASTIGLIONI O. 1971 'La cultura di Remedello. Problematica ergologica di una facies dell'Eneolitico Padano. ' *Memorie della Società Italiana di Scienze Naturali e del Museo Civ. St. Nat. di Milano* Vol. 20 No. 1, 1-79 (20 tavv. fuori testo).

CORNAGGIA CASTIGLIONI O. 1971a 'La 'Cultura di Civate': una nuova 'facies' arcaica della 'Civiltà eneolitica' della Lombardia.' *Natura. Soc. It. Sc. Nat.* Vol. 62 No. 1, 101-125.

ČOVIĆ B. 1961 'Rezultati sondiranja na preistorijskom naselju u Gornjoj Tuzli.' *Godišnjak Zemaljskog Muzeja Sarajevo 1960-1961*, 79 sq.

CREMONESI G. AND VIGLIARDI A. 1988 'L'età del rame nell'Italia peninsulare: problemi generali.' *Rassegna di Archeologia* 7, 307-314.

CROWFOOD G.M. 1931 *Methods of Hand Spinning in Egypt and Sudan*. Halifax.

D'AMICO C., GHEDINI M., MICHELI R. AND MONTAGNARI KOKELJ E. 1996 'Le asce forate del Friuli-Venezia Giulia.' In: *Le vie della pietra verde*. L'industria litica levigata nella preistoria dell'Italia settentrionale. Torino, 229-38.

DAVERPORT W.H. 1986 'Two Kinds of Value in the Eastern Solomon Islands.' In: APPADURAI A. (ED.), *The Social Life of Things: Commodities in Cultural Perspectives*. Cambridge University Press. Cambridge.

DE GARIS DAVIES N. 1913 *Five Theban Tombs*. London.

DE GARIS DAVIES N. 1929 'The Town House in Ancient Egypt.' *Metropolitan Museum Studies* 1.

DELEV P. 1982 'Problemi na trakijskite megalitni pametnitsi.' In: FOL A. (ED.), *Trakijski pametnitsi*, 398-427.

DELEV P. 1984 'Megalithic Thracian Tombs in South-Eastern Bulgaria.' *Anatolica* 11, 17-45.

DELEV P. 1993 'Megalitna kultura.' In: *Trakijska drevnost. Kratka entsiklopediya*. Sofia, 173 sq.

DE MARINIS R.C. 1992 'La più antica metallurgia nell'Italia settentrionale.' In: *Der Mann im Eis*. Band I. Bericht über das Internationale Symposium 1992 in Innsbruck. Innsbruck, 389-409.

DE MARINIS R.C. 1995 'Le stele antropomorfe di Aosta.' *Notizie archeologiche Bergomensi* 3, 213-220.

DE MARINIS R.C. AND PEDROTTI A. 1997 'L'età del rame nel versante italiano della Alpi centro-occidentali.' In: *La Valle d'Aosta nel quadro della preistoria e protostoria dell'arco alpino centro-occidentale*. Atti della XXX Riunione Scientifica IIPP, Courmayeur 2-5 giugno 1994, 247-300.

DEMOULE J. P. 1989 'La tradition du néolithique au Bronze Ancien dans le Nord de l'Egée: les données de Dikili Tash.' *Archaeometry* 25, 687-96.

DEMOULE J.-P. AND LICHARDUS-ITTEN M. 1994 'Fouilles Franco-Bulgares du site Néolithique ancien de Kovačevo (Bulgarie du Sud-Ouest).' *BCH* 118, 561-618.

DEMOULE J.-P. AND PERLÈS C. 1993 'The Greek Neolithic.' *Journal of World Prehistory* Vol. 7 No. 4, 355-416.

DERGACHOV V. 1986 *Moldaviya i sosednye territorii v epokhu bronzy (analiz i kharakteristika kul'turnykh grup)*. Shtiintsa. Kishinev.

DESHAYES J. 1970. 'Dikili Tash.' *BCH*, Vol. 94, No.2, 799-808.

DESHAYES J. 1973. 'Les fouiles de Dikili Tash et l'arhéologie Yugoslave.' *ZNM* 6, 21-41

DETEV P. 1950a 'Selishtnata mogila pri Kapitan Dimitrievo.' *GNAMPl* 2, 1-21.

DETEV P. 1950b 'Selishtnata mogila Razkopanitsa.' *Izvestiya na Arkheologicheskiya Institut* 17, 171-90.

DETEV P. 1954 'Selishtna mogila pri s. Bikovo.' *GMPO* 1, 157-97.

DETEV P. 1960 'Praistoricheski selishta v basejna na r. Mechka.' *GMP* 3, 58, fig. 20.

DETEV P. 1962 'Razkopki na selishtnata mogila pri Lenovo, Plovdivsko.' *Arkheologiya* 4, 49-53.

DETEV P. 1963 'Razkopki na Nebettepe v Plovdiv.' *INAMPl* 5, 61-78.

DETEV P. 1968a 'Materiali za praistoriyata na Trakiya.' *GNAMPl* 6, 139-50.

DETEV P. 1968b 'Traces de la civilisation de Razkopanitza en Transylvanie.' *Apulum* Vol. 7, No. 1, 61-91.

DETEV P. 1971 'Données archéologiques de la continuité de la culture du tell Razkopanitza près du village Manole, départament de Plovdiv.' *Studia Balcanica* 5, 93–105.

DETEV P. 1981'Le Tell Razkopanica.' *IAI* 36, 141-88.

DETEV P. AND MATSANOVA V. 1977 'Praistoricheskoto selishte pri s. Ognyanovo.' *IMYUB* 3, 45-86.

DOMARADZKI M. 1994 'Les lieux de culte thraces (deuxieme moitie du II-e – I-er mil. Av. J.-C.).' *Helis* Vol. 3 part 1, 69-108.

DOMARADZKI M. WITH COLLECTIVE. 1999 *Pametnitsi na trakijskata kultura po gornoto techenie na reka Mesta*. Razkopki i prouchvaniya. Vol. 26. Sofia.

DOTHAN TH. 1963 'Spinning-bowls.' *Israel Exploration Journal* 13, 97-112.

DUMIRESCU V. 1961 *Necropole de incinerație din epoca bronzului de la Cîrna*. EARSR. București.

DUMITROAIA GH. 1987 'La station de Lunca – Poiana Slatini.' In: *La Civilisastion de Cucuteni en contexte Europeen*. Iaşi 1987.

DUNĂREANU-VULPE E. 1938 'Le nécropole de l'âge du bronze de Poiana.' *Dacia* 5-6, 151-67.

DUNBABIN T. 1957 *The Greeks and Their Eastern Neighbours*. London.

DZHAMBAZOV N. 1963 'Loveshkite peshteri.' *Izvestiya na Arkheologicheskiya Institut* 26, 195-237.

ERDOĞU B. 1993 *The Early Prehistoric Sites in Eastern Thrace: Neolithic and Chalcolithic Ages*. Unpublished M.A. Thesis. University College. London.

ERDOĞU B. 1995 'Prehistoric Surveys in Eastern Thrace (Turkey).' *Orient Oxpress* 3, 97-98.

ERICH R. (ED.) 1965 *Chronology in Old World Archaeology*. Chicago.

ERICH R. (ED.) 1992 *Chronology in Old World Archaeology*. Chicago.

ERICH R. AND BANKOFF H. 1992 'Geographical and Chronological Patterns in East Central Europe.' In: ERICH R. (ED.), *Chronology*, 375-94.

EVANS R. 1986 'The Pottery from Phase III.' In: RENFREW C. ET AL. (EDS.), *Sitagroi*, 345-428.

FALKENSTEIN F. 1998 *Feudvar II. Die Siedlungsgeschichte des Titeler Plateaus*. Prähistorische Archäologie in Südosteuropa 14.

FASANI L. 1990 (1988) 'La sepoltura e il forno di fusione de La Vela di Valbusa (Trento).' *Preistoria Alpina* 24, 165-81.

FILIPOV T. 1978 Idolna plastika ot kusnobronzoviya nekropol pri s. Orsoya, Mikhajlovgradski okrug. *Muzei i pametnitsi na kulturata* 2, 9-17.

FITZGERALD G.M.F. 1930 *The Four Caananite Temples of Beth Shean*. Part II: *The Pottery*. Philadelphia.

FOL A. (ED.) 1982 *Trakijski pametnitsi*, Vol. 3. *Megalitite v Trakiya. Trakiya Pontica,* Part 2. Nauka i izkustvo. Sofiya, Fig. 225.

FOL A., KATINČAROV R., LICHARDUS J., BERTEMES F. AND ILIEV I.K. 1989 'Bericht über die bulgarishdeutschen Ausgrabungen in Drama (1983-88).' *BRGK* 70, 5-128.

FOL A. AND LICHARDUS J. (EDS.) 1988 *Macht, Herrschaft und Gold. Das Gräberfeld von Varna (Bulgarien) und die Anfänge einer neuen europäischen Zivilization.* Moderne Galerie des Saarland-Museums. Saarbrücken.

FOTAKIEVA E., MILCHEVA M., ANDONOV T., GRUEV TS. AND BABEV I. 1976 *Prouchvane na pochvite v Bulgariya*. Vol. 3. Nauka i Izkustvo. Sofia, p. 9.

FRENCH D. 1961 'Late Chalcolithic Pottery in Northwest Turkey and Anatolia.' *Anatolian Studies* 11, 99-142.

FRENCH E. AND RUTTER J.B. 1977 'The Handmade Burnished Ware of the Late Helladic IIIC Period: Ist Modern Historical Context.' *American Journal of Archaeology* 81, 111-112.

GALE N.H., STOS-GALE Z.A., LILOV F., DIMITROV M. AND TODOROV T. 1991 'Recent Studies of Eneolithic Copper Ores and Artefacts in Bulgaria.' In: *La découverte du métal. Amis du Musée des Antiquités nationales. Millénaires. Collection dirigée par J.-P. Mohen. Dossier* 2, 49-75.

GALLAY A. 1995 'Les stèles anthropomorphes du site mégalithique du Petit-Chasseur à Sion (Valais, Suisse).' *Notizie Archeologiche Bergomensi* 3, 167-94.

GAMBARI F.M. AND VENTURINO GAMBARI M. 1985-1986 'La ceramica a fori passanti nel quadro dell'eneolitico dell'Italia nord-occidentale.' *Sibrium* 18, 61-79.

GARAŠANIN M. 1954 'Quelques elements datant la civilisation de Bubanj-Hum.' *Archaeologia Yugoslavica* 1, 19-24.

GARAŠANIN M. 1958 'Kontrollgrabung in Bubanj bei Niš.' *Praehistorische Zeitschrift* 36, 223-44.

GARAŠANIN M. 1973 *Praistorja na tlu Srbije*. Vol. 1. Srpska Kn'iževa Zadruga. Beograd.

GARAŠANIN M 1996 'Zur chronologischen und kulturellen Stellung von Junacite in der Bronzezeit.' *Starinar*, 99-104.

GARAŠANIN M. 1998 'Zur Verbreitung des Sălcuţa-Krivodol-Bubanj-Komplexes auf dem Mittleren Balkan.' In: *Man and the Animal World.* Studies in Archaeozoology, Archaeology and Palaeolinguistics in memoriam Sandor Bökönyi, 1998, 266-74.

GARAŠANIN M. AND SIMOSKA D. 1976 'Kontrolni iskopuvanja na Šuplevac i nekoi problemi na grupata Šuplevac-Bakarno Gumno.' *MAA* 2, 9-41.

GATSOV I. 1984. 'Technology and Typology of Cores from the Collection "Pobiti Kamani", Dikilitash, Bulgaria.' *Varia* CCI, *Archaeologia Interregionalis*, 135–50.

GATSOV I. 1985. 'Early Holocene Flint Assemblages form the Bulgarian Black Sea Coast.' BONSAL C. (ED.), *The Mesolithic in Europe*. Edinburgh, 471-74.

GATSOV I. 1992 'Kharakter i spetsifika na kremuchnoto proizvodstvo prez neolita v dneshna Zapadna Bulgariya.' *Arkheologiya* 2, 1-11.

GATSOV I. 1993 *Neolithic Chipped Stone Assemblages in Western Bulgaria*. Krakow.

GATSOV I. 1998 'Ağačli Pinar – Techno-Typological Analysis of the Chipped Stone Material.' *Archaeologia Bulgarica* 3, 1 – 6.

GATSOV I. 1999a 'Chipped stone assemblages from Neolithic and Chalcolithic Settlement "Kapitan Dimitrievo".' In: NIKOLOV V. ET AL., *Selisna mogila " Kapitan Dimitrievo",* 115-123.

GATSOV I. 1999b 'Neolithic Chipped Stone Assemblages from NW Turkey.' Paper presented at VIII Flint Symposium, Bohum, Germany.

GATSOV I. AND GUROVA M. 1998 'Izsledvane na kremachnite artefakti ot rannoneolithnia plast na Azmashkata selishna mogila kraj Stara Zagora. ' *Izvestiya na Muzeite ot Jugoiztochna Bulgaria* 19, 7 –21.

GATSOV I. AND KURČATOV V. 1997 'Neolithische Feuersteinartefacte. Mineralogische Untersuchung und technisch-typologische Charakteristik.' In: HILLER ST. AND NIKOLOV V. (HGS.), *Osterreichisch-Bulgarische Ausgrabungen und Forschungen in Karanovo*, 213-227.

GATSOV I., KURCHATOV V. AND GUROVA M. 1999(IN PRINT) 'State of Researches of the Neolithic and Chalcolithic Chipped Stone Assemblages in South Bulgaria.' Papers presented at Conference " From Rocks to Rouble", 19th –21st of March 1999, Durham, England.

GATSOV I. AND ÖZDOGAN M. 1994 'Some Epi–Palaeolithic Sites from NW Turkey. Ağačli, Domali, Gümüşdere.' *Anatolica* 20, 97-120.

GEORGIEV G. IL. 1948 'Eneolitnoto selishte do s. Kolena, Starozagorsko.' *Razkopki i prouchvaniya* 1, 63-74.

GEORGIEV G. IL. 1961 'Kulturgruppen der Jungstein- und der Kupferzeit in der Ebene von Tharzien (Südbulgarien).' In: BÖHM J. AND DE LAET S.J. (EDS.), *L'Europe à la fin de l'âge de la pierre*. Prague, 45-100.

GEORGIEV G. IL. 1963 'Glavni rezultati ot razkopkite na Azmashkata selishtna mogila prez 1961.' *Izvestiya na Arkheologicheskiya Institut* 26, 157-76.

GEORGIEV G. IL. 1966 'Mnogoslojnoe poselenie Azmashka mogila bliz Staroj Zagory (Bolgariya).' *KSIA* 106, 3-10.

GEORGIEV G. IL. 1972a 'Das Neolithikum und Chalkolithikum in der thrakischen Tiefebene (Südbulgarien). Probleme des heutigen Forschungsstandes.' *Thracia* 1, 5-27.

GEORGIEV G. IL. 1972b 'Die Gemeinsamkeit der Kulturen Südbulgarien und Nordgriechenlands wärend des Neolithikums und Frühbronzezeit.' In: *Actes of IInd Colloquim on Aegean Prehistory*. Athens, 115-127.

GEORGIEV G. IL. 1981 'Die neolithische Siedlung bei Čavdar, Bezirk Sofia.' *Izvestiya na Arkheologicheskiya Institut* 36, 63-109.

GEORGIEV G.IL., MERPERT N. YA., KATINCHAROV R.V. AND DIMITROV D.G. (EDS.) 1979 *Ezero. Rannobronzovoto selishte*. BAN. Sofiya.

GEORGIEV G.IL., NIKOLOV V., NIKOLOVA V. AND ČOHADŽIEV S. 108 'Die neolithische Siedlung bei Sapareva Banja, Bezirk Kjustendil.' *Studia Praehistorica* 8, 108-151.

GEORGIEVA P. 1988 'Die prähistorische Sidlung in der Gegend Čukata bein Dorf Galatin bei Vraca (Bulgarien).' *Studia Praehistorica* 9, 11-46.

GEORGIEVA P. 1994 'Prouchvane na selischte ot prehodniya period ot eneolita kum bronzovata epokha pri s. Reburkovo, Vrachansko.' *Arkheologiya* 1, 9-26.

GEORGIEVA R. 1977 'Sakralniyat kharakter na ornitoplastikata ot rannozhelyaznata epokha v basejna na Vardar i Haliakmon.' *Thracia Antiqua* 2, 63-72.

GEORGIEVA R. 1991 'Prouchvaniya na zhelyaznata epokha v rayona na "Maritsa-Iztok".' In: PANAYOTOV ET AL. (EDS.), *Maritsa-Iztok*, 91-104.

GEORGIEVSKI B. 1993 'Praistoriski gradishta od Kumasko.' *MAA* 13, 61-68.

GERGOV V. 1987 'Medni nakhodki ot praistoricheskoto selishte v m. Redutite pri s. Telish, Plevenski okrug.' *Arkheologiya* 4, 44-54.

GERGOV V. 1992 'Doistoricheskoe poselenie Telish – Redutite.' *Studia Praehistorica* 11-12, 347-57.

GERGOV V. 1995 'A Dwelling of Telich IV.' In: NIKOLOVA L. (ED.), *Early Bronze Age Settlement Patterns*, 309-322.

GERGOVA D. 1982 'Proizhod i datirovka na sukrovishteto ot s. Burzitsa (b. Sheremet), Varnenski okrug.' *Arkheologiya* 3-4, 6-17.

GERGOVA D. 1987 *Früh- und ältereisenzeilliche Fibel in Bulgarien*. PBF 14-7. München.

GIFFORD D. P. 1981 'Taphonomy and Paleoecology: A Critical review of Archaeology's Sister Disciplines.' *Recent Advances in Archaeological Method and Theory* 4, 365-438.

GILLI E. AND MONTAGNARI KOKELJ E. 1993 'La grotta dei Ciclami nel Carso Triestino (materiali degli scavi 1959-1961).' In: *Atti della Società per la Preistoria e Protostoria della regione Friuli-Venezia Giulia* VII (1992), 65-162.

GILLI E. AND MONTAGNARI KOKELJ E. 1994 (1993) 'La grotta delle Gallerie nel Carso Triestino. ' In: *Atti della Società per la Preistoria e Protostoria della regione Friuli-Venezia Giulia* VIII, 121-94.

GIMBUTAS M. 1991 *The Civilization of the Goddess. The World of Old Europe*. San Francisco.

GIMBUTAS M., WINN SH. AND SHIMABUKU D. 1989 *Achilleion. A Neolthic Settlement in Thessaly, Greece, 6400-5600 BC*. University of California. Los Angeles. Monumenta archaeologica 14.

GLUMAC P. AND TRINGHAM R. 1990 'The exploitation of copper minerals.' In: TRINGHAM R. AND KRSTIĆ D. (EDS.), *Selevac: A Neolithic Village in Yugoslavia*. Institute of Archaeology, UCLA. Los Angelos, 549-66.

GOGÂLTAN F. 1993 'Foeni, eine frühbronzezeitliche Siedlung.' *Thraco-Dacica* 14, 51-64.

GOGÂLTAN F. 1995 'Die Frühe Bronzezeit im Südwesten Rumäniens. Stand der Forschung.' *Thraco-Dacica* 16, 55-79.

GOGÂLTAN F. 1996 'About the Early Bronze Age in the Romanian Banat.' In: N. TASIĆ (ED.), *The Yugoslav Danube Basin and the Neighbouring Regions,* 43-67.

GRAHAM A. 1990 'Pre-colonial Contacts: Question and Problems.' In: *Preccedings of the First Australian Congress of Classical Archaeology*. Oxford and New York, 45-60.

GREENFIELD H. J. 1986a 'Fauna: Kičmenjaci.' In: BOGDANOVIĆ M. (ED.), *Ljuljaci: Naselje Protovatinske i Vatinske Kulture*. Narodni Muzej. Kragujevac, 75-78.

GREENFIELD H. J. 1986b *The Paleoeconomy of the Central Balkans (Serbia): A Zooarchaeological Perspective on the Late Neolithic and Bronze Age (ca. 4500-1000 B.C.)*. Vols. 1-2. BAR. Oxford. British Archaeological Reports, International Series 304.

GREENFIELD H. J. 1986c 'Summary Report on the Vertebrate Fauna from Novačka Ćuprija, Serbia (Eneolithic-Late Bronze Age).' *Zbornik Radova Narodnog Muzeja* 12, 63-74.

GREENFIELD H. J. 1991 'Fauna from the Late Neolithic of the Central Balkans: Issues in Subsistence and Land Use.' *Journal of Field Archaeology* 18, 161-86.

GREENFIELD H. J. 1996 'Sarina Medja, Vrbica, and Večina Mala: The Zooarchaeology of Three Late Bronze Age/Early Iron Age Transition Localities near Jagodina, Serbia.' *Starinar* 45-46,133-42.

GREENFIELD H. J. 1999 'The Origins of Metallurgy: Distinguishing Stone from Metal Cut Marks on Bones from Archaeological Sites.' *Journal of Archaeological Science* 26 (7), 797-808.

GREENFIELD H. J. 2000a 'The Application of Integrated Surface and Subsurface Reconnaissance Techniques to Later Prehistoric SE European Sites: Blagotin Serbia.' *Geoarchaeology* 15 (2), 20 pages.

GREENFIELD H. J. 2000a 'Monitoring the Origins of Metallurgy: an Application of Cut Mark Analysis on Animal Bones from the Central Balkans.' *Environmental Archaeology*, 30 pages. Accepted for publication April 1999.

GREENFIELD H. J. n.d. a 'Distinguishing Metal (Steel and Low-Tin Bronze) from Stone (Flint and Obsidian) Tool Cut Marks on Bone: an Experimental Approach.' In: MATHIEU J. (ED.), *Recent Advances in Experimental Archaeology*. Altamira Press. Accepted for publication - June 1999.

GREENFIELD H. J. n.d. b (IN PRESS) 'Origins of Metallurgy: A Zooarchaeological Perspective from the Central Balkans.' In: HARRISON R. AND GILLESPIE M. (EDS.), *Eureka: The Archaeology of Innovation. Proceedings of the 27th Annual Chacmool Conference*. The Archaeological Association of the University of Calgary Calgary. Accepted for publication 1996. (R)

GREENFIELD H. J.. n.d. c *Vertebrate Fauna from Blagotin*. Unpublished manuscript on file at the University of Manitoba.

GREENFIELD H. J. n.d. d *Vertebrate fauna from Stragari*. Unpublished manuscript on file at the University of Manitoba.

GREENFIELD H. J. n.d. e *Vertebrate fauna from Kadica Brdo*. Unpublished manuscript on file at the University of Manitoba.

GRIFONI CREMONESI R. 1987 'Storia delle teorie relative al Neolitico in Italia.' In: *Il Neolitico in Italia*. Atti della XXVI Riunione Scientifica IIPP, Firenze 7-10 novembre 1985. Vol. 1, 11-20.

GRIFONI CREMONESI R. 1996 'Le grotte e la loro funzione. Premessa metodologica.' In: *L'Antica età del bronzo in Italia*. Atti del Congresso di Viareggio, 9-12 gennaio 1995. Firenze, 305-312.

GUERRESCHI A. 1995 'Museo della Val Fiorentina (Selva di Cadore, Belluno).' In: BROGLIO A. (ED.), *Guide archeologiche. Preistoria e Protostoria in Italia 4. Paleolitico, Mesolitico e Neolitico dell'Italia nord-orientale*, 198-205.

GUMĂ M. 1995 'The End of the Bronze Age and the Beginning of the Early Iron Age in South-western Romania. A Short Review.' *Thraco-Dacica* 16, 99-117.

GUMĂ M. 1996 *Epoca bronzului în Banat. The Bronze Age in Banat*. Mirton. Timişoara.

HAGERSTRAND T. 1967 *Innovation Diffusion as a Spatial Process*. University of Chicago Press. Chicago.

HÄNSEL B. 1969 'Plastik der jungeren Bronzezeit und der alteren Eisenzeit aus Bulgarien.' *Germania* 17, 62-86.

HÄNSEL B. 1976 *Beiträge zur regionalen und chronologischen Gliederung der Alteren Hallstattzeit an der unteren Donau*. Bonn. Beiträge zur ur- und frühgeschichtlichen Archäologie des Mittelmeer-Kulturraumes 16.

HANSCHMANN E. AND MILOJČIĆ V. 1976 *Die deutsche Ausgrabungen auf der Argissa-Magula in Thessalien*. Vol. 3. Dr. Rudolf Habelt GmbH. Bonn.

HARDING A.F. 1984 *The Myceneans and Europe*. London.

HAUPTMANN H. 1967 'Zum Neolithikum in Makedonien.' *Istanbuler Mitteilungen* 17, fig. 6: 7-9.

HEURTLEY W.A. 1939 *Prehistoric Makedonia*. Cambridge University Press. Cambridge.

HIGHAM CH.F.W. 1968 'Faunal Sampling and Economic Prehistory.' *Zeischrift für Säugetierkunde* Vol. 33, No. 5, 297-305.

HILL J.D. 1996 'The Identification of Ritual Deposits of Animal Bones. A General Perspective from a Specific Study of 'Special Animal Deposits' from the Southern English Iron Age.' In: ANDERSON S. AND BOYLE K. (EDS.), *Ritual Treatment of Hu-*

man and Animal Remains: Preceedings of the First Meeting of the Osteoarchaeological Research Group Held in Cambridge on 8th October 1994. Oxbow. Oxford, 17-32.

HILLER S. AND NIKOLOV V. (HGS.) 1997 *Karanovo. Die Ausgrabungen im Südsektor 1984 – 1992*. Ferdinand Berger u. Söhne. Salzburg – Sofia.

ILCHEVA V. 1982 'Prouchvane na bronzovata epokha po basejna na reka Yantra.' *GMSB* 8, 17-24.

ILCHEVA V. 1996 'Technologiya i klasifikatsiya na keramikata ot Hotnitsa-Vodopada – prekhoden period ot eneolita kum bronzovata epokha.' *IIMVT* 11, 87-139.

ILCHEVA V. 1997 'Technologiya i klasifikatsiya na keramikata na selishta ot prekhodniya period ot eneolita kum bronzovata epokha v sredniya basejn na reka Yantra.' *IIMVT* 12, 91-114.

IVANOV I. AND AVRAMOVA M. 1997 *Varnenski nekropol*. Biblioteka "Virtus" 1. Agatho. Sofia.

JOVANOVIĆ B. 1976 'Tumuli stepske kulture grobova jama u Podunavl'u.' *Starinar* 26, 9-24.

JOVANOVIĆ B. 1980 'The Origins of Copper Mining in Europe.' *Scientific American* Vol. 242, No. 5, 152-67.

JOVANOVIĆ B. 1991 'La métallurgie énéolithique du cuivre dans les Balkans.' In: *La découverte du métal*. Amis du Musée des Antiquités nationales. Millénaires. Collection dirigée par J.-P. Mohen. Dossier 2, 93-102.

KAISER T. AND VOYTEK B. 1983 'Sedentism and Economic Change in the Balkan Neolithic.' *Journal of Anthropological Archaeology* 2, 323-53.

KĂNČEV M. AND KĂNČEVA-RUSSEVA T. 1996 'Architektur der prähistorischen Siedlung Ovcaritza II.' In: NIKOLOVA L. (ED.), *Early Bronze Age Settlement Patterns*, 227-38.

KARMANSKI S. 1991 *Donja Branjevina*. Muzejna edinica pri narodnom universitetu "Odžaci". Odžaci.

KATINČAROV R. 1972a 'Habitations de l'âge du bronze moyen du tell de Nova Zagora (Bulgarie de Sud).' *Thracia* 1, 43-54.

KATINČAROV R. 1972b 'Osnovnye rezult'taty raskopok pervobytnogo poseleniya v g. Nova Zagora (Yuzhnaya Bulgariya) in 1968 and 1970.' *Sovetskaya arkheologiya* 1, 242-47.

KATINČAROV R. 1973. 'Trouvailles de l'âge du bronze moyen a Nova Zagora (Bulgarie de Sud).' In: *Actes du VIIIe CISPP*. Beograd. Vol. 3, 66-75.

KATINČAROV R. 1975 'Traits caractérisques de la civilization de l'âge du bronze ancien et moyen en Bulgarie.' *AAC* 15, 85-110.

KATINČAROV R. 1975a. 'Prouchvaniya na bronzovata epokha v Bulgariya.' *Arkheologiya* 2, 1-17.

KATINČAROV R. 1981 'État des recherches sur l'âge du bronze en Bulgarie du Sud-Est.' *Izvestoya na Arkheologicheskiya Institut* 36, 117-40.

KATINČAROV R. 1982 'Sur la syncronisation de l'âge du bronze ancien en Thrace et la région d'Egéen et d'Anatolie.' *Thracia Praehistorica. Supplementum Pulpudeva* 3, 117-40.

KATINČAROV R. AND DIMITROV M. 1978 'Razkopki na nadgobna mogila pri s. Glavan, Starozagorski okrug.' *AOR prez 1977*. Vidin, p. 44.

KATINČAROV R. [KATINCHAROV R.] AND MATSANOVA V. 1993 'Razkopki na selishtanata mogila pri s. Yunatsite, Pazardzhishko.' In: NIKOLOV V. (ED.), *Praistoricheski nakhodki*, 151-73.

KATINČAROV R. [KATINCHAROV R.], MERPERT N. YA., MATSANOVA V. AND AVILOVA L. I. *1995 Selishtnata mogila pri selo Yunatsite.Istoriya and prouchvaniyata. Obshta stratigraphiya. Plast A*. Dios and Agatho. Sofia.

KITANOVSKI B. 1971 'Bakarno gumno à Cepigovo près de Prilep.' In: *Époque préhistorique et protohistorique en Yougoslavie. Recherches et resultats*. Beograd, 139-41.

KITOV G. AND PAVLOV P. 1973 'Grobna nakhodka ot Lovech.' *Muzei i pametnitsi na kulturata* 2, 4-9.

KITOV G., PANAYOTOV I. AND PAVLOV P. 1991 *Mogilni nekropoli v Loveshkiya kraj. Ranna bronzova epokha (nekropolut Goran-Slatina)*. BAN. Sofiya. Razkopki i prouchvaniya 23.

KOEN, V. 1992 *Final'nyi paleolit gornogo Krima*. Kiev. Avtoreferat.

KOHL PH.L. 1987 'The Use and Abuse of World Systems Theory: The Case of the Pristine West Asian State.' *Advances in Archaeological Method and Theory* 11, 1-36.

KOLEVA B. 1988 'Praistoricheskoto selishte v m. Izvora pri Perushtitsa.' *IMYUB* 14, 75-97.

KOLEVA B. 1993 'Kusnokhalkolitni askosi, Plovdivsko.' In: NIKOLOV V. (ED.), *Praistoricheski nakhodki*, 121-26.

KOLIŠTIKORSKA-NASTEVA I. 1993 'Eneolitsko naogalište na Kostoperska kapra.' *MAA* 13, 43-49.

KOPPENHÖFER D. 1997 'Troia VII – Versuch einer Zusammenschau einschliesslich der Ergebnisse des Jahres 1995.' *Studia Troica* 7, 295-353.

KOROŠEC P. AND KOROŠEC J. 1969 *Najdbe s kolišcarskih naselbin pri Igu na Ljuljanskem Barju*. Ljubljana.

KOSSAK G. 1954 *Studien zum Symbolgut der Urnenfelder-und Halstattzeit Mitteleuropa*. Berlin.

KOUKOULI-CHRYSANTHAKI CH. 1982 'Late Bronze Age in Eastern Macedonia.' *Thracia Praehistorica. Suppl. Pulpudeva* 3, 231-58.

KOUKOULI-CHRYSANTHAKI CH. 1987 'Early Bronze Age Settlement at Skala Soteros in the Island of Thasos.' *To Arhailogiko Ergo eth Makedonia kai Thrake* 1, 289-406, p. 536.

KOUKOULI-CHRYSANTHAKI CH. 1989 'The excavations at Skala Soteros.' *To Arhailogiko Ergo eth Makedonia kai Thrake* 3, 507-520.

KOVACHEVA M., JORDANOVA D. AND JORDANOVA D. 1995 Archaeomagnetic Study of the Early Bronze Age Settlement Sarovka Near Dubene, Karlovo District. In: NIKOLOVA L. (ed.), *Early Bronze Age Settlement Patterns*, 32-43.

KOVACHEVA M. AND TOSHKOV A. 1991 'Arheomagnitno datirane na nyakolko neolitni selishta v Bulgariya.' *Arkheologiya* 33, 4, 9-13.

KOZLOWSKI J. K. AND KOZLOWSKI S. K. 1984. 'Chipped-stone industries from Lepenski Vir, Yugoslavia.' *Preistoria Alpina* 19, 259-293.

KRAJCHEV N. 1970 'Predi 7000 godini. Novootkrito selishte v Karlovo.' *Karlovska tribuna* 148, 12.06.1970.

KRAMER C. 1982 *Village Ethnoarchaeology: Rural Iran in Archaeological Perspective*. Academic Press. New York.

KRUMOV G. 1984 'Osnovni etapi v morfostrukturnoto razvitie na Sakar Planina.' *GSU GGF* book 2-Geography Vol. 76.

KULL. B. 1988 *Demircihüyük*. Band 5. *Mittelbronzezeitliche Siedlung*. Mainz am Rhein.

KUNCHEV K. 1995 'Stratigrafsko razpredelenie na risuvanata keramika v mnogoslojnoto rannoneolitno selishte Chavdar, Pirdopsko.' *Arkheologiya* 2, 1-4.

KUNCHEV M. 1984 'Selishta i nekropoli ot kusnata bronzova i rannozhelyaznata epokha v Novozagorsko.' *Thracia* 6, 134-59.

KUNCHEV M. 1991 'Nekropoli ot bronzovata epokha v rayona and "Maritsa-Iztok".' In: PANAYOTOV I. ET AL. (EDS.), *"Maritsa-Iztok"*, Vol. 1, 41-70.

KUNCHEVA-RUSSEVA T. 1994 'Yamni pogrebeniya v mogila No. 2 ot mogilen nekropol v zemlishteto na s. Pet Mogili, Novozagorsko.' In: PANAYOTOV I. ET AL. (EDS.), *Maritsa-Iztok*, 91-94.

KUZMAN P. 1996 'Kutlina.' *MAA* 14, 9-22.

LARSEN H. 1941 'Vorbericht über die Schweidischen Grabungen in Abu Ghalib 1936/1937.' *MDIK* 10, 30-31.

LÁSZLÓ A. 1997 'Legături între regiunea Dunării de Jos şi spaţiul egeo-anatolian la sfârşitul epocii bronzului şi începutul epocii fierului.' *Memoria Antiquitatis* 21, 115-130.

LÁSZLÓ A. 1998 'Some Connections between the Northern Thrace and Asia Minor during the Late Bronze and Early Iron Age.' In: TUNA N., AKTÜRE Z. AND LYNCH M. (EDS.), *Thracians and Phrygians. Problems of Parallelism. Proceedings of an International Symposium on the Archaeology, History and Ancient Languages of Thrace and Phrygia. Ankara, 3-4 June 1995*. Ankara, 41-44.

LAZAROVICI GH. 1979 *Neoliticul Banatului*. Muzeul de istorie al Transilvaniei. Cluj-Napoca.

LAZAROVICI GH. 1998 'Once again about the Ceramics from Gornea-Vodneac of the Early Bronze in Banat.' In: SCHUSTER C. (ED.), *Die Kulturen der Bronzezeit in dem Gebiet des Eisernen Tores. Kolloquium in Drobeta-Turnu Severin (22.-24. November 1997), Rumänisch-Jugoslawische Kommission für die Erforschung der Region des Eisernen Tores, Archäologische Abteilung* II, Bukarest, 47-70.

LAZIĆ M. (ED.) 1997 *Arkheologija istochne Srbije*. Centre for Archaeological Research. Belgrade.

LAZIĆ M. AND SLADIĆ M. 1997 'Eneolitsko naselje u Škodrenom polju kod Knjaževca.' In: LAZIĆ M. (ED.), *Arkheologija istochne Srbije*, 211-23.

LEAHU V. 1966 *Cultura Tei*. Bucureşti.

LEAHU V. 1978 'Cu privire la raporturile culturii Tej cu unele ale epocii bronzului din Balcani.' *SCIV* Vol. 29 No. 1, 43-75.

LEAHU V. 1995 'Cultura Tei. Tei Culture.' In: PETRESCU-DÎMBOVIŢA M. ET AL. (EDS.), *Comori ale epocii bronzului din Romania*, 253-57.

LEGGE A.J. 1989 'Milking and Evidence: A Reply to Entwistle and Grant.' In: MILLES A., WILLIAMS D. AND GARDNER N. (EDS.), *The Beginning of Agriculture*. BAR. Oxford. BAR International Series 496.

LEKOVIĆ V. 1990 'The Vinčanization of Starčevo culture.' In: SREJOVIĆ D. AND TASIĆ N. (EDS.), *Vinča and Its World*, 67-74.

LEIGHTON R. 1996 'From Chiefdom to Tribe? Social Organization and Change in Later Prehistory.' In: LEIGHTON R. (ED.), *Early Societies in Sicily. New Developments and Archaeological Research*. Accordia Research Center, University of London. Accordia Specialist Studies on Italy. Vol. 5, 101-115.

LESHTAKOV K. 1988 'Osnovni klasifikatsionni printsipi za keramikata or rannata i srednata bronzova epokha v Trakiya.' *Arkheologiya* 3, 1-15.

LESHTAKOV K. 1992 'Izsledvaniya vurkhu bronzovata epokha v Bulgariya. I. Sravnitelna stratigraphiya na selishtnite mogili ot RBE v Yugoiztochna Bulgariya.' *GSU IF*, Vol. 84-85, 5-119.

LESHTAKOV K. 1993 'The End of the Early Bronze Age in Thrace.' In: PAVUK J. (ED.), *Actes du XIIe Congrès International des SPP*. IAASS, Nitra. Bratislava, 556-60.

LESHTAKOV K. 1994 'Opit za datirane na I-IV str. h. na selishtnata mogila do s. Dyadovo.' In: PANAYOTOV I. ET AL. (EDS.), *Maritsa-Iztok*, 95-103.

LESHTAKOV K. 1996. 'Trade Centres in the Early Bronze Age III and Middle Bronze Age in Upper Thrace

(On the Interpretation of Baa dere - Konstantia - Gulubovo).' In: NIKOLOVA L. (ED.), *Early Bronze Age Settlement Paterns*, 239-87.

LESHTAKOV K. 1997 'Tsiganova Mogila, Soundage Excavations, 1993'; 'Rescue Archaeological Excavations at the Luda Reka Site, 1993'; 'Preliminary Report on the Dervishov Odzhak Rescue Excavations, 1993.' In: Leshtakov K. (ED.), *Maritsa Project*, 15-28, 55-146.

LESHTAKOV K (ED.) 1997 *Maritsa Project*. Vol. 1. Roads Agency Ltd. Sofia.

LEŠTAKOV K. 1993 'Die mittelbronzezeitliche Besiedlung des Siedlungshügels von Gălăbovo in Südbulgarien.' *Saarbrücken Studien und Materialien zur Altertumskunde* 2, 191-222.

LEŠTAKOV K 1997 'Typologie und vergleichende Chronologie der mittelbronzezeitliche Keramik aus Gălăbovo - Südostbulgarien.' *Saarbrücken Studien und Materialien zur Altertumskunde* 6 (in print).

LEWTHWAITE J. 1987 'Three steps to leaven: applicazione del modello di disponibilità al neolitico italiano.' In: *Il Neolitico in Italia.* Atti della XXVI Riunione Scientifica IIPP, Firenze 7-10 novembre 1985. Vol. 1, 89-102.

LICHARDUS J. (ED.) 1991 *Die Kupferzeit als historische Epoche. Symposium Saarbrücken und Otzenhausen, 6-13.11.*1988. Dr. Rudolf Habelt GmbH. Bonn, 287-97.

LOUD G. 1948 *Megiddo*. Vol. 4. Chicago.

LYMAN R. L. 1984 'Bone density and differential survivorship of fossil classes.' *Journal of Anthropological Archaeology* 3, 259-99.

LYMAN L. L. 1994 *Vertebrate Taphononomy*. Cambridge University Press. Cambridge.

MAGGI R. 1996 'Le grotte e la loro funzione. L'Italia settentrionale.' In: *L'Antica età del bronzo in Italia.* Atti del Congresso di Viareggio, 9-12 gennaio 1995. Firenze, 313-22.

MAGGI R. (ED.) 1997 *Arene Candide: a Functional and Environmental Assessment of the Holocene Sequence (Excavations Bernabò Brea-Cardini 1940-50)*. Memorie dell'Istituto Italiano di Paleontologia Umana. Nuova Serie 5.

MAGGI R. AND DEL LUCCHESE 1988 'L'età del rame nell'Italia centrale. Aspects of the Copper Age in Liguria.' In: *L'età del Rame in Europa*. Congresso Internazionale, Viareggio 15-18 ottobre 1987. *Rassegna di Archeologia* 7, 331-38.

MALTBY J.M. 1985 'Patterns in faunal assemblage variability.' In: BARKER G. AND GAMBLE C. (EDS.), *Beyond Domestication in Prehistoric Europe*. Academic Press. New York, 33-74.

MANZURA I. 1993 'The East-West Interactions in the Mirror of the Eneolithic and Early Bronze Age Cultures in the North-West Pontic.' *Revista Arheologica* 1, 23-53.

MARAN J. 1989 'Die Schaftlochaxt aus dem Depotfund von Theben (Mittelgriechenland) und ihre Stellung im Rahmen der Bronzezeitlichen Axte Südosteuropas.' *Archäologisches Korrespondenzblatt* 19, 129-136.

MARAN J. 1997 'Neue Ansätze für die Beurteilung der Balkanisch-Ägäischen Beziehungen im 3 Jahrtausend. v. Chr.' In: ROMAN P. (ED.), *The Thracian World*, 171-92.

MARAN J. 1998 *Kulturwandel auf dem griechischen Festland und den Kykladen. Studien zur kulturellen Verhältnissen in Südosteuropa und dem zentralen sowie östlichen Mittelmeerraum in der später Kupfer- und frühen Bronzezeit*. 1-2. Dr. Rudolf Habelt GmbH. Bonn.

MARCHESETTI C. 1889 'Ricerche preistoriche nelle caverne di S. Canziano presso Trieste.' *Bollettino della Società Adriatica di Scienze Naturali in Trieste* 1, 1-19 (and 2 tables).

MATEESCU C.N. 1978 'Contributions to the Study of Neolithic Dwellings in Romania: A Dwelling of the Second Phase of the Vădastra Culture (Middle Neolithic).' *Dacia* N.S., 22, 65-71.

MATSANOVA V. 1994 'Krustati idoli ot selishtnata mogila pri s. Yunatsite, obshtina Pazardzhik.' In: PANAYOTOV I. ET AL. (EDS.), *Maritsa-Iztok*, 53-63.

MATSANOVA V. AND RADUNCHEVA A. 1995 'Neolitho seloshte pri grad Rakitovo.' *Slaveevi gori* I, 114 –125.

MATSAS D. 1991 'Samothrace and the Northeastern Aegean: The Minoan Connections.' In: *Studia Troica*, Vol. 1, 159-79.

MAZANOVA V. 1992 'Tellsiedlung Junazite die Spätkupferzeit.' *Studia Praehistorica 11-12, 248- 61.*

MAZANOVA V. 1996 'Cult Objects from the Early Neolithic Site at the Town of Rakitovo.' *Poročilo* 23, 105-128.

MAZAR B. 1951 (1950-1951) 'The Excavations at Tell Oasile. Preliminary Report.' *Israel Exploration Journal* 1, pp. 128 and 139.

MEADOW R.H. 1975 'Mammal remains from Hajji Firuz: A study in methodology.' In: CLASON A.T. (ED.), *Archaeozoological Studies*. North Holland. Amsterdam, 265-83.

MEADOW R. H. 1978 'Effects of context on the interpretation of faunal remains: A Case Study.' In: MEADOW R. H. AND MELINDA A. Z. (EDS.), *Approaches to Faunal Analysis in the Middle East.* Peabody Museum and Harvard University. Cambridge. Peabody Museum Bulletin no. 2, 15-21.

MEADOW R.H. 1983 'The vertebrate faunal remains from Hasanlu period X at Hajji Firuz.' In: VOIGT M.M. (ED.), *Hajji Firuz Tepe, Iran: The Neolithic Set-*

tlement. Hasanlu Excavation Reports Vol. 1. University of Pennsylvania, University Museum Monographs No. 50, 369-422.

MEDVEDSKAYA I. 1986 'A Study on the Chronological Parallels between the Greek Geometric Style and Sialk B Pottery.' *Iranica Antiqua* 21, 92-115.

MELLAART J. 1971 'Prehistory of Anatolia and its Relations with the Balkans.' *Studia Balcanica* 5, 119–47.

MEZZENA F. 1997 'La Valle d'Aosta nel Neolitico e nell'Eneolitico.' In: *La Valle d'Aosta nel quadro della preistoria e protostoria dell'arco alpino centro-occidentale*. Atti della XXX Riunione Scientifica IIPP, Courmayeur 2-5 giugno 1994, 17-138.

MIKIĆ Ž. 1990 'Lepenksi Vir and das Neolithisationproblem in der Antropologie.' In: SREJOVIĆ D. AND TASIĆ N. (EDS.), *Vinča and Its World*, 61-65.

MIKOV V. 1926-1927 'Selishtnata mogila pri Balbunar.' *IBAI* 4, 236 sq.

MIKOV V. 1926-1931 'Selishtanata mogila v s. Gabarevo (Kazanlushka okoliya).' *GNM* 5, 84-113.

MIKOV V. 1928-1929 'Prinos kum predistoricheskiata kultura u nas.' *IBAI* 1928-1929, 316-317.

MIKOV V. 1933 *Predistoricheski selishta i hakhodki v Bulgariya*. BAN. Sofiya.

MIKOV V. 1934 'Posledni mogilni nakhodki.' In: *Sb. "Madara"* Vol. 1. Sofia, 429-38.

MIKOV V. 1936 'Stari mogilni pogrebeniya pri Kalugeritsa i Kyulevcha.' In: *Sb. Madara* Vol. 2, 3 sq.

MIKOV V. 1940 'Selishtnata mogila pri s. Yunatsite (Pazardzhishko).' *GPNBM za 1937-1938*, 55-84.

MIKOV V. 1948 'Praistoricheskoto selishte do Krivodol, Vrachansko.' *Razkopki i prouchvaniya* 1, 26-62.

MIKOV V. 1961 'Selishtnata mogila do s. Zavet, Burgasko.' *Izvestiya na Arkheologicheskiya Institut* 24, 269-83.

MIKOV V. 1970 'Materiali ot posledniya etap na bronzovata epokha ot Severozapadna Bulgariya.' *Arkheologiya* 3, 48-63.

MIKOV V. 1971 'La Bulgarie a l'âge du bronze.' *Studia Balcanica* 5, 51-61.

MIKOV V. AND DZHAMBAZOV N. 1960 *Devetashkata peshtera*. BAN. Sofia.

MILOJČIĆ V. 1949 *Chronologie der Steinzeit Mittel- und Südosteuropas*. Berlin.

MIRACLE P.T. 1986 *Preliminary Analysis of the Distribution of Archaeological Materials at Opovo-Ugar Bajbuk, Yugoslavia*. Unpublished B.A. Honor thesis, University of California, Berkeley.

MOHEN J.-P. 1991 'Les sépultures de métallurgistes du début des Ages des Métaux en Europe.' In: La découverte du métal. Amis du Musée des Antiquités nationales. Millénaires. Collection dirigée par J.-P. Mohen. *Dossier* 2, 131-42.

MONTAGNARI KOKELJ E. 1990 'Industrie litiche della Bassa Friulana inquadrabili nel III e nella prima metà del II millennio a.C.' *Aquileia Nostra* 60, Col. 9-28.

MONTAGNARI KOKELJ E. 1998 (1997) 'Alcune osservazioni sui dati archeologici relativi alla Grotta Gigante nel Carso triestino.' In: *Atti e Memorie della Commissione Grotte "E. Boegan"* 35, 27-42.

MONTAGNARI KOKELJ E. AND CRISMANI A. 1997 (1996) 'La grotta del Mitreo nel Carso Triestino.' In: *Atti della Società per la Preistoria e Protostoria della regione Friuli-Venezia Giulia* 10, 7-98.

MOORE H.L. 1981 'Bone refuse – Possibilities for the future.' In: SHERIDAN A. AND BAILEY G. (EDS.), *Economic Archaeology: Towards an Integration of Ecological and Social Approaches*. BAR. Oxford. BAR International Series 96, 87-94.

MOORE H.L. 1982 'The interpretation of spatial patterning in settlement residues.' In: HODDER I. (ED.), *Symbolic and Structural Archaeology*. Cambridge University Press. Cambridge, 74-79.

MOORE H.L. 1986 'Space, Text, and Gender: An Anthropological Study of the Marakwet of Kenya.' Cambridge University Press. Cambridge.

MORETTI M. 1993 'Aspetti della metallurgia dell'età del bronzo antico e medio in Friuli.' In: *Preistoria del Caput Adriae*. Catalogo della mostra Trieste – Castello di S. Giusto 1983, 69-74.

MORINTZ S. AND ROMAN P. 1968 'Aspekte des Ausgangs des Äneolithikums und der Übergangsstufe zur Bronzezeit im Raum der Niederdonau.' *Dacia* N. S. 12, 45-128.

MOTTES E. 1996 'Considerazioni sulle lame di pugnale litiche del territorio veronese nel quadro dell'Eneolitico dell'Italia Settentrionale.' In: BELLUZZO G. AND SALZANI L. (EDS.), *Dalla terra al museo*. Forlì, 33-56.

MOTZOI-CHICIDEANU I. 1995 'Fremdgüter im Monteoru.-Kulturraum.' In: B. HÄNSEL (ED.), *Handel, Tausch und Verkehr im bronze- und frühzeitlichen Südosteuropa*. Südosteuropa-Schriften Bd. 17, Prähistorische Archäologie in Südosteuropa Bd. 11, 219-42.

MULTHAUF R.P. 1978 *Neptune's Gift A history of common salt*. Baltimore and London.

NACSN1983 'North American Stratigraphic Code.' *Amer. Assoc. Petrol. Geol. Bull.* 67, 841-75.

NAGEL G. 1938 *La Céramique du Nouvel Empire à Deir el-Medineh*. Vol. 1. Le Caire.

NAKAU G. 1997 'The Role of the North Aegean in the Development of the Aegean Metallurgy.' In: DOUMAS CHR. AND LA ROSA V. (EDS.), *Poliochni e l'Antica eta del Bronzo nell'Egeo setentrionale*.

Convegno Internationale Atene, 22-25 Aprile 1996. Atene, 634-48.

NEKHRIZOV G. 1996 'Mogilnite nekropoli v Iztochnite Rodopi.' In: ZAREV K, KITOV G., GOCHEVA Z., VASILEVA M. AND LILOVA B. (EDS.), *Purvi mezhdunaroden simpozium 'Sevtopolic' 'Nadgrobnite mogili v Yugoiztochna Bulgariya.* PIK. V. Turnovo, 7-18.

NEKHRIZOV G. 1999a 'Pogrebalni praktiki v Iztochnite Rodopi prez purvoto hilyadoletie pr. Hr.' *Rhodopica* Vol. 2, No. 1, 21-34.

NEKHRIZOV G. 1999b (IN PRINT) 'Kultovi mesta na trakite v Iztochnite Rodopi.' In: *Mezhdunaroden simpozium in memoriam M. Domaradzki 'Trakite i tehnite susedi'. Kazanluk, 30.09.-03.10.1999.*

NENQUIN J. 1961 *Salt. A Study in Economic Prehistory.* De Tempel. Brugge.

NEWBERRY P.E. 1893 *Beni Hasan.* Vol. 2. London.

NEWBERRY P.E. 1895 (1894-1895) *El* Bershe. Vol. 1. London.

NICA M. 1970 'Asupra origini si dezvoltarii culturii Vădastra de la Fărcaşele (Jud. Olt).' *Historica* 1, 31-52.

NICA M. 1982 'Locuirea preistorică de la Sucidava-Celej din perioadă de trecere de la neolitic la epoca bronzului. Oltenia.' *Studii şi Comunicari. Istorie* 4, fig. 6.

NICA M. 1996 'Date noi cu privire la generza şi ervoluţia culturii Verbicioara.' *Drobeta* 7, 18-34.

NICOLIS F. 1996 'Strutture e riti funebri. L'Italia settentrionale.' In: *L'Antica età del bronzo in Italia.* Atti del Congresso di Viareggio, 9-12 gennaio 1995. Firenze, 337-44.

NIKOLIĆ D. 1997 'Eneolitska nasel'a u okolini Majdanpeka.' In: LAZIĆ M. (ED.), *Arkheologija istochne Srbije,* 197-208.

NIKOLOV B. 1964 'Selishta i nekropoli ot bronzovata epokha vuv Vrachansko.' *Arkheolohiya* 2, 69-77.

NIKOLOV B. 1968 'Praistoricheskoto selishte pri Ohoden, Vrachanski okrug.' *Arkheologiya* 1, 65-75

NIKOLOV B. 1974 *Gradeshnitsa.* Nauka i izkustvo. Sofia.

NIKOLOV B. 1975 *Zaminets. Praistorichesko selishte.* Nauka i izkustvo. Sofia.

NIKOLOV B. 1976 'Mogilni pogrebeniya ot rannobronzovata epokha pri s. Turnava i Knezha, Vrachanski okrug.' *Arkheologiya* 3, 38-51.

NIKOLOV B. 1982 'Epokha rannej bronzy v Severo-Zapadnoj Bolgarii i nekropol' u sely Tyrnava.' *Thracia Praehistorica. Supplement Pulpudeva* 3, 192-200.

NIKOLOV B. 1984 *Krivodol-drevni kulturi.* Nauka i izkustvo. Sofia.

NIKOLOV B. 1978 'Développment du Chalcolithique en Bulgarie de l'Ouest et du Nord-Ouest.' *Studia Praehistorica* 1-2, 121-29.

NIKOLOV V. 1983 'Trakiya i Tesaliya prez bronzovata epokha.' *Vekove* 1, 12-19.

NIKOLOV V. 1986 'Vruzki na dolinata na r. Mesta s Anatoliya prez ranniya neolit.' *Arkheologiya* 2, 5-10.

NIKOLOV V. 1987 'Neolitni keramichni sudove za prigotvyane na khrana (Po materiali ot Yugozapadna Bulgariya).' *Bulgarska etnografiya* 2, 29-35.

NIKOLOV V 1988 'Liste der erforschten Siedlungen und Gräberfelder der Karanovo VI-Periode in Bulgarien.' In: FOL A. AND LICHARDUS J. (EDS.), *Macht, Herrschaft und Gold,* 267-72.

NIKOLOV V. 1992 *Rannoneolitno zhilishte ot Slatina (Sofia).*.BAN. Sofia. Razkopki i prouchvaniya 25.

NIKOLOV V. (ED.) 1993 *Praistoricheski nakhodki i izsledvaniya v pamet na prof. G. Il. Georgiev.* Archaeological Institute. Sofiya.

NIKOLOV V. 1994 'Risuvanata ornamentatsiya vurhu rannoneolitni keramichni sudove ot Chavdar.' *GDA-NBU* 1, 185-208.

NIKOLOV V. 1996 'Jugozapaden variant na kultura Karanovo I.' *Arkheologiya* 2-3, 1 –8.

NIKOLOV V. 1998a *Prouchvaniya vurkhu neolitnata keramika v Trakiya.* Agatho. Sofia.

NIKOLOV V. 1998b 'The Circumpontic Cultural Zone during the 6th Millennium BC.' *Documenta Praehistorica* 25, 81-89.

NIKOLOV V., MATSANOVA V., STEFANOVA T., BOZHILOV V., BUCHVAROV K., GATSOV I., MARINOVA E. AND NINOV L. 1999 *Selishtnata mogila Kapitan Dimitrievo.* Arkheologicheski Institut s muzej – BAN and Istoricheski muzej – Peshtera. Sofia and Peshtera.

NIKOLOV V. AND MASLAROV K. 1987 *Ancient settlement near Eleshnitsa.* Sofia Press. Sofia.

NIKOLOV V. AND RADEVA M. 1992 'Sondazhno prouchvane na rannoneolitnoto selishte v Dobrinishte, Razlozhko ' *Arkheologiya* 1, 1-14.

NIKOLOV V. AND STANEV S. 1996 'Sondazhni razkopki na praistoricheskoto selishte kraj Pomoshtitsa, Popovsko.' *AOR prez 1995.* Sofia, 11-12.

NIKOLOVA L. 1985 'Semantichnata transformatsiya na ogneniya kult v indoevropejskata obrednokultova traditsiya (po pismeni danni).' In: TACHEVA M. (ED.), *Vtori vissh seminar "Drevnost i suvremie".* SU "Kliment Okhridski" – Istoricheski fakultet. Sofia, 106-116.

NIKOLOVA L. 1994 'Novootkriti selishta v gornoto techenie na r. Stryama (Severozapadna Trakiya).' *Anali* 2-3, 5-11.

NIKOLOVA L. 1995 'Sarovka Tell near the Village of Dubene, Karlovo District.' In: NIKOLOVA L. (ED.),

Early Bronze Age Settlement Patterns, 15-23.

NIKOLOVA L (ED.) 1995 *Early Bronze Age Settlement Patterns in the Balkans (Ca. 3500-2000 BC, Calibrated Dates)*. Part 1. RPRP Vol. 1, No. 1.

NIKOLOVA L. 1996a 'Settlements and Ceramics: the Experience of Early Bronze Age in Bulgaria.' In: NIKOLOVA L (ED.), *Early Bronze Age Settlement Patterns*, 145-213.

NIKOLOVA L. 1996b 'On the Incrusted Ceramics from the Early Bronze Age Settlement of Dubene. In the light of the 1992-1995 Investigations.' In: TASIĆ N. (ED.), *The Yugoslav Danube and the Neighbouring Regions*, 81-93.

NIKOLOVA L. (ED.) 1996 *Early Bronze Age Settlement Patterns in the Balkans (Ca. 3500-2000 BC, Calibrated Dates)*. Parts 2-3. RPRP Vol. 1, Nos. 2-4.

NIKOLOVA L. 1998a 'Neolithic Sequence: The Upper Stryama Valley in Western Thrace (With Appendix: Radiocarbon Dating of the Balkan Neolithic).' *Documenta Praehistorica* 25, 99-131.

NIKOLOVA L. 1998b 'The Yunatsite Culture I: Manifestation and Chronology during Early Bronze I.' In: P. ROMAN (ED.), *The Thracian World*, 431-62.

NIKOLOVA L. 1999 *The Balkans in Later Prehistory. Periodization, Chronology and Cultural Development in the Final Copper and Early Bronze Age (Fourth and Third Millennia BC)*. BAR. Oxford. BAR International Series 791.

NIKOLOVA L. 2000a (1999) *The Yunatsite culture: Periodization, chronology and synchronizations*. Reports of Prehistoric Research Projects 2-3, 33 sq.

NIKOLOVA L. 2000b 'Social Transformation and Evolution in the Balkans in the Fourth and Third Millennia BC.' In: NIKOLOVA L (ED.), *Analyzing the Bronze Age*, 1-8.

NIKOLOVA L (ED.) 2000 *Analyzing the Bronze Age. Contributions to the Prehistoric Cultural Pattern in the Balkans*. Reports of Prehistoric Research Projects 4.

NIKOLOVA L. AND MADJEV N. 1993 *Prähistorische Funde aus Bulgarien (das Strjama – Tal)*. MKM. Sofia.

NIKOLOVA L. AND MADZHEV N. 1992 'Sondazhno prouchvane na rannoneolitni mnogoslojni selishta pri s. Kliment and s. Dubene, Karlovsko.' In: NIKOLOV V. (ED.), *Arheologicheski razkopki na praistoricheski obekti prez 1992*. Arheologicheski institut s muzej – Sektsiya za praistoriya. Sofia, 10-11.

NIKOLOVA L. AND ZAKHARIEVA T. 1994 'Selishte ot Karanovo VI pri s. Dubene, Karlovsko.' In: PANAYOTOV I. ET AL. (EDS.), *Maritsa-Iztok*, 63-72.

NIKOLOVA YA. AND ANGELOV N. 1961 'Razkopki na Emenskata peshtera.' *Izvestiya na Arkheologicheskiya Institut* 24, 297-316.

NIKOV K. 1996 'Ukrasa na keramikata ot Rannozhelyaznata epokha ot mestnostta "Gradishteto" kraj selo Glavan.' In: PANAYOTOV ET AL. (EDS.), *Maritsa-Iztok* Vol. 3, p. 120, fig. 2d.

NIKOV K. (IN PRINT) 'Relief Pithoi with Stamped and Insized Decoration from South-Eastern Thrace.' In: *Sbornik v chest na prof. A. Milchev*. Sofia.

OLSEN S. L. 1988 'The Identification of Stone and Metal Tool Marks on Bone Artifacts.' In: OLSEN S. L. (ED.), *Scanning Electron Microscopy in Archaeology*. British Archaeological Reports, I.S.S. 452, 337-60.

OTTAWAY B.S. 1996 'Technological and Social Basis for Early Prehistoric Copper. Procurement and Production in Europe.' In: *XIII International Congress of Prehistoric and Protohistoric Sciences Forlì – Italia – 8/14 September 1996*. 10. *The Copper Age in the Near East and Europe*. Colloquium XIX. *Metallurgy: Origins and Technology*. Forlì, 51-56.

ÖZDOĞAN M. 1982a 'Tikiburnu: A Late Chalcolithic Site in Eastern Thrace.' *Anatolica* 9, 1-26.

ÖZDOĞAN M. 1982b 'Dogu Marmara ve Trakya arastirmalari.' *Türk Arkeoloji Derfisi* Vol. 26 No. 1, 37-49.

ÖZDOĞAN M. 1983 'Pendik. A Neolithic site of Fikir tepe culture in the Marmara region.' In: BOEHMER R. M. AND HAUPTMAN H. (HGS.), *Beiträge zur Altertumskunde Kleinsasien*. Festschrift fur Kurt Bittel. Mainz, 401-411.

ÖZDOĞAN M. 1985 'Surface Survey for Prehistoric and Early Historic Sites in Northwestern Turkey.' *National Georgraphic Research Projects for 1979* (1985) 20, p. 532.

ÖZDOĞAN M. 1989 'Neolithic Cultures of Northwestern Turkey.' *Varia Archaeologica Hungarica* 2, 201-15.

ÖZDOĞAN M. 1994. 'Obsidian in Anatolia: An Archaeological Perspective on Status of Research.' *Archaeometry*, 423-31.

ÖZDOĞAN M. 1995 'Neolithic in Turkey.' In: *Reading in Prehistory. Studies presented to Halet Cambel*. Istanbul, 53-55.

ÖZDOĞAN M. 1998 'Recent Excavations in Eastern Thrace and Contact between the Prehistoric Cultures of Anatolia and Balkans.' /Abstract in English/ *TUBA – AR* 1, 63-92.

ÖZDOĞAN M. AND GATSOV I. 1998 'The Aceramic Neolithic period in Western Turkey and in the Eagean.' *Anatolica* 24, 209-232.

ÖZDOĞAN M., PARZINGER H. AND KARUL N. 1997 'Kirklareli Kazilari (Aşağipinar ve Kanliğecit Hoyükleri).' *Arkeoloji ve sanat* 19. 77: 2 - 11.

PANAJOTOV I. 1989 'Zur Chronologie und Periodisierung der Bronzezeit in den bulgarischen Territorien.' *Thracia* 9, 74-103.

PANAYOTOV I. 1975 'Za chronologoyata i periodizatsiyata

na bronzovata epokha v bulgarskite zemi.' *Istoricheski pregled* 6, 24-46.

PANAYOTOV I. 1984 'Za formiraneto na rannobronzovite kulturi v bulgarskite zemi.' *Arkheologiya* 2-3, 7-15.

PANAYOTOV I. 1988 'Studies on the Bronze Age in Bulgarian Lands (Historiographic Notes).' *Thracia* 8, 157-73.

PANAYOTOV I. 1989 *Yamnata kultura v bulgarskite zemi.* BAN. Sofiya. Razkopki i prouchvaniya 21.

PANAYOTOV I. 1995 'The Bronze Age in Bulgaria: Studies and Problems.' In: BAILEY D. AND PANAYOTOV I. (EDS.) *Prehistoric Bulgaria*, 243-52.

PANAYOTOV I. AND ALEXANDROV S. 1988 'Za kultura MAGURA-Coţofeni v bulgarskite zemi.' *Arkheologiya* 2, 1-14.

PANAYOTOV I. AND ALEXANDROV S. 1995 'Mogilen nekropol ot rannata bronzova epokha v zemlishtata na selata Mednikarovo I Iskritsa.' In: PANAYOTOV I., LESHTAKOV K., GEORGIEVA R., ALEXANDROV S. AND BORISOV B. (EDS.), *"Maritsa-Iztok". Arkheologicheski prouchvaniya.* Vol. 3. Ekspeditsiya "Maritsa-Iztok". Radnevo, 87-113.

PANAYOTOV I., LESHTAKOV K., GEORGIEVA R., ALEXANDROV S. AND BORISOV B. (EDS.) 1991 *"Maritsa-Iztok". Arkheologicheski prouchvaniya.* Vol. 1. Ekspeditsiya "Maritsa-Iztok". Sofia.

PANAYOTOV I., LESHTAKOV K., GEORGIEVA R., ALEXANDROV S. AND BORISOV B. (EDS.) 1994 *"Maritsa-Iztok". Arkheologicheski prouchvaniya.* Vol. 2. Ekspeditsiya "Maritsa-Iztok". Sofia.

PAPAEFTHYMIOU-PAPANTHIMOU A. AND PILALI-PAPASTERIOU A. 1997 *Prehistoric Macedonia: An Itinerary*. Paratiritis. Thessaloniki.

PAPAZOGLU F. 1969 *Srednjobalkanska plemena u predrimsko doba*. Djela, Book 30. Sarajevo.

PARZINGER H. 1993 *Studien zur Chronologie und Kulturgeschichte der Jungstein-, Kupfer- und Frühbronzezeit zwischen Karpaten und Mittlerem Taurus*. Philipp von Zabern. Mainz an Rhein.

PARZINGER H. AND ÖZDOĞAN M. 1995 'Die Ausgrabungen in Kirklareli (Turkish–Thrakien) und ihre Bedeutung für die Kulturbeziehungen zwischen Anatolien und dem Balkan vom Neolithikum bis zur Frühbronzezeit.' *Bericht der Römisch-Germanischen Kommission* 76, 5 – 29.

PĂUNESCU A. 1970 *Evoluţia uneltelor si armelor de piatră ciopită descoperite pe teritoriul României*. Academia R.S.R. Bucureşti, figs. 17, 18, 19, 21, 22.

PEDROTTI A.L. 1990 'L'insediamento di Kanzianiberg: rapporti culturali fra Carinzia ed Italia settentrionale durante il neolitico.' In: BIAGI P. (ED.), *The Neolithisation of the Alpine Region*. Monografie di Natura Bresciana 13, 213-26.

PEDROTTI A.L. 1995 'Le statue-stele e le stele antropomorfe del Trentino Alto Adige e del Veneto occidentale. Gruppo atesino, gruppo di Brentonico, gruppo della Lessinia.' *Notizie Archeologiche Bergomensi* 3, 259-80.

PEEF T.E. AND WOOLY C.L. 1923 *The City of Akhenaton.* Vol. 1. London.

PEJKOV A. 1995 'Stratigrafiya v krepostta na drevnotrakijskiya grad Evmolpiya.' *IMYUB* 21, 23-42.

PERNICHEVA L. 1983 'Prehistory of the Strumesnica Valley.' In: ŚLIWA J. AND DOMARADZKI M. (EDS.), *The Lower Strumesnica Valley*, 11-34.

PERNICHEVA L. 1992 'Issledovanie khalkolita v srednem porech'e reki Struma.' *Studia Praehistorica* 11-12, 221-35.

PERNICHEVA L. 1993 'Zoomorfni skulpturni izpbrazheniya ot praistoricheski selishta v Sofijsko.' *Godishnik na Natsionalniya Arkheologicheski Muzej* 9, 39-44.

PERNICHEVA L. 1995 'Prehistoric Cultures in the Middle Struma Valley: Neolithic and Eneolithic.' In: BAILEY D. AND PANAYOTOV I. (EDS.), *Prehistoric Bulgaria*, 99-147.

PERONI R. 1962-1963 'La Romita di Asciano (Pisa). Riparo sotto roccia utilizzato dall'età Neolitica alla Barbarica.' *Bullettino di Paletnologia Italiana* n.s. 14, 71-72, 251-452.

PETKOV N. 1950 'Selishtnata mogila Okol-Glava.' *Izvestoya na Arkheologicheskiya Institute* 17, 157-66.

PETKOV N. 1963 'Peklyuk, praistoricheskoto selishte pri s. Gulubovtsi, Sofijsko.' *Izvestoya na Arkheologicheskiya Institute* 26, 177-94.

PETKOV N. 1964 'Gruncharska pesht ot eneolitnoto selishte pri s. Gulubovtsi, Sofijsko.' *Arkheologiya* 1, 48-59.

PETRE-GOVORA GH. 1988 'Descoperiri arheologice din Oltenia privind epoca timpurie a bronzului.' *Thraco-Dacica* 9, 137-47.

PETRE-GOVORA GH. 1995 *O preistorie a nord-estului Olteniei.* Râmnicu Vâlcea.

PETRESCU-DÎMBOVIŢA ET AL. (EDS.) 1995 *Comori ale epocii bronzului din Romania. Treasures of the Bronze Age in Romania.* Bucureşti.

PETRIE W.M.F. 1890 *Kahum, Gugar and Hawara.* London.

PETRIE W.M.F. 1928 *Gerar.* London.

PETRIE W.M.F. 1932 *Ancient Gaza.* Vol. 2. London.

PETROVIĆ E. 1984 *Gomolava. Arheološko nalazište.* Zavičajni muzej u Rumi / Vojvođanski Muzej u Novom Sadu. Novi Sad.

POGGIANI KELLER R. 1988. 'L'età del rame nell'Italia settentrionale. Gli aspetti sepolcrali dell'area alpina centrale. ' In: *L'età del Rame in Europa.* Congresso Internazionale, Viareggio 15-18 ottobre 1987. Rassegna di Archeologia 7, 401-411.

POPESCU D. 1956 'Cercetari archeologice in Transilvania.' *MCA* 2, 203 and fig. 119.

POPOV R. 1918 'Kodzha-Dermenskata mogila pri gr. Shumen.' *IBAD* 6, 71-155.

POPOV R. 1922-1925 'Mogilata Deve-Bargan.' *GNM* 4, 72-115.

POPOV R. 1930-1931 'Mogilni grobove pri s. Endzhe.' *IBAI* 6, 89 sq.

POPOV V. 1994 'Periodizatsiya i hronologiya na neolitnite kulturi ot porechieto na Rusenski Lom.' *GDA NBU* 1, 294-302.

POPOV V. 1996 *Periodizatsiya i hronologiya na neolitnite i halkolitnite kulturi ot porechieto na r. Rusenski Lom*. Russe.

POPOV V AND VAJSOV I. 1992 'Spasitelni razkopki na praistoricheskoto selitshte kraj s. Koprivets, obshtina Balya' *AOR prez 1991*. Sofia, 9-11.

POPOV V AND VAJSOV I. 1996 'Arkheologichesko prouchvane na neolitnoto selitshe pri s. Koprivets.' *AOR prez 1995*. Sofia, p. 15.

PRIZER K.M. 1994 *Food Sharing in the Neolithic: Refitting Red Deer Skeletons from Opovo, Serbia*. Unpublished Senior Honor thesis. University of California, Berkeley. Berkeley.

PUPIN J-P. 1980 'Zircon and granite petrology.' *Contrib. Mineral. Petrol.* 73, 207-220.

RACSKY P. 1980 'A Körös kultúra újabb figurális ábrázolásai a Közép-Tiszavidékről és történeti összefüggéseik.' *SZMMÉ 1979-1980*, 5-33.

RADUNCHEVA A. 1981 'Praistoricheski selishta /V-II khil. pr. e./.' In: IVANOV T. (ED.), *Pernik*. Vol. 1. BAN. Sofia, 11-51.

RADUNCHEVA A. 1986 'Obshtestveno-ikonomicheskiya zhivot na Dobrudzha i Zapadnoto Chernomorie prez eneolita.' *Vekove* 1.

RADUNCEVA A. 1991 'Kurzer vorläugiger Bericht über ddie Ausgrabungen in Dolnoslav.' In: *Lichardus J. (Hrg), Die Kupferzeit als historische Epoche, 107-110.*

RADUNCHEVA A. 1992 'Kum vuprosa za nyakoi ukrepitelni suoruzheniya ot neolita i eneolita.' *Prinosi kum bulgarskata arkheologiya* 1.

RADUNCHEVA A. 1994 'Otnovo za Kukova Mogila i okolnostite na s. Duvanlij, Plovdivsko.' *GDA-NBU* 1, 178-83.

RADUNCHEVA A. 1999a *Kusnoeneolitnoto obshtestvo v bulgarskite zemi*. Avtoreferat na disertatsiya za poluchavane na stepen "doktor na istoricheskite nauki". Sofia.

RADUNCHEVA A. 1999b 'Organizatsiya na rudodobiva i metalurgiyata v bulgarskite zemi prez eneolita.' *GNBU-DA* 4 (in print).

RENFREW C. 1969 'The autonomy of the southeast European Copper Age.' *Proceedings of the Prehistoric Society* 35, 12-47.

RENFREW C. 1971 'Sitagroi, Radiocarbon and Prehistory of South-east Europe.' *Antiquity* 45, 275-82.

RENFREW C. 1979 'The Tree-Ring Calibration of Radiocarbon'; 'The Autonomy of the South-East European Copper Age.' In: *Problems in European Prehistory. Edinburgh, 138-78.*

RICHARDS C. AND THOMAS J. 1984 'Ritual Activity and Structural Deposition in Later Neolithic Wessex.' In: BRADLEY R. AND GARDINER J. (EDS.), *Neolithic Studies: A Review of Some Current Research*. BAR. Oxford. BAR British Series 133.

RIGAUD J.-PH. AND SIMEK J.F. 1991 'Interpreting Spatial Pattern at the Grotte XV: A Multiple-Method Approach.' In: KROLL E.M. AND PRICE T.D. (EDS.), *The Interpretation of Archaeological Spatial Patterning*. Plenum. New York, 199-220.

ROGER J.-M. 1995 'A propos de l'abandon des concepts de Chalcolithique et de Néolithique final.' In: *Chronologies néolithiques. De 6000 à 2000 avant notre ère dans le Bassin rhodanien.* Actes du Colloque d'Ambérieu-en-Bugey, 19 et 20 septembre 1992, sous la direction de J.-L.Voruz. Documents du Département d'anthropologie et d'écologie de l'université de Genève 20, 304-305.

ROMAN P. 1971 'Strukturänderungen des Endäneolithicums im Donau-Karpaten-Raum.' *Dacia* N.S. 15, p. 31 sq., Abb. 2, 6-8.

ROMAN P. 1976 *Cultura Coţofeni*. Academia R. S. R. Bucureşti.

ROMAN P. 1981 'Zur rumänischen Frühbronzezeit (der Forschungsstand).' In: KALICZ N. AND KALICZ-SCHREIBER (EDS.) R., *Die Frühbronzezeit im Karpatenbecken und in den Nachbargebieten*. Internationaler Symposium 1977. Budapest-Velem, 157-69.

ROMAN P. 1984 'Probleme în legătură cu perioada timpurie a epocii bronzului şi începuturile culturii Otomani.' *SCIVA* Vol. 35, No. 4, 266-74.

ROMAN P. 1985 'Cercetări la Govora Sat-Runcuri în 1977.' *SCIVA* Vol. 36 No. 4, 279-97.

ROMAN P. 1986 'Perioada timpurie a epocii bronzului pe teritoriul României.' *SCIVA* Vol. 37 No.1, 29-55.

ROMAN P. 1988 'Ostrovul Corbului (Rumänien) und Malé Kosihy (Tschechoslowakei).' *Slovenská Archeológia* 36, 217-24.

ROMAN P. (ED.) 1998 *The Thracian World at the Crossroads of Civilizations*. Vol. 2. Institutul Român de Tracologie. Bucharest.

ROMAN P. (ED.) 1997 *The Thracian World at the Crossroads of Civilizations*. Vol. 1. Vavila Edinf SRL. Bucarest.

ROMAN P. AND NÉMETI I. 1978 *Cultura Baden în România*. Academia Republicii Socialiste România. Bucareşti.

ROMAN P. AND NÉMETI J. 1990 'Epoca bronzului în nord-vestul României.' *Symposia Thracologica* 8. Satu Mare-Carei, 34-41.

ROODENBERG J., THISSEN L. AND BUITENHUIS H. 1989-1990 'Preliminary Report on the Archeological Investigations at Illipinar in NW Anatolia.' *Anatolica* 16, 61 – 144.

ROODENBERG J. 1995. *The Illipinar Excavations I. Five Seasons on Fieldwork in NW Anatolia, 1987 – 91*. Istanbul.

ROSEN S. A. 1984 'The Adoption of Metallurgy in the Levant: A Lithic Perspective.' *Current Anthropology* 25, 504-505.

ROSEN S. A. 1997 *Lithics after the Stone Age: A Handbook of Stone Tools from the Levant*. Altamira Press. Walnut Creek, CA.

ROWLETT R.M. 1984 'Archaeological Evidence for Early Indo-European Chieftains.' *JIES* Vol. 12 No. 1-2, 191-219.

RUNNELS C AND VAN ANDEL T. 1988 'Trade and the Origins of Agriculture in the Eastern Mediterranean.' *Journal of Mediterranean Archaeology* 1.

RUSSEL N. 1993 'Hunting, Herding and Feasting: Human Use of Animals in Neolithic Southeast Europe.' Unpublished Ph.D. thesis. University of California, Berkeley. Berkeley.

RUSSEL N. 1993 (IN PRESS) 'Symbolic dimentions of animals and meat at Opovo, Yugoslavia.' In: ROBB J. (ED.), *Material Symbols*. Southern Illinois University, Center for Archaeological Investigations, Occasional Papers.

RYNDINA N. 1971 *Drevnejshie metalloobrabotywashtee proizvodstvo Vostochnoj Evropy*. Moskva.

RYNDINA N. 1993 'Novye dannye o drevnejshej metallurgii Bolgarii.' *Pulpudeva* 6, 138-51.

SAMS K. 1974 'Phrygian Painted Animals: Anatolian Orientalizing Art.' *Anatolian Studies* 24, 169-195.

SAMS K. 1978 'Schools of Geometric Painting in Early Iron Age Anatolia.' In: *The Proceedings of the 10th International Congress of Classical Archaeology*. Ankara. Vol. 2, p.227.

SANEV B., SIMOSKA D., KITANOSKI B. AND SARZHOSKI S. 1975 *Praistorija vo Makedonija*. Arheoloshko drushtvo na SR Makedonija. Skopije.

SCHLOR I.1997 'Die chalkolitische Keramik aus Grube Z. ' In: HILLER ST. AND NIKOLOV V. (HGS.), *Karanovo, 285-321.*

SCHNEIDER G. 1989 'Naturwissenschaftliche Kriterien und Verfahren zur Beschreibing der Keramik.' *Acta Prehistorica et Archaeologica 21,7-39.*

SCHORTMAN E.M. AND URBAN P.A. 1987 'Modeling Interregional Interactions in Prehistory.' *Advances in Archaeological Method and Theory* 11, 37-97.

SCHUSTER C. 1992a 'Aşezări Glina pe cursul inferior al Argeşului şi Valea Cîlniştei (I). Mihăileşti-Tufa.' *Thraco-Dacica* 13, 35-41.

SCHUSTER C. 1992b 'Cîteva precizări privind faza a III-a a culturii Glina în lumina celor mai noi cercetări.' *Litua* 5, 7-10.

SCHUSTER C. 1994 'Aria de răspândire a culturii Tei.' *Analele Banatului*, S.N., 3, 171-78.

SCHUSTER C. 1995 'Cercetări arheologice în aşezarea culturii Glina de la Varlaam, jud. Giurgiu.' *Cultură şi Civilizaţie la Dunărea de Jos* 13-14, 53-63.

SCHUSTER C. 1996 'Zu den hallstattzeitlichen Funden im südlichen Teil Mittelmunteniens.' In: GARAŠANIN M. AND ROMAN P. (EDS), *Der Basarabi-Komplex in Mittel- und Südosteuropa. Kolloquium in Drobeta-Turnu Severin (7.-9. November 1996).* Rumänisch-Jugoslawische Kommission für die Erforschung der Region des Eisernen Tores, Archäologische Abteilung I, Bukarest, 147-163.

SCHUSTER C. 1997a *Perioada timpurie a epocii bronzului în bazinele Argeşului şi Ialomiţei Superioare. Bibliotheca Thracologica* 20, Bucureşti.

SCHUSTER C. 1997b 'Cultura Tei şi sud-estul Transilvaniei.' *Angustia* 2, 131-35.

SCHUSTER C. 1998a (1996-1998) 'Unele probleme ale Bronzului Mijlociu (BM) şi Final (BF) din centrul Munteniei (I).' *Giurgiu* Vol. 2-4, No. 2-4, 141-53.

SCHUSTER C. 1998b 'Die Frühbronzezeit in der Walachei und in Südostsiebenbürgen – kulturelle Verbindungen, Beziehungen und Einflüsse und ethnische Bewegungen.' *Acta Musei Napocensis* Vol. 35, No. 1, 25-35.

SCHUSTER C. 2000 'Zur Besiedlungen der West- und Mitttelwalachei (Rumänien) in der Frühbronzezeit.' In: NIKOLOVA L (ED.), *Analyzing the Bronze Age*, 9-19.

SCHUSTER C. AND POPA T. 1995 'Cercetări privind epoca bronzului în judeţul Giurgiu (investigaţiile din anii 1986-1994).' *Giurgiu* Vol. 1, No. 1, 20-54.

SÉFÉRIADÈS M. 1983. 'Dikili Tash: Introduction á la Préhistoire de la Macédoine Orientale.' *BCH* 107, 635-77.

SÉFÉRIADÈS M. 1991 'Pierre taillée et métallurgie.' In: *La découverte du Métal*. Amis du Musée des Antiquités nationales. Millénaires. Collection dirigée par J.-P. Mohen. Dossier 2, 325-330.

SÉFÉRIADÈS M. 1996 'Deshayes' Excavations at Dikili Tash: the Early Bronze Age Levels.' In: NIKOLOVA L. (ED.), *Early Bronze Age Settlement Patterns*, 95-128.

ŞERBĂNESCU D. AND BĂLTEANU C. 1998 'Considérations sur le rite, le rituel funeraire et l'etude anthropologique des squelettes de la civilisation Tei.' In: ROMAN P. (ED.), *The Thracian World*, 482-96.

SEREBRYANNYI L. 1980 *Laboratornyi analiz v geomorphologii*. VINITI. Moskva.

SEURE G. AND DEGRAND A.1906 'Explorationde quelques telles de la Thrace.' *BCH 30*, 359-432.

SHALGANOVA T. 1993 *Razvitie na kulturite prez kusnobronzovata i nachaloto na starozhelyaznata epoha v Severozapadna Bulgariya*. Avt. ... za k.i.n. Archaeological Institute with Museum. Sofia.

SHALGANOVA T. 1995 The Lower Danube Incrusted Pottery Culture. In: BAILEY D.W. AND PANAYOTOV I. (EDS.), *Prehistoric Bulgaria*, 291-308.

SHERRATT A. 1997 'Plough and Pastoralism: Aspects of the Secondary Products Revolution (1981).' In: SHERRATT A., *Economy and Society in Prehistoric Europe. Changing Perspectives*. Edinburgh University Press. 158-98.

SHKORPIL K. AND SHKORPIL H. 1888 *Pametnitsi iz Bulgarsko. Part 1. Trakiya*. Plovdiv.

SHKORPIL K. AND SHKORPIL H. 1892 'Severoiztochna Bulgariya v geografsko i arkheologichesko otnoshenie, I.' *SbNUNK* 7, 68-70.

SHKORPIL K. AND SHKORPIL H. 1896 *Purvobitnite lyudi v Bulgariya*. Plovdiv.

SHKORPIL K. AND SHKORPIL H. 1898 *Mogilite. Pametnitsi ot Bulgarsko*. Plovdiv.

SHMAGLIJ N.M. AND CHERNYAKOV I.T. 1970 'Kurgany stepnoj chasti mezhdurech'ya Dunaya i Dnestra (1964-1969).' *MASP* 6, 5-115.

SIMEONOVA H. 1968 (1970) 'Die vorgeschichtliche Funde von Sadovec (Nordbulgarien).' *Jahrbuch des römisch-germanischen Zentralmuseums Mainz* 15, 15-63.

SIMOSKA D., KITANOSKI B. AND TODOROVIĆ J. 1976 'Naselbata Crnobuki i problemot na istoimenata kultura vo svetlinata na novite arkheoloshki istrazhuvani'a.' *MAA* 2, 43-83.

SIMOSKA D., KITANOSKI B. AND TODOROVIĆ J. 1977 'Praistoriska naselba Tumba vo selo Karamani kaj Bitola.' *MAA* 3, 9-25.

SIMOSKA D. AND SANEV V. 1976 *Praistorija na Tsentralna Pelagonija*. Bitola.

ŠINKOVEC I. 1995 'Katalog posameznih kovinskih najdb bakrene in bronaste dobe / Catalogue of Individual Metal Finds from the Eneolithic and Bronze Ages.' In: TERŽAN B. (ED.), *Depojske in posamezne kovinske najdbe bakrene in bronaste dobe na Slovenskem*. Vol. 1, 13-127. Vol. II, 125-63.

SIRAKOV N., SIRAKOVA S., IVANOVA S., GATSOV I. AND TSONEV TS. 1994 'The Epigravettian Sequence.' In: KOZLOWSKI J. K., LAVILLE H. AND GINTER B. (EDS.), *Temnata Cave, Excavations in Karlukovo Karst Area*. Vol. 1.2. Jagellonian University Press. Krakow, 169-315.

SIRAKOV, N. AND SKAKUN N. 1983 'Stone Artefacts connected with Chipping Technique.' In: ŚLIWA J. AND DOMARADZKI M. (EDS.), *The Lower Strumesnica Valley*, 35-42.

SIRAKOV, N. AND TSONEV TS. 1995. 'Chipped-Stone Assemblage of Hotnitsa-Vodopada (Eneolithic/Early Bronze Age Transition in Northern Bulgaria) and the Problem of the Earliest "Steppe Invasion" in Balkans.' *Preistoire Europeenne* 7, 241-64.

SKEATES R. 1994 'Radiocarbon Date-List for Prehistoric Italy (46,400 – 2450 BP/400 cal. BC).' In: SKEATES R. AND WHITEHOUSE R.D. (EDS.), *Radiocarbon Dating and Italian Prehistory*. Accordia Research Centre and the British School at Rome. London.

ŚLIWA J. AND DOMARADZKI M. (EDS.) 1983 *The Lower Strumesnica Valley in Prehistoric, Ancient and Early Medieval Times*. Institute of Archaeology Cracow and Institute of Thracology Sofia. Krakow.

SOLOMON K. AND VOIGT M. 1986 'Bread and Beer: The Early Use of Cereals in Human Diet.' *Expedition* Vol. 28, No. 2, 23-34.

SPINDLER K., RASTBICHLER-ZISSERNIG E., WILFING H., ZUR NEDDEN D. AND NORTHDURFTER H. (EDS.) 1984 *Der Mann Im Eis. Neue Funde und Ergebnisse*. Wien-New York.

SREJOVIĆ D. AND TASIĆ N. (EDS.) 1990 *Vinča and Its World. International Symposium*. Serbian Academy of Science and Arts. Belgrade. Symposia Vol. 51, Book 15.

STALIO B. 1972 *Gradac. Praistorisko nasel'e*. Narodni muzej. Beograd.

STALIO B. 1992 'Grupno sakhran'ivan'e na Ajmani-Mala Vrbica.' *ZNM* Vol. 14, No. 1, 65-76.

STANEV P. 1995. 'Topografiya i stratigraphiya na neoliten kompleks Orlovets, zhilishtna arhitektura na rannoneolitnoto selishte.' *Izvestiya na Istoricheski muzej Veliko Turnovo* 10, 58-60.

STANEV P., NAJDENOVA E. AND ELENSKA N. 1994 'Arkheologicheski prouchvaniya v semlitshteto na s. Orlovets, Polskotrumbeshka obshtina.' *Izvestiya na Istoricheski muzej Veliko Turnovo* 9, p. 90.

STEFANOVA M. 2000 'Control Excavations at Mihalich in 1998-1999 (Preliminary Communication).' In: NIKOLOVA L (ED.), *Analyzing the Bronze Age*, 21-31.

STEFANOVA T. 1991 'Selishtna mogila Gulubovo – Khalkolitna epokha.' In: PANAYOTOV ET AL. (EDS.), *Maritsa-Iztok*, 141-53.

STEFANOVICH M. 1999 *Arkheologiya, etnichnost i sotsialna praistoriya*. Avtoreferat na disertatsiya za prisuzhdane na nauchnata stepen 'doktor na istoricheskite nauki'. Sofiya.

STEFANOVICH M. AND BANKOFF H.A. 1998 'Kamenska

Chuka 1993-1995 - Preliminary Report' In: STEFANOVICH M. ET AL. (EDS.), *In the Steps of James Harvey Gaul*, 255-338.

STEFANOVICH M., TODOROVA H. AND HAUPTMANN H. (EDS.) 1998 *In the Steps of James Harvey Gaul*. Vol. 1. *James Harvey Gaul - In Memoriam*. James Harvey Gaul Foundation. Sofia.

STEVANOVIĆ M. 1996 *The Age of Clay: The Social Dynamics of House Destruction*. Unpublished Ph.D. Thesis. Univesity of California, Berkeley. Berkeley.

STOICA C. 1989 'Complexe de la sfîrşitul epocii bronzului descoperite la Cătunu (jud. Dîmboviţa).' *Symposia Thracologica* 7. Tulcea, 258-59.

STOJIĆ M. 1996 'L'Énéolithique final dans le centre des Balkans.' In: NIKOLOVA L (ED.), *Early Bronze Age Settlement Patterns*, 353-72.

STOYANOV M. 1991 'Sondazhno prouchvane na praistorichesko selishte ot ranniya eneolit kraj s. Umarevtsi, Loveshka obshtina.' *AOR prez 1990*. Lovech, 18-19.

STOYANOV M. 1992a 'Prouchvane na rannoeneolitnoto selishte kraj s. Umarevtsi, Loveshka obshtina (Predvaritelno suobshtenie).' *GMSB* 18, 7-16.

STOYANOV M. 1992b 'Spasitelni arheologicheski razkopki v Devetashkata peshtera.' *AOR prez 1992*. Sofia, 14-16.

STOYANOV M. 1993 'Spasitelni arkheologicheski razkopki v Devetashkata peshtera.' *GMSB* 19, 7-20.

STRAHM C. 1988 'Chalkolithikum und Metallikum: Kupferzeit und frühe Bronzezeit in Südwestdeutschland und der Schweiz.' In: *L'età del Rame in Europa*. Congresso Internazionale, Viareggio 15-18 ottobre 1987. Rassegna di Archeologia 7, 175-92.

STRAHM C. 1991 'Les débuts de la métallurgie dans l'Ouest de l'Europe centrale.' In: *La découverte du Métal*. Amis du Musée des Antiquités nationales. Millénaires. Collection dirigée par J.-P. Mohen. Dossier 2, 143-56.

SZÉKELY ZS. 1997 *Perioada timpurie şi începutul celei mijlocii în sud-estul Transilvaniei. Bibliotheca Thracologica* 21, Bucureşti.

TASIĆ N. 1983 *Iugoslavensko podunavl'e od indoevropske seobe do prodora skita*. Institute for Balkan Studies. Novi Sad. Beograd.

TASIĆ N. (ED.) 1996 *The Yugoslav Danube Basin and the Neighbouring Regions in the 2 nd Millenium B.C.* Institute for Balkan Studies. Beograd abd Vršac.

TASIC N. 1995 *Eneolithic Cultures of Central and West Balkans*. Draganich. Belgrade.

TASIĆ N. 1997 'Eneolit i bronzano doga severoistočne Srbije.' In: LAZIĆ M. (ED.), *Arkheologija istochne Srbije*, 79-89.

TASIĆ N. N. 1992 'A Review of the C14 Series of Dates from Đerdap.' *Balcanica* 23, 89-98.

TASIĆ N.N. 1997 *Hronologja starčevačke kulture*. Typescript of Doctorate thesis. Beograd.

THEOCHARIS D. 1972 *Prehistory of Eastern Macedonia and Thrace*. Ancient Greek Sites 9. Athens.

THOMAS H. 1967 *Near Eastern, Mediterranean and European Chronology*. SIMA 7, vol. 1–2, Lund.

THOMAS H. 1992 'Summary.' In: EHRICH R. (ED.), *Chronologies*, 504–515.

THORPE O.W., WARREN S.E. AND BARFIELD L.H. 1979 'The Sources and Distribution of Archaeological Obsidian in Northern Italy.' *Preistoria Alpina* 15, 73-92.

TILLEY CH. 1981 'Economy and Society: What Relationship?' In: SHERIDAN A. AND BAILEY G. (EDS.), *Economic Archaeology: Towards an Integration of Ecological and Social Approaches*. BAR. Oxford. BAR International Series 96.

TITOV V. 1995 'V. Milojčić za myastoto i znachenieto na selishtnata mogila.' In: KATINCHAROV ET AL., *Selishtnata mogila*, 14-17.

TONČEVA G. 1984 'Nécropole tumulaire de l'âge du bronze ancien près du village Belogradec.' *Studia Praehisrica* 1-2, 228-37.

TONCHEVA G. 1984 'Monumentalni skulpturni pametnitsi ot kusnobronzovata i ot rannozhelyaznata epokha v Bulgariya.' *Thracia* 6, fig. 17.

TODOROVA H. 1978 *Eneolit Bolgarii*. Sofia Press. Sofia.

TODOROVA H. 1982 'Nyakoi aktualni problemi na statisticheskata obrabotka na keramikata ot arkheologicheski obekti.' *Istoricheski pregled* 9, 11-16.

TODOROVA H. 1986 *Kamenno-mednata epokha v Bulgariya*. Nauka i izkustvo. Sofia.

TODOROVA H. 1990 'Ein mitteläneolithisches Haus aus Poduene (Sofia) aus der Vinča D_2 Zeit.' In: SREJOVIĆ D. AND TASIĆ N. (EDS.), *Vinča and Its World*, 156-65.

TODOROVA H. 1992 'Neuauswertung der prähistorischen Funde von 1935-1937.' *In:* ÜNZE S., *Die Spätantiken Befestigungen von Sadovec. Ergebnisse der Deutsch - Bulgarisch - Österreichischen Ausgrabungen 1934-1937*. C.H. Beck. München, 370-80.

TODOROVA H. 1998 'Der Balkan-Anatolische Kulturbereich vom Neolithikum bis zur Frühbronzezeit (Stand der Forschung).' In: STEFANOVICH M. ET AL. (EDS.), *In the Steps of James Harvey Gaul*, 27-54.

TODOROVA N. 2000 *Sravnitelen analysis na kusnohalkolitnata keramika ot Yunatsite*. PhD Thesis. Sofia University.

TORBOV N. 1994 'Rezultati ot arkheologicheskoto prouchvane na Ninovskata mogila pri s. Turnak, Beloslatinsko.' *IMSZB* 22, 11-22.

TREUIL R. 1983 *Le Néolithique et le Bronze Ancien égéens.* Bibliotheque des écoles françaises d'Athenes et de Rome, fasc. 248. Paris.

TRIANDAPHYLLOS D. 1983 'Les monuments mégalithiques en Thrace Occidentale.' *Pulpudeva* 4, 145-63.

ΤΡΙΑΥΤΑΦΓΛΛΟΣ Δ. 1987 'Δοκιμαστικη αυασκαφη στα Ριζια του Εβρου.' *Αρχαιολογικοα ΕΡΓ Οασιασ Μαεδουιακατ* Θρακη 1, 492-93.

TRINGHAM R., BRUKNER B., KAISER T., BOROJEVIĆ K., BUKVIĆ L., ŠTELI P., RUSSEL N., STEVANOVIĆ M. AND VOYTEK B. 1992 'Excavations at Opovo, 1985-1987: Socioeconomic Change in the Balkan Neolithic.' *Journal of Field Archaeology* Vol. 19, No. 3, 351-86.

TRINGHAM R., BRUKNER B. AND VOYTEK B. 1985 'The Opovo Project: A Study of Socioeconomic Change in the Balkan Neolithic.' *Journal of Field Archaeology* Vol. 12, No. 4, 425-44.

TSANKOV V. 1946 *Petrografiya na Gornooryahovsko.* Geofond na GUGOZN. Sofia.

TSONCHEV D. 1938 *Prinosi kum starata istoriya na Plovdiv.* Sofiya.

TSONCHEV D. 1940 'Starinite okolo Haskovskite toplomineralni izvori.' *GPlNBM* 1937/1939, 85-104.

TSONEV TS. 1995 'Collection of Flint Artefacts from the Upper Valley of the Stryama River, Karlovo Region.' In: NIKOLOVA L. (ED.), *Early Bronze Age Settlement Patterns*, 57-63.

TRINKAUS K.M. 1987 'On Boundaries and Social Complexity.' In: TRINKAUS K.M. (ED.), *Polities and Partitions*, 1-28.

TRINKAUS K.M. (ED.) 1987 *Polities and Partitions: Human Boundaries and the Growth of Complex Societies.* Arizona State University. Anthropological Research Paper No. 37.

TUDOR E. 1982 'Neue Angaben zur frühen Bronzezeit in Südrumänien.' *Dacia*, N.S., 26, 59-75.

TUDOR E. 1983 'Săpăturile arheologice de la Odaia Turcului (jud. Dîmbovița).' *Materiale și Cercetări Arheologice.* Brașov, 108-111.

TYLECOTE R. F. 1987 *The Early History of the Metallurgy in Europe.* Longsmans. London.

TYLECOTE R. F. 1992 *A History of Metallurgy.* The Institute of Metals. London.

VAJSOV I. 1984 'Antropomorfnaya plastika iz praistoricheskogo poseleniya Kurilo-Kremenitsa, Sofijskogo okruga.' *Studia Praehistorica* 7, 33-63.

VAJSOV I. 1992a 'Anthropomorphe Plastik aus dem prähistorischen Gräberfeld bei Durankulak.' *Studia Praehistorica* 11-12, 95-113.

VAJSOV I. 1992b 'Antropomorfnata plastika na kulturata Hamangia.' *Dobrudzha* 9, 35-70.

VAJSOV I. 1993a 'Duhoven zhivot na neolitnoto naselenie v Bulgariya.' In: TODOROVA H. AND VAJSOV I., *Novokamennata epoha v Bulgariya.* Nauka i izkustvo. Sofia, 183-240.

VAJSOV I. 1993b 'Die frühesten Metalldolche Südost- und Mitteleuropas.' *Praehistorische Zeitschrift* 68, 103-145.

VAJSOV I. 1998 'Studies in Memoriam of James Harvey Gaul on the Typology of the Anthropomorphic Figurines from Northeastern Bulgaria.' In: STEFANOVICH M. ET AL. (EDS.), *In the Steps of James Harvey Gaul*, 107-141.

VELKOV I. 1942 (1940-1942) 'Predistoricheskoto selishte do s. Dyakovo.' *IBAI* 14, 208-209.

VELUŠCEK A. (IN PRESS) 'Presentazione del pugnale rinvenuto nella Grotta Andreolotti. ' *Atti e Memorie della Commissione Grotte "E. Boegan".*

VENEDIKOV I. 1960 *Trakijskata kolesnitsa.* Sofiya.

VITRI S. 1995 'Palù di Livenza (Pordenone).' In: ASPES A. AND FASANI L. (EDS.), *Guide archeologiche.* Preistoria e Protostoria in Italia 7. Veneto e Friuli-Venezia Giulia, 182-93.

VOYTEK B. 1990 'Resource Use and the Inference of a Pastoral Economy.' *Rivista di Studi Liguri* A. LVI, 1-4, 47 –58.

VOSS J. 1987 'Prehistoric Tribalization in Northwestern Europe.' In: TRINKAUS K.M. (ED.), *Polities and Partitions*, 29-60.

VOSSEN R. 1970 'Klassifikationsprobleme und Klassifikationssysteme in der Amerikanischen Archäologie.' *Acta Praehistorica et Arhaeologica 1.*

VULPE A. 1970 *Die Äxte und Beile in Rumänien.* Vol. 1. C. H. Beck. München. PBF 9-2.

VULPE A. 1983 'Pe marginea unor noi descoperiri din bronzul timpuriu.' *Studii și Cercetări de Istorie Veche și Arheologie* Vol. 34 No.1, 71-75.

VULPE A. 1990 'Neue Beiträge zur Chronologie und kulturellen Gliederung der Frühbronzezeit im unteren Donaugebiet.' *Starinar* 40-41, 105-111.

VULPE A. 1997a 'Considerații privind începutul și definirea perioadei timpurii a epocii bronzului în România.' In: NISTOR V. AND ZAHARIA D. (EDS.), *In honorem Ligiae Bârzu.* Memorie și patrimoniu I, București, 37-50.

VULPE A. 1997b 'Săpăturile de la Popești. Prezentarea campaniilor 1988-1993.' *Cercetări Arheologice* 10, 163-72.

VULPE A. AND DRÂMBOCEANU V. 1981 'Săpăturile arheologice în raza comunei Năeni (Buzău).' *Studii și Cercetări de Istorie Veche și Arheologie* Vol. 32, No. 3, 171-93.

WARREN P. AND HANKEY V. 1989 *Aegean Bronze Age Chronology.* Bristol Classical Press. Bristol.

WARSON J.P.N. 1979 'The estimation of the relative frequencies of mammalian species: Khirokitia 1972.' *Journal of Archaeological Science* 6, 127-37.

WEINBERG S. 1965 'The Relative Chronology of the Aegean in the Stone and Early Bronze Age.' In: ERICH R. (ED.), *Chronology*, 285-320.

WEISSHAAR, H.–J. 1989 *Die Deutschen Ausgrabungen auf der Pevkakia-Magula in Thessalien. Das späte Neolithikum und das Chalkolithikum*. Dr. Rudolf Habelt. Bonn. BAM 28.

WELLS P.S. 1987 'Sociopolitical Change and Core-Periphery Interactions: An Example from Early Iron Age Europe.' In: TRINKAUS K.M. (ED.), *Polities and Partitions*, 141-55.

WILLIS K. AND BARNETT K.D. 1994 'The Neolithic Transition – Fact or Fiction? Palaeoecological Evidence from the Balkans.' *The Holocene* Vol. 4, No. 3, 326-30.

WINLOCK H.E. 1955 Models of Daily Life in Ancient Egypt from the Tomb of Meket-Reʿ at Thebesh. Harvard, 1955.

YAKAR J. 1985 *The Later Prehistory of Anatolia. The Late Chalcolithic and the Early Bronze Age*. BAR. International Studies 268.

YAROVOJ E. 1985. *Drevnejshie skotovodchieskie plemena Yugo-Zapada SSSR*. Shtiintsa. Kishinev.

ZAGORCHEV I. 1984 'Sostav i stroezh dokambrijskikh metamorfitov Ograzhdenskogo bloka (Serbo-Makedonskij massiv) i Vlakhinskogo bloka (Kraishtidy) v yugo-zapadnoj Bolgarii.' In: *Osobenosti stanovleniya Zemnoj kory v Dokambrij Yuzhnoj Bolgarii*. BAN. PK 9 AISS, 31-43.

ZAHARIA E. 1987 'La culture Monteoru. L'étape des débuts à la lumière des fouilles de Sărata Monteoru.' *Dacia*, N.S., 31, 21-49.

ZAHARIA E. 1995 'Cultura Zăbala. Zăbala Culture.' In: PETRESCU-DÎMBOVIȚA ET AL. M. (EDS.), *Comori ale epocii bronzului în România*, 285-87.

ZIRRA V. 1960 'Kul'tury pogrebenij s okhroj v Zakarpatskikh oblastyakh RNR.' In: *Materialy i issledovaniya po arkheologii Yugozapada SSSR i RNR*. Kishinev, 97-123.

ZLATEVA R. 1997 *Rannoholotsenski kremuchni anambli ot porechieto na reka Rusenski Lom*. MSc Dissertation. New Bulgarian University. Sofia Unpublished).

ZMEJKOVA I. 1990 'Kum problematikata na keramikata ot bronzovata epokha otkrita v Tabashkata peshtera.' *Ekspeditsionen godishnik na SU "Kl. Okhridski"* Vol. 5, A-1989, 105-107.